Coral Reefs of the World

Volume 24

Coral Reefs of the World is a series presenting the status of knowledge of the world's coral reefs authored by leading scientists. The volumes are organized according to political or regional oceanographic boundaries. Emphasis is put on providing authoritative overviews of biology and geology, explaining the origins and peculiarities of coral reefs in each region. The information is so organized that it is up to date and can be used as a general reference and entry-point for further study. The series will cover all recent and many of the fossil coral reefs of the world.

Prospective authors and/or editors should consult the Series Editors B.M. Riegl and R.E. Dodge for more details. Any comments or suggestions for future volumes are welcomed:

Dr. Bernhard M. Riegl/Dr. Richard E. Dodge
Nova Southeastern University
Dania Beach, FL 33004, USA
e-mail: rieglb@nova.edu and dodge@nova.edu

Marcelo O. Soares • Tyler B. Smith
Emma F. Camp • Verena Schoepf
Editors

Extreme and Marginal Coral Reefs and Communities

Living at the Edge of Environmental Limits

Springer

Editors
Marcelo O. Soares
Instituto de Ciências do Mar
Universidade Federal do Ceará
Fortaleza, Ceará, Brazil

Emma F. Camp
Climate Change Cluster
Faculty of Science
University of Technology Sydney
Sydney, NSW, Australia

Tyler B. Smith
Center for Marine and Environmental Studies
University of the Virgin Islands
Saint Thomas, VI, USA

Verena Schoepf
Department of Freshwater and Marine Ecology
Institute for Biodiversity and Ecosystem
Dynamics
University of Amsterdam
Amsterdam, The Netherlands

ISSN 2213-719X ISSN 2213-7203 (electronic)
Coral Reefs of the World
ISBN 978-3-032-28034-3 ISBN 978-3-032-28035-0 (eBook)
https://doi.org/10.1007/978-3-032-28035-0

This Springer imprint is published by the registered company Springer Nature Switzerland AG
The registered company address is: Gewerbestrasse 11, 6330 Cham, Switzerland

Marcelo O. Soares:
I dedicate this book to my family, who gave me wings—or fins—to fly, dive, and explore the world.

Tyler B. Smith:
I dedicate this book to my children, Liam and Hazel. May you both make the world a better place for us all.

Emma F. Camp:
I dedicate this book to my parents for fostering my passion for the ocean and for always believing in me.

Verena Schoepf:
I dedicate this book to my mother and my husband, who have always believed in my dreams and supported me in pursuing them.

Preface

> The most dangerous worldview is the worldview of those who have not viewed the world.
>
> —Alexander von Humboldt (1769–1859)

The initial idea for this book emerged from research we have been conducting in different reef environments across continents around the world. Sometimes, diving into deeper, more turbid, warmer, or more acidic waters, we encounter organisms such as corals, fish, and microorganisms with remarkable adaptations. Despite the exponential increase in research over the past few decades and the publication of numerous articles, there was still a gap for a book that would gather and synthesize this valuable information about the changing oceans we face today.

We know that it is impossible (perhaps even in many lifetimes) for a scientist to know the entire world—or all the oceans. However, as the great naturalist Alexander von Humboldt reminds us at the beginning of this preface, by bringing together more than 90 researchers from around the world, along with thousands of references produced over the last century, we aim to provide a truly global perspective on extreme and marginal reefs and their communities. At a time when many countries find it increasingly difficult to interact with other nations or cultures (or even individuals), we seek to demonstrate that we are indeed one humanity, and that local perspectives can offer a powerful and essential global vision.

As scientists, we are often traveling and participating in expeditions across the world's oceans. In 2022, the idea for this book began to take shape. Marcelo Soares, a Brazilian researcher, was on a train in Bremen, Germany, speaking with Italian scientist Sergio Rossi (author of Chap. 14), when he asked, "Sergio, you have already led two books—what do you think about a new one on marginal and extreme reefs around the world? Do you have any advice?" Rossi responded enthusiastically, "Marcelo, I think it's a great idea! Talk to Eva Loerinczi at Springer Nature—there's huge potential there!"

With this idea in mind, and taking advantage of his time as a visiting professor at ZMT (Leibniz Centre for Tropical Marine Research) in 2022, Marcelo took another train to the beautiful city of Amsterdam, the Netherlands. It was a quick two-day trip to meet Dra. Verena Schoepf for the first time. Verena has extensive experience in this field, having worked with tolerant corals in locations such as the Caribbean and Australia. After a fantastic and energetic day full of inspiring discussions, we had already secured two editors for the book. Soon after, we invited Dra. Emma F. Camp from Sydney, who has built a remarkable career as a researcher, diver, and science communicator.

Finally, in 2023, during a three-month research trip to the United States Virgin Islands and the Caribbean's blue waters, Marcelo met American researcher Dr. Tyler B. Smith, completing the team of editors. Two men and two women from different regions: South America (Marcelo O. Soares), North America/Caribbean (Tyler B. Smith), Europe (Verena Schoepf), and Oceania (Emma F. Camp). This configuration helped shape our vision for the book. It had to be global, with authors from every continent, ensuring meaningful gender representation and diversity in age—in other words, a balance between emerging researchers and those with extensive experience.

With the editorial team in place, we drafted the proposal in late 2023 and submitted it to Springer Nature. The reception was excellent, and we were thrilled to have the book accepted as part of the globally renowned Coral Reefs of the World series. During the United Nations Decade of Ocean Science (2021–2030), our collective effort to better understand our seas, their reefs, and how to preserve them served as a central motivation throughout the 3 years spent developing this volume.

At the same time, the years during which the book was written (2023–2026) were extremely challenging for coral reef researchers. The effects of global climate change have become more rapid, frequent, and intense, and reef ecosystems have been among the primary—and most visible—victims. Many of the authors found themselves in the field trying to understand the impacts of the severe heatwaves of 2023 and 2024. There were moments when we knew contributors were returning from dives with deep sadness after witnessing underwater mortality events. Yet, as authors and resilient human beings—much like the corals described in this book that thrive in extreme environments—we sought inspiration even while watching parts of the ocean seemingly "on fire."

The accelerating impacts of climate change have only heightened interest in the reefs and organisms presented here, given their potential role as refuges under current conditions and/or as analogs of what we may face in the near future. The lessons offered by reef organisms living in extreme environments are invaluable for shaping conservation policies and expanding human knowledge.

To achieve this, we aimed to create a book with a holistic and multidisciplinary perspective. The volume begins with foundational concepts, followed by chapters on key reef organisms, various types of extreme and marginal environments, and concludes with discussions on refugia, ecosystem services, persistence, and conservation strategies. In this way, the book serves not only marine scientists but also managers, entrepreneurs, rights holders, and nongovernmental organizations working to protect the world's oceans.

We invited contributors and received an enthusiastic response. This first edition, *Extreme and Marginal Coral Reefs and Communities: Living at the Edge of Environmental Limits*, includes 17 chapters authored by researchers from the Americas, Africa, Asia, Europe, and Oceania.

The opening chapter provides a conceptual framework: what marginal and extreme reefs are, where they occur, why they matter, and why they are attracting increasing attention. It also revisits the legacy of Charles Darwin, who was captivated by coral reefs during his voyage aboard the HMS *Beagle* between 1832 and 1836.

Chapter 2, authored by Dr. Miguel Mies from São Paulo (Brazil), examines corals in extreme and marginal environments: their adaptations, resilience, and ecological roles. Chapter 3, led by Dr. Juliano Morais from Australian Institute of Marine Sciences (Australia), explores reef fish diversity, biogeography, human impacts, and conservation.

Following this, Chaps. 2, 3, and 4 focus on key reef organisms: corals, fishes, and microorganisms. Chapter 4 (led by Dr. Matteo Monti) highlights the often unseen but fundamental role of microorganisms, presented by five researchers from KAUST (King Abdullah University of Science and Technology, Saudi Arabia), part of Dr. Raquel Peixoto's team. Their amazing work explores not only microbial ecology but also innovative restoration strategies. As classic French literature reminds us, "What is essential is invisible to the eye"—yet in this book, microorganisms are made visible.

From Chap. 5 onward, we examine different types of extreme and marginal environments. In Chap. 5, Dr. Bernhard Riegl Jr. presents the Arabian Gulf, a region characterized by naturally high and variable temperatures where "super corals" thrive. Chapter 6, developed by Dr. Verena Schoepf and collaborators from South Africa, Australia, and California, focus on intertidal reefs—environments exposed to dramatic daily fluctuations that may offer insights into future warming scenarios. Chapters 7 and 8 examine naturally acidified environments, including submarine CO_2 vents in places such as Papua New Guinea. Dr. Riccardo Rodolfo-Metalpa and Dr. Sylvain Agostini present ecosystem-level perspectives in Chap. 7, while a team of researchers led by Dr. Timothy Ravassi and Dr. Michael Izumiyama (OIST, Japan) provide genetic and transcriptomic insights in Chap. 8. Chapter 9, by Dr. Shannon Dee and Dr. Adi Zweifler (Australia), synthesizes research on turbid reefs, which may function as potential climate refuges. Chapter 10, by Dr. Marlene Wall and Dr. Claudio Richter (Germany), explores upwelling systems and their complex role as either stressors or buffers.

Chapter 11, by Dr. Heather Stewart (USA) and colleagues, examines corals in mangroves and seagrass beds. Chapter 12, authored by Dr. Tyler B. Smith and Dr. Gal Eyal, explores mesophotic reefs in the twilight zone (30–150 m depth). Chapter 13, by Dr. Ciara Baines and Dr. Maria Beger (UK) and an amazing international team, addresses high-latitude coral communities and the phenomenon of tropicalization.

Dr. Alex Bastos from Espírito Santo (Brazil) leads Chap. 15 on the geological foundations of these reefs—how they formed and how Earth's history shaped these extraordinary environments. This geological perspective is essential for understanding how habitats for countless organisms developed into some of the most remarkable underwater structures on the planet. Darwin himself was deeply engaged with both biology and geology, reflecting the integrated scientific worldview championed decades earlier by Alexander von Humboldt.

The Chaps. 14, 16, and 17 focus on ecosystem processes, services, persistence, and conservation. Chapter 14, by Dr. Sergio Rossi, examines trophic ecology. Chapter 16, by Dr. Nicola Browne, Dr. Kyle Morgan, and Dr. Lorenzo Alvarez-Filip, analyzes carbonate budgets and geological functions. Chapter 17, by Dr. Emma F. Camp, Dr. Ariel Pezner, and Dr. Ian Enochs, discusses persistence and resilience mechanisms in stress-tolerant corals.

We hope this international collaboration serves as a strong foundation of both basic and applied knowledge for coral reefs and the oceans as a whole during this Decade of Ocean Science for Sustainable Development. While we cannot cover every organism or environment, we offer this book as a meaningful contribution toward understanding diverse ecosystems and promoting the science needed for the ocean we want—a guiding motto for the challenging times in which we live.

Fortaleza, Ceará, Brazil — Marcelo O. Soares
Saint Thomas, VI, USA — Tyler B. Smith
Sydney, NSW, Australia — Emma F. Camp
Amsterdam, The Netherlands — Verena Schoepf

Acknowledgments

We wish to acknowledge the community of researchers whose dedication, time, and effort underpin the data presented throughout these chapters. We are grateful for their contributions, which have advanced understanding in this field.

We also pay our respects to the Traditional Owners and custodians of the Sea Country visited as part of the research for this book. We acknowledge their enduring connection to land and waters, and recognize the importance of Indigenous knowledge systems.

We value all knowledge streams and their contributions to the ongoing advancement of this field.

We would like to take this opportunity to express our sincere gratitude to the reviewers who generously devoted their time, expertise, and thoughtful insights to the evaluation of chapters in the book *Extreme and Marginal Coral Reefs and Communities*, part of the Coral Reefs of the World series by Springer Nature.

High-quality academic publishing relies fundamentally on the rigor, dedication, and collegial spirit of peer reviewers. The authors of this volume have consistently expressed their appreciation for the constructive, detailed, and insightful comments received, which have significantly strengthened the scientific quality and clarity of the chapters. Even in cases of critical evaluation, the feedback was delivered with professionalism and a genuine commitment to advancing the field.

We are therefore pleased to formally acknowledge and thank the following reviewers for their invaluable contributions:

Prof. Dr. Sergio Rossi
Profa. Dra. Oscarina Viana
Prof. Dr. Amorim Reis
Dr. Pedro Pereira
Dra. Jessica Bleuel
Prof. Dr. Antonio Ximenes Neto
Dra. Tatiane Martins Garcia
Dra. Tallita Cruz Lopes Tavares

Their efforts have been essential in shaping this novel volume, and we deeply appreciate their role in maintaining the high standards expected of this important book series.

Contents

Extreme and Marginal Coral Reefs and Communities: A Global Synthesis

1

Marcelo O. Soares, Tyler B. Smith, Emma F. Camp, Verena Schoepf, Anne L. Gurgel, Mariany Oliveira, Alexandre C. Siqueira, Juliano Morais, Guilherme O. Longo, Mariana Bender, Bráulio A. Santos, Jessica Bleuel, and Emanuelle F. Rabelo

Abstract

Corals are foundation organisms that thrive in a wide range of environmental conditions. While much of the research on tropical coral reefs has focused on areas conducive to carbonate growth, such as clear, low-nutrient, shallow waters, there is a significant gap in studying coral communities and reefs that thrive at the edge of their environmental limits. Recent theoretical advancements have provided a novel framework for studying these extreme and marginal reefs, re-defining coral reef extremeness and marginality and providing a broader perspective on their ecological significance. Extreme coral environments are characterised by large deviations from optimal abiotic conditions, with extreme in this case meaning: elevated siliciclastic sedimentation, turbid waters, high nutrient levels (e.g., nitrogen and phosphorus), substantial depths, and fluctuating or extreme temperature, pH, salinity, and dissolved oxygen conditions. In some cases, but certainly not all, these extreme conditions can lead to altered coral cover, community composition, and ecosystem functioning characteristic of marginal reefs and coral communities. Thus, areas with extreme environmental conditions are often mistakenly considered unsuitable for sustaining biogenic growth and biodiversity. However, extreme environments may serve as refugia and natural laboratories in the face of ongoing climate change, and can potentially offer insights into how coral ecosystems might respond to future oceanic conditions. This chapter serves as a comprehensive introduction to the global study of extreme and marginal reef environments, consolidating the knowledge accumulated to date. It provides an overview of the unique coral communities found in extreme conditions, such as those in the hottest waters, turbid-zone reefs, intertidal zones, low pH areas around submarine springs and CO_2 vents, regions affected by upwelling and internal waves, as well as seagrass and mangrove-coral ecosystems, semi-enclosed bays, mesophotic reefs, and high-latitude reefs. Given the increasing impacts of climate change and human activities on global coral reefs, research into these important habitats offers critical insights for conservation and the future resilience of marine ecosystems.

M. O. Soares (✉) · A. L. Gurgel · M. Oliveira · J. Bleuel
Instituto de Ciências do Mar (LABOMAR), Universidade Federal do Ceará (UFC), Fortaleza, Ceará, Brazil
e-mail: marcelosoares@ufc.br

T. B. Smith
Center for Marine and Environmental Studies (CMES), University of the Virgin Islands, Saint Thomas, US Virgin Islands, USA

E. F. Camp
Faculty of Science, Climate Change Cluster, University of Technology Sydney, Sydney, NSW, Australia

V. Schoepf
Department of Freshwater and Marine Ecology, Institute for Biodiversity and Ecosystem Dynamics, University of Amsterdam, Amsterdam, The Netherlands

A. C. Siqueira
Centre for Marine Ecosystems Research, School of Science, Edith Cowan University, Perth, WA, Australia

J. Morais
Research Hub for Coral Reef Ecosystem Functions and College of Science and Engineering, James Cook University, Townsville, QLD, Australia

G. O. Longo
Departamento de Oceanografia e Limnologia (DOL), Universidade Federal do Rio Grande do Norte (UFRN), Natal, Rio Grande do Norte, Brazil

M. Bender
Departamento de Ecologia e Evolução, Universidade Federal de Santa Maria (UFSM), Santa Maria, Rio Grande do Sul, Brazil

B. A. Santos
Departamento de Sistemática e Ecologia, Universidade Federal da Paraíba (UFPB), João Pessoa, Paraíba, Brazil

E. F. Rabelo
Universidade Federal Rural do Semiárido (UFERSA), Mossoró, Rio Grande do Norte, Brazil

M. O. Soares et al. (eds.), *Extreme and Marginal Coral Reefs and Communities*, Coral Reefs of the World 24,
https://doi.org/10.1007/978-3-032-28035-0_1

Keywords

Reef-building corals · Mesophotic reefs · Intertidal and back-reef coral environments · Turbid-zone reefs · Semi-enclosed bays · Low pH habitats · Upwelling · Hot-reefs · Mangroves · Seagrasses · High-latitude reefs · Cold-water communities · Climate change

1.1 A Journey to Understand Coral Reefs and Communities

> *12th. [April 1836] In the morning we stood out of the Lagoon. I am glad we have visited these Islands; such formations surely rank high amongst the wonderful objects of this world. It is not a wonder which at first strikes the eye of the body, but rather after reflection, the eye of reason. We feel surprised when travellers relate accounts of the vast piles & extent of some ancient ruins; but how insignificant are the greatest of these, when compared to the matter here accumulated by various small animals.*
> (Charles Darwin, Beagle Diary)

During his voyage on the Beagle, the famous naturalist Charles Darwin had the opportunity to visit and study coral reefs. He was particularly fascinated by the reefs around the islands of Mauritius and Cocos (Keeling) in the Indian Ocean in 1836. Darwin was intrigued by how corals, which usually thrive in clear, shallow, and nutrient-poor waters, grew around underwater mountains and formed carbonate structures. This subject became the focus of his famous book "The Voyage of the Beagle", published upon his return to England in 1839. Since then, numerous scientists, equally fascinated by coral reefs, have dedicated their lives to understanding these complex and intriguing systems and how they change over time.

Coral reefs, despite occupying a small area of the world's oceans, have always captivated humans due to their rich biodiversity and resources such as fisheries. Though they cover less than 1% of the seafloor, these unique habitats support about 25% of all marine species, including over 4000 species of fish, 800 species of hard corals and a sum of 830,000 (95% credible limits: 550,000–1,330,000) multicellular species (Dubinsky and Stambler 2011; Fisher et al. 2015; Hoegh-Guldberg et al. 2017). Estimates indicate that the global economic value of coral reefs is around $375 billion per year (Costanza et al. 2014). This significant value arises from various ecosystem goods and services, including tourism, fisheries, and coastal protection (Beck et al. 2018). For example, reef-related tourism generates over $36 billion annually, and an estimated 1 billion people worldwide depend on coral reefs for their livelihoods, food, and income from fisheries (Hughes et al. 2010; Spalding et al. 2017; Sing Wong et al. 2022). In terms of coastal protection, coral reefs act as natural barriers, shielding coastlines from sea-level rise, storms, hurricanes, and coastal erosion. Their rigid carbonate structures can absorb up to 97% of wave energy, reducing the impact on shorelines and saving billions in annual infrastructure damages and human livelihoods (Elliff and Silva 2017).

The remarkable ecosystem services and biodiversity of coral reefs are vitally important to the world's oceans and humankind. These seascapes are among the most diverse ecosystems on Earth (Fisher et al. 2015), often compared to tropical rainforests for their richness and variety (Rossi et al. 2017). These rainforests of the sea typically grow under optimal conditions of light, temperature, and nutrient-poor waters, as observed by Charles Darwin. However, are shallow and nutrient-poor waters the only habitats where we find these marine forests with reef-building corals? Just as we have various types of tropical land forests in different climates, wouldn't we also have different reefs and communities in diverse ocean conditions? For example, biodiverse tropical forests occur naturally in extreme climates, such as areas with frequent droughts or high temperatures (Stan and Sanchez-Azofeifa 2019). So, which are the extreme marine environments where corals can thrive, persist, and sometimes even build unique carbonate structures?

Corals are living, adaptable and foundation (or bioengineering) organisms that can flourish in a range of environmental conditions, much like trees in land forests. Their trophic ecology (see Chap. 14), which combines feeding and photosynthesis, growth rates, and ecological roles, can differ significantly between those in optimal conditions (like those Darwin observed) and those in more extreme or suboptimal environmental conditions (Perry and Larcombe 2003; Camp et al. 2018a, b; Wang et al. 2018). For reef-building shallow corals, optimal conditions for growth and survival typically refer to waters shallower than 20 m deep, temperatures between 23 and 29 °C, pH levels around 8.1–8.4, clear waters with low turbidity, salinity between 32 and 42 ppt, and high sunlight availability (Kleypas et al. 1999; Hoegh-Guldberg et al. 2017). However, these optimal conditions (e.g., nutrients) have been discussed in the recent literature (Morais et al. 2025).

These environmental characteristics can be found, for example, in the Caribbean Sea and the Indo-Pacific, a reef region visited by Charles Darwin and described in his letters, as seen in an excerpt taken from these letters that begins this chapter. Most scientific research on coral reefs has been carried out in these shallow, low-latitude environments (Morais et al. 2018, 2022) due to the ease of access and the presence of colonial and/or developed countries. Under these optimal conditions, and with low-level of current day human impacts, we find high species diversity (taxonomic, functional, and/or phylogenetic) and high coral cover in reefs, such as in the Coral Triangle in the Indo-Pacific (Schoepf et al. 2023). However, research discussed throughout this book and in this opening chapter indicates that corals can grow significantly and thrive outside these environmental conditions, forming important and unique marine forests (e.g., Chap 10 and 11).

Numerous studies have focused on the conservation, ecology, and ecosystem services of coral reefs in areas with optimal conditions for coral growth in shallow waters (Morais et al. 2018). However, there is a lack of studies on coral communities that form marine forests occurring in extreme or suboptimal environmental conditions (Camp et al. 2018a, b; Soares et al. 2020; Schoepf et al. 2023). These understudied conditions are often wrongly considered unsuitable for coral growth. A similar misunderstanding occurred with dry forests in very hot climates with little rainfall, where it was once thought there was little biodiversity or low tree growth (Sunderland et al. 2015; Dexter et al. 2015; Araujo et al. 2023). Today, we know that these land forests in extreme climates are crucial for the global biodiversity of plants and animals (Blackie et al. 2014; Sunderland et al. 2015; Silva et al. 2017). Similarly, we now understand the potential of extreme coral reefs and communities (see Chaps. 3 and 4) as resilience hotspots, natural laboratories, and potential refugia habitats, such as high-latitude and intertidal reefs (Beger et al. 2014; Chaps. 6 and 13), mesophotic reefs (Bongaerts et al. 2017; Chap. 12), and turbid-zone reefs (Chap. 9). But how have scientific advances in marine technology enabled this change in thinking?

1.2 Better Understanding the Distribution of Corals and All Types of Reefs

When performing research in the nineteenth century, such as that conducted by Charles Darwin with the support of Captain FitzRoy, technological limitations meant that indirect marine sampling methods were used to indicate the presence of corals, such as seafloor sampling. FitzRoy's investigations at the Cocos (Keeling) Islands in Australia suggested a depth limit for living corals of about 37 m. Based on numerous observations by other scientists and marine workers, Darwin estimated a probable limit of 55 m (Darwin 1889). This led to the traditional belief that the optimal environmental conditions for reef-building corals were in shallow, clear waters with high light incidence. In other words, it was thought that corals could not thrive in low-light conditions, such as those in mangroves (Chap. 11), turbid waters, or at greater depths (considered suboptimal or extreme environmental conditions). However, today, with advancements in Scuba diving, underwater cameras and various modern marine sampling methods, it is known that extensive mesophotic reefs, ranging usually from 30 m to 150 m deep, can harbor reef-building corals (Chap. 12).

With the advancement of methods and global research in the twentieth and twenty-first centuries (Kleypas et al. 1999; Bak and Meesters 2000; Potts and Jacobs 2000; Perry and Larcombe 2003; Camp et al. 2018a, b), the presence of reefs and coral communities in these understudied or "suboptimal" conditions was confirmed. For example, coral communities have been found in uncommon environments such as mangrove areas (177 coral species) and seagrass beds (39 species) (Chap. 11), not necessarily on hard reef bottoms. In this context, various terms have been used to define these marine forests, with marginal and extreme being the most common ones (Schoepf et al. 2023). Only recently has there been a clearer definition of what constitutes an extreme and/or marginal reef. A group of scientists led by Verena Schoepf published an important perspective article with definitions and core concepts to help classify these environments and communities (Schoepf et al. 2023) (Fig. 1.1).

The research group began by conceptualizing what constitutes an extreme reef. On one hand, extreme reefs are characterized by large deviations from optimal abiotic environmental conditions, such as temperature, pH, turbidity, salinity, depth, nutrients, and oxygen (Schoepf et al. 2023). These reefs can serve as adaptive refugia and/or natural laboratories for future ocean conditions (Maggioni et al. 2021), which explains the increasing interest in these unique environments (Camp et al. 2018a, b; Burt et al. 2020; Soares et al. 2020). On the other hand, marginal reefs, which exhibit altered coral cover, community composition, and ecosystem functioning (Schoepf et al. 2023), are becoming more common due to increase of environmental variations resulting from intense human impacts and fast climate change.

In the theoretical framework proposed recently, coral reef extremeness is based on environmental conditions, defined by large deviations from optimal ocean conditions in terms of mean and/or variance. Marginality, on the other hand, is based solely on ecological criteria, such as altered reef community composition and/or ecosystem functioning (Schoepf et al. 2023). This integrated but independent evaluation of both environmental and ecological criteria is crucial to avoid the misconception that coral communities existing outside the presumed optimal conditions for reef development are inferior to coral reefs in more traditional conditions (Schoepf et al. 2023), much like land forests in extreme climates. For instance, some extreme hot-reefs (e.g., the Persian Arabian Gulf) (see Chap. 5) exhibit robust carbonate production due to the high saturation states of calcium carbonate in warm

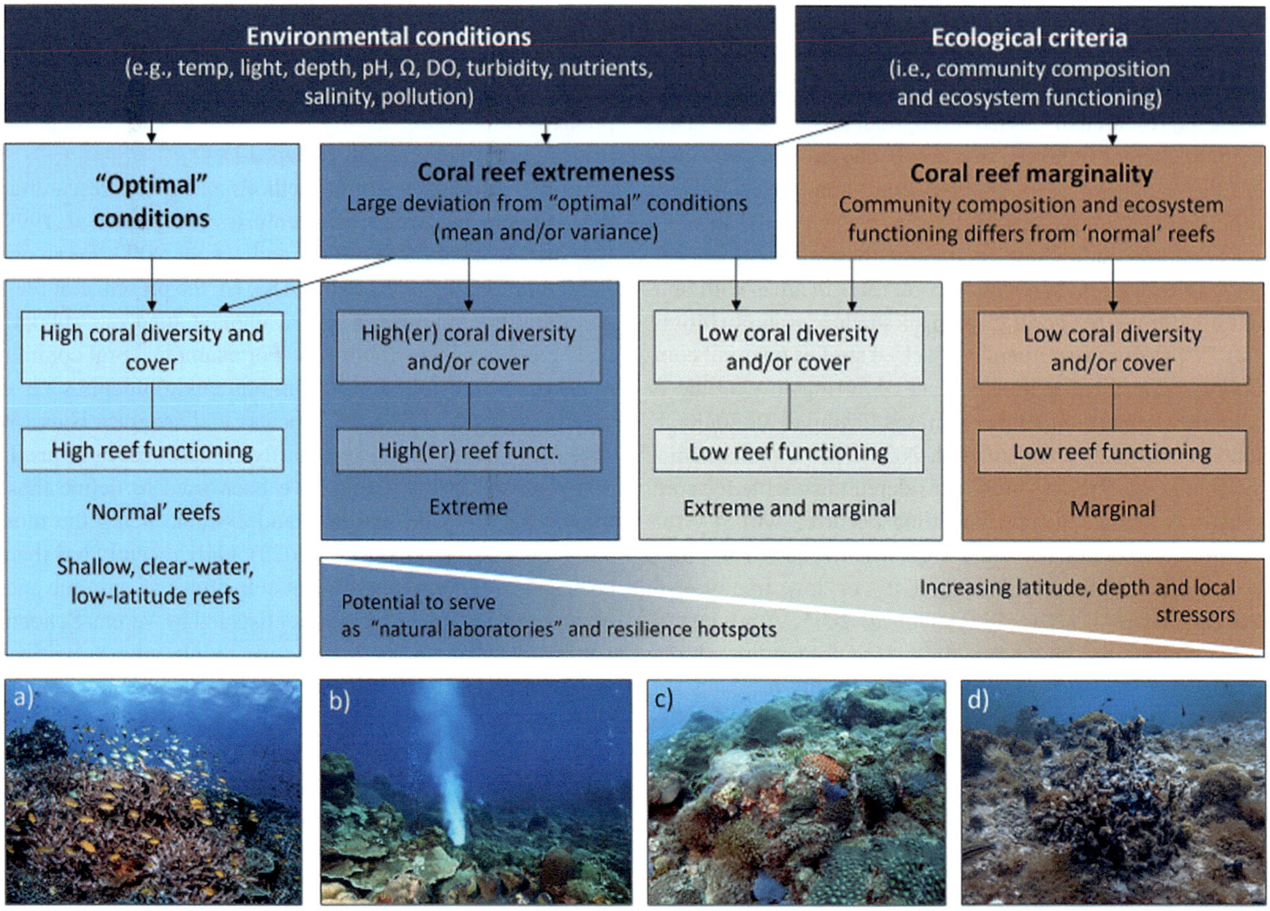

Fig. 1.1 Concepts and theoretical framework on extreme and/or marginal reefs and communities. (Source: Modified from Schoepf et al. (2023) by the authors)

and saline waters. Similarly, higher carbonate accretion rates occur in inshore turbid-zone reefs of the Great Barrier Reef, Singapore reefs, and high-latitude Japan reefs (see Chap. 16).

In marginal reefs, community changes can occur in the types of organisms dominating the marine forest architecture. Reefs previously dominated by coral builders may experience, for example, a phase-shift, with a greater occurrence of fleshy algae, sponges, or soft corals (such as octocorals or zoantharians) (Dudgeon et al. 2010a, b; Bell et al. 2021). In such cases, the maintenance of the reef carbonate structure, an important coral reef function, is compromised (see Chap. 16). These environmental changes in marine ecosystems are often driven by human-induced impacts, such as coastal development (deforestation and dredging), which increase turbidity, or the mismanagement of sewage, which contributes to coastal pollution in reef areas (Soares 2020; Bell et al. 2021).

Similarly to marine forests, we observe a comparable situation with land forests. There are important forests that are not limited to low-latitude tropical rainforests like the Amazon. These forests can occur in different conditions, such as high altitudes, regions with frequent droughts, or areas with significant temperature variations like high-latitude regions (Medina and Klinge 1983). Forests in extreme climates are crucial for global biodiversity (Blackie et al. 2014; Sunderland et al. 2015; Silva et al. 2017). However, recent impacts, including local human activities such as deforestation and fires, as well as global climate change effects like high temperatures, wildfires, and increased droughts, have led to deep changes on forest ecosystems (e.g., ongoing savannization of the Amazon forest, "desertification" of South American seasonally dry tropical forests) (Lapola et al. 2023; Mau et al. 2023; Moura et al. 2023; Norlen et al. 2024).

This chapter provides an overview of extreme and marginal reefs and coral communities that occur in a wider range of environmental settings worldwide. We aim to discuss and answer intriguing scientific questions about the occurrence of these reefs and the coral communities associated with them or other coastal habitats, such as mangroves or seagrass beds: (1) Where do these environments occur? (2) Why is there so much interest today in these reefs and the communities that inhabit them? (3) What have scientists studied most extensively in these fascinating ecosystems in recent years? (4) What are the main threats faced by these unique environments? (5) What still needs to be studied and understood in these environments?

1.3 Scientific Questions About Extreme and Marginal Reefs and Communities

1.3.1 Where Do These Extreme and Marginal Environments Occur?

The distribution of these reefs and coral communities is global, including tropical and subtropical areas in the South Atlantic, Caribbean, Indo-Pacific, Japan, Red Sea, Persian Gulf, among other areas (Figs. 1.2 and 1.3). The corals in extreme environments can be found under the influence of cold spells (Tuckett and Wernberg 2018; Chap. 10) or warm waters (often above 30 °C) (Bento et al. 2016; see Chap. 5), waters with lower and variable pH levels (often below 8.0) near CO_2 vents (see Chaps. 7 and 8), and turbid waters with moderate or high levels of suspended particles (between 0.080 to 0127 K_d490), reducing light penetration (Chap. 9). The coral communities can be found even in estuarine waters (e.g., mangrove habitats) with variable salinity, high levels of nutrients, and organic matter (Chap. 11). In fact, the vast majority of coral reefs occur under more productive conditions, with approximately 80% found in mesotrophic and eutrophic waters rather than oligotrophic (nutrient-poor) environments (Morais et al. 2025).

These extreme reefs and communities (Fig. 1.3) exist naturally under these unique environmental conditions and living near the maximum environmental threshold. In other words, on the edge of environmental limits for coral survival and growth. Red Sea reefs (e.g., central and southern part) endure temperatures ranging from 20 °C to 34 °C, which is significantly higher than the global average for coral reefs (Grottoli et al. 2017). Moreover, they are found under extremely high salinity levels of around 40–42 ppt, compared to the typical 32–35 found in most reef areas worldwide (Fine et al. 2019). These Red Sea corals, such as those in the Gulf of Aqaba, have adapted to these high temperatures and salinity, showing partial resilience to bleaching events that devastate other reefs globally. Under a similar hot-reef situation are the Persian—Arabian Gulf Reefs (Chap. 5). These are one of the hottest marine environments, with summer water temperatures exceeding 35 °C and salinity levels reaching up to 45 due to high evaporation rates and limited freshwater influx. Reef-building coral species such as *Acropora* and *Porites* have adapted to these extreme conditions, though they exhibit slower growth rates (see Chap. 5).

Contrasting with the Red Sea, the southernmost reefs in Australia (e.g., Lord Howe Island) are usually exposed to cooler water temperatures, ranging from 18 °C to 24 °C, which is at the lower end of the thermal tolerance range for most coral species (Kleypas et al. 1999; Harriott and Banks

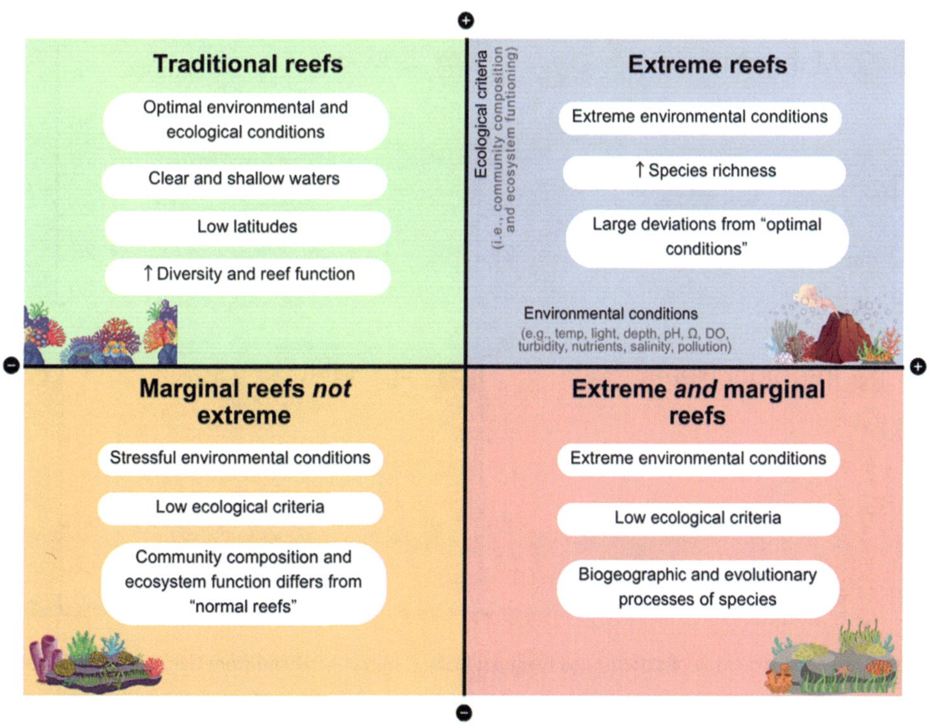

Fig. 1.2 Coral communities and founding organisms in extreme environments or with sub-optimal conditions. (Modified from Schoepf et al. (2023))

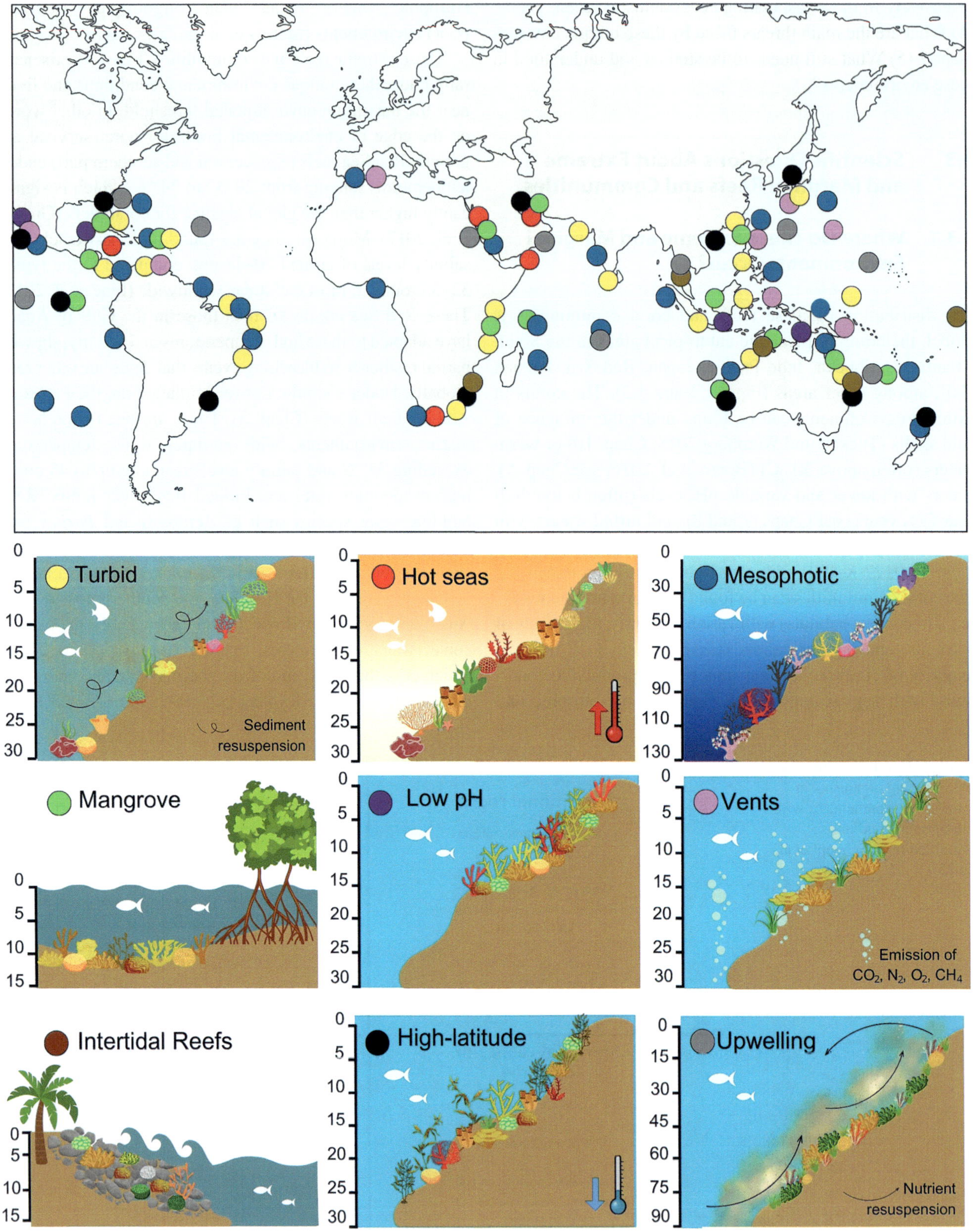

Fig. 1.3 Global distribution of extreme and marginal reefs. (Source: Modified from Camp et al. (2018a, b), Soares et al. (2020), this book)

2002). These reefs host unique species and high levels of endemism (Beger et al. 2014). Some coral species in the genera *Pocillopora* and *Stylophora* have adapted to cooler temperatures and lower light conditions compared to tropical reefs. In Kermadec Islands (New Zealand), some reef habitats are also exposed to cooler waters, typically around 15–25 °C, and seasonal temperature fluctuations (see Chap. 13).

In some areas of the Western Atlantic reefs (Florida, Caribbean and Brazil) (Fig. 1.3), reefs are found under high turbidity levels due to river runoff, coastal development, and periodic upwelling events (Hughes et al. 2003; Suggett et al. 2012; Cruz et al. 2018). Moreover, they are influenced by freshwater influx from rivers and estuaries, leading to fluctuating salinity levels. Brazilian shallow-water reefs are usually dominated by coral species tolerant to moderate sedimentation and turbidity, such as *Siderastrea* spp., *Madracis*, *Montastrea* and the endemic genus *Mussismilia* spp. (Suggett et al. 2012; Leão et al. 2016). The most abundant scleractinian coral in Brazil (*Montastraea cavernosa*), largely dominates (71%–93% of total coral cover) both the shallow turbid and deeper clear-water reefs (Santana et al. 2023). The Amazon mesophotic reefs exhibit high turbidity and significant depths (>80 m deep) under the influence of the Amazon River; one of the world's largest freshwater sources, leading to low light penetration. The reef community is dominated by calcareous algae, sponges and few coral species (Francini-Filho et al. 2018). Corals in this equatorial region have adapted to low light, mesophotic depths, and varying salinity, relying more on heterotrophy than photosymbiosis such as *Madracis decactis* (Francini-Filho et al. 2018).

1.3.2 Reasons why there is Growing Interest in Reefs and Coral Communities at the Edge of Their Environmental Limits

Studies of communities in extreme and marginal coral environments have grown significantly, especially in the last decade, as detected in recent analyses (Camp et al. 2018a, b; Burt et al. 2020; Schoepf et al. 2023). What is the reason for the increase in these studies by the scientific community?

Various research studies have shown that corals in certain extreme environments (such as warmer, turbid, deeper waters or under the influence of seasonal cooler water) have lower bleaching and/or mortality rates after the onset of thermal stress (Tuckett and Wernberg 2018; Fine et al. 2019; Soares et al. 2020; see Chap. 2). While traditional clear-water shallow reefs showed mass coral bleaching and high mortality after thermal stress, there are cases in which extreme reefs showed remarkable resistance or resilience to marine heatwaves and other disturbances (Portugal et al. 2016; Cruz et al. 2018; Freitas et al. 2019). Some of these reef types may bleach less because abiotic stressors are mitigated (e.g. turbid reefs having lower light levels which mitigates heat stress) while others have extreme conditions that enhance heat tolerance of the corals themselves (e.g. highly variable temperature environments). This is the difference between spatial and adaptive refugia, with only the latter being able to serve as potential resilience hotspots (see Chap. 17).

Similar processes occur in arid and semiarid land forests, which have been called "last forests" because they supposedly are more resistant to heatwaves, fires, and global warming than rainforests (De Keersmaecker et al. 2015; Mau et al. 2023; Norlen et al. 2024). This reasoning comes from a human logic in which what "what doesn't kill you makes you stronger" (i.e., stress-hardening). However, recent studies in tropical dry forests indicate that most of their native plant and animal species are close to their limit of thermal tolerance; by 2060, more than 99% of plant assemblages are projected to lose native species (Moura et al. 2023) and 85% of the non-volant mammal species will lose suitable habitats, with one quarter of species projected to completely lose suitable habitats (Moura et al. 2023).

The same rationale has been used to recognize coral species already adapted to extreme conditions as "tough corals" (Camp et al. 2018a, b), which could potentially cope with the new coming conditions imposed by global climate change. Moreover, even "super corals" can be overwhelmed in their tolerance when e.g. a marine heatwave is super-imposed on the already extreme thermal environment (Schoepf et al. 2019; Klepac and Barshis 2020). In this regard, this hypothesis of resistance is flawed, mainly because the speed, frequency and intensity of anthropogenic climate change events might be beyond their capacity to resist.

A recent study (Dimitrijević et al. 2024) examines tropical coral reefs over the last 275 million years and four major climate warming events, finding that hyperthermal events coincided with global reef crises. After these events, coral reefs shifted to deeper mesophotic environments, and expanded into higher latitudes and turbid waters, suggesting these areas served as refuges for corals escaping past warming seas. However, while these shifts offer some hope for coral survival in the face of current climate change, the recovery of full reef ecosystems typically took millions of years (Dimitrijević et al. 2024); which is clearly different from the present anthropogenic climate change. In addition, coral species living in extreme environments are already at the edge of abiotic limits or the limit of their resistance (e.g. Schoepf et al. 2019; Klepac and Barshis 2020). In other words, "what doesn't make you weak makes you stronger, until it kills you" (Schoepf et al. 2023).

Global-scale and rapid human-induced climate change and local impacts are significantly altering marine environmental conditions (Halpern et al. 2019; Smale et al. 2019), affecting: (1) seawater temperature, (2) turbidity, (3) acidity, (4) salinity and (5) oxygen (Pezner et al. 2024). These changing environmental conditions pose a challenge to the conservation of coral reef communities globally (Hughes et al. 2010; Hoegh-Guldberg et al. 2017).

Firstly, the ongoing global warming effects show us that the average global air temperatures have increased ~1.5 °C (WMO 2024) and the oceans heat ~1.0 °C (Collins et al. 2013) since the late nineteenth century (pre-industrial period). The January–September 2024 global mean surface air temperature was 1.54 °C (with a margin of uncertainty of ±0.13 °C) above the pre-industrial average (Fig. 1.4). 2015–2024 will be the warmest 10 years on record and the loss of ice from glaciers, sea-level rise, and ocean heating are accelerating (Fig. 1.4).

Marine heatwaves (MHWs) have intensified globally, with a 54% rise in annual MHW days between 1925 and 2016. Since 2024, the world has faced the most severe thermal stress on record, triggering an ongoing, large-scale coral bleaching event—the fourth globally and the second this decade (Glynn 1996; Hughes et al. 2018; Reimer et al. 2024). By 2100, sea surface temperatures may rise 1–4 °C depending on greenhouse gas emissions, and MHWs could become 50 times more frequent under a 3.5 °C global warming scenario (Smale et al. 2019). Longer and more intense MHWs will severely impact coral reefs worldwide, including extreme environments (Skirving et al. 2019; Sully et al. 2019; Chap. 17). The fourth global coral bleaching (2023-present) represent the most extensive and severe period of coral heat stress ever recorded. Between 2018 and 2025, an uninterrupted episode of global heat stress affected 87% of the world's coral reef areas, with bleaching likely impacting reefs in 98 of the 102 coral reef-bearing countries (Spady et al. 2026).

Coastal development, deforestation, and poor land use are increasing sediment runoff and turbidity, degrading coral habitats (Cruz et al. 2018). Elevated turbidity alters coral community structure and threatens species sensitive to siltation (Schoepf et al. 2023). Even turbid-adapted reefs can shift ecologically when exposed to excessive sedimentation (Leão et al. 2016). Agricultural runoff and wastewater further elevate nutrient loads, promoting harmful algal blooms, while intensified rainfall and storms under climate change may exacerbate sediment and nutrient inputs.

The ocean has absorbed about 30% of anthropogenic CO_2 since the industrial revolution, lowering surface pH from 8.2 to 8.1. This acidification impairs calcifying organisms such as corals, mollusks, and echinoderms. Under high-emission scenarios, pH may drop another 0.3–0.4 units by 2100, further reducing calcification and threatening reef ecosystem stability. See Chaps 7 and 8 for the effects on corals and fish in naturally acidified extreme environments.

Finally, global changes in the water cycle due to climate emergencies are causing some regions to become saltier and others dryer. High evaporation rates increase salinity in some coastal areas, while increased precipitation decreases it in others, including floods and droughts. Freshwater inputs from river diversions, damming, and agricultural runoff can significantly alter local salinity levels, particularly in interconnected estuarine and coastal environments, affecting coral reefs and communities (Soares 2020).

As global and local impacts increasingly alter reef function and reef structure worldwide, research into these habitats and coral communities is extremely relevant, as it can provide useful insights into the future trajectories of traditional, extreme and marginal coral reefs. Extreme reefs and coral communities are unique and distinct in many ways from traditional clear-water reefs, and they are therefore also important in their own right, including conservation strate-

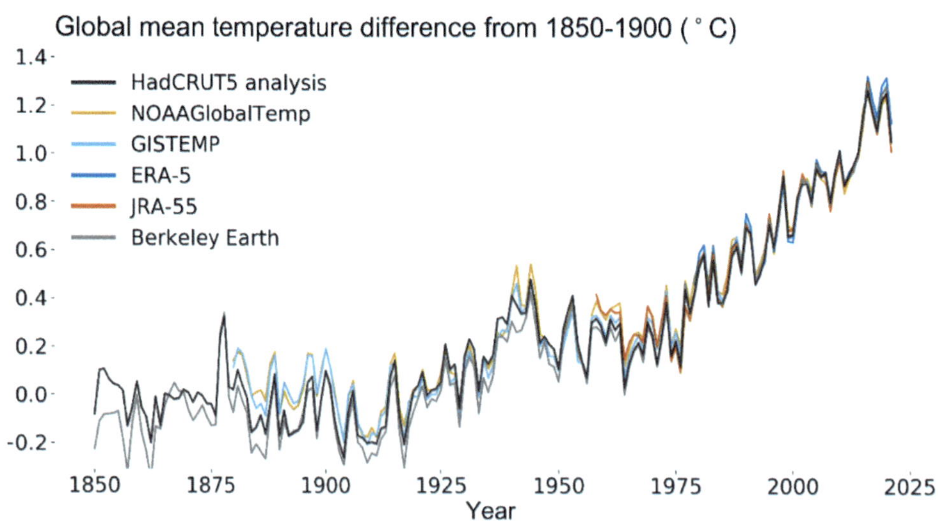

Fig. 1.4 Global annual mean temperature difference from preindustrial conditions (1850–1900) for six international datasets. (Source: WMO (2024))

gies (Camp et al. 2018a, b; Soares 2020; Schoepf et al. 2023). We understand the reasons behind the increase in studies on extreme and/or marginal coral communities over the past decade, but what aspects of these environments have received the most attention?

1.3.3 What Has Been Studied Most in Extreme and Marginal Coral Environments?

Studies on extreme and/or marginal reefs have mainly been conducted at the level of individuals, especially with corals adapted to these unique conditions. Most of the research topics mainly involve: (1) the structural properties of coral skeletons; (2) coral physiology and reproduction; (3) aspects involving symbiosis with microalgae and coral health; and (4) extreme environmental conditions. The smallest number of studies involve aspects of higher levels of biological organization, such as reef communities and ecosystems.

The focus on the structural properties of coral skeletons (e.g., dissolution, density, mineralogy, calcification, cementation, etc.) were investigated in extreme environments such as mangroves, areas with high acidity (*ojos* or close to CO_2 vents), turbid waters and intertidal areas. These field or laboratory studies aim to understand how corals that live and grow in these environments are adapted to greater acidity, for example (Chap. 7). In this case, as a natural analog for future ocean acidification conditions due climate change (Chap. 8). Corals in harsh environmental conditions showcase remarkable adaptations that allow them to survive and maintain ecosystem functions (e.g., carbonate growth and net accretion), albeit living at the edge of abiotic limits (Camp et al. 2018a, b; see Chap. 2).

Studies on coral reproduction usually investigated topics such as spawning, recruitment, and fecundity. Moreover, topics of coral physiology are one of the most researched areas in extreme and marginal reefs such as respiration, proteins, photosynthesis, pH homeostasis, lipids, feeding rates and chlorophyll a (see review of Camp et al. (2018a, b) for some studies). These aspects of population biology and physiology are important for understanding how corals maintain their populations despite significant fluctuation in marine environmental factors such as turbidity, pH, temperature or nutrients. Research topics involving symbiosis with microalgae and coral health are widely studied such as zooxanthellae types, population genetics, diseases, and bacterial communities. These corals are often referred to as "supercorals" because they might exhibit higher tolerance to temperature extremes (Smith et al. 2017), often by hosting heat-resistant (Hume et al. 2015) and/or generalist strains of zooxanthellae, and/or skeletal structures to cope with lower pH (Kurihara et al. 2019), and/or have larger polyps to capture more food particles from the water (Chap. 14). In other words, they may rely more on heterotrophy than photosynthesis (Mies et al. 2018; Conti-Jerpe et al. 2020). Finally, estuarine corals are adapted to fluctuating salinity levels and might have mechanisms to regulate internal ion concentration (Chap. 11). Understanding these differences is crucial for conservation efforts and resilience mechanisms, particularly in the face of climate change and other cumulative anthropogenic stressors (e.g., pollution and deforestation) (Soares 2020).

Environmental studies have been conducted in extreme or marginal coral communities to characterize abiotic factors (e.g. Maggioni et al. 2021; De Jong et al. 2025). In this case, field studies analyze factors such as water movement, turbidity, total alkalinity, temperature, conductivity or salinity, pH, oxygen, nutrients, light etc. Understanding the seasonal fluctuation and spatial variation of these environmental factors allows us to understand the limits of coral adaptation and survival. The greatest scientific interest has been to compare the environmental factors naturally found in these extreme coral ecosystems with forecast climate scenarios based on greenhouse gas emissions.

We used VOSviewer 1.6.19 to visualize and analyse trends in the form of bibliometric maps about the extreme and marginal reefs and communities. We used this software to conduct data mining, mapping, and grouping of topics that were taken from the Web of Science database (van Eck and Waltman 2010). The number of publications was determined based on a Web of Science search (topic = coral AND marginal OR peripheral OR subtropical OR extreme OR turbid OR mesophotic OR vent OR intertidal); year = 1965–2024), with each publication examined for relevance. However, these search terms do not cover all types of extreme reef environments explored in this book (e.g. mangrove and seagrass environments, semi-enclosed bays). We selected 172 publications for analysis on VOSviewer. The minimum number of relationships with terms in the use of VOSviewer was set to 12 terms.

The bibliometric analysis revealed (Fig. 1.5) three clusters (indicated by the green, blue and red colors) highlighting the relationships among studies on extreme and marginal coral communities. The font size and circles in the network represent the frequency of occurrences, with larger circles representing more frequently cited topics in the abstracts. There is a close relationship between the keywords *adaptation, marginal habitat, refugia, population, shallow, structure,* and *mesophotic ecosystem,* which are grouped in the same green cluster area, this reflects an increase in the number of research studies comparing the population structure of shallow and mesophotic coral ecosystems, mainly seeking to learn about their adaptations to greater depths and their capacity to provide refuge for mesophotic coral ecosystems (MCEs) (Chap. 12) in view of the degradation of shallow coral reefs by local and global impacts.

Similar relationships can be seen for the other two clusters, red and blue, on the left side of the figure (Fig. 1.5). The red cluster includes keywords such as *coral species, marginal environment, degree celsius, temperature, condition, coral growth, sediment,* and *turbidity.* This reflects the increase in the number of studies analyzing turbid reefs with significant sedimentation, in particular aiming to understand how different coral species adapt to these extreme environmental conditions. These turbid-zones, like the MCEs, have been much discussed in terms of their climate change refugia potential due to the hypothesis that turbidity can reduce bleaching and heat-stress through shading (see Chap. 9).

The blue cluster addresses keywords such as *bleaching, resilience, stress, climate change, response, effect, climate change* and *future* (Fig. 1.5). This cluster is connected to other themes in the red and green clusters. In other words, the main research focus has been on understanding the resilience of different extreme coral systems to climate change (see Chaps. 2 and 17), including turbid and MCEs, and what the effect and responses to these stresses have been, especially in terms of understanding future scenarios of global change. The density figure shows that the majority of published papers cite these keywords and goals (Fig. 1.5).

1.3.4 What Are the Remaining Knowledge Gaps in Extreme and Marginal Coral Environments?

As we have seen in the previous sections of this chapter, most research studies on extreme and marginal coral communities have been carried out due their current-day resistance to marine heatwaves or potential climate-change refugia. In addition, studies of hottest reefs such as the Persian-Arabian Gulf (see Chap. 5) and the Red Sea, have been increasing (Burt et al. 2011, 2013, 2019), mainly as analogues of future global warming and hotter waters. Our analysis in the previous section also showed us that the research focus has been on experimental, physiological and reproductive studies, coral-algae symbiosis, or on understanding the environmental conditions in which these coral communities live. Research carried out at higher levels of biological hierarchy, such as communities and ecosystems, or other reef organisms (e.g., microorganisms and fishes; Chaps. 3 and 4) are still rare in these unique ecosystems.

Since the first research studies on coral reefs, including those by Charles Darwin on the Beagle, technological and scientific advances have enabled new approaches to be conducted on extreme and marginal reefs and their coral

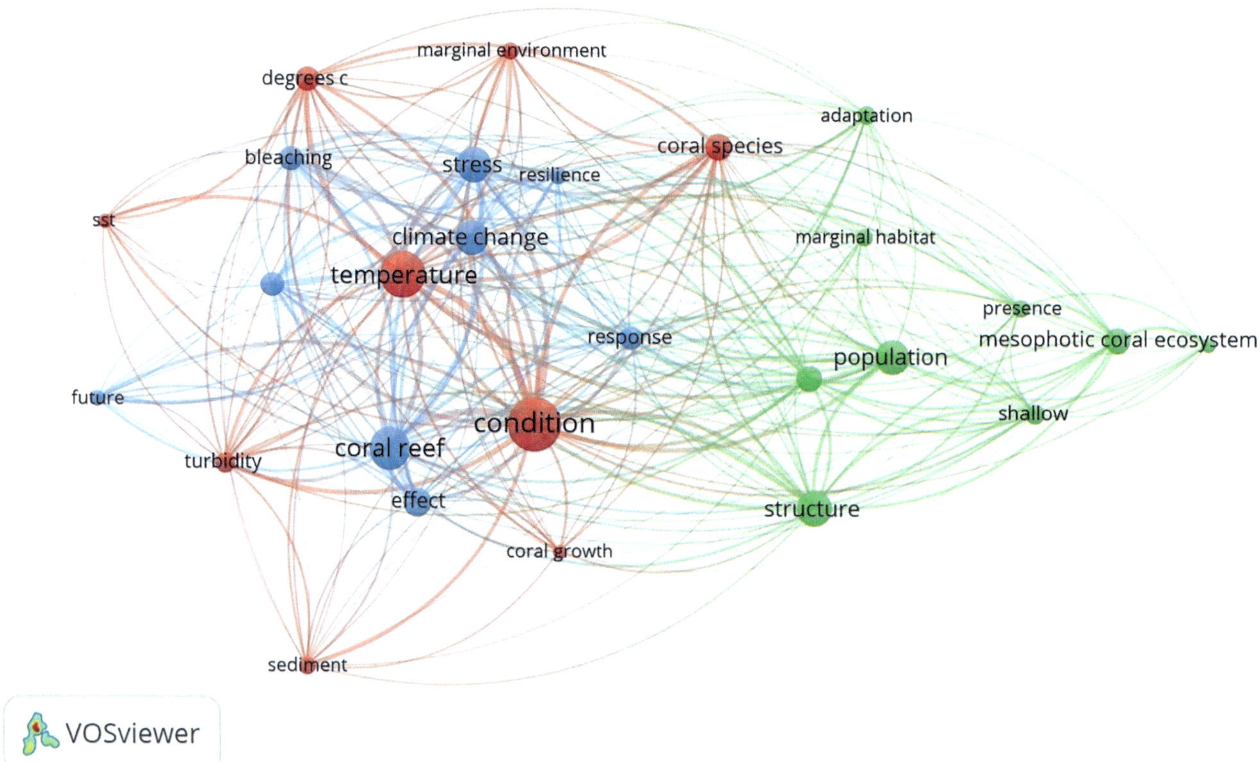

Fig. 1.5 Bibliometric network visualization of research topics related to some extreme and marginal reefs and communities, generated with VOSviewer

communities. This hot topic was discussed in the recent reviews led by Emma Camp et al. (2018a, b), John Burt et al. (2020) and Verena Schoepf et al. (2023). Thus, this section aims to compile what has been discussed previously and complement these suggestions for future research (Table 1.1).

The study of taxonomic, functional and phylogenetic diversity on marginal and extreme reefs is a topic that needs to be better understood (Table 1.1). These studies have advanced on traditional clear-water reefs, while studies on non-traditional reefs have focused primarily on taxonomic diversity (Soares et al. 2020) with only a few exceptions addressing functional diversity (Bender et al. 2017 Chaps. 2 and 3). Functional studies are important because they can help us understand the balance among fundamental ecosystem processes, such as herbivory/ primary production, nutrient uptake/release, some of the key processes highlighted by Brandl et al. (2019). These processes are carried out by species from different taxa and are therefore influenced by species composition and abundance.

Table 1.1 Priorities for future research on extreme and marginal coral communities

Future research priorities on extreme and marginal coral communities
Individual to organismal level
Investigate eco-physiological mechanisms underlying variability in stress responses of reef organisms under global and local environmental changes.
Examine the time scales and molecular mechanisms that enable enhanced resistance and long-term survival, and determine how long favorable traits persist in novel environments.
Experimentally test whether organisms from marginal and/or extreme reefs exhibit elevated resistance to sustained (chronic) climate stressors.
Assess trophic ecology, reproduction, and growth across a wider range of reef species beyond the commonly studied genera.
Population to community level
Study interactions among key community members (e.g., microorganisms, sponges, corals, crustose coralline algae, macroalgae, fishes) and evaluate how these relationships shift under changing abiotic conditions.
Evaluate the role of non-coral taxa in shaping ecological dynamics and influencing future reef trajectories.
Analyze taxonomic, functional, and phylogenetic diversity within marginal and extreme reef communities.
Ecosystem to landscape/seascape level
Determine multiple aspects of reef functioning in marginal and extreme coral communities, including productivity, nutrient cycling, and resilience metrics.
Assess reef geology and geo-ecological functions, including reef accretion and erosion through reef budget analysis.
Conduct long-term monitoring integrating high-resolution environmental and ecological data to track the structure and dynamics of extreme and marginal reefs.
Improve understanding of spatial and adaptive refugia, connectivity patterns, and tropicalisation processes across reef networks.

Sources: Camp et al. (2018a, b), Burt et al. (2020); Soares et al. (2020); Schoepf et al. (2023), the authors

To better understand the flow of energy and matter within ecosystems and across environments over time and space, it is necessary to access species traits related to these processes, focusing on analogous traits and functions that can be expressed in a common currency to enable comparison and integration. Similarly, phylogenetic studies can provide important information about the diversity of these ecosystems. However, in order to make progress in this research agenda, long-term monitoring is needed at important sites at each ocean basin, as well as collecting analogous cross-taxa species traits that are related to key ecosystem processes. Species traits should be collected across different taxa (e.g., from algae to fishes) in a way that allows us to integrate these traits (using analogous/common trait metrics) and along a common currency to account for important ecological processes in reef systems (e.g., rate-based measures of the core ecological processes). Monitoring site networks and online free-access information repositories have advanced for corals (https://www.coraltraits.org/) and mesophotic reefs (such as mesophotic.org) but are uncommon for other non-traditional reef systems. Therefore, long-term monitoring programs are needed to collect environmental and ecological data in order to understand the structure and dynamics of marginal and extreme reefs (Table 1.1).

Long-term programs can also involve aspects of reef geology (see Chap. 15). Reefs and communities in extreme or marginal environments are located in a vast geological diversity (i.e., geodiversity). This ranges from active biogenic reefs, drowned reefs in deeper waters or even the occurrence of coral communities in CO_2 vents, rocky shores, sandstone reefs and estuarine areas. Understanding the geological history in the Holocene and in different geological times is important for predicting the future trajectories of these extreme coral communities in the Anthropocene. In addition, understanding the geo-ecological functions in these reef environments such as reef framework production, sediment generation, maintenance of reef habitat and reef growth potential needs to be better explored in these unique reefs (Chap. 16). Topics such as reef budgets have been explored worldwide, but are still little explored in marginal and extreme reefs. Data has shown that reef growth in marginal settings can rapidly turn-on or -off according to regional environmental changes such as studies under high-latitude reefs (Pereira-Filho et al. 2021).

Reef carbonate accretion can be determined by the growth of reef-building corals (see Chap. 16), but also by other builders such as calcareous algae and bryozoans, which have already been identified as important in extreme and/or marginal reefs. In this context, physiological research into the growth of corals under natural stress conditions in extreme environments and under the effect of ongoing climate emergency is a relevant topic. Understanding trophic ecology such as spatial and seasonal changes in heterotrophy and autotrophy rates, as well as coral growth and reproduction processes, can help us understand reef functioning. Currently, most studies have focused on a few genera such as *Acropora* and *Porites*. This demonstrates the need to expand to other coral genera that are important carbonate builders worldwide (Table 1.1).

In extreme and marginal reefs, the main benthic groups studied to date are corals, crustose coralline algae, sponges and macroalgae. Studies on traditional shallow and clear-water reefs seek to understand the phase-shift from corals to macroalgae, which leads to a drastic change in the reef structure and functioning (Mumby 2009; Bell et al. 2021). However, in extreme reefs, sponges, zoantharians, and/or coralline algae, as the main engineering organisms, naturally exhibit high coverage. Also, in marginal reefs, there are ecological shifts among the dominant benthic groups (Cruz et al. 2018). Therefore, research on the role of non-coral organisms is fundamental to the understanding of the ecological dynamics and future trajectories of these unique reefs. Thus, baseline ecological studies with a long-term approach are essential to understand these patterns, the interactions of these benthic groups and how changes in their dominance affect reef-associated organisms such as fishes and small invertebrates (Table 1.1).

Extreme environments may function as important resilience hotspots (see Chap. 17) in the face of climate change or even host coral species through the tropicalization phenomenon (see Chap. 13) that is occurring on high latitude reefs (Figueira and Booth 2010; Tuckett et al. 2017; Tuckett and Wernberg 2018; González-Delgado et al. 2018) (Table 1.1). In both cases, connectivity plays an important role as these reefs can be important stepping stone habitats for range-shift species distribution and recovering populations affected by local impacts or extreme climate events. Thus, better understanding ecological and genetic connectivity and the role of extreme and marginal reefs is a relevant research topic.

1.4 Conclusions

This initial chapter opens the first book on extreme and marginal environments and their associated coral communities. Since the time of the first naturalists, including Charles Darwin (as referred to at the start of the chapter), scientists have been fascinated by coral reefs. Much remains to be understood about these reef systems, which are threatened by global climate change and local impacts in tropical and subtropical regions. Understanding the extreme and marginal ecosystems and their ecological and socio-economic relevance, as well as traditional reefs, can provide important insights for public policy. In addition to corals, it is essential to understand how other reef organisms such as fishes and microorganisms live and act in the dynamics of these reef ecosystems.

The resistance (or resilience) of organisms to extreme environmental conditions on these reefs awakened the worldwide interest in the subject. It used to be thought that these abiotic conditions were suboptimal, but the presence and abundance of organisms such as corals suggests otherwise. Living on the edge of environmental conditions has led to the occupation of a specific niche and the formation of unique habitats. However, living on the edge of the environmental limits comes with high risks, especially in today's rapidly changing ocean. We can no longer neglect these organisms and ecosystems in extreme habitats or see them just as a lifeboat for traditional reefs. Evaluating them only as reflections in a mirror of the future of traditional reefs is also inadequate. The diversity of reef forms and the current distribution of reef organisms must be understood as a whole, including these unique extreme reefs and communities, with a science-based view to conservation strategies for these singular marine forests.

Acknowledgments This book chapter celebrates the United Nations (UN) Decade of Ocean Science for Sustainable Development (2021–2030) and is the first book on extreme and marginal reefs and coral communities of the series *Coral Reefs of the World* (Springer-Nature). We hope that this decade will provide a 'once in a lifetime' opportunity to create a science-based foundation to society, across the science-policy interface, to strengthen the governance of our oceans for the benefit of humankind and all coral reefs. We wish to acknowledge the two reviewers for comments on our draft, which significantly improved this book chapter. MOS thanks the financial support provided by the Conselho Nacional de Desenvolvimento Científico e Tecnológico—CNPq (Grants 442337/2020-5, 445974/2024-9, 444172/2024-6, 448511/2025-8), FUNCAP Chief-Scientist Program, FUNCAP PELD (PS1-0186-00295.01.00/2), CAPES-GLOBAL Program, INTEGRAMAR, Conhecimento Brasil, CAPES-COFECUB, and PELD Costa Semiárida do Brasil. MOS, GL, MB, and BAS thanks CNPq Research Productivity Fellowship.

References

Araujo HFP, Canassa NF, Machado CCC (2023) Human disturbance is the major driver of vegetation changes in the Caatinga dry forest region. Sci Rep 13:18440. https://doi.org/10.1038/s41598-023-45571-9

Bak RPM, Meesters EH (2000) Acclimatization/adaptation of coral reefs in a marginal environment. In: Proceedings 9th international coral reef symposium, Bali, Indonesia. http://coremap.or.id/downloads/ICRS9th-definitiefJPEGfigs.pdf

Beck MW, Losada IJ, Menéndez P, Reguero BG, Díaz-Simal P, Fernández F (2018) The global flood protection savings provided by coral reefs. Nat Commun 9(1):1–7. https://doi.org/10.1038/s41467-018-04568-z

Beger M, Sommer B, Harrison PL, Smith SD, Pandolfi JM (2014) Conserving potential coral reef refuges at high latitudes. Divers Distrib 20(3):245–257. https://doi.org/10.1111/ddi.12140

Bell JJ, Micaroni V, Strano F (2021) Regime shifts on tropical coral reef ecosystems: future trajectories to animal-dominated states in response to anthropogenic stressors. Emerg Top Life Sci 6(1):95–106. https://doi.org/10.1042/ETLS20210231

Bender MG, leprieur F, Mouillot D, Kulbicki M, Parravicini V, Pie MR, Barneche DR, Oliveira- Santos LGR, Floeter SR (2017) Isolation drives taxonomic and functional nestedness in tropical reef fish faunas. Ecography 40(3):425–435. https://doi.org/10.1111/ecog.02293

Bento R, Hoey AS, Bauman AG, Feary DA, Burt JA (2016) The implications of recurrent disturbances within the world's hottest coral reef. Mar Pollut Bull 105(2):466–472. https://doi.org/10.1016/j.marpolbul.2015.10.006

Blackie R, Baldauf C, Gautier D, Gumbo D, Kassa H, Parthasarathy N, Paumgarten F, Sola P, Pulla S, Waeber P, Sunderland T (2014) Tropical dry forests: The state of global knowledge and recommendations for future research. Discussion Paper 2. CIFOR, Bogor. http://www.cifor.org/publications/pdf_files/WPapers/DPBlackie1401.pdf

Bongaerts P, Riginos C, Brunner R, Englebert N, Smith SR, Hoegh-Guldberg O (2017) Deep reefs are not universal refuges: reseeding potential varies among coral species. Sci Adv 3(2):e1602373. https://doi.org/10.1126/sciadv.1602373

Brandl SJ, Rasher DB, Côté IM, Casey JM, Darling ES, Lefcheck JS, Duffy JE (2019) Coral reef ecosystem functioning: eight core processes and the role of biodiversity. Front Ecol Environ 17(8):445–454. https://doi.org/10.1002/fee.2088

Burt J, Al-Harthi S, Al-Cibahy A (2011) Long-term impacts of coral bleaching events on the world's warmest reefs. Mar Environ Res 72(4):225–229. https://doi.org/10.1016/j.marenvres.2011.08.005

Burt JA, Al-Khalifa K, Khalaf E, AlShuwaikh B, Abdulwahab A (2013) The continuing decline of coral reefs in Bahrain. Mar Pollut Bull 72(2):357–363. https://doi.org/10.1016/j.marpolbul.2012.08.022

Burt J, Paparella F, Al-Mansoori N, Al-Mansoori A, Al-Jailani H (2019) Causes and consequences of the 2017 coral bleaching event in the southern Persian/Arabian Gulf. Coral Reefs 38(4):567–589. https://doi.org/10.1007/s00338-019-01767-y

Burt JA, Camp EF, Enochs IC, Johansen JL, Morgan KM, Riegl B, Hoey AS (2020) Insights from extreme coral reefs in a changing world. Coral Reefs 39:495–507. https://doi.org/10.1007/s00338-020-01966-y

Camp EF, Schoepf V, Mumby PJ, Hardtke LA, Rodolfo-Metalpa R, Smith DJ, Suggett DJ (2018a) The future of coral reefs subject to rapid climate change: lessons from natural extreme environments. Front Mar Sci 5(4):1–15. https://doi.org/10.3389/fmars.2018.00004

Camp EF, Schoepf V, Suggett DJ (2018b) How can "super corals" facilitate global coral reef survival under rapid environmental and climatic change? Glob Chang Biol 24(7):2755–2757

Collins M, Knutti R, Arblaster J, Dufresne JL, Fichefet T, Friedlingstein P et al (2013) Long-term climate change: projections, commitments and irreversibility. In: Climate change 2013: the physical science basis: contribution of Working Group I to the fifth assessment report of the Intergovernmental Panel On Climate Change. Cambridge University Press, pp 1029–1136

Conti-Jerpe I, Thompson PD, Wong CWM, Oliveira NL, Duprey NN, Moynihan MA, Baker DM (2020) Trophic strategies and bleaching resistance in reef-building corals. Sci Adv 6(15):eaaz5443. https://doi.org/10.1126/sciadv.aaz5443

Costanza R, de Groot R, Sutton P, der Ploeg SV, Anderson SJ, Kubiszewiski I, Farber S, Turner RK (2014) Changes in the global value of ecosystem services. Glob Environ Chang 26:152–158. https://doi.org/10.1016/j.gloenvcha.2014.04.022

Cruz ICS, Waters LG, Kikuchi RKP, Leão ZMAN, Turra A (2018) Marginal coral reefs show high susceptibility to phase shift. Mar Pollut Bull 135:551–561. https://doi.org/10.1016/j.marpolbul.2018.07.043

Darwin C (1889) The structure and distribution of coral reefs. University of California Press, Berkeley

De Jong C, Van Os I, Sepúlveda-Rodríguez G, de Baat ML, Schoepf V (2025) High-resolution temporal assessment of physicochemical variability and water quality in tropical semi-enclosed bays and coral reefs. Sci Total Environ 968:178810. https://doi.org/10.1016/j.scitotenv.2025.178810

De Keersmaecker W, Lhermitte S, Tits L, Honnay O, Somers B, Coppin P (2015) A model quantifying global vegetation resistance and resilience to short-term climate anomalies and their relationship with vegetation cover. Glob Ecol Biogeogr 24:539–548. https://doi.org/10.1111/geb.12279

Dexter KG, Smart B, Baldauf C, Baker TR, Balinga MPB, Brienen RJW, Fauset S, Feldpausch TR, Silva LF-D, Muledi JI, Lewis SL, Lopez-Gonzalez G, Marimon-Junior BH, Marimon BS, Meerts P, Page N, Parthasarathy N, Phillips OL, Sunderland TCH et al (2015) Floristics and biogeography of vegetation in seasonally dry tropical regions. Int For Rev 17(Supplement 2):10–32. https://doi.org/10.1505/146554815815834859

Dimitrijević D, Santodomingo N, Kiessling W (2024) Reef refugia in the aftermath of past episodes of global warming. Coral Reefs 43:1431–1442. https://doi.org/10.1007/s00338-024-02548-y

Dubinsky Z, Stambler N (2011) Coral reefs: an ecosystem in transition, vol 552. Springer, Dordrecht. https://doi.org/10.1007/978-94-007-0114-4

Dudgeon SR, Aronson RB, Bruno JF, Precht WF (2010a) Phase shifts and stable states on coral reefs. Mar Ecol Prog Ser 413:201–216. https://doi.org/10.3354/meps08751

Dudgeon SR, Aronson RB, Bruno JF, Precht WF (2010b) Phase shifts and stable states on coral reefs. Mar Ecol Prog Ser 413:201–2016. https://doi.org/10.3354/meps08751

Elliff CI, Silva IR (2017) Coral reefs as the first line of defense: shoreline protection in face of climate change. Mar Environ Chang 127:148–154. https://doi.org/10.1016/j.marenvres.2017.03.007

Figueira WF, Booth DJ (2010) Increasing ocean temperatures allow tropical fishes to survive overwinter in temperate waters. Glob Chang Biol 16(2):506–516. https://doi.org/10.1111/j.1365-2486.2009.01934.x

Fine M, Cinar M, Voolstra CR, Safa A, Rinkevich B, Laffoley D, Hilmi N, Allemand D (2019) Coral reefs of the Red Sea—challenges and potential solutions. Reg Stud Mar Sci 25:100498. https://doi.org/10.1016/j.rsma.2018.100498

Fisher R, O'Leary RA, Low-Choy S, Mengersen K, Knowlton N, Brainard RE, Caley MJ (2015) Species richness on coral reefs and the pursuit of convergent global estimates. Curr Biol 25(4):500–505. https://doi.org/10.1016/j.cub.2014.12.022

Francini-Filho R, Asp NE, Siegle E, Hocevar J, Lowyck K, D'Avila N, Vasconcelos AA, Baitelo R, Rezende CE, Omachi CY, Thompson CC, Thompson FL (2018) Perspectives on the great Amazon reef: extension, biodiversity, and threats. Front Mar Sci 5:142. https://doi.org/10.3389/fmars.2018.00142

Freitas LM, Oliveira MDM, Leão ZMAN, Kikuchi RKP (2019) Effects of turbidity and depth on the bioconstruction of the Abrolhos reefs. Coral Reefs 38(2):241–253. https://doi.org/10.1007/s00338-019-01770-3

Glynn PW (1996) Coral reef bleaching: facts, hypotheses and implications. Glob Chang Biol 2:495–509. https://doi.org/10.1111/j.1365-2486.1996.tb00063.x

González-Delgado S, López C, Brito A, Clemente S (2018) Marine community effects of two colonial zoanthids in intertidal habitats of the Canary Islands. Reg Stud Mar Sci 23:23–31. https://doi.org/10.1016/j.rsma.2018.03.006

Grottoli AG, Tchernov D, Winters G (2017) Physiological and biogeochemical responses of super-corals to thermal stress from the northern Gulf of Aqaba, Red Sea. Front Mar Sci 4:1–10. https://doi.org/10.3389/fmars.2017.00215

Halpern BS, Frazier M, Afflerbach J, Lowndes JS, Micheli F, O'Hara C, Scarborough C, Selkoe KA (2019) Recent pace of change in human impact on the world's ocean. Sci Rep 9:11609. https://doi.org/10.1038/s41598-019-47201-9

Harriott VJ, Banks SA (2002) Latitudinal variation in coral communities in eastern Australia: a qualitative biophysical model of factors regulating coral reefs. Coral Reefs 21:83–94. https://doi.org/10.1007/s00338-001-0201-x

Hoegh-Guldberg O, Poloczanska ES, Skirving W, Dove S (2017) Coral reef ecosystems under climate change and ocean acidification. Front Mar Sci 4:158. https://doi.org/10.3389/fmars.2017.00158

Hughes TP et al (2003) Climate change, human impacts, and the resilience of coral reefs. Science 301:929–933

Hughes TP, Graham NAJ, Jackson JBC, Mumby PJ, Steneck RS (2010) Rising to the challenge of sustaining coral reef resilience. Trends Ecol Evol 25(11):633–642. https://doi.org/10.1016/j.tree.2010.07.011

Hughes TP, Anderson KD, Connolly SR, Heron SF, Kerry JT, Lough JM, Baird AH, Baum JK, Berumen ML, Bridge TC, Claar DC, Eakin CM, Gilmour JP, Graham NAJ, Harrison H, Hobbs JPA, Hoey AS, Hoogenboom M, Lowe RJ, McCulloch MT, Pandolfi JM, Pratchett M, Schoepf V, Torda G, Wilson SK (2018) Spatial and temporal patterns of mass bleaching of corals in the Anthropocene. Science 359(6371):80–83. https://doi.org/10.1126/science.aan8048

Hume BCC, D'Angelo C, Smith EG, Stevens SR, Burt J, Wiedenmann J (2015) *Symbiodinium thermophilum* sp. nov., a thermotolerant symbiotic alga prevalent in corals of the world's hottest sea, the Persian/Arabian Gulf. Sci Rep 5:8562. https://doi.org/10.1038/srep08562

Klepac CN, Barshis DJ (2020) Reduced thermal tolerance of massive coral species in a highly variable environment. Proc Biol Sci 287(1933):20201379. https://doi.org/10.1098/rspb.2020.1379

Kleypas JA, McManus JW, Menez LAB (1999) Environmental limits to coral reef development: where do we draw the line? Am Zool 39:146–159. https://doi.org/10.1093/icb/39.1.146

Kurihara H, Wouters J, Yasuda N (2019) Seasonal calcification of the coral Acropora digitifera from a subtropical marginal Okinawa reef under ocean acidification. Coral Reefs 38(3):443–454. https://doi.org/10.1007/s00338-019-01794-9

Lapola DM, Pinho P, Barlow J, Aragão LE, Berenguer E, Carmenta R et al (2023) The drivers and impacts of Amazon forest degradation. Science 379(6630):eabp8622

Leão ZMAN, Kikuchi RKP, Ferreira P, Neves EG, Sovieroski HH, Oliveira MD, Maida M, Correia MD, Johnsson R (2016) Brazilian coral reefs in a period of global change: a synthesis. Braz J Oceanogr 64(2):97–116. https://doi.org/10.1590/S1679-875920160916064sp2

Maggioni F, Pujo-Pay M, Aucan J, Cerrano C, Calcinai B, Payri C, Benzoni F, Letourneur Y, Rodolfo-Metalpa R (2021) The Bouraké semi-enclosed lagoon (New Caledonia)–a natural laboratory to study the lifelong adaptation of a coral reef ecosystem to extreme environmental conditions. Biogeosciences 18(18):5117–5140

Mau Y, Feuer E, Preisler Y (2023) Tree resilience to extreme heatwave events in a semi-arid forest. EGU General Assembly 2023, Vienna, EGU23-12091. https://doi.org/10.5194/egusphere-egu23-12091

Medina E, Klinge H (1983) Produtividade de florestas tropicais e bosques tropicais. In: Lange OL, Nobel PS, Osmond CB, Ziegler H (eds) Physiological plant ecology IV. Encyclopedia of plant physiology, vol 12/D. Springer, Berlin. https://doi.org/10.1007/978-3-642-68156-1_10

Mies M, Güth AZ, Tenório AA, Banha TNS, Waters LG, Polito PS, Taniguchi S, Bícego MC, Sumida PYG (2018) In situ shifts of predominance between autotrophic and heterotrophic feeding in the reef-building coral *Mussismilia hispida*: an approach using fatty acid trophic markers. Coral Reefs 37(6):677–689. https://doi.org/10.1007/s00338-018-1692-z

Morais RA, Patricio-Valerio L, Narvaez P, Parravicini V, Brandl SJ (2025) Rethinking Darwin's coral reef paradox and the ubiquity of "marine oases". Curr Biol 35(13):3241–3250.e6. https://doi.org/10.1016/j.cub.2025.05.033

Morais J, Medeiros AP, Santos BA (2018) Research gaps of coral ecology in a changing world. Mar Environ Res 140:243–250. https://doi.org/10.1016/j.marenvres.2018.06.021

Morais J, Cardoso APL, Santos BA (2022) A global synthesis of the current knowledge on the taxonomic and geographic distribution of major coral diseases. Environ Adv 8:100231

Moura MR, Oliveira GA, Paglia AP, Pires MM, Santos BA (2023) Climate change should drive mammal defaunation in tropical dry forests. Glob Chang Biol 29(24):6931–6944. https://doi.org/10.1111/gcb.16979

Mumby PJ (2009) Phase shifts and the stability of macroalgal communities on Caribbean coral reefs. Coral Reefs 28(3):761–773. https://doi.org/10.1007/s00338-009-0506-8

Norlen CA, Hemes KS, Wang JA, Randerson JT, Battles JJ, Tubbesing CL, Goulden ML (2024) Recent fire history enhances semi-arid conifer forest drought resistance. For Ecol Manag 573:122331. https://doi.org/10.1016/j.foreco.2024.122331

Pereira-Filho GH, Mendes VR, Perry CT, Shintate GI, Niz WC, Sawakuchi AO, Bastos AC, Giannini PCF, Motta FS, Millo C, Paula-Santos GM, Moura RL (2021) Growing at the limit: reef growth sensitivity to climate and oceanographic changes in the South Western Atlantic. Glob Planet Change 201:103479. https://doi.org/10.1016/j.gloplacha.2021.103479

Perry CT, Larcombe P (2003) Marginal and non-reef-building coral environments. Coral Reefs 22(4):427–432. https://doi.org/10.1007/s00338-003-0330-5

Pezner AK, Courtney TA, Chou WC, Chu HC, Frable BW, Kekuewa SA, Soong K, Wei Y, Andersson AJ (2024) Coral growth along a natural gradient of seawater temperature, pH, and oxygen in a nearshore seagrass bed on Dongsha atoll, Taiwan. PLoS One 19(10):e0312263

Portugal AB, Carvalho FL, PBM C, Rossi S, Soares MO (2016) Increased anthropogenic pressure decreases species richness in tropical intertidal reefs. Mar Environ Res 120:44–54. https://doi.org/10.1016/j.marenvres.2016.07.005

Potts DC, Jacobs JR (2000) Evolution of reef-building scleractinian corals in turbid environments: a paleo-ecological hypothesis. In: Proceedings 9th International Coral Reef Symposium (Bali), pp 249–254

Reimer JD, Peixoto RS, Davies SW, Traylor-Knowles N, Short ML, Cabral-Tena RA, Voolstra CR (2024) The fourth global coral bleaching event: where do we go from here? Coral Reefs 43(4):1121–1125. https://doi.org/10.1007/s00338-024-02504-w

Rossi S, Bramanti L, Gori A, Orejas C (2017) Animal forests of the world: an overview. In: Rossi S, Bramanti L, Gori A, Orejas C (eds) Marine animal forests. Springer, Cham. https://doi.org/10.1007/978-3-319-21012-4_1

Santana EFC, Mies M, Longo GO, Menezes R, Aued AW, Luza AL (2023) Turbidity shapes shallow southwestern Atlantic benthic reef communities. Mar Environ Res 183:105807. https://doi.org/10.1016/j.marenvres.2022.105807

Schoepf V, Carrion SA, Pfeifer SM, Naugle M, Dugal L, Bruyn J, McCulloch MT (2019) Stress-resistant corals may not acclimatize to ocean warming but maintain heat tolerance under cooler temperatures. Nat Commun 10:4031. https://doi.org/10.1038/s41467-019-12065-0

Schoepf V, Baumann JH, Barshis DJ, Browne NK, Camp EF, Comeau S et al (2023) Corals at the edge of environmental limits: a new conceptual framework to re-define marginal and extreme coral communities. Sci Total Environ 884:163688. https://doi.org/10.1016/j.scitotenv.2023.163688

Silva JMC, Leal IR, Tabarelli M (2017) Caatinga: the largest tropical dry Forest region in South America. Springer, Cham, 482p. https://doi.org/10.1007/978-3-319-68339-3

Sing Wong A, Vrontos S, Taylor ML (2022) An assessment of people living by coral reefs over space and time. Glob Chang Biol 28(23):7139–7153. https://doi.org/10.1111/gcb.16391

Skirving WJ, Heron SF, Marsh BL, Liu G, De La Cour JL, Geiger EF, Eakin CM (2019) The relentless march of mass coral bleaching: a global perspective of changing heat stress. Coral Reefs 38(4):547–557. https://doi.org/10.1007/s00338-019-01799-4

Smale DA, Wernberg T, Oliver ECJ, Thomsen MS, Harvey BP, Straub SC, Burrows M, Alexander LV, Benthuysen J, Donat M, Hobday AJ, Holbrook NJ, Perkins-Kirkpatrick SE, Scannell H, Gupta AS, Payne BL, Moore PJ (2019) Marine heatwaves threaten global biodiversity and the provision of ecosystem services. Nat Clim Chang 9(4):306–312. https://doi.org/10.1038/s41558-019-0412-1

Smith EG, Hume BCC, Delaney P, Wiedenmann J, Burt JA (2017) Genetic structure of coral-*Symbiodinium* symbioses on the world's harmest reefs. PLoS One 12(6):e0180169. https://doi.org/10.1371/journal.pone.0180169

Soares MO (2020) Marginal reef paradox: a possible refuge from environmental changes? Ocean Coast Manag 185:105063. https://doi.org/10.1016/j.ocecoaman.2019.105063

Soares MO, Araújo JT, Ferreira SMC, Santos BA, Boavida JR, Costantini F, Rossi S (2020) Why do mesophotic coral ecosystems have to be protected? Sci Total Environ 726:138456. https://doi.org/10.1016/j.scitotenv.2020.138456

Spady BL, Skirving WJ, De La Cour JL, Geiger EF, Liu G, Hoegh-Guldberg O, Norrie A, Heron SF, Pomeroy MW, Kolodziej G, Brown KT, Manzello DP (2026) The 4th global coral bleaching event: ushering in an era of near-annual bleaching. Coral Reefs. https://doi.org/10.1007/s00338-025-02810-x

Spalding M, Burke L, Wood SA, Ashpole J, Hutchinson J, Ermgassen PZ (2017) Mapping the global value and distribution of coral reef tourism. Mar Policy 82:104–113. https://doi.org/10.1016/j.marpol.2017.05.014

Stan K, Sanchez-Azofeifa A (2019) Tropical dry Forest diversity, climatic response, and resilience in a changing climate. Forests 10(5):443. https://doi.org/10.3390/f10050443

Suggett DJ, Kikuchi RKP, Oliveira MDM, Spanó S, Carvalho R, Smith DJ (2012) Photobiology of corals from Brazil's nearshore marginal reefs of Abrolhos. Mar Biol 159:1461–1473. https://doi.org/10.1007/s00227-012-1925-6

Sully S, Burkepile DE, Donovan MK, Hodgson G, Woesik RV (2019) A global analysis of coral bleaching over the past two decades. Nat Commun 10:1264. https://doi.org/10.1038/s41467-019-09238-2

Sunderland T, Guimarães Apgaua DM, Baldauf C, Blackie R, Colfer CJP, Colfer CJP, Cunningham AB, Dexter KG, Dexter KG, Djoudi H, Gautier D, Gautier D, Gumbo D, Ickowitz A, Kassa H, Parthasarathy N, Pennington RT, Paumgarten F et al (2015) Global dry forests: a prologue. Int For Rev 17(2):1–9. https://doi.org/10.1505/146554815815834813

Tuckett CA, Wernberg T (2018) High latitude corals tolerate severe cold spell. Front Mar Sci 5:1–5. https://doi.org/10.3389/fmars.2018.00014

Tuckett CA, de Bettignies T, Fromont J, Wernberg T (2017) Expansion of corals on temperate reefs: direct and indirect effects of marine heatwaves. Coral Reefs 36(3):947–956. https://doi.org/10.1007/s00338-017-1586-5

van Eck NJ, Waltman L (2010) Software survey: VOSviewer, a computer program for bibliometric mapping. Scientometrics 84(2):523–538

Wang L, Shantz AA, Payet JP, Sharpton TJ, Foster A, Burkepile DE, Thurber RV (2018) Corals and their microbiomes are differently affected by exposure to elevated nutrients and a natural thermal anomaly. Front Mar Sci 5:1–14. https://doi.org/10.3389/fmars.2018.00101

WMO (2024) 2024 is on track to be hottest year on record as warming temporarily hits 1.5°C. WMO. Available at https://wmo.int/news/media-centre/2024-track-be-hottest-year-record-warming-temporarily-hits-15degc

The Ecology of Corals in Extreme and Marginal Settings

2

Miguel Mies

Abstract

Corals are reef-building cnidarians that support diverse marine life. In extreme environmental settings, characterized by suboptimal conditions like high turbidity, extreme salinity, or low light, corals face unique physiological challenges. Although species richness is often lower, many exhibit remarkable resilience. Studying these marginal species is vital for anticipating coral responses to climate change. Oceanographic factors including currents, wind, stratification, river discharge, and tides shape extreme coral environments by influencing temperature, light, salinity, and sedimentation. Processes like upwelling, freshwater input, and thermocline formation create challenging habitats. These dynamics limit coral diversity and define global biogeographic coral boundaries. Corals in suboptimal environments such as turbid, deep, cold, or warm waters exhibit specialized adaptations enabling survival beyond typical reef conditions. These include stress–tolerant morphologies (*e.g.*, massive, plating forms), flexible symbioses with specialized Symbiodiniaceae, and enhanced heterotrophic feeding abilities. Light–limited corals often show morphological flattening and physiological photoacclimation, while corals in thermally extreme conditions rely on resistant symbionts and higher metabolic rates. Ecologically, such reefs host fewer species, with dominance by slow-growing, stress-tolerant taxa. Nonetheless, extreme reefs contribute to coral biodiversity and may serve as reservoirs for adaptation under changing climate conditions. However, the future of corals in extreme and/or marginal reefs is uncertain as key stress responses remain poorly understood, and rising pressures may exceed their physiological and adaptive limits.

Keywords

Light · Oceanography · Nutrients · River · Temperature · Turbidity

2.1 Introduction

Corals are benthic and sessile animals that can form large colonies made up of multiple individual polyps. Belonging to the Phylum Cnidaria, a basal clade within Eumetazoa, these organisms first emerged around 700 million years ago, in the late Proterozoic (Stanley and Fautin 2001; Park et al. 2012; McFadden et al. 2021). Corals, especially those in the Scleractinia order, are keystone species because of their role in reef-building and provision of habitat complexity for millions of eukaryotic and prokaryotic species (Knowlton et al. 2010; Graham and Nash 2013). The calcium carbonate framework produced by scleractinians provides shelter, feeding grounds, and breeding sites for these organisms (Kobluk 1988; Idjadi and Edmunds 2006; Kerry and Bellwood 2012). Currently, there are 838 extant species of photosymbiotic (*i.e.*, associated with photosynthetic Symbiodiniaceae dinoflagellates—see LaJeunesse et al. 2018) scleractinian corals (Veron et al. 2016; WoRMS 2024). Most of these species are found in shallow and tropical areas of the Indo-Pacific and Atlantic, particularly in biodiversity hotspots such as the Coral Triangle, the Great Barrier Reef, and the Caribbean. However, in extreme settings, where physicochemical conditions typically differ from those in a traditional reef environment, coral diversity tends to decrease considerably.

Because of their basal position in the evolutionary tree of Animalia, corals are simple organisms from anatomical and morphological perspectives. Therefore, they developed simple and occasionally limited mechanisms for coping with suboptimal conditions, such as those in extreme reefs.

M. Mies (✉)
Instituto Oceanográfico, Universidade de São Paulo, Praça do Oceanográfico, 191, São Paulo, Brazil

Instituto Coral Vivo, Rua dos Coqueiros, 87, Santa Cruz Cabrália, Brazil
e-mail: miguel.mies@usp.br

M. O. Soares et al. (eds.), *Extreme and Marginal Coral Reefs and Communities*, Coral Reefs of the World 24,
https://doi.org/10.1007/978-3-032-28035-0_2

According to the definition in Chap. 1, extreme reefs (previously called "marginal reefs") display large deviations in physicochemical conditions compared to optimal levels (Kleypas et al. 1999; Perry and Larcombe 2003; Schoepf et al. 2023). This means that reefs that often deviate from ideal conditions of warm temperatures, high irradiance, higher and stable salinity, and low nutrient levels may be considered extreme. As such, deviations from this "optimal" scenario pose a challenge to most photosymbiotic corals and other reef-dwelling organisms, requiring adaptive traits tailored specifically to these extreme conditions (Buddemeier and Smith 1999; Hennige et al. 2010; Morgan et al. 2016). Consequently, extreme reefs function in a much different manner compared to more typical reefs, usually supporting fewer coral species that are generally more robust (Harriott and Banks 2002; Suzuki et al. 2016; Soares et al. 2020).

Although marginal coral assemblages are often understudied in comparison to those under optimal or near-optimal conditions, they are widely distributed across the oceans and the globe. These include the highly turbid reefs on the African coast in the Western Indian Ocean (McClanahan and Obura 1997; Reuter et al. 2019), the turbid and nutrient-enriched reefs in the Southwestern Atlantic (Leão et al. 2019; Mies et al. 2020), the high-temperature and high-salinity reefs of the Red Sea (Fine et al. 2019; Hereher et al. 2022), and high-latitude reefs such as those found in Bermuda and the Southern Great Barrier Reef, which experience intense seasonal fluctuations in irradiance and temperature (Cook et al. 1990; van Woesik and Done 1997; Abrego et al. 2021). All of these have a unique and distinctive coral fauna, often with high endemism. Furthermore, the equally understudied mesophotic reefs, which are also globally distributed, harbor coral assemblages adapted to reduced irradiance and may also be considered extreme environments (Lesser and Slattery 2009; Kahng et al. 2019). Therefore, marginal coral assemblages are certainly not rare and require increased attention.

The responses of coral assemblages to extreme conditions have been critical in shaping modern-day extreme and non-extreme reefs. Corals have faced widespread extreme conditions on several occasions in the past, influenced by changes in climate, sea level and ocean chemistry (Stanley Jr 2003; Pandolfi and Kiessling 2014; Vasseur et al. 2021). These include low oxygen and high sedimentation during the Ordovician (Munnecke et al. 2010; Cocks and Torsvik 2021), anoxia and sea level changes during the Late Devonian extinction (Bond et al. 2004; Kabanov et al. 2023), fluctuating salinity and nutrient concentrations in the Triassic (Kidder and Worsley 2004; Sun 2024), intense variations in sea level and temperatures during the Neogene (Valentine 1984; Lear et al. 2015), among several other occasions. These physicochemical challenges drove coral evolution across multiple temporal and spatial scales.

With the advance of climate change and other anthropogenic stressors, reefs are currently being pushed towards extreme conditions (Guinotte et al. 2003). In this context, investigating how corals specifically managed to adapt to some of these extreme conditions is critical for understanding their evolutionary physiology and generating insights on how they may cope under future changing climate (see Burt et al. 2020). Therefore, if the ecosystem services (*e.g.*, fisheries, tourism, coastal protection) provided by corals and reef environments to millions of people across the world (Moberg and Folke 1999; Costanza et al. 2014; Giglio et al. 2024) are to be preserved, further effort on understanding and protecting marginal coral assemblages is warranted.

2.2 The Role of Oceanography in Shaping Extreme Conditions for Corals

Modern oceanography may be defined by the integrated and interdisciplinary study of biological, chemical, physical, and geological processes in the oceans (Lalli and Parsons 1997). It is subdivided into four main categories that investigate ocean processes related to (i) marine life, (ii) dissolved elements in seawater and their cycles, (iii) fluid dynamics, and (iv) geological features and their formation and transformation. The interaction between these processes has a profound role in shaping coral assemblages and their distribution, and often determine where optimal and extreme conditions manifest across ocean basins. For example, suboptimal conditions along the Western Atlantic equatorial margin restrict gene flow between the Caribbean and the South Atlantic, acting as a biogeographic filter (Floeter et al. 2008; Araújo et al. 2022). The Amazon-Orinoco plume carries freshwater, high sediment loads, and elevated nutrient levels across extensive distances, creating conditions unfavorable for coral survival (Nittrouer and DeMaster 1996; Leão et al. 2003; Drake et al. 2021). Combined with the northward-flowing North Brazil Current (Silveira et al. 1994), coral dispersal, especially towards the south, is significantly constrained in this region. Thus, the elements of the oceanographic context, namely the geological setting, dissolved elements, and hydrodynamic patterns, interact to produce pronounced extreme effects on local coral communities, reducing their richness, cover, and/or functional diversity. Therefore, the oceanographic context is pivotal in producing the extreme setting.

2.2.1 Wind: Key Driver of Thermally-extreme Conditions for Corals

The distribution of reef corals is largely dictated by the presence of warm (~26 °C) temperatures (Kleypas et al. 1999). However, the temperature of currents bathing shallow continental shelf environments is often defined by the regional wind pattern. The Ekman Theory shows that large-scale ocean surface circulation is driven by both the wind and the acceleration of Coriolis, a consequence of the Earth's rotating reference frame (Ekman 1905; Price et al. 1987). As such, a given water parcel is transported 90° to the left of the wind direction (in the Southern Hemisphere; 90° to the right for the Northern Hemisphere). This means that water will be deflected away from the shore in western coasts where the trade wind pattern is directed towards the Equator. As the surface water is displaced, colder and nutrient-rich water from deeper layers of the ocean is drawn up to replace it—resulting in an upwelling episode. Consequently, local currents tend to be colder and thus extreme for coral development (Fig. 2.1). These processes explain why coral diversity is impoverished in the southwestern Africa coast (subject to the colder Benguela Current in the Atlantic Ocean), northwestern Africa (Canary Current in the Atlantic Ocean), western.

South America (Peru Current in the Pacific Ocean), and western Australia (West Australia Current in the Indian Ocean). All these tropical areas became thermally-extreme for coral development precisely because of the presence of upwelling-influenced colder currents. Thus, global wind patterns largely define the distribution of both warm, temperature-optimal coral reefs and cooler, marginal coral assemblages.

Wind also plays an important role in producing extreme conditions at smaller scales. Intense wind shear produces resuspension in extreme reefs by generating strong surface currents and turbulent forces that disturb the seafloor (Ogston et al. 2004; Browne et al. 2013). These forces are strong enough to lift and suspend fine sediments and organic matter from the seafloor into the water column. In reef environments, where water depth is often limited, the effects of wind-driven turbulence are amplified (see Pond and Pickard 1983). This resuspension process greatly increases water turbidity and reduces light availability for corals and other photosynthetic organisms (Hennige et al. 2008; Bessell-Browne et al. 2017; López-Londoño et al. 2021). Examples of naturally turbid reefs due to wind-driven resuspension include those at Kenya (Western Indian Ocean), Puerto Rico (Caribbean), and Paluma Shoals at the Great Barrier Reef (van Katwijk et al. 1993; Larcombe et al. 2001; Otero and Carbery 2005).

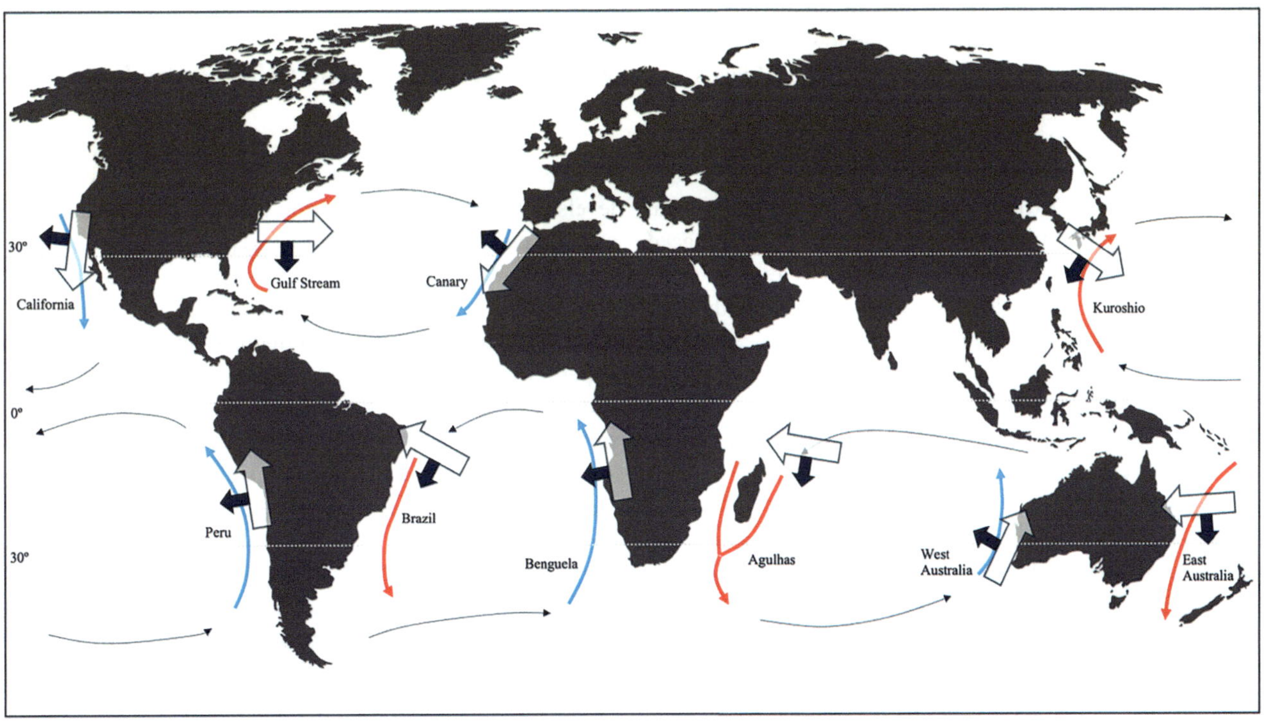

Fig. 2.1 The main ocean surface currents in tropical regions. Thick white arrows indicate the direction of persistent wind patterns, while adjacent thick black arrows represent the Ekman Transport, which moves water 90° to the left of the wind direction in the Southern Hemisphere and 90° to the right in the Northern Hemisphere. When winds blow parallel to the coast and toward the equator, water parcels are transported offshore, leading to upwelling of colder water parcels (Ekman Pumping). Consequently, colder currents (thin blue arrows) are formed and a highly extreme setting for corals is developed, as opposed to the more optimal conditions found in areas dominated by warm currents (thin red arrows)

2.2.2 Ocean Stratification: Establishing Thermal Limits for Corals

Ocean stratification is the vertical layering of ocean water parcels based on differences in density, which is primarily influenced by temperature and salinity. Stratification prevents the vertical mixing of water, which would otherwise occur in a well-mixed water column. Usually, stratification in reef environments is caused by the sun warming the upper layer. Occasionally, this causes for surface waters to heat more rapidly during warm periods, which intensifies coral bleaching (Wyatt et al. 2023). However, some reefs may be affected by seasonal stratification caused by the intrusion of colder water masses. In this case, although corals below the thermocline may be, theoretically, more protected from bleaching, they also face considerably lower (and extreme) temperatures. In insular rocky reefs of the Southwestern Atlantic, the intrusion of the South Atlantic Central Water (SACW) in the spring summer produces an intense thermocline (Silva et al. 2023; Sasaki et al. 2024). Coral colonies below the thermocline and under the influence of the SACW face temperatures as low as 13 °C (Banha et al. 2020), which is considerably lower than the lethal limit for several coral species (Sheppard et al. 2018). Other notable coral assemblages subject to colder temperatures below a marked thermocline include those in the Gulf of Panama, Galápagos, and the Chagos Archipelago (D'Croz and O'Dea 2007; Sheppard 2009; Glynn et al. 2018). The global increase in ocean stratification, as evidenced by the rise in the Brunt-Väisälä frequency (Li et al. 2020), is likely to further reduce vertical mixing, thus promoting the development of pronounced thermoclines in reef environments. This may expose some coral assemblages to colder temperatures below thermoclines, potentially making their habitat extreme.

2.2.3 River Discharges: Major Boundaries for Corals

River discharges promote dramatic changes to local oceanographic dynamics and serve as some of the most relevant biogeographic filters for corals (McLaughlin et al. 2003). The freshwater inputs from rivers reduce the salinity of adjacent coastal waters. This salinity change also often creates stratified layers, where less dense freshwater overlays denser seawater. Because corals are intolerant of freshwater and also sensitive to reduced salinity, areas near river discharges have few, if any, coral species (Coles and Jokiel 1992).

Besides freshwater, rivers also transport nutrients and sediment into the ocean. Nutrients frequently stimulate phytoplankton blooms, leading to eutrophication and sometimes creating hypoxic conditions (Rabalais et al. 2014). More importantly, elevated nutrient levels drive algal overgrowth, and therefore opportunistic turf algae quickly outcompete corals for sunlight and space in the benthos (O'Brien and Scheibling 2018). Rivers also deliver suspended terrigenous sediments, which increase turbidity and reduce light penetration in coastal waters (Zweifler et al. 2021). Although it also blocks heat (Cacciapaglia and van Woesik 2016), high turbidity decreases the amount of sunlight reaching the corals, thus limiting the Symbiodiniaceae photosynthetic output (Bessell-Browne et al. 2017). This comes at an energetic cost for corals and reduces their growth rates (Anthony and Connolly 2004; Crabbe and Smith 2005).

Large river plumes may also influence coastal circulation patterns. The freshwater flow and associated sediment and nutrient plumes alter local current dynamics, which can affect larval transport, sedimentation rates, among numerous other processes (Kourafalou et al. 1996; Bolle et al. 2009). Therefore, coral assemblages subject to river discharge influence are typically regarded as marginal and are widely distributed across the globe. These include those in the Great Amazon Reef System, Rosario Islands in Colombia, and those in Kenya, Tanzania, and Mozambique in Southeastern Africa, among several others (Obura et al. 2000; Restrepo et al. 2016; Francini-Filho et al. 2018).

2.2.4 Tides, Waves, and Other Hydrodynamic Processes

Waves and tidal cycles cause regular fluctuations in water levels, which affect exposure to sunlight, temperature, and salinity, particularly in shallow areas (Hoitink 2004). Internal waves, formed at the boundary between stratified water layers, can alter local temperature and nutrient availability (Wall et al. 2012). Corals exposed to frequent internal waves often experience conditions considered extreme, due to the sudden influxes of cooler, nutrient-rich water that challenge their physiological stability. Other hydrodynamic forces such as surface and longshore currents, as well as monsoon and climatic events, may also produce suboptimal physicochemical conditions for corals.

2.3 Diversity, Physiology, and Ecology of Corals Under Extreme Conditions

Marginal coral species occupy challenging reef environments, often on the edges of their biogeographic ranges, where conditions such as turbidity, temperature extremes, and nutrient variability limit traditional (*i.e.*, "optimum") reef development (Fig. 2.2). These coral species display a broad range of functional traits for adapting to the unique oceanographic conditions. Physiologically, they are characterized by robust mechanisms to cope with stress, such as

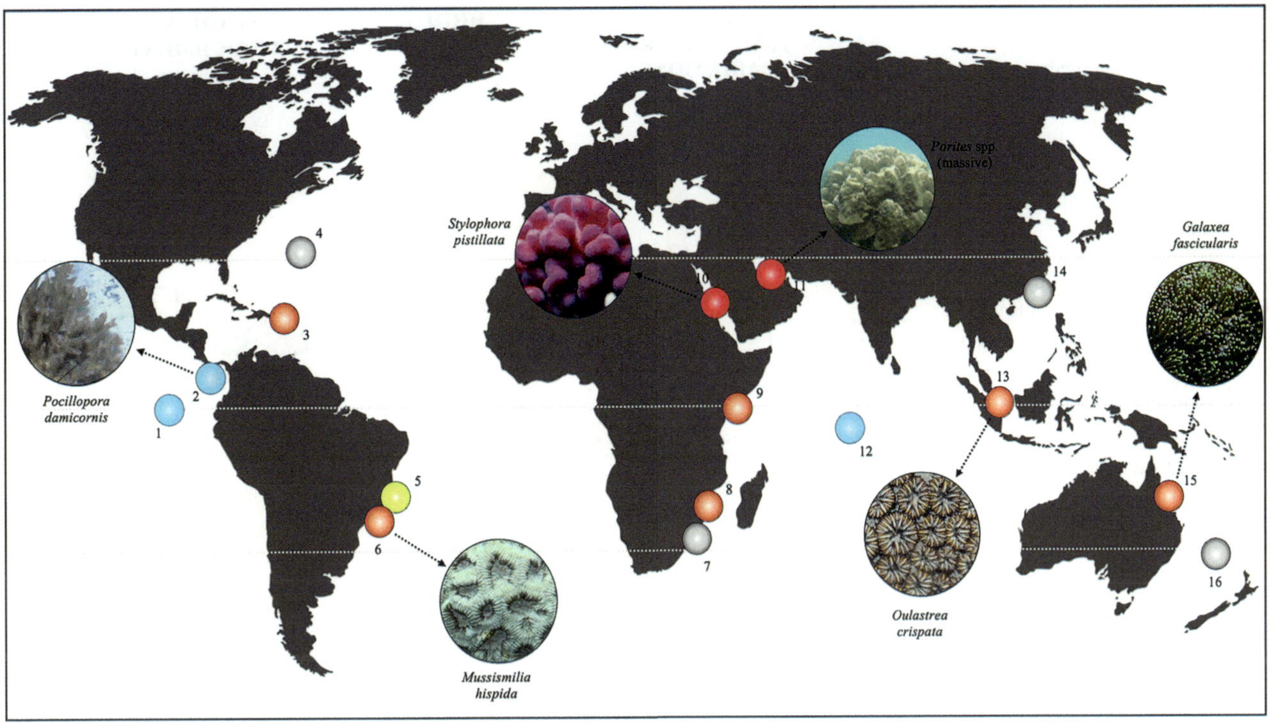

Fig. 2.2 Representative coral species commonly found in some of the main extreme reef systems around the world. (1) Galápagos Islands, (2) Western Panama coast, (3) Puerto Rico, (4) Bermuda, (5) Abrolhos Bank in Brazil, (6) São Paulo coast in Brazil, (7) Sodwana Bay in South Africa, (8) Mozambique, (9) Kenya, (10) Red Sea, (11) Persian Gulf, (12) Chagos Archipelago, (13) Singapore, (14) Hong Kong, (15) Paluma Shoals in the Great Barrier Reef, and (16) Lord Howe Island. Blue, red, brown, yellow, and grey circles represent reef settings that are colder, warmer, more turbid, nutrient-enriched, and located at high latitudes, respectively

flexible symbioses and enhanced energy acquisition strategies (Fig. 2.3). Therefore, in this section, their main adaptive traits to some of the most common extreme conditions are presented and discussed.

2.3.1 Light: Corals in Deeper and Darker Waters

Light plays a key role in reef formation, driving essential processes such as photosynthesis and calcification. Reefs typically develop under light intensities ranging from 300 to 600 µmol photons m^{-2} s^{-1}, sustained over a daily photoperiod of approximately 12 h (Kleypas 1997; Gattuso et al. 2006). Consequently, the distribution of reef communities is often restricted to the upper depth strata where light is more abundantly present, particularly photosynthetically active radiation, which encompasses the 400–700 nm spectrum (Sheppard et al. 2018). However, many coral species thrive under much lower light availability, even at 50 µmol photons m^{-2} s^{-1} (Lesser and Slattery 2009; Carpenter et al. 2022), a condition commonly resulting from increased turbidity ('turbid reefs') or greater depth ('mesophotic reefs')—two key factors that reduce light availability in reef environments.

The diversity of corals adapted to low-light environments is considerable. Of all 838 known photosymbiotic reef-building coral species, 15% (134 species) are adapted to turbid conditions, while 33% (280 species) occur in mesophotic zones (Madin et al. 2016; Mies et al. 2020). A key distinction from well-lit environments is the near or complete absence of branching morphology species such as acroporids. Nonetheless, low-light habitats function as extensions of shallow reefs and may act as genetic reservoirs for multiple coral species (Hinderstein et al. 2010; Kahng et al. 2010; Semmler et al. 2017). Among recognized low-light specialists are *Montastraea cavernosa*, a depth-adapted species in the Caribbean and Southwestern Atlantic (Lesser et al. 2010; Soares et al. 2018; Santana et al. 2023), recorded at depths reaching 180 m (Castro et al. 2006), and *Leptoseris* spp. and *Porites* spp. in the Indo-Pacific, both frequently occurring across broad mesophotic depth ranges (Pochon et al. 2015; Englebert et al. 2017).

Light presence also largely dictates the morphology of corals. Light-limited corals often exhibit flattened, plating morphologies that maximize surface area for light capture (Kramer et al. 2020; Lesser et al. 2021). For example, *Leptoseris* spp. in depths >60 m can grow in ultra-thin sheets less than 5 mm thick, oriented horizontally to maximize photon absorption (Kahng et al. 2020). In turbid zones, massive

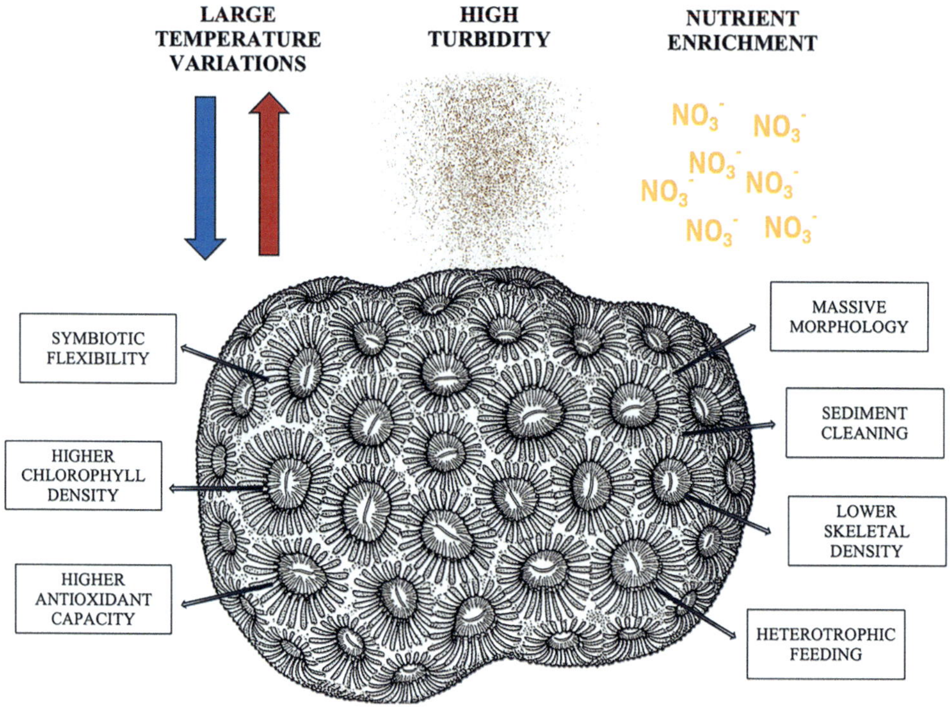

Fig. 2.3 Corals inhabiting extreme reef environments, characterized by elevated or reduced temperatures, high turbidity, and nutrient enrichment, exhibit a range of ecophysiological adaptations including (i) a stress-tolerant massive growth form, (ii) active sediment removal via tentacle movement and polyp inflation, (iii) reduced skeletal density to conserve energy, (iv) enhanced heterotrophic feeding to increase carbon acquisition, (v) elevated total antioxidant capacity to mitigate oxidative stress, (vi) increased chlorophyll concentrations to optimize photosynthesis, and (vii) a generalist symbiotic strategy that facilitates association with stress-tolerant Symbiodiniaceae phylotypes

or encrusting forms dominate, possibly due to mechanical resistance to sediment load and high particle flux (Zweifler et al. 2021). Further, skeletal density is often lower in corals under lower irradiance, reducing metabolic cost while maintaining growth (Carricart-Ganivet and Merino 2001).

Likewise, coral physiology and photosynthetic efficiency are critical under reduced light conditions. Corals in low-light environments tend to have higher chlorophyll-*a* concentrations per symbiont cell and greater maximum quantum yields of photosystem II (Lesser et al. 2010; Ziegler et al. 2015). Complementing these physiological adjustments, corals under low irradiance often host distinct shade-adapted Symbiodiniaceae phylotypes (Chan et al. 2009; Cooper et al. 2011; Smith et al. 2020; Hussain et al. 2025). In mesophotic zones, *Cladocopium* phylotypes often dominate, particularly those with enhanced photoacclimation capacity (LaJeunesse, 2005; Bongaerts et al. 2011). This symbiotic flexibility of some coral species in turbid and deeper settings also allows for symbiont shuffling, which enables change towards a more effective symbiotic assemblage from a photosynthetic output perspective (see Baker 2003; Baker et al. 2004). Additionally, under reduced autotrophic nutrition, heterotrophic capacity may be enhanced, especially in turbid reefs where particulate organic matter is abundant (see Sect. 2.4).

Low-light reefs exhibit distinct dynamics from an ecological standpoint. Coral growth rates are generally slower due to limited light availability (Kahng et al. 2019), and in mesophotic reefs, sea-level rise poses a particular threat—if it outpaces vertical reef accretion, these systems risk 'drowning' (Wood 1999). Limited light also constrains the abundance of frondose and turf algae, which in turn reduces food availability for herbivores. Consequently, herbivore biomass is often lower (Kahng et al. 2010), and the naturally reduced coral cover diminishes competitive interactions both among corals and between corals and frondose or turf algae (Kahng and Kelley 2007). Nonetheless, these conditions frequently favor the proliferation of sponges and crustose coralline algae, which can dominate benthic space and further reshape community structure and ecological interactions (Kahng et al. 2010; Slattery et al. 2011).

2.3.2 Temperature: Corals in Warmer and Colder Waters

Temperature is widely regarded as the primary driver shaping the biogeographic distribution of reef-building organisms, as it regulates key biological processes including calcification, reproduction, and metabolic performance. Most reef communities are confined to waters ranging from approximately 18 to 30 °C, which restricts their global distribution to tropical and subtropical latitudes (Sheppard et al. 2018). Prolonged exposure to temperatures at or below 18 °C generally induces physiological stress (Kleypas et al. 1999). On the upper end, temperatures approaching 30 °C frequently trigger coral bleaching. However, these thermal thresholds vary greatly according to context. Globally, the average sea surface temperature on coral reefs is 27.6 °C, with an average maximum of 29.5 °C (Sheppard et al. 2018). Importantly, thermal variability in tropical reef systems has historically been low, meaning that many coral taxa have evolved near their upper thermal tolerance limits.

The diversity of reef-building corals declines sharply at thermal extremes. While Indo-Pacific hotspots can support over 500 photosymbiotic scleractinian species, thermally-extreme reefs often host far fewer. For example, Lord Howe Island (~31 °S), exposed to higher-frequency thermal variability, supports around 60 species (Harriott et al. 1995). The Galápagos Islands, although at lower latitudes, experience strong upwelling and harbor less than 20 species (Riegl et al. 2019). Some extreme reefs support even fewer—the São Paulo State (23 °S) rocky reefs, at the tropical-subtropical boundary of the Southwestern Atlantic and influenced by the SACW with temperatures dropping to 13 °C, sustain only two species (*Madracis decactis* and *Mussismilia hispida*—see Leão et al. 2003, 2019; Banha et al. 2020). In the warmer end of the spectrum, in the reef environments of the Persian Gulf, summer sea surface temperatures regularly exceed 34 °C (Samiei et al. 2015). Despite these extreme conditions, coral assemblages persist, but they are dominated by a few thermally-tolerant genera like *Porites* spp., *Platygyra* spp., and *Cyphastrea* spp. (Burt et al. 2019).

At thermal extremes, coral morphology and physiology often shift toward traits that enhance survival under stress. In warmer environments, massive species generally perform better, largely due to their slower growth and higher metabolic rates, which promote greater protein turnover and facilitate acclimation (Gates and Edmunds 1999). In addition, their thicker tissues further enhance resilience by providing increased photoprotection through self-shading, storing higher lipid reserves, and displaying elevated antioxidant capacity (Loya et al. 2001; Maor-Landaw and Levy 2016; Liu et al. 2022). In contrast, the effects of cold stress on coral morphology are less well understood. Although massive and encrusting forms are often dominant (and branching morphologies notably scarce) in high-latitude reefs, a cold anomaly in Florida in 2010 caused disproportionately high mortality among massive species (Lirman et al. 2011). Thus, the role of coral morphology in cold tolerance remains underexplored. Nonetheless, the apparent prevalence of massive, slow-growing morphologies at high-latitude marginal reefs reflects a convergence toward structural forms that minimize thermal damage and enable persistence under suboptimal conditions. This pattern is evident in the dominance of massive and encrusting species such as *Goniastrea* spp. at Hall Bank, Western Australia (Thomson and Frisch 2010), *Mussismilia hispida* in southern Brazil (Oliveira et al. 2025), and *Oulastrea crispata* in Hong Kong (Lam 2000).

To further endure suboptimal temperatures, corals may associate with Symbiodiniaceae phylotypes that are specifically adapted to those thermal conditions. Symbiont identity is a major predictor of thermal tolerance because there is a remarkable diversity of functional responses to thermal stress across Symbiodiniaceae (Swain et al. 2017; Nitschke et al. 2022). In the warmer reefs of the Persian Gulf, thermally-tolerant *Durusdinium* spp. and *Cladocopium thermophilum* are dominant, increasing the overall thermal tolerance of the coral host (Mostafavi et al. 2007; Oladi et al. 2019). Similarly, the colder conditions in upwelling regions promote coral symbioses with Symbiodiniaceae phylotypes that are specifically adapted to low-temperature environments (Zhou et al. 2012; Picciani et al. 2016).

Coral communities in thermally extreme settings, whether warmer or colder, operate under distinct ecological constraints compared to typical tropical reefs. In warmer systems, such as the Red Sea or Persian Gulf, elevated temperatures and frequent thermal anomalies drive a shift toward heat-specialist taxa (Gonzalez et al. 2024). These environments often host fewer coral species, but those present may exhibit high local abundance and dominance (Sawall et al. 2014). In colder reefs, such as those influenced by seasonal upwelling or located at higher latitudes, coral assemblages are typically even more depauperate in species richness. Cold-tolerant, slow-growing, and often massive or encrusting species tend to prevail, forming low-diversity assemblages adapted to high frequency thermal variability and periodic stress (Cartwright et al. 2023). Both warmer and colder systems may also exhibit enhanced seasonal or temporal fluctuations in community composition, shaped by episodic mortality and selective recovery (Beger et al. 2014; Schoepf et al. 2023). Despite their lower taxonomic richness, thermally extreme reefs often display high ecological resilience, driven by species with broad physiological tolerances, plastic life-history traits, and efficient resource use (Morgan et al. 2016; Banha et al. 2020; Browne and Bauman 2023).

2.3.3 Turbidity, Nutrients, and Freshwater: Corals Under River Influence

Rivers typically serve as biogeographic filters for coral species due to elevated freshwater influx, nutrient enrichment, and turbidity, which are conditions generally considered extreme for coral growth. Corals have limited capacity to regulate osmotic balance and are typically intolerant to prolonged salinity fluctuations (Mayfield and Gates 2007). Consequently, coral communities rarely prosper near river plumes and typically thrive at higher salinities ranging from 33 to 36 ppt (Sheppard et al. 2018). In terms of nutrients, coral reefs are generally associated with oligotrophic waters characterized by low concentrations of chlorophyll-a (~0.5 μg L^{-1}), nitrate (~0.6 μM), phosphate (~0.2 μM), and total organic carbon (~1.0 mg L^{-1}) (Pagès et al. 1997; Yahel et al. 2003; Fabricius et al. 2005; Sheppard et al. 2018). Regarding turbidity, reefs distant from rivers usually exhibit sedimentation rates below 10 mg cm^{-2} day^{-1} (Rogers 1990;

Nugues and Roberts 2003; Wolanski et al. 2005; Sofonia and Anthony 2008). Another measure of turbidity, the diffuse attenuation coefficient K_d490, quantifies light attenuation at 490 nm caused by suspended particles (Wang et al. 2009), with typical values ranging from 0.01 to 0.07 m^{-1} (Yee and Barron 2010; García-Sais et al. 2017; Ceccarelli et al. 2020).

Despite this adverse scenario for reef development, some coral species and even entire reef systems can persist and occasionally thrive near major river discharges. Examples include *Siderastrea siderea* occurring near several river mouths in Puerto Rico (Torres and Morelock 2002; Larsen and Webb 2009), *Galaxea fascicularis*-dominated reefs at Paluma Shoals on the Great Barrier Reef (Larcombe et al. 2001), *Favia gravida* thriving near the Buranhém River in Brazil (Pereira et al. 2020), and a diverse coral assemblage comprising dozens of species in the river-influenced waters of Ngermeduu Bay, Palau (Golbuu et al. 2011). In general, reef systems influenced by rivers display depauperate but resilient coral assemblages.

In these high-turbidity, low-light environments, coral morphology often shifts toward massive, submassive, or plating forms that are better suited to resist sedimentation and maximize light capture (Jones et al. 2019; Lesser et al. 2021). Massive corals typically avoid sediment smothering and resuspend deposited particles through mucus shedding and polyp inflation (Brown and Bythell 2005; Duckworth et al. 2017). This sediment rejection capacity, although energetically costly, prevents smothering and maintains polyp functionality. On the other hand, branching species are rarer due to their higher sediment entrapment and sensitivity. Physiologically, corals under river influence are often more tolerant to nutrification, such as the case of *Mussismilia hispida* in the Southwestern Atlantic, which can tolerate up to 30 μM of nitrate (Patrocinio et al. 2025). However, the osmotic mechanisms developed to cope with reduced salinities remain largely underassessed for corals. Nonetheless, collectively, these physiological traits allow a subset of coral species to persist in estuarine and plume-influenced reef environments.

Although the role of Symbiodiniaceae in corals exposed to riverine influence remains quite unexplored, it is well established that corals thriving under high turbidity typically host Symbiodiniaceae assemblages adapted to low-light conditions (LaJeunesse et al. 2010; Cooper et al. 2011). Likewise, nutrient concentration also plays a role in defining symbiont assemblages and overall coral health. Excessive enrichment with dissolved inorganic nitrogen and phosphate can trigger rapid symbiont proliferation, intensifying competition for resources and reducing both photosynthetic performance and carbon transfer per symbiont cell (Ezzat et al. 2015; Morris et al. 2019). When nitrate levels rise or phosphate becomes limiting, photosynthesis is impaired, which greatly reduces the total carbon transferred to the host and negatively affects coral health (Wiedenmann et al. 2013; Ezzat et al. 2015; Rosset et al. 2017; Morris et al. 2019). However, these processes are highly dependent on the identity and functional aspects of the symbiont lineages (McIlroy et al. 2022; Nitschke et al. 2022; Huang et al. 2024). Therefore, corals under nutrient enrichment likely harbor symbionts adapted to such conditions.

Ecologically, reefs influenced by riverine inputs are shaped by strong environmental filtering, where only a subset of coral species can tolerate the combined pressures of high turbidity, elevated nutrients, and salinity fluctuations. These conditions limit light availability, increase sediment deposition, and periodically expose corals to osmotic stress, narrowing the ecological niche space for many reef builders. Consequently, coral assemblages in these settings often consist of a small number of tolerant species, yet these can establish persistent and resilient populations. Instead of the hundreds of coral species found in clear, oligotrophic systems like the Coral Triangle, river-influenced reefs may support fewer than 20 photosymbiotic scleractinian species (Leão et al. 2003). Diversity is often associated with environmental gradients across distance from shore and river plume exposure (Golbuu et al. 2008). This creates mosaic-like reefscapes, where different species dominate depending on small-scale hydrodynamic and sedimentation regimes. Thus, although overall diversity is lower than on offshore reefs, the coral communities near river discharges are not necessarily impoverished in ecological function (Luza et al. 2023; Mies et al. 2024). Instead, they represent a form of specialized, stress-adapted biodiversity that may offer key insights into coral resilience under future coastal stress scenarios.

2.4 The Importance of Heterotrophic Feeding

Corals are mixotrophic organisms (Fig. 2.4), obtaining energy through both autotrophic and heterotrophic pathways. Autotrophy relies on the translocation of photosynthetically fixed carbon from their intracellular endosymbiotic dinoflagellates to the coral host (Tremblay et al. 2012). In contrast, heterotrophy involves the active capture and ingestion of organic carbon sources, including plankton and detritus (Houlbrèque and Ferrier-Pagès 2009). Additionally, corals can absorb dissolved organic matter directly from the water column, in a process known as osmotrophy (Sorokin 1973; Al-Moghrabi et al. 1993).

In extreme reef settings, environmental stressors such as lower light availability, elevated temperatures, and nutrient enrichment can impair the photosynthetic efficiency of Symbiodiniaceae (Perry and Larcombe 2003), thereby compromising autotrophic energy acquisition. Increased turbidity reduces light penetration, limiting the energy available

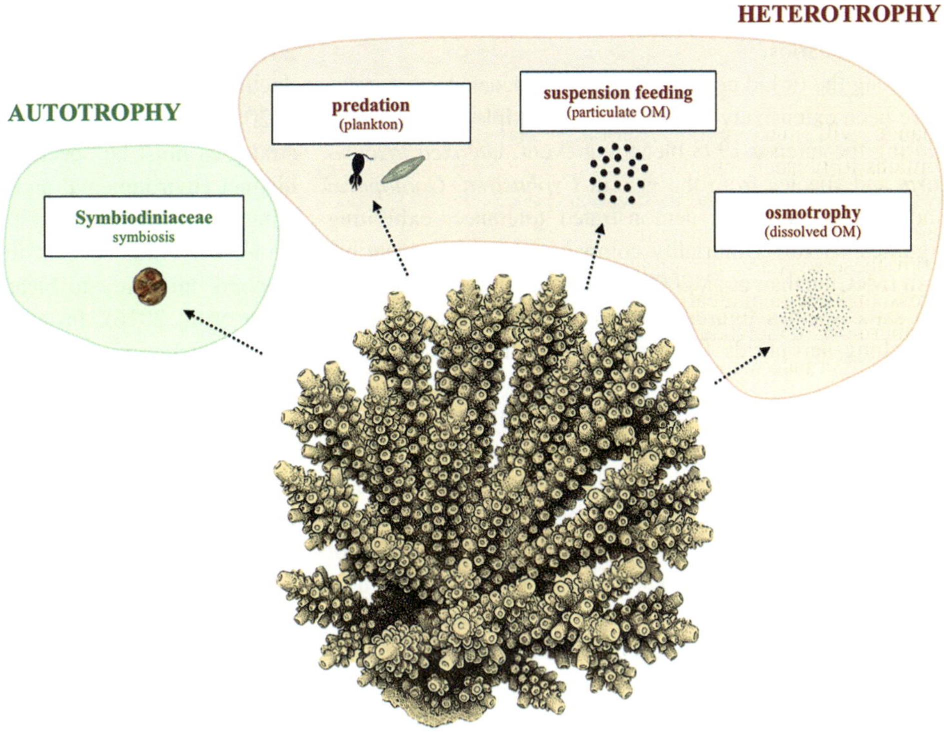

Fig. 2.4 The trophic ecology of reef-building corals is mixotrophic and quadripartite, encompassing the autotrophic input of (i) photosynthetically derived carbon produced by the intracellular Symbiodiniaceae dinoflagellates, as well as the heterotrophic carbon acquisition through (ii) predation on plankton and the uptake of (iii) particulate and (iv) dissolved organic matter (OM). In extreme reef settings, the heterotrophic behavior is enhanced for the compensation of reduced autotrophy

for photosynthesis. Additionally, excessive nutrients can disrupt the delicate balance of the coral-algal symbiosis, leading to decreased photosynthetic performance and increased susceptibility to bleaching (Wiedenmann et al. 2013). To adapt to these challenging conditions, many coral species have evolved enhanced heterotrophic capabilities as a compensation mechanism (Anthony and Fabricius 2000; Hughes 2013). This includes morphological adaptations such as larger polyps and increased tentacle expansion, which facilitate the capture of plankton and particulate organic matter (Houlbrèque and Ferrier-Pagès 2009). In the rocky reefs off São Paulo, Brazil, colonies of *Mussismilia hispida* exposed to prolonged periods of extreme turbidity exhibited a marked decline in fatty acids typically supplied via autotrophic translocation from their Symbiodiniaceae symbionts. This deficit, however, was offset by an increase in zooplankton predation, demonstrating a compensatory shift toward heterotrophy (Mies et al. 2018). Similarly, soft corals inhabiting mesophotic zones of the Red Sea showed reduced concentrations of autotrophy-associated fatty acids, while simultaneously upregulating their consumption of zooplankton (Pupier et al. 2021). These examples highlight the capacity of corals to adjust their nutritional strategy under suboptimal conditions and therefore reduce their susceptibility to bleaching and mortality.

Heterotrophy also plays a fundamental role in coral reproduction, particularly in turbid and other extreme environments. Evidence suggests that in such settings, the energy shortfall resulting from impaired photosynthetic efficiency is counterbalanced by increased heterotrophic feeding, which may supply the necessary reserves for gametogenesis (Loures et al. 2026). Corals inhabiting these challenging environments can thus maintain or even enhance reproductive output by relying more heavily on heterotrophy (Pires et al. 2011; Godoy et al. 2021). Thus, the heterotrophic strategy enhances the resilience of coral populations in extreme habitats, ensuring their persistence despite adverse environmental conditions. In summary, heterotrophy serves as a vital adaptive mechanism for corals inhabiting extreme reef environments. By supplementing energy intake through heterotrophic feeding, corals can withstand the detrimental effects of reduced light availability, thermal stress, and nutrient enrichment. This trophic flexibility not only supports individual coral health and growth but also contributes to the overall resilience and stability of reef ecosystems facing environmental challenges.

2.5 Specific Responses of Coral Species to Climate Change in Extreme Settings

Coral reefs situated in extreme environments host coral species that display varied responses to bleaching events. These responses are shaped by local environmental pressures, species-specific traits, and the historical exposure of coral populations to stress. Understanding how specific species respond to bleaching in these settings provides insights into coral resilience and the potential for survival under future climate scenarios.

Along the turbid coastline of Kenya, bleaching dynamics have been extensively studied across multiple coral species. During the intense 1998 bleaching event, *Galaxea fascicularis* and species from the genera *Cyphastrea*, *Goniopora*, and *Porites* (massive) demonstrated tolerance, exhibiting significantly lower mortality compared to more susceptible taxa (McClanahan et al. 2004). Likewise, in the 2010 event, the same species figured among the most resistant, while branching acroporids faced the highest declines (Mwaura et al. 2017).

In the central Red Sea, especially within the Gulf of Aqaba, massive *Porites* species and the branching coral *Stylophora pistillata* have consistently exhibited high thermal tolerance across multiple bleaching events (Monroe et al. 2018; Ziegler et al. 2019). While these species have not been entirely unaffected, experiencing some mortality during extreme thermal anomalies, they remain among the most resilient corals in a region where temperatures often exceed bleaching thresholds for many Indo-Pacific taxa. Their persistence has been linked to a suite of ecological, genetic, and life-history traits that promote resistance and recovery under stress (Meziere et al. 2022). In contrast, other branching *Acropora* species have suffered substantial declines over time, with repeated bleaching events leading to widespread loss of cover and diversity (Furby et al. 2013).

In Brazil, species of the endemic genus *Mussismilia* have shown contrasting responses to recent bleaching events. During the severe episodes of 2019 and 2024, the massive species *Mussismilia hispida* and *M. braziliensis* exhibited high resilience, with extensive bleaching but minimal mortality (Banha et al. 2020; Ferreira et al. 2021). In contrast, the phaceloid species *M. harttii* experienced significant population declines, with losses ranging from 10% to 50% (Braz et al. 2022; Corazza et al. 2024; Pereira et al. 2022).

Urban reefs in Singapore offer another compelling example in extreme settings. These reefs, persist under conditions of chronic turbidity, high nutrient input, and intense human disturbance (Guest et al. 2016). Remarkably, during the 2016 global bleaching event, which caused widespread mortality across tropical reefs, coral mortality in Singapore remained low. Notably, species such as *Galaxea fascicularis*, *Diploastrea* spp., and *Turbinaria* spp. exhibited no mortality despite exposure to elevated thermal stress (Bauman et al. 2022).

This context may highlight the elevated resilience of some corals inhabiting extreme reef environments (Osman et al. 2018; Mies et al. 2020; Mies et al. 2025a, b and Mariano et al. 2026). However, these ecosystems are increasingly exposed to intensified thermal stress events (Genevier et al. 2019; Destri et al. 2025), and rising mortality rates have been documented even in these relatively tolerant settings (Soares 2020; Mo et al. 2022; De et al. 2023). As such, conservation strategies must be specifically designed to account for the distinct environmental and ecological conditions characterizing each extreme reef system. Furthermore, these examples show that coral species in extreme environments do not respond uniformly to bleaching (see Sommer et al. 2021; Faria et al. 2026). Instead, their responses are shaped by complex interactions among physiology, environment, and anthropogenic stress. Some show remarkable flexibility in energy acquisition or thermal tolerance, while others succumb to bleaching despite living in historically challenging environments. These contrasting outcomes underscore the importance of species-level investigations when assessing reef vulnerability and guiding conservation efforts.

2.6 Research Gaps and Perspectives

Several important research gaps remain regarding the biology and responses of corals in extreme reef settings. One major gap is the limited understanding of how specific coral species maintain metabolic function under chronic environmental stressors such as turbidity, low irradiance, thermal extremes, and nutrient enrichment. While some corals survive or even thrive under such conditions, the physiological processes, particularly those related to energy balance, respiration, and tissue maintenance, remain poorly characterized. Another key gap concerns the role and variability of heterotrophic feeding strategies. Although heterotrophy is widely assumed to buffer corals against low photosynthetic input, there is little quantitative data on heterotrophic rates across species. It also remains unclear how energy derived from heterotrophy supports specific functions such as gametogenesis, regeneration, or calcification in extreme environments. The plasticity and stability of coral–Symbiodiniaceae associations in marginal settings are also insufficiently studied.

Coral reefs are changing and facing more extreme conditions very quickly (Guinotte et al. 2003). The future of coral species in extreme reefs will depend on their ability to withstand intensifying climate stress and local pressures. While many host resilient coral species with unique adaptations, the increasing frequency of heatwaves, and other local scale stressors may exceed their adaptive capacity. Targeted research, protection tailored to their specific stress regimes, and inclusion in global conservation frameworks are essential to ensure these often-overlooked species and systems persist.

Acknowledgements I would like to thank the editorial team for kindly inviting me to take part in this edition, and the Coral Vivo Project and its sponsors, Petrobras, through the Petrobras Socioenvironmental Program, and Arraial d'Ajuda Eco Parque. I also acknowledge the support of Fundação de Amparo à Pesquisa do Estado de São Paulo grant #23/09180-3. I also thank Prof. Ilson Silveira for his kind assistance and Jessica Bleuel and Pedro Henrique Pereira for their reviews.

References

Abrego D, Howells EJ, Smith SD, Madin JS, Sommer B, Schmidt-Roach S, Cumbo VR, Thomson DP, Rosser NL, Baird AH (2021) Factors limiting the range extension of corals into high-latitude reef regions. Diversity 13(12):632

Al-Moghrabi S, Allemand D, Jaubert J (1993) Valine uptake by the scleractinian coral *Galaxea fascicularis*: characterization and effect of light and nutritional status. J Comp Physiol B 163:355–362

Anthony KR, Connolly SR (2004) Environmental limits to growth: physiological niche boundaries of corals along turbidity–light gradients. Oecologia 141:373–384

Anthony KR, Fabricius KE (2000) Shifting roles of heterotrophy and autotrophy in coral energetics under varying turbidity. J Exp Mar Biol Ecol 252:221–253

Araújo GS, Rocha LA, Lastrucci NS, Luiz OJ, Di Dario F, Floeter SR (2022) The Amazon-Orinoco barrier as a driver of reef-fish speciation in the Western Atlantic through time. J Biogeogr 49:1407–1419

Baker AC (2003) Flexibility and specificity in coral-algal symbiosis: diversity, ecology, and biogeography of *Symbiodinium*. Annu Rev Ecol Evol Syst 34:661–689

Baker AC, Starger CJ, TR MC, Glynn PW (2004) Corals' adaptive response to climate change. Nature 430:741

Banha TN, Capel KC, Kitahara MV, Francini-Filho RB, Francini CL, Sumida PY, Mies M (2020) Low coral mortality during the most intense bleaching event ever recorded in subtropical southwestern Atlantic reefs. Coral Reefs 39:515–521

Bauman AG, Januchowski–Hartley FA, Teo A, Todd PA (2022) Further flattening of a degraded, turbid reef system following a severe coral bleaching event. Front Mar Sci 9:910085

Beger M, Sommer B, Harrison PL, Smith SD, Pandolfi JM (2014) Conserving potential coral reef refuges at high latitudes. Divers Distrib 20:245–257

Bessell-Browne P, Negri AP, Fisher R, Clode PL, Duckworth A, Jones R (2017) Impacts of turbidity on corals: the relative importance of light limitation and suspended sediments. Mar Pollut Bull 117:161–170

Bolle LJ, Dickey-Collas M, van Beek JK, Erftemeijer PL, Witte JI, van der Veer HW, Rijnsdorp AD (2009) Variability in transport of fish eggs and larvae. III. Effects of hydrodynamics and larval behaviour on recruitment in plaice. Mar Ecol Prog Ser 390:195–211

Bond D, Wignall PB, Racki G (2004) Extent and duration of marine anoxia during the Frasnian–Famennian (Late Devonian) mass extinction in Poland, Germany, Austria and France. Geol Mag 141:173–193

Bongaerts P, Sampayo EM, Bridge TCL, Ridgway T, Vermeulen F, Englebert N, Webster JM, Hoegh-Guldberg O (2011) Symbiodinium/ppdiversity in mesophotic coral communities on the great Barrier reef: a first assessment. Mar Ecol Prog Ser 439:117–126

Braz GB, Lacerda CH, Evangelista H, Güth AZ, Rumbelsperger AM, Capel KC, PL D'O, Mies M (2022) Unprecedented erosion of *Mussismilia harttii*, a major reef-building species in the southwestern Atlantic, after the 2019 bleaching event. Coral Reefs 41:1537–1548

Brown BE, Bythell JC (2005) Perspectives on mucus secretion in reef corals. Mar Ecol Prog Ser 296:291–309

Browne NK, Bauman AG (2023) Marginal reef systems: resilience in a rapidly changing world. Diversity 15:703

Browne NK, Smithers SG, Perry CT (2013) Spatial and temporal variations in turbidity on two inshore turbid reefs on the great barrier reef, Australia. Coral Reefs 32:195–210

Buddemeier RW, Smith SV (1999) Coral adaptation and acclimatization: a most ingenious paradox. Am Zool 39:1–9

Burt JA, Paparella F, Al-Mansoori N, Al-Mansoori A, Al-Jailani H (2019) Causes and consequences of the 2017 coral bleaching event in the southern Persian/Arabian gulf. Coral Reefs 38:567–589

Burt JA, Camp EF, Enochs IC, Johansen JL, Morgan KM, Riegl B, Hoey AS (2020) Insights from extreme coral reefs in a changing world. Coral Reefs 39:495–507

Cacciapaglia C, van Woesik R (2016) Climate-change refugia: shading reef corals by turbidity. Glob Chang Biol 22:1145–1154

Carpenter GE, Chequer AD, Weber S, Mass T, Goodbody-Gringley G (2022) Light and photoacclimatization drive distinct differences between shallow and mesophotic coral communities. Ecosphere 13:e4200

Carricart-Ganivet JP, Merino M (2001) Growth responses of the reef-building coral *Montastraea annularis* along a gradient of continental influence in the southern Gulf of Mexico. Bull Mar Sci 68:133–146

Cartwright PJ, Browne NK, Belton D, Parnum I, O'Leary M, Valckenaere J, Fearns P, Lowe R (2023) Long-term spatial variations in turbidity and temperature provide new insights into coral-algal states on extreme/marginal reefs. Coral Reefs 42:859–872

Castro CB, Pires DO, Medeiros MS, Loiola LL, Arantes RCM, Thiago CM, Berman E (2006) Filo Cnidaria. In: Lavrado HP, Ignacio BL (eds) Biodiversidade bentônica da região central da Zona Econômica Exclusiva brasileira. Museu Nacional, Rio de Janeiro, pp 147–192

Ceccarelli DM, Evans RD, Logan M, Mantel P, Puotinen M, Petus C, Russ GR, Williamson DH (2020) Long-term dynamics and drivers of coral and macroalgal cover on inshore reefs of the great barrier reef Marine Park. Ecol Appl 30:e02008

Chan YL, Pochon X, Fisher MA, Wagner D, Concepcion GT, Kahng SE, Toonen RJ, Gates RD (2009) Generalist dinoflagellate endosymbionts and host genotype diversity detected from mesophotic (67–100 m depths) coral *Leptoseris*. BMC Ecol 9:1–7

Cocks LR, Torsvik TH (2021) Ordovician palaeogeography and climate change. Gondwana Res 100:53–72

Coles SL, Jokiel PL (1992) Effects of salinity on coral reefs. In: Pollution in tropical aquatic systems. CRC Press, pp 147–166

Cook CB, Logan A, Ward J, Luckhurst B, Berg CJ (1990) Elevated temperatures and bleaching on a high latitude coral reef: the 1988 Bermuda event. Coral Reefs 9:45–49

Cooper TF, Ulstrup KE, Dandan SS, Heyward AJ, Kühl M, Muirhead A, O'Leary RA, Ziersen BE, Van Oppen MJ (2011) Niche specialization of reef-building corals in the mesophotic zone: metabolic trade-offs between divergent *Symbiodinium* types. Proc R Soc B 278:1840–1850

Corazza BM, Lacerda CHF, Güth AZ, Marcançoli RKM, Bianchini A, Calderon EN, Capel KCC, Conceição E, Faria SC, Francini-Filho RB, Garrido AG, Guebert FM, Kitahara MV, Longo GO, Lotufo TMC, Salvi KP, Segal B, Sumida PYG, Zilberberg C, Mies M (2024) No coral recovery three years after a major bleaching event in reefs in the southwestern Atlantic refugium. Mar Biol 171:114

Costanza R, De Groot R, Sutton P, Van der Ploeg S, Anderson SJ, Kubiszewski I, Farber S, Turner RK (2014) Changes in the global value of ecosystem services. Glob Environ Chang 26:152–158

Crabbe MJ, Smith DJ (2005) Sediment impacts on growth rates of *Acropora* and *Porites* corals from fringing reefs of Sulawesi, Indonesia. Coral Reefs 24:437–441

D'Croz L, O'Dea A (2007) Variability in upwelling along the Pacific shelf of Panama and implications for the distribution of nutrients and chlorophyll. Estuar Coast Shelf Sci 73:325–340

De K, Nanajkar M, Mote S, Ingole B (2023) Reef on the edge: resilience failure of marginal patch coral reefs in eastern Arabian Sea under recurrent coral bleaching, coral diseases, and local stressors. Environ Sci Pollut Res 30:7288–7302

Destri G, Güth AZ, Luza AL, Ibanhez JY, Dottori M, Silveira IC, Braz GB, De La Cour JL, Manzello DP, Skirving WJ, Mies M (2025) The thermal stress history of South Atlantic reefs reveals increasing intensity, duration, frequency, and likely undocumented bleaching episodes. Glob Chang Biol 31:e70162

Drake TW, Hemingway JD, Kurek MR, Peucker-Ehrenbrink B, Brown KA, Holmes RM, Galy V, Moura JM, Mitsuya M, Wassenaar LI, Six J (2021) The pulse of the Amazon: fluxes of dissolved organic carbon, nutrients, and ions from the world's largest river. Glob Biogeochem Cycles 4:e2020GB006895

Duckworth A, Giofre N, Jones R (2017) Coral morphology and sedimentation. Mar Pollut Bull 125(1-2):289–300

Ekman VW (1905) On the influence of the earth's rotation on ocean-currents. Arch Math Astron Phys 2:1–52

Englebert N, Bongaerts P, Muir PR, Hay KB, Pichon M, Hoegh-Guldberg O (2017) Lower mesophotic coral communities (60–125 m depth) of the northern great barrier reef and Coral Sea. PLoS One 2:e0170336

Ezzat L, Maguer JF, Grover R, Ferrier-Pagès C (2015) New insights into carbon acquisition and exchanges within the coral–dinoflagellate symbiosis under NH_4^+ and NO_3^- supply. Proc R Soc B Biol Sci 1812:20150610

Fabricius K, De'ath G, McCook L, Turak E, Williams DM (2005) Changes in algal, coral and fish assemblages along water quality gradients on the inshore great barrier reef. Mar Pollut Bull 51:384–398

Faria SC, Goodbody-Gringley G, Kitahara MV, Marangoni LFB, Silva CA, Neto RR, Bianchini A, Maraschi AC, Seiblitz IGL, Solms B, Pereira CM, Mies M, Garland T, Navas CA (2026) Thermal refugia indicate physiological evolution in Atlantic reef corals. Biol J Linn Soc 147:blaf13

Ferreira LC, Grillo AC, Repinaldo Filho FP, Souza FN, Longo GO (2021) Different responses of massive and branching corals to a major heatwave at the largest and richest reef complex in South Atlantic. Mar Biol 168:54

Fine M, Cinar M, Voolstra CR, Safa A, Rinkevich B, Laffoley D, Hilmi N, Allemand D (2019) Coral reefs of the Red Sea—challenges and potential solutions. Reg Stud Mar Sci 25:100498

Floeter SR, Rocha LA, Robertson DR, Joyeux JC, Smith-Vaniz WF, Wirtz P, Edwards AJ, Barreiros JP, Ferreira CEL, Gasparini JL, Brito A, Falcón JM, Bowen BW, Bernardi G (2008) Atlantic reef fish biogeography and evolution. J Biogeogr 35:22–47

Francini-Filho RB, Asp NE, Siegle E, Hocevar J, Lowyck K, D'Avila N, Vasconcelos AA, Baitelo R, Rezende CE, Omachi CY, Thompson CC, Thompson FL (2018) Perspectives on the great Amazon reef: extension, biodiversity, and threats. Front Mar Sci 5:142

Furby KA, Bouwmeester J, Berumen ML (2013) Susceptibility of Central Red Sea corals during a major bleaching event. Coral Reefs 32:505–513

García-Sais JR, Williams SM, Amirrezvani A (2017) Mortality, recovery, and community shifts of scleractinian corals in Puerto Rico one decade after the 2005 regional bleaching event. PeerJ 5:e3611

Gates RD, Edmunds PJ (1999) The physiological mechanisms of acclimatization in tropical reef corals. Am Zool 39:30–43

Gattuso JP, Gentili B, Duarte CM, Kleypas JA, Middelburg JJ, Antoine D (2006) Light availability in the coastal ocean: impact on the distribution of benthic photosynthetic organisms and their contribution to primary production. Biogeosciences 3(4):489–513

Genevier LG, Jamil T, Raitsos DE, Krokos G, Hoteit I (2019) Marine heatwaves reveal coral reef zones susceptible to bleaching in the Red Sea. Glob Chang Biol 25(7):2338–2351

Giglio VJ, Aued AW, Cordeiro CA, Eggertsen L, Ferrari DS, Gonçalves LR, Hanazaki N, Luiz OJ, Luza AL, Mendes TC, Pinheiro HT (2024) A global systematic literature review of ecosystem services in reef environments. Environ Manag 73:634–645

Glynn PW, Feingold JS, Baker A, Banks S, Baums IB, Cole J, Colgan MW, Fong P, Glynn PJ, Keith I, Manzello D (2018) State of corals and coral reefs of the Galápagos Islands (Ecuador): past, present and future. Mar Pollut Bull 133:717–733

Godoy L, Mies M, Zilberberg C, Pastrana Y, Amaral A, Cruz N, Pereira CM, Garrido AG, Paris A, Santos LF, Pires DO (2021) Southwestern Atlantic reef-building corals *Mussismilia* spp. are able to spawn while fully bleached. Mar Biol 168:1–8

Golbuu Y, Fabricius K, Victor S, Richmond RH (2008) Gradients in coral reef communities exposed to muddy river discharge in Pohnpei, Micronesia. Estuar Coast Shelf Sci 1:14–20

Golbuu Y, van Woesik R, Richmond RH, Harrison P, Fabricius KE (2011) River discharge reduces reef coral diversity in Palau. Mar Pollut Bull 62(4):824–831

Gonzalez K, Daraghmeh N, Lozano-Cortés D, Benzoni F, Berumen ML, Carvalho S (2024) Differential spatio-temporal responses of Red Sea coral reef benthic communities to a mass bleaching event. Sci Rep 14(1):24229

Graham NA, Nash KL (2013) The importance of structural complexity in coral reef ecosystems. Coral Reefs 32:315–326

Guest JR, Tun K, Low J, Vergés A, Marzinelli EM, Campbell AH, Bauman AG, Feary DA, Chou LM, Steinberg PD (2016) 27 years of benthic and coral community dynamics on turbid, highly urbanised reefs off Singapore. Sci Rep 6:36260

Guinotte JM, Buddemeier RW, Kleypas JA (2003) Future coral reef habitat marginality: temporal and spatial effects of climate change in the Pacific basin. Coral Reefs 22:551–558

Harriott V, Banks S (2002) Latitudinal variation in coral communities in eastern Australia: a qualitative biophysical model of factors regulating coral reefs. Coral Reefs 21:83–94

Harriott VJ, Harrison PL, Banks SA (1995) The coral communities of Lord Howe Island. Mar Freshw Res 2:457–465

Hennige SJ, Smith DJ, Perkins R, Consalvey M, Paterson DM, Suggett DJ (2008) Photoacclimation, growth and distribution of massive coral species in clear and turbid waters. Mar Ecol Prog Ser 369:77–88

Hennige SJ, Smith DJ, Walsh SJ, McGinley MP, Warner ME, Suggett DJ (2010) Acclimation and adaptation of scleractinian coral communities along environmental gradients within an Indonesian reef system. J Exp Mar Biol Ecol 391(1-2):143–152

Hereher M, Bantan R, Gheith A, El-Kenawy A (2022) Spatio-temporal variability of sea surface temperatures in the Red Sea and their implications on Saudi Arabia coral reefs. Geocarto Int 19:5636–5652

Hinderstein LM, Marr JC, Martinez FA, Dowgiallo MJ, Puglise KA, Pyle RL, Zawada DG, Appeldoorn R (2010) Theme section on "Mesophotic coral ecosystems: characterization, ecology, and management". Coral Reefs 29:247–251

Hoitink AJ (2004) Tidally-induced clouds of suspended sediment connected to shallow-water coral reefs. Mar Geol 208:13–31

Houlbrèque F, Ferrier-Pagès C (2009) Heterotrophy in tropical scleractinian corals. Biol Rev 1:1–7

Huang Y, Gao B, Long C, Wang Y, Long L, Yang F (2024) The growth and nitrogen utilization strategies in two dominant Symbiodiniaceae species facing nitrogen deficiency and enrichment. Algal Res 84:103772

Hughes AD, Grottoli AG (2013) Heterotrophic compensation: a possible mechanism for resilience of coral reefs to global warming or a sign of prolonged stress? PLoS One 11:e81172

Hussain A, Kumar SH, Kumar AA, Prathiviraj R, Renjith K, Kiran GS, Selvin J (2025) Delineating the emergence of thermally tolerant Symbiodiniaceae genotypes across the dominant coral species of a turbid reef: adaptation and resilience strategy. Sci Total Environ 963:178255

Idjadi JA, Edmunds PJ (2006) Scleractinian corals as facilitators for other invertebrates on a Caribbean reef. Mar Ecol Prog Ser 319:117–127

Jones R, Fisher R, Bessell-Browne P (2019) Sediment deposition and coral smothering. PLoS One 6:e0216248

Kabanov P, Hauck TE, Gouwy SA, Grasby SE, van der Boon A (2023) Oceanic anoxic events, photic-zone euxinia, and controversy of sea-level fluctuations during the middle-late Devonian. Earth Sci Rev 241:104415

Kahng SE, Kelley C (2007) Vertical zonation of habitat forming benthic species on a deep photosynthetic reef (50–140 m) in the Au'au channel, Hawaii. Coral Reefs 26:679–687

Kahng SE, Garcia-Sais JR, Spalding HL, Brokovich E, Wagner D, Weil E, Hinderstein L, Toonen RJ (2010) Community ecology of mesophotic coral reef ecosystems. Coral Reefs 29:255–275

Kahng SE, Akkaynak D, Shlesinger T, Hochberg EJ, Wiedenmann J, Tamir R, Tchernov D (2019) Light, temperature, photosynthesis, heterotrophy, and the lower depth limits of mesophotic coral ecosystems. In: Loya Y, Puglise K, Bridge T (eds) Mesophotic coral ecosystems. Springer, Cham

Kahng SE, Watanabe TK, Hu HM, Watanabe T, Shen CC (2020) Moderate zooxanthellate coral growth rates in the lower photic zone. Coral Reefs 39:1273–1284

Kerry JT, Bellwood DR (2012) The effect of coral morphology on shelter selection by coral reef fishes. Coral Reefs 31:415–424

Kidder DL, Worsley TR (2004) Causes and consequences of extreme Permo-Triassic warming to globally equable climate and relation to the Permo-Triassic extinction and recovery. Palaeogeogr Palaeoclimatol Palaeoecol 203:207–237

Kleypas JA (1997) Modeled estimates of global reef habitat and carbonate production since the last glacial maximum. Paleoceanography 12:533–545

Kleypas JA, McManus JW, Meñez LAB (1999) Environmental limits to coral reef development: where do we draw the line? Am Zool 39:146–159

Knowlton N, Brainard RE, Fisher R, Moews M, Plaisance L, Caley MJ (2010) Coral reef biodiversity. In: McIntyre AD (ed) Life in the world's oceans: diversity, distribution, and abundance. Blackwell Publishing, Oxford, pp 65–77

Kobluk DR (1988) Cryptic faunas in reefs: ecology and geologic importance. PALAIOS 3:379–390

Kourafalou VH, Oey LY, Wang JD, Lee TN (1996) The fate of river discharge on the continental shelf: 1. Modeling the river plume and the inner shelf coastal current. J Geophys Res Oceans 101:3415–3434

Kramer N, Tamir R, Eyal G, Loya Y (2020) Coral morphology portrays the spatial distribution and population size-structure along a 5–100 m depth gradient. Front Mar Sci 7:615

LaJeunesse TC (2005) "Species" radiations of symbiotic dinoflagellates in the Atlantic and Indo-Pacific since the Miocene-Pliocene transition. Mol Biol Evol 22:570–581

LaJeunesse TC, Pettay DT, Sampayo EM, Phongsuwan N, Brown B, Obura DO, Hoegh-Guldberg O, Fitt WK (2010) Long-standing environmental conditions, geographic isolation and host–symbiont specificity influence the relative ecological dominance and genetic diversification of coral endosymbionts in the genus Symbiodinium. J Biogeogr 37:785–800

LaJeunesse TC, Parkinson JE, Gabrielson PW, Jeong HJ, Reimer JD, Voolstra CR, Santos SR (2018) Systematic revision of Symbiodiniaceae highlights the antiquity and diversity of coral endosymbionts. Curr Biol 28:2570–2580

Lalli C, Parsons TR (1997) Biological oceanography: an introduction. Elsevier

Lam KK (2000) Sexual reproduction of a low-temperature tolerant coral Oulastrea crispata (Scleractinia, Faviidae) in Hong Kong, China. Mar Ecol Prog Ser 205:101–111

Larcombe P, Costen A, Woolfe KJ (2001) The hydrodynamic and sedimentary setting of nearshore coral reefs, central great barrier reef shelf, Australia: Paluma shoals, a case study. Sedimentology 48:811–835

Larsen MC, Webb RM (2009) Potential effects of runoff, fluvial sediment, and nutrient discharges on the coral reefs of Puerto Rico. J Coast Res 25:189–208

Leão ZMAN, Kikuchi RKP, Testa V (2003) Corals and coral reefs of Brazil. In: Cortés J (ed) Latin American coral reefs. Elsevier, Amsterdam, pp 9–52

Leão ZMAN, Kikuchi RKP, Oliveira MDM (2019) The coral reef province of Brazil. In: Sheppard C (ed) World seas: an environmental evaluation: volume I: Europe, the Americas and West Africa. Academic Press, pp 813–833

Lear CH, Coxall HK, Foster GL, Lunt DJ, Mawbey EM, Rosenthal Y, Sosdian SM, Thomas E, Wilson PA (2015) Neogene ice volume and ocean temperatures: insights from infaunal foraminiferal Mg/Ca paleothermometry. Paleoceanography 30:1437–1454

Lesser MP, Slattery M (2009) Ecology of mesophotic coral reefs. J Exp Mar Biol Ecol 375:1–8

Lesser MP, Slattery M, Stat M, Ojimi M, Gates RD, Grottoli A (2010) Photoacclimatization by the coral Montastraea cavernosa in the mesophotic zone: light, food, and genetics. Ecology 91:990–1003

Lesser MP, Mobley CD, Hedley JD, Slattery M (2021) Incident light on mesophotic corals is constrained by reef topography and colony morphology. Mar Ecol Prog Ser 670:49–60

Li G, Cheng L, Zhu J, Trenberth KE, Mann ME, Abraham JP (2020) Increasing ocean stratification over the past half-century. Nat Clim Chang 10:1116–1123

Lirman D, Schopmeyer S, Manzello D, Gramer LJ, Precht WF, Muller-Karger F, Banks K, Barnes B, Bartels E, Bourque A, Byrne J (2011) Severe 2010 cold-water event caused unprecedented mortality to corals of the Florida reef tract and reversed previous survivorship patterns. PLoS One 6:e23047

Liu C, Zhang Y, Huang L, Yu X, Luo Y, Jiang L, Sun Y, Liu S, Huang H (2022) Differences in fatty acids and lipids of massive and branching reef-building corals and response to environmental changes. Front Mar Sci 9:882663

López-Londoño T, Galindo-Martínez CT, Gómez-Campo K, González-Guerrero LA, Roitman S, Pollock FJ, Pizarro V, López-Victoria M, Medina M, Iglesias-Prieto R (2021) Physiological and ecological consequences of the water optical properties degradation on reef corals. Coral Reefs 40:1243–1256

Loures A, Rädecker N, Voolstra CR, Garrido AG, Zilberberg C, Faria SC, Sumida PYG, Mies M (2026) The larva-Symbiodiniaceae association at risk: putative impacts of climate change on reproduction, dispersal, and recruitment in coral reefs. Coral Reefs 45:1015–1037. https://doi.org/10.1007/s00338-025-02777-9

Loya Y, Sakai K, Yamazato K, Nakano Y, Sambali H, Van Woesik R (2001) Coral bleaching: the winners and the losers. Ecol Lett 4:122–131

Luza AL, Aued AW, Barneche DR, Dias MS, Ferreira CEL, Floeter SR, Francini-Filho RB, Longo GO, Quimbayo JP, Bender MG (2023) Functional diversity patterns of reef fish, corals and algae in the Brazilian biogeographical province. J Biogeogr 50:1163–1176

Madin JS, Anderson KD, Andreasen MH, Bridge TC, Cairns SD, Connolly SR, Darling ES, Diaz M, Falster DS, Franklin EC, Gates RD et al (2016) The coral trait database, a curated database of trait information for coral species from the global oceans. Sci Data 3:1–22

Maor-Landaw K, Levy O (2016) Gene expression profiles during short-term heat stress; branching vs. massive Scleractinian corals of the Red Sea. PeerJ 4:e1814

Mariano AEB, Mies M, Abessa DMS (2026) Reducing local stressors enhances the resistance of the coral Mussismilia hispida and the zoantharian Palythoa caribaeorum to global stressors in island reefs under marginal conditions. Coral Reefs. https://doi.org/10.1007/s00338-026-02879-y

Mayfield AB, Gates RD (2007) Osmoregulation in anthozoan–dinoflagellate symbiosis. Comp Biochem Physiol A Mol Integr Physiol 147:1–10
McClanahan TR, Obura D (1997) Sedimentation effects on shallow coral communities in Kenya. J Exp Mar Biol Ecol 209:103–122
McClanahan TR, Baird AH, Marshall PA, Toscano MA (2004) Comparing bleaching and mortality responses of hard corals between southern Kenya and the great barrier reef, Australia. Mar Pollut Bull 48:327–335
McFadden CS, Quattrini AM, Brugler MR, Cowman PF, Dueñas LF, Kitahara MV, Paz-García DA, Reimer JD, Rodríguez E (2021) Phylogenomics, origin, and diversification of Anthozoans (Phylum Cnidaria). Syst Biol 70:635–647
McIlroy SE, TerHorst CP, Teece M, Coffroth MA (2022) Nutrient dynamics in coral symbiosis depend on both the relative and absolute abundance of Symbiodiniaceae species. Microbiome 10:192
McLaughlin CJ, Smith CA, Buddemeier RW, Bartley JD, Maxwell BA (2003) Rivers, runoff, and reefs. Glob Planet Chang 39:191–199
Meziere Z, Rich WA, Carvalho S, Benzoni F, Morán XA, Berumen ML (2022) Stylophora under stress: a review of research trends and impacts of stressors on a model coral species. Sci Total Environ 816:151639
Mies M, Güth AZ, Tenório AA, Banha TNS, Waters LG, Polito PS, Taniguchi S, Bícego MC, Sumida PYG (2018) In situ shifts of predominance between autotrophic and heterotrophic feeding in the reef-building coral Mussismilia hispida: an approach using fatty acid trophic markers. Coral Reefs 37:677–689
Mies M, Francini-Filho RB, Zilberberg C, Garrido AG, Longo GO, Laurentino E, Güth AZ, Sumida PYG, Banha TNS (2020) South Atlantic coral reefs are major global warming refugia and less susceptible to bleaching. Front Mar Sci 7:514
Mies M, Banha TNS, Castro CB, Cordeiro RTS, Lotufo TMC, Bianchini A (2024) Biology of Brazilian benthic reef builders and dwellers. In: Kikuchi RKP, Leão ZMAN, Araújo ME, Lotufo TMC (eds) Brazilian coral reefs: a multidisciplinary approach. Springer-Verlag, pp 81–112
Mies M, Destri G, Lacerda CHF, Carvalho JLM, Ibanhez JY, Güth AZ, Luza AL, Campos LP, Shimada AC, Omaki KNT, Jahn BE, Abreu YA, Amario M, Andrade MG, Angonese MS, Banha T, Barros F, Berrettini AB, Bianchini A, Bleuel J, Braz GB et al (2025a) Coral bleaching and mortality across a 24° latitudinal range in the Southwestern Atlantic during the fourth global bleaching event. Coral Reefs. https://doi.org/10.1007/s00338-025-02743-5
Mies M, Longo GO, Bianchini A, Calderon EN, Castro CB, Faria SC, Francini-Filho RB, Guebert FM, Kitahara MV, Lacerda CHF, Lotufo TMC, Marangoni LFB, Pires DO, Cordeiro RTS (2025b) Challenges for coral restoration in Southwestern Atlantic reefs: guidelines for ethical and sustainable practices. Biodivers Conserv 34:2287–2313
Mo S, Chen T, Chen Z, Zhang W, Li S (2022) Marine heatwaves impair the thermal refugia potential of marginal reefs in the northern South China Sea. Sci Total Environ 825:154100
Moberg F, Folke C (1999) Ecological goods and services of coral reef ecosystems. Ecol Econ 29:215–233
Monroe AA, Ziegler M, Roik A, Röthig T, Hardenstine RS, Emms MA, Jensen T, Voolstra CR, Berumen ML (2018) In situ observations of coral bleaching in the central Saudi Arabian Red Sea during the 2015/2016 global coral bleaching event. PLoS One 13:e0195814
Morgan KM, Perry CT, Smithers SG, Johnson JA, Daniell JJ (2016) Evidence of extensive reef development and high coral cover in nearshore environments: implications for understanding coral adaptation in turbid settings. Sci Rep 6:29616
Morris LA, Voolstra CR, Quigley KM, Bourne DG, Bay LK (2019) Nutrient availability and metabolism affect the stability of coral–Symbiodiniaceae symbioses. Trends Microbiol 27:678–689
Mostafavi PG, Fatemi SM, Shahhosseiny MH, Hoegh-Guldberg O, Loh WK (2007) Predominance of clade D Symbiodinium in shallow-water reef-building corals off Kish and Larak Islands (Persian Gulf, Iran). Mar Biol 153:25–34
Munnecke A, Calner M, Harper DA, Servais T (2010) Ordovician and Silurian sea–water chemistry, sea level, and climate: a synopsis. Palaeogeogr Palaeoclimatol Palaeoecol 296:389–413
Mwaura J, Umezawa Y, Furaha J, Kimeli A, Kamau J, Aura CM (2017) Spatial variability of scleractinian coral bleaching susceptibility in 2010 El Niño–Southern Oscillation between northern and southern reefs, Kenya. Coast Mar Sci 40:17–27
Nitschke MR, Rosset SL, Oakley CA, Gardner SG, Camp EF, Suggett DJ, Davy SK (2022) The diversity and ecology of Symbiodiniaceae: a traits-based review. Adv Mar Biol 92:55–127
Nittrouer CA, DeMaster DJ (1996) The Amazon shelf setting: tropical, energetic, and influenced by a large river. Cont Shelf Res 16:553–573
Nugues MM, Roberts CM (2003) Coral mortality and interaction with algae in relation to sedimentation. Coral Reefs 22:507–516
O'Brien JM, Scheibling RE (2018) Turf wars: competition between foundation and turf-forming species on temperate and tropical reefs and its role in regime shifts. Mar Ecol Prog Ser 590:1–7
Obura DA, Suleiman MO, Motta HE, Schleyer MI (2000) Status of coral reefs in East Africa: Kenya, Mozambique, South Africa and Tanzania. In: Status of coral reefs of the world. Australian Institute of Marine Science (AIMS), pp 65–76
Ogston AS, Storlazzi CD, Field ME, Presto MK (2004) Sediment resuspension and transport patterns on a fringing reef flat, Molokai, Hawaii. Coral Reefs 23:559–569
Oladi M, Shokri MR, Rajabi-Maham H (2019) Extremophile symbionts in extreme environments; a contribution to the diversity of Symbiodiniaceae across the northern Persian Gulf and Gulf of Oman. J Sea Res 144:105–111
Oliveira LS, Sanches FH, Silva MA, PAS L, Laurino IR, Motta FS, Pereira-Filho GH (2025) Calcium carbonate production by the massive coral Mussismilia hispida in subtropical reefs of the southwestern Atlantic. Mar Environ Res 209:107218
Osman EO, Smith DJ, Ziegler M, Kürten B, Conrad C, El-Haddad KM, Voolstra CR, Suggett DJ (2018) Thermal refugia against coral bleaching throughout the northern Red Sea. Glob Chang Biol 24:e474–e484
Otero E, Carbery KK (2005) Chlorophyll a and turbidity patterns over coral reefs systems of La Parguera natural reserve, Puerto Rico. Rev Biol Trop 53:25–32
Pagès J, Torreton JP, Sempéré R (1997) Dissolved organic carbon in coral-reef lagoons, by high temperature catalytic oxidation and UV spectrometry. C R Acad Sci Ser IIA Earth Planet Sci 324:915–922
Pandolfi JM, Kiessling W (2014) Gaining insights from past reefs to inform understanding of coral reef response to global climate change. Curr Opin Environ Sustain 7:52–58
Park E, Hwang DS, Lee JS, Song JI, Seo TK, Won YJ (2012) Estimation of divergence times in cnidarian evolution based on mitochondrial protein-coding genes and the fossil record. Mol Phylogenet Evol 62:329–345
Patrocinio GT, Lopes FC, Ciotti ÁM, Faria SC, Martins CD (2025) The southwestern Atlantic coral Mussismilia hispida exhibit remarkable tolerance to nitrate enrichment. Coral Reefs 44(2):1–4
Pereira CM, Fonseca JS, Paiva ES, Costa PG, Mies M, Silva AG, Calderon EN, Bianchini A, Castro CB (2020) Larvae of the South Atlantic coral Favia gravida are tolerant to salinity and nutrient concentrations associated with river discharges. Mar Environ Res 161:105118
Pereira PH, Lima GV, Pontes AV, Côrtes LG, Gomes E, Sampaio CL, Seoane JCS (2022) Unprecedented coral mortality on Southwestern Atlantic coral reefs following major thermal stress. Front Mar Sci 9:725778

Perry CT, Larcombe P (2003) Marginal and non-reef-building coral environments. Coral Reefs 22:427–432

Picciani N, Seiblitz IGL, Paiva PC, Castro CB, Zilberberg C (2016) Geographic patterns of Symbiodinium diversity associated with the coral Mussismilia hispida (Cnidaria, Scleractinia) correlate with major reef regions in the southwestern Atlantic Ocean. Mar Biol 163:236

Pires DO, Segal B, Caparelli AC (2011) Reproductive effort of an endemic major reef builder along an inshore–offshore gradient in South-Western Atlantic. J Mar Biol Assoc U K 91:1613–1616

Pochon X, Forsman ZH, Spalding HL, Padilla-Gamiño JL, Smith CM, Gates RD (2015) Depth specialization in mesophotic corals (Leptoseris spp.) and associated algal symbionts in Hawai'i. R Soc Open Sci 2:140351

Pond S, Pickard GL (1983) Introductory dynamical oceanography. Pergamon Press Ltd

Price JF, Weller RA, Schudlich RR (1987) Wind-driven ocean currents and Ekman transport. Science 238:1534–1538

Pupier CA, Mies M, Fine M, Bastos Francini-Filho R, Pereira Brandini F, Zambotti-Villela L, Colepicolo P, Ferrier-Pagès C (2021) Lipid biomarkers reveal the trophic plasticity of octocorals along a depth gradient. Limnol Oceanogr 66:2078–2087

Rabalais NN, Cai WJ, Carstensen J, Conley DJ, Fry B, Hu X, Quinones-Rivera Z, Rosenberg R, Slomp CP, Turner RE, Voss M (2014) Eutrophication-driven deoxygenation in the coastal ocean. Oceanography 27:172–183

Restrepo JD, Park E, Aquino S, Latrubesse EM (2016) Coral reefs chronically exposed to river sediment plumes in the southwestern Caribbean: Rosario Islands, Colombia. Sci Total Environ 553:316–329

Reuter M, Bosellini FR, Budd AF, Ćorić S, Piller WE, Harzhauser M (2019) High coral reef connectivity across the Indian Ocean is revealed 6–7 ma ago by a turbid-water scleractinian assemblage from Tanzania (Eastern Africa). Coral Reefs 38:1023–1037

Riegl B, Johnston M, Glynn PW, Keith I, Rivera F, Vera-Zambrano M, Banks S, Feingold J, Glynn PJ (2019) Some environmental and biological determinants of coral richness, resilience and reef building in Galápagos (Ecuador). Sci Rep 9:10322

Rogers CS (1990) Responses of coral reefs and reef organisms to sedimentation. Mar Ecol Prog Ser 62:185–202

Rosset S, Wiedenmann J, Reed AJ, D'Angelo C (2017) Phosphate deficiency promotes coral bleaching and is reflected by the ultrastructure of symbiotic dinoflagellates. Mar Pollut Bull 118:180–187

Samiei JV, Saleh A, Mehdinia A, Shirvani A, Kayal M (2015) Photosynthetic response of Persian Gulf acroporid corals to summer versus winter temperature deviations. PeerJ 3:e1062

Santana EFC, Mies M, Longo GO, Menezes R, Aued AW, Luza AL, Bender MG, Segal B, Floeter SR, Francini-Filho RB (2023) Turbidity shapes shallow southwestern Atlantic benthic reef communities. Mar Environ Res 183:105807

Sasaki DK, Silva D, Del Giovannino SR Jr, ICA S, Belo WC, Martins RP, Dottori M (2024) Hydrographic climatology of the South Brazil bight continental shelf and slope. Theor Appl Climatol 155(10):1–9

Sawall Y, Al-Sofyani A, Kürten B, Al-Aidaroos AM, Hoang Xuan B, Marimuthu N, Khomayis HS, Sommer U, Gharbawi WY, Wahl M (2014) Coral communities, in contrast to fish communities, maintain a high assembly similarity along the large latitudinal gradient along the Saudi Red Sea coast. J Ecosyst Ecography 4(003):1–7

Schoepf V, Baumann JH, Barshis DJ, Browne NK, Camp EF, Comeau S, Cornwall CE, Guzmán HM, Riegl B, Rodolfo-Metalpa R, Sommer B (2023) Corals at the edge of environmental limits: a new conceptual framework to re-define marginal and extreme coral communities. Sci Total Environ 884:163688

Semmler RF, Hoot WC, Reaka ML (2017) Are mesophotic coral ecosystems distinct communities and can they serve as refugia for shallow reefs? Coral Reefs 36:433–444

Sheppard C (2009) Large temperature plunges recorded by data loggers at different depths on an Indian Ocean atoll: comparison with satellite data and relevance to coral refuges. Coral Reefs 28:399–403

Sheppard CRC, Davy SK, Pilling GM, Graham NAJ (2018) The biology of coral reefs. Oxford University Press, Oxford

Silva DA, Sasaki DK, Dottori M, Silveira ICA, Belo WC, Martins RP (2023) An inventory of salty and warm subsurface intrusions in the South Brazil bight. Cont Shelf Res 259:104984

Silveira ICA, Miranda LBM, Brown WS (1994) On the origins of the North Brazil current. J Geophys Res Oceans 99:22501–22512

Slattery M, Lesser MP, Brazeau D, Stokes MD, Leichter JJ (2011) Connectivity and stability of mesophotic coral reefs. J Exp Mar Biol Ecol 408:32–41

Smith EG, Gurskaya A, Hume BC, Voolstra CR, Todd PA, Bauman AG, Burt JA (2020) Low Symbiodiniaceae diversity in a turbid marginal reef environment. Coral Reefs 39:545–553

Soares MO (2020) Marginal reef paradox: a possible refuge from environmental changes? Ocean Coastal Manage 185:105063

Soares MO, Davis M, Paiva CC, PBM C (2018) Mesophotic ecosystems: coral and fish assemblages in a tropical marginal reef (northeastern Brazil). Mar Biodivers 48:1631–1636

Soares MO, Cruz IC, Santos BA, Tavares TC, Garcia TM, Menezes N, Lopes BD, de Araújo JT, Gurgel AL, Rossi S (2020) Marginal reefs in the Anthropocene: they are not Noah's ark. In: Perspectives on the marine animal forests of the world. Springer, pp 87–128

Sofonia JJ, Anthony KR (2008) High-sediment tolerance in the reef coral Turbinaria mesenterina from the inner great barrier reef lagoon (Australia). Estuar Coast Shelf Sci 78:748–752

Sommer B, Butler IR, Pandolfi JM (2021) Trait-based approach reveals how marginal reefs respond to acute and chronic disturbance. Coral Reefs 40:735–749

Sorokin YI (1973) On the feeding of some scleractinian corals with bacteria and dissolved organic matter. Limnol Oceanogr 18:380–386

Stanley GD, Fautin DG (2001) The origins of modern corals. Science 291:1913–1914

Stanley GD Jr (2003) The evolution of modern corals and their early history. Earth Sci Rev 60:195–225

Sun Y (2024) Dynamics of nutrient cycles in the Permian–Triassic oceans. Earth Sci Rev 258:104914

Suzuki G, Keshavmurthy S, Hayashibara T, Wallace CC, Shirayama Y, Chen CA, Fukami H (2016) Genetic evidence of peripheral isolation and low diversity in marginal populations of the Acropora hyacinthus complex. Coral Reefs 35:1419–1432

Swain TD, Chandler J, Backman V, Marcelino L (2017) Consensus thermotolerance ranking for 110 Symbiodinium phylotypes: an exemplar utilization of a novel iterative partial-rank aggregation tool with broad application potential. Funct Ecol 31(1):172–183

Thomson DP, Frisch AJ (2010) Extraordinarily high coral cover on a nearshore, high-latitude reef in south-West Australia. Coral Reefs 29(4):923–927

Torres JL, Morelock JA (2002) Effect of terrigenous sediment influx on coral cover and linear extension rates of three Caribbean massive coral species. Caribb J Sci 38(3/4):222–229

Tremblay P, Grover R, Maguer JF, Legendre L, Ferrier-Pagès C (2012) Autotrophic carbon budget in coral tissue: a new 13C-based model of photosynthate translocation. J Exp Biol 215(8):1384–1393

Valentine JW (1984) Neogene marine climate trends: implications for biogeography and evolution of the shallow-sea biota. Geology 12(11):647–650

van Katwijk MM, Meier NF, Van Loon R, Van Hove EM, Giesen WB, Van der Velde G, Den Hartog C (1993) Sabaki River sediment load and coral stress: correlation between sediments and condition of the Malindi-Watamu reefs in Kenya (Indian Ocean). Mar Biol 117:675–683

van Woesik R, Done TJ (1997) Coral communities and reef growth in the southern great barrier reef. Coral Reefs 16:103–115
Vasseur R, Lathuilière B, Lazăr I, Martindale RC, Bodin S, Durlet C (2021) Major coral extinctions during the early Toarcian global warming event. Glob Planet Chang 207:103647
Veron JE, Stafford-Smith MG, Turak E, DeVantier LM (2016) Corals of the World. Australian Institute of Marine Science (AIMS). Available online at: http://www.coralsoftheworld.org/page/home/. Accessed 20 May 2025
Wall M, Schmidt GM, Janjang P, Khokiattiwong S, Richter C (2012) Differential impact of monsoon and large amplitude internal waves on coral reef development in the Andaman Sea. PLoS One 7(11):e50207
Wang M, Son S, Harding LW Jr (2009) Retrieval of diffuse attenuation coefficient in the Chesapeake Bay and turbid ocean regions for satellite ocean color applications. J Geophys Res Oceans 114:C10011
Wiedenmann J, D'Angelo C, Smith EG, Hunt AN, Legiret FE, Postle AD, Achterberg EP (2013) Nutrient enrichment can increase the susceptibility of reef corals to bleaching. Nat Clim Chang 3:160–164
Wolanski E, Fabricius K, Spagnol S, Brinkman R (2005) Fine sediment budget on an inner-shelf coral-fringed Island, great barrier reef of Australia. Estuar Coast Shelf Sci 65:153–158
Wood R (1999) Reef evolution. Oxford University Press, Oxford
World Register of Marine Species (WoRMS) (2024). Available online at: https://www.marinespecies.org. Accessed 20 Oct 2024
Wyatt AS, Leichter JJ, Washburn L, Kui L, Edmunds PJ, Burgess SC (2023) Hidden heatwaves and severe coral bleaching linked to mesoscale eddies and thermocline dynamics. Nat Commun 14(1):25
Yahel G, Sharp JH, Marie D, Häse C, Genin A (2003) In situ feeding and element removal in the symbiont-bearing sponge Theonella swinhoei: bulk DOC is the major source for carbon. Limnol Oceanogr 48:141–149
Yee SH, Barron MG (2010) Predicting coral bleaching in response to environmental stressors using 8 years of global-scale data. Environ Monit Assess 161:423–438
Zhou G, Huang H, Lian J, Zhang C, Li X (2012) Habitat correlation of Symbiodinium diversity in two reef-building coral species in an upwelling region, eastern Hainan Island, China. J Mar Biol Assoc U K 92:1309–1316
Ziegler M, Roder CM, Büchel C, Voolstra CR (2015) Mesophotic coral depth acclimatization is a function of host-specific symbiont physiology. Front Mar Sci 2:4
Ziegler M, Arif C, Voolstra CR (2019) Symbiodiniaceae diversity in Red Sea coral reefs & coral bleaching. In: Coral Reefs of the Red Sea. Springer, pp 69–89
Zweifler A, O'Leary M, Morgan K, Browne NK (2021) Turbid coral reefs: past, present and future—a review. Diversity 13:251

Diversity and Vulnerability of Reef Fishes in Marginal and Extreme Habitats

3

Juliano Morais, Alexandre C. Siqueira, Luiz A. Rocha, Aline P. M. Medeiros, Aiara P. L. R. Cardoso, and Bráulio A. Santos

Abstract

In this book chapter, we review and discuss the diversity and vulnerability of reef fishes in marginal and extreme habitats. These habitats, characterised by elevated rates of siliciclastic sedimentation, extreme levels of turbidity, significant depths, elevated nutrient levels, and/or substantial temperature variability, among other factors, present unique conditions for reef fish assemblages. By synthesising current literature, we explore the evolutionary, biogeographic, and ecological aspects of reef fish diversity in these habitats. We also examine the impacts of human-induced stressors such as global warming, and plastic pollution, on these assemblages. The chapter emphasizes the importance of integrating historical and geographical perspectives in research to better understand and protect these unique ecosystems. Additionally, we highlight the need for targeted conservation and management strategies, such as marine spatial planning, zoning for fisheries, oil and gas exploitation, and other large-scale economic activities, to preserve the diversity and ecological functions of reef fish in marginal and extreme habitats. Our review underscores the urgency of adopting resilience-based management actions at local and global scales to mitigate the adverse effects of environmental changes on these vulnerable ecosystems. By understanding the dynamics of reef fish in these non-optimal environments, we can better inform conservation efforts and ensure the long-term sustainability of coral reef ecosystems under the growing pressures of the Anthropocene.

Keywords

Coral reefs · Reef fish diversity · Marginal habitats · Extreme habitats · Conservation strategies.

3.1 Introduction

Coral reefs harbour the highest biodiversity of all marine ecosystems globally (Connell 1978; Roberts et al. 2002). These ecosystems hold immense ecological and economic importance through services like carbon storage, tourism, materials, coastal protection, and valuable biochemicals, making it essential for human societies and industries (Souter and Lindén 2000; Chen et al. 2015; Woodhead et al. 2019; Eddy et al. 2021; Carlot et al. 2023). More importantly, coral reefs are a lifeline for millions of small-scale fishers (Newton et al. 2007; Teh et al. 2013), ensuring a continuous food source for coastal communities throughout the tropics (Teh et al. 2013), even when facing the challenges of fishing-induced biomass depletion (Donner and Potere 2007; Morais et al. 2023).

Beyond their socio-economic importance, coral reefs maintain extensive biological and genetic diversity, serving as a cornerstone for important ecological and evolutionary processes, such as species diversification, habitat provision, nutrient recycling, and the maintenance of complex food

J. Morais (✉)
College of Science and Engineering, James Cook University, Townsville, QLD, Australia

Centro de Ciências exatas e da Natureza, Departamento de Sistemática e ecologia, Universidade Federal da Paraíba, Cidade Universitária, Castelo Branco, João Pessoa, PB, Brazil
e-mail: juliano.morais@my.jcu.edu.au

A. C. Siqueira
Research Hub for Coral Reef Ecosystem Functions and College of Science and Engineering, James Cook University, Townsville, QLD, Australia

Centre for Marine Ecosystems Research, School of Science, Edith Cowan University, Perth, WA, Australia

L. A. Rocha
Department of Ichthyology, California Academy of Sciences, San Francisco, CA, USA

A. P. M. Medeiros · Aiara P. L. R. Cardoso · B. A. Santos
Centro de Ciências exatas e da Natureza, Departamento de Sistemática e ecologia, Universidade Federal da Paraíba, Cidade Universitária, Castelo Branco, João Pessoa, PB, Brazil

M. O. Soares et al. (eds.), *Extreme and Marginal Coral Reefs and Communities*, Coral Reefs of the World 24,
https://doi.org/10.1007/978-3-032-28035-0_3

webs (Moberg and Folke 1999; Graham and Nash 2013; Kerry and Bellwood 2015). Coral reef-associated fishes are arguably among the most important groups in driving some of these processes (Sale 2015). Reef fishes deliver critical functions, facilitating coral reef processes including biomass production, reef framework erosion, sediment reworking, macroalgae regulation, and predation (Bellwood 1995; Morais and Bellwood 2019; Mihalitsis et al. 2021; Schlaefer et al. 2022; Tebbett et al. 2023). However, coral reef fishes have been severely affected by overexploitation, habitat loss, and climate change, especially during thermal anomalies (Eddy et al. 2021).

Traditionally, most of the research exploring the diversity and ecology of coral reef fishes predominantly focused on areas typified by optimal conditions, or "classic" coral reefs (Berumen et al. 2013), especially in the Caribbean and Australia. These conditions generally imply shallow waters in low-latitude settings, featuring low nutrient levels, suitable pH levels, high water transparency, and low temperature fluctuations. Combined, these conditions enhance carbonate production, promote high benthic productivity, and support biodiversity (Cabioch et al. 2011; Hughes et al. 2018b; Morais et al. 2024b). Indeed, while most marine processes, including reef fish-driven functions, are well-studied in shallow "classic" reefs worldwide, research in marginal and extreme environments is still in its infancy (Morais et al. 2018; Rocha et al. 2018; Burt et al. 2020; Soares et al. 2021; Sommer 2022; Schoepf et al. 2023).

Marginal and extreme coral reefs are generally defined as ecosystems that exist in conditions that depart from the optimal for coral reef growth/functioning (Schoepf et al. 2023). These conditions can include elevated rates of siliciclastic sedimentation, extreme levels of turbidity, significant depths, elevated nutrient levels, heightened primary productivity, and/or substantial temperature variability, among other factors (Kleypas et al. 1999; Chow et al. 2019; Soares et al. 2021; Zweifler et al. 2021; Browne and Bauman 2023). Such characteristics bring substantial challenges to directly surveying fish assemblages, likely explaining the low number of studies performed in marginal and extreme coral reefs (Rocha et al. 2018; Pyle 2019; Medeiros et al. 2021; Pinheiro et al. 2023a). Yet, research interest on this topic has been increasing globally (Burt et al. 2020), and is revealing that marginal and extreme reef ecosystems are indeed unique and should be the focus of more research and conservation efforts (Soares 2020; Slattery et al. 2024).

In a recent review focused on coral communities, Schoepf et al. (2023) proposed a definition that differentiates between extreme reefs (i.e. reefs characterised by large deviations from optimal abiotic conditions) and marginal reefs (i.e. reefs with altered coral cover, community composition, and ecosystem functioning relative to reefs developing under optimal environmental conditions, such as those in shallow, clear-water tropical settings). Here, we apply these definitions in conjunction to explore the evolutionary, biogeographic, and ecological aspects of fish diversity in both marginal and extreme coral reefs. We identify potential challenges and opportunities for further research, and discuss the vulnerability of assemblages in marginal and extreme reefs at local and global scales. Finally, we highlight the main human-induced pressures that threaten these unique environments, and suggest conservation and management actions that can help protect their biodiversity. Our chapter draws on the recent literature to stimulate a better understanding of marginal and extreme reef fish assemblages.

3.2 Diversity of Reef Fishes in Marginal and Extreme Habitats

3.2.1 Global Patterns of Diversity

3.2.1.1 Evolution and Diversity of Reef Fishes in Marginal and Extreme Habitats

Studies exploring the evolution and diversity of reef fishes in marginal and extreme habitats generally focus on comparing macroevolutionary processes (i.e. diversification and dispersal) and assemblage composition between typical coral reef habitats and deep reefs (e.g. Tornabene et al. 2016; Baldwin et al. 2018; Rocha et al. 2018; Pinheiro et al. 2023a), high-latitude reefs (e.g. Floeter et al. 2004; Siqueira et al. 2016; Stuart-Smith et al. 2021), and thermally extreme reefs (e.g. Brandl et al. 2020; Bouwmeester et al. 2022). A common finding across these studies is that marginal and extreme reefs are less diverse and have simplified fish assemblage structures compared to shallow-water coral reefs (Fig. 3.1). For example, Stuart-Smith et al. (2021) found that being a generalist was one of the most important traits in predicting the poleward range extent of reef fish species across a large latitudinal gradient, from tropical coral reefs to temperate rocky reefs. Similarly, Pinheiro et al. (2023a) identified that deeper reefs are generally composed of fewer trophic guilds, regardless of biogeographic location. In conjunction, these studies highlight the ubiquitous role of environmental filters (see Cavender-Bares et al. 2009) in shaping the composition of reef fishes across abiotic gradients.

From an evolutionary perspective, understanding reef fish community assemblages over time involves recognizing that the diversity gradient from core coral reef areas towards marginal or extreme reefs is defined by species tolerances to environmental parameters. Marginal reefs tend to contain a subset of lineages derived from high-diversity areas (Robertson 1998; Siqueira et al. 2016; Stuart-Smith et al. 2021). However, this pattern shifts when certain thresholds of environmental conditions are reached, leading to a turnover in fish assemblage composition. The latitudinal distri-

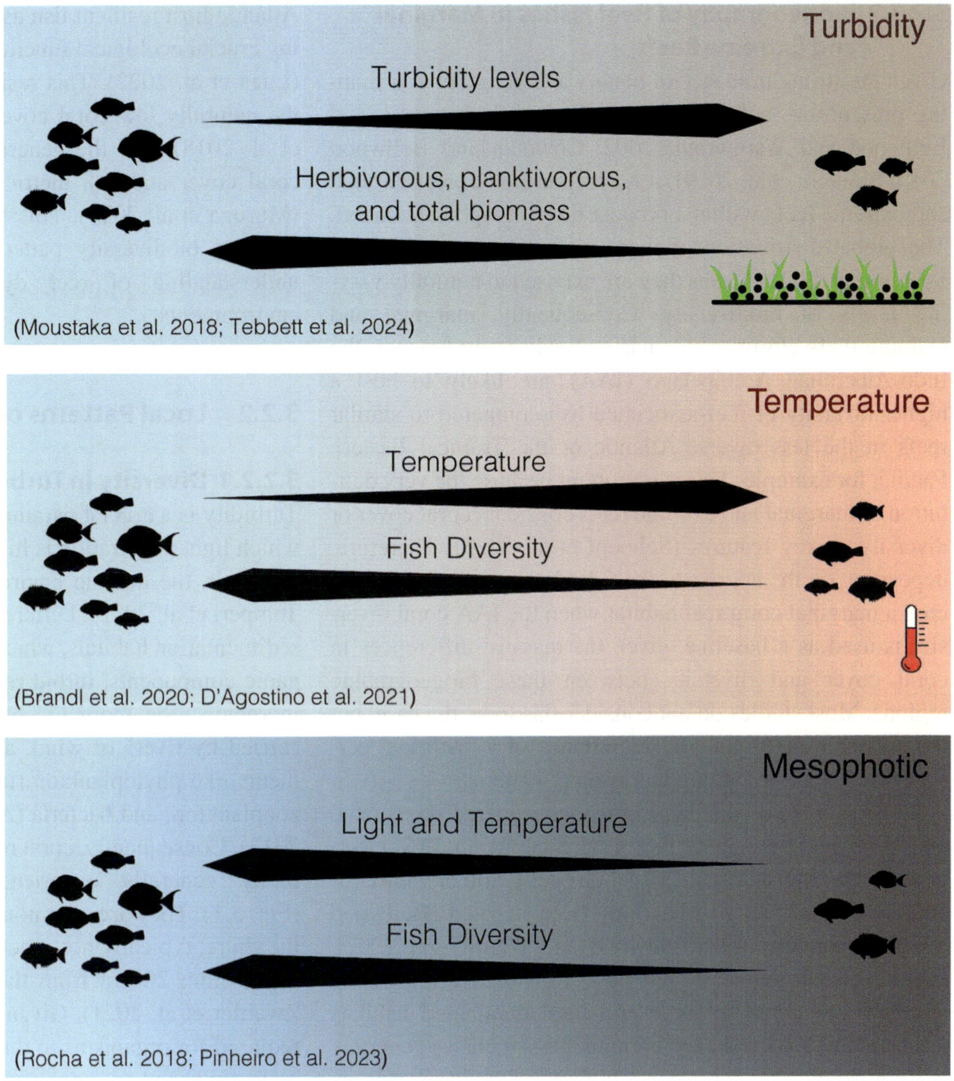

Fig. 3.1 Conceptual diagram showing the effects of various marginal and extreme habitats on reef fish assemblages. (Modified from Tebbett et al. 2024)

bution of reef fish biodiversity illustrates this point well. In many high-latitude reefs, such as those in subtropical regions (e.g. Floeter et al. 2004; Quimbayo et al. 2023), fish assemblages represent a subset of their tropical counterparts, with typical coral reef fish families (e.g. Pomacentridae, Labridae, Chaetodontidae) still present (Siqueira et al. 2016). However, at the extreme latitudes of temperate reefs, the composition becomes highly dissimilar (Bellwood and Wainwright 2002), and extra-tropical taxa with smaller range sizes (Pie et al. 2021) start to dominate assemblages.

This transition from subtropical to extreme high-latitude temperate reefs thus indicates a fundamental shift from environmental tolerances to ecological opportunity. For instance, classic coral reef-associated families show higher diversification rates in tropical waters (Siqueira et al. 2016). However, in extremely marginal habitats, including polar and deep-ocean regions, diversification rates may surpass those of tropical lineages (Rabosky et al. 2018; Friedman and Muñoz 2023). This suggests that the level of habitat extremism significantly influences the evolutionary dynamics of marine fish lineages. In marginal zones, tropical, coral reef-associated lineages with broader tolerances tend to be the source of dispersal, while in very extreme zones, ecological opportunity drives evolutionary rates. Essentially, if lineages can cross the most extreme environmental filters, they tend to radiate and become specialised, given the opportunities offered by newly colonized habitats.

These very extreme examples (i.e. polar and aphotic marine zones; Baldwin et al. 2018; Rabosky et al. 2018) provide valuable insights into the evolutionary and ecological processes determining species compositions. However, our focus here lies on the less extreme and/or marginal reefs, where fish assemblages resemble, at least in higher taxonomic levels, "classic" coral reefs. Therefore, in the next sections, we will describe patterns related to fish lineages with tropical reef affinities.

3.2.1.2 Biogeography of Reef Fishes in Marginal and Extreme Reefs

Given the strong influence of history and geography in shaping present-day patterns of global reef fish diversity (e.g. Bellwood and Wainwright 2002; Cowman and Bellwood 2013; Siqueira et al. 2019), it is crucial to interpret marginal and extreme reefs within a broader biogeographical context. The global distribution of these reefs (Camp et al. 2018; Soares et al. 2021) means they are exposed to naturally varying levels of biodiversity. Consequently, marginal and extreme reefs closer to the marine biodiversity hotspot, the Indo-Australian Archipelago (IAA), are likely to host a higher diversity of reef-associated fish compared to similar reefs in the less diverse Atlantic or the Tropical Eastern Pacific, for example. This is important because the very definition of marginal and extreme reefs considers coral cover or diversity as key features (Schoepf et al. 2023). Therefore, depending on the approach, the Caribbean might be considered a marginal coral reef habitat when the IAA coral diversity is used as a baseline, given the massive differences in coral cover and diversity between these biogeographic regions. Most studies do not consider this to be the case, but this example highlights the importance of a historical perspective when interpreting biodiversity patterns.

In this context, distinguishing between reef-associated and coral reef-associated fishes becomes crucial (Siqueira et al. 2023a; Muruga et al. 2024). The definition of what constitutes a coral reef fish has long been debated (Bellwood 1996; Robertson 1998; Bellwood and Wainwright 2002). Coral reef fish faunas do not have a distinctive taxonomic structure compared to some non-coral dominated habitats (Robertson 1998), although coral reef assemblages generally present predictable patterns at higher taxonomic levels on regional and global scales (Bellwood and Hughes 2001; Bellwood and Wainwright 2002). Thus, while coral-dominated reefs might represent the epitome of marine fish diversity, this diversity is not exclusive to coral reefs and often extends to other non-biogenic reefs (Robertson 1998) and even non-reef habitats (Sambrook et al. 2019). Within a biogeographical context, this means that fish distributions naturally extend beyond those of coral species, suggesting that marginal and extreme habitats may influence coral diversity patterns more strongly than fish diversity (Mies 2025 in this book).

Considering these nuances, we propose that research on marginal and extreme reef fish biodiversity should also integrate historical and geographical perspectives to provide a more comprehensive understanding of these ecosystems. While marginal and extreme reefs have been suggested as natural laboratories for understanding resilience and adaptation mechanisms in reef ecosystems (Camp et al. 2018; Burt et al. 2020; Schoepf et al. 2023), this will likely be context dependent. For instance, marginal reefs in the Southwestern Atlantic host resilient fish assemblages capable of maintaining crucial ecological functions despite significant coral loss (Luza et al. 2022). This resilience is likely associated with the naturally low coral cover/diversity of the region (Aued et al. 2018), and the generally weak correlations between coral cover and fish metrics for that biogeographic realm (Muruga et al. 2024). This approach will refine our predictions of biodiversity patterns and enhance our scientific understanding of reef dynamics within these unique environments.

3.2.2 Local Patterns of Diversity

3.2.2.1 Diversity in Turbid Waters

Turbidity is a crucial parameter that represents the degree to which light penetration is hindered by suspended particulate matter in the aquatic environment (Van Duin et al. 2001; Tomperi et al. 2022). Different from the elevated siliciclastic sedimentation habitats, which are mainly influenced by inorganic components, turbid reefs consist of both organic and inorganic—the inorganic often being terrestrial sediments carried by rivers or wind, and the organic comprising elements like phytoplankton (indicated by chlorophyll levels), zooplankton, and bacteria (Anthony 1999; Macdonald et al. 2013). Consequently, coral reef habitats affected by high turbidity generally experience reduced light availability (Fig. 3.1). They are commonly found in shallow waters near the shore, typically not exceeding a depth of 10 m, and situated within 20 km from the coastline (Obura et al. 2019; Zweifler et al. 2021). Given the challenges and adaptations required for organisms to thrive in turbid reef habitats, coral reefs subjected to these conditions are considered extreme habitats. Indeed, as sediment inputs to reefs are predicted to increase due to anthropogenic impacts, understanding the challenges organisms face in these ecosystems can provide valuable insights into the future functioning of coral reefs (Suárez-Castro et al. 2021; Andrello et al. 2022). Notably, some coral reefs most vulnerable to increased turbidity are often crucial to local communities. Their proximity to land makes them highly accessible for resource extraction, such as fishing, thus playing a significant role in supporting livelihoods (Fig. 3.2) (Maire et al. 2016).

Despite occupying 30% of reefs in the Coral Triangle and 12% globally (Sully and van Woesik 2020), turbid coral reefs are relatively less explored. Due to the logistical challenges associated with working in low visibility conditions both directly (i.e. *in situ*) and indirectly (i.e. using remote sensing technologies), there is a large gap in our understanding of these habitats (Hedley et al. 2016; Obura et al. 2019). Consequently, our knowledge of how turbid coral reefs function is way more limited when compared to their clear-water counterparts (Browne et al. 2012; Zweifler et al. 2021;

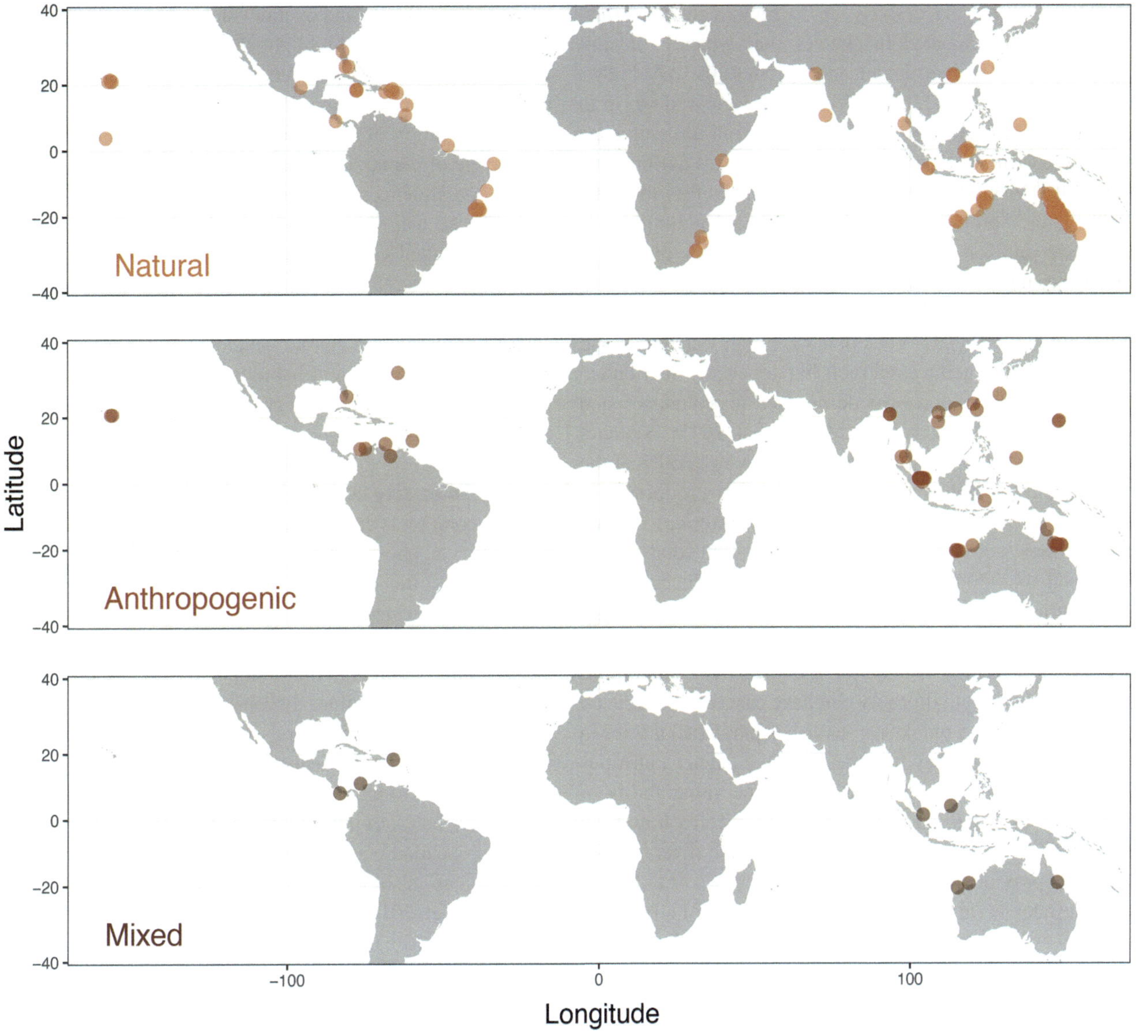

Fig. 3.2 Global distribution of the studies on turbid coral reefs reviewed by Zweifler et al. (2021). Maps show the locations of studies performed on Natural (i.e. river runoff and hydrodynamics), Anthropogenic (i.e. dredging and land use), and Mixed (i.e. hydrodynamics/dredging, hydrodynamics/land use, and runoff/land use) turbid reefs

Cartwright et al. 2023). Moreover, while studies investigating sessile benthic organisms like corals and sponges in these habitats are challenging (Teague et al. 2022; Morais et al. 2024a), the complexity increases significantly when the focus shifts to mobile organisms like reef fishes (Travers et al. 2018; Figueroa-Pico et al. 2020), especially concerning functions such as plankton consumption and herbivory (Moustaka et al. 2018; Tebbett et al. 2024).

In a study examining the impact of suspended sediment on coral reef fish assemblages in north-west Australia, Moustaka et al. (2018) showed that elevated turbidity levels significantly affect some functional groups of reef fishes. Specifically, planktivores showed a marked decrease in abundance at sites with high turbidity. This study used both remotely sensed and in situ turbidity data to characterise the turbidity regime and found that the abundance of these fish groups was negatively correlated with long-term levels of suspended sediment (Fig. 3.1). The reduction in planktivores was attributed to decreased foraging success due to limited visibility, reduced prey availability, and lower structural complexity of the habitat (Moustaka et al. 2018). Indeed, the impact of sediment-induced turbidity on the foraging performance and prey choice of planktivorous coral reef fishes is profound. Elevated turbidity levels significantly impair the ability of these visual predators to effectively forage (Johansen and Jones 2013).

On the Great Barrier Reef, even low turbidity levels of 4 NTU (nephelometric turbidity units) reduced the average planktivore attack success by up to 56%, with the effect being more pronounced in species that typically inhabit clearer offshore waters (Johansen and Jones 2013). At higher turbidity levels (8 NTU), attack success decreased by up to 69%, leading to a 21–24% reduction in overall foraging rates for mid- to outer-shelf species. This decline was attributed to diminished visual range, which affected the ability to detect and capture prey, particularly mobile planktonic prey (Johansen and Jones 2013). In turbid conditions, planktivorous fishes tended to rely more on immobile prey, which still resulted in lower foraging efficiency, especially at higher flow velocities. Planktivorous fishes are highly abundant and diverse on Indo-Pacific coral reefs but not evenly distributed, leading to a disproportional dominance in abundance and species richness in that region (Morais et al. 2021b; Siqueira et al. 2021). This fish group can constitute up to 50% of the fish standing biomass and 41% of biomass production in some areas, playing a significant role in nutrient cycling within reef ecosystems (Morais and Bellwood 2019; Campbell et al. 2020). However, as turbidity reduces the foraging efficiency of these fishes, it leads to lower energy intake and higher mortality rates, ultimately resulting in a decline in both their abundance and diversity. This diminished abundance and diversity can have cascading effects on reef ecosystems, as planktivorous fishes play crucial roles in nutrient cycling as prey and through their highly nutritious feces (Schiettekatte et al. 2023). Thus, the vulnerability of planktivorous fishes on turbid reefs underscores how such extreme habitat conditions directly affect the diversity and functions provided by this trophic group.

Moustaka et al. (2018) also showed that herbivorous scrapers were negatively affected by increased sediment deposition on the benthos, which diminishes the quality and palatability of the available algae (Moustaka et al. 2018). Similarly, Tebbett et al. (2024), demonstrated the near absence of herbivorous fish on high-turbidity reefs, prompting an investigation into the factors influencing this distribution. They explored the relationship between turbidity levels and the abundance of herbivorous fishes, considering the structure and productivity of the epilithic algal matrix (EAM) as a potential contributing factor. Results revealed that areas with higher sediment loads were correlated with lower herbivorous fish presence, suggesting that the quality and availability of nutritional resources in these environments were constrained, possibly due to decreased light penetration caused by the turbidity and sediment coverage affecting EAM productivity. This highlights the complexity of bottom-up influences on reef fish assemblages in turbid conditions, but may also suggest that reduced herbivory on turbid reefs can lead to algal overgrowth (Bosch et al. 2022). Additionally, Moustaka et al. (2018) reported a significant decline in species richness and total biomass with increasing turbidity, suggesting that elevated sediment levels can lead to a loss of biodiversity and overall fish biomass in reef ecosystems. These findings highlight the vulnerability of certain fish groups to sediment-induced turbidity and underscore the importance of considering these impacts in the management and conservation of coral reef ecosystems. The decrease in abundance of planktivores and herbivorous species can have severe consequences for the resilience and recovery of coral reefs. These fish play crucial roles in nutrient cycling and controlling algal growth, which is essential for maintaining reef health. Therefore, there is a need for careful management of anthropogenic activities that increase sedimentation, such as dredging and coastal development, to prevent long-term damage to coral reef habitats. Identifying and mitigating the sources of sediment runoff can help protect vulnerable fish populations and support reef conservation efforts.

3.2.2.2 Diversity Under High Nutrient Contents

Coastal coral reefs constantly receive substantial amounts of nutrients (i.e., phosphorus and nitrogen) (Brodie et al. 2012). This influx, often originated from agricultural runoff and urban sewage, can result in significant alterations reef ecosystems, including stimulating algal blooms, reducing water clarity, and changing the composition of the habitats (Fabricius 2005; Martinez-Escobar and Mallela 2019). As nutrient levels increase, coral reef communities transition from dominance by nutrient-recycling symbiotic organisms, such as corals, to increasing proportions of macroalgae and further to heterotrophic filter feeders (Birkeland 1987). However, beyond benthic communities, these environmental changes can also have profound effects on reef fishes (Allgeier et al. 2014). Species adapted to clear, nutrient-poor waters may struggle to survive in such conditions, while some species may thrive, leading to shifts in community structure (Fabricius 2005).

High nutrient content and sewage pollution can have sublethal effects on fish reproduction (Kingsford et al. 1996). Elevated nutrient levels can disrupt endocrine functions, leading to reproductive anomalies and reduced spawning success. For example, Kingsford et al. (1996) showed that exposure to sewage plumes can have detrimental effects on fish larvae. In their study, conducted along the coast of New South Wales, they found a variety of deformities in fish larvae exposed to different levels of pollution. The study documented several types of serious deformities, including notochord issues such as 'dorso-ventral-curl' and 'lateral zig-zag', as well as eye deformities like 'bubble eye' and 'deformed eye'. These deformities were found in fish larvae across all sampled locations, indicating that the high nutrient levels and pollution may have a pervasive impact on fish health (Kingsford et al. 1996). Similarly, Azzurro et al. (2010) demonstrated the significant impact of sewage pollution on Mediterranean rocky-reef fish assemblages. They revealed that untreated sewage led to notable changes at both the assemblage and species levels. Specifically, the total abundance of fish was higher in polluted areas, with an increase in opportunistic species such as *Gobius bucchichii* and *Parablennius rouxi*. Conversely, species sensitive to pollution, such as those from the genus *Symphodus*, showed a decrease in abundance near sewage outfalls. This study underscores the utility of fish assemblages as biological indicators of marine coastal water quality and highlights how sewage pollution can disrupt the ecological balance of these environments (Azzurro et al. 2010). Furthermore, evidence suggests that nutrient enrichment can lead to hypoxic (low oxygen) conditions in the water, especially in areas with poor water circulation. Hypoxia disrupts the migratory patterns of benthic and demersal species, and reduces growth and recruitment rates (Pollock et al. 2007; Lau et al. 2021). Additionally, fish exposed to high-nutrient environments often suffer from physiological stress and increased susceptibility to diseases. The altered water chemistry and increased organic matter may affect the health of fish, making them more vulnerable to infections and reducing their overall fitness (Wear and Thurber 2015).

It is important to note that while most studies focus on anthropogenic nutrient enrichment, some reef systems are naturally influenced by nutrient inputs, such as those from groundwater discharge or oceanic upwelling. Studying these naturally enriched systems is essential for distinguishing between baseline variability and pollution-driven impacts on reef fish assemblages (Moore 2010). For example, Lilkendey et al. (2019) demonstrated that juvenile *Chrysiptera glauca* exposed to fresh submarine groundwater discharge (SGD) exhibited significantly higher growth rates and earlier settlement compared to those in purely marine environments, suggesting that SGD can enhance physiological conditions for fish development. Similarly, Eisele et al. (2021) examined reef fish assemblages in an upwelling region of the Eastern Tropical Pacific and found that seasonal upwelling strongly influenced fish assemblage structure, with effects modulated by habitat complexity and depth. These studies highlight how naturally elevated nutrient levels can also shape fish assemblages, much like pollution-driven eutrophication. However, unlike naturally occurring processes, nutrient enrichment from human activities—such as urban expansion and agricultural development—is expected to intensify in the coming decades.

In an increasingly urbanized world, not only will turbidity negatively impact reef fishes, but so will chemical pollutants and high nutrient levels. Despite the critical importance of this issue, there is limited information on the vulnerability of reef fish in these environments. Thus, understanding the impact of high nutrient content and sewage pollution on reef fish diversity is crucial, particularly because these reefs are often located near areas where local communities rely on fisheries as essential protein sources (Maire et al. 2016). Comprehensive studies are needed to fill the knowledge gap on the effects of elevated nutrient levels on reef fishes and to inform effective conservation and management strategies to mitigate the impacts of nutrient pollution on reef ecosystems.

3.2.2.3 Diversity in Mesophotic Depths

Mesophotic coral ecosystems (MCEs) are found at depths of approximately 30–150 m, functioning as an extension of shallow reefs (Kahng et al. 2010; Lesser et al. 2018). These ecosystems are characterized by steep gradients in irradiance and temperature, which significantly influences their productivity and biodiversity (Soares et al. 2018; Slattery et al. 2024). The unique conditions of MCEs create distinct habitats that support a wide variety of marine life, including depth generalists and specialists (Rocha et al. 2018; Medeiros et al. 2021). The upper mesophotic zone (30–60 m) represents a transition zone with biodiversity similar to shallow reefs, while the lower mesophotic zone (60–150 m) supports more endemic species adapted to low-light/temperature environments (Englebert et al. 2017; Gould et al. 2021). MCEs are estimated to encompass approximately 60–80% of the coral reef area globally, which is considerably more than the area covered by shallow reefs (Pyle and Copus 2019). The structural complexity of mesophotic reefs, with their diverse topographies like fore-reef banks (Smith et al. 2016a), buttress slopes (Sherman et al. 2010), and vertical walls (Slattery 2012), also provides essential habitats for fish. However, despite being relatively unexplored (Morais et al. 2018; Eyal et al. 2021), these ecosystems are not immune to the stressors affecting shallow reefs. As shallow reefs, they also face threats from climate change, overfishing, and pollution, necessitating specific conservation strategies to protect their unique biodiversity (Lesser et al. 2018; Pinheiro et al. 2023b).

Beyond the threats that may limit the occurrence of some species, physical factors correlated with depth, such as light availability, temperature, and pressure also play a critical role in shaping the fish assemblages in MCEs (Lesser et al. 2018; Rocha et al. 2018) (Fig. 3.3). For example, the significant reduction in light at depths beyond 30 m limits the growth of photosynthetic organisms like algae, which are a primary food source for herbivorous fish (Brokovich et al. 2008; Loiseau et al. 2023). Therefore, herbivorous species are less prevalent in these environments, shifting the assemblage composition towards fish that rely on other food sources (Rocha et al. 2018). The cooler, more stable temperatures found in MCEs also influence the types of species that can thrive there. These conditions favour species adapted to mesophotic conditions over those requiring the warmer, more variable temperatures of shallow reefs (Kahng et al. 2017). Additionally, the increased pressure at mesophotic depths poses physiological challenges that many shallow-water species cannot tolerate (Baker et al. 2016). Combined, these factors create an environment that selects for certain fish families, such as the Anthiadidae (basslets), while limiting the presence of others, such as Acanthuridae (surgeonfishes) (Medeiros et al. 2021; Loiseau et al. 2023) (Fig. 3.3). This selective pressure results in unique and specialised assemblages that are distinct from those found in shallower reef areas (Pinheiro et al. 2023a). For example, a study showed that 71% of fish species in mesophotic settings in Bermuda and Curacao were restricted to depths below 60 m, 8% occurred only in the lower mesophotic, and only 21% were found in both upper and lower mesophotic zones (Pinheiro et al. 2016). This suggests that many species are confined to the lower mesophotic depth zone, likely because of its distinct environmental conditions.

Deep reefs, unlike their shallow counterparts (less than 30 m in depth), were believed to be less affected by ocean warming and human-induced stresses such as coastal pollu-

Fig. 3.3 Typical reef fishes on mesophotic habitats. (**a**) *Genicanthus watanabei*. (**b**) *Cephalopholis polleni*. (**c**) *Chaetodon flavocoronatus*. (**d**) *Pseudanthias pulcherrimus*. (**e**) *Cirrhilabrus finifenmaa*. (**f**) *Pseudanthias fasciatus*. (All photographs taken by Luiz A. Rocha)

tion and overfishing (Hoegh-Guldberg and Bruno 2010; Mies et al. 2020; Soares et al. 2020a). This relative protection has led scientists to hypothesize that deep reefs could serve as refuges, offering resilience for reef organisms in the event of significant losses in shallow reef habitats ("deep reef refugia hypothesis" (*sensu* Bongaerts et al. (2010), also see Glynn (1996), Kahng et al. (2010)). This hypothesis suggests that deeper reefs may provide both short-term and long-term shelter for various species and biological communities against specific or multiple disturbances. In this context, *depth refuge* refers to a depth range that offers temporary protection during a disturbance event, *depth refugium* indicates a depth range providing long-term shelter from various disturbances, and *resilience area* refers to a depth range that supports species or communities resilient to multiple types of long-term disturbances (Bongaerts and Smith 2019). Since the original proposition of this hypothesis, however, a substantial body of evidence highlighted that deep reefs rarely serve as refuges for shallow-water fish assemblages, as these assemblages are composed of different species (Pereira et al. 2018; Rocha et al. 2018; Medeiros et al. 2021; Pinheiro et al. 2023a).

In a recent study, Pinheiro et al. (2023a) highlighted the complex patterns of fish diversity along the depth gradient within mesophotic ecosystems. They revealed a trend where the well-recognized biodiversity drivers from the Indo-Pacific and other regions diminish in importance with the increase in depth. Through extensive global surveys, they revealed that mesophotic coral ecosystems harbour fish assemblages that are shaped by distinct ecological forces, diverging markedly from their shallower counterparts. These deep-reef assemblages are characterized by lower taxonomic and functional richness compared to shallow reefs, exhibiting higher local saturation and convergence across different ocean realms. This suggests that the distinct ecological and evolutionary processes governing mesophotic reefs lead to inherently different assemblages between deep and shallow reefs, supporting the view that deep reefs do not serve as effective refuges for shallow-water fish assemblages. All these findings emphasise the need for conservation efforts that consider the distinct biodiversity and ecological dynamics at these depths and also reinforce the distinctiveness of mesophotic communities (Pereira et al. 2018; Rocha et al. 2018; Medeiros et al. 2021; Soares et al. 2021; Loiseau et al. 2023).

3.2.2.4 Diversity Under High and Highly Variable Temperatures

Additionally to the other marginal and extreme conditions already discussed here, coral reefs under constant high and highly variable temperature conditions are unique. Such variations can impose physiological challenges to many organisms, limiting their occurrence (Camp et al. 2018). Scleractinian corals, the foundational habitat forming organisms on tropical reefs, are highly vulnerable to thermal extremes (Riegl Jr. et al. 2025 in this book). This sensitivity has contributed to the accelerated decline of coral reefs on local and global scales (Hughes et al. 2018a; Sully et al. 2019; Morais et al. 2021a). In the current scenario of habitat loss due to high coral mortality and rapid erosion (Leggat et al. 2019; Morais et al. 2022), fish assemblages also decline or change in composition, directly impacting the resources available to people dependent on reef fisheries (Bellwood et al. 2006; Pratchett et al. 2011; Robinson et al. 2019). Although there is evidence suggesting a weak association between fish and coral metrics, and showing that some fish species may adapt to or even benefit from the decrease in live coral cover (Robinson et al. 2019; Wismer et al. 2019; MacDonald et al. 2021; Siqueira et al. 2023a; Muruga et al. 2024), tropical reef fishes are typically adapted to a relatively narrow range of environmental conditions. Thus, reef fishes may also be vulnerable to the direct effects of climate change, such as increased sea surface temperatures (Pörtner and Knust 2007; Deutsch et al. 2015). Extremes temperatures can disrupt metabolic processes, affect reproductive success, and alter species interactions, ultimately influencing the structure and resilience of reef fish populations (Comte and Olden 2017).

Although other environmental factors, such as salinity and oxygen saturation, significantly affect reef fish physiology (Ern et al. 2014), temperature remains the most frequently studied environmental stressor for reef fishes (Nagelkerken et al. 2023). Despite notable variations in species-specific tolerances to elevated temperatures (Johansen and Jones 2011; Bernal et al. 2020), most reef fish species experience non-lethal adverse physiological, developmental, or behavioural responses when exposed to temperatures beyond their normal range (Mora and Ospína 2001). Currently, our general understanding indicates a potential long-term harmful effect on reef fish populations (Munday et al. 2012), however, there are few documented cases of direct temperature-induced population declines in situ for reef fish assemblages (Feary et al. 2014). Transgenerational plasticity, which can improve the performance of offspring at higher temperatures through developmental, genetic, or epigenetic pathways (Donelson et al. 2012; Bernal et al. 2018), may mitigate some adverse effects of rising temperatures in the wild. This phenomenon has been observed in a few species (Donelson et al. 2012; Grenchik et al. 2013) but requires increased energetic investments (Bernal et al. 2018; Miller et al. 2018). Nevertheless, it remains uncertain whether this process can genuinely enhance the survival of reef fishes in competitive, uncontrolled environments, and how differences in species-specific temperature tolerance might influence coexistence within ecological communities.

The Arabian Gulf is a prime example of a habitat with extremely high and variable temperatures (Riegl Jr. et al. 2025 in this book). Situated in a region with harsh climatic conditions, it experiences some of the highest sea surface temperatures (SST) in the world (Alosairi et al. 2020; Brandl et al. 2020; Howells et al. 2020). The SST in the Gulf often exceeds 36.0 °C during major heat waves, with temperatures in specific areas, like Kuwait Bay, reaching as high as 37.6 °C (Alosairi et al. 2020). This thermally extreme environment significantly impacts marine life, influencing everything from coral bleaching events to the survival and composition of fish assemblages (Hereher 2020; Vaughan et al. 2021). The Gulf's unique hydrodynamic characteristics, combined with regional climatic factors such as the Shamal and Kous winds, contribute to these temperature extremes. The SST in this region has been increasing at a rate of 0.4 °C per decade, more than double the global average, highlighting the significant local impact of global climate change (Al-Rashidi et al. 2009). A recent study by Brandl et al. (2020) provides a comprehensive examination of the effects of extreme temperatures on cryptobenthic reef fish assemblages. They compared reefs in the world's hottest coral reefs in the southern Arabian Gulf with those in the Gulf of Oman. They found that cryptobenthic fish assemblages in the Arabian Gulf were markedly less diverse and abundant, with species richness averaging 1.62 species per m^2 compared to 3.40 species per m^2 in the Gulf of Oman. Additionally, the abundance of cryptobenthic fishes was dramatically lower, averaging 6.12 individuals per m^2 in the Arabian Gulf versus 31.94 individuals per m^2 in the Gulf of Oman. Despite similar benthic composition and coral cover, the extreme temperatures in the Arabian Gulf resulted in energetically challenging conditions, reducing fish biomass production to 0.038 grams per day per m^2, compared to 0.231 grams per day per m^2 in the Gulf of Oman. These findings illustrate how thermally extreme environments, expected to become more common with climate change, can severely disrupt the diversity, abundance, and productivity of reef fish assemblages. Similar to the small-body size cryptobenthic fishes, larger reef fish species also responded negatively to the extreme temperatures of the Arabian Gulf. D'Agostino et al. (2021) compared growth patterns of two coral reef fish species, *Lutjanus ehrenbergii* and *Pomacanthus maculosus*, in the same environmentally extreme Arabian Gulf and the relatively cooler Oman Sea. They found that both species showed smaller sizes at age and lower maximum sizes in the Arabian Gulf, suggesting a significant impact of environmental parameters such as salinity and temperature. Furthermore, experimental studies, models, and meta-analyses on temperature-size responses indicate that the body size of fish might further reduce by 14–24% per degree of warming by 2050 under a higher greenhouse gas emission scenario (Cheung et al. 2013; Pauly and Cheung 2018). This effect has even been recently demonstrated in evolutionary timescales (Siqueira et al. 2023b). Consequently, the likely decrease in body size will impact the prey-predator interactions (Barnes et al. 2010), fecundity and fisheries productivity (Baudron et al. 2014; Barneche et al. 2018), and ecosystem functioning (Bellwood et al. 2012; Cheung et al. 2013).

In addition to environments that experience persistently high temperatures, those with variable temperature regimes can also select and limit fish species occurrence, providing valuable insights into the heat tolerance and acclimation/adaptation potential of reef fish (Sponaugle and Grorud-Colvert 2006). Variations in temperature regimes can occur due to localized conditions such as tide pools and lagoons, which are characterized by highly variable thermal environments, or can result from broader temporal differences in high-latitude environments where corals are subject to significant seasonal temperature fluctuations (Koweek et al. 2015; Camp et al. 2018). In back reef tide pools, temperature fluctuations can be extreme, with water temperatures reaching up to 35 °C during noontime low tides, exceeding the local critical bleaching temperature for corals, and dropping to significantly low temperatures during the night (Palumbi et al. 2014; Schoepf et al. 2015). For reef fishes, these pools create energetically challenging conditions, yet some species have adapted to these environments (Andrades et al. 2023). For example, Davis and Nimbs (2022) found that cryptobenthic fish in smaller high-variance pools exhibit significantly lower diversity and abundance compared to those in larger, more stable pools. Larger pools supported a greater number of species and larger fish populations, while the depth of the pool did not significantly affect these factors. This indicates that the area of the pool plays a more important role in maintaining higher diversity and abundance of cryptobenthic fish (Davis and Nimbs 2022). Therefore, even within a highly variable temperature system, different characteristics, such as pool size, can significantly impact abundance and diversity on a small scale.

Another example is the high-latitude reef environments, which are characterized by cooler, more seasonably variable temperatures compared to tropical reefs (Abdo et al. 2012; Abrego et al. 2021). These areas have also gained attention as potential refuges from climate change and ocean warming (Beger et al. 2014). Fish in high-latitude environments often exhibit traits that enable them to survive in highly variable and marginal and extreme conditions (Pie et al. 2021). For example, fish populations in these regions may show different growth rates and maximum sizes, reflecting adaptation to the local temperature regimes (Holland et al. 2020). These adaptations highlight the species-specific mechanisms influenced by local environmental factors, including temperature and productivity gradients (Holland et al. 2020).

In summary, high and variable temperature conditions pose significant challenges to coral reef ecosystems, particularly affecting fish diversity. These thermal stresses can limit the occurrence and abundance of various species by disrupting physiological processes and altering assemblage structures. As global climate change continues to exacerbate these conditions, understanding their impact on fish diversity is crucial for developing effective conservation strategies and ensuring the resilience of reef ecosystems.

3.3 Vulnerability of Reef Fishes in Marginal and Extreme Habitats

3.3.1 Global Human Stressors

3.3.1.1 Global Warming

Global human-induced stressors, such as global warming, are well known for their impact on marine biodiversity at all levels. Climate change is rapidly altering the functioning of shallow "classic" and marginal reef ecosystems (Bellwood et al. 2004; Smith et al. 2016b; Hughes et al. 2019; Duarte et al. 2020; Cornwall et al. 2021), placing their future at risk. For reef fishes living in suboptimal habitats, global warming exacerbates the suboptimal conditions of marginal and extreme reefs.

Globally, from 1950 to 2011, sea surface temperatures (SST) increased by approximately 0.08 °C per decade (Lough 2012). Additionally, while the 2023 report by the Intergovernmental Panel on Climate Change (IPCC) suggested that there was more than a 50% probability of global temperatures reaching or exceeding 1.5 °C between 2021 and 2040, observations now confirm that this threshold was already surpassed for the first time in 2024. According to the World Meteorological Organization (2024), the global average temperature in 2024 reached approximately 1.55 °C above pre-industrial levels. Without significant emissions reductions, this level of warming is projected to become increasingly frequent—and may only represent an early stage in a more severe trajectory. For instance, if carbon emissions continue to rise at a high rate, global temperatures could increase between 3.3 °C and 5.7 °C by the end of the century (IPCC 2023). Although some ecosystems may exhibit delayed responses to climate change, coral reefs are already demonstrating significant impacts through recurrent bleaching events. This highlights a close connection between these ecosystems and climate extremes. Indeed, the frequency and severity of coral mortality due to extensive ocean-wide bleaching events have escalated in recent decades (Eakin et al. 2010; Hughes et al. 2018a, b; Sully et al. 2019; Vercelloni et al. 2020). Coral reefs have already faced four global bleaching events (1998, 2010, 2014–2017, and 2024), which may have a severe impact on reef fishes, especially in marginal habitats (Richardson et al. 2018; Skirving et al. 2019; Reimer et al. 2024; Morais et al. 2026). Global warming indirectly impacts reef fishes through coral bleaching, affecting the corals and the entire reef ecosystem, including those fishes that rely on them for shelter and food. Additionally, warmer waters can directly alter fish metabolism, reproduction, and distribution (Pörtner and Knust 2007; Johansen and Jones 2011; Deutsch et al. 2015). Species that are unable to adapt or migrate to cooler waters may face population declines or even extinction. Therefore, even the distribution and occurrence of reef fish species already adapted to the marginal condition may shift poleward in response to warming waters, disrupting the existing ecological conditions in these habitats.

Climate change may exacerbate the challenges faced by reef fishes in different marginal and extreme conditions. For example, although elevated levels of suspended particles in turbid reefs can benefit corals by reducing incident light and subsequently mitigating coral bleaching during thermal-stress events (Morgan et al. 2017; Sully and van Woesik 2020), this combination offers no advantage to reef fishes. Elevated temperatures caused by global warming exacerbate these conditions by increasing the metabolic demands and reducing the swimming ability of some reef fish groups (Johansen and Jones 2011; Deutsch et al. 2015; Comte and Olden 2017), making it harder for them to find food in an already challenging turbid environment (Newport et al. 2021). Similarly, global warming can also negatively impact mesophotic coral ecosystems (Diaz et al. 2023; McWhorter et al. 2024). These ecosystems were once thought to be refuges from the impacts of climate change, but recent research indicates that they are also vulnerable (Medeiros et al. 2021; Soares et al. 2021). This vulnerability is harder to assess in reef fishes, as physiological impacts are the most likely cause of change. However, these effects are more evident in corals, which show physical signs of stress and damage. For example, there is evidence of mesophotic coral bleaching in the central Pacific (Venegas et al. 2019), the Great Barrier Reef (Frade et al. 2018), the eastern Caribbean Sea (Smith et al. 2016b), and the western and central Indian Ocean (Diaz et al. 2023), which contradict the idea of thermal refugia. The reduced light and cooler temperatures of mesophotic reefs mean that many species are already living near their thermal tolerance limits. Therefore, changes in temperature at these depths will likely result in severe thermal stress. The same is expected to happen in highly variable temperature habitats: reefs that experience these conditions, such as those in high-latitude, tide pools, or upwelling zones, face unique challenges (Camp et al. 2018; Davis and Nimbs 2022). Fish in these regions must constantly adjust to rapid temperature fluctuations, which can be energetically costly. Global warming can amplify these temperature swings, pushing fish beyond their thermal tolerance limits. Increased temperatures lead to metabolic stress and lower reproductive success, further threatening fish populations already adapted to extreme conditions (Comte and Olden 2017; Holland et al. 2020).

3.3.1.2 Plastic Pollution

Ocean plastic pollution has become a global issue due to the widespread use of plastic materials in various aspects of human life. Reports indicate that plastic pollution affects every ocean and sea on Earth (Eriksen et al. 2014), posing a significant threat to marine life and the economies of coastal nations (Derraik 2002). These plastics often end up in the marine environment through improper waste management, littering, and runoff from urban areas. Once in the ocean, plastics break down into smaller particles known as microplastics, which can persist in the environment for hundreds of years (Andrady 2011). Studies have highlighted the alarming extent of plastic pollution and its increasing trend. For example, a study by Jambeck et al. (2015) estimated that between 4.8 and 12.7 million metric tons of plastic waste entered the ocean in 2010 alone, with predictions that this amount could increase tenfold by 2025 if no significant measures are taken. Furthermore, Lebreton et al. (2017) has mapped the distribution of plastic debris in the world's oceans, revealing vast accumulations in ocean gyres, such as the Great Pacific Garbage Patch.

The impact of plastic pollution on reef fishes is multifaceted. Ingested plastics can cause physical harm, such as blockages in the digestive system, and can lead to malnutrition or starvation. Recent findings from Lizard Island, Australia, showed that reef fishes were among the most frequently contaminated organisms sampled, exhibiting higher levels of microplastic contamination than other taxa, with concentrations 1.6 to 2 times greater than those found in sponges, corals, and sea cucumbers (Santana et al. 2025). Furthermore, microplastics can act as vectors for harmful pathogens, introducing new diseases to reef ecosystems (Lamb et al. 2018). Critchell and Hoogenboom (2018) found that microplastics affect the body condition and behaviour of planktivorous reef fish (*Acanthochromis polyacanthus*) differently depending on the size and concentration of the plastic pieces. When exposed to microplastics of the same size as their natural food, there was no significant impact on fish growth, body condition, or behaviour. However, when exposed to smaller microplastic particles (approximately one-quarter the size of natural food), the fish ingested significantly more plastics, which led to negative effects on growth and body condition when plastics replaced their food. Indeed, as a global problem (Eriksen et al. 2014), microplastics can be particularly damaging in marginal habitats, such as turbid reefs (Utami et al. 2021) or even on seafloor (e.g., submarine canyons, (Sanchez-Vidal et al. 2018) and seamounts (Woodall et al. 2014).

Recent studies have expanded our understanding of the impacts of plastic pollution on deeper marginal reef ecosystems (Fig. 3.4a–d). According to Pinheiro et al. (2023b), mesophotic reefs (30–150 m) are significantly affected by plastic debris, particularly macroplastics. The study surveyed 84 coral ecosystems across the Pacific, Atlantic, and Indian Oceans, finding anthropogenic debris in 77 of the surveyed reefs, with macroplastics comprising 88% of this debris. Notably, deeper reefs showed higher densities of plastic debris compared to shallower reefs, with fishing activities identified as a primary source. This finding contrasts with other nearshore marine ecosystems where macroplastic densities typically decrease with depth and are dominated by consumer items. Additionally, microplastics were found in the diet of the mesophotic *Prognathodes olbiquus* in St. Paul's Rocks, one of the most remote islands in the world (Nunes et al. 2019). Overall, both microplastic and macroplastic pollution not only contribute to habitat degradation but also present significant physical and physiological hazards for all marine life, including reef fish, in marginal habitats.

3.3.2 Regional and Local Human Stressors

Regional and local human stressors may involve overfishing, non-regulated tourism, sewage disposal, biological invasion, oil and gas exploitation, aquaculture farms, artificial reefs and beaches, and other human activities. They usually act synergistically with global stressors and amplify the negative impacts on reef fish assemblages. An example is the opening of new exploration frontiers for petroleum in environmentally and socially sensitive regions such as the Brazilian Amazon Continental Margin, where extensive mesophotic reefs occur (Moura et al. 2016). Vilardo and Barbosa (2024) argue that such drilling should not be allowed because the region has poorly understood, recently discovered reef ecosystems; transboundary contingency plans were not developed; and coastal ecosystems are highly sensitive and vulnerable to oil spills. Other political, administrative reasons are given to avoid environmental disasters such as the 2019 spill that tarnished Brazil. In early September of that year, dense crude oil began to wash the beaches of Brazil's tropical coast, where marginal reefs that sustain tourism and artisanal fishing are common (Soares et al. 2020b). Four months later, the oil was found along >3000 km of the Brazilian coastline on >980 beaches, including the Amazon coast. It was one of the most extensive and severe environmental disasters ever recorded in Brazilian history and in tropical coastal regions worldwide (Soares et al. 2020b).

The spill affected echinoderms, coral symbionts, polychaetes, and sponges, besides at least 34 threatened species, mostly vertebrates (Soares et al. 2022). Socioeconomic impacts were detected in food security, public health, lodging, gender equality, tourism, and fishing, with reduced sales,

Human stressors

1. Global warming
2. Plastic pollution
3. Overfishing
4. Non-regulated tourism
5. Agricultural and urban runoff
6. Sewage disposal
7. Biological invasion
8. Mining, oil and gas exploitation
9. Offshore wind energy
10. Maritime transport
11. Aquaculture farms
12. Artificial structures

Conservation and management priorities

1. Reduce greenhouse emissions
2. Marine spatial planning
3. Marine environmental licensing
4. Marine protected areas
5. Ocean zoning
6. Pollution control

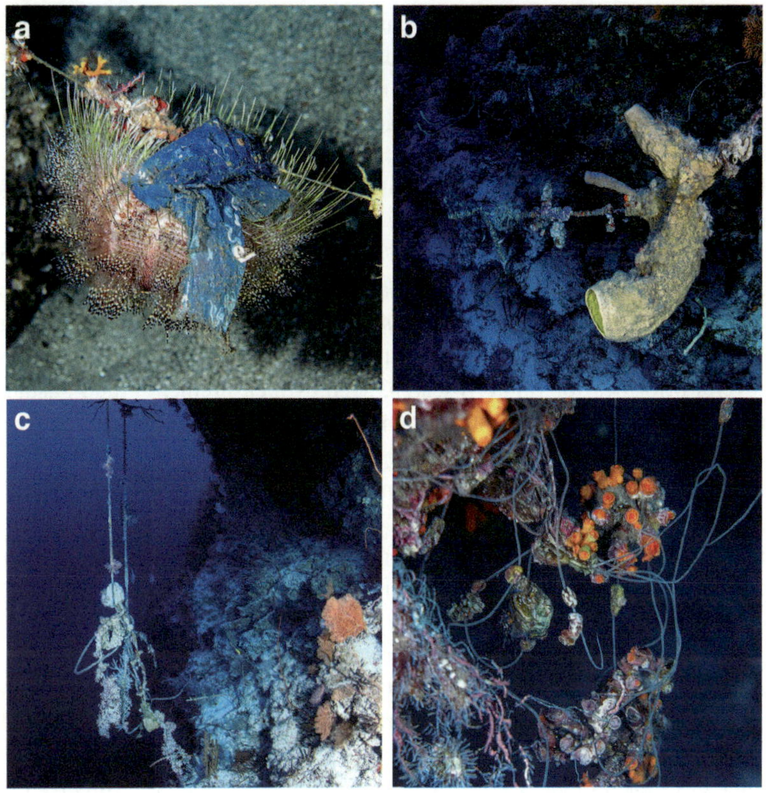

Fig. 3.4 Marginal and extreme habitats: Human stressors and conservation management priorities that must be addressed to preserve these unique environments. Photographs showing anthropogenic debris (i.e. macroplastics) on mesophotic coral reefs of the world. (**a**) Sea urchin, *Asthenosoma varium*, entangled with fishing line while camouflaging itself with a plastic bag at 130-m depth in the Philippines. (**b**) Broken anchor line observed at a depth of 92 m in Honduras. (**c**) Anchor line at 100-m depth in Palau. (**d**) Fishing line wrapped around coral at 70-m depth in Cape Verde (Pinheiro et al. 2023b)

prices, tourist attractiveness, gross domestic product, and employment (Soares et al. 2022). Chemical contamination was detected by toxic metals (Hg, As, Cd, Pb, and Zn) and polycyclic aromatic hydrocarbons (acenaphthalene, fluoranthene, fluorene, naphthalene, and phenanthrene). A total of 57 protected areas were affected by more than 5300 tons of oil that were collected. Ten ecosystems were impacted, with potentially more severe effects in mangroves and seagrasses, but the negative effects on rhodolith beds and marginal reefs are still understudied (Soares et al. 2022). In fact, recovery from this disaster is still ongoing and we do not know to what extent the fish assemblages of marginal reefs have been affected, nor how other human stressors such as overfishing are hindering recovery.

Another important regional impact is biological invasion, which poses a growing threat to reef ecosystems and their associated fish assemblages (Ahmed et al. 2022; Albins and Hixon 2013). A notable example is the lionfish (*Pterois* spp.), an invasive predator that has recently established populations along the Brazilian coast (Ferreira et al. 2015; Soares et al. 2023). The introduction of lionfish into the Atlantic is most commonly attributed to aquarium releases, but other hypotheses, such as ballast water discharge from ships, have also been proposed (Schofield 2009). Their recent detection in Brazil may reflect a combination of southward larval dispersal from the Caribbean and human-mediated transport. Between 2020 and 2023, 352 individuals were recorded along 2766 km of coastline, across eight Brazilian states, twelve marine protected areas, and multiple habitats, including mangroves, seagrass beds, and mesophotic reefs (1–110 m depth) (Soares et al. 2023). Recent modeling work further shows that suitable habitat for lionfish already encompasses approximately 60% of Brazil's marine protected areas, with 18 MPAs having confirmed records of the species (Soares et al. 2025). These areas include biodiversity hotspots such as the Fernando de Noronha Archipelago, where high endemism rates raise particular concern (Soares et al. 2023, 2025). Lionfish are voracious predators with wide dietary breadth and high fecundity, capable of triggering cascading ecological and socioeconomic impacts through intense predation on native juvenile fish and competition for resources (Albins and Hixon 2013; Côté and Smith 2018). The invasion threatens both biodiversity and key economic sectors, including artisanal fisheries and tourism, particularly in northeastern Brazil (Bumbeer et al. 2018; Soares et al. 2023). To address this, Soares et al. (2023) calls for urgent, integrated management including early detection with eDNA, citizen science, inter-institutional coordination, and regionally harmonized regulation, highlighting the importance of functional eradication and long-term monitoring strategies.

Certainly, there is a vast field for scientific investigation on fish responses to regional and local stressors in marginal and extreme reefs. Research on corals has made recent progress in this issue. For instance, Camp et al. (2018) outlined several priorities for future research, such as measurement time frames for environmental parameters and investigation on the fitness trade-offs underlying survival under sub-optimal conditions. These priorities may be addressed in rare pristine marginal, and extreme reefs, but also in the increasingly common disturbed marginal and extreme reefs that are currently facing multiple human stressors. Schoepf et al. (2023) complement Camp et al. (2018) priorities and state that we need a collection of longer-term, high-resolution abiotic and ecological baseline data sets that include multiple parameters across different temporal and spatial scales. Understanding the way local, regional, and global stressors interact to each other over a range of temporal and spatial scales is probably the most important priority, given its high potential for application in reef management and conservation.

3.4 Conservation and Management Efforts

As stated by Schoepf et al. (2023), the large diversity and heterogeneity of marginal and extreme communities make it difficult to recommend conservation and management actions that are universally applicable across locations. The potential solutions for protecting and/or recovering marginal and extreme reefs are under debate (e.g. Graham et al. 2014; Bates et al. 2019; Darling et al. 2019; Vergés et al. 2019), but there is a consensus that human stressors are jeopardizing (Edenhofer et al. 2011; Ogar et al. 2020). As discussed above, human impacts involve local, regional and global stressors that act synergistically and require integrated and coordinated conservation and management actions. Because decision-making is not effective if scientific data are lacking (Oliveira et al. 2024), the research agenda of marginal and extreme reefs should be aligned with the current and future uses of reef ecosystems (Soares et al. 2020a). We outline six priorities for the long-term protection of fish assemblages in marginal and extreme reefs: (1) drastic reductions in greenhouse gas emissions; (2) marine spatial planning; (3) strict marine environmental licensing processes for mining, oil and gas exploitation, and other large-scale economic activities; (4) marine protected areas; (5) ocean zoning for fisheries, tourism, maritime transport, renewable energy, and artificial structures; and (6) pollution control (Fig. 3.4).

While drastic cuts in CO_2 emissions represent the most critical global action to mitigate climate-related threats to marine biodiversity (see Sect. 3.1.1), their implementation depends largely on international agreements and national policies beyond the scope of local management. Therefore, complementary strategies become essential to enhance resilience at regional and local scales. Among these, marine spatial planning emerges as a key tool to balance conservation and sustainable use in marginal and extreme reef environments. According to UNESCO, marine spatial planning is a public process of analysing and allocating the spatial and temporal distribution of human activities in marine areas to achieve ecological, economic and social objectives that have been specified through a political process (Iglesias-Campos et al. 2021). It is not an end in itself but a practical way to establish a more rational use of marine space, to balance demands for development with the need to protect the environment and to deliver social and economic outcomes in an open and planned way (Iglesias-Campos et al. 2021). Human populations, particularly traditional communities that rely directly on reef fish for subsistence, should be involved to ensure that their perspectives are adequately represented and that the social benefits of spatial planning are equally distributed. Given that many marginal and extreme reefs are understudied and largely unknown (Schoepf et al. 2023), the precautionary principle of environmental law should be followed rigorously. This will avoid habitat degradation and biodiversity loss during the process of making informed and coordinated decisions about the use of marine resources (Smythe 2017; Johnson et al. 2018). Also, very few efforts in marine spatial planning currently consider the ongoing effects of climate change (Frazão Santos et al. 2024). Thus, using climate-smart marine spatial planning will also be important for marginal and extreme reefs and their fish assemblages.

Ocean zoning is also a critical component of the marine spatial planning that divides marine areas into specific zones and regulates their activities and uses (Iglesias-Campos et al. 2021). While marine spatial planning aims to balance ecological, economic, and social objectives, zoning ensures that specific activities are conducted in appropriate areas to minimize conflicts and environmental impacts; it is the separation of uses that are thought to be incompatible (Iglesias-Campos et al. 2021). In the context of marginal and extreme reefs, ocean zoning is particularly important for regulating the activities that have direct or indirect effects on these environments, such as fisheries, tourism, shipping lines, renewable energy, and other artificial structures such as oil platforms, man-made reefs, submarine cables, and shipwrecks (Rocha et al. 2018; Soares et al. 2021; Jia et al. 2022; Medeiros et al.

2022). In the case of fish assemblages, regulations within designated areas for marginal and extreme reef conservation should include fishing and tourism closures with a solid scientific basis and participation of local communities. Because the value of fishing and tourism varies across species and localities, actions should be carefully planned and conducted considering the socioecological context of each locality (Ogar et al. 2020). The same holds true for shipping lines, deployment of submarine cables and decommissioned vessels, artificial reefs, offshore wind farms, oil platforms among others. Ultimately, a co-location strategy may be required to implement multi-use zoning, reduce conflicts, and increase the long-term benefits of the rational use of marginal and extreme reefs (e.g. Jia et al. 2022).

Marine protected areas are also linked to the process of marine spatial planning and its implementation via zoning, but have their own legal and operational framework (Agardy et al. 2011; Silva and Santos 2023). There is no doubt that we have to expand our network of protected areas in many regions (Graham et al. 2014; Mouillot et al. 2016), but the protection of marginal and extreme reefs has rarely been the main motivation for protected area inception. To correct this asymmetry, adapting the existing network of marine protected areas to embrace these environments is needed (Soares 2020). We can achieve this by adjusting management objectives when marginal and extreme reefs are already included in existing protected areas, or by creating new areas or expanding existing ones to include them. Importantly, the implementation of marine protected areas designated to conserve marginal and extreme reefs should be conducted along with integrated actions aimed at recovering the populations of vulnerable, endangered and threatened species (Silva and Santos 2024), as well as controlling invasive ones (Soares et al. 2023).

Rigorous marine licensing, particularly for mining, oil and gas exploitation, but also offshore wind farms and other large-scale activities, is another priority for marginal and extreme reef conservation (Silva et al. 2016; Frometa et al. 2017; Soares et al. 2020). In fact, any economic activity being developed in the sea should pass through a comprehensive analysis of its potential environmental impacts. Marine licensing is fundamental for the protection of marine ecosystems and their biodiversity, sustainable resource management, community and economic benefits, international compliance and cooperation, climate change mitigation, and pollution control. The two last points—climate change and pollution—are possibly the most complex ones because they are transboundary in nature; their sources and effects are globally widespread. In many cases, the reduction of greenhouse emissions and pollution requires compliance with international agreements and cooperation on global efforts (Edenhofer et al. 2011; IPCC 2023), which are rarely implemented (Soares et al. 2021). As a result, fish species in marginal and extreme reefs are facing the challenge of living in mostly unprotected, and increasingly warmer and polluted habitats. Continued business-as-usual greenhouse gas emissions will likely further disrupt their ecological interactions. Thus, it is urgent to decarbonize the economy, enhance carbon sinks, and ensure the effective conservation of these irreplaceable ecosystems.

3.5 Concluding Remarks

Reefs under marginal and extreme conditions host their unique biotas that have evolved and colonized marine ecosystems in many ways. They are essential for the functioning of the global ocean and should be the focus of studies and protection efforts in a similar extent to more "classic" clear water shallow reefs. Research agendas should be aligned with management and conservation priorities, which require integrated and coordinated actions to solve problems and share socioeconomic benefits at each locality. To effectively mitigate human stressors at both local and global scales, reduce impacts on marginal reefs, and strengthen their long-term protection, these unique ecosystems — and the reef fish species they support — must be fully integrated into ocean conservation efforts. We are in the middle of the United Nations Decade of Ocean Science for Sustainable Development (2021–2030), as well as the Decade of Ecosystem Restoration. Both initiatives focus heavily on reefs exposed to various environmental and ecological conditions, and that includes "classic", marginal and extreme reefs. Protecting these irreplaceable environments should extend beyond academic interest and international boundaries, and become a global priority.

Acknowledgements JM acknowledges the Ian Potter Foundation and the Australian Museum's Lizard Island Research Station for financial support. BAS is grateful to Brazilian Council for Scientific and Technological Development (CNPq, grant number 307260/2022-4). ACS acknowledges funding from a Vice Chancellor's Research fellowship from Edith Cowan University.

References

Abdo DA, Bellchambers LM, Evans SN (2012) Turning up the heat: increasing temperature and coral bleaching at the high latitude coral reefs of the Houtman Abrolhos Islands. PLoS One 7:e43878

Abrego D, Howells EJ, Smith SDA, Madin JS, Sommer B, Schmidt-Roach S, Cumbo VR, Thomson DP, Rosser NL, Baird AH (2021) Factors limiting the range extension of corals into high-latitude reef regions. Diversity 13:1–16

Agardy T, di Sciara GN, Christie P (2011) Mind the gap: addressing the shortcomings of marine protected areas through large scale marine spatial planning. Mar Policy 35:226–232

Ahmed DA, Hudgins EJ, Cuthbert RN, Kourantidou M, Diagne C, Haubrock PJ, Leung B, Liu C, Leroy B, Petrovskii S, Beidas A, Courchamp F (2022) Managing biological invasions: the cost of inaction. Biol Invasions 24:1927–1946. https://doi.org/10.1007/s10530-022-02755-0

Albins MA, Hixon MA (2013) Worst case scenario: potential long-term effects of invasive predatory lionfish (Pterois volitans) on Atlantic and Caribbean coral-reef communities. Environ Biol Fish 96:1151–1157. https://doi.org/10.1007/s10641-011-9795-1

Allgeier JE, Layman CA, Mumby PJ, Rosemond AD (2014) Consistent nutrient storage and supply mediated by diverse fish communities in coral reef ecosystems. Glob Chang Biol 20:2459–2472

Alosairi Y, Alsulaiman N, Rashed A, Al-Houti D (2020) World record extreme sea surface temperatures in the northwestern Arabian/Persian Gulf verified by in situ measurements. Mar Pollut Bull 161:111766

Al-Rashidi TB, El-Gamily HI, Amos CL, Rakha KA (2009) Sea surface temperature trends in Kuwait Bay, Arabian Gulf. Nat Hazards 50:73–82

Andrades R, González-Murcia S, Buser TJ, Macieira RM, Andrade JM, Pinheiro HT, Vilar CC, Pimentel CR, Gasparini JL, Quintão TL, Machado FS, Castellanos-Galindo G, Ruiz-Campos G, Ojeda FP, Martin KL, Giarrizzo T, Joyeux J-C (2023) Ecology, evolution and conservation of tidepool fishes of the Americas. Rev Fish Biol Fish 33:1263–1290

Andrady AL (2011) Microplastics in the marine environment. Mar Pollut Bull 62:1596–1605

Andrello M, Darling ES, Wenger A, Suárez-Castro AF, Gelfand S, Ahmadia GN (2022) A global map of human pressures on tropical coral reefs. Conserv Lett 15:1–12

Anthony KR (1999) Coral suspension feeding on fine particulate matter. J Exp Mar Bio Ecol 232:85–106

Aued AW, Smith F, Quimbayo JP, Cândido DV, Longo GO, Ferreira CEL, Witman JD, Floeter SR, Segal B (2018) Large-scale patterns of benthic marine communities in the Brazilian Province. PLoS One 13:e0198452

Azzurro E, Matiddi M, Fanelli E, Guidetti P, La Mesa G, Scarpato A, Axiak V (2010) Sewage pollution impact on Mediterranean rocky-reef fish assemblages. Mar Environ Res 69:390–397

Baker E, Thygesen K, Harris P, Andradi-Brown D, Appeldoorn RS, Ballantine D, Bejarano I, Bridge TCL, Colin PL, Eyal G (2016) Mesophotic coral ecosystems: a lifeboat for coral reefs? United Nations Environ Program GRID-Arendal, Nairobi Arendal, p 98

Baldwin CC, Tornabene L, Robertson DR (2018) Below the Mesophotic. Sci Rep 8:4920

Barneche DR, Robertson DR, White CR, Marshall DJ (2018) Fish reproductive-energy output increases disproportionately with body size. Science (80-) 360:642–645

Barnes C, Maxwell D, Reuman DC, Jennings S (2010) Global patterns in predator–prey size relationships reveal size dependency of trophic transfer efficiency. Ecology 91:222–232

Bates AE, Cooke RSC, Duncan MI, Edgar GJ, Bruno JF, Benedetti-Cecchi L, Côté IM, Lefcheck JS, Costello MJ, Barrett N, Bird TJ, Fenberg PB, Stuart-Smith RD (2019) Climate resilience in marine protected areas and the 'Protection Paradox'. Biol Conserv 236:305–314

Baudron AR, Needle CL, Rijnsdorp AD, Tara Marshall C (2014) Warming temperatures and smaller body sizes: synchronous changes in growth of North Sea fishes. Glob Chang Biol 20:1023–1031

Beger M, Sommer B, Harrison PL, Smith SDA, Pandolfi JM (2014) Conserving potential coral reef refuges at high latitudes. Divers Distrib 20:245–257

Bellwood DR (1995) Direct estimate of bioerosion by two parrotfish species, Chlorurus gibbus and C. sordidus, on the Great Barrier Reef, Australia. Mar Biol 121:419–429

Bellwood DR (1996) The Eocene fishes of Monte Bolca: the earliest coral reef fish assemblage. Coral Reefs 15:11–19

Bellwood DR, Hughes TP (2001) Regional-scale assembly rules and biodiversity of coral reefs. Science (80-) 292:1532–1535

Bellwood DR, Wainwright PC (2002) The history and biogeography of fishes on coral reefs. In: Coral reef fishes: dynamics and diversity in a complex ecosystem, Elsevier, San Diego, CA, USA, pp. 5–32

Bellwood DR, Hughes TP, Folke C, Nyström M (2004) Confronting the coral reef crisis. Nature 429:827–833

Bellwood DR, Hoey AS, Ackerman JL, Depczynsky M (2006) Coral bleaching, reef fish community phase shifts and the resilience of coral reefs. Glob Chang Biol 12:1587–1594

Bellwood DR, Hoey AS, Hughes TP (2012) Human activity selectively impacts the ecosystem roles of parrotfishes on coral reefs. Proc R Soc B Biol Sci 279:1621–1629

Bernal MA, Donelson JM, Veilleux HD, Ryu T, Munday PL, Ravasi T (2018) Phenotypic and molecular consequences of stepwise temperature increase across generations in a coral reef fish. Mol Ecol 27:4516–4528

Bernal MA, Schunter C, Lehmann R, Lightfoot DJ, Allan BJM, Veilleux HD, Rummer JL, Munday PL, Ravasi T (2020) Species-specific molecular responses of wild coral reef fishes during a marine heatwave. Sci Adv 6:eaay3423

Berumen ML, Hoey AS, Bass WH, Bouwmeester J, Catania D, Cochran JEM, Khalil MT, Miyake S, Mughal MR, Spaet JLY, Saenz-Agudelo P (2013) The status of coral reef ecology research in the Red Sea. Coral Reefs 32:737–748

Birkeland C (1987) Nutrient availability as a major determinant of differences among coastal hard-substratum communities in different regions of the tropics. Comparison between Atlantic and Pacific tropical marine coastal ecosystems: community structure, ecological processes, and productivity. UNESCO Reports in Marine Science. pp 45–97

Bongaerts P, Smith TB (2019) Beyond the "Deep Reef Refuge" hypothesis: a conceptual framework to characterize persistence at depth. In: Loya Y, Puglise KA, Bridge TCL (eds) Mesophotic coral ecosystems. Coral reefs of the world. Springer International Publishing, Cham, pp 881–895

Bongaerts P, Ridgway T, Sampayo EM, Hoegh-Guldberg O (2010) Assessing the 'deep reef refugia' hypothesis: focus on Caribbean reefs. Coral Reefs 29:309–327

Bosch NE, Pessarrodona A, Filbee-Dexter K, Tuya F, Mulders Y, Bell S, Langlois T, Wernberg T (2022) Habitat configurations shape the trophic and energetic dynamics of reef fishes in a tropical–temperate transition zone: implications under a warming future. Oecologia 200:455–470

Bouwmeester J, Ben-Hamadou R, Range P, Al Jamali F, Burt JA (2022) Spatial patterns of reef fishes and corals in the thermally extreme waters of Qatar. Front Mar Sci 9:1–14

Brandl SJ, Johansen JL, Casey JM, Tornabene L, Morais RA, Burt JA (2020) Extreme environmental conditions reduce coral reef fish biodiversity and productivity. Nat Commun 11:1–14

Brodie JE, Kroon FJ, Schaffelke B, Wolanski EC, Lewis SE, Devlin MJ, Bohnet IC, Bainbridge ZT, Waterhouse J, Davis AM (2012) Terrestrial pollutant runoff to the Great Barrier Reef: an update of issues, priorities and management responses. Mar Pollut Bull 65:81–100

Brokovich E, Einbinder S, Shashar N, Kiflawi M, Kark S (2008) Descending to the twilight-zone: changes in coral reef fish assemblages along a depth gradient down to 65 m. Mar Ecol Prog Ser 371:253–262

Browne NK, Bauman AG (2023) Marginal reef systems: resilience in a rapidly changing world. Diversity 15:703

Browne NK, Smithers SG, Perry CT (2012) Earth-science reviews coral reefs of the turbid inner-shelf of the Great Barrier Reef, Australia: an environmental and geomorphic perspective on their occurrence , composition and growth. Earth Sci Rev 115:1–20

Bumbeer J, da Rocha RM, Bornatowski H, de Robert MC, Ainsworth C (2018) Predicting impacts of lionfish (Pterois volitans) invasion in a coastal ecosystem of southern Brazil. Biol Invasions 20:1257–1274. https://doi.org/10.1007/s10530-017-1625-8

Burt JA, Camp EF, Enochs IC, Johansen JL, Morgan KM, Riegl B, Hoey AS (2020) Insights from extreme coral reefs in a changing world. Coral Reefs 39:495–507

Cabioch G, Davies P, Done T, Gischler E, Macintyre IG, Wood R, Woodroffe C (2011) Encyclopedia of modern coral reefs. Springer Netherlands, Dordrecht

Camp EF, Schoepf V, Mumby PJ, Hardtke LA, Rodolfo-Metalpa R, Smith DJ, Suggett DJ (2018) The future of coral reefs subject to rapid climate change: lessons from natural extreme environments. Front Mar Sci 5:1–21

Campbell SJ, Darling ES, Pardede S, Ahmadia G, Mangubhai S, Amkieltiela E, Maire E (2020) Fishing restrictions and remoteness deliver conservation outcomes for Indonesia's coral reef fisheries. Conserv Lett 13:e12698

Carlot J, Vousdoukas M, Rovere A, Karambas T, Lenihan HS, Kayal M, Adjeroud M, Pérez-Rosales G, Hedouin L, Parravicini V (2023) Coral reef structural complexity loss exposes coastlines to waves. Sci Rep 13:1683

Cartwright PJ, Browne NK, Belton D, Parnum I, O'Leary M, Valckenaere J, Fearns P, Lowe R (2023) Long-term spatial variations in turbidity and temperature provide new insights into coral-algal states on extreme/marginal reefs. Coral Reefs 42:859–872

Cavender-Bares J, Kozak KH, Fine PVA, Kembel SW (2009) The merging of community ecology and phylogenetic biology. Ecol Lett 12:693–715

Chen PY, Chen CC, Chu L, McCarl B (2015) Evaluating the economic damage of climate change on global coral reefs. Glob Environ Chang 30:12–20

Cheung WWL, Sarmiento JL, Dunne J, Frölicher TL, Lam VWY, Deng Palomares ML, Watson R, Pauly D (2013) Shrinking of fishes exacerbates impacts of global ocean changes on marine ecosystems. Nat Clim Chang 3:254–258

Chow GSE, Chan YKS, Jain SS, Huang D (2019) Light limitation selects for depth generalists in urbanised reef coral communities. Mar Environ Res 147:101–112

Comte L, Olden JD (2017) Climatic vulnerability of the world's freshwater and marine fishes. Nat Clim Chang 7:718–722

Connell JH (1978) Diversity in tropical rain forests and coral reefs. Science (80-) 199:1302 LP–1310

Cornwall CE, Comeau S, Kornder NA, Perry CT, van Hooidonk R, DeCarlo TM, Pratchett MS, Anderson KD, Browne N, Carpenter R, Diaz-Pulido G, D'Olivo JP, Doo SS, Figueiredo J, Fortunato SAV, Kennedy E, Lantz CA, McCulloch MT, González-Rivero M, Schoepf V, Smithers SG, Lowe RJ (2021) Global declines in coral reef calcium carbonate production under ocean acidification and warming. Proc Natl Acad Sci 118:e2015265118

Côté IM, Smith NS (2018) The lionfish Pterois sp. invasion: has the worst-case scenario come to pass? J Fish Biol 92:660–689. https://doi.org/10.1111/jfb.13544

Cowman PF, Bellwood DR (2013) The historical biogeography of coral reef fishes: global patterns of origination and dispersal. J Biogeogr 40:209–224

Critchell K, Hoogenboom MO (2018) Effects of microplastic exposure on the body condition and behaviour of planktivorous reef fish (Acanthochromis polyacanthus). PLoS One 13:1–19

D'Agostino D, Burt JA, Santinelli V, Vaughan GO, Fowler AM, Reader T, Taylor BM, Hoey AS, Cavalcante GH, Bauman AG, Feary DA (2021) Growth impacts in a changing ocean: insights from two coral reef fishes in an extreme environment. Coral Reefs 40:433–446

Darling ES, McClanahan TR, Maina J, Gurney GG, Graham NAJ, Januchowski-Hartley F, Cinner JE, Mora C, Hicks CC, Maire E, Puotinen M, Skirving WJ, Adjeroud M, Ahmadia G, Arthur R, Bauman AG, Beger M, Berumen ML, Bigot L, Bouwmeester J, Brenier A, Bridge TCL, Brown E, Campbell SJ, Cannon S, Cauvin B, Chen CA, Claudet J, Denis V, Donner S, Estradivari FN, Feary DA, Fenner D, Fox H, Franklin EC, Friedlander A, Gilmour J, Goiran C, Guest J, Hobbs J-PA, Hoey AS, Houk P, Johnson S, Jupiter SD, Kayal M, Kuo C, Lamb J, Lee MAC, Low J, Muthiga N, Muttaqin E, Nand Y, Nash KL, Nedlic O, Pandolfi JM, Pardede S, Patankar V, Penin L, Ribas-Deulofeu L, Richards Z, Roberts TE, Rodgers KS, Safuan CDM, Sala E, Shedrawi G, Sin TM, Smallhorn-West P, Smith JE, Sommer B, Steinberg PD, Sutthacheep M, Tan CHJ, Williams GJ, Wilson S, Yeemin T, Bruno JF, Fortin M-J, Krkosek M, Mouillot D (2019) Social–environmental drivers inform strategic management of coral reefs in the Anthropocene. Nat Ecol Evol 3:1341–1350

Davis TR, Nimbs M (2022) Susceptibility of Tidal Pool fish assemblages to climate change. Ecologies 3:510–520

Soares MO (2020) Marginal reef paradox: a possible refuge from environmental changes? Ocean Coast Manag 185:105063

Soares MO, de Araújo JT, Ferreira SMC, Santos BA, Boavida JRH, Costantini F, Rossi S (2020a) Why do mesophotic coral ecosystems have to be protected? Sci Total Environ 726:138456

Soares MO, Teixeira CEP, Bezerra LEA, Paiva SV, Tavares TCL, Garcia TM, de Araújo JT, Campos CC, Ferreira SMC, Matthews-Cascon H, Frota A, Mont'Alverne TCF, Silva ST, Rabelo EF, Barroso CX, de Freitas JEP, de Melo Júnior M, de Campelo RPS, de Santana CS, de Carneiro PBM, Meirelles AJ, Santos BA, de Oliveira AHB, Horta P, Cavalcante RM (2020b) Oil spill in South Atlantic (Brazil): environmental and governmental disaster. Mar Policy 115:103879

Derraik JG (2002) The pollution of the marine environment by plastic debris: a review. Mar Pollut Bull 44:842–852

Deutsch C, Ferrel A, Seibel B, Pörtner H-O, Huey RB (2015) Climate change tightens a metabolic constraint on marine habitats. Science (80-) 348:1132–1135

Diaz C, Foster NL, Attrill MJ, Bolton A, Ganderton P, Howell KL, Robinson E, Hosegood P (2023) Mesophotic coral bleaching associated with changes in thermocline depth. Nat Commun 14:1–11

Donelson JM, Munday PL, McCormick MI, Pitcher CR (2012) Rapid transgenerational acclimation of a tropical reef fish to climate change. Nat Clim Chang 2:30–32

Donner SD, Potere D (2007) The inequity of the global threat to coral reefs. Bioscience 57:214–215

Duarte CM, Agusti S, Barbier E, Britten GL, Castilla JC, Gattuso JP, Fulweiler RW, Hughes TP, Knowlton N, Lovelock CE, Lotze HK, Predragovic M, Poloczanska E, Roberts C, Worm B (2020) Rebuilding marine life. Nature 580:39–51

Eakin CM, Morgan JA, Heron SF, Smith TB, Liu G, Alvarez-Filip L, Baca B, Bartels E, Bastidas C, Bouchon C, Brandt M, Bruckner AW, Bunkley-Williams L, Cameron A, Causey BD, Chiappone M, Christensen TRL, Crabbe MJC, Day O, de la Guardia E, Díaz-Pulido G, DiResta D, Gil-Agudelo DL, Gilliam DS, Ginsburg RN, Gore S, Guzmán HM, Hendee JC, Hernández-Delgado EA, Husain E, Jeffrey CFG, Jones RJ, Jordán-Dahlgren E, Kaufman LS, Kline DI, Kramer PA, Lang JC, Lirman D, Mallela J, Manfrino C, Maréchal J-P, Marks K, Mihaly J, Miller WJ, Mueller EM, Muller EM, Orozco Toro CA, Oxenford HA, Ponce-Taylor D, Quinn N, Ritchie KB, Rodríguez S, Ramírez AR, Romano S, Samhouri JF, Sánchez JA, Schmahl GP, Shank BV, Skirving WJ, Steiner SCC, Villamizar E, Walsh SM, Walter C, Weil E, Williams EH, Roberson KW, Yusuf Y (2010) Caribbean corals in crisis: record thermal stress, bleaching, and mortality in 2005. PLoS One 5:e13969

Eddy TD, Lam VWY, Reygondeau G, Cisneros-Montemayor AM, Greer K, Palomares MLD, Bruno JF, Ota Y, Cheung WWL (2021) Global decline in capacity of coral reefs to provide ecosystem services. One Earth 4:1278–1285

Edenhofer O, Pichs-Madruga R, Sokona Y, Seyboth K, Arvizu D, Bruckner T, Christensen JM, Devarnay J-M, Faaij A, Fischedick M (2011) Summary for policymakers. In: Edenhofer O, Pichs-Madruga R, Sokona Y, Seyboth K, Matschoss P, Kadner S, Zwickel T, Eickemeier P, Hansen G, Schlömer S, von Stechow C (eds) IPCC Special Report on Renewable Energy Sources and Climate Change Mitigation. Cambridge University Press, Cambridge, United Kingdom and New York, NY, USA

Eisele MH, Madrigal-Mora S, Espinoza M (2021) Drivers of reef fish assemblages in an upwelling region from the Eastern Tropical Pacific Ocean. J Fish Biol 98:1074–1090. https://doi.org/10.1111/jfb.14639

Englebert N, Bongaerts P, Muir PR, Hay KB, Pichon M, Hoegh-Guldberg O (2017) Lower Mesophotic Coral Communities (60-125 m depth) of the Northern Great Barrier Reef and Coral Sea. PLoS One 12:e0170336

Eriksen M, Lebreton LCM, Carson HS, Thiel M, Moore CJ, Borerro JC, Galgani F, Ryan PG, Reisser J (2014) Plastic pollution in the world's oceans: more than 5 trillion plastic pieces weighing over 250,000 tons afloat at sea. PLoS One 9:e111913

Ern R, Huong DTT, Cong NV, Bayley M, Wang T (2014) Effect of salinity on oxygen consumption in fishes: a review. J Fish Biol 84:1210–1220

Eyal G, Laverick JH, Bongaerts P, Levy O, Pandolfi JM (2021) Mesophotic coral ecosystems of the great barrier reef are under-studied and underexplored. Front Mar Sci 8:1–5

Fabricius KE (2005) Effects of terrestrial runoff on the ecology of corals and coral reefs: review and synthesis. Mar Pollut Bull 50:125–146

Feary DA, Pratchett MS, Emslie JM, Fowler AM, Figueira WF, Luiz OJ, Nakamura Y, Booth DJ (2014) Latitudinal shifts in coral reef fishes: why some species do and others do not shift. Fish Fish 15:593–615

Ferreira CEL, Luiz OJ, Floeter SR, Lucena MB, Barbosa MC, Rocha CR, Rocha LA (2015) First record of invasive Lionfish (Pterois volitans) for the Brazilian Coast. PLoS One 10:e0123002

Figueroa-Pico J, Carpio AJ, Tortosa FS (2020) Turbidity: a key factor in the estimation of fish species richness and abundance in the rocky reefs of Ecuador. Ecol Indic 111:106021

Floeter SR, Ferreira CEL, Dominici-Arosemena A, Zalmon IR (2004) Latitudinal gradients in Atlantic reef fish communities: trophic structure and spatial use patterns. J Fish Biol 64:1680–1699

Frade PR, Bongaerts P, Englebert N, Rogers A, Gonzalez-Rivero M, Hoegh-Guldberg O (2018) Deep reefs of the Great Barrier Reef offer limited thermal refuge during mass coral bleaching. Nat Commun 9:3447

Frazão Santos C, Agardy T, Crowder LB, Day JC, Pinsky ML, Himes-Cornell A, Reimer JM, García-Morales S, Bennett NJ, Lombard AT, Calado H, Scherer M, Flannery W, Wedding LM, Gissi E (2024) Key components of sustainable climate-smart ocean planning. npj Ocean Sustain 3:1–10. https://doi.org/10.1038/s44183-024-00045-x

Friedman ST, Muñoz MM (2023) A latitudinal gradient of deep-sea invasions for marine fishes. Nat Commun 14:765

Frometa J, DeLorenzo ME, Pisarski EC, Etnoyer PJ (2017) Toxicity of oil and dispersant on the deep water gorgonian octocoral Swiftia exserta, with implications for the effects of the Deepwater Horizon oil spill. Mar Pollut Bull 122:91–99

Glynn PW (1996) Coral reef bleaching: facts, hypotheses and implications. Glob Chang Biol 2:495–509

Gould K, Bruno JF, Ju R, Goodbody-Gringley G (2021) Upper-mesophotic and shallow reef corals exhibit similar thermal tolerance, sensitivity and optima. Coral Reefs 40:907–920

Graham NAJ, Nash KL (2013) The importance of structural complexity in coral reef ecosystems. Coral Reefs 32:315–326

Graham NAJ, Cinner JE, Norström AV, Nyström M (2014) Coral reefs as novel ecosystems: EMBRACING new futures. Curr Opin Environ Sustain 7:9–14

Grenchik MK, Donelson JM, Munday PL (2013) Evidence for developmental thermal acclimation in the damselfish, Pomacentrus moluccensis. Coral Reefs 32:85–90

Hedley JD, Roelfsema CM, Chollett I, Harborne AR, Heron SF, Weeks SJ, Skirving WJ, Strong AE, Mark Eakin C, Christensen TRL, Ticzon V, Bejarano S, Mumby PJ (2016) Remote sensing of coral reefs for monitoring and management: a review. Remote Sens 8:118

Hereher ME (2020) Assessment of climate change impacts on sea surface temperatures and sea level rise-The Arabian Gulf. Climate 8:50

Hoegh-Guldberg O, Bruno JF (2010) The impact of climate change on the world's marine ecosystems. Science (80-) 328:1523–1528

Holland MM, Smith JA, Everett JD, Vergés A, Suthers IM (2020) Latitudinal patterns in trophic structure of temperate reef-associated fishes and predicted consequences of climate change. Fish Fish 21:1092–1108

Howells EJ, Bauman AG, Vaughan GO, Hume BCC, Voolstra CR, Burt JA (2020) Corals in the hottest reefs in the world exhibit symbiont fidelity not flexibility. Mol Ecol 29:899–911

Hughes TP, Anderson KD, Connolly SR, Heron SF, Kerry JT, Lough JM, Baird AH, Baum JK, Berumen ML, Bridge TC, Claar DC, Eakin CM, Gilmour JP, Graham NAJ, Harrison H, Hobbs JPA, Hoey AS, Hoogenboom M, Lowe RJ, McCulloch MT, Pandolfi JM, Pratchett M, Schoepf V, Torda G, Wilson SK (2018a) Spatial and temporal patterns of mass bleaching of corals in the Anthropocene. Science (80-) 359:80–83

Hughes TP, Kerry JT, Baird AH, Connolly SR, Dietzel A, Eakin CM, Heron SF, Hoey AS, Hoogenboom MO, Liu G, McWilliam MJ, Pears RJ, Pratchett MS, Skirving WJ, Stella JS, Torda G (2018b) Global warming transforms coral reef assemblages. Nature 556:492–496

Hughes TP, Kerry JT, Baird AH, Connolly SR, Chase TJ, Dietzel A, Hill T, Hoey AS, Hoogenboom MO, Jacobson M, Kerswell A, Madin JS, Mieog A, Paley AS, Pratchett MS, Torda G, Woods RM (2019) Global warming impairs stock–recruitment dynamics of corals. Nature 568:387–390

Iglesias-Campos A, Rubeck J, Sanmiguel-Esteban D, Schwarz G, Ansong JO, Isaksson I, da Quesada Silva M, Smith J, Suárez de Vivero JL, Varjopuro R (2021) MSPglobal: international guide on marine/maritime spatial planning

IPCC (2023) IPCC, 2023: Climate Change 2023: synthesis report. Contribution of Working Groups I, II and III to the Sixth Assessment Report of the Intergovernmental Panel on Climate Change [Core Writing Team, H. Lee and J. Romero (eds.)]. IPCC, Geneva, Switzerland

Jambeck JR, Geyer R, Wilcox C, Siegler TR, Perryman M, Andrady A, Narayan R, Law KL (2015) Plastic waste inputs from land into the ocean. Science (80-) 347:768–771

Jia R, Gao J, Gao F (2022) Robust Ocean zoning for conservation, fishery and marine renewable energy with co-location strategy. Appl Energy 328:120166

Johansen JL, Jones GP (2011) Increasing ocean temperature reduces the metabolic performance and swimming ability of coral reef damselfishes. Glob Chang Biol 17:2971–2979

Johansen JL, Jones GP (2013) Sediment-induced turbidity impairs foraging performance and prey choice of planktivorous coral reef fishes. Ecol Appl 23:1504–1517

Johnson DE, Barrio Froján C, Turner PJ, Weaver P, Gunn V, Dunn DC, Halpin P, Bax NJ, Dunstan PK (2018) Reviewing the EBSA process: improving on success. Mar Policy 88:75–85

Kahng SE, Garcia-Sais JR, Spalding HL, Brokovich E, Wagner D, Weil E, Hinderstein L, Toonen RJ (2010) Community ecology of mesophotic coral reef ecosystems. Coral Reefs 29:1–21

Kahng S, Copus JM, Wagner D (2017) Mesophotic coral ecosystems. In: Rossi S, Bramanti L, Gori A, Orejas C (eds) Marine animal forests. Springer International Publishing, Cham, pp 1–22

Kerry JT, Bellwood DR (2015) Do tabular corals constitute keystone structures for fishes on coral reefs? Coral Reefs 34:41–50

Kingsford MJ, Suthers IM, Gray CA (1996) Exposure to sewage plumes and the incidence of deformities in larval fishes. Mar Pollut Bull 33:201–212

Kleypas JA, McMaus JW, Menez LAB (1999) Environmental limits to coral reef development: where do we draw the line? Am Zool 39:146–159

Koweek DA, Dunbar RB, Monismith SG, Mucciarone DA, Woodson CB, Samuel L (2015) High-resolution physical and biogeochemical variability from a shallow back reef on Ofu, American Samoa: an end-member perspective. Coral Reefs 34:979–991

Lamb JB, Willis BL, Fiorenza EA, Couch CS, Howard R, Rader DN, True JD, Kelly LA, Ahmad A, Jompa J, Harvell CD (2018)

Plastic waste associated with disease on coral reefs. Science (80-) 359:460–462

Lau SC, Mehdi H, Bragg LM, Servos MR, Balshine S, Scott GR (2021) Exposure to wastewater effluent disrupts hypoxia responses in killifish (Fundulus heteroclitus). Environ Pollut 284:117373

Lebreton LCM, van der Zwet J, Damsteeg J-W, Slat B, Andrady A, Reisser J (2017) River plastic emissions to the world's oceans. Nat Commun 8:15611

Leggat WP, Camp EF, Suggett DJ, Heron SF, Fordyce AJ, Gardner S, Deakin L, Turner M, Beeching LJ, Kuzhiumparambil U, Eakin CM, Ainsworth TD (2019) Rapid coral decay is associated with marine heatwave mortality events on reefs. Curr Biol 29:2723–2730.e4

Lesser MP, Slattery M, Mobley CD (2018) Biodiversity and functional ecology of mesophotic coral reefs. Annu Rev Ecol Evol Syst 49:49–71

Lilkendey J, Pisternick T, Neumann SI, Dumur Neelayya D, Bröhl S, Neehaul Y, Moosdorf N (2019) Fresh submarine groundwater discharge augments growth in a reef fish. Front Mar Sci 6:1–11. https://doi.org/10.3389/fmars.2019.00613

Loiseau N, Villéger S, Le Bozec C, Gimenez M, Kawahara SL, Claverie T (2023) Mesophotic reefs are not refugia for neither taxonomic nor functional diversity of reef fishes. Coral Reefs 42:63–75

Lough JM (2012) Small change, big difference: sea surface temperature distributions for tropical coral reef ecosystems, 1950–2011. J Geophys Res Ocean 117:C09018

Luza AL, Quimbayo JP, Ferreira CEL, Floeter SR, Francini-Filho RB, Bender MG, Longo GO (2022) Low functional vulnerability of fish assemblages to coral loss in Southwestern Atlantic marginal reefs. Sci Rep 12:1–12

Macdonald RK, Ridd PV, Whinney JC, Larcombe P, Neil DT (2013) Towards environmental management of water turbidity within open coastal waters of the Great Barrier Reef. Mar Pollut Bull 74:82–94

MacDonald C, Pinheiro HT, Shepherd B, Phelps TAY, Rocha LA (2021) Disturbance and distribution gradients influence resource availability and feeding behaviours in corallivore fishes following a warm-water anomaly. Sci Rep 11:1–12

Maire E, Cinner J, Velez L, Huchery C, Mora C, Dagata S, Vigliola L, Wantiez L, Kulbicki M, Mouillot D (2016) How accessible are coral reefs to people? A global assessment based on travel time. Ecol Lett 19:351–360

Martinez-Escobar DF, Mallela J (2019) Assessing the impacts of phosphate mining on coral reef communities and reef development. Sci Total Environ 692:1257–1266

McWhorter JK, Halloran PR, Roff G, Mumby PJ (2024) Climate change impacts on mesophotic regions of the Great Barrier Reef. Proc Natl Acad Sci 121:e2303336121

Medeiros APM, Ferreira BP, Alvarado F, Betancur-R R, Soares MO, Santos BA (2021) Deep reefs are not refugium for shallow-water fish communities in the southwestern Atlantic. Ecol Evol 11:4413–4427

Medeiros APM, Ferreira BP, Betancur-R R, Cardoso APLR, Matos MRSBC, Santos BA (2022) Centenary shipwrecks reveal the limits of artificial habitats in protecting regional reef fish diversity. J Appl Ecol 59:286–299

Mies M (2025) Diversity and vulnerability of reef-building corals in extreme and marginal habitats. In: Extreme and marginal coral reefs and communities

Mies M, Francini-Filho RB, Zilberberg C, Garrido AG, Longo GO, Laurentino E, Güth AZ, Sumida PYG, Banha TNS (2020) South Atlantic Coral Reefs are major global warming Refugia and less susceptible to bleaching. Front Mar Sci 7:1–13

Mihalitsis M, Hemingson CR, Goatley CHR, Bellwood DR (2021) The role of fishes as food: a functional perspective on predator–prey interactions. Funct Ecol 35:1109–1119

Miller DD, Ota Y, Sumaila UR, Cisneros-Montemayor AM, Cheung WWL (2018) Adaptation strategies to climate change in marine systems. Glob Chang Biol 24:e1–e14

Moberg FF, Folke C (1999) Ecological goods and services of coral reef ecosystems. Ecol Econ 29:215–233

Moore WS (2010) The effect of submarine groundwater discharge on the ocean. Annu Rev Mar Sci 2:59–88. https://doi.org/10.1146/annurev-marine-120308-081019

Mora C, Ospína A (2001) Tolerance to high temperatures and potential impact of sea warming on reef fishes of Gorgona Island (tropical eastern Pacific). Mar Biol 139:765–769

Morais RA, Bellwood DR (2019) Pelagic subsidies underpin fish productivity on a degraded coral reef. Curr Biol 29:1521–1527.e6

Morais J, Medeiros APM, Santos BA (2018) Research gaps of coral ecology in a changing world. Mar Environ Res 140:243–250

Morais J, Morais RA, Tebbett SB, Pratchett MS, Bellwood DR (2021a) Dangerous demographics in post-bleach corals reveal boom-bust versus protracted declines. Sci Rep 11:18787

Morais RA, Siqueira AC, Smallhorn-West PF, Bellwood DR (2021b) Spatial subsidies drive sweet spots of tropical marine biomass production. PLoS Biol 19:4–6. https://doi.org/10.1371/journal.pbio.3001435

Morais J, Morais R, Tebbett SB, Bellwood DR (2022) On the fate of dead coral colonies. Funct Ecol 36:3148–3160

Morais RA, Smallhorn-West P, Connolly SR, Ngaluafe PF, Malimali S, Halafihi T, Bellwood DR (2023) Sustained productivity and the persistence of coral reef fisheries. Nat Sustain 6:1199

Morais J, Cordeiro IL, Medeiros APM, Santos GG, Santos BA (2024a) Exploring depth-related patterns of sponge diversity and abundance in marginal reefs. Ecol Evol 14:1–13

Morais J, Tebbett SB, Morais RA, Bellwood DR (2024b) Natural recovery of corals after severe disturbance. Ecol Lett 27:1–12

Morais J, Almeida PM, Avelino CD, Souza MCA, Cardoso APLR, Gurgel ALAR, Soares MO, Santos BA (2026) Risk or relief? The influence of river plumes on coral bleaching and disease in marginal and extreme reefs. Mar Pollut Bull 222:118741

Morgan KM, Perry CT, Johnson JA, Smithers SG (2017) Nearshore turbid-zone corals exhibit high bleaching tolerance on the Great Barrier Reef following the 2016 ocean warming event. Front Mar Sci 4:1–13

Mouillot D, Parravicini V, Bellwood DR, Leprieur F, Huang D, Cowman PF, Albouy C, Hughes TP, Thuiller W, Guilhaumon F (2016) Global marine protected areas do not secure the evolutionary history of tropical corals and fishes. Nat Commun 7:10359

Moura RL, Amado-Filho GM, Moraes FC, Brasileiro PS, Salomon PS, Mahiques MM, Bastos AC, Almeida MG, Silva JM, Araujo BF, Brito FP, Rangel TP, Oliveira BCV, Bahia RG, Paranhos RP, Dias RJS, Siegle E, Figueiredo AG, Pereira RC, Leal CV, Hajdu E, Asp NE, Gregoracci GB, Neumann-Leitao S, Yager PL, Francini-Filho RB, Froes A, Campeao M, Silva BS, Moreira APB, Oliveira L, Soares AC, Araujo L, Oliveira NL, Teixeira JB, Valle RAB, Thompson CC, Rezende CE, Thompson FL (2016) An extensive reef system at the Amazon River mouth. Sci Adv 2:e1501252–e1501252

Moustaka M, Langlois TJ, McLean D, Bond T, Fisher R, Fearns P, Dorji P, Evans RD (2018) The effects of suspended sediment on coral reef fish assemblages and feeding guilds of north-west Australia. Coral Reefs 37:659–673

Munday PL, McCormick MI, Nilsson GE (2012) Commentary impact of global warming and rising CO2 levels on coral reef fishes: what hope for the future? J Exp Biol 215:3865–3873

Muruga P, Siqueira AC, Bellwood DR (2024) Meta-analysis reveals weak associations between reef fishes and corals. Nat Ecol Evol 8:676–685

Nagelkerken I, Allan BJM, Booth DJ, Donelson JM, Edgar GJ, Ravasi T, Rummer JL, Vergés A, Mellin C (2023) The effects of climate change on the ecology of fishes. PLOS Clim 2:e0000258
Newport C, Padget O, de Perera TB (2021) High turbidity levels alter coral reef fish movement in a foraging task. Sci Rep 11:1–10
Newton K, Côté IM, Pilling GM, Jennings S, Dulvy NK (2007) Current and future sustainability of Island Coral Reef Fisheries. Curr Biol 17:655–658
Nunes LT, Cord I, Francini-Filho RB, Stampar SN, Pinheiro HT, Rocha LA, Floeter SR, Ferreira CEL (2019) Ecology of Prognathodes obliquus, a butterflyfish endemic to mesophotic ecosystems of St. Peter and St. Paul's Archipelago. Coral Reefs 38:955–960
Obura DO, Aeby G, Amornthammarong N, Appeltans W, Bax N, Bishop J, Brainard RE, Chan S, Fletcher P, Gordon TAC, Gramer L, Gudka M, Halas J, Hendee J, Hodgson G, Huang D, Jankulak M, Jones A, Kimura T, Levy J, Miloslavich P, Chou LM, Muller-Karger F, Osuka K, Samoilys M, Simpson SD, Tun K, Wongbusarakum S (2019) Coral reef monitoring, reef assessment technologies, and ecosystem-based management. Front Mar Sci 6:1–21
Ogar E, Pecl G, Mustonen T (2020) Science must embrace traditional and indigenous knowledge to solve our biodiversity crisis. One Earth 3:162–165
Oliveira BSSP, Santos RG, Santos BA (2024) Improving the knowledge management of marine megafauna strandings. J Environ Manag 351:119815
Palumbi SR, Barshis DJ, Traylor-Knowles N, Bay RA (2014) Mechanisms of reef coral resistance to environmental stress,making its relative ability to acclimate or adapt extremely important to the to future climate change. Science (80-) 344:895–898
Pauly D, Cheung WWL (2018) Sound physiological knowledge and principles in modeling shrinking of fishes under climate change. Glob Chang Biol 24:e15–e26
Pereira PHC, Macedo CH, de Nunes JACC, de Marangoni LFB, Bianchini A (2018) Effects of depth on reef fish communities: insights of a "deep refuge hypothesis" from Southwestern Atlantic reefs. PLoS One 13:1–20
Pie MR, Divieso R, Caron FS, Siqueira AC, Barneche DR, Luiz OJ (2021) The evolution of latitudinal ranges in reef-associated fishes: heritability, limits and inverse Rapoport's rule. J Biogeogr 48:2121–2132
Pinheiro H, Goodbody-Gringley G, Jessup ME, Shepherd B, Chequer A, Rocha L (2016) Upper and lower mesophotic coral reef fish communities evaluated by underwater visual censuses in two Caribbean locations. Coral Reefs 35:139
Pinheiro HT, MacDonald C, Quimbayo JP, Shepherd B, Phelps TA, Loss AC, Teixeira JB, Rocha LA (2023a) Assembly rules of coral reef fish communities along the depth gradient. Curr Biol 33:1421–1430.e4
Pinheiro HT, MacDonald C, Santos RG, Ali R, Bobat A, Cresswell BJ, Francini-Filho R, Freitas R, Galbraith GF, Musembi P, Phelps TA, Quimbayo JP, Quiros TEAL, Shepherd B, Stefanoudis PV, Talma S, Teixeira JB, Woodall LC, Rocha LA (2023b) Plastic pollution on the world's coral reefs. Nature 619:311–316
Pollock MS, Clarke LMJ, Dubé MG (2007) The effects of hypoxia on fishes: from ecological relevance to physiological effects. Environ Rev 15:1–14
Pörtner HO, Knust R (2007) Climate change affects marine fishes through the oxygen limitation of thermal tolerance. Science (80-) 315:95–97
Pratchett MS, Hoey AS, Wilson SK, Messmer V, Graham NAJ (2011) Changes in biodiversity and functioning of reef fish assemblages following coral bleaching and coral loss. Diversity 3:424–452
Pyle RL (2019) Advanced technical diving. In: Loya Y, Puglise KA, Bridge TCL (eds) Mesophotic Coral Ecosystems. Coral Reefs of the World, vol 12. Springer, Cham, Switzerland, pp 959–972. https://doi.org/10.1007/978-3-319-92735-0_50
Pyle RL, Copus JM (2019) Mesophotic coral ecosystems: introduction and overview. In: Loya Y, Puglise KA, Bridge TCL (eds) Mesophotic coral ecosystems. Springer International Publishing, Cham, pp 3–27
Quimbayo JP, Nunes LT, Silva FC, Anderson AB, Barneche DR, Canterle AM, Cord I, Dalben A, Ferrari DS, Fontoura L, Fiuza TMJ, Liedke AMR, Longo GO, Morais RA, Siqueira AC, Floeter SR (2023) TimeFISH: long-term assessment of reef fish assemblages in a transition zone in the Southwestern Atlantic. Ecology 104:e3966
Rabosky DL, Chang J, Title PO, Cowman PF, Sallan L, Friedman M, Kaschner K, Garilao C, Near TJ, Coll M, Alfaro ME (2018) An inverse latitudinal gradient in speciation rate for marine fishes. Nature 559:392–395
Reimer JD, Peixoto RS, Davies SW, Traylor-Knowles N, Short ML, Cabral-Tena RA, Burt JA, Pessoa I, Banaszak AT, Winters RS, Moore T, Schoepf V, Kaullysing D, Calderon-Aguilera LE, Wörheide G, Harding S, Munbodhe V, Mayfield A, Ainsworth T, Vardi T, Eakin CM, Pratchett MS, Voolstra CR (2024) The fourth global coral bleaching event: where do we go from here? Coral Reefs 43:1121
Richardson LE, Graham NAJ, Pratchett MS, Eurich JG, Hoey AS (2018) Mass coral bleaching causes biotic homogenization of reef fish assemblages. Glob Chang Biol 24:3117–3129
Riegl B Jr, Cavalcante G, Bento R, Burt J, Riegl B, D'Angelo C, Wiedenmann J (2025) Marginal reefs of the Persian/Arabian Gulf: patterns and processes. In: Extreme and marginal coral reefs and communities
Roberts CM, McClean CJ, Veron JEN, Hawkins JP, Allen GR, McAllister DE, Mittermeier CG, Schueler FW, Spalding M, Wells F, Vynne C, Werner TB (2002) Marine biodiversity hotspots and conservation priorities for tropical reefs. Science 80-) 295:1280–1284
Robertson DR (1998) Do coral-reef fish faunas have a distinctive taxonomic structure? Coral Reefs 17:179–186
Robinson JPW, Wilson SK, Robinson J, Gerry C, Lucas J, Assan C, Govinden R, Jennings S, Graham NAJ (2019) Productive instability of coral reef fisheries after climate-driven regime shifts. Nat Ecol Evol 3:183–190
Rocha LA, Pinheiro HT, Shepherd B, Papastamatiou YP, Luiz OJ, Pyle RL, Bongaerts P (2018) Mesophotic coral ecosystems are threatened and ecologically distinct from shallow water reefs. Science (80-) 361:281–284
Sale PF (2015) The future for coral reef fishes. In: Mora C (ed) Ecology of Fishes on Coral Reefs. Cambridge University Press, Cambridge, UK, pp 283–288. https://doi.org/10.1017/CBO9781316105412.037
Sambrook K, Hoey AS, Andréfouët S, Cumming GS, Duce S, Bonin MC (2019) Beyond the reef: the widespread use of non-reef habitats by coral reef fishes. Fish Fish 20:903–920
Sanchez-Vidal A, Thompson RC, Canals M, de Haan WP (2018) The imprint of microfibres in southern European deep seas. PLoS One 13:e0207033
Santana MFM, Tonin H, Vamvounis G, Van Herwerden L, Motti CA, Kroon FJ (2025) Predicting microplastic dynamics in coral reefs : presence , distribution , and bioavailability through field data and numerical simulation analysis. Environ Sci Pollut Res. https://doi.org/10.1007/s11356-025-36234-5
Schiettekatte NMD, Casey JM, Brandl SJ, Mercière A, Degregori S, Burkepile D, Van Wert JC, Ghilardi M, Villéger S, Parravicini V (2023) The role of fish feces for nutrient cycling on coral reefs. Oikos 2023:e09914
Schlaefer JA, Tebbett SB, Bowden CL, Collins WP, Duce S, Hemingson CR, Huertas V, Mihalitsis M, Morais J, Morais RA, Siqueira AC, Streit RP, Swan S, Valenzuela J, Bellwood DR (2022) A snapshot of sediment dynamics on an inshore coral reef. Mar Environ Res 181:105763
Schoepf V, Stat M, Falter JL, McCulloch MT (2015) Limits to the thermal tolerance of corals adapted to a highly fluctuating, naturally extreme temperature environment. Sci Rep 5:1–14

Schoepf V, Baumann JH, Barshis DJ, Browne NK, Camp EF, Comeau S, Cornwall CE, Guzmán HM, Riegl B, Rodolfo-Metalpa R, Sommer B (2023) Corals at the edge of environmental limits: a new conceptual framework to re-define marginal and extreme coral communities. Sci Total Environ 884:163688

Schofield PJ (2009) Geographic extent and chronology of the invasion of non-native lionfish (Pterois volitans [Linnaeus 1758] and P. miles [Bennett 1828]) in the Western North Atlantic and Caribbean Sea. Aquat Invasions 4:473–479. https://doi.org/10.3391/ai.2009.4.3.5

Sherman C, Nemeth M, Ruíz H, Bejarano I, Appeldoorn R, Pagán F, Schärer M, Weil E (2010) Geomorphology and benthic cover of mesophotic coral ecosystems of the upper insular slope of southwest Puerto Rico. Coral Reefs 29:347–360

Silva LCC, Santos BA (2023) Evolution of Brazilian legislation regarding the marine fauna: advances and shortcomings between 1960 and 2020. Mar Policy 153:105638

Silva LCC, Santos BA (2024) Are Brazilian marine protected areas repositories of threatened species? Biodivers Conserv 33:793–805

Silva M, Etnoyer PJ, MacDonald IR (2016) Coral injuries observed at Mesophotic Reefs after the Deepwater Horizon oil discharge. Deep Sea Res Part II Top Stud Oceanogr 129:96–107

Siqueira AC, Oliveira-Santos LGR, Cowman PF, Floeter SR (2016) Evolutionary processes underlying latitudinal differences in reef fish biodiversity. Glob Ecol Biogeogr 25:1466–1476

Siqueira AC, Bellwood DR, Cowman PF (2019) Historical biogeography of herbivorous coral reef fishes: the formation of an Atlantic fauna. J Biogeogr 46:1611–1624

Siqueira AC, Morais RA, Bellwood DR, Cowman PF (2021) Planktivores as trophic drivers of global coral reef fish diversity patterns. Proc Natl Acad Sci USA 118:e2019404118

Siqueira AC, Muruga P, Bellwood DR (2023a) On the evolution of fish–coral interactions. Ecol Lett 26:1348–1358

Siqueira AC, Yan HF, Morais RA, Bellwood DR (2023b) The evolution of fast-growing coral reef fishes. Nature 618:322–327

Skirving WJ, Heron SF, Marsh BL, Liu G, De La Cour JL, Geiger EF, Eakin CM (2019) The relentless march of mass coral bleaching: a global perspective of changing heat stress. Coral Reefs 38:547–557

Slattery M (2012) Mesophotic coral reefs: a global model of community structure and function. Proc 12th Int Coral Reef Symp, pp 9–13

Slattery M, Lesser MP, Rocha LA, Spalding HL, Smith TB (2024) Function and stability of mesophotic coral reefs. Trends Ecol Evol 39:585–598

Smith TB, Brandtneris VW, Canals M, Brandt ME, Martens J, Brewer RS, Kadison E, Kammann M, Keller J, Holstein DM (2016a) Potential structuring forces on a shelf edge upper mesophotic coral ecosystem in the US Virgin Islands. Front Mar Sci 3:1

Smith TB, Gyory J, Brandt ME, Miller WJ, Jossart J, Nemeth RS (2016b) Caribbean mesophotic coral ecosystems are unlikely climate change refugia. Glob Chang Biol 22:2756–2765

Smythe TC (2017) Marine spatial planning as a tool for regional ocean governance?: an analysis of the New England Ocean planning network. Ocean Coast Manag 135:11–24

Soares M, Davis M, de Paiva CC, de Macêdo Carneiro PB (2018) Mesophotic ecosystems: coral and fish assemblages in a tropical marginal reef (northeastern Brazil). Mar Biodivers 48:1631–1636

Soares MO, Teixeira CEP, Bezerra LEA, Rossi S, Tavares T, Cavalcante RM (2020) Brazil oil spill response: time for coordination. Science (80-) 367:155–155

Soares M, Cruz ICS, Santos BA, Tavares TCL, Garcia TM, Menezes N, Lopes BD, de Araújo JT, Gurgel ALAR, Rossi S (2021) Marginal reefs in the Anthropocene: they are not Noah's Ark. Perspectives on the marine animal forests of the world. Springer International Publishing, Cham, pp 87–128

Soares MO, Teixeira CEP, Bezerra LEA, Rabelo EF, Castro IB, Cavalcante RM (2022) The most extensive oil spill registered in tropical oceans (Brazil): the balance sheet of a disaster. Environ Sci Pollut Res 29:19869–19877

Soares MO, Pereira PHC, Feitosa CV, Maggioni R, Rocha RS, Bezerra LEA, Duarte OS, Paiva SV, Noleto-Filho E, Silva MQM, Csapo-Thomaz M, Garcia TM, Arruda Júnior JPV, Cottens KF, Vinicius B, Araújo R, do Eirado CB, Santos LPS, Guimarães TCS, Targino CH, Amorim-Reis Filho J, dos Santos WCR, de Klautau AGCM, de Gurjão LM, Machado DAN, Maia RC, Santos ES, Sabry R, Asp N, Carneiro PBM, Rabelo EF, Tavares TCL, de Lima GV, Sampaio CLS, Rocha LA, Ferreira CEL, Giarrizzo T (2023) Lessons from the invasion front: integration of research and management of the lionfish invasion in Brazil. J Environ Manag 340:117954

Soares MO, Pereira PHC, Rabelo EF, Sampaio CLS, de Xavier AL, Mendes LF, Nunes J, Bleuel J, Reis-Filho JA, Giarrizzo T (2025) Invasive lionfish spread through southwestern Atlantic marine protected areas. Mar Environ Res 208. https://doi.org/10.1016/j.marenvres.2025.107099

Sommer B (2022) Marginal reefs: distinct ecosystems of extraordinarily high conservation value. In: Hamylton SM, Hutchings P, Hoegh-Guldberg O (eds) Coral Reefs of Australia: Perspectives from Beyond the Water's Edge. CSIRO Publishing, Collingwood, Australia, pp 139–143

Souter DW, Lindén O (2000) The health and future of coral reef systems. Ocean Coast Manag 43:657–688

Sponaugle S, Grorud-Colvert K (2006) Environmental variability, early life-history traits, and survival of new coral reef fish recruits. Integr Comp Biol 46:623–633

Stuart-Smith RD, Mellin C, Bates AE, Edgar GJ (2021) Habitat loss and range shifts contribute to ecological generalization among reef fishes. Nat Ecol Evol 5:656–662

Suárez-Castro AF, Beyer HL, Kuempel CD, Linke S, Borrelli P, Hoegh-Guldberg O (2021) Global forest restoration opportunities to foster coral reef conservation. Glob Chang Biol 27:5238–5252

Sully S, van Woesik R (2020) Turbid reefs moderate coral bleaching under climate-related temperature stress. Glob Chang Biol 26:1367–1373

Sully S, Burkepile DE, Donovan MK, Hodgson G, van Woesik R (2019) A global analysis of coral bleaching over the past two decades. Nat Commun 10:1264

Teague J, Megson-Smith DA, Allen MJ, Day JCC, Scott TB (2022) A review of current and new optical techniques for coral monitoring. Oceans 3:30–45

Tebbett SB, Schlaefer JA, Bowden CL, Collins WP, Hemingson CR, Ling SD, Morais J, Morais RA, Siqueira AC, Streit RP, Swan S, Bellwood DR (2023) Bio-physical determinants of sediment accumulation on an offshore coral reef: a snapshot study. Sci Total Environ 895:165188

Tebbett SB, Bellwood DR, Bassett T, Cuttler MVW, Moustaka M, Wilson SK, Yan HF, Evans RD (2024) The limited role of herbivorous fishes and turf-based trophic pathways in the functioning of turbid coral reefs. Rev Fish Biol Fish 34:439–460

Teh LSL, Teh LCL, Sumaila UR (2013) A global estimate of the number of coral reef fishers. PLoS One 8:e65397

Tomperi J, Isokangas A, Tuuttila T, Paavola M (2022) Functionality of turbidity measurement under changing water quality and environmental conditions. Environ Technol 43:1093–1101

Tornabene L, Van Tassell JL, Robertson DR, Baldwin CC (2016) Repeated invasions into the twilight zone: evolutionary origins of a novel assemblage of fishes from deep Caribbean reefs. Mol Ecol 25:3662–3682

Travers MJ, Clarke KR, Newman SJ, Hall NG, Potter IC (2018) To what extents are species richness and abundance of reef fishes along a tropical coast related to latitude and other factors? Cont Shelf Res 167:99–110

Utami DA, Reuning L, Konechnaya O, Schwarzbauer J (2021) Microplastics as a sedimentary component in reef systems: a case study from the Java Sea. Sedimentology 68:2270–2292

Van Duin EHS, Blom G, Los FJ, Maffione R, Zimmerman R, Cerco CF, Dortch M, Best EPH (2001) Modeling underwater light climate in relation to sedimentation, resuspension, water quality and autotrophic growth. Hydrobiologia 444:25–42

Vaughan GO, Shiels HA, Burt JA (2021) Seasonal variation in reef fish assemblages in the environmentally extreme southern Persian/Arabian Gulf. Coral Reefs 40:405–416

Venegas RM, Oliver T, Liu G, Heron SF, Clark SJ, Pomeroy N, Young C, Eakin CM, Brainard RE (2019) The rarity of depth refugia from coral bleaching heat stress in the Western and Central Pacific Islands. Sci Rep 9:19710

Vercelloni J, Liquet B, Kennedy EV, González-Rivero M, Caley MJ, Peterson EE, Puotinen M, Hoegh-Guldberg O, Mengersen K (2020) Forecasting intensifying disturbance effects on coral reefs. Glob Chang Biol 26:2785–2797

Vergés A, McCosker E, Mayer-Pinto M, Coleman MA, Wernberg T, Ainsworth T, Steinberg PD (2019) Tropicalisation of temperate reefs: implications for ecosystem functions and management actions. Funct Ecol 33:1000–1013

Vilardo C, Barbosa AF (2024) Oil and Gas in the Brazilian Amazon Continental Margin: to drill or not to drill? Available at SSRN: https://doi.org/10.2139/ssrn4559105

Wear SL, Thurber RV (2015) Sewage pollution: mitigation is key for coral reef stewardship. Ann N Y Acad Sci 1355:15–30

Wismer S, Tebbett SB, Streit RP, Bellwood DR (2019) Young fishes persist despite coral loss on the Great Barrier Reef. Commun Biol 2:1–7

WMO (2024) World Meteorological Organization, 2024: WMO confirms 2024 warmest year on record, about 1.55 °C above pre-industrial level. WMO, Geneva. Available at: https://wmo.int/news/media-centre/wmo-confirms-2024-warmest-year-record-about-155degc-above-pre-industrial-level

Woodall LC, Sanchez-Vidal A, Canals M, Paterson GLJ, Coppock R, Sleight V, Calafat A, Rogers AD, Narayanaswamy BE, Thompson RC (2014) The deep sea is a major sink for microplastic debris. R Soc Open Sci 1:140317

Woodhead AJ, Hicks CC, Norström AV, Williams GJ, Graham NAJ (2019) Coral reef ecosystem services in the Anthropocene. Funct Ecol 33:1023–1034

Zweifler A, O'Leary M, Morgan K, Browne NK (2021) Turbid Coral Reefs: past, present and future—a review. Diversity 13:251

Microbial Communities in Extreme and Marginal Coral Reefs

4

Matteo Monti, Marco Casartelli, Laura Beenham, Yusuf C. El-Khaled, and Raquel S. Peixoto

Abstract

This chapter explores the complex microbial communities associated with coral reefs, particularly those in extreme and marginal environments, including critical temperature regimes, mesophotic depths, low pH conditions, mangroves ecosystems, turbid environments and anthropogenically disturbed reefs, highlighting their crucial role in reef resilience and response to environmental stressors. Microbial associates of corals, known collectively as the coral microbiome, are integral to coral health, influencing nutrient cycling, disease resistance, and, consequently, overall ecosystem stability. Extreme coral communities, found naturally in suboptimal conditions, feature stress-tolerant microbiomes that resist disturbances and may underlie the holobiont (i.e., the coral host and its associated microbiome) resistance. Here, we detail the dynamics of these communities across different coral compartments and their response to a range of environmental "extremes", which serve as natural laboratories for studying microbial adaptations to harsh conditions. The study of these microbiomes is critical in the context of the current climate change scenario, as they may provide key insights into mechanisms of coral resistance, as well as help identify potential microbial interventions, and novel conservation approaches to enhance coral reefs resilience in a rapidly changing world.

Keywords

Microbiome · Bleaching · Mesophotic reefs · Acidification · Mangroves · Anthropogenic impact

M. Monti · M. Casartelli · L. Beenham · Y. C. El-Khaled · R. S. Peixoto (✉)
Biological and Environmental Sciences and Engineering Division (BESE), King Abdullah University of Science and Technology (KAUST), Thuwal, Saudi Arabia
e-mail: matteo.monti@kaust.edu.sa; marco.casartelli@kaust.edu.sa; laura.beenham@kaust.edu.sa; yusuf.khaled@kaust.edu.sa; raquel.peixoto@kaust.edu.sa

4.1 Guardians of the Reef: The Pivotal Role of Microbiomes in Coral Health and Resilience

Coral reefs are among the most biodiverse and ecologically significant ecosystems on our planet, offering habitat (Rolim et al. 2022; Kochan et al. 2023) and coastal protection (Ferrario et al. 2014; Beck et al. 2018; Reguero et al. 2021), as well as resources for millions of people across the globe (Burke et al. 2011a; Sobha et al. 2023). However, natural and anthropogenically driven climate change is currently exerting unprecedented pressure on the organisms inhabiting these environments. Recent estimates suggest that coral reefs are experiencing a worldwide decline, with an estimated 30%–50% of these ecosystems already lost. Projections indicate an additional 70%–90% decline under the scenario of a 1.5 °C increase in global warming (Voolstra et al. 2023). Addressing the potential of reef-inhabiting organisms to withstand the challenges posed by these environmental changes requires a broad approach that explores stress adaptation and response at the organismal, community, and ecosystem levels. Within this framework, reef-inhabiting microorganisms, and especially host-associated microbes, have gained increasing interest among scientists in the last three decades (Rohwer et al. 2002; Rosenberg et al. 2007; Bourne et al. 2016; Apprill 2017; O'Brien et al. 2019).

The microbial community associated with a certain environment and/or host is collectively referred to as "microbiome" (Whipps et al. 1988), including a plethora of eukaryotic and prokaryotic microorganisms such as bacteria, archaea, fungi, algae, protists, and viruses (Berg et al. 2020) (Fig. 4.1). Members of the microbiome interact with one another (Santoro et al. 2025) and with their surrounding environment, defining and occupying specific ecological niches. When residing within a living, multicellular eukaryotic host, these microbial assemblages engage in interactions that can significantly influence the host's health and fitness (Santoro et al. 2025). Selective environmental and physiological

M. O. Soares et al. (eds.), *Extreme and Marginal Coral Reefs and Communities*, Coral Reefs of the World 24,
https://doi.org/10.1007/978-3-032-28035-0_4

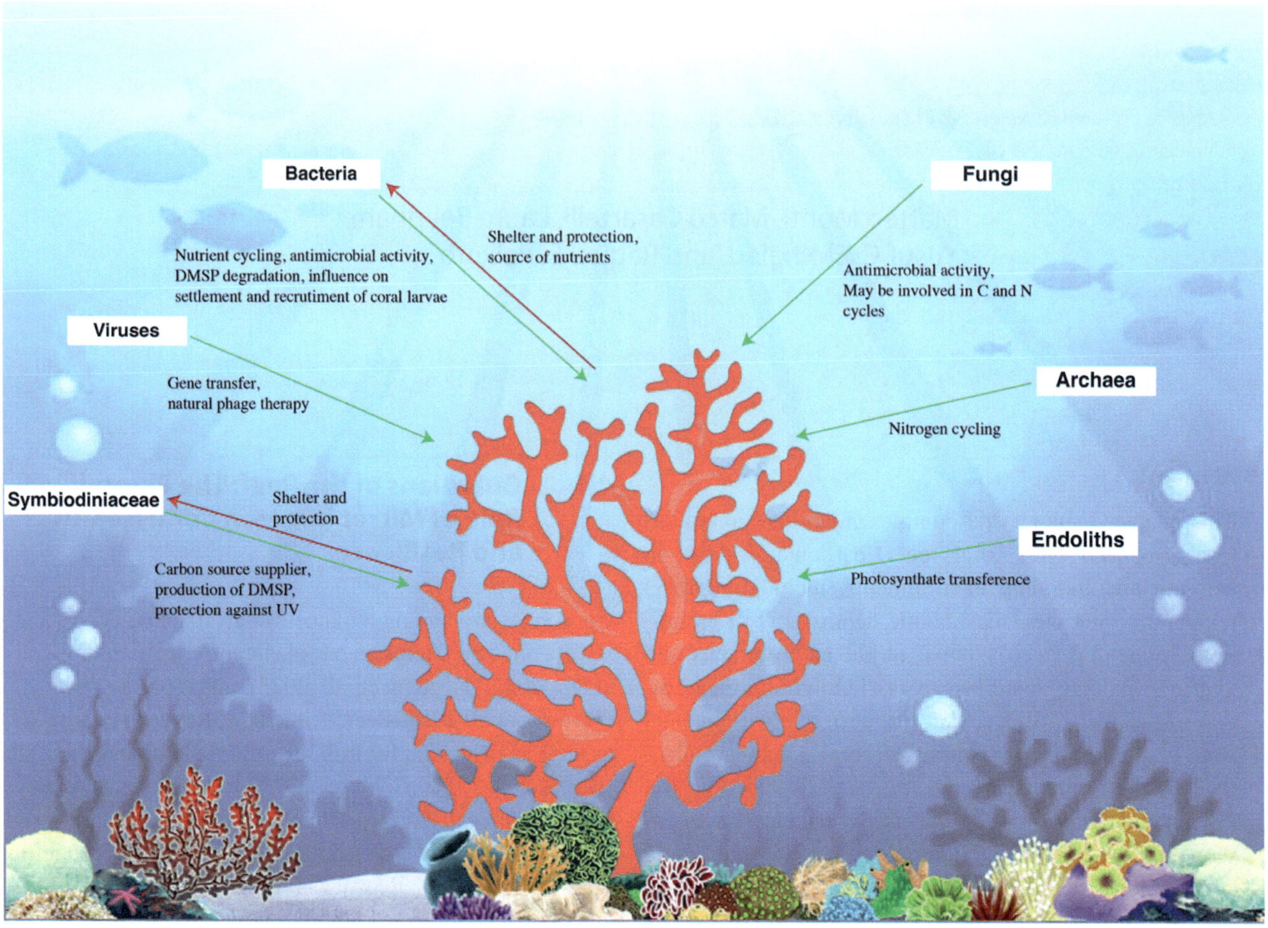

Fig. 4.1 Coral microbiome composition highlighting the principal players and their putative roles in maintaining holobiont health

forces act on each member of the holobiont individually, as well as on the phenotypic variations generated by the interaction of host, associated microbiome, and environment, shaping the holobiont phenotype (Bordenstein and Theis 2015; Moran and Sloan 2015; O'Brien et al. 2019; Voolstra et al. 2024; Santoro et al. 2025).

Like all metazoans, corals are characterized by an extremely complex microbiome, with members inhabiting different host's compartments, including the surface mucopolysaccharide layer (SML), the epidermis and gastrodermis, the mesoglea, the gastric cavity, and the calcium carbonate skeleton in scleractinians (Bourne et al. 2016; Ainsworth et al. 2017; van Oppen and Blackall 2019; Mohamed et al. 2023). Members of the coral microbiome include symbiotic dinoflagellate algae of the family Symbiodiniaceae, endolithic algae, fungi, bacteria, archaea, and viruses (Bourne et al. 2016; LaJeunesse et al. 2018; van Oppen and Blackall 2019; Voolstra et al. 2021). Despite a growing interest in neglected members of this community (Raghukumar and Ravindran 2012; Siboni et al. 2012; Vega Thurber et al. 2017; Galindo-Martínez et al. 2022; Modolon et al. 2023; Voolstra et al. 2024), the vast majority of the research on coral microbiome focused on Symbiodiniaceae and bacteria (e.g., Bourne et al. 2016; Neave et al. 2017; Ziegler et al. 2019). Accordingly, unless specified otherwise, the information presented in this chapter refers to these two groups.

Host phylogeny appears to be the main driver for coral microbiome composition, but other environmental variables such as geographical location, habitat, depth, temperature, light, and nutrient availability have been identified as important drivers dictating differences in the microbial makeup of various coral taxa, which can also vary between different compartments of the same coral colony (Bourne et al. 2016; van Oppen and Blackall 2019; Vega Thurber et al. 2020; Delgadillo-Ordoñez et al. 2022; Lewis et al. 2022). Moreover, colony age and size, development stage (larva, recruit, and adult), as well as biological interactions with other organisms can directly or indirectly influence the coral microbiome composition (Leite et al. 2017; Pollock et al. 2018; Epstein et al. 2019). Additionally, different species of Symbiodiniacea produce specific metabolites that contribute for the selection of specific assemblages of the remaining members of the microbiome in the "phycosphere" (Santoro et al. 2025), which is the halo surrounding the algae cells (Bell and Mitchell 1972; Garrido et al. 2021). These specific microbiome assemblages are correlated with different thermal resilience rates (Santoro et al. 2025).

The composition and diversity of coral-inhabiting microbial communities are generally significantly different from the one characterizing the surrounding seawater (Frias-Lopez et al. 2002; Sunagawa et al. 2010; Neave et al. 2017). This divergence led scientists to hypothesize the presence of control mechanisms, mediated by the host or by members of the associated microbial consortia, which could select and consequently shape the coral microbiome composition (Vidal-Dupiol et al. 2011; Vega Thurber et al. 2017; Welsh et al. 2017), as already observed for other marine organisms (Nyholm and McFall-Ngai 2004; Deines et al. 2017; Pita et al. 2018).

Bacterial assemblages within corals can be species-specific (Rohwer et al. 2002), whereby colonies of a given coral species can share small groups of persistent symbionts (Roder et al. 2015; Wessels et al. 2017). Defining core members of the coral microbiome is challenging and may vary across hosts and environmental conditions (Voolstra et al. 2024). Overall, the coral microbiome is composed of (1) transient microorganisms, usually associated with the coral mucus and highly controlled by environmental conditions (Roder et al. 2015; Hernandez-Agreda et al. 2016, 2017; Marchioro et al. 2020); (2) a resident host-specific microbial community (Delgadillo-Ordoñez et al. 2022; Buitrago-López et al. 2023; Longley et al. 2024); and (3) near universal members that are associated with multiple coral species across different environmental conditions and locations (Hernandez-Agreda et al. 2017).

Members of the core community are thought to be persistent within their host even under changes in environmental conditions, and they are proposed to exert a crucial part in host physiology (Sweet and Bulling 2017). For example, the bacterial genus *Endozoicomonas* (Gammaproteobacteria, Oceanospirillales, Endozoicomomnadaceae) has been proposed as an important constituent of the bacterial core microbiome of different coral species, despite its function still remains mostly unresolved (Peixoto et al. 2017; Pogoreutz and Ziegler 2024). This common coral symbiont exhibits co-phylogeny with its host (Pollock et al. 2018). Moreover, it has been observed forming aggregates in the coral tissue, or cell-associated microbial aggregates (CAMAs) (Neave et al. 2017; Maire et al. 2023), eventually adjacent to CAMAs containing other bacterial groups (Maire et al. 2023), and differences in its phylogenetic composition among different coral species have been related to different reproduction modes (i.e. brooders vs spawners) in the examined corals (Neave et al. 2017). Contrarily to *Endozoicomonas*, which often composes a high proportion of the entire bacterial community of some coral species, rare and low-abundance bacterial taxa have also been found associated with corals. For example, members of the orders Actinomycetales and Burkholderiales have been proposed as key core microbiome taxa despite being represented in low proportions in the coral bacterial community (Ainsworth et al. 2015).

Numerous studies have examined the potential contribution of scleractinian coral-associated microbial communities in holobiont health and physiology (Reshef et al. 2006; Ziegler et al. 2017, 2019). The active stewardship of the coral microbiome (Peixoto et al. 2022), for example through the use of beneficial microorganisms for corals (BMCs) (Peixoto et al. 2017, 2021), the so-called coral probiotics, has been proven to be an efficient tool to enhance coral health (Rosado et al. 2019; Tang et al. 2020; Li et al. 2021; Santoro et al. 2021; Ushijima et al. 2023; Moradi et al. 2023; Cardoso et al. 2024) and growth (Zhang et al. 2021).

The role of Symbiodiniaceae in coral host metabolism has long been documented, with the algae releasing photosynthetically fixed carbon compounds, as well as other key metabolites including amino acids, sugars and fatty acids, that the coral uses as a source of nutrition in exchange for protection and a supply of inorganic nutrients from the host (Stat et al. 2006; Matthews et al. 2017). Furthermore, through photosynthesis, the dinoflagellates shift the equilibrium of carbonate species in seawater towards calcium carbonate deposition, favoring coral calcification (Pearse and Muscatine 1971) and producing reactive oxygen species (ROS) that can act as a mechanism against pathogen infection (Diaz et al. 2016; Szabó et al. 2020). However, stressful conditions, and especially temperature stress, can disrupt this finely regulated relationship, leading to a competition for nutrients (Rädecker et al. 2021) and the toxic overproduction of ROS (Suggett and Smith 2020; Szabó et al. 2020), followed by the rejection or expulsion of the symbiotic algae from the coral tissue, a phenomenon known as coral bleaching, which can ultimately lead to the host's death (Glynn 2003; Brown 1997; Wilkinson 1999).

Besides Symbiodiniaceae, other microbiome members also play crucial roles in coral holobiont health by mitigating toxic compounds, providing nutrients, regulating the pathogenic populations and supporting coral health (Peixoto et al. 2017). More specifically, BMCs can drive carbon (C), nitrogen (N), and sulfur (S) cycling, which are beneficial for both the host and its symbiotic partners (Raina et al. 2009; Kimes et al. 2010). Nitrogen (N), essential for corals, can be sourced through heterotrophic feeding from the water column or internal recycling (Muscatine et al. 1998; Yellowlees et al. 2008). Microbially-performed nitrogen cycling is vital (O'Neil and Capone 2008), where prokaryotic N_2 fixers convert N_2 into bioavailable ammonium, thus alleviating nitrogen limitation for coral reef primary producers (Cardini et al. 2016).

Conversely, microbial denitrification processes can remove excess nitrogen during high environmental availability, such as in eutrophication (see below) (Capone et al. 1992; Koop et al. 2001; Tilstra et al. 2019; El-Khaled et al. 2021). The endolithic microbial community within coral skeletons notably contributes about half of the host's nitrogen needs, especially during bleaching events, with cyanobacteria providing organic compounds via photosynthesis, similar to mechanisms observed in marine sponges (Fine and Loya 2002). Furthermore, bacteria residing in the surface mucus layer represent the first line of defense against possible pathogen invasions through both competitive exclusion and production of antimicrobial compounds providing protection against disease onset and transmission (Ritchie 2006; Rosenberg et al. 2007; Shnit-Orland and Kushmaro 2009; Teplitski and Ritchie 2009; Rosado et al. 2019; Tang et al. 2020; Monti et al. 2022; Ushijima et al. 2023). Notably, bacterial strains of the genera *Pseudoalteromonas*, *Halomonas*, and *Cobetia* were selected based on their beneficial traits (including nitrogen cycling, ROS scavenging, and dimethylsulfoniopropionate degradation) to assemble probiotic consortia effectively employed to mitigate pathogen and temperature stressors in *Pocillopora damicornis* corals (Rosado et al. 2019). Similarly, a strain of *Pseudaolteromonas*, namely McH1–7, was reported to produce korormicin and tetrabromopyrrole antibacterial compounds, and its inoculation on corals affected by Stony Coral Tissue Loss Disease (SCTLD) was successful in slowing and arresting disease progression (Ushijima et al. 2023).

Both biotic and abiotic stressors, as well as a combination of the two, can lead to a disruption in the host-microbiome interactions (from here on referred to as dysbiosis), altering the structure of the resident bacterial community, possibly leading to a deterioration of host health due, for example, to the proliferation of opportunistic pathogens which can thrive during altered homeostatic conditions (Sweet and Bulling 2017; Peixoto et al. 2017; Lachnit et al. 2019; Moriarty et al. 2020). Understanding the specific role of microbes in influencing coral holobiont fitness is, nevertheless, a challenging task, and many of the functions identified have been proposed without being properly experimentally proven. Despite recent advances in cultivation techniques (Schultz et al. 2022; Modolon et al. 2023), a great part of the coral microbial community still remains unculturable, hampering our ability to univocally assign functions to specific microbial strains. Modern sequencing technologies and multi-omics analyses (e.g., metagenomic, metatranscriptomic, metaproteomic, metabolomic) provide scientists with tools to infer metabolic processes and potential roles of coral-associated bacteria (Burke et al. 2011b; Jiao et al. 2021; Robbins et al. 2019; Santoro et al. 2021; Reddy et al. 2023; Mannochio-Russo et al. 2023). Nevertheless, the understating of the functions played by host-associated microbial assemblages in contributing to the overall holobiont health is still in its infancy.

Coral reefs are usually found in clear, warm and oligotrophic areas, although their rare but natural occurrence in marginal (or "extreme") conditions (e.g., highly turbid coral reefs in Brazil - Lucas et al. 2023) may expand as a consequence of climate change (Camp et al. 2018). Extreme coral reef systems may therefore offer, in this sense, a unique setup to study the response of reef organisms and their associated microbiomes to unabated environmental stressors (Perry and Larcombe 2003; Burt et al. 2013; Camp et al. 2018; Schoepf et al. 2023).

This knowledge can be very useful for the selection of key microbial players to be used for different microbial therapies and microbiome stewardship (Peixoto et al. 2019, 2022; Maire and van Oppen 2022; Garcias-Bonet et al. 2023). Reefs occurring in thermally extreme habitats and/or thermally variable environments (e.g., Howells et al. 2016; Camp et al. 2018), highly turbid or urban areas (e.g., Loya et al. 2016; Morgan et al. 2016; Lucas et al. 2023) and mesophotic depths (e.g., Loya et al. 2016; Semmler et al. 2017), among others offers the opportunity to study holobiont exposure to conditions close to tolerance thresholds for a prolonged time. The study of coral-associated microbial communities in these environments is of utmost importance as they can provide a window into the near future for other areas of the world, incorporating high degrees of environmental complexity, naturally co-occurring stressors while involving diverse community interactions, and complicated ecological processes which are difficult or even impossible to simulate *in-vitro* in experimental setups.

4.2 Coral-Associated Microbial Communities in Thermally Extreme Reefs

Corals possess a thermal optimum in which the photosynthetic performance of their symbionts is at its maximum. Currently, the most pressing issue for these organisms all over the world is the increasing frequency of prolonged summer heat waves in which coral's upper thermal optimum is often exceeded, triggering symbiosis dysfunction (Hoegh-Guldberg 1999; Oliver et al. 2009; Hughes et al. 2017; Silbiger et al. 2019). Many researchers have examined the microbiome differences in coral during bleaching events, with some finding that the functional communities stay relatively stable in different geographical locations (e.g., Bourne et al. 2008; Hadaidi et al. 2017; Morrow et al. 2018; van Oppen and Blackall 2019; Meenatchi et al. 2020; Bergman et al. 2023). However, results also showed differences in carbon-, sulfur-, and nitrogen-cycling processes, as well as loss of beneficial bacteria being replaced with more opportunistic/potentially pathogenic bacteria (such as Vibrionaceae) (Hadaidi et al. 2017; Pootakham et al. 2018, 2019; Chavanich et al. 2022; Delgadillo-Ordoñez et al. 2022).

For example, Rädecker et al. (2021) experimentally demonstrated that heat stress destabilizes symbiotic carbon translocation well before the breakdown of the coral-algal symbiosis and prior to any visible signs of bleaching, leading to severe carbon limitations for the coral host. As such, the observed mass mortality of corals during global heat anomalies may not be the direct consequence of the loss of algal symbionts per se, but more fundamentally reflect the starvation of the coral host due to the collapse of symbiotic nutrient cycling.

Another intriguing study analyzing coral transcriptional footprint noted that corals exposed to prolonged heat stress conditions may experience post-heat stress disorder (PHSD). This disorder is marked by a contrast between the corals' responses and their efforts to recover from the heat-stress while still deteriorating due to the cellular, immune, and metabolic impacts of such stress (Santoro et al. 2021). Because of the current increase in the occurrence, frequency, and severity of bleaching events, researchers have started to investigate potential approaches to enhance coral resistance and resilience to this global threat (van Oppen et al. 2015; Peixoto et al. 2017; Voolstra et al. 2021). These approaches include the stewardship or manipulation of microbiomes through the use of different microbial therapies, such as the addition of beneficial microorganisms (Garcias-Bonet et al. 2023). Encouraging results have been obtained in studies on scleractinian corals, which show the ability of a selected consortium of BMCs to increase coral resistance to bleaching and other impacts (Rosado et al. 2019; Tang et al. 2020; Santoro et al. 2021; Silva et al. 2021; Li et al. 2023; Ushijima et al. 2023; Cardoso et al. 2024). Particularly, Rosado et al. (2019) isolated a consortium of native bacteria from the scleractinian *Pocillpora damicornis*, including members of the genera *Pseudoalteromonas*, *Halomonas*, and *Cobetia*, which demonstrated to mitigate coral bleaching in aquarium experiments. Similar results were obtained in a more recent study by Santoro et al. (2021), in which heat-stressed samples of *Mussismilia hispida* inoculated with selected bacteria from healthy colonies of the same coral species had a better recovery once temperatures were reduced to ambient. In this work, the authors proved the concept that the inoculation of BMCs resulted in a reconstructed microbiome, suggesting a signal cascade from the microbes to the host, which resulted in the induction of thermal stress protection and blockage of PHSD by instigating the reconstruction of the host cell transcriptional network and cellular homeostasis. Moreover, the authors highlighted the correspondence between desirable traits of selected probiotics, such as dimethylsulfoniopropionate (DMSP) degradation, and observed changes in metabolomic profiles in BMC-treated corals dampening heat stress.

The inoculation of BMCs yields promises not only in laboratory settings but also for *in situ* applications. In an impressive collective effort, Delgadillo-Ordoñez et al. (2024) inoculated colonies of *Pocillopora verrucosa* 3-times per week for 3 months with a consortium of BMCs *in situ* in the Red Sea and revealed that BMCs successfully temporarily reshaped the host microbiome without causing changes in the microbial communities of the surrounding environment providing supporting evidence of the potentially safe application of coral probiotics *in situ*.

In addition to manipulative studies that help understand key mechanisms and processes related to heat stress and explore novel methods to enhance coral resistance and resilience, tracking these mechanisms and the associated microbes in corals that are naturally exposed to extreme conditions can provide valuable insights. Corals inhabiting particular geographical locations such as the Red Sea, the Arabian Gulf, and the back reef pools of American Samoa (Fig. 4.2) are characterized by higher-than-average (compared to most tropical coral reefs) sea temperatures, salinity levels, and other physicochemical parameters that would usually stress and kill other corals in less extreme environments (Berumen et al. 2019; Bouwmeester et al. 2020; Birkeland et al. 2021). These extreme reefs represent a unique opportunity for studying microbial adaptations to thermal stress and for understanding how coral-associated microorganisms could enhance resistance, resilience, and adaptation of their hosts.

Particularly, Red Sea corals are known to be well-adapted to withstand high seawater temperatures by developing a higher thermal tolerance when compared to corals from other parts of the world, with research highlighting maximum photosynthesis rates at temperatures ranging between 25–32 °C (Evensen et al. 2021, 2022; Savary et al. 2021; Blanckaert et al. 2022; Eladawy et al. 2022). Even more extreme thermal conditions are encountered in the Arabian Gulf, where summer sea surface temperatures can reach up to 37.6 °C (Alosairi et al. 2020). Corals in this region have been noted to withstand high temperatures up to 35–37.5 °C (Howells et al. 2016; Hume et al. 2016; Smith et al. 2017a), which would be lethal to similar species in other regions.

Coral thermal tolerance not only reflects the ability of these organisms to withstand extreme temperatures but also to adapt to particularly large fluctuations of this parameter. Although corals are generally highly susceptible to changes in their environmental conditions, key reef-building coral genera such as *Acropora*, *Pocillopora*, and *Porites* are seen thriving in the backreef pools of the National Park of American Samoa (NPSA), which experience extreme temperature fluctuations with studies reporting up to 6 °C daily variation, with highs of 34.5 °C (Craig et al. 2001; Smith and Birkeland 2003; Barker 2018; Vargas-Ángel and Huntington 2020). The Red Sea basin is particularly interesting, covering

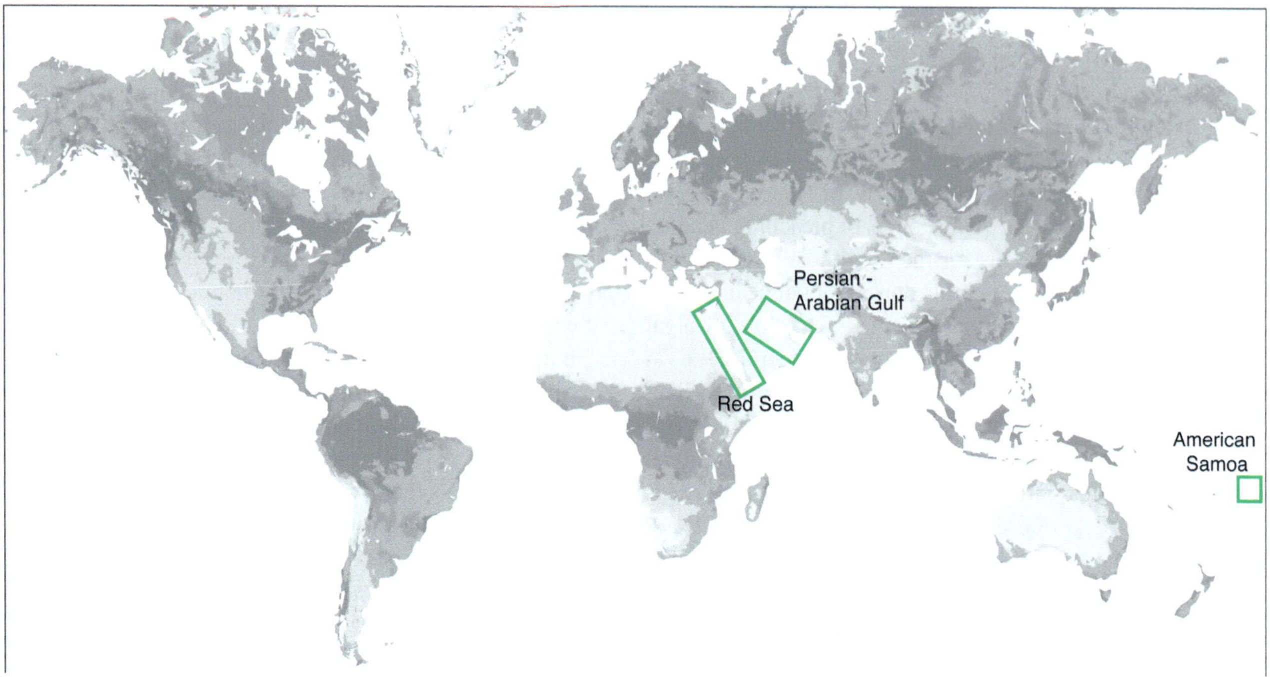

Fig. 4.2 Location of the thermally extreme coral reefs discussed in this chapter

15° latitude in which corals in the northern area experience high thermal fluctuations of up to 15 degree heating weeks (DHW) without visible bleaching in comparison to their central and southern conspecifics (Osman et al. 2018). This potential resistance to extreme temperature is determined by the holobiont as a whole, including the host expression of genes that promote heat stress tolerance, as well as the Symbiodinaceae community genotype, and host-associated bacterial microbiome (Avila-Magaña et al. 2021; Díaz-Almeyda et al. 2022; Voolstra et al. 2024).

4.2.1 Coral-Associated Algal Photosymbionts in Thermally Extreme Reefs

Regarding coral-algal symbionts, the genera *Symbiodinium*, *Cladocopium*, and *Durusdinium* (LaJeunesse et al. 2018) have been highlighted in many coral studies, with *Cladocopium* being noted to be particularly dominant in the Red Sea coral (Hume et al. 2020; Osman et al. 2020; Terraneo et al. 2023; Gardner et al. 2023). It has been proposed that these genotypes are host- and environment-specific, suggesting that the selection of specific genotypes may confer resistance or resilience to their coral hosts (Quigley et al. 2018; Hume et al. 2020; Martinez et al. 2022; Davies et al. 2023; Terraneo et al. 2023). Interestingly, the presence of *Durusdinium* spp. that are typically associated with stress tolerance (Glynn et al. 2001; Berkelmans and van Oppen 2006) was found to be limited in the northern Red Sea and not encountered in the Gulf of Aqaba (Terraneo et al. 2023), further suggesting that the coral-host specific symbionts are niche-adapted, rather than generalist as theorized in Hume et al. 2020. Similarly, corals in the Arabian Gulf appear to establish stable associations (despite temperatures spiking to 35 °C) with photosymbionts belonging to *Cladocopium*, *Durusdinium*, *Symbiodinium* clade C3, and a recently described thermotolerant dinoflagellate, *Cladocopium thermophilum*, which has been recorded as particularly prevalent locally and likely playing a key factor in the survival of coral during extreme stress (Hume et al. 2013, 2016, 2018; LaJeunesse et al. 2014; D'Angelo et al. 2015; Silverstein et al. 2015).

Conversely, little research has been conducted on the Symbiodinaceae communities in the back pools of American Samoa. Studies on *Acropora* species have found that *Durusdinium* spp. symbionts dominate habitats with the highest temperatures, representing up to 83% of the dinoflagellate community. In contrast, both genera *Cladocopium* and *Durusdinium* are common in areas with lower maximum temperatures (Oliver and Palumbi 2009). Similarly, species of the coral genus *Pavona* were noted to have 'hundredfold' more *Durusdinium* spp. in warmer, more variable pools than that of *Cladocopium* spp. (Cunning et al. 2015), whilst other coral genera exhibited variable results, suggesting that the photosymbiosis can depend on biotic and abiotic factors other than temperature, such as light, water flow, host genetic identity and surrounding metacommunity composition.

The isolation and cultivation of Symbiodiniaceae from coral hosts experiencing extreme thermal regimes have potential applications in the context of human-assisted evolution of corals (van Oppen et al. 2015). Active manipulation of symbiotic algal communities through the inoculation of naturally resistant and laboratory-selected Symbiodiniaceae strains has been proposed as a tool to promote coral holobiont tolerance to thermal stress (van Oppen et al. 2015; Torda et al. 2017) and has been attempted with various degrees of success in tank experiments (Howells et al. 2016; Chakravarti et al. 2017; Morgans et al. 2020). Improving our understanding of the genetic, physiological, and ecological factors determining the development of thermal resistance in Symbiodiniaceae species in these challenging environments, as well as the mechanisms underlying the establishment of the symbiosis with the coral host, could greatly benefit the research on active human intervention practices to assist coral reef adaptation and resilience to climate changes.

4.2.2 Coral Bacterial Communities in Thermally Extreme Reefs

The bacterial communities associated with corals inhabiting the Red Sea have also been proposed to contribute to their thermal tolerance despite the few studies addressing this topic (Osman et al. 2020; Delgadillo-Ordoñez et al. 2022). Similar to corals from other locations (Neave et al. 2019), the main coral-associated bacteria observed in Red Sea studies include those belonging to the phyla Proteobacteria, Actinobacteria, Cyanobacteria and Chloroflexi, with the more frequently occurring families belonging to Endozoicomonadaceae, Rhodobacteraceae and Vibrionaceae (Lampert et al. 2008; Bayer et al. 2013; Delgadillo-Ordoñez et al. 2022). Despite this, the Red Sea coral bacteriomes exhibit, alongside conserved endosymbiotic communities, a dynamic bacterial composition able to potentially contribute to the holobiont wellbeing and resilience in these challenging environments (Röthig et al. 2017; Osman et al. 2020). Osman et al. (2020) examined the SML bacterial communities associated with six coral species across 4° of latitude in sites that varied in mean summer temperatures within the northern Red Sea. Results from this research highlighted that, for a given host, bacterial community composition and diversity were site-specific and varied significantly along the latitudinal gradient, with several prokaryotic taxa, including the genera *Pseudoalteromonas*, *Alteromonas*, *Ruegeria*, and *Pseudomonas* being determined as indicator taxa representative of specific sites. Interestingly, the photosynthetic bacterium *Erythrobacter* sp. constituted the dominant indicator taxon in the warmest site analyzed. As this microorganism contains large amounts of carotenoids, it has been proposed to play a fundamental role in antioxidant activities during heat stress, but such functional implications have yet to be fully assessed. Overall, this compositional variability may reflect the potential capacity for local adaptation, suggesting a selection of beneficial bacteria that may be critical to promoting coral survival under extreme thermal fluctuations and regimes.

4.3 Coral-Associated Microbial Communities in Mesophotic Reefs

Mesophotic coral ecosystems (MCEs) provide an extensive, although still relatively unexplored habitat, ranging from 30 to 150 m depth (Lesser et al. 2009; Hinderstein et al. 2010). These ecosystems are usually divided into an upper mesophotic zone (30–60 m) and a lower mesophotic zone (60–150 m) (Bongaerts et al. 2015) (Fig. 4.3). However, this definition is not based on any ecological or environmental characteristics. There is ongoing debate within the scientific community about whether to establish a new definition for MCEs based on the underwater light environment and its effects on MCE communities (Lesser et al. 2009, 2018). MCEs have been historically considered to possess stable environmental conditions, rarely affected by anthropogenic stressors compared to shallow coral reefs (Pinheiro et al. 2019). Their populations are thought to act as 'seed banks' for degraded shallow reefs through larval connectivity (Slattery et al. 2011) according to the Deep Reef Refuge Hypothesis (DRRH) (reviewed by Bongaerts and Smith 2019). Nevertheless, this hypothesis is still under debate, as highlighted in a recent review by Slattery et al. (2024), as several studies examined the degree of genetic connectivity between shallow and mesophotic reef taxa with broad distribution. Some findings indicate, for example, that some corals present a stable microbial community, being depth-specialists, whereas other species, associated with a more dynamic microbiome are depth-generalists (Glasl et al. 2017). Ultimately, the interpretation of DRRH might be context-dependent, based on several variables such as geographic locations, sea bottom topography, depth range of the study, and potential ecological disturbances. In support of this statement, different research highlighted that coral species inhabiting MCEs could represent distinct populations with unique genetic and physiological adaptations, contrasting the popular thought about MCEs being a refuge for coral taxa with broad-depth distribution and a potential source for reef restoration efforts (Baker et al. 2016; Bongaerts et al. 2017). For example, in a study from 2013 in the Bahamas, the reef-building *Montastrea cavernosa* exhibited limited vertical genetic connectivity to both shallow reef and upper MCE communities from lower MCE ones, revealing a distinct genetic structure across its depth distribution (Brazeau et al. 2013).

In addition, recent data indicate that despite mesophotic octocorals present high tolerance to temperatures ranging from 12 to 21 °C, a decrease in the abundance of certain

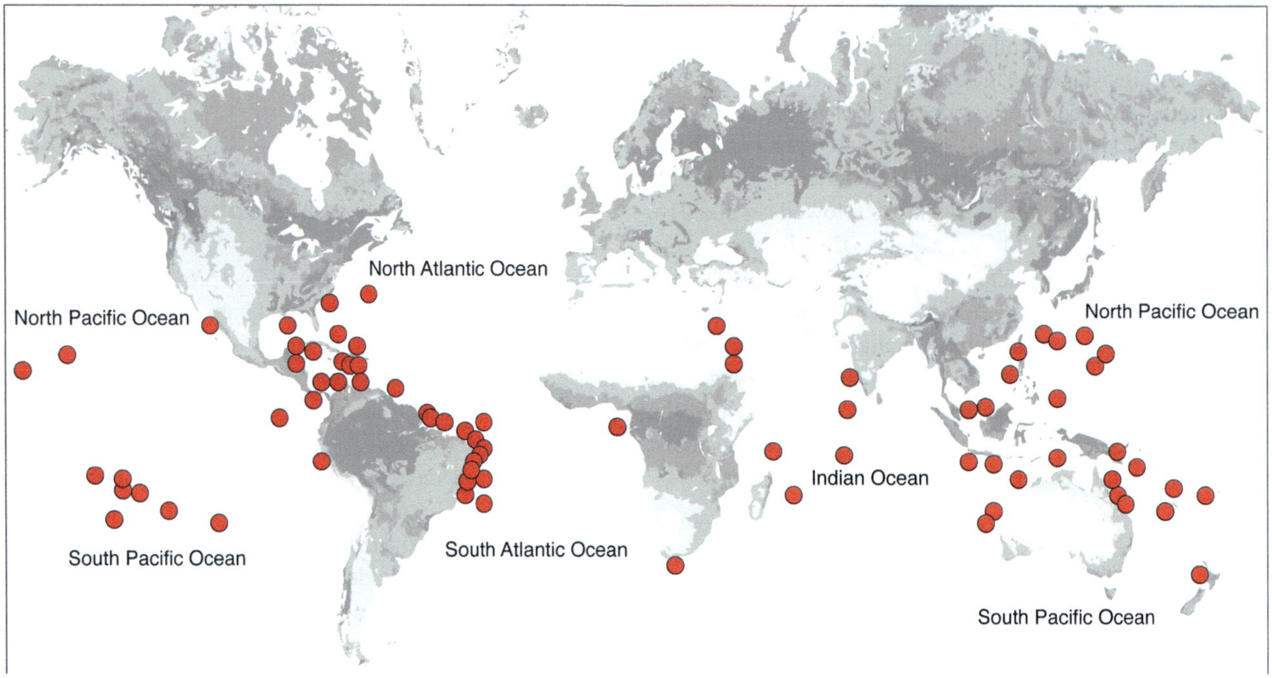

Fig. 4.3 Map of Mesophotic coral ecosystems (MCEs) worldwide. Source: www.mesophotic.org

bacterial symbionts, an increase in others, associated with tissue loss and gene overexpression was observed when these holobionts were exposed to 24 °C, indicating a certain level of vulnerability to warmer temperatures, with implications for understanding the effects of marine heatwaves and underlying resistance in mesophotic coral reefs (Tignat-Perrier et al. 2023).

Although research on mesophotic reef ecosystems is currently advancing, most of the current understanding of the coral holobiont still comes from research conducted in shallow waters. Nevertheless, corals are extensively found in the mesophotic environment, where the relationships and interactions between the host and its associated microbial communities may be fundamentally different from the ones characterizing shallow-water corals.

4.3.1 Coral-Associated Algal Photosymbionts in Mesophotic Reefs

Corals inhabiting shallow-water environments rely up to 95% of their energy requirement on photosynthetically derived products from their symbiotic dinoflagellates living in association with the gastrodermal cell layer of the coral polyps (Davy et al. 2012; LaJeunesse et al. 2018) providing their host with metabolic carbon through translocation of photosynthate, while producing amino acids and organosulfur compounds, e.g. dimethylsulfoniopropionate (DMSP) (Muscatine and Porter 1977; Durham et al. 2015). Conversely, these symbiotic microalgae acquire nutrients both from the host cells as well as from photosynthetically active sun radiation (Gordon and Leggat 2010; Roth 2014).

As depth increases, changes in the wavelength of solar radiation, both in irradiance and spectral quality useful for algae photosynthesis (400–700 nm), represents the primary abiotic factor constraining the distribution and productivity of Symbiodiniaceae-associated corals (Gattuso et al. 2006; Frade et al. 2008; Kahng et al. 2010) (Fig. 4.4). To date, scleractinian corals have been occasionally found in the lower mesophotic zones, with the deepest zooxanthellate scleractinians recorded as colonies of *Agaricia* spp. growing at 119 m in the Bahamas (Reed 1985) and, more recently, colonies of *Leptoseris hawaiiensis* reported to occur as deep as 172 m in the Gambier archipelago (French Polynesia), representing the deepest scleractinian corals associated with Symbionianceae and other photosymbionts (Rouzé et al. 2021).

Despite the paucity of studies focusing on MCEs, several revealed that, moving from shallow reefs to mesophotic depths, the Symbionianceae composition associated with scleractinian species can change with depth specialist strains replacing the ones not suited to lower light levels encountered at MCEs depths (Pochon et al. 2015; Glasl et al. 2017). For example, Bongaerts et al. (2015) demonstrated through the analysis of conserved mitochondrial markers that corals belonging to the genera *Agaricia* and *Madracis* found at depths comprised between 80 and 90 meters in Curaçao host the same photosymbionts genotypes, which are rarely encountered in shallow reef communities and likely reflect specialization to deep-water conditions (Bongaerts et al.

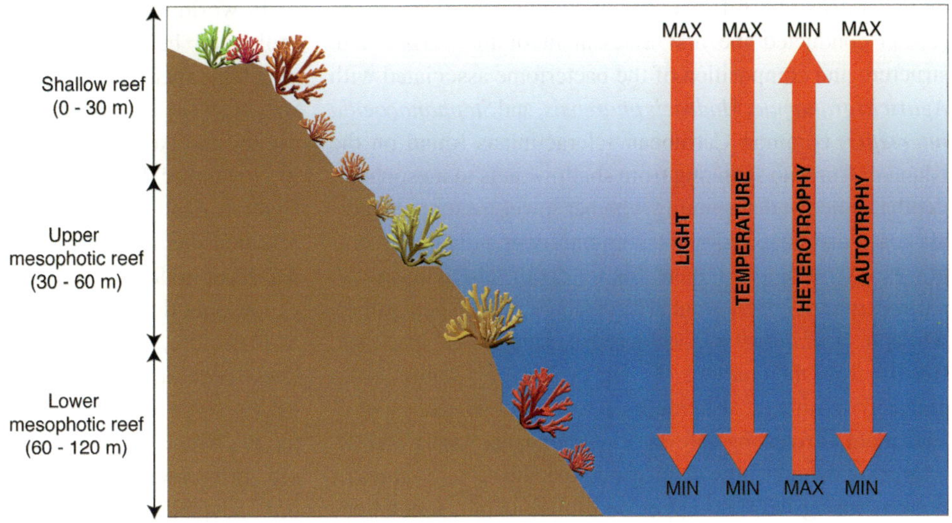

Fig. 4.4 Reef slope zonation from shallow reefs (0–30 m), upper mesophotic reef (30–60 m), and lower mesophotic reef (60–120 m). The arrows highlight generalized trends across depth gradients

2015). Accordingly, as highlighted in an extensive review of Symbiodiniaceae genetic diversity by Goulet et al. (2019), the photosymbiont genus *Cladocopium* dominates the dinoflagellate communities associated with different scleractinians such as the genera *Agaricia*, *Leptoseris*, *Fungia*, *Echinophyllia*, *Montastrea*, *Galaxea*, *Seriatopora* and *Stylophora* found in both the upper and lower mesophotic environments (reviewed in Goulet et al. 2019).

Nevertheless, hosting peculiar Symbiodiniaceae genotypes is not the only adaptation of the coral holobiont to deep-sea environments. The potential of photoacclimation of scleractinians to the low light irradiance of MCEs can involve hosting the same photosymbiont genotypes across depth gradients while expressing host-associated accessory pigments (such as fluorescent pigments), specifically increasing the photoconvertible red fluorescent proteins which can convert non-photochemical light wavelengths to those that can be used by the host-associated Symbiodiniaceae, facilitating greater photosynthetic efficiency in the endosymbiotic dinoflagellates (Schlichter and Fricke 1991; Smith et al. 2017b).

4.3.2 Coral Bacterial Communities in Mesophotic Reefs

Despite the prokaryotic component of the coral microbiome has gained a lot of interest recently due to their important role in mediating metabolic needs and protecting the host from the infections of potential pathogens (Ainsworth et al. 2015; Thompson et al. 2015), the vast majority of studies present to date in the scientific literature focus on shallow-water coral microbial communities with only sparse information on the bacteriome associated with coral inhabiting the upper-mesophotic available, and even less known about bacteriomes associated in lower MCEs organisms (Olson and Kellogg 2010). Many species of corals inhabiting mesophotic depths appear to shift their trophic reliance from primarily being autotrophic to a relatively greater dependence on heterotrophy, thanks to the increased availability of Particulate Organic Matter (POM), zooplankton, and picoplankton confirming their mixotrophy strategy for carbon requirements (Padilla-Gamiño et al. 2019; Lesser et al. 2022). Nevertheless, the trophic ecology of corals in MCEs is not a topic that can be generalized or simplified as different patterns occur in different taxa.

In this context, bacteria, compared to the symbiotic Symbiodiniaceae, may have a central role in contributing to the metabolic and energetic balance of the coral holobiont in MCEs as light becomes a limiting factor for dinoflagellates growth. Particularly, the emerging role of nitrogen fixation by symbiotic diazotrophic bacteria supports the importance of particular members of the bacteriome associated with deep-water corals (Benavides et al. 2017). For example, it has been shown that *Montastrea cavernosa* colonies from the Caribbean basin with a depth distribution spanning from approximately 3 to 100 m (Reed 1985) can host nitrogen-fixing Cyanobacteria in their tissue (Lesser et al. 2007). Interestingly, as depth increased, the prevalence of *M. cavernosa* colonies containing symbiotic Cyanobacteria increased from 0% in shallow sites to 33% in mesophotic depths (Lesser et al. 2007). Similarly, Bednarz et al. (2017) showed that colonies of *Stylophora pistillata* found in the upper mesophotic relied for up to 15% of their nitrogen budget on diazotroph bacterial symbionts compared to 1% from their conspecific inhabiting shallow reefs.

Since it is well established that coral-associated prokaryotic communities can vary depending on their host phylogeny as well as surrounding environmental conditions, an essential aspect of the research about MCEs coral microbiology is to investigate if coral species found across depth gradients (shallow to mesophotic) host variable or conserved bacteriomes depending on whether they are depth specialist or generalists. To fill this gap of knowledge, Glasl et al. (2017) conducted the first assessment of the variation in structure and composition of the bacteriome associated with *Agaricia grahamae*, *Madracis pharensis*, and *Stephanocoenia intersepta*, common Caribbean scleractinians found on the island of Curaçao spanning from shallow reefs to mesophotic depths, aimed at determining whether lower mesophotic corals host a distinct prokaryotic community or indicator assemblages. According to their broad depth distribution, *M. pharensis* and *S. intersepta* are considered depth generalists, while *A. grahamae* is considered a deep-water specialist. Results from the analyses of their prokaryotic communities revealed significant differences between depth zones in *S. intersepta* (driven by rare or low-abundance taxa) and *M. pharensis*, while in the deep-water specialist, *A. grahamae*, no structuring between upper and lower depth populations were found. These findings suggest that depth-specialist hosts maintain a specific associated microbial community while depth-generalists can shuffle their prokaryotic composition to better adapt to the surrounding environment (Glasl et al. 2017). In an attempt to clarify the intrinsic complexity of coral-associated microbial communities, the concept of the core microbiome framework was introduced to better understand the dynamics of the coral microbiome. This concept aims to identify potentially fundamental microorganisms within the host-associated microbial community based on their persistence despite differences in host phylogeny and/or spatio-temporal boundaries (Ainsworth et al. 2015; Hernandez-Agreda et al. 2016, 2017). Using this concept, Hernandez-Agreda et al. (2016) examined the prokaryotic community of 123 individuals of the depth generalist *Pachyseris speciosa* on reefs separated by over 700 km in the Great Barrier Reef and Coral Sea to the mesophotic depth of 80 m and found that 80% of the core microbiome across sampling depths was constituted principally of eight highly persistent bacterial phylotypes, providing evidence for stable associations that could be maintained through different environmental conditions (e.g., light availability and temperature). Moreover, the same study highlighted the presence of an additional 14 phylotypes belonging to the classes Alpha-, Beta-, Delta-, and Gammaproteobacteria, Actinobacteria, Bacilli, Flavobacteriia, and Synechococcophycideae that were consistently associated with *P. speciosa* samples from the deepest depth range analyzed (60–80 m), suggesting habitat partitioning and ecological diversification in core bacteriome related to depth, which members likely contribute to their host success in these extreme environments.

Overall, MCE coral microbiology remains an understudied topic to date. Nevertheless, the microbial partnerships mesophotic corals establish with their symbionts represent a key component to understand and underpin biological changes happening in the context of the current climate change scenario (Ainsworth and Gates 2016). As shallow reefs worldwide continue to be impacted by frequent and prolonged heat waves, episodes of thermal anomalies in which the thermal thresholds of mesophotic coral holobionts will be exceeded, ultimately leading to deep water bleaching events, are likely to happen. Two research cruises undertaken in November 2019 and March 2020 in two sites within the Egmont Atoll (Chagos Archipelago, Central Indian Ocean) revealed the susceptibility of MCEs to bleaching events at depths up to 90 m, demonstrating the potential vulnerability of mesophotic corals to thermal stress (Diaz et al. 2023). It is, therefore, of fundamental importance to advance our knowledge on the unique partnerships that corals inhabiting MCEs settle with specific microorganisms, which may aid the resilience of these corals to thermal stress by, for example, providing enhanced nutritional flexibility when low-light adapted photosymbionts are lost.

4.4 Coral-Associated Microbial Communities in Low pH Environments

Changes in ocean pH are expected to impact marine ecosystems and cause shifts in community structure (Ishimatsu et al. 2005; Bibby et al. 2007; Hoegh-Guldberg et al. 2007; Kuffner et al. 2008; Martin et al. 2008), especially among calcifying organisms (Gattuso et al. 1998; Orr et al. 2005). Since the industrial revolution, oceans have absorbed ~30% of the anthropogenically derived CO_2 emission, which led to a reduction of seawater pH by an average of 0.1 units, while current emission models predict a further decline by 0.2–0.4 units by 2100 (Doney et al. 2009; Kapsenberg et al. 2015). Despite this alarming trend, the effects of ocean acidification on host-associated microbiomes are still relatively understudied (reviewed in O'Brien et al. 2016). Because the coral holobiont has emerged as a model system to explore host-microbial interactions, most of the research focusing on the potential effects of low pH on the holobiont microbial communities has been focused on this system (Bourne and Webster 2013).

Different research examined the response of the coral bacteriome to decreased pH conditions, reporting significant shifts in their equilibrium towards enrichment in bacterial taxa frequently found in higher relative abundances in stress and/or disease state (Vega Thurber et al. 2009; Meron et al. 2011, 2012; Webster et al. 2016). For example, studies conducted on the reef-building *Acropora eurystoma* exposed for several weeks to pH conditions lower than 8.0 reported an increase in the bacteria families Vibrionaceae, Alteromonadaceae, and Rhodobacteraceae which members have been previously isolated from diseased, injured, or stressed corals (Ritchie 2006; Arboleda and Reichardt 2009; Sunagawa et al. 2010; Meron et al. 2011; Ushijima et al.

2020). In addition, the same studies revealed an increase in the expression of genes encoding for resistance to toxins and secondary metabolites, as well as the loss of sequences belonging to the genus *Endozoicomonas* known to be associated with healthy corals (Vega Thurber et al. 2009; Meron et al. 2011; Webster et al. 2016).

Similarly, Anthony et al. 2008 revealed that the rise of CO_2 concentrations in seawater, acting in synergy with temperature, affects coral - Symbiodiniaceae relationships leading to higher bleaching susceptibility in *Acropora intermedia* and *Porites lobata* potentially hindering the preferred pathway for CO_2 accumulation at the site for rubisco within the intracellular dinoflagellates, or interfering directly with the electron transport chain by destabilizing thylakoid proton gradients and, accordingly, the ability of individual symbionts to fix carbon.

Although interesting and useful to expand our knowledge on coral microbiomes under future ocean acidification conditions, these results were all obtained from studies performed under controlled laboratory conditions. Thus, care must be taken when interpreting their conclusions, as those may not be representative of real-world scenarios.

Conversely, volcanic CO_2 seeps can provide insights into the adaptation of holobionts to ocean acidification. In 2019, Wee and colleagues conducted one of the few field-based studies on microbiomes in CO_2 seeps. Particularly, this study investigated the diversity of Symbioniniaceae associated with the zoantharian *Palythoa tubercolosa* in acidified and non-acidified locations at Iwotorishima, a volcanic island in southern Japan. Through the analysis of ITS2, ITS4 and psbA sequences, this research revealed a reduction in the Symbioniaceae genotype diversity in acidified locations. This result could indicate that the partial pressure of CO_2 may drive the selection and/or dominance of photosymbionts in the coral host (Wee et al. 2019).

Furthermore, in 2021, Shore and colleagues explored the bacterial communities associated with three species of scleractinians, *Pocillopora eydouxi*, *Porites lobata*, and *Porites rus*, present the reefs of Maug, located in the Northern Mariana Islands, characterized by submarine volcanic vents that bubble CO_2 creating a localized gradient of reduced pH. The results revealed that the bacteriome of *P. eydouxi*, which is considered to be relatively inflexible as in all members of the genus *Pocillopora*, experienced a higher dispersion in the site characterized by lower pH, suggesting that ocean acidification may represent a chronic environmental stress able to shape even intransigent associations between the coral host and their microbiomes. In contrast, the bacterial communities associated with the two species of *Porites* analyzed across the same pH gradient were revealed to be more homogeneous in their composition, highlighting a potential process of acclimation and tolerance to ocean acidification (Shore et al. 2021). This study is another example demonstrating coral species-specific responses to challenging conditions, in which the host and its phylogeny can drive the selection of the most advantageous associated microbial community in adapting to extreme environments.

4.5 Coral-Associated Microbial Communities in Mangrove Ecosystems

Mangrove ecosystems represent one of the most important coastal habitats providing a plethora of ecological services including, but not limited to, carbon sequestration, protection of the coast from solar UV-B radiation, wave action, coastal erosion, sea level rise, floods, and extreme events such as hurricanes (Akram et al. 2023). Mangroves also act as sediment traps and sink for nutrients, while their root system contributes to the stability of the substrate. Moreover, these ecosystems provide food, breeding grounds, and nurseries for an incredible variety of marine organisms, including many species of fish, representing crucial biodiversity hotspots (Macintosh and Ashton 2002; Kathiresan 2012; Rog et al. 2017).

Nevertheless, mangroves are considered one of the most extreme coastal marine habitats, being characterized by elevated temperatures, acidic and deoxygenated waters (Rogers and Herlan 2012; Maggioni et al. 2021), and versatile microbiomes (Allard et al. 2020). Particularly, compared to adjacent coral reefs, these environments are, on average, 2 °C warmer and 0.3–0.4 pH levels more acidic, with daily drops of almost 1.0 pH level as recorded by Camp et al. (2017, 2020). Despite these extreme conditions, corals have been observed to thrive in mangrove habitats. For example, Camp et al. (2017) described a semi-enclosed lagoon system surrounded by mangroves in New Caledonia, inhabited by diverse coral populations that persist under the combined action of high temperature, low pH, and low dissolved oxygen stressors, periodically fluctuating according to tidal and diurnal cycles. These corals exhibited different physiological and metabolic signatures compared to conspecifics inhabiting surrounding reefs, such as enhanced respiration, which likely facilitated their persistence in the mangrove extreme environments (Camp et al. 2017). An interesting follow-up study investigated the holobiont physiology and the coral-associated microbiome of three dominant scleractinian species (*Acropora muricata*, *Acropora pulchra* and *Porites lutea*) in the same mangrove lagoon (Bouraké, New Caledonia), revealing that coral species exhibited metabolic adjustments of reduced photosynthesis-to-respiration ratios as well as highly divergent host-specific microbial associations (Camp et al. 2020). Particularly, compared with colonies of the same species from adjacent reefs, *A. muricata* exhibited a conserved Symbiodiniaceae ITS2-type profile and an increased abundance of the bacterial family Propionibacteriaceae and several rare taxa. *Porites lutea* was instead characterized by an association with different genotypes of the algal symbiont *Cladocopium* of the C15 radiation across sampled habitats while increasing low abundant bacterial taxa in the mangrove environment, similarly to *A. muricata* but by a much larger magnitude. In contrast, *A. pulchra* was associated with a divergent algal symbiotic relationship between mangrove lagoon and sampled reefs. Additionally, the bacteriome of this coral species inhabiting Bouraké lagoon was dominated by prokaryotes identified taxonomically as belonging to the genera *Fritschea* (family Simkaniaceae) and/or *Psychrobacter* (family Moraxellaceae), with the latter being postulated to be highly adaptable to continuously changing environmental conditions typical of the mangroves ecosystems (Badhai et al. 2016). However, it is unclear whether these differences in coral microbiomes can be a direct consequence of the extreme environmental conditions characterizing mangroves. Accordingly, to underpin the coral microbial communities adaptability and plasticity in these ecosystems, Haydon et al. (2021) conducted a 9-months transplantation experiment in which colonies of the scleractinian *Pocillopora acuta* were translocated between adjacent mangrove and reef sites on the northern Great Barrier Reef. Results of this study revealed that the bacterial communities of *P. acuta* living in mangroves are highly diverse and reactive to changing surrounding conditions by rapidly changing after transplantation, which suggests strong selective pressures from the environment, whereas the Symbiodiniaceae communities remained highly stable.

Altogether these results reinforce that corals living in extreme conditions, such as mangrove environments, host more transient bacteriomes compared to the ones in more stable environments. This characteristic may allow them to cope with environmental variability (Ziegler et al. 2017, 2019) while yielding valuable insights into the future of the same coral species in different habitats under the current global change scenario.

4.6 Coral-Associated Microbial Communities in Anthropogenically Disturbed Reefs

4.6.1 The Anthropocene

During the past decades, the nature and frequency of anthropogenic disturbances have changed. The scale of human intervention is increasing through a growing population and the globalization of markets, which promotes developments in agriculture, industries, fisheries, and international commerce (Vitousek et al. 1997). Because these activities also became significant biological and geological forces, the current geological epoch was termed "Anthropocene" (Crutzen 2006). In the Anthropocene, human activity has negatively impacted almost all marine ecosystems, including coral reefs (Halpern et al. 2007).

4.6.2 Ecosystem Resilience, Phase Shifts and Consequences on Coral Reef Microbiomes

The combination of anthropogenic stressors and various competitive mechanisms among major space occupiers on coral reefs causes persistent alterations, which are commonly referred to as 'phase shifts', 'regime shifts', or 'movements between alternate states' from complex coral-dominated communities to simpler communities with a predominance of other benthic organisms (Done 1992; Bellwood et al. 2004; Hughes et al. 2007, 2010, 2018; Graham et al. 2015; Bryant et al. 2017). A 'resilient' coral reef can absorb recurrent disturbances from climate change and human activities without fundamentally switching to an alternative state (Nyström et al. 2008; Hughes et al. 2010). However, in the light of intensifying threats, the resilience of coral reefs has been eroded to the extent that they are unable to recover after recurrent disturbances.

Phase-shift studies for coral reefs traditionally focused on the shift from coral- to algal- dominance, describing a range of consequences for the reef ecosystem, which all lead to a substantial loss of ecosystem services that cannot be provided anymore compared to a stable, close-to-pristine environment (McManus and Polsenberg 2004; Crisp et al. 2022).

Nevertheless, long-lasting shifts from corals to assemblages other than benthic algae (e.g., sponges, tunicates, zoanthids) have also been widely reported (reviewed in Norström et al. 2009). Commonly, however, changes on an ecosystem level are intrinsically linked to other levels as well. As such, we will outline the extent to which these phase shifts alter the reef's stability on the microbial scale.

Substrates are limited in coral reef environments, one of the aspects that drives a constant competition of various functional groups like coral holobionts or algal turfs. In an unperturbed scenario, corals outcompete algal turfs due to their efficient nutrient recycling and utilization in naturally oligotrophic waters. However, external nutrient input, i.e., eutrophication, can lead to the proliferation of algal turfs (Karcher et al. 2020). At the interface of corals and algal assemblages, this can have dramatic consequences. For example, certain species of fleshy algae (*Halimeda opuntia*) have been shown to serve as a reservoir for the bacterium *Aurantimonas coralicida*, which itself is a causative agent of the white plague type II coral disease (Nugues et al. 2004). Physical contact between *H. opuntia* and adjacent hard coral species can trigger the outbreak of this disease, suggesting that its spread in the Caribbean was stimulated by increasing *H. opuntia* prevalence (Nugues et al. 2004). In an ecological context, several other studies (Goreau et al. 1998; Hayes and Goreau 1998; Harvell et al. 1999, 2004; Smith et al. 2006) suggested that an increase of algae on coral reefs can induce a positive feedback loop whereby compounds released by algae could enhance microbial activity on live coral surfaces causing mortality and further algal growth (Smith et al. 2006). Accordingly, together with the pressure of further anthropogenic factors that often co-occur, such as the absence of herbivores due to overfishing, chemically rich seaweeds and algal turfs will have competitive advantage over calcifying corals and coralline algae (Haas et al. 2016) while they may also poison corals when increasing in abundance as a consequence of uncontrolled growth (Rasher and Hay 2010).

Besides the release of pathogens and the related promotion of disease outbreaks, further indirect effects of algae, particularly in the context of increasing algal abundances on reefs, have been less considered. For example, it has become clear that algal exudates can also negatively impact the coral-microbe interaction. Increased releases of organic carbon from abundant algal assemblages have been shown to alter microbial processes in hard corals by inducing coral-symbiont breakdown (Pogoreutz et al. 2017). Subsequently, altered stoichiometric ratios between the coral host and its symbionts, i.e., increased nitrogen:phosphorus ratios, indicate the disruption of the crucial N limitation in coral holobionts with potential cascading effects in warming oceans (Pogoreutz et al. 2017). In other words, imbalanced nutrient availability, i.e., elevated inorganic N concentrations in combination with phosphate depletion, can destabilize the host-symbiont symbiosis (Wiedenmann et al. 2013) due to a shift from N-limitation to phosphate limitation/starvation (Rädecker et al. 2015). However, eutrophication commonly occurs by the introduction of mixed elements rather than inorganic or organic nitrogen alone, respectively. As such, nutrient enrichment, i.e., nitrogenous and phosphorous compounds together, may at least be tolerated by coral reefs and their microbial communities (Bongiorni et al. 2003; McClanahan et al. 2003; Becker et al. 2021b). Additionally, nutrient enrichment has been shown to stimulate N_2 fixation in coral reefs that may increase the amount of bioavailable N (Koop et al. 2001; El-Khaled et al. 2020) or organic carbon (Pogoreutz et al. 2017). However, studies indicated that microbially performed metabolic processes, such as N_2 fixation or its counteracting pathway denitrification, can be moderated by the amount of available nutrients (El-Khaled et al. 2020, 2021). The extent to which denitrification can relieve coral reefs from excess N availability largely depends on its dominant benthic functional groups and its associated microbial communities (El-Khaled et al. 2021). In addition, the type of N that corals are exposed to, i.e., ammonia or nitrate, has been indicated to uniquely alter the microbiome of coral holobionts and speed the recovery from other stressors such as corallivory (Rice et al. 2019).

Phase shifts from initially coral-dominated to algae-dominated states have been also demonstrated to facilitate the release of significant amounts of bioavailable, algal-derived dissolved organic carbon (DOC; Haas et al. 2010, 2011; Wild et al. 2010) stimulating copiotrophic bacterial growth (Haas et al. 2011; Morrow et al. 2011; Nelson et al. 2013). Copiotrophic bacteria are known to possess a competitive advantage over native bacteria in remineralizing available organic nutrients (Haas et al. 2013) and can facilitate the encoding of virulence factor genes. Further, studies observed that these prokaryotes create zones of prolonged hypoxia that, together with pathogenic mechanisms, may harm or kill hard corals (Kuntz et al. 2005; Kline et al. 2006; Smith et al. 2006; Nelson et al. 2013). As dead coral skeletons are likely to be overgrown by algal assemblages, particularly under elevated nutrient levels, this positive feedback loop is called DDAM (DOC, disease, algae, microorganism). This critical feedback loop, characterized by the ecosystem shift towards more abundant microbial biomass and energy use, is defined as global microbialization of coral reefs and threatens coral reefs around the world (Haas et al. 2016).

Despite the outlined range of indirect effects on the microbiome of coral reefs and its functional groups, nutrient concentrations in surrounding seawater may induce changes within the coral microbial communities themselves (McDevitt-Irwin et al. 2017). Particularly, due to additional nutrient inputs within coral reefs, altered microbial communities often seem to favor diseases (Dinsdale et al. 2008). For example, human sewage inputs may shape the coral-associated microbiome (Leite et al. 2018), and even led to an introduction of *Serratia marcescens*, a common fecal enterobacterium, into the reef ecosystem that induced white-pox diseases in coral holobionts (Sutherland et al. 2010). Moreover, eutrophication may also have direct effects on the microbiome by altering its composition by promoting opportunistic and pathogenic microbial taxa such as Vibrionales, Flavobacteriales, Rhodobacterales, Alteromonadales, Rhizobiales, Rhodospirillales and Desulfovibrionales (Mcdevitt-Irwin et al. 2017) and Enterobacteriaceae (Leite et al. 2018). Particularly, members of the order Rhodobacterales are fast-growing and opportunistic bacteria native to both healthy and stressed corals (Meron et al. 2011; Sharp et al. 2012). However, due to their opportunistic characteristics, they may boom under stress conditions, thereby potentially revealing pathogenic functions.

At the same time, elevated nutrient levels have been linked to a decrease in the abundance of the bacterial symbiont of the genus *Endozoicomonas*. Reduced abundance of this proposed key player within coral tissues is of significant importance and can have profound impacts due to its multiple putative functions and roles: despite also being potentially associated with stress (Pogoreutz and Ziegler 2024), some *Endozoicomonas* species may help corals under stress by preventing mitochondrial dysfunction and by stimulating gluconeogenesis (Ding et al. 2016). Further, *Endozoicomonas* is proposed to be crucial for sulfur cycling (Raina et al. 2009), an important partner within the nutritional and metabolism (La Rivière et al. 2013; Neave et al. 2017) and Pantos et al. (2015) hypothesize that it may protect Symbiodiniaceae from bleaching pathogens.

In conclusion, nutrient enrichment on reefs and their functional groups has a range of direct and indirect effects during the Anthropocene. Direct effects can be described as alterations in the microbiome of hard corals by a reduction of beneficial bacteria and an increase in pathogens, and subsequently resulting in the outbreak of diseases, increased susceptibility to bleaching or loss of certain functions. Indirect effects are associated with changes in the ecosystem's structure as a whole as a result of eutrophication, for example, the proliferation of algal assemblages that may have a competitive advantage over hard corals (Karcher et al. 2020). In this context, microbially performed pathways, such as those associated with the N cycle, may be altered on an ecosystem level due to a change in benthic abundance in key taxa (El-Khaled et al. 2021) or due to a direct stimulation or suppression of these metabolic processes (El-Khaled et al. 2020). However, few studies have also indicated the potential beneficial effects of moderately elevated nutrient levels on the performance and resistance of corals (Rice et al. 2019). These ambiguous effects of nutrient enrichment on reefs and their main ecosystem engineers suggest that effects may be species-specific and dependent on further additional stressors (Wang et al. 2018).

4.7 Coral-Associated Microbial Communities in Turbid Reefs

Turbid coral reefs, contrasting with their clear-water counterparts, endure higher sedimentation and, currently, elevated temperatures due to climate change, potentially resulting in the loss of stress-sensitive corals and ecosystem decline (Restrepo et al. 2016; Camp et al. 2018; Lucas et al. 2023). Yet, these turbid habitats boast resilient coral holobionts, equipped with stress-tolerant species that defy disturbances and may function as climate change refugees, despite persistent threats. Recent data indicate the protective role of moderate turbidity in shielding corals from temperature stress, exemplified notably in Brazilian nearshore reefs (Lucas et al. 2023). Understanding the resilience of holobionts thriving in turbid reefs may provide crucial insights into the impacts of changing environmental conditions.

4.7.1 Coral-Associated Algal Photosymbionts in Turbid Reefs

The diversity of Symbiodiniaceae species associated with the endemic coral *Mussismilia hispida* along the Brazilian coast seems to be influenced by temperature and turbidity (Picciani et al. 2016). Similarly, *Favia gravida*, a native South Atlantic coral species, spans a wide geographic range, including the Mid-Atlantic islands, despite its preference for shallower waters. *F. gravida* thrives in environments with extreme conditions like high irradiance, temperature, and turbidity, such as very shallow tide pools, and their associated symbiont diversity varies across biogeographic regions, with different types found in the Tropical Eastern Atlantic, Mid-Atlantic, and Southwestern Atlantic, influenced by local environmental factors (Teschima et al. 2019). These findings suggest ecological processes shaping both reef communities and host-symbiont associations in turbid coral reefs, highlighting the coral's ability to associate with multiple symbiont species, potentially contributing to their wide distribution in marginal reefs.

4.7.2 Coral-Associated Bacteria in Turbid Reefs

Similar to other coral species, prokaryotic communities associated with corals from turbid reef environments encompass transient, near-universal, and resident members (Reis et al. 2009; Lins-de-Barros et al. 2013; Leite et al. 2017, 2018), with overall shifts influenced by environmental conditions (Leite et al. 2018). These communities often include species commonly found in areas impacted by human activities, such as *Acinetobacter*, *Alteromonas*, *Marinomonas*, *Neptuniibacter*, *Bacillus*, *Pseudomonas* and *Vibrio* (Castro et al. 2013; Leite et al. 2017, 2018; Santoro et al. 2021), some of which seem to be vertically transferred from coral to their offspring (Leite et al. 2017). Additionally, comparative studies have highlighted fine-scale adaptations and significant bacterial composition changes correlated with morphological differences, such as the absence of soft tissue connections between polyps observed in *Mussismilia harttii* compared to other species within the same coral genus, namely *M. hispida* and *M. braziliensis* (Castro et al. 2013). Changes in the microbial community of Brazilian corals in response to thermal stress indicate shifts in both prokaryotic and eukaryotic phototrophic plastids compared to healthy corals (Lins-de-Barros et al. 2013), as well as specific microbiome and metabolic restructuring when these corals are treated with probiotics (Santoro et al. 2021) that seem to prevent PHSD.

High sediment loads in environments such as reefs near river mouths can dramatically impact the health of corals by altering their associated bacteriome composition towards the enrichment of potentially harmful bacteria (Erftemeijer et al. 2012; Rosales et al. 2020). For example, the bacterial communities associated with the scleractinian *Porites lobata* inhabiting Fouha Bay in Guam in proximity of the river mouth, where suspended sediments can reach 1000 mg/ml during periods of heavy rain, have been shown to be enriched in ASVs belonging to *Catenococcus* and *Cohaesibacter*. These genera have been recognized as potentially pathogenic and observed to be enriched in corals exhibiting signs of stony coral tissue loss disease (SCTLD) and white-plague infection (Sunagawa et al. 2009; Roder et al. 2014; Rosales et al. 2020; Becker et al. 2021a). This evidence suggest that corals living in high sedimentation and turbid environments might be at higher risk for coral diseases and may have developed strategies to minimize their susceptibility.

4.8 Closing Remarks

The intricate microbial communities associated with coral reefs play fundamental roles in maintaining coral health and ecological resilience, including in extreme and marginal environments. These microbial communities encompass a diverse array of microorganisms, including bacteria, archaea, fungi, algae, protists, and viruses, each of them interacting and contributing uniquely to the coral holobiont's biology. These microbial communities also seem to be pivotal in facilitating coral adaptation to varying and often harsh environmental conditions, from turbid shallow reefs in Brazil to marginal reefs near mangroves to the warm Red Sea and Arabian Gulf or mesophotic reefs. Overall, the microbial assemblages associated with corals in extreme reefs are shaped by harsh environmental conditions such as high temperature, low pH, or fluctuating salinity, which may drive the selection for stress-tolerant symbionts that provide critical adaptive advantage to their host, bolstering coral survival under critical abiotic characteristics. Coral microbiomes associated with corals inhabiting marginal reefs often reflect ecological shifts, responding to altered dynamics such as nutrient loading by prioritizing functions such as organic matter degradation, aligning with ecosystem needs. Although these peculiar communities are key factors in underpinning coral resistance and resilience in a changing world, the mechanisms underlying coral-microbiome interactions and adaptations to extreme conditions are only beginning to be explored. Polyphasic approaches that include manipulative and field studies, combining innovative microbial culturing techniques and multi-omics, are necessary to fill these gaps. Moreover, expanding the geographical scope of studies to include under-researched coral ecosystems, particularly those in extreme environmental conditions, may provide crucial insights to dissect the complex interactions within the microbiome under various climate change scenarios. Understanding microbial interactions in coral holobionts adapted to extreme conditions, compared to most tropical coral reefs, particularly under climate stressors, is vital for developing targeted conservation strategies and enhancing coral resilience through microbiome stewardship.

References

Ainsworth TD, Gates RD (2016) Corals' microbial sentinels. Science 352:1518–1519. https://doi.org/10.1126/science.aad9957

Ainsworth TD, Krause L, Bridge T et al (2015) The coral core microbiome identifies rare bacterial taxa as ubiquitous endosymbionts. ISME J 9:2261–2274. https://doi.org/10.1038/ismej.2015.39

Ainsworth TD, Fordyce AJ, Camp EF (2017) The other microeukaryotes of the coral reef microbiome. Trends Microbiol 25:980–991. https://doi.org/10.1016/j.tim.2017.06.007

Akram H, Hussain S, Mazumdar P et al (2023) Mangrove health: a review of functions, threats, and challenges associated with mangrove management practices. Forests 14:1698. https://doi.org/10.3390/f14091698

Allard SM, Costa MT, Bulseco AN et al (2020) Introducing the mangrove microbiome initiative: identifying microbial research priorities and approaches to better understand, protect, and rehabilitate mangrove ecosystems. mSystems 5:e00658–e00620. https://doi.org/10.1128/mSystems.00658-20

Alosairi Y, Alsulaiman N, Rashed A, Al-Houti D (2020) World record extreme sea surface temperatures in the northwestern Arabian/Persian Gulf verified by in situ measurements. Mar Pollut Bull 161:111766. https://doi.org/10.1016/j.marpolbul.2020.111766

Anthony KRN, Kline DI, Diaz-Pulido G et al (2008) Ocean acidification causes bleaching and productivity loss in coral reef builders. PNAS 105:17442–17446. https://doi.org/10.1073/pnas.0804478105

Apprill A (2017) Marine animal microbiomes: toward understanding host–microbiome interactions in a changing ocean. Front Mar Sci 4:222. https://doi.org/10.3389/fmars.2017.00222

Arboleda M, Reichardt W (2009) Epizoic communities of prokaryotes on healthy and diseased Scleractinian corals in Lingayen Gulf, Philippines. Microb Ecol 57:117–128. https://doi.org/10.1007/s00248-008-9400-0

Avila-Magaña V, Kamel B, DeSalvo M et al (2021) Elucidating gene expression adaptation of phylogenetically divergent coral holobionts under heat stress. Nat Commun 12:5731. https://doi.org/10.1038/s41467-021-25950-4

Badhai J, Ghosh TS, Das SK (2016) Composition and functional characterization of microbiome associated with mucus of the coral *Fungia echinata* collected from Andaman Sea. Front Microbiol 7:936. https://doi.org/10.3389/fmicb.2016.00936

Baker E, Thygesen K, Harris P et al (2016) Mesophotic coral ecosystems: a lifeboat for coral reefs? United Nations Environment Programme (UNEP) and GRID-Arendal

Barker V (2018) Exceptional thermal tolerance of coral reefs in American Samoa: a review. Curr Clim Chang Rep 4:417–427. https://doi.org/10.1007/s40641-018-0112-3

Bayer T, Arif C, Ferrier-Pagès C et al (2013) Bacteria of the genus *Endozoicomonas* dominate the microbiome of the Mediterranean gorgonian coral *Eunicella cavolini*. Mar Ecol Prog Ser 479:75–84. https://doi.org/10.3354/meps10197

Beck MW, Losada IJ, Menéndez P et al (2018) The global flood protection savings provided by coral reefs. Nat Commun 9:2186. https://doi.org/10.1038/s41467-018-04568-z

Becker CC, Brandt M, Miller CA, Apprill A (2021a) Stony coral tissue loss disease biomarker bacteria identified in corals and overlying waters using a rapid field-based sequencing approach. bioRxiv. https://doi.org/10.1101/2021.02.17.431614
Becker DM, Putnam HM, Burkepile DE et al (2021b) Chronic low-level nutrient enrichment benefits coral thermal performance in a fore reef habitat. Coral Reefs 40:1637–1655. https://doi.org/10.1007/s00338-021-02138-2
Bednarz VN, Grover R, Maguer J-F et al (2017) The assimilation of diazotroph-derived nitrogen by Scleractinian corals depends on their metabolic status. MBio 8:e02058-16. https://doi.org/10.1128/mbio.02058-16
Bell W, Mitchell R (1972) Chemotactic and growth responses of marine bacteria to algal extracellular products. Biol Bull 143:265–277. https://doi.org/10.2307/1540052
Bellwood DR, Hughes TP, Folke C, Nyström M (2004) Confronting the coral reef crisis. Nature 429:827–833. https://doi.org/10.1038/nature02691
Benavides M, Bednarz VN, Ferrier-Pagès C (2017) Diazotrophs: overlooked key players within the coral symbiosis and tropical reef ecosystems? Front Mar Sci 4:428–445. https://doi.org/10.3389/fmars.2017.00010
Berg G, Rybakova D, Fischer D et al (2020) Microbiome definition re-visited: old concepts and new challenges. Microbiome 8:103. https://doi.org/10.1186/s40168-020-00875-0
Bergman JL, Ricci F, Leggat W, Ainsworth TD (2023) Characteristics of the bleached microbiome of the generalist coral *Pocillopora damicornis* from two distinct reef habitats. Integr Org Biol 5:obad012. https://doi.org/10.1093/iob/obad012
Berkelmans R, Van Oppen MJH (2006) The role of zooxanthellae in the thermal tolerance of corals: a 'nugget of hope' for coral reefs in an era of climate change. Proc R Soc B 273:2305–2312. https://doi.org/10.1098/rspb.2006.3567
Berumen ML, Voolstra CR, Daffonchio D et al (2019) The Red Sea: environmental gradients shape a Natural Laboratory in a Nascent Ocean. In: Voolstra CR, Berumen ML (eds) Coral reefs of the Red Sea. Springer International Publishing, Cham, pp 1–10
Bibby R, Cleall-Harding P, Rundle S et al (2007) Ocean acidification disrupts induced defenses in the intertidal gastropod *Littorina littorea*. Biol Lett 3:699–701. https://doi.org/10.1098/rsbl.2007.0457
Birkeland C, Green A, Lawrence A et al (2021) Different resiliencies in coral communities over ecological and geological time scales in American Samoa. Mar Ecol Prog Ser 673:55–68
Blanckaert ACA, Omanović D, Fine M et al (2022) Desert dust deposition supplies essential bioelements to Red Sea corals. Glob Chang Biol 28:2341–2359. https://doi.org/10.1111/gcb.16074
Bongaerts P, Smith TB (2019) Beyond the "deep reef refuge" hypothesis: a conceptual framework to characterize persistence at depth. In: Loya Y, Puglise KA, Bridge TCL (eds) Mesophotic coral ecosystems. Springer International Publishing, Cham, pp 881–895
Bongaerts P, Frade PR, Hay KB et al (2015) Deep down on a Caribbean reef: lower mesophotic depths harbor a specialized coral-endosymbiont community. Sci Rep 5:7652. https://doi.org/10.1038/srep07652
Bongaerts P, Riginos C, Brunner R et al (2017) Deep reefs are not universal refuges: reseeding potential varies among coral species. Sci Adv 3:e1602373. https://doi.org/10.1126/sciadv.1602373
Bongiorni L, Shafir S, Angel D, Rinkevich B (2003) Survival, growth and gonad development of two hermatypic corals subjected to *in situ* fish-farm nutrient enrichment. Mar Ecol Prog Ser 253:137–144
Bordenstein SR, Theis KR (2015) Host biology in light of the microbiome: ten principles of holobionts and hologenomes. PLoS Biol 13:e1002226. https://doi.org/10.1371/journal.pbio.1002226
Bourne DG, Webster NS (2013) Coral reef bacterial communities. Prokaryotes: Prokaryot Commun Ecophysiol 4:163–187
Bourne DG, Iida Y, Uthicke S et al (2008) Changes in coral-associated microbial communities during a bleaching event. ISME J 2:350–363. https://doi.org/10.1038/ismej.2007.112
Bourne DG, Morrow KM, Webster NS (2016) Insights into the coral microbiome: underpinning the health and resilience of reef ecosystems. Ann Rev Microbiol 70:317–340. https://doi.org/10.1146/annurev-micro-102215-095440
Bouwmeester J, Riera R, Range P et al (2020) Coral and reef fish communities in the thermally extreme Persian/Arabian Gulf: insights into potential climate change effects. In: Rossi S, Bramanti L (eds) Perspectives on the marine animal forests of the world. Springer International Publishing, Cham, pp 63–86
Brazeau DA, Lesser MP, Slattery M (2013) Genetic structure in the coral, *Montastraea cavernosa*: assessing genetic differentiation among and within mesophotic reefs. PLoS One 8:1–11. https://doi.org/10.1371/journal.pone.0065845
Brown BE (1997) Coral bleaching: causes and consequences. Coral Reefs 16:129–138. https://doi.org/10.1007/s003380050249
Bryant DEP, Rodriguez-Ramirez A, Phinn S et al (2017) Comparison of two photographic methodologies for collecting and analyzing the condition of coral reef ecosystems. Ecosphere 8:e01971. https://doi.org/10.1002/ecs2.1971
Buitrago-López C, Cárdenas A, Hume BCC et al (2023) Disparate population and holobiont structure of pocilloporid corals across the Red Sea gradient demonstrate species-specific evolutionary trajectories. Mol Ecol 32:2151–2173. https://doi.org/10.1111/mec.16871
Burke L, Reytar K, Spalding M, Perry A (2011a) Reefs at risk revisited. World Resources Institute (WRI), Washington, DC
Burke C, Steinberg P, Rusch D et al (2011b) Bacterial community assembly based on functional genes rather than species. Proc Natl Acad Sci USA 108:14288–14293. https://doi.org/10.1073/pnas.1101591108
Burt JA, Feary DA, Van Lavieren H (2013) Persian Gulf reefs: an important asset for climate science in urgent need of protection. Ocean Challenge 20:49–56
Camp EF, Nitschke MR, Rodolfo-Metalpa R et al (2017) Reef-building corals thrive within hot-acidified and deoxygenated waters. Sci Rep 7:2434. https://doi.org/10.1038/s41598-017-02383-y
Camp EF, Schoepf V, Mumby PJ et al (2018) The future of coral reefs subject to rapid climate change: lessons from natural extreme environments. Front Mar Sci 5:4. https://doi.org/10.3389/fmars.2018.00004
Camp EF, Suggett DJ, Pogoreutz C et al (2020) Corals exhibit distinct patterns of microbial reorganization to thrive in an extreme inshore environment. Coral Reefs 39:701–716. https://doi.org/10.1007/s00338-019-01889-3
Capone DG, Dunham SE, Horrigan SG, Duguay LE (1992) Microbial nitrogen transformations in unconsolidated coral reef sediments. Marine ecology progress series. Inter-Research, pp 75–88
Cardini U, Bednarz VN, van Hoytema N et al (2016) Budget of primary production and dinitrogen fixation in a highly seasonal Red Sea coral reef. Ecosystems 19:771–785. https://doi.org/10.1007/s10021-016-9966-1
Cardoso PM, Hill LJ, Villela HDM et al (2024) Localization and symbiotic status of probiotics in the coral holobiont. mSystems 9:e00261-24. https://doi.org/10.1128/msystems.00261-24
Castro APD, Araújo SD, Reis AMM et al (2013) Bacterial communities associated with three Brazilian endemic reef corals (Mussismilia spp.) in a coastal reef of the Abrolhos shelf. Cont Shelf Res 70:135–139. https://doi.org/10.1016/j.csr.2013.05.009
Chakravarti LJ, Beltran VH, Van Oppen MJH (2017) Rapid thermal adaptation in photosymbionts of reef-building corals. Glob Chang Biol 23:4675–4688. https://doi.org/10.1111/gcb.13702
Chavanich S, Kusdianto H, Kullapanich C et al (2022) Microbiomes of healthy and bleached corals during a 2016 thermal bleaching event in the Andaman Sea of Thailand. Front Mar Sci 9:763421. https://doi.org/10.3389/fmars.2022.763421

Craig P, Birkeland C, Belliveau S (2001) High temperatures tolerated by a diverse assemblage of shallow-water corals in American Samoa. Coral Reefs 20:185–189. https://doi.org/10.1007/s003380100159

Crisp SK, Tebbett SB, Bellwood DR (2022) A critical evaluation of benthic phase shift studies on coral reefs. Mar Environ Res 178:105667. https://doi.org/10.1016/j.marenvres.2022.105667

Crutzen PJ (2006) The "Anthropocene". In: Ehlers E, Krafft T (eds) Earth system science in the Anthropocene. Springer, Berlin/Heidelberg, pp 13–18

Cunning R, Yost DM, Guarinello ML et al (2015) Variability of Symbiodinium communities in waters, sediments, and corals of thermally distinct reef pools in American Samoa. PLoS One 10:e0145099. https://doi.org/10.1371/journal.pone.0145099

D'Angelo C, Hume BC, Burt J et al (2015) Local adaptation constrains the distribution potential of heat-tolerant Symbiodinium from the Persian/Arabian Gulf. ISME J 9:2551–2560. https://doi.org/10.1038/ismej.2015.80

Davies SW, Gamache MH, Howe-Kerr LI et al (2023) Building consensus around the assessment and interpretation of Symbiodiniaceae diversity. PeerJ 11:e15023

Davy SK, Allemand D, Weis VM (2012) Cell biology of cnidarian-dinoflagellate symbiosis. Microbiol Mol Biol Rev 76:229–261. https://doi.org/10.1128/mmbr.05014-11

Deines P, Lachnit T, Bosch TCG (2017) Competing forces maintain the *Hydra* metaorganism. Immunol Rev 279:123–136. https://doi.org/10.1111/imr.12564

Delgadillo-Ordoñez N, Raimundo I, Barno AR et al (2022) Red Sea atlas of coral-associated bacteria highlights common microbiome members and their distribution across environmental gradients—a systematic review. Microorganisms 10:2340. https://doi.org/10.3390/microorganisms10122340

Delgadillo-Ordoñez N, Garcias-Bonet N, Raimundo I et al (2024) Probiotics reshape the coral microbiome in situ without detectable off-target effects in the surrounding environment. Commun Biol 7:434. https://doi.org/10.1038/s42003-024-06135-3

Diaz JM, Hansel CM, Apprill A et al (2016) Species-specific control of external superoxide levels by the coral holobiont during a natural bleaching event. Nat Commun 7:13801. https://doi.org/10.1038/ncomms13801

Diaz C, Foster NL, Attrill MJ et al (2023) Mesophotic coral bleaching associated with changes in thermocline depth. Nat Commun 14:6528. https://doi.org/10.1038/s41467-023-42279-2

Díaz-Almeyda EM, Ryba T, Ohdera AH et al (2022) Thermal stress has minimal effects on bacterial communities of thermotolerant Symbiodinium cultures. Front Ecol Evol 10:764086. https://doi.org/10.3389/fevo.2022.764086

Ding J-Y, Shiu J-H, Chen W-M et al (2016) Genomic insight into the host–endosymbiont relationship of *Endozoicomonas montiporae* CL-33T with its coral host. Front Microbiol 7:251. https://doi.org/10.3389/fmicb.2016.00251

Dinsdale EA, Pantos O, Smriga S et al (2008) Microbial ecology of four coral atolls in the Northern Line Islands. PLoS One 3:1–17. https://doi.org/10.1371/journal.pone.0001584

Done TJ (1992) Phase shifts in coral reef communities and their ecological significance. Hydrobiologia 247:121–132. https://doi.org/10.1007/BF00008211

Doney SC, Fabry VJ, Feely RA, Kleypas JA (2009) Ocean acidification: the other CO2 problem. Annu Rev Mar Sci 1:169–192. https://doi.org/10.1146/annurev.marine.010908.163834

Durham BP, Sharma S, Luo H et al (2015) Cryptic carbon and sulfur cycling between surface ocean plankton. Proc Natl Acad Sci 112:453–457. https://doi.org/10.1073/pnas.1413137112

Eladawy A, Nakamura T, Shaltout M et al (2022) Appraisal of coral bleaching thresholds and thermal projections for the northern Red Sea refugia. Front Mar Sci 9:938454

El-Khaled Y, Roth F, Tilstra A et al (2020) In situ eutrophication stimulates dinitrogen fixation, denitrification, and productivity in Red Sea coral reefs. Mar Ecol Prog Ser 645:55–66

El-Khaled YC, Roth F, Rädecker N et al (2021) Nitrogen fixation and denitrification activity differ between coral- and algae-dominated Red Sea reefs. Sci Rep 11:11820. https://doi.org/10.1038/s41598-021-90204-8

Epstein HE, Torda G, Munday PL, Van Oppen MJH (2019) Parental and early life stage environments drive establishment of bacterial and dinoflagellate communities in a common coral. ISME J 13:1635–1638. https://doi.org/10.1038/s41396-019-0358-3

Erftemeijer PLA, Riegl B, Hoeksema BW, Todd PA (2012) Environmental impacts of dredging and other sediment disturbances on corals: a review. Mar Pollut Bull 64:1737–1765. https://doi.org/10.1016/j.marpolbul.2012.05.008

Evensen NR, Fine M, Perna G et al (2021) Remarkably high and consistent tolerance of a Red Sea coral to acute and chronic thermal stress exposures. Limnol Oceanogr 66:1718–1729. https://doi.org/10.1002/lno.11715

Evensen NR, Voolstra CR, Fine M et al (2022) Empirically derived thermal thresholds of four coral species along the Red Sea using a portable and standardized experimental approach. Coral Reefs 41:239–252. https://doi.org/10.1007/s00338-022-02233-y

Ferrario F, Beck MW, Storlazzi CD et al (2014) The effectiveness of coral reefs for coastal hazard risk reduction and adaptation. Nat Commun 5:1–9

Fine M, Loya Y (2002) Endolithic algae: an alternative source of photoassimilates during coral bleaching. Proc R Soc Lond B 269:1205–1210. https://doi.org/10.1098/rspb.2002.1983

Frade P, Englebert N, Faria J et al (2008) Distribution and photobiology of Symbiodinium types in different light environments for three color morphs of the coral *Madracis pharensis*: is there more to it than total irradiance? Coral Reefs 27:913–925. https://doi.org/10.1007/s00338-008-0406-3

Frias-Lopez J, Zerkle AL, Bonheyo GT, Fouke BW (2002) Partitioning of bacterial communities between seawater and healthy, black band diseased, and dead coral surfaces. Appl Environ Microbiol 68:2214–2228. https://doi.org/10.1128/AEM.68.5.2214-2228.2002

Galindo-Martínez CT, Weber M, Avila-Magaña V et al (2022) The role of the endolithic alga *Ostreobium* spp. during coral bleaching recovery. Sci Rep 12:2977. https://doi.org/10.1038/s41598-022-07017-6

Garcias-Bonet N, Roik A, Tierney B et al (2023) Horizon scanning the application of probiotics for wildlife. Trends Microbiol 32:252–269. https://doi.org/10.1016/j.tim.2023.08.012

Gardner SG, Leggat W, Ainsworth TD (2023) The microbiome of the endosymbiotic Symbiodiniaceae in corals exposed to thermal stress. Hydrobiologia 850:3685–3704. https://doi.org/10.1007/s10750-023-05221-7

Garrido AG, Machado LF, Zilberberg C, Leite DCA (2021) Insights into 'Symbiodiniaceae phycosphere' in a coral holobiont. Symbiosis 83:25–39. https://doi.org/10.1007/s13199-020-00735-3

Gattuso J-P, Frankignoulle M, Bourge I et al (1998) Effect of calcium carbonate saturation of seawater on coral calcification. Glob Planet Chang 18:37–46. https://doi.org/10.1016/S0921-8181(98)00035-6

Gattuso J-P, Gentili B, Duarte CM et al (2006) Light availability in the coastal ocean: impact on the distribution of benthic photosynthetic organisms and their contribution to primary production. Biogeosciences 3:489–513. https://doi.org/10.5194/bg-3-489-2006

Glasl B, Bongaerts P, Elisabeth NH et al (2017) Microbiome variation in corals with distinct depth distribution ranges across a shallow–mesophotic gradient (15–85 m). Coral Reefs 36:447–452. https://doi.org/10.1007/s00338-016-1517-x

Glynn PW (2003) Coral bleaching: ecological perspectives. Coral Reefs 12:1–17. https://doi.org/10.1007/BF00303779

Glynn PW, Maté JL, Baker AC, Calderón MO (2001) Coral bleaching and mortality in Panama and Ecuador during the 1997–1998

El Niño–Southern Oscillation event: spatial/temporal patterns and comparisons with the 1982–1983 event. Bull Mar Sci 69:79–109
Gordon BR, Leggat W (2010) Symbiodinium—invertebrate symbioses and the role of metabolomics. Mar Drugs 8:2546–2568. https://doi.org/10.3390/md8102546
Goreau TJ, Cervino J, Goreau M et al (1998) Rapid spread of diseases in Caribbean coral reefs. Rev Biol Trop 46:157–171
Goulet TL, Lucas MQ, Schizas NV (2019) Symbiodiniaceae genetic diversity and symbioses with hosts from shallow to mesophotic coral ecosystems. In: Mesophotic coral ecosystems. Springer International Publishing, pp 537–551. https://doi.org/10.1007/978-3-319-92735-0_30
Graham NAJ, Jennings S, MacNeil MA et al (2015) Predicting climate-driven regime shifts versus rebound potential in coral reefs. Nature 518:94–97. https://doi.org/10.1038/nature14140
Haas A, Jantzen C, Naumann M et al (2010) Organic matter release by the dominant primary producers in a Caribbean reef lagoon: implication for in situ O2 availability. Mar Ecol Prog Ser 409:27–39
Haas AF, Nelson CE, Wegley Kelly L et al (2011) Effects of coral reef benthic primary producers on dissolved organic carbon and microbial activity. PLoS One 6:1–12. https://doi.org/10.1371/journal.pone.0027973
Haas AF, Nelson CE, Rohwer F et al (2013) Influence of coral and algal exudates on microbially mediated reef metabolism. PeerJ 1:e108. https://doi.org/10.7717/peerj.108
Haas AF, Fairoz MFM, Kelly LW et al (2016) Global microbialization of coral reefs. Nat Microbiol 1:16042. https://doi.org/10.1038/nmicrobiol.2016.42
Hadaidi G, Röthig T, Yum LK et al (2017) Stable mucus-associated bacterial communities in bleached and healthy corals of *Porites lobata* from the Arabian seas. Sci Rep 7:45362. https://doi.org/10.1038/srep45362
Halpern BS, Selkoe KA, Micheli F, Kappel CV (2007) Evaluating and ranking the vulnerability of global marine ecosystems to anthropogenic threats. Conserv Biol 21:1301–1315. https://doi.org/10.1111/j.1523-1739.2007.00752.x
Harvell CD, Kim K, Burkholder JM et al (1999) Emerging marine diseases–climate links and anthropogenic factors. Science 285:1505–1510. https://doi.org/10.1126/science.285.5433.1505
Harvell D, Aronson R, Baron N et al (2004) The rising tide of ocean diseases: unsolved problems and research priorities. Front Ecol Environ 2:375–382. https://doi.org/10.1890/1540-9295(2004)002[0375:TRTOOD]2.0.CO;2
Haydon TD, Seymour JR, Raina J-B et al (2021) Rapid shifts in bacterial communities and homogeneity of Symbiodiniaceae in colonies of *Pocillopora acuta* transplanted between reef and mangrove environments. Front Microbiol 12:756091. https://doi.org/10.3389/fmicb.2021.756091
Hayes RL, Goreau NI (1998) The significance of emerging diseases in the tropical coral reef ecosystem. Rev Biol Trop 46:173–185
Hernandez-Agreda A, Leggat W, Bongaerts P, Ainsworth TD (2016) The microbial signature provides insight into the mechanistic basis of coral success across reef habitats. mBio 7:e00560-16. https://doi.org/10.1128/mBio.00560-16
Hernandez-Agreda A, Gates RD, Ainsworth TD (2017) Defining the core microbiome in corals' microbial soup. Trends Microbiol 25:125–140. https://doi.org/10.1016/j.tim.2016.11.003
Hinderstein LM, Marr JCA, Martinez FA et al (2010) Theme section on "Mesophotic coral ecosystems: characterization, ecology, and management". Coral Reefs 29:247–251. https://doi.org/10.1007/s00338-010-0614-5
Hoegh-Guldberg O (1999) Climate change, coral bleaching and the future of the world's coral reefs. Mar Freshw Res 50:839–866
Hoegh-Guldberg O, Mumby PJ, Hooten AJ et al (2007) Coral reefs under rapid climate change and ocean acidification. Science 318:1737–1742. https://doi.org/10.1126/science.1152509
Howells EJ, Abrego D, Meyer E et al (2016) Host adaptation and unexpected symbiont partners enable reef-building corals to tolerate extreme temperatures. Glob Chang Biol 22:2702–2714. https://doi.org/10.1111/gcb.13250
Hughes TP, Rodrigues MJ, Bellwood DR et al (2007) Phase shifts, herbivory, and the resilience of coral reefs to climate change. Curr Biol 17:360–365. https://doi.org/10.1016/j.cub.2006.12.049
Hughes TP, Graham NA, Jackson JB et al (2010) Rising to the challenge of sustaining coral reef resilience. Trends Ecol Evol 25:633–642. https://doi.org/10.1016/j.tree.2010.07.011
Hughes TP, Kerry JT, Álvarez-Noriega M et al (2017) Global warming and recurrent mass bleaching of corals. Nature 543:373–377. https://doi.org/10.1038/nature21707
Hughes TP, Anderson KD, Connolly SR et al (2018) Spatial and temporal patterns of mass bleaching of corals in the Anthropocene. Science 359:80–83. https://doi.org/10.1126/science.aan8048
Hume B, D'Angelo C, Burt J et al (2013) Corals from the Persian/Arabian Gulf as models for thermotolerant reef-builders: prevalence of clade C3 *Symbiodinium*, host fluorescence and ex situ temperature tolerance. Mar Pollut Bull 72:313–322. https://doi.org/10.1016/j.marpolbul.2012.11.032
Hume BCC, Voolstra CR, Arif C et al (2016) Ancestral genetic diversity associated with the rapid spread of stress-tolerant coral symbionts in response to Holocene climate change. Proc Natl Acad Sci 113:4416–4421. https://doi.org/10.1073/pnas.1601910113
Hume BC, D'Angelo C, Burt JA, Wiedenmann J (2018) Fine-scale biogeographical boundary delineation and sub-population resolution in the *Symbiodinium thermophilum* coral symbiont group from the Persian/Arabian Gulf and Gulf of Oman. Front Mar Sci 5:138
Hume BCC, Mejia-Restrepo A, Voolstra CR, Berumen ML (2020) Fine-scale delineation of Symbiodiniaceae genotypes on a previously bleached Central Red Sea reef system demonstrates a prevalence of coral host-specific associations. Coral Reefs 39:583–601. https://doi.org/10.1007/s00338-020-01917-7
Ishimatsu A, Hayashi M, Lee K-S et al (2005) Physiological effects on fishes in a high-CO2 world. J Geophys Res Oceans 110:C09S09. https://doi.org/10.1029/2004JC002564
Jiao J-Y, Liu L, Hua Z-S et al (2021) Microbial dark matter coming to light: challenges and opportunities. Natl Sci Rev 8:nwaa280. https://doi.org/10.1093/nsr/nwaa280
Kahng S, Garcia-Sais J, Spalding H et al (2010) Community ecology of mesophotic coral reef ecosystems. Coral Reefs 29:255–275. https://doi.org/10.1007/s00338-010-0593-6
Kapsenberg L, Kelley AL, Francis LA, Raskin SB (2015) Exploring the complexity of ocean acidification: an ecosystem comparison of coastal pH variability. Sci Scope 39:51
Karcher DB, Roth F, Carvalho S et al (2020) Nitrogen eutrophication particularly promotes turf algae in coral reefs of the Central Red Sea. PeerJ 8:e8737. https://doi.org/10.7717/peerj.8737
Kathiresan K (2012) Importance of mangrove ecosystem. Int J Mar Sci 2:70–89
Kimes NE, Van Nostrand JD, Weil E et al (2010) Microbial functional structure of *Montastraea faveolata*, an important Caribbean reef-building coral, differs between healthy and yellow-band diseased colonies. Environ Microbiol 12:541–556. https://doi.org/10.1111/j.1462-2920.2009.02113.x
Kline DI, Kuntz NM, Breitbart M et al (2006) Role of elevated organic carbon levels and microbial activity in coral mortality. Mar Ecol Prog Ser 314:119–125
Kochan DP, Mitchell MD, Zuercher R, Harborne AR (2023) Winners and losers of reef flattening: an assessment of coral reef fish species and traits. Oikos 2023:e10011. https://doi.org/10.1111/oik.10011
Koop K, Booth D, Broadbent A et al (2001) ENCORE: the effect of nutrient enrichment on coral reefs. Synthesis of results and conclusions. Mar Pollut Bull 42:91–120. https://doi.org/10.1016/S0025-326X(00)00181-8

Kuffner IB, Andersson AJ, Jokiel PL et al (2008) Decreased abundance of crustose coralline algae due to ocean acidification. Nat Geosci 1:114–117. https://doi.org/10.1038/ngeo100

Kuntz NM, Kline DI, Sandin SA, Rohwer F (2005) Pathologies and mortality rates caused by organic carbon and nutrient stressors in three Caribbean coral species. Mar Ecol Prog Ser 294:173–180

La Rivière M, Roumagnac M, Garrabou J, Bally M (2013) Transient shifts in bacterial communities associated with the temperate gorgonian *Paramuricea clavata* in the Northwestern Mediterranean Sea. PLoS One 8:1–14. https://doi.org/10.1371/journal.pone.0057385

Lachnit T, Bosch TCG, Deines P (2019) Exposure of the host-associated microbiome to nutrient-rich conditions may lead to dysbiosis and disease development—an evolutionary perspective. MBio 10:1–8. https://doi.org/10.1128/mbio.00355-19

LaJeunesse TC, Wham DC, Pettay DT et al (2014) Ecologically differentiated stress-tolerant endosymbionts in the dinoflagellate genus *Symbiodinium* (Dinophyceae) Clade D are different species. Phycologia 53:305–319. https://doi.org/10.2216/13-186.1

LaJeunesse TC, Parkinson JE, Gabrielson PW et al (2018) Systematic revision of Symbiodiniaceae highlights the antiquity and diversity of coral endosymbionts. Curr Biol 28:2570–2580.e6. https://doi.org/10.1016/j.cub.2018.07.008

Lampert Y, Kelman D, Nitzan Y et al (2008) Phylogenetic diversity of bacteria associated with the mucus of Red Sea corals. FEMS Microbiol Ecol 64:187–198. https://doi.org/10.1111/j.1574-6941.2008.00458.x

Leite DCA, Leão P, Garrido AG et al (2017) Broadcast spawning coral *Mussismilia hispida* can vertically transfer its associated bacterial core. Front Microbiol 8:176. https://doi.org/10.3389/fmicb.2017.00176

Leite DCA, Salles JF, Calderon EN et al (2018) Coral bacterial-core abundance and network complexity as proxies for anthropogenic pollution. Front Microbiol 9:833. https://doi.org/10.3389/fmicb.2018.00833

Lesser MP, Falcón LI, Rodríguez-Román A et al (2007) Nitrogen fixation by symbiotic cyanobacteria provides a source of nitrogen for the scleractinian coral *Montastraea cavernosa*. Mar Ecol Prog Ser 346:143–152. https://doi.org/10.3354/meps0700

Lesser MP, Slattery M, Leichter JJ (2009) Ecology of mesophotic coral reefs. J Exp Mar Biol Ecol 375:1–8. https://doi.org/10.1016/j.jembe.2009.05.009

Lesser MP, Slattery M, Mobley CD (2018) Biodiversity and functional ecology of mesophotic coral reefs. Annu Rev Ecol Evol Syst 49:49–71. https://doi.org/10.1146/annurev-ecolsys-110617-062423

Lesser MP, Slattery M, Macartney KJ (2022) Using stable isotope analyses to assess the trophic ecology of Scleractinian corals. Oceans 3:527–546. https://doi.org/10.3390/oceans3040035

Lewis RE, Davy SK, Gardner SG et al (2022) Colony self-shading facilitates Symbiodiniaceae cohabitation in a South Pacific coral community. Coral Reefs 41:1433–1447. https://doi.org/10.1007/s00338-022-02292-1

Li H-Y, Zhou D-D, Gan R-Y et al (2021) Effects and mechanisms of probiotics, prebiotics, synbiotics, and postbiotics on metabolic diseases targeting gut microbiota: a narrative review. Nutrients 13:3211. https://doi.org/10.3390/nu13093211

Li J, Zou Y, Li Q et al (2023) A coral-associated actinobacterium mitigates coral bleaching under heat stress. Environ Microbiome 18:83. https://doi.org/10.1186/s40793-023-00540-7

Lins-de-Barros MM, Cardoso AM, Silveira CB et al (2013) Microbial community compositional shifts in bleached colonies of the Brazilian reef-building coral *Siderastrea stellata*. Microb Ecol 65:205–213. https://doi.org/10.1007/s00248-012-0095-x

Longley R, Benucci GMN, Pochon X et al (2024) Species-specific coral microbiome assemblages support host bleaching resistance during an extreme marine heatwave. Sci Total Environ 906:167803. https://doi.org/10.1016/j.scitotenv.2023.167803

Loya Y, Eyal G, Treibitz T et al (2016) Theme section on mesophotic coral ecosystems: advances in knowledge and future perspectives. Coral Reefs 35:1–9. https://doi.org/10.1007/s00338-016-1410-7

Lucas CC, Lima IC, Garcia TM et al (2023) Turbidity buffers coral bleaching under extreme wind and rainfall conditions. Mar Environ Res 192:106215. https://doi.org/10.1016/j.marenvres.2023.106215

Macintosh DJ, Ashton EC (2002) A review of mangrove biodiversity conservation and management. Centre for Tropical Ecosystems Research, University of Aarhus, Aarhus

Maggioni F, Pujo-Pay M, Aucan J et al (2021) The Bouraké semi-enclosed lagoon (New Caledonia)–a natural laboratory to study the lifelong adaptation of a coral reef ecosystem to extreme environmental conditions. Biogeosciences 18:5117–5140. https://doi.org/10.5194/bg-18-5117-2021

Maire J, van Oppen MJH (2022) A role for bacterial experimental evolution in coral bleaching mitigation? Trends Microbiol 30:217–228. https://doi.org/10.1016/j.tim.2021.07.006

Maire J, Tandon K, Collingro A et al (2023) Colocalization and potential interactions of *Endozoicomonas* and chlamydiae in microbial aggregates of the coral *Pocillopora acuta*. Science. Advances 9:eadg0773. https://doi.org/10.1126/sciadv.adg0773

Mannochio-Russo H, Swift SOI, Nakayama KK et al (2023) Microbiomes and metabolomes of dominant coral reef primary producers illustrate a potential role for immunolipids in marine symbioses. Commun Biol 6:896. https://doi.org/10.1038/s42003-023-05230-1

Marchioro GM, Glasl B, Engelen AH et al (2020) Microbiome dynamics in the tissue and mucus of acroporid corals differ in relation to host and environmental parameters. PeerJ 8:e9644. https://doi.org/10.7717/peerj.9644

Martin S, Rodolfo-Metalpa R, Ransome E et al (2008) Effects of naturally acidified seawater on seagrass calcareous epibionts. Biol Lett 4:689–692. https://doi.org/10.1098/rsbl.2008.041

Martinez S, Grover R, Baker DM, Ferrier-Pagès C (2022) Symbiodiniaceae are the first site of heterotrophic nitrogen assimilation in reef-building corals. MBio 13:e01601–e01622. https://doi.org/10.1128/mbio.01601-22

Matthews JL, Crowder CM, Oakley CA et al (2017) Optimal nutrient exchange and immune responses operate in partner specificity in the cnidarian-dinoflagellate symbiosis. PNAS 114:13194–13199. https://doi.org/10.1073/pnas.1710733114

McClanahan TR, Sala E, Stickels PA et al (2003) Interaction between nutrients and herbivory in controlling algal communities and coral condition on Glovers Reef, Belize. Mar Ecol Prog Ser 261:135–147

McDevitt-Irwin JM, Baum JK, Garren M, Vega Thurber RL (2017) Responses of coral-associated bacterial communities to local and global stressors. Front Mar Sci 4:286253. https://doi.org/10.3389/fmars.2017.00262

McManus JW, Polsenberg JF (2004) Coral–algal phase shifts on coral reefs: ecological and environmental aspects. Prog Oceanogr 60:263–279. https://doi.org/10.1016/j.pocean.2004.02.014

Meenatchi R, Thinesh T, Brindangnanam P et al (2020) Revealing the impact of global mass bleaching on coral microbiome through 16S rRNA gene-based metagenomic analysis. Microbiol Res 233:126408. https://doi.org/10.1016/j.micres.2019.126408

Meron D, Atias E, Iasur Kruh L et al (2011) The impact of reduced pH on the microbial community of the coral *Acropora eurystoma*. ISME J 5:51–60. https://doi.org/10.1038/ismej.2010.102

Meron D, Rodolfo-Metalpa R, Cunning R et al (2012) Changes in coral microbial communities in response to a natural pH gradient. ISME J 6:1775–1785. https://doi.org/10.1038/ismej.2012.19

Modolon F, Schultz J, Duarte G et al (2023) In situ devices can culture the microbial dark matter of corals. iScience 26:108374. https://doi.org/10.1016/j.isci.2023.108374

Mohamed AR, Ochsenkühn MA, Kazlak AM et al (2023) The coral microbiome: towards an understanding of the molecular mecha-

nisms of coral–microbiota interactions. FEMS Microbiol Rev 47:fuad005. https://doi.org/10.1093/femsre/fuad005
Monti M, Giorgi A, Kemp DW, Olson JB (2022) Cultivable bacteria associated with Caribbean octocorals are active against coral pathogens but exhibit variable bioactivity when grown under different temperature conditions. Coral Reefs 41:1365–1377. https://doi.org/10.1007/s00338-022-02285-0
Moradi M, Magalhaes PR, Peixoto RS et al (2023) Probiotics mitigate thermal stress- and pathogen-driven impacts on coral skeleton. Front Mar Sci 10:1212690. https://doi.org/10.3389/fmars.2023.1212690
Moran NA, Sloan DB (2015) The hologenome concept: helpful or hollow? PLoS Biol 13:e1002311. https://doi.org/10.1371/journal.pbio.1002311
Morgan KM, Perry CT, Smithers SG et al (2016) Evidence of extensive reef development and high coral cover in nearshore environments: implications for understanding coral adaptation in turbid settings. Sci Rep 6:29616. https://doi.org/10.1038/srep29616
Morgans CA, Hung JY, Bourne DG, Quigley KM (2020) Symbiodiniaceae probiotics for use in bleaching recovery. Restor Ecol 28:282–288. https://doi.org/10.1111/rec.13069
Moriarty T, Leggat W, Huggett MJ, Ainsworth TD (2020) Coral disease causes, consequences, and risk within coral restoration. Trends Microbiol 28:793–807. https://doi.org/10.1016/j.tim.2020.06.002
Morrow KM, Paul VJ, Liles MR, Chadwick NE (2011) Allelochemicals produced by Caribbean macroalgae and cyanobacteria have species-specific effects on reef coral microorganisms. Coral Reefs 30:309–320. https://doi.org/10.1007/s00338-011-0747-1
Morrow KM, Muller E, Lesser MP (2018) How does the coral microbiome cause, respond to, or modulate the bleaching process? In: van Oppen M, Lough J (eds) Coral bleaching. Ecological studies, vol 233. Springer, Cham. https://doi.org/10.1007/978-3-319-75393-5_7
Muscatine L, Porter JW (1977) Reef corals: mutualistic symbioses adapted to nutrient-poor environments. Bioscience 27:454–460. https://doi.org/10.2307/1297526
Muscatine L, Ferrier-Pagès C, Blackburn A et al (1998) Cell-specific density of symbiotic dinoflagellates in tropical anthozoans. Coral Reefs 17:329–337. https://doi.org/10.1007/s003380050133
Neave MJ, Rachmawati R, Xun L et al (2017) Differential specificity between closely related corals and abundant *Endozoicomonas* endosymbionts across global scales. ISME J 11:186–200. https://doi.org/10.1038/ismej.2016.95
Neave MJ, Apprill A, Aeby G et al (2019) Microbial communities of Red Sea coral reefs. In: Coral reefs of the Red Sea. Springer, pp 53–68. https://doi.org/10.1007/978-3-030-05802-9_4
Nelson CE, Goldberg SJ, Wegley Kelly L et al (2013) Coral and macroalgal exudates vary in neutral sugar composition and differentially enrich reef bacterioplankton lineages. ISME J 7:962–979. https://doi.org/10.1038/ismej.2012.161
Norström A, Nyström M, Lokrantz J, Folke C (2009) Alternative states on coral reefs: beyond coral–macroalgal phase shifts. Mar Ecol Prog Ser 376:295–306
Nugues MM, Smith GW, van Hooidonk RJ et al (2004) Algal contact as a trigger for coral disease. Ecol Lett 7:919–923. https://doi.org/10.1111/j.1461-0248.2004.00651.x
Nyholm SV, McFall-Ngai M (2004) The winnowing: establishing the squid–vibrio symbiosis. Nat Rev Microbiol 2:632–642. https://doi.org/10.1038/nrmicro957
Nyström M, Graham NAJ, Lokrantz J, Norström AV (2008) Capturing the cornerstones of coral reef resilience: linking theory to practice. Coral Reefs 27:795–809. https://doi.org/10.1007/s00338-008-0426-z
O'Brien PA, Morrow KM, Willis BL, Bourne DG (2016) Implications of ocean acidification for marine microorganisms from the free-living to the host-associated. Front Mar Sci 3:47. https://doi.org/10.3389/fmars.2016.00047
O'Brien PA, Webster NS, Miller DJ, Bourne DG (2019) Host-microbe coevolution: applying evidence from model systems to complex marine invertebrate holobionts. mBio 10:e02241-18. https://doi.org/10.1128/mBio.02241-18
O'Neil JM, Capone DG (2008) Nitrogen cycling in coral reef environments. In: Nitrogen in the marine environment, vol 10. Springer, pp 948–989. https://doi.org/10.1016/B978-0-12-372522-6.00021-9
Oliver TA, Palumbi SR (2009) Distributions of stress-resistant coral symbionts match environmental patterns at local but not regional scales. Mar Ecol Prog Ser 378:93–103. https://doi.org/10.3354/meps07871
Oliver J, Berkelmans R, Eakin C (2009) Coral bleaching in space and time. In: Van Oppen M, Lough J (eds) Coral bleaching: patterns and processes, causes and consequences. Springer, pp 21–40
Olson JB, Kellogg CA (2010) Microbial ecology of corals, sponges, and algae in mesophotic coral environments. FEMS Microbiol Ecol 73:17–30. https://doi.org/10.1111/j.1574-6941.2010.00862.x
Orr JC, Fabry VJ, Aumont O et al (2005) Anthropogenic ocean acidification over the twenty-first century and its impact on calcifying organisms. Nature 437:681–686. https://doi.org/10.1038/nature04095
Osman EO, Smith DJ, Ziegler M et al (2018) Thermal refugia against coral bleaching throughout the northern Red Sea. Glob Chang Biol 24:e474–e484. https://doi.org/10.1111/gcb.13895
Osman EO, Suggett DJ, Voolstra CR et al (2020) Coral microbiome composition along the northern Red Sea suggests high plasticity of bacterial and specificity of endosymbiotic dinoflagellate communities. Microbiome 8:8. https://doi.org/10.1186/s40168-019-0776-5
Padilla-Gamiño JL, Roth MS, Rodrigues LJ et al (2019) Ecophysiology of mesophotic reef-building corals in Hawai'i is influenced by symbiont–host associations, photoacclimatization, trophic plasticity, and adaptation. Limnol Oceanogr 64:1980–1995. https://doi.org/10.1002/lno.11164
Pantos O, Bongaerts P, Dennis PG et al (2015) Habitat-specific environmental conditions primarily control the microbiomes of the coral *Seriatopora hystrix*. ISME J 9:1916–1927. https://doi.org/10.1038/ismej.2015.3
Pearse VB, Muscatine L (1971) Role of symbiotic algae (zooxanthellae) in coral calcification. Biol Bull 141:350–363
Peixoto RS, Rosado PM, Leite DCDA et al (2017) Beneficial microorganisms for corals (BMC): proposed mechanisms for coral health and resilience. Front Microbiol 8:341. https://doi.org/10.3389/fmicb.2017.00341
Peixoto RS, Sweet M, Bourne DG (2019) Customized medicine for corals. Front Mar Sci 6:686. https://doi.org/10.3389/fmars.2019.00686
Peixoto RS, Sweet M, Villela HDM et al (2021) Coral probiotics: premise, promise, prospects. Annu Rev Anim Biosci 9:265–288. https://doi.org/10.1146/annurev-animal-090120-115444
Peixoto RS, Voolstra CR, Sweet M et al (2022) Harnessing the microbiome to prevent global biodiversity loss. Nat Microbiol 7:1726–1735. https://doi.org/10.1038/s41564-022-01173-1
Perry CT, Larcombe P (2003) Marginal and non-reef-building coral environments. Coral Reefs 22:427–432. https://doi.org/10.1007/s00338-003-0330-5
Picciani N, de Lossio e Seiblitz IG, de Paiva PC et al (2016) Geographic patterns of *Symbiodinium* diversity associated with the coral *Mussismilia hispida* (Cnidaria, Scleractinia) correlate with major reef regions in the Southwestern Atlantic Ocean. Mar Biol 163:236. https://doi.org/10.1007/s00227-016-3010-z
Pinheiro HT, Eyal G, Shepherd B, Rocha LA (2019) Ecological insights from environmental disturbances in mesophotic coral ecosystems. Ecosphere 10:e02666. https://doi.org/10.1002/ecs2.2666
Pita L, Rix L, Slaby BM et al (2018) The sponge holobiont in a changing ocean: from microbes to ecosystems. Microbiome 6:46. https://doi.org/10.1186/s40168-018-0428-1
Pochon X, Forsman Z, Spalding H et al (2015) Depth specialization in mesophotic corals (*Leptoseris* spp.) and associ-

ated algal symbionts in Hawai'i. R Soc Open Sci 2:140351. https://doi.org/10.1098/rsos.140351

Pogoreutz C, Ziegler M (2024) Frenemies on the reef? Resolving the coral–*Endozoicomonas* association. Trends Microbiol 32:422–434. https://doi.org/10.1016/j.tim.2023.11.006

Pogoreutz C, Rädecker N, Cárdenas A et al (2017) Sugar enrichment provides evidence for a role of nitrogen fixation in coral bleaching. Glob Chang Biol 23:3838–3848. https://doi.org/10.1111/gcb.13695

Pollock FJ, McMinds R, Smith S et al (2018) Coral-associated bacteria demonstrate phylosymbiosis and cophylogeny. Nat Commun 9:4921. https://doi.org/10.1038/s41467-018-07275-x

Pootakham W, Mhuantong W, Putchim L et al (2018) Dynamics of coral-associated microbiomes during a thermal bleaching event. Microbiol Open 7:e00604. https://doi.org/10.1002/mbo3.604

Pootakham W, Mhuantong W, Yoocha T et al (2019) Heat-induced shift in coral microbiome reveals several members of the Rhodobacteraceae family as indicator species for thermal stress in *Porites lutea*. MicrobiologyOpen 8:e935. https://doi.org/10.1002/mbo3.935

Quigley KM, Baker AC, Coffroth MA et al (2018) Bleaching resistance and the role of algal endosymbionts. In: van Oppen MJH, Lough JM (eds) Coral bleaching: patterns, processes, causes and consequences. Springer International Publishing, Cham, pp 111–151

Rädecker N, Pogoreutz C, Voolstra CR et al (2015) Nitrogen cycling in corals: the key to understanding holobiont functioning? Trends Microbiol 23:490–497. https://doi.org/10.1016/j.tim.2015.03.008

Rädecker N, Pogoreutz C, Gegner HM et al (2021) Heat stress destabilizes symbiotic nutrient cycling in corals. Proc Natl Acad Sci 118:e2022653118. https://doi.org/10.1073/pnas.2022653118

Raghukumar C, Ravindran J (2012) Fungi and their role in corals and coral reef ecosystems. In: Raghukumar C (ed) Biology of marine fungi. Springer, Berlin/Heidelberg, pp 89–113

Raina J-B, Tapiolas D, Willis BL, Bourne DG (2009) Coral-associated bacteria and their role in the biogeochemical cycling of sulfur. Appl Environ Microbiol 75:3492–3501. https://doi.org/10.1128/AEM.02567-08

Rasher DB, Hay ME (2010) Chemically rich seaweeds poison corals when not controlled by herbivores. Proc Natl Acad Sci 107:9683–9688. https://doi.org/10.1073/pnas.0912095107

Reddy MM, Goossens C, Zhou Y et al (2023) Multi-omics determination of metabolome diversity in natural coral populations in the Pacific Ocean. Commun Earth Environ 4:281. https://doi.org/10.1038/s43247-023-00942-y

Reed J (1985) Deepest distribution of Atlantic hermatypic corals discovered in The Bahamas. In: Proc 5th Int Coral Reef symp. Antenne Museum-EPHE, pp 249–254

Reguero BG, Storlazzi CD, Gibbs AE et al (2021) The value of US coral reefs for flood risk reduction. Nat Sustainability 4:688–698. https://doi.org/10.1038/s41893-021-00706-6

Reis AMM, Araújo SD Jr, Moura RL et al (2009) Bacterial diversity associated with the Brazilian endemic reef coral *Mussismilia braziliensis*. J Appl Microbiol 106:1378–1387. https://doi.org/10.1111/j.1365-2672.2008.04106.x

Reshef L, Koren O, Loya Y et al (2006) The coral probiotic hypothesis. Environ Microbiol 8:2068–2073. https://doi.org/10.1111/j.1462-2920.2006.01148.x

Restrepo JD, Park E, Aquino S, Latrubesse EM (2016) Coral reefs chronically exposed to river sediment plumes in the southwestern Caribbean: Rosario Islands, Colombia. Sci Total Environ 553:316–329. https://doi.org/10.1016/j.scitotenv.2016.02.140

Rice MM, Maher RL, Thurber RV, Burkepile DE (2019) Different nitrogen sources speed recovery from corallivory and uniquely alter the microbiome of a reef-building coral. PeerJ 7:e8056. https://doi.org/10.7717/peerj.8056

Ritchie KB (2006) Regulation of microbial populations by coral surface mucus and mucus-associated bacteria. Mar Ecol Prog Ser 322:1–14

Robbins SJ, Singleton CM, Chan CX et al (2019) A genomic view of the reef-building coral *Porites lutea* and its microbial symbionts. Nat Microbiol4:2090–2100.https://doi.org/10.1038/s41564-019-0532-4

Roder C, Arif C, Bayer T et al (2014) Bacterial profiling of white plague disease in a comparative coral species framework. ISME J 8:31–39. https://doi.org/10.1038/ismej.2013.127

Roder C, Bayer T, Aranda M et al (2015) Microbiome structure of the fungid coral *Ctenactis echinata* aligns with environmental differences. Mol Ecol 24:3501–3511. https://doi.org/10.1111/mec.13251

Rog SM, Clarke RH, Cook CN (2017) More than marine: revealing the critical importance of mangrove ecosystems for terrestrial vertebrates. Divers Distrib 23:221–230. https://doi.org/10.1111/ddi.12514

Rogers CS, Herlan JJ (2012) Life on the edge: corals in mangroves and climate change. In: Proceedings of the 12th international coral reef symposium, Cairns, Australia. National Coral Reef Institute, pp 9–13

Rohwer F, Seguritan V, Azam F, Knowlton N (2002) Diversity and distribution of coral-associated bacteria. Mar Ecol Prog Ser 243:1–10. https://doi.org/10.3354/meps243001

Rolim FA, Langlois T, Motta FDS et al (2022) Habitat and marine reserve status drive reef fish biomass and functional diversity in the largest South Atlantic coral reef system (Abrolhos, Brazil). Front Mar Sci 9:701244. https://doi.org/10.3389/fmars.2022.701244

Rosado PM, Leite DCA, Duarte GAS et al (2019) Marine probiotics: increasing coral resistance to bleaching through microbiome manipulation. ISME J 13:921–936. https://doi.org/10.1038/s41396-018-0323-6

Rosales SM, Clark AS, Huebner LK et al (2020) Rhodobacterales and Rhizobiales are associated with stony coral tissue loss disease and its suspected sources of transmission. Front Microbiol 11:681. https://doi.org/10.3389/fmicb.2020.00681

Rosenberg E, Koren O, Reshef L et al (2007) The role of microorganisms in coral health, disease and evolution. Nat Rev Microbiol 5:355–362. https://doi.org/10.1038/nrmicro1635

Roth MS (2014) The engine of the reef: photobiology of the coral–algal symbiosis. Front Microbiol 5:97199. https://doi.org/10.3389/fmicb.2014.00422

Röthig T, Yum LK, Kremb SG et al (2017) Microbial community composition of deep-sea corals from the Red Sea provides insight into functional adaption to a unique environment. Sci Rep 7:44714. https://doi.org/10.1038/srep44714

Rouzé H, Galand PE, Medina M et al (2021) Symbiotic associations of the deepest recorded photosynthetic scleractinian coral (172 m depth). ISME J 15:1564–1568. https://doi.org/10.1038/s41396-020-00857-y

Santoro EP, Borges RM, Espinoza JL et al (2021) Coral microbiome manipulation elicits metabolic and genetic restructuring to mitigate heat stress and evade mortality. Sci Adv 7:eabg3088. https://doi.org/10.1126/sciadv.abg3088

Santoro EP, Cardenas A, Villela HMD et al (2025) Inherent differential microbial assemblages and functions associated with corals exhibiting different thermal phenotypes. Sci Adv 11:eadq2583. https://doi.org/10.1126/sciadv.adq2583

Savary R, Barshis DJ, Voolstra CR et al (2021) Fast and pervasive transcriptomic resilience and acclimation of extremely heat-tolerant coral holobionts from the northern Red Sea. Proc Natl Acad Sci 118:e2023298118. https://doi.org/10.1073/pnas.2023298118

Schlichter D, Fricke H (1991) Mechanisms of amplification of photosynthetically active radiation in the symbiotic deep-water coral *Leptoseris fragilis*. Hydrobiologia 216:389–394. https://doi.org/10.1007/BF00026491

Schoepf V, Baumann JH, Barshis DJ et al (2023) Corals at the edge of environmental limits: a new conceptual framework to re-define marginal and extreme coral communities. Sci Total Environ 884:163688. https://doi.org/10.1016/j.scitotenv.2023.163688

Schultz J, Modolon F, Rosado AS et al (2022) Methods and strategies to uncover coral-associated microbial dark matter. mSystems 7:e00367-22. https://doi.org/10.1128/msystems.00367-22

Semmler RF, Hoot WC, Reaka ML (2017) Are mesophotic coral ecosystems distinct communities and can they serve as refugia for shallow reefs? Coral Reefs 36:433–444. https://doi.org/10.1007/s00338-016-1530-0

Sharp KH, Distel D, Paul VJ (2012) Diversity and dynamics of bacterial communities in early life stages of the Caribbean coral *Porites astreoides*. ISME J 6:790–801. https://doi.org/10.1038/ismej.2011.144

Shnit-Orland M, Kushmaro A (2009) Coral mucus-associated bacteria: a possible first line of defense. FEMS Microbiol Ecol 67:371–380. https://doi.org/10.1111/j.1574-6941.2008.00644.x

Shore A, Day R, Stewart J, Burge C (2021) Dichotomy between regulation of coral bacterial communities and calcification physiology under ocean acidification conditions. Appl Environ Microbiol 87:e02189–e02120. https://doi.org/10.1128/aem.02189-20

Siboni N, Ben-Dov E, Sivan A, Kushmaro A (2012) Geographic specific coral-associated ammonia-oxidizing archaea in the Northern Gulf of Eilat (Red Sea). Microb Ecol 64:18–24. https://doi.org/10.1007/s00248-011-0006-6

Silbiger NJ, Goodbody-Gringley G, Bruno JF, Putnam HM (2019) Comparative thermal performance of the reef-building coral *Orbicella franksi* at its latitudinal range limits. Mar Biol 166:126. https://doi.org/10.1007/s00227-019-3573-6

Silva DP, Villela HDM, Santos HF et al (2021) Multi-domain probiotic consortium as an alternative to chemical remediation of oil spills at coral reefs and adjacent sites. Microbiome 9:118. https://doi.org/10.1186/s40168-021-01041-w

Silverstein RN, Cunning R, Baker AC (2015) Change in algal symbiont communities after bleaching, not prior heat exposure, increases heat tolerance of reef corals. Glob Chang Biol 21:236–249. https://doi.org/10.1111/gcb.12706

Slattery M, Lesser M, Brazeau D et al (2011) Connectivity and stability of mesophotic coral reefs. J Exp Mar Biol Ecol 408:32–41. https://doi.org/10.1016/j.jembe.2011.07.024

Slattery M, Lesser MP, Rocha LA et al (2024) Function and stability of mesophotic coral reefs. Trends Ecol Evol 39:585–594. https://doi.org/10.1016/j.tree.2024.01.011

Smith L, Birkeland C (2003) Managing NPSA's coral reefs in the face of global warming: research project report for year 1. Hawai'i Cooperative Fishery Research Unit, Zoology Department, University of Hawai'i at Manoa, Honolulu

Smith JE, Shaw M, Edwards RA et al (2006) Indirect effects of algae on coral: algae-mediated, microbe-induced coral mortality. Ecol Lett 9:835–845. https://doi.org/10.1111/j.1461-0248.2006.00937.x

Smith E, Vaughan G, Ketchum R et al (2017a) Symbiont community stability through severe coral bleaching in a thermally extreme lagoon. Sci Rep 7:2428. https://doi.org/10.1038/s41598-017-01569-8

Smith EG, D'angelo C, Sharon Y et al (2017b) Acclimatization of symbiotic corals to mesophotic light environments through wavelength transformation by fluorescent protein pigments. Proc R Soc B Biol Sci 284:20170320. https://doi.org/10.1098/rspb.2017.0320

Sobha TR, Vibija CP, Fahima P (2023) Coral reef: a hot spot of marine biodiversity. In: Sukumaran ST, Ramakrishnan KT (eds) Conservation and sustainable utilization of bioresources. Springer Nature Singapore, Singapore, pp 171–194

Stat M, Carter D, Hoegh-Guldberg O (2006) The evolutionary history of *Symbiodinium* and scleractinian hosts—symbiosis, diversity, and the effect of climate change. Perspect Plant Ecol Evol Syst 8:23–43. https://doi.org/10.1016/j.ppees.2006.04.001

Suggett DJ, Smith DJ (2020) Coral bleaching patterns are the outcome of complex biological and environmental networking. Glob Chang Biol 26:68–79. https://doi.org/10.1111/gcb.14871

Sunagawa S, DeSantis TZ, Piceno YM et al (2009) Bacterial diversity and white plague disease-associated community changes in the Caribbean coral Montastraea faveolata. ISME J 3:512–521. https://doi.org/10.1038/ismej.2008.131

Sunagawa S, Woodley CM, Medina M (2010) Threatened corals provide underexplored microbial habitats. PLoS One 5:e9554. https://doi.org/10.1371/journal.pone.0009554

Sutherland KP, Porter JW, Turner JW et al (2010) Human sewage identified as likely source of white pox disease of the threatened Caribbean Elkhorn coral, *Acropora palmata*. Environ Microbiol 12:1122–1131. https://doi.org/10.1111/j.1462-2920.2010.02152.x

Sweet MJ, Bulling MT (2017) On the importance of the microbiome and pathobiome in coral health and disease. Front Mar Sci 4. https://doi.org/10.3389/fmars.2017.00009

Szabó M, Larkum AWD, Vass I (2020) A review: the role of reactive oxygen species in mass coral bleaching. In: Larkum AWD, Grossman AR, Raven JA (eds) Photosynthesis in algae: biochemical and physiological mechanisms. Springer International Publishing, Cham, pp 459–488

Tang K, Zhan W, Zhou Y et al (2020) Antagonism between coral pathogen *vibrio coralliilyticus* and other bacteria in the gastric cavity of scleractinian coral *Galaxea fascicularis*. Sci China Earth Sci 63:157–166. https://doi.org/10.1007/s11430-019-9388-3

Teplitski M, Ritchie KB (2009) How feasible is the biological control of coral diseases? Trends Ecol Evol 24:378–385. https://doi.org/10.1016/j.tree.2009.02.008

Terraneo TI, Ouhssain M, Castano CB et al (2023) From the shallow to the mesophotic: a characterization of Symbiodiniaceae diversity in the Red Sea NEOM region. Front Mar Sci 10:1077805. https://doi.org/10.3389/fmars.2023.1077805

Teschima MM, Garrido A, Paris A et al (2019) Biogeography of the endosymbiotic dinoflagellates (Symbiodiniaceae) community associated with the brooding coral *Favia gravida* in the Atlantic Ocean. PLoS One 14:e0213519. https://doi.org/10.1371/journal.pone.0213519

Thompson JR, Rivera HE, Closek CJ, Medina M (2015) Microbes in the coral holobiont: partners through evolution, development, and ecological interactions. Front Cell Infect Microbiol 4:176. https://doi.org/10.3389/fcimb.2014.00176

Tignat-Perrier R, van de Water JAJM, Allemand D, Ferrier-Pagès C (2023) Holobiont responses of mesophotic precious red coral Corallium rubrum to thermal anomalies. Environ Microbiome 18:70. https://doi.org/10.1186/s40793-023-00525-6

Tilstra A, El-Khaled YC, Roth F et al (2019) Denitrification aligns with N2 fixation in Red Sea corals. Sci Rep 9:19460. https://doi.org/10.1038/s41598-019-55408-z

Torda G, Donelson JM, Aranda M et al (2017) Rapid adaptive responses to climate change in corals. Nat Clim Change 7:627–636. https://doi.org/10.1038/nclimate3374

Ushijima B, Meyer JL, Thompson S et al (2020) Disease diagnostics and potential coinfections by *Vibrio coralliilyticus* during an ongoing coral disease outbreak in Florida. Front Microbiol 11:569354. https://doi.org/10.3389/fmicb.2020.569354

Ushijima B, Gunasekera SP, Meyer JL et al (2023) Chemical and genomic characterization of a potential probiotic treatment for stony coral tissue loss disease. Commun Biol 6:248. https://doi.org/10.1038/s42003-023-04590-y

van Oppen MJH, Blackall LL (2019) Coral microbiome dynamics, functions and design in a changing world. Nat Rev Microbiol 17:557–567. https://doi.org/10.1038/s41579-019-0223-4

van Oppen MJH, Oliver JK, Putnam HM, Gates RD (2015) Building coral reef resilience through assisted evolution. Proc Natl Acad Sci 112:2307–2313. https://doi.org/10.1073/pnas.1422301112

Vargas-Ángel B, Huntington B (2020) Status and trends assessment of benthic coral reef communities in Vatia Bay, American Samoa, 2015–2020. NOAA. https://doi.org/10.25923/r7dd-wg50

Vega Thurber R, Willner-Hall D, Rodriguez-Mueller B et al (2009) Metagenomic analysis of stressed coral holobionts. Environ Microbiol 11:2148–2163. https://doi.org/10.1111/j.1462-2920.2009.01935.x

Vega Thurber R, Payet JP, Thurber AR, Correa AMS (2017) Virus–host interactions and their roles in coral reef health and disease. Nat Rev Microbiol 15:205–216. https://doi.org/10.1038/nrmicro.2016.176

Vega Thurber R, Mydlarz LD, Brandt M et al (2020) Deciphering coral disease dynamics: integrating host, microbiome, and the changing environment. Front Ecol Evol 8:575927. https://doi.org/10.3389/fevo.2020.575927

Vidal-Dupiol J, Ladrière O, Meistertzheim A-L et al (2011) Physiological responses of the scleractinian coral *Pocillopora damicornis* to bacterial stress from vibrio coralliilyticus. J Exp Biol 214:1533–1545. https://doi.org/10.1242/jeb.053165

Vitousek PM, Aber JD, Howarth RW et al (1997) Human alteration of the global nitrogen cycle: sources and consequences. Ecol Appl 7:737–750. https://doi.org/10.1890/1051-0761(1997)007[0737:HAOTGN]2.0.CO;2

Voolstra CR, Suggett DJ, Peixoto RS et al (2021) Extending the natural adaptive capacity of coral holobionts. Nat Rev Earth Environ 2:747–762. https://doi.org/10.1038/s43017-021-00214-3

Voolstra CR, Peixoto RS, Ferrier-Pagès C (2023) Mitigating the ecological collapse of coral reef ecosystems: effective strategies to preserve coral reef ecosystems. EMBO Rep. https://doi.org/10.15252/embr.202356826

Voolstra CR, Raina J-B, Dörr M et al (2024) The coral microbiome in sickness, in health and in a changing world. Nat Rev Microbiol. https://doi.org/10.1038/s41579-024-01015-3

Wang L, Shantz AA, Payet JP et al (2018) Corals and their microbiomes are differentially affected by exposure to elevated nutrients and a natural thermal anomaly. Front Mar Sci 5:101. https://doi.org/10.3389/fmars.2018.00101

Webster N, Negri A, Botté E et al (2016) Host-associated coral reef microbes respond to the cumulative pressures of ocean warming and ocean acidification. Sci Rep 6:19324. https://doi.org/10.1038/srep19324

Wee HB, Kurihara H, Reimer JD (2019) Reduced Symbiodiniaceae diversity in *Palythoa tubercolosa* at a heavily acidified coral reef. Coral Reefs 38:311–319. https://doi.org/10.1007/s00338-019-01776-x

Welsh RM, Rosales SM, Zaneveld JR et al (2017) Alien vs. predator: bacterial challenge alters coral microbiomes unless controlled by *Halobacteriovorax* predators. PeerJ 5:e3315. https://doi.org/10.7717/peerj.3315

Wessels W, Sprungala S, Watson S-A et al (2017) The microbiome of the octocoral *Lobophytum pauciflorum*: minor differences between sexes and resilience to short-term stress. FEMS Microbiol Ecol 93:fix013. https://doi.org/10.1093/femsec/fix013

Whipps JM, Lewis K, Cooke RC (1988) Mycoparasitism and plant disease control. Manchester University Press, Manchester

Wiedenmann J, D'Angelo C, Smith EG et al (2013) Nutrient enrichment can increase the susceptibility of reef corals to bleaching. Nat Clim Chang 3:160–164. https://doi.org/10.1038/nclimate1661

Wild C, Niggl W, Naumann MS, Haas AF (2010) Organic matter release by Red Sea coral reef organisms—potential effects on microbial activity and in situ O2 availability. Mar Ecol Prog Ser 411:61–71

Wilkinson CR (1999) Global and local threats to coral reef functioning and existence: reviews and predictions. Mar Freshw Res 50:867–878. https://doi.org/10.1071/MF99121

Yellowlees D, Rees TAV, Leggat W (2008) Metabolic interactions between algal symbionts and invertebrate hosts. Plant Cell Environ 31:679–694

Zhang Y, Yang Q, Ling J et al (2021) Shifting the microbiome of a coral holobiont and improving host physiology by inoculation with a potentially beneficial bacterial consortium. BMC Microbiol 21:130. https://doi.org/10.1186/s12866-021-02167-5

Ziegler M, Seneca FO, Yum LK et al (2017) Bacterial community dynamics are linked to patterns of coral heat tolerance. Nat Commun 8:14213. https://doi.org/10.1038/ncomms14213

Ziegler M, Grupstra CGB, Barreto MM et al (2019) Coral bacterial community structure responds to environmental change in a host-specific manner. Nat Commun 10:3092. https://doi.org/10.1038/s41467-019-10969-5

Marginal Reefs of the Persian/Arabian Gulf: Patterns and Processes

5

B. Riegl Jr, G. Cavalcante, R. Bento, J. Burt, B. Riegl, C. D'Angelo, and J. Wiedenmann

Abstract

The Persian/Arabian Gulf (PAG) is a high-latitude, peripheral sea with a extreme temperature regime for reef-builders. The climatological and oceanographic setting cause repetitive coral mass mortality that stunt reef-building. Reef-like structures classify as "marginal" and "extreme". During the formation of PAG during Holocene flooding (~12 k ybp), shorelines that could have supported reefs moved too rapidly for significant frame building, but corals did establish early on. Presently, the PAG is home to a coral fauna of ~70 species near the Straits of Hormuz and ~ 30 in its inner basin, many of wide Indo-Pacific distribution with a subset of regional endemics (The PAG and Arabian region). The PAG corals thermo-tolerance ranges from ~10 °C–>35 °C. Bleaching onset is after several days above 35 °C and mortality above 35.7 °C. Corals from the hottest, south-eastern part of the PAG, are predominantly associated with *Cladocopium thermophilum*, an extremely heat tolerant dinoflagellate symbiont species that is present in the wider Arabian region and invaded the PAG during Holocene flooding. Along a gradient of environmental stress factors including increasing salinity, seasonal temperature variation and annual minimal temperature, *C. thermophilum* becomes increasingly dominant over the other heat tolerant symbiont that can be found in the PAG, *Durusdinium* sp. This extreme environment makes the PAG a natural laboratory for understanding how corals and their symbionts respond to severe and rapidly changing environments.

Keywords

Persian/Arabian Gulf · Hottest sea · Marginal reefs · Highest bleaching threshold · *Cladocopium thermophilum* · *Durusdinium* · *Symbionts* · *Coral*

B. Riegl Jr (✉)
College of Science and Engineering, James Cook University, Townsville, QLD, Australia
e-mail: rieglb@nova.edu

G. Cavalcante
Instituto de Ciências Atmosféricas (ICAT), Universidade Federal de Alagoas, Maceió, Brazil

R. Bento
Instituto de Ciências Atmosféricas (ICAT), Universidade Federal de Alagoas, Maceió, Brazil

Mubadala Arabian Center for Climate and Environmental Sciences (ACCESS) and Water Research Center (WRC), New York University Abu Dhabi, Abu Dhabi, UAE

J. Burt
Mubadala Arabian Center for Climate and Environmental Sciences (ACCESS) and Water Research Center (WRC), New York University Abu Dhabi, Abu Dhabi, UAE

B. Riegl
National Coral Reef Institute, Halmos College of Arts and Sciences, Nova Southeastern University, Dania Beach, FL, USA

C. D'Angelo · J. Wiedenmann
Coral Reef Laboratory, National Oceanography Centre Southampton, University of Southampton, Southampton, UK

5.1 Introduction

The Persian/Arabian Gulf (the PAG) is a peripheral, high latitude sea with marginal reef building (Riegl and Purkis 2012; Schoepf et al. 2023). It is situated at a similar latitudinal span as the northern Red Sea, the Ryukyu Islands, the northern Hawaiian Islands, and Florida/Bahamas. Among these peers, only Gulf coral reefs are truly marginal throughout, and the reef associated fauna is significantly less diverse and abundant.

The PAG's oceanographic characteristics, forced by its geographic and tectonic setting, are extreme for corals and other reef-building organisms and coral growth in this region persists despite some of the coldest winter temperatures (15 degrees: Burt and Paparella 2023; in extreme cases 10 °C: Shinn 1976), highest summer temperatures (maxima reach-

M. O. Soares et al. (eds.), *Extreme and Marginal Coral Reefs and Communities*, Coral Reefs of the World 24,
https://doi.org/10.1007/978-3-032-28035-0_5

ing 35.5 to 36 °C; Riegl et al. 2011; Burt and Paparella 2023), variable alkalinity and nutrient levels among different parts of the basin (Purser 1973; Riegl and Purkis 2012; Burt and Paparella 2023), and recurrent hypoxia on reefs in summer (de Verneil et al. 2021). A narrow entrance to the Gulf via the Straits of Hormuz (39 to 96 km wide) and the influence of the Arabian Sea upwelling zone driven by the Indian monsoon clearly separates this reef province from the greater Indo-Pacific reef belt (Paparella and Burt 2023). The PAG coral fauna declines from the Straits of Hormuz (~73 species) to <30 species in Kuwait, and the soft coral fauna attenuates in a similar pattern (Sheppard and Sheppard 1991; Coles 2003; Riegl et al. 2012; Samimi-Namin and van Ofwegen 2012). Reefal areas are intermittently situated along the shorelines of the Gulf (Purser 1973; Sheppard and Sheppard 1991; Sheppard et al. 2010; Riegl and Purkis 2012; Pohl et al. 2014). No unequivocal evidence (by cores, etc.) for true frame building throughout the Holocene exists for the area (Riegl 1999). Similarly, while Pleistocene deposits exist in Iran (Kish Island: Preusser et al. 2003; Mossadegh et al. 2012; Larak Island: Samimi-Namin and Riegl 2012), these do not represent true reef-building episodes. Therefore, it is certainly correct to consider the PAG reefs both "extreme" and "marginal" (Burt et al. 2020a, b; Schoepf et al. 2023).

5.2 What Makes Persian/Arabian Gulf Reefs "Marginal"?

Coral reefs, in the strict sense of the word, are defined as consisting of *in-situ* frameworks of reef builders in growth position (James 1983; Fluegel 2004). The term "coral framework" refers to the rigid, interlocking structure of skeletons made mostly by corals in the modern ocean, or by calcareous sponges or other taxa that have varied throughout Earth history, but are primary contributors to the construction of reefs. "Framestones" (sensu James 1983; Fagerstrom 1987) are limestone rocks composed largely of these original skeletal frameworks, showing minimal post-mortem transportation or alteration. They tend to form distinctive, massive, convex-up and convex-down, lenticular structures (referred to as bioherms) often surrounded by bedded peripheral sediments (James 1983). Other framework types that occur frequently throughout Earth history and also in today's ocean are biostromes, laterally continuous frameworks formed by reef-builders (like corals, calcareous sponges, calcareous algae, etc.) that usually occur at middle depths (not near the surface but above the mesophotic) (James 1983; Fagerstrom 1987). While previously mentioned framework structures are common throughout Earth history and in today's ocean, reef-forming organisms are in no way bound to these framework-forming locales or time-periods, where they occur in such density that they become rock-forming. Just as frequently, and in sum throughout Earth history maybe even more frequently, otherwise frame-building organisms can occur in sparse assemblages where they do not form framestones and, hence, no reef structures (Fagerstrom 1987; Riegl and Piller 2000). In the fossil record, such situations are evidenced by coral rudstones or floatstones (James 1983) and in today's ocean are often referred to as "coral communities" (the definition as used in the geological literature should not be confused with a somewhat different meaning of the term in the ecological literature). What is within this book chapter referred to as "marginal reefs" is encompassed within the continuum from biostrome to coral community but mostly excluding the typical bioherms. Thus, from a rock-forming perspective, the PAG reefs represent a situation in which frame-building organisms can be dominant but only rarely, if ever, occur in such a density that they are rock-forming as framestones. These are small and potentially ephemeral in structure. Such situations are referred to as "marginal" with regards to frame-building potential, which is caused by an "extreme" environmental situation (Schoepf et al. 2023). The PAG has been a prime example of such environments since the Pleistocene and therefore serves to illustrate the dynamics of reef-building and the potential for framework deposition in extreme environments in fossil and today's oceans. This is not only of ecological, but also of actuo-paleontological (using the present to interpret the past) interest.

Within the PAG, coral areas have experienced much change due to increased heat stress over the past decades (Burt et al. 2020a, b). Community structure was relatively uniform with corals occurring in a gradient of dense (>50% cover, which is now rarely seen) to very sparse (~1% cover) (Fig. 5.2). Only mild depth gradients existed in assemblage structure (dense coral usually occurring in a depth band of about 5–10 m) and horizontal gradients (for example along substratum quality) were often very similar and involved the same species. From northern Iran and Kuwait to Musandam, patches of very dense *Acropora* cover occurred prior to the repeated mass mortality events of the twenty-first century. These formerly dense assemblages have now been severely degraded and space cover overall has declined in all assemblages over the past decades from around 60% to below 10% space cover (unpubl. data).

Where dense cover of *Acropora* tables existed, they harbored an understorey of smaller massive colonies that were a mixed assemblage of merulinids, siderastreids and poritids (Riegl and Purkis 2012). Large massive colonies (30–50 cm diameter) were able to persist despite being surrounded by a mostly closed canopy of *Acropora* if they were in a size-class that made overtopping by *Acropora* impossible. Dense assemblages dominated by stands of the locally endemic *Porites harrisoni* occurred within the same habitat and often adjacent to massive *Porites* or tabular *Acropora* assemblages without any notable environmental gradient (i.e. either different salinities or depths; Sheppard and Sheppard 1991). It is believed that many dense *Porites* assemblages arose after previous mortality events eradicated *Acropora* since *P. harrisoni* and *P. nodifera* usually had lower mortality in bleaching events than *Acropora*. In western Abu Dhabi, eastern Qatar and Bahrain, *P. harrisoni* dominated in the hottest and most saline regions with dense coral growth (43–45 ppt, Sheppard and Sheppard 1991) but even these previously highly resilient assemblages have been significantly degraded with live cover of <10% even a decade ago (Burt et al. 2011, 2013, 2016).

Sparse coral assemblages still cover wide areas and are an important repository of coral biomass. The dominant corals are mostly *Porites* or *Dipsastraea* species, in the sandiest areas, isolated colonies of *Siderastrea* occur. The above-described community structure is unique to the PAG and already strongly modified in the adjacent Gulf of Oman/Arabian Sea which have much greater biodiversity and thus opportunities for ecological differentiation of assemblages. The relative paucity of species in the PAG, the absence of strong community gradients (largely driven by the absence of strong gradients in the substrate, given the Gulf's nature as a shallow ramp), and the general absence of well-developed and coherent reefal frameworks are all hallmarks of marginal reefs in an extreme environment. The following sections will explain why the PAG represents an extreme environment for corals and how this influences reefal processes. We will also describe what mechanisms allow these marginal "reefs" to nonetheless persist.

5.3 Geological History of the Persian/Arabian Gulf

While most other reef areas at comparable latitudes had much of the Holocene to initiate, develop and backstep and therewith keep pace with sea level (e.g. Florida; Banks et al. 2007; Toth et al. 2021), PAG reefs, except maybe on islands closer to the Gulf's axis, had under 6000 years to develop. During the Pleistocene, the Gulf did not even exist or was a much smaller body of water. Between 110,000 yr BP (before present) and 30,000 yr BP sea-level fluctuated from 30 and 60 m below present (inset Fig. 5.1). During the Last Glacial Maximum (LGM), sea-level dropped to 120–150 m below present, and the PAG dried up completely. PAG sea-level history is likely comparable to that of surrounding basins, like the Red Sea. The transgression after the LGM was driven by sea-level fluctuations in the Indian Ocean and allows understanding present-day reef morphology (Purser 1973; Purkis et al. 2010; Purkis and Riegl 2012). The Gulf began to fill prior to 12,000 yr BP as Indian Ocean waters flooded the basin via the Strait of Hormuz (Lambeck 1996; Teller et al. 2000). It then witnessed rapid rise in sea-level during the period 12,000 yr BP to 9000 yr BP from <90 m to >30 m

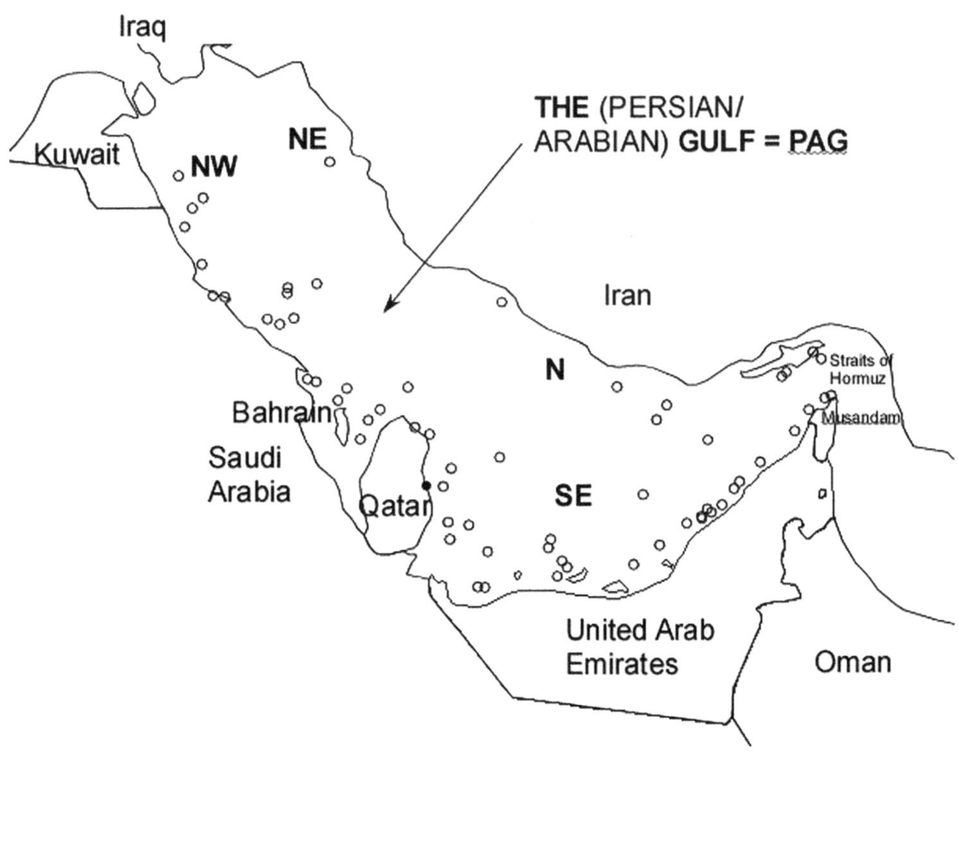

Fig. 5.1 The Persian/Arabian Gulf (The PAG) is an area with much coral growth, however few (if any) well-developed biogenic reefs. It qualifies best as an area of marginal reef growth in an extreme environment. Areas with significant coral occurrence reported in the literature are shown as circles. (From Riegl et al. 2018)

below the present level. Today's sea-level was only slowly reached between 3000 to 6000 yr BP. A potential overshoot of 1 m above present sea level may have occurred at ~6000 yr BP prior to a retreat to today's level (Chappell 1983). The shallow morphology without a strong depth-cline caused rapid landward displacement of shorelines and therewith rapid and strong lateral displacement of the ecological zone where reefs could have otherwise developed. Shoreline position may have changed by up to 5 km per century and maybe even faster as the gently sloping central shelf was inundated (Purkis and Riegl 2012). Here, over a distance of ~500 km exists only an 18 m rise in elevation (Teller et al. 2000). Thus, even if conditions had otherwise been perfect, there was hardly any time for reef initiation, let alone frame-building in any potential fringing reef settings.

Smith et al. (2022) used genomics to infer the colonization patterns in *Platygyra*. From the model, the presence of an ancestral PAG population was inferred that likely had established itself along the Iranian coastline because of the initial PAG flooding to the south of Iran between 12,000 and 10,000 yr BP. Then, populations split into those establishing reefs to the north, and in the southern/central PAG. After this split, populations were established at locations in the southern/central PAG consistent with the expansion of available habitats ~8000 yr BP. In the southern PAG, new subpopulations were formed in a step wise manner away from the pioneer population, after the establishment of the PAG's modern coastlines ~6000 yr BP.

Given the very dynamic nature of sea level stands in the PAG basin during the Holocene, very little time was available for significant framework accretion at most localities, but most particularly so on the wide homoclinal ramp constituting its southern half. Thus, geology alone, by forcing sea level dynamics and thus acting in concert with it, has been a powerful factor in shaping the marginal nature of the PAG reefs (Purkis and Riegl 2012). Added to this,oceanographic and climatic factors were also forced by the regions geological history that defines reef development (Burt and Paparella 2023). In concert, the PAG's geological and climatological history has not only created an extreme environment for reef-building organisms, it has also severely restricted the time available for adaptation to take place (Smith et al. 2022). Sea level can only be considered stable for ~4000 yr which coincides with hyperarid and hot climate conditions settling in over the Gulf region and persisting since 2400 yr (Purkis et al. 2010). Thus, the PAG corals have been subjected to extreme conditions almost since their establishment in the region and are now suffering even more severe conditions in an anthropogenically heated climate.

5.4 Oceanography of the Persian/Arabian Gulf Makes the Region Extreme for Reefs

The PAG is a semi-enclosed sea, situated in a subtropical arid region with distinct hydrographic characteristics. High temperature variability, extreme salinity, and limited exchange with the open ocean contribute to a challenging environment for marine life. The PAG experiences significant seasonal variations, with summer water temperatures exceeding 35 °C and winter temperatures dropping to around 15 °C (Sheppard et al. 2010) (Fig. 5.3). Thus, the upper and lower temperatures of the waterbody push hard against the physiological limits of reef-building corals. The PAG salinity levels are among the highest globally, reaching up to 44 PSU close to the United Arab Emirates (UAE) (Burt and Paparella 2023), and regularly exceed 50 PSU in coastal lagoons and embayments (Evans et al. 1973).

Surface salinity is higher in early winter (February) than summer (August) (Fig. 5.2) because winds are slightly stronger. The cold, dry-air emanating from the northwest in winter leads to excessive evaporation which results in the formation of the saltier and denser Gulf water (Thoppil and Hogan 2010) (Fig. 5.4).

The PAG's climatological conditions and seasonal circulation patterns have been extensively studied (Reynolds 1993; Kämpf and Sadrinasab 2006; Thoppil and Hogan 2010; Alosairi et al. 2011; Pous et al. 2015; Al Azhar et al. 2016; Campos et al. 2020a). The primary drivers of its oceanography are evaporation, freshwater input from rivers, and the seasonal monsoon winds. This semi-enclosed marginal sea, connected to the Indian Ocean through the Strait of Hormuz and the Gulf of Oman, experiences a reverse estuarine circulation due to greater evaporation than freshwater inflow (Swift and Bower 2003; Alosairi et al. 2011; Al Azhar et al. 2016). The resulting density currents import Indian Ocean Surface Waters (IOSW) in the top layers while exporting higher salinity waters in the bottom layers through the Strait of Hormuz, crucial for the dispersion and removal of salt and suspended material (Thoppil and Hogan 2009).

The mean circulation in the Gulf follows the general representation of a barotropic cyclonic gyre (counter-clockwise), with less saline Indian Ocean Surface Waters (IOSW) entering through the Straits of Hormuz (Campos et al. 2020b). Two branches form: one flows along the Iranian coast, creating a cyclonic circulation to the west of Qatar, and a returning flow represented by a coastal current along the United Arab Emirates coastlines, while the other turns southward, driven mainly by Ekman drift due to the predominant northwesterly winds (Reynolds 1993; Pous et al. 2015). The circulation pattern exhibits marked seasonal variation, with the northern

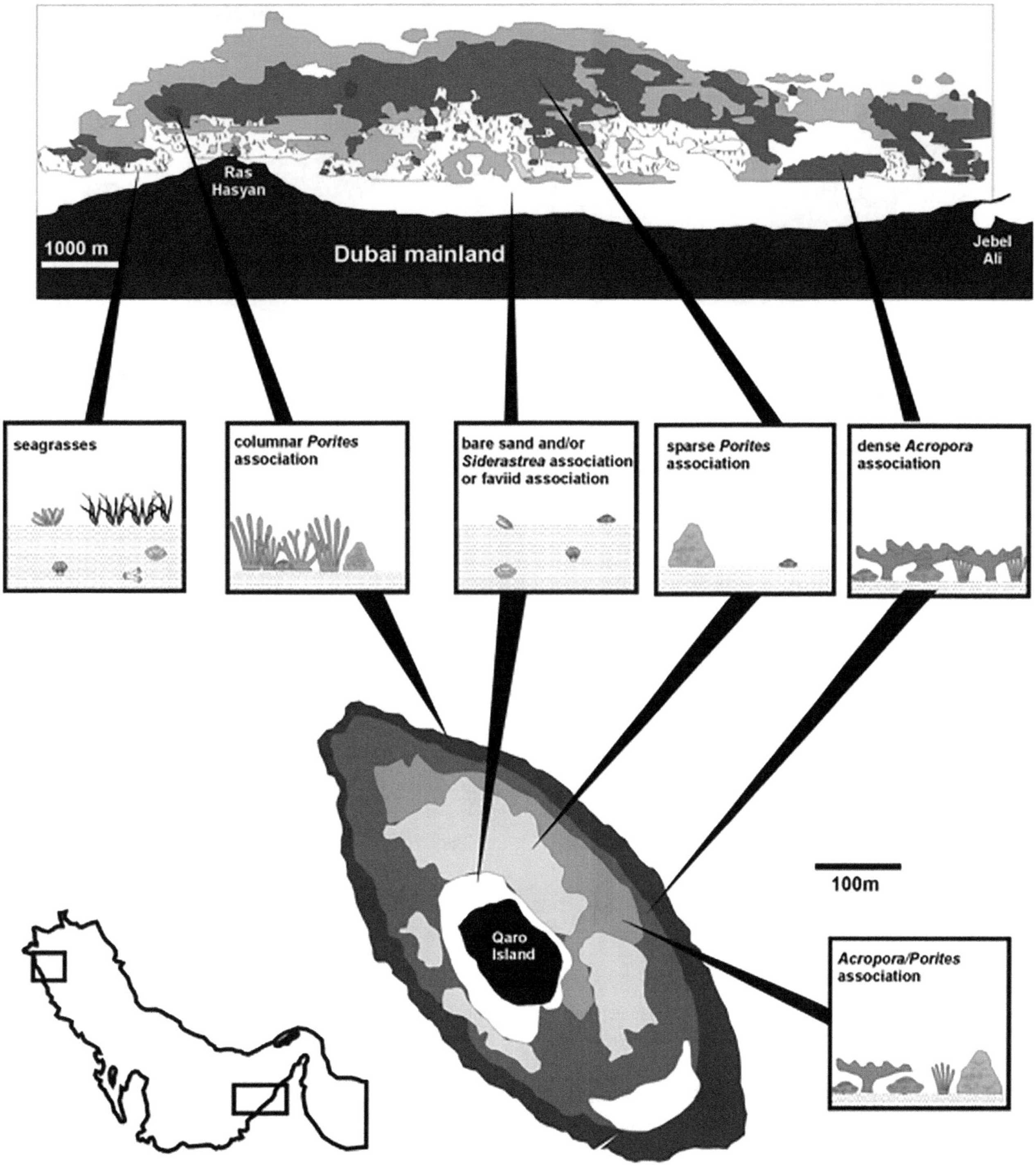

Fig. 5.2 Community structure on the PAG marginal reefs in the 1990s was roughly comparable across the basin (examples shown from Jebel Ali, UAE, and Qaro Island, Kuwait). A limited number of species showed similar ecological and successional dynamics on most dense assemblages that occurred within a limited and uniform depth window. In deeper water, only sparse assemblages were found (from Riegl and Purkis 2012)

branch reaching its northernmost extent in the summer, and the southern branch being strongest during the winter (Johns et al. 2003; Yao and Johns 2010). The IOSW undergoes significant salinity increase by evaporation and mixing with ambient saltier water as it spreads into the Gulf (Campos et al. 2022). These circulation patterns persist throughout the year, influenced by seasonal winds and thermocline variations (Thoppil and Hogan 2010; Pous et al. 2015).

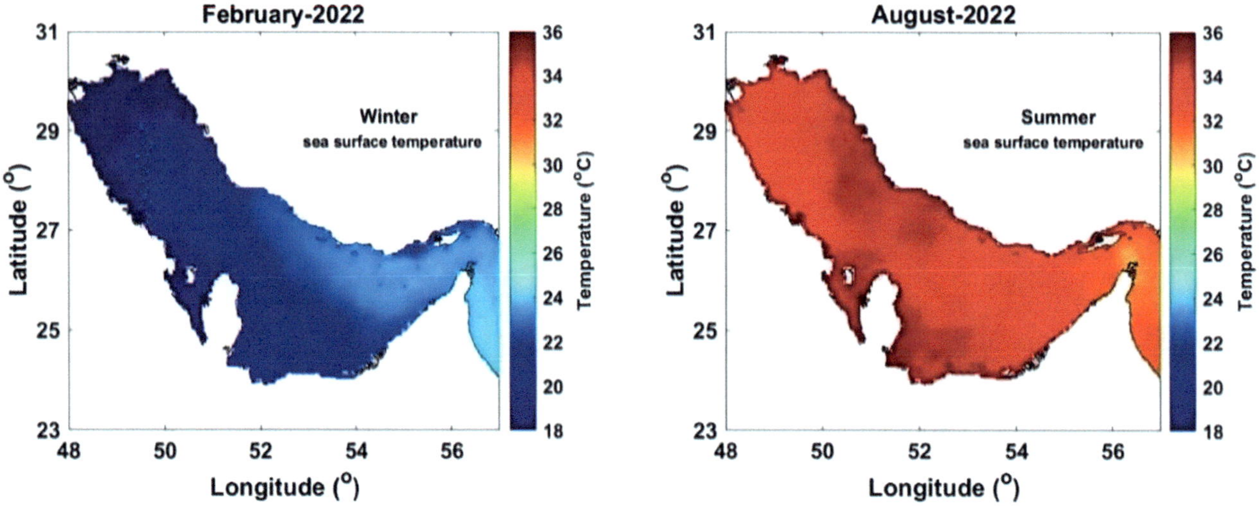

Fig. 5.3 Monthly averaged sea surface temperature in February (left) and August (right), 2021. Level 3 Standard Mapped Image (SMI) products, for sea surface temperature (SST) at 4 km resolution from the NASA Ocean Biology Distributed Active Archive Center (OB.DAAC). NASA Ocean Biology Processing Group. (2024). Ancillary Data Sources. Retrieved from https://oceandata.sci.gsfc.nasa.gov/odps/. Accessed on 18/02/2024

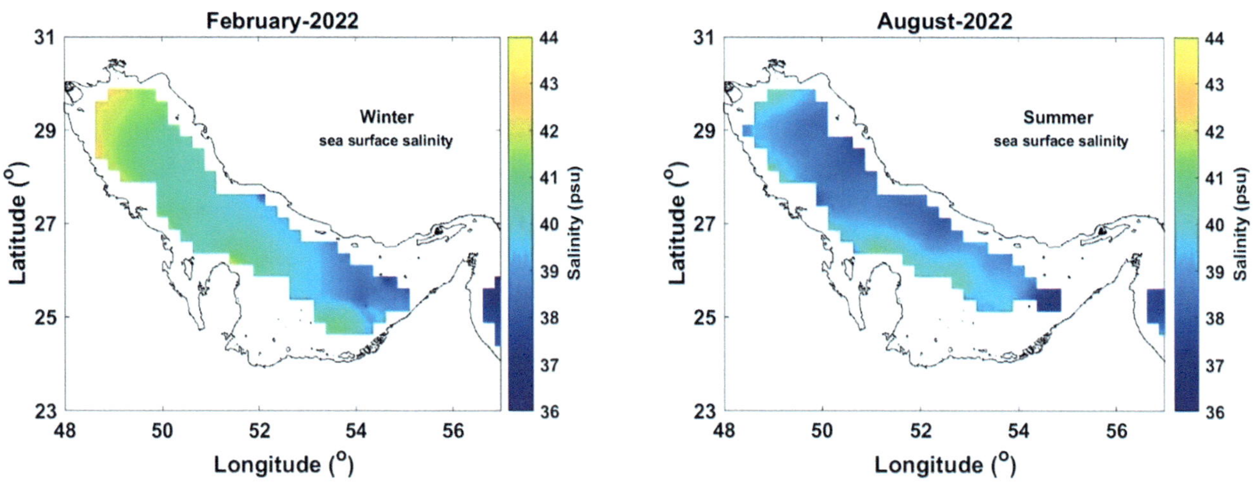

Fig. 5.4 Monthly averaged sea surface salinity (SSS) in February (left) and August (right), 2021. PI-produced JPL SMAP-SSS V5.0 CAP, level 3, monthly mapped sea surface salinity (SSS) product from the NASA Soil Moisture Active Passive (SMAP) observatory (Nasa/Jpl 2020)

The shallow nature of the Gulf amplifies the impact of these factors, leading to variations in temperature, salinity, and nutrient concentrations. The monsoon-driven upwelling during the southwest monsoon enhances nutrient availability, influencing primary productivity and subsequently affecting the entire marine ecosystem (Burt and Paparella 2023). Extreme conditions, including periodic temperature spikes and limited nutrient availability, pose significant threats to coral health (Riegl and Purkis 2012; Rosset et al. 2017). The elevated temperatures contribute to coral bleaching events (Burt et al. 2019a, b). The high summer temperatures have also been implicated in enhanced biological oxygen demand which can lead to recurrent hours-long periods of hypoxia on reefs at night in the summer, as well as a large and growing oxygen minimum zone in the central Gulf that is considered a threat to fisheries (de Verneil et al. 2021; Lachkar et al. 2022). Tidal forcing minimally impacts circulation, affecting smaller spatial and temporal scales (Pous et al. 2004; Pous et al. 2015). Wind forcing, particularly the Shamal phenomenon (a strong, low-level wind moving from the Anatolian plateau into the Gulf basin and propagating roughly along its axis), plays a significant role. The northwesterly Shamal winds, occurring mostly in winter but occasionally also in summer, drive mid-Gulf eddies and transport water to the southern basin (Reynolds 1993; Thoppil and Hogan 2009; Campos et al. 2022). These winds have eco-

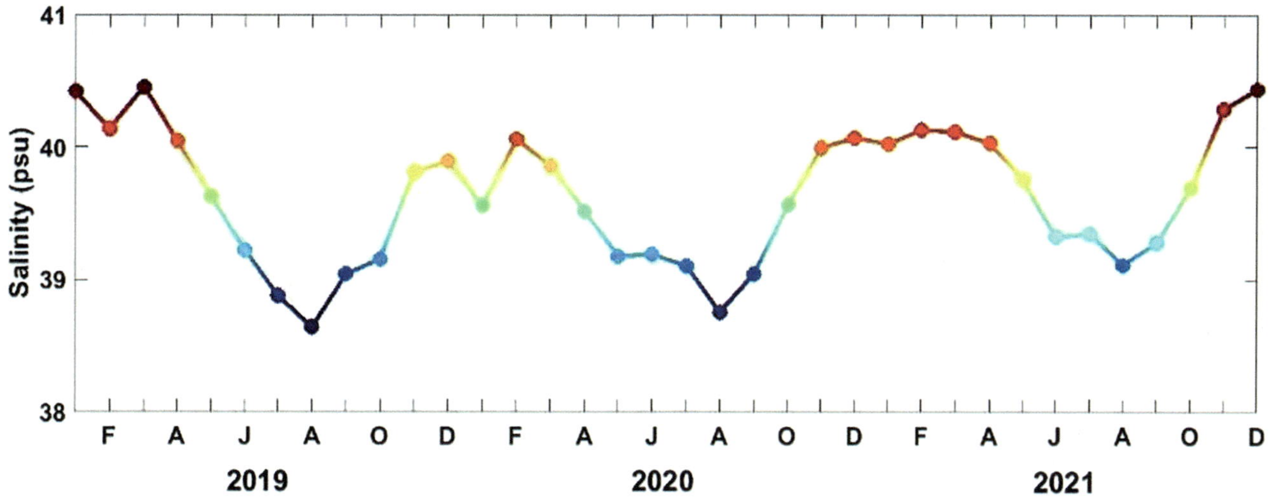

Fig. 5.5 Time series of basin-averaged (west of 56°E) monthly mean of SSS (sea surface salinity in psu) for 3 years (2019–21) of the PI-produced JPL SMAP-SSS V5.0 CAP, level 3, monthly mapped sea surface salinity (SSS) product from the NASA Soil Moisture Active Passive (SMAP) observatory (Nasa/Jpl 2020). The colors represent levels of salinity, red being high or extreme salinity levels and dark blue means lower and less extreme salinity levels

logical importance, facilitating larval transport and providing evaporative cooling during extremely hot summer sea temperatures (Cavalcante et al. 2016; Riegl et al. 2018; Paparella et al. 2019; Cavalcante et al. 2020).

Small-scale cyclonic eddies can form and travel at ~0.3–0.4 m/s as a result of summer Shamal (northerly winds) events (Cavalcante et al. 2016; Cavalcante et al. 2020). Such eddies may play a crucial role in creating internal salinity and temperature instability along the UAE coast, shaping the health and dynamics of coral reefs (Sheppard et al. 2010), and may be important for the long-distance transport of coral larvae from Iran to various parts of the UAE (Burt and Paparella 2023). Wind-driven circulation shifts in temperature can have dramatic implications for coral survival, possibly inducing hot- or cold-water coral bleaching or reducing physiological stress and limiting mortality if temperatures become more benign during seasonal temperature extremes (Riegl et al. 2011; Paparella et al. 2019; Burt and Paparella 2023). Such small-scale variability of conditions within this overall, for corals, extreme environment is important. The distribution of ocean surface heat during bleaching events is not uniform (Fig. 5.6 illustrating those of 2010/11/12) making the probability of any part of the coral metapopulation to be exposed to lethal temperatures variable (Riegl et al. 2018). Geographical variability in the bleaching response during stressful years is indeed recorded across the PAG (Riegl et al. 2012; Shuail et al. 2016). The Iranian, upstream part of the PAG coral metapopulation usually experiences more benign conditions. However, virtually the entire trajectory that coral larvae must take to reach the populations on the shores of the southern Gulf experiences heat. Therefore, larval competence periods will be shortened in hot years, no matter what the exact heat distribution looks like (Figueiredo et al. 2014). Such dynamics has important implications for connectivity (Riegl et al. 2018; Cavalcante et al. 2020).

In conclusion, the PAG oceanography significantly influences the health and dynamics of coral reefs in the region. The extreme environmental conditions, coupled with seasonal variations and nutrient limitations, create a unique set of challenges for coral ecosystems.

5.5 Extreme Environment and Absence of Reef Framework Building Suggests Repetitive Mass Mortality of Reef Builders

Many PAG coral assemblages formed part of a successional cycle, primarily influenced by temperature-related coral death and probably also coral diseases (Riegl 1999, 2002; Purkis and Riegl 2005; Riegl and Purkis 2009, Aeby et al. 2024). This dynamic has recently become obscured due to the widespread coral death over the last decades. Frequent recurrence of disturbances, at roughly decadal scale, kept reefs at least geologically-speaking in an early successional, non-reefal, i.e. "marginal" state. Many coral assemblages showed a halo-like structure with patches of dense dead coral surrounded by sparse live coral and then by hardground with only occasional corals (Fig. 5.7). After corals are killed, coral skeletons were soon settled either by *Chama reflexa* bivalves or coralline red algae *Lithophyton kotschyanum*

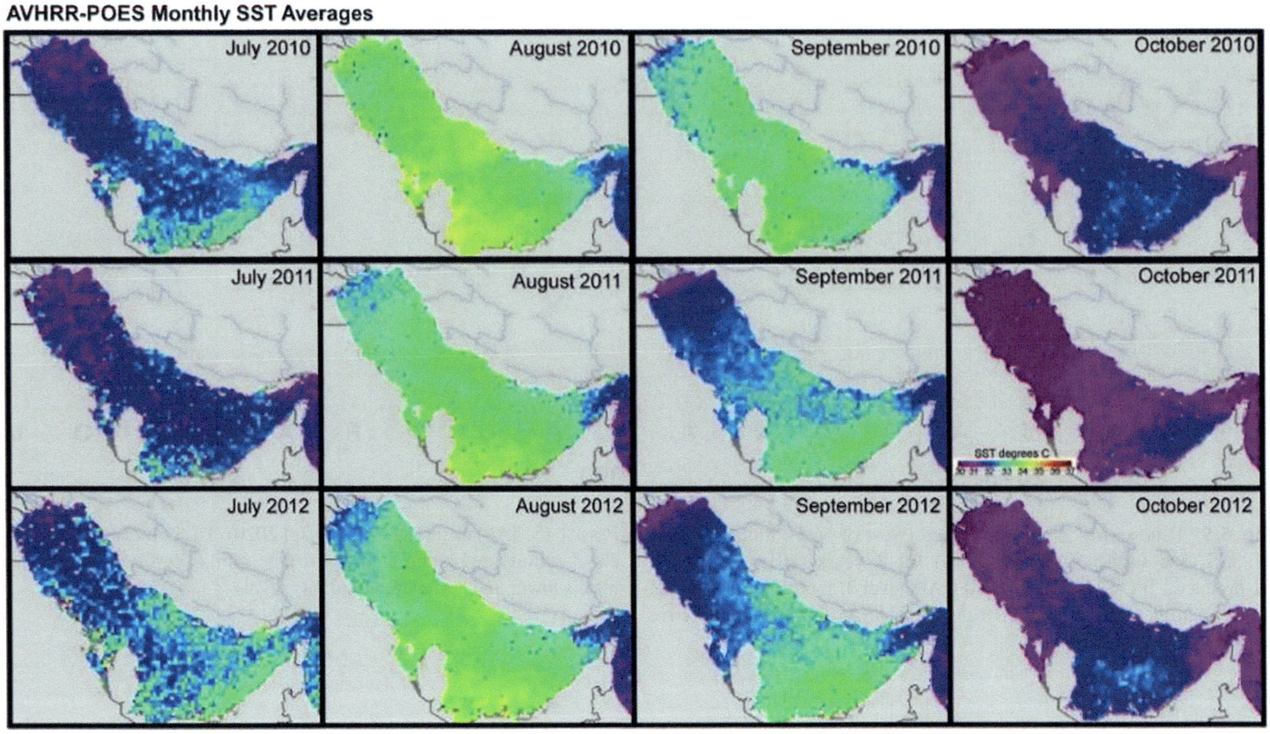

Fig. 5.6 Spatial distribution of heat during the 2010/11/12 bleaching events. Bleaching occurs in late August in the areas with yellow/orange pixels. It abates quickly over September, when disease outbreaks further reduce the weakened corals. Data downloaded from http://coastwatch.pfeg.noaa.gov/erddap. (From Riegl et al. 2018).

(Fig. 5.7). A variety of green and brown algae also settled on the skeletons. While in the late 1990s and early 2000s these were still macro-algae, the latter have disappeared from reefs and now most dead substrata are covered by algal turfs (Riegl Jr et al. 2024). Increasing encrustation and also boring by lithophagid bivalves and clionid sponges weaken the dead coral skeletons, (Al-Mansoori et al. 2019) eventually causing their disintegration after about 6 years (Fig. 5.7). Rubble then disperses quickly on the flat seafloor (Rasser and Riegl 2002). Where dense rubble beds are encountered, they tend to cement rapidly into hardground (within only a few decades; Shinn 2006). Observations over 25 years and evidence in the literature (Riegl 1999, 2002; Purkis and Riegl 2005, Riegl and Purkis 2009) show that these successions were, prior to the upheaval of the last decade, recurrent roughly every 15–20 years and cyclical.

5.6 Symbionts as Key to Coral Survival in the Extreme PAG Environment

The coral communities of the PAG consist mostly of species that live in symbiotic association with photosynthetic dinoflagellates (Sheppard and Sheppard 1991; Riegl and Purkis 2012). The symbionts provide the coral hosts with carbon-rich photosynthates for energy production, facilitating for instance calcification required for the formation of the coral skeleton (Davy et al. 2012). The symbionts also take up dissolved inorganic N- and P-containing nutrients (ammonium, nitrate, phosphate) from the water that are not accessible to the coral host and turn them into bioavailable N and P rich organic compounds (Davy et al. 2012; Ferrier-Pagès et al. 2016; Rädecker et al. 2015). Through the regular digestion of parts of the symbiont population, the host can acquire nitrogen and phosphorus (Wiedenmann et al. 2023) essential for its tissue growth and complement its nutrient acquisition through the uptake of dissolved and particulate organic N- and P-rich compounds including particulate food (Rosset et al. 2015). Thereby, the symbionts enable the corals to exploit two independent N- and P pools, offering the symbiotic association a competitive advantage in oligotrophic waters over non-symbiotic organisms which rely on the N and P uptake either in organic (animals) or inorganic (photosynthetic organisms) forms (Wiedenmann et al. 2023). The symbionts further underpin this selective advantage by recycling N- and P-rich endproducts of the host metabolism and retaining them in the symbiotic association (Ferrier-Pagès et al. 2016; Rädecker et al. 2015).

Various forms of environmental stress including temperature, light and nutrient stress can result in a breakdown of the symbiotic association, the loss of the vital symbionts from the host tissue and a bleached appearance of the corals

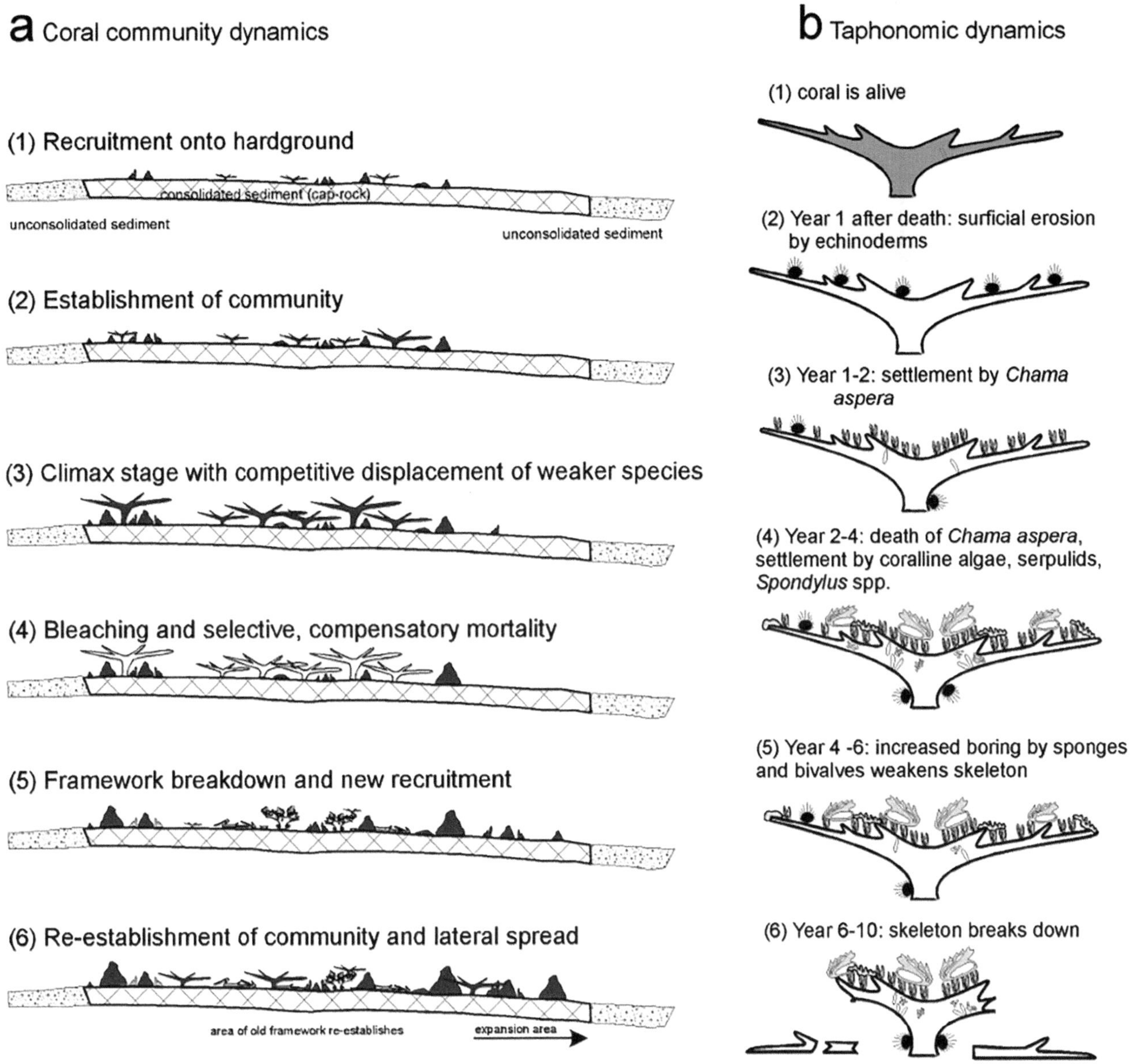

Fig. 5.7 Ecological (**a**) and taphonomic (**b**) dynamics within the coral communities that make up the marginal reefs of the PAG. (Modified from Riegl and Purkis 2009)

(Douglas 2003; Davy et al. 2012; Wiedenmann et al. 2013, 2023; Rosset et al. 2017). The regional bleaching thresholds of symbiotic PAG corals can range up to ~35 ° C (Shuail et al. 2016), facilitating their survival at temperatures that would kill corals anywhere else on the planet (Riegl et al. 2011). Still, over recent decades bleaching events have resulted in severe coral mortality and reduction of coral cover also in the PAG (Riegl 1999; Riegl et al. 2018; Burt et al. 2019a, b). Accordingly, the symbionts must be considered as a major driver of the adaptation of the PAG corals to their extreme environment and can provide important insights how coral elsewhere may respond and adapt to rising ocean temperatures (Burt et al. 2020a, b) (Fig. 5.8).

Molecular biological techniques advanced the understanding of the taxonomy of coral symbionts and led to the realization that corals are not just associated with "clades" of a single, dominant species, but that these clades usually represent genera within the taxon Symbiodiniaceae, many of which are comprised of a multitude of species (LaJeunesse et al. 2018). Accordingly, functional traits such as thermal tolerance previously attributed to certain "clades" can now be associated with distinct taxonomic units (Stat and Gates 2011; LaJeunesse et al. 2018). Research on symbionts of the PAG corals progressed over the same period as the development of the molecular systematics of Symbiodiniaceae. Accordingly, the corresponding scientific literature reflects

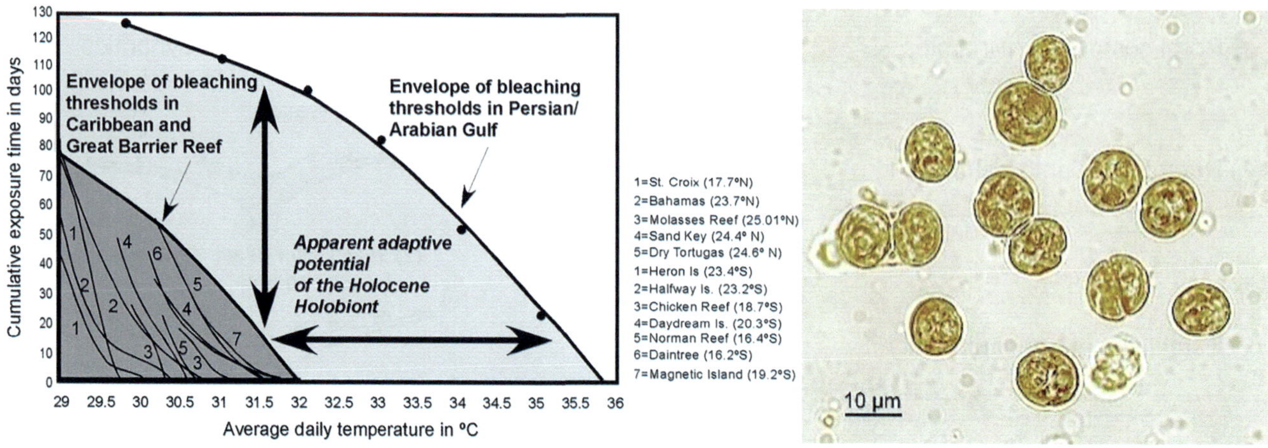

Fig. 5.8 The unusually high bleaching thresholds on the PAG reefs. (From Coles and Riegl 2013) are associated with the presence of an unusually well adapted dinoflagellate symbiont *Cladocopium thermophilum*

the transition from the clade- to species-specific taxonomy. An early study identified *Symbiodinium* Clade D in 62% of coral samples in the western PAG off the Saudi coast (Jana and Karan Islands, Ras Abu Ali) (Baker et al. 2004). Representatives of this clade were also found in samples from the eastern PAG along the Iranian coast (Kish, Hengam and Larak Islands) (Ghavam Mostafavi et al. 2007; Shahhosseiny et al. 2011). Since clade D symbionts are associated with increased coral thermotolerance at the global scale (Baker et al. 2004), the stress tolerance of the PAG corals from the Gulf has been attributed to the association with this clade, which has subsequently been described as genus *Durusdinium* (LaJeunesse et al. 2018). However, the presence of a clade C symbiont in *Acropora clathrata* from the hottest, south-eastern part of the PAG off the coast of Abu Dhabi (Hartle-Mougiou et al. 2012) initiated further surveys. A subsequent study of this region demonstrated that clade C3 symbionts are year-round the prevalent partner in some of the most abundant corals on Abu Dhabi reefs (*Porites lobata, P. lutea, P. harrisoni, Cyphastrea microphthalma, Favia pallida, Platygyra daedalea* and *A. clathrata*) (Hume et al. 2013; Hume et al. 2015). Clade D (*Durusdinium* sp.) was essentially absent from this region, being detected only in one single sample collected during the winter months (Hume et al. 2015). Since clade C3 symbionts were thought to be a cosmopolitan, thermally sensitive and host generalist symbiont (LaJeunesse et al. 2003), their prevalence in the hottest reef waters came as a surprise. Yet, through a detailed analysis of a suite of molecular markers that provide resolution beyond the clade level, a distinct taxonomic status of the "Gulf C3 clade" could be demonstrated, leading to the description of a new symbiont species, *Symbiodinium thermophilum,* with a holotype derived from a *P. lobata* specimen from Saadiyat reef in Abu Dhabi, UAE (Hume et al. 2015. 2018a). The later systematic revision of Symbiodiniaceae led to the adoption of clade C symbionts in the genus *Cladocopium* (LaJeunesse et al. 2018). Accordingly, it is now referred to as *Cladocopium thermophilum* (Hume et al. 2015; LaJeunesse et al. 2018). Using a species-specific mutation ("C3-Gulf" / "S. thermo" indel) within the clade C3-defining internal transcribed spacer 2 (ITS2) sequence (Hume et al. 2015) along with other markers, the presence of *Cladocopium (Symbiodinium) thermophilum* has been confirmed for the wider PAG, the Gulf of Oman and the Red Sea (D'Angelo et al. 2015; Hume et al. 2016; Varasteh et al. 2018; Oladi et al. 2019, 2021) (Fig. 5.9). Recently, next-generation sequencing-based phylogenetic methods (Symportal, Hume et al. 2019) have been also used to identify *C. thermophilum* (Howells et al. 2020). Across the middle eastern region, *C. thermophilum* associations were detected not only in 21 species from 13 families of scleractinian corals, but also in symbiotic soft corals, sea anemones and zoanthids (Hume et al. 2016). Other symbiont genera ("clades") detected in the south-eastern PAG in mixed populations with *C. thermophilum* include for example *Cladocopium* (C1, C15), *Symbiodinium* (A1), *Durusdinum* (clade D) and *Geraklaium* (clade G) (Hume et al. 2015; LaJeunesse et al. 2018). *Symbiodinium* sp. (Clade A) was found also in samples from the western PAG (Baker et al. 2004; Hume et al. 2016).

The distribution of symbiont types around the Middle Eastern region, specifically of the heat tolerant species *C. thermophilum* and *Durusdinium* sp., is now well studied (Fig. 5.10). While differences in sampling scale across different studies put some limitations to the comparability between locations, there are clear trends in symbiont community composition that indicate that the distribution of *C. thermophilum* and *Durusdinium* sp. in the PAG is strongly correlated with key environmental drivers. Specifically, there is a strong correlation between salinity and temperature-related parameters and the prevalence of these species (Fig. 5.10). Across the PAG and the adjacent Gulf

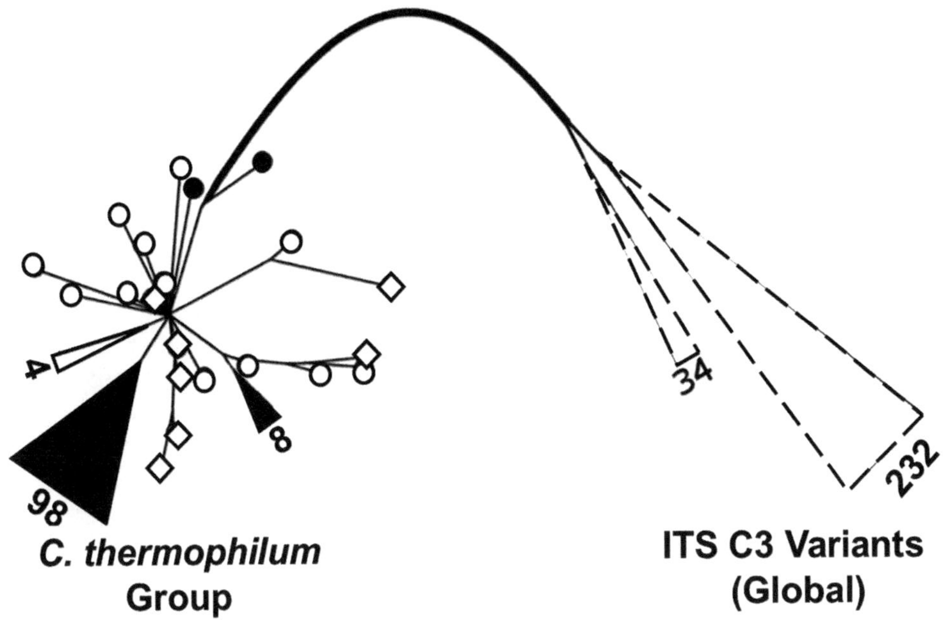

Fig. 5.9 PsbAncr (noncoding region of the chloroplastic psbA gene) estimated phylogeny indicates taxonomic separation of *Cladocopium thermophilum* from the Middle Eastern region (left sequence cluster) and *Cladocopium* sp. elsewhere (dashed wedges/right sequence cluster). The genetic uniformity of the PAG *C. thermophilum* (black spheres and black wedges) indicates selection of a low number of variants from a diverse group present at low abundance in the Middle Eastern region (white spheres & white wedge: Gulf of Oman; diamonds: Red Sea). Wedges represent collapsed groups. The number of contained sequences is given next to the wedges. Re-drawn from Hume et al. 2016

of Oman, the dominance of *C. thermophilum* increases with increasing salinity and temperature. Contrastingly, the contribution of *Durusdinium* to the symbiont communities decreases along the same salinity and temperature gradients, indicating that *C. thermophilum* can tolerate harsher environmental conditions compared to *Durusdinium* sp. This is further underpinned by the correlation of the distribution of the two symbiont taxa with the annual minimal temperature and the seasonal temperature variability (Fig. 5.11).

D'Angelo et al. (2015) first suggested that the Strait of Hormuz represents a biogeographical boundary for symbiont distribution based on the gradual decrease of the dominance of *C. thermophilum* along the coast of the western part of the Strait of Hormuz and the Gulf of Oman. Subsequent fine-scale genetic population analysis could clearly distinguish between a *C. thermophilum* population inside the PAG and a related, yet distinct, population outside (Hume et al. 2016, 2018b). The biogeographical boundary could be located within 20 km distance between sampling sites on the western and eastern side of the Musandam peninsula in the Strait of Hormuz (Hume et al. 2018b).

Using PsbA^{ncr} sequence-based phylogeny as indicator of the genetic diversity of symbiont populations, *C. thermophilum* inside the PAG appears to be genetically very homogenous (Hume et al. 2016) (Fig. 5.10). In the hottest and most saline waters of the southern PAG, representatives of this group can be found with high frequency in a broad range of coral species (Hume et al. 2015). Contrastingly, outside the PAG, genetic diversity among *C. thermophilum* populations is high, but their representatives are found only in low abundance in a limited number of coral species (Hume et al. 2016).

5.7 Physiological Mechanisms of Stress Tolerance and Local Adaptation and Associated Trade-Offs

The distribution of the heat tolerant symbiont *C. thermophilum* and *Durusdinium* sp. in the PAG suggests their increased temperature stress tolerance is influenced by adaptations to other local environmental factors, prominently, to elevated salinity levels. Indeed, an experimental study found that *Porites lobata* corals hosting *C. thermophilum* were not affected by experimental heat stress when they were kept at the elevated salinity (42 PSU) characteristic for the waters of the Southern PAG from which they originate (D'Angelo et al. 2015). In contrast, they lost their superior heat stress tolerance when they were cultured in water with ambient oceanic salinity (36.5 PSU) (D'Angelo et al. 2015). Mass spectrometric analyses revealed that *C. thermophilum* from *P. lobata* kept at the PAG salinity levels contained elevated levels of the osmolyte 2-*O*-glycerol-α-d-galactopyranoside (floridoside), presumably to mitigate the osmotic stress resulting from the high salinity (Ochsenkühn et al. 2017).

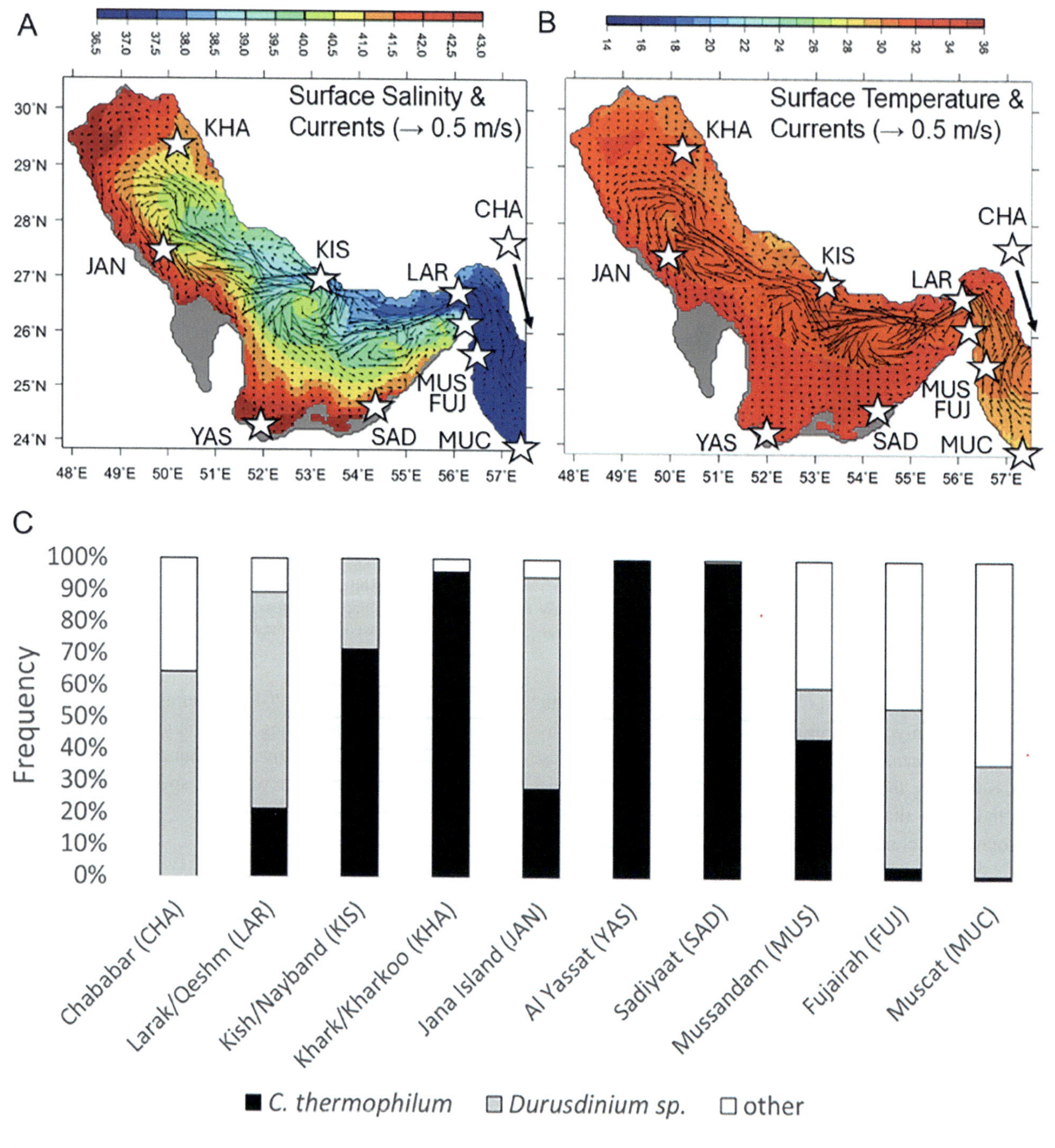

Fig. 5.10 Environmental parameters and the distribution of symbiont species in the PAG and the Gulf of Oman. Maps show model outputs for major currents and (**a**) salinity and (**b**) temperature during summer in the PAG and Gulf of Oman. Colour scales indicate salinity and temperature (°C), respectively (Modified with permission from Yao and Johns 2010). Regions for which taxonomic information for symbionts are available are highlighted by asterisks. The names of regions are given as 3-letter abbreviations defined in the labels of the horizontal axis in (**c**). Frequency of *C. thermophilum*, *Durusdinium* sp. and other symbiont species in host corals in regions with deviating salinity and temperature environments (**c**). Frequency estimates of symbiont species were compiled for CHA=Chahbahar (Oladi et al. 2019, 2021); LAR=Larak (Varasteh et al. 2018; Oladi et al. 2019, 2021); KIS=Kish (Varasteh et al. 2018; Oladi et al. 2019); KHA=Kharg and Khargoo (Varasteh et al. 2018; Oladi et al. 2019, 2021); JAN=Jana (Baker et al. 2004; Hume et al. 2016); YAS=Yas (Hume et al. 2016); SAD=Saadiyat (Hume et al. 2013, 2015, 2016); MU=MusandamS (D'Angelo et al. 2015); FUJ=Fujairah (D'Angelo et al. 2015; Hume et al. 2016) and MUC=Muscat (D'Angelo et al. 2015; Hume et al. 2016). Limitations of the deduced set of estimates arising from differences in analytical methods and sampling depth between the original studies are acknowledged

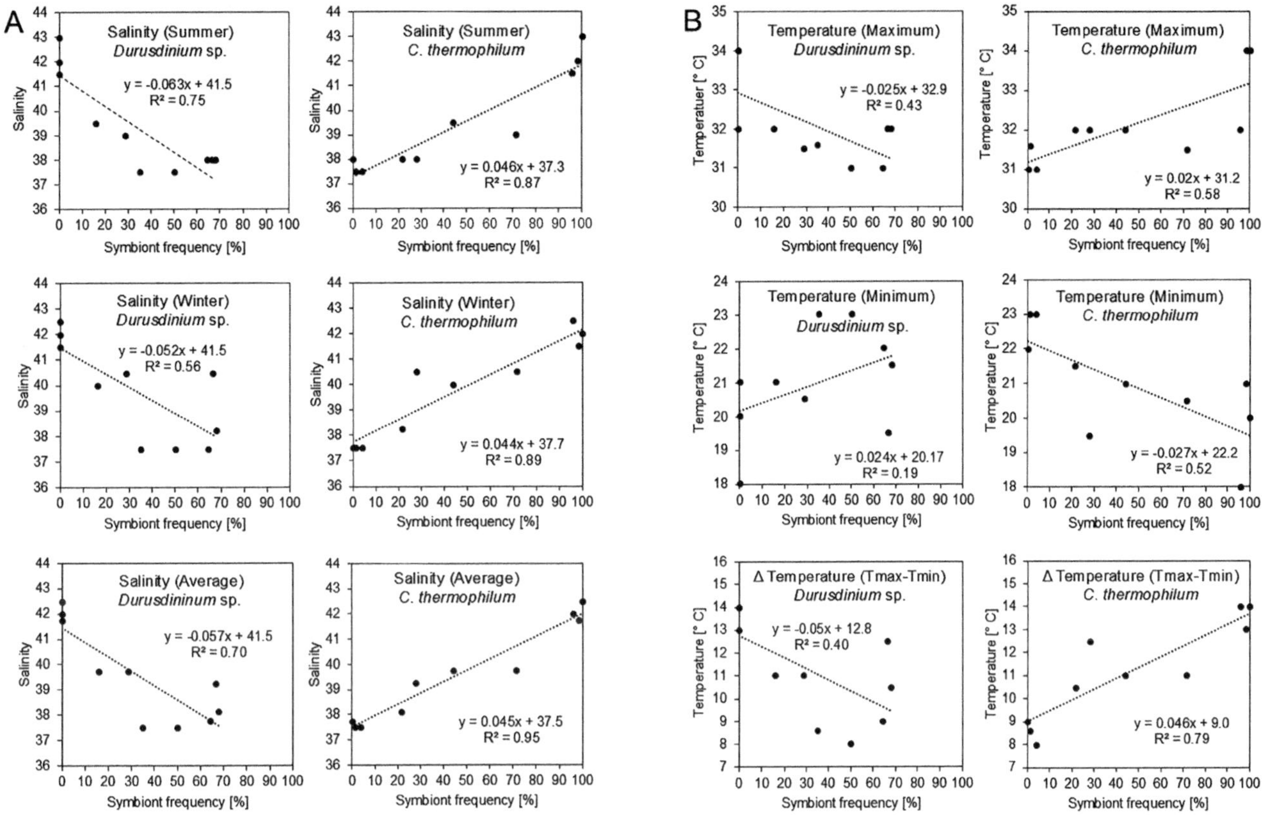

Fig. 5.11 Prevalence of *Durusdinium* sp. and *C. thermophilum* among the symbiont communities from regions of the PAG and the adjacent Gulf of Oman with deviating temperature and salinity conditions. Scatterplots show the relationship between salinity (**a**) and temperature (**b**) and the symbiont type frequency in coral hosts from the different sampling regions. Symbiont frequency estimate data were taken from Fig. 5.10c. Salinity and temperature data were compiled from the models of Yao and Johns 2010, and from Varasteh et al. 2018 and Oladi et al. 2019. Salinity data for Jana Island was deduced from Burt and Paparella 2023. Muscat minimal and maximal temperature values are from https://www.seatemperature.org/middle-east/oman/muscat.htm

Since floridoside has antioxidant properties, it may function not only as an organic osmolyte, facilitating survival at elevated salinities, but also to scavenge reactive oxygen species, thereby mitigating the impact of temperature stress for the symbiont and, subsequently, its host. Such a mechanistic coupling between high salinity and high temperature adaptations could help to explain why *C. thermophilum* becomes increasingly dominant with increasing levels of salinity and temperature stress across the PAG. Ongoing research will likely reveal further mechanisms involved in the stress tolerance of this symbiont species.

Durusdinium sp. show increased temperature stress tolerance compared to other symbiont taxa from reef regions across the globe (Jones and Berkelmans 2010; Stat and Gates 2011; Pettay et al. 2015). However, this symbiont taxon tends to become less abundant in the PAG corals along with increasing salinity levels and temperature extremes (Fig. 5.5). Therefore, the mechanisms involved in the adaptation to environmental conditions are potentially different in this symbiont genus compared to *C. thermophilum*. The results of comparative analyses of *Durusdinium trenchii* and the thermosensitive taxon *Cladocopium* sp. (C3) in corals from the South Pacific indicate that specific types of lipids could mediate the thermotolerance of *D. trenchii* by stabilising thylakoid membranes and photosystem II interactions and by providing chaperone-like functions (Rosset et al. 2019) (Fig. 5.12).

The partnership with thermotolerant *Durusdinium* symbionts can be associated with trade-offs such as reduced calcification or reduced growth (Pettay et al. 2015; Jones and Berkelmans 2010). While the differences in calcification rates can persist over a range of temperatures, the growth-related trade-offs may become less pronounced at elevated temperatures (Pettay et al. 2015; Cunning et al. 2015). The corals *Platygyra daedalea* and *Cyphastrea microphthalma* from the Southern PAG are typically associated with *C. thermophilum* (Hume et al. 2015; Howells et al. 2020). Their linear extension and calcification rates measured in field experiments where broadly comparable to those of the same species from the Gulf of Oman where they are usually found

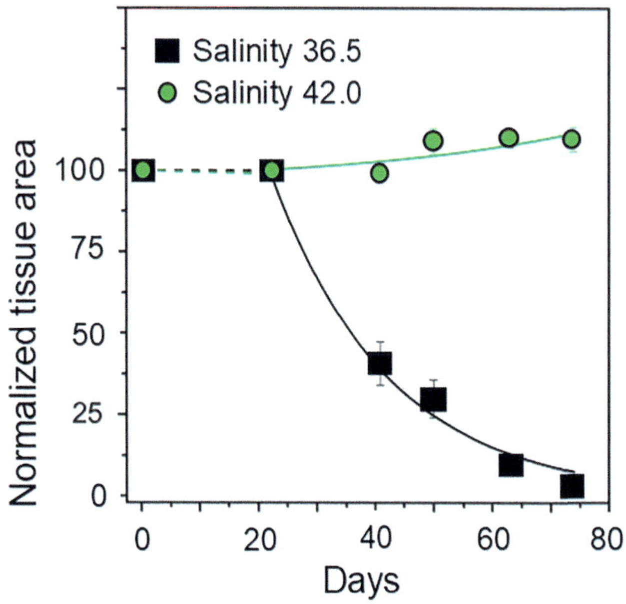

Fig. 5.12 Survival of *Porites lobata* associated with *C. thermophilum* at 32° Celsius at salinity levels characteristic for the southern PAG (42) and the Gulf of Oman (36.5). Data points represent averages, error bars indicate S.E.M. Re-drawn with data from D'Angelo et al. 2015

in partnership with *Durusdinium* sp. (Howells et al. 2018, 2020). Considering that the environmental stress levels are higher in the PAG compared to the Gulf of Oman, it can be concluded that the increased stress tolerance gained by the association with *C. thermophilum* is not reflected in an equal increase in growth and calcification-related trade-offs compared to the less tolerant *Durusdinium* sp.

5.8 Origin of *C. thermophilum* and Implications for Climate Change Adaptations

The PAG represents an ideal natural laboratory to study the adaptation of coral symbiont communities to rapid climate change. This is facilitated by the well-defined oceanographic and climate history, allowing to put the adaptation process in a temporal context. The PAG did not exist as a marine waterbody until ~12 k years ago, when the sea level rise that followed the end of the last ice age started to fill up its basin (Sheppard et al. 1992; Purkis et al. 2010). The temperatures started to increase only ~5 k years ago to reach present day's levels around ~2.4 k years ago (Purkis et al. 2010). Finally, as a landlocked sea, the only way for colonization is through the narrow Strait of Hormuz which facilitates the identification of source populations (Sheppard et al. 1992). Historically, during the marine transgression, the waters filling up the PAG could have brought in symbionts from the Gulf of Oman and potentially the wider region either in their free-living, motile stage or associated with free-swimming coral larvae. Later, the overturning circulation of the PAG with its prevalent surface influx of oceanic waters from adjacent waters (Sheppard et al. 1992; Yao and Johns 2010) can be expected to import symbionts continuously into the PAG.

There are two major mechanisms that could explain the exceptional stress tolerance of the PAG symbionts, specifically of *C. thermophilum*: 1) A rapid evolution of stress tolerant variants by natural selection from the PAG population along with the rise of temperature and salinity over the past ~12 k year or 2) the selection of pre-adapted variants that are able to tolerate the harsh environment from the pool of symbionts continuously introduced into the PAG.

Molecular dating analyses of *C. thermophilum* populations suggest that this group divided from related taxa already about 13 million years ago and that the genetic diversity within this group has evolved several million years ago (Hume et al. 2016, 2018b). Therefore, the genetic diversity of the *C. thermophilum* group existed well before the PAG started to form ~12 k years ago, let alone before it reached present day's temperatures ~2.4 k years ago. Therefore, the genetic homogeneity of the *C. thermophilum* population in the PAG makes the scenario of a rapid evolution of a heat tolerant strain within the PAG not likely. The presence of substantial genetic diversity of *C. thermophilum* in the eastern Strait of Hormuz and the coast of Oman, the region from which genetic material can be assumed to be continuously imported into the PAG, makes it equally unlikely that the lack of diversity within the PAG population is the result of genetic drift or bottleneck or founder effects.

Instead, the genetic uniformity of the PAG *C. thermophilum* rather supports the hypothesis that this population represents the descendants of very few pre-adapted variants that have been selected from representatives of a genetically diverse group outside the PAG (Hume et al. 2016). This is clearly supported by the non-random distribution of *C. thermophilum* and *Durusdinium* along temperature and salinity gradients established in this chapter which indicates that the PAG environment indeed favours variants that are able to tolerate the harsh conditions. The variants that can thrive in this environment face limited competition with other symbionts and can therefore rapidly become the dominant partner in anthozoa associations. The recent anthropogenic introduction and subsequent spread of *Durusdinium trenchii* across the Caribbean also gave rise to a population with exceptionally low genetic diversity (Pettay et al. 2015), suggesting that the spread of one or very few stress-tolerant strains through mostly asexual reproduction may represent a common response of reef ecosystems adapting to climate change. Undoubtedly, the stress tolerant *C. thermophilum* can be considered a key factor for the rapid adaptation of coral communities to the hottest and most saline water of the world during Holocene climate change in the middle eastern region (Hume et al. 2016).

5.9 Implications for Reef Conservation and Climate Change Adaptation Elsewhere

The rapid decline of coral reefs under the pressure of climate change has led to the consideration of a broad range of new conservation approaches including assisted migration of heat tolerant species (Coles and Riegl 2013; Hoegh-Guldberg et al. 2008). The PAG corals and their symbionts offer themselves as potential source populations due to their exceptional heat tolerance. However, the corals of the PAG show clear signs of local adaptation to other environmental conditions apart from the maximal temperature. Specifically, the thermotolerance of *C. thermophilum* appears to be tightly coupled to the adaptation to the high salinity environment (D'Angelo et al. 2015) and its capability to compete against other symbiont types may also depend on other environmental parameters such as annual minimal temperature or seasonal temperature variability (Fig. 5.5). While the oceanographic setting of the Strait of Hormuz favour the import of organisms into the Gulf over the export, this boundary does not seem to be absolute (D'Angelo et al. 2015; Ketchum et al. 2020). Therefore, the strict genetic separation of the PAG *C. thermophilum* over only 20 km from the outside population (Hume et al. 2018b), suggest that the adaptations to the local environment of the PAG, more than the physical separation by the oceanographic settings alone, may impose a distribution barrier to the high-temperature tolerant *C. thermophilum* from the PAG. Accordingly, the direct introduction of *C. thermophilum* into reefs systems with normal oceanic salinity may result in a loss of the thermotolerance and does not offer itself as a straightforward solution to transition reefs elsewhere to a more thermotolerant state (D'Angelo et al. 2015). Future research will need to establish if assisted evolution (Chan et al. 2023) or other manipulations can generate symbiont variants that combine the unusual heat tolerance with the capacity to succeed in environments with otherwise ambient conditions.

Clearly, the colonization of the PAG with pre-adapted symbionts indicates that natural populations exist under overall less extreme environmental conditions that include pre-adapted variants that are considerably more tolerant than seemingly "required" by the prevailing environmental parameters (Hume et al. 2016). If conditions change, these variants can become dominant. Therefore, conservation approaches proven to be effective to preserve the existing genetic diversity of symbiont populations should help to facilitate the transition of coral communities to more tolerant states. This applies equally for symbionts and hosts. Establishing methods to pinpoint these tolerant variants, e.g. by molecular markers or physiological tests (Hume et al. 2015; Voolstra et al. 2020) offer avenues to assist this natural adaptation processes through targeted conservation methods including propagation programs for tolerant variants. Taking the PAG coral communities as a model, it can be expected that the selection processes and the transition to more heat tolerant systems driven by climate change elsewhere will likely go along with a reduction of diversity both at species and genotype level (Pettay et al. 2015; Hume et al. 2016). Over the recent decades, episodes of elevated sea water temperatures caused bleaching and mortality also among the most stress tolerant coral communities of the PAG, resulting in a dramatic reduction of coral cover (Riegl et al. 2018; Burt et al. 2019a, b). Therefore, the bleaching threshold of ~35 °C of corals in the southern PAG (Shuail et al. 2016) may well represent the limit at which symbiotic corals can exist at the global scale. Accordingly, this temperature limit can facilitate predictions of coral survival under various future warming scenarios. While the PAG supports corals with remarkable heat tolerance, it should not be considered a refuge from climate change, as extreme temperature events still regularly lead to bleaching and mortality. Their specialized adaptations–particularly to high salinity, which underscores much of their heat tolerance– may not be easily transferable to other regions with different environmental conditions. Nonetheless, safeguarding the PAG's genetic diversity could inform broader coral resilience strategies. Ultimately, the PAG serves more as a natural test-bed for adaptation than a guaranteed refuge against climate change.

Acknowledgements: Cecilia D'Angelo acknowledges funding from the Leverhulme Trust (Research Leadership Award).

References

Aeby GS, Ben-Hamadou R, Burt JA, Padierna ML, Range PAG, Riegl B, Salimi MA, Torquato FO (2024) Coral reefs in the world's warmest sea: the paradoxical vulnerability of the world's most robust corals. In: Coral reefs and associated marine fauna around the Arabian Peninsula. CRC Press

Al Azhar M, Temimi M, Zhao J, Ghedira H (2016) Modeling of circulation in the Arabian Gulf and the Sea of Oman: skill assessment and seasonal thermohaline structure. J Geophys Res Oceans 121(3):1700–1720

Al-Mansoori N, McParland D, Howells E, Bauman A, Burt JA (2019) Coral bioerosion on the marginal reefs of northeastern Arabia. Aquat Ecosyst Health Manag 22:22–30

Alosairi Y, Imberger J, Falconer R (2011) Mixing and flushing in the Persian Gulf (Arabian Gulf). J Geophys Res Oceans 116:C03029

Baker AC, Starger CJ, McClanahan TR, Glynn PW (2004) Corals' adaptive response to climate change. Nature 430(7001):741

Banks KW, Riegl BM, Shinn EA, Piller WE, Dodge RE (2007) Geomorphology of the Southeast Florida continental reef tract (Miami-Dade, Broward, and Palm Beach counties, USA). Coral Reefs 26:617–640

Burt JA, Paparella F (2023) The marine environment of the Emirates. In: Burt JA (ed) A natural history of the Emirates. Springer Nature, Cham, pp 95–117

Burt JA, Al-Harthi S, Al-Cibahy A (2011) Long-term impacts of bleaching events on the world's warmest reefs. Mar Env Res 72(4):225–229

Burt JA, Al-Khalifa K, Khalaf E, AlShuwaik B, Abdulwahab A (2013) The continuing decline of coral reefs in Bahrain. Marine Pollut Bull 72(2):357–363

Burt JA, Smith EG, Warren C, Dupont J (2016) An assessment of Qatar's coral communities in a regional context. Mar Pollut Bull 105(2):473–479

Burt JA, Paparella F, Al-Mansoori N, Al-Mansoori A, Al-Jailani H (2019a) Causes and consequences of the 2017 coral bleaching event in the southern Persian/Arabian Gulf. Coral Reefs 38:567–589

Burt JA, Paparella F, Al-Mansoori N, Al-Mansoori A, Richlen M (2019b) Sudden shifts in summer wind regimes lead to severe heat stress and widespread mortality on a Gulf coral reef. Glob Chang Biol 25(10):3831–3846

Burt JA, Camp E, Enochs IC, Johansen JL, Morgan KM, Riegl B, Hoey AS (2020a) Insights from extreme coral reefs in a changing world. Coral Reefs 39(3):495–507

Burt JA, Paparella F, Al-Mansoori N, Al-Mansoori A, Al-Jailani H (2020b) Long-term impacts of bleaching events on the world's warmest reefs. Mar Environ Res 154:104836

Campos EJD, Vieira F, Cavalcante G, Kjerfve B, Abouleish MY, Shahriar S et al (2020a) Impacts of brine disposal from water desalination plants on the physical environment in the Persian/Arabian Gulf. Environ Res Commun 2:125003. https://doi.org/10.1088/2515-7620/abd0ed

Campos EJD, Gordon AL, Kjerfve B, Vieira F, Cavalcante G (2020b) Freshwater budget in the Persian (Arabian) Gulf and exchanges at the Strait of Hormuz. PLoS One 15(5):e0233090

Campos EJD, Kjerfve B, Cavalcante G, Vieira F, Abouleish M (2022) Water exchange across the Strait of Hormuz. Effects of tides and rivers runoff. Reg Stud Mar Sci 52:102336

Cavalcante GH, Feary DA, Burt JA (2016) The influence of extreme winds on coastal oceanography and its implications for coral population connectivity in the southern Arabian Gulf. Mar Pollut Bull 105(2):489–497

Cavalcante G, Vieira F, Mortensen J, Ben-Hamadou R, Range P, Goergen EA et al (2020) Biophysical model of coral population connectivity in the Arabian/Persian Gulf. In: Riegl B (ed) Population dynamics of the reef crisis, Advances in marine biology. Academic, pp 193–221

Chan WY, Meyers L, Rudd D, Topa SH, van Oppen MJ (2023) Heat-evolved algal symbionts enhance bleaching tolerance of adult corals without trade-off against growth. Glob Chang Biol 29(24):6945–6968

Chappell J (1983) Evidence for smoothly falling sea level relative to North Queensland, Australia, during the last 6,000 yr. Nature 302:406–408

Coles SL (2003) Coral species diversity and environmental factors in the Arabian Gulf and the Gulf of Oman: a comparison to the Indo-Pacific region. Atoll Res Bull 507:1–19

Coles SL, Riegl BM (2013) Thermal tolerances of reef corals in the Gulf: a review of the potential for increasing coral survival and adaptation to climate change through assisted translocation. Mar Pollut Bull 72(2):323–332

Cunning R, Gillette P, Capo T, Galvez K, Baker AC (2015) Growth tradeoffs associated with thermotolerant symbionts in the coral Pocillopora damicornis are lost in warmer oceans. Coral Reefs 34:155–160

D'Angelo C, Hume BC, Burt J, Smith EG, Achterberg EP, Wiedenmann J (2015) Local adaptation constrains the distribution potential of heat-tolerant Symbiodinium from the Persian/Arabian Gulf. The ISME J 9(12):2551–2560

Davy SK, Allemand D, Weis VM (2012) Cell biology of cnidarian-dinoflagellate symbiosis. Microbiol Mol Biol Rev 76:229–261

de Verneil A, Burt JA, Mitchell M, Paparella F (2021) Summer oxygen dynamics on a southern Arabian gulf coral reef. Front Mar Sci 8(1676):1676. https://doi.org/10.3389/fmars.2021.781428

Douglas AE (2003) Coral bleaching—how and why? Mar Pollut Bull 46(4):385–392

Evans G, Murray JW, Biggs HE, Bate R, Bush PR (1973) The oceanography, ecology, sedimentology, and geomorphology of parts of the Trucial Coast barrier Island complex, Persian Gulf. In: Purser BH (ed) The Persian Gulf. Springer, Berlin, pp 211–233

Fagerstrom AJ (1987) The evolution of reef communities. Wiley, New York, p 600

Ferrier-Pagès C, Godinot C, D'Angelo C, Wiedenmann J, Grover R (2016) Phosphorus metabolism of reef organisms with algal symbionts. Ecol Monogr 86(3):262–277

Fluegel E (2004) Microfacies of carbonate rocks. Analysis, interpretation and application. Springer, Berlin, p 975

Figueiredo J, Baird AH, Harii S, Connolly SR (2014) Increased local retention of reef coral larvae as a result of ocean warming. Nature Climate Change 4(6):498–502

Ghavam Mostafavi P, Fatemi SMR, Shahhosseiny MH, Hoegh-Guldberg O, Loh WKW (2007) Predominance of clade D Symbiodinium in shallow-water reef-building corals off Kish and Larak Islands (Persian Gulf, Iran). Mar Biol 153:25–34

Hartle-Mougiou K, D'Angelo C, Smith EG, Burt JA, West P, Wiedenmann J (2012) Diversity of zooxanthellae from corals and sea anemones after long-term aquarium culture. J Mar Biol Assoc UK 92(4):687–691

Hoegh-Guldberg O, Hughes L, McIntyre S, Lindenmayer DB, Parmesan C, Possingham HP, Thomas CD (2008) Assisted colonization and rapid climate change. Science 321(5887):345–346

Howells EJ, Dunshea G, McParland D, Vaughan GO, Heron SF, Pratchett MS, Burt JA, Bauman AG (2018) Species-specific coral calcification responses to the extreme environment of the southern Persian Gulf. Frontiers Mar Sci 5:56

Howells EJ, Bauman AG, Vaughan GO, Hume BC, Voolstra CR, Burt JA (2020) Corals in the hottest reefs in the world exhibit symbiont fidelity not flexibility. Molecular Ecol 29(5):899–911

Hume B, D'Angelo C, Burt JA, Baker AC, Riegl B, Wiedenmann J (2013) Corals from the Persian/Arabian Gulf as models for thermotolerant reef-builders: prevalence of clade C3 Symbiodinium, host fluorescence and ex situ temperature tolerance. Marine Pollut Bull 72(2):313–322

Hume BC, D'Angelo C, Smith EG, Stevens JR, Burt JA, Wiedenmann J (2015) *Symbiodinium thermophilum* sp. nov., a thermotolerant symbiotic alga prevalent in corals of the world's hottest sea, the Persian/Arabian Gulf. Sci Rep. 5(1):8562

Hume BC, Voolstra CR, Arif C, D'Angelo C, Burt JA, Eyal G, Loya Y, Wiedenmann J (2016) Ancestral genetic diversity associated with the rapid spread of stress-tolerant coral symbionts in response to Holocene climate change. Proc Nat Acad Sci 113(16):4416–4421

Hume BC, D'Angelo C, Burt JA, Wiedenmann J (2018a) Fine-scale biogeographical boundary delineation and sub-population resolution in the *Symbiodinium thermophilum* coral symbiont group from the Persian/Arabian Gulf and Gulf of Oman. Front Mar Sci 5:138

Hume BC, D'Angelo C, Smith EG, Stevens JR, Burt JA, Wiedenmann J (2018b) Validation of the binary designation *Symbiodinium thermophilum* (Dinophyceae). J Phycol 54(5):762–764

Hume BC, Smith EG, Ziegler M, Warrington HJ, Burt JA, La Jeunesse TC, Wiedenmann J, Voolstra CR (2019) SymPortal: a novel analytical framework and platform for coral algal symbiont next-generation sequencing ITS2 profiling. Mol Ecol Resour 19(4):1063–1080

James NP (1983) Reef environment. In: Scholle PA, Bebout DG, Moore CA (eds) Carbonate depositional environments, AAPG Memoir 33. American Association of Petroleum Geologists, Tulsa, pp 345–462

Johns WE, Yao F, Olson DB, Josey SA, Grist JP, Smeed DA (2003) Observations of seasonal exchange through the Straits of Hormuz and the inferred heat and freshwater budgets of the Persian Gulf. J Geophys Res Oceans 108:3391

Jones A, Berkelmans R (2010) Potential costs of acclimatization to a warmer climate: growth of a reef coral with heat tolerant vs. sensitive symbiont types. PLoS One 5(5):e10437

Kämpf J, Sadrinasab M (2006) The circulation of the Persian Gulf: a numerical study. Ocean Sci 2(1):27–41

Ketchum RN, Smith EG, DeBiasse MB, Vaughan GO, McParland D, Leach WB, Al-Mansoori N, Ryan JF, Burt JA, Reitzel AM (2020) Population genomic analyses of the sea urchin *Echinometra* sp. EZ across an extreme environmental gradient. Genome Biol Evol 12(10):1819–1829

Lachkar Z, Mehari M, Lévy M, Paparella F, Burt JA (2022) Recent expansion and intensification of hypoxia in the Arabian Gulf and its drivers. Front Mar Sci 9:892334

LaJeunesse TC, Loh W, van Woesik R, Hoegh-Guldberg O, Schmidt GW, Fitt WK (2003) Low symbiont diversity in southern Great Barrier Reef corals, relative to those of the Caribbean. Limnol Oceanogr 48:2046–2054

LaJeunesse TC, Parkinson JE, Gabrielson PW, Jeong HJ, Reimer JD, Voolstra CR, Santos SR (2018) Systematic revision of Symbiodiniaceae highlights the antiquity and diversity of coral endosymbionts. Curr Biol 28(16):2570–2580

Lambeck K (1996) Shoreline reconstructions for the Persian Gulf since the last glacial maximum. Earth Planet Sci Lett 142:43–57

Mossadegh ZK, Parker J, Gischler E, Oschmann W, Blakeway D, Eisenhauer A (2012) Paleoecology of well-preserved coral communities in a siliciclastic environment from the late Pleistocene (MIS 7), Kish Island, Persian Gulf (Iran): the development of low-relief frameworks (biostromes) in increasingly restricted environments. Int J Earth Sci 102:545–570

Ochsenkühn MA, Röthig T, D'Angelo C, Wiedenmann J, Voolstra CR (2017) The role of floridoside in osmoadaptation of coral-associated algal endosymbionts to high-salinity conditions. Sci Adv 3(8):e1602047

Oladi M, Shokri MR, Rajabi-Maham H (2019) Extremophile symbionts in extreme environments; a contribution to the diversity of Symbiodiniaceae across the northern Persian Gulf and Gulf of Oman. J Sea Res 144:105–111

Oladi M, Rouzbehani S, Ahmadzadeh F, Ghazilou A (2021) Dynamics of Dipsastraea pallida-symbiont association following bleaching events across the northern Persian Gulf and Gulf of Oman. Symbiosis 84(2):141–149

Paparella F, Burt JA (2023) Climate of The United Arab Emirates: present, past and impacts on life. In: Burt JA (ed) A natural history of the Emirates. Springer, Cham, pp 65–94

Paparella F, Xu C, Vaughan GO, Burt JA (2019) Coral bleaching in the Persian/Arabian gulf is modulated by summer winds. Front Mar Sci 6:10.3389

Pettay DT, Wham DC, Smith RT, Iglesias-Prieto R, LaJeunesse TC (2015) Microbial invasion of the Caribbean by an Indo-Pacific coral zooxanthella. Proc Nat Acad Sci 112(24):7513–7518

Pohl T, Al-Muqdadi SW, Ali MH, Fawzi NA-M, Ehrlich H, Merkel B (2014) Discovery of a living coral reef in the coastal waters of Iraq. Sci Rep 4:4250

Pous SP, Carton X, Lazure P (2004) Hydrology and circulation in the Strait of Hormuz and the Gulf of Oman—results from the GOGP99 experiment: 2. Gulf of Oman. J Geophys Res Oceans 109:C12S08

Pous S, Lazure P, Carton X (2015) A model of the general circulation in the Persian Gulf and in the Strait of Hormuz: Intraseasonal to interannual variability. Cont Shelf Res 94:55–70

Preusser F, Radke U, Fontugne M, Haghipour A, Hilgers A, Kasper HU, Nazari H, Pirazzoli PA (2003) ESR dating of rised coral reefs from Kish Island, Persian Gulf. Quat Sci Rev 22(10–13):1317–1322

Purkis SJ, Riegl B (2005) Spatial and temporal dynamics of Arabian Gulf coral assemblages quantified from remote-sensing and in situ monitoring data. Mar Ecol Prog Ser 287:99–113

Purkis S, Riegl B (2012) Geomorphology and reef building. In: Riegl B, Purkis S (eds) Reefs of the Gulf. Adaptation to environmental extremes. Springer, Dordrecht, pp 33–50

Purkis SJ, Rowlands GP, Riegl B, Renaud PG (2010) The paradox of tropical karst morphology in the coral reefs of the arid Middle East. Geology 38(3):227–230

Purser BH (1973) The Persian Gulf. Springer, Berlin, p 471

Rädecker N, Pogoreutz C, Voolstra CR, Wiedenmann J, Wild C (2015) Nitrogen cycling in corals: the key to understanding holobiont functioning? Trends in Microbiol 23(8):490–497

Rasser M, Riegl B (2002) Holocene coral reef rubble and its binding agents. Coral Reefs 21:57–72

Reynolds MR (1993) Physical oceanography of the Gulf, Strait of Hormuz, and the Gulf of Oman—results from the Mt Mitchell expedition. Mar Pollut Bull 27:35–59

Riegl B (1999) Corals in a non-reef setting in the southern Arabian Gulf (Dubai, UAE): fauna and community structure in response to recurring mass mortality. Coral Reefs 18:63–73

Riegl B (2002) Effects of the 1996 and 1998 positive sea-surface temperature anomalies on corals, coral diseases and fish in the Arabian Gulf (Dubai, UAE). Mar Biol 140(1):29–40

Riegl B, Piller WE (2000) Biostromal coral facies—a Miocene example from the Leitha limestone (Austria) and its actualistic interpretation. PALAIOS 15:399–413

Riegl B, Purkis SJ (2009) Markov models for linking environments and facies in space and time (recent Arabian Gulf, Miocene Paratethys). Int Assoc Sedimentol Spec Publ 41:337–360

Riegl BM, Purkis SJ (2012) Dynamics of Gulf coral communities: observations and models from the world's hottest coral sea. In: Riegl B (ed) Coral reefs of the Gulf: adaptation to climatic extremes. Springer, Dordrecht, pp 71–93

Riegl BM, Purkis SJ, Al-Cibahy AS, Abdel-Moati MA, Hoegh-Guldberg O (2011) Present limits to heat-adaptability in corals and population-level responses to climate extremes. PLoS One 6(9):e24802

Riegl B, Benzoni F, Samimi-Namin K, Sheppard CRC (2012) The hermatypic (hard) coral fauna of the Gulf. In: Riegl B (ed) Coral reefs of the Gulf: adaptation to climatic extremes. Springer, Dordrecht, pp 187–224

Riegl B, Johnston M, Purkis S, Howells E, Burt JA, Steiner SC, Sheppard CR, Bauman A (2018) Population collapse dynamics in *Acropora downingi*, an Arabian/Persian Gulf ecosystem-engineering coral, linked to rising temperature. Glob Chang Biol 24(6):2447–2462

Riegl Jr B, Bauman A, Hadj-Hamou J, Burt JA (2024) A decade of benthic changes on coral reefs in the Southern Persian/Arabian Gulf (2010–2020). Coral Reefs 43(6):1647–1657

Rosset S, D'Angelo C, Wiedenmann J (2015) Ultrastructural biomarkers in symbiotic algae reflect the availability of dissolved inorganic nutrients and particulate food to the reef coral holobiont. Front Mar Sci 2:103

Rosset S et al (2017) Phosphate deficiency promotes coral bleaching and is reflected by the ultrastructure of symbiotic dinoflagellates. Mar Pollut Bull 118(1–2):180–187

Rosset S, Koster G, Brandsma J, Hunt AN, Postle AD, D'Angelo C (2019) Lipidome analysis of Symbiodiniaceae reveals possible mechanisms of heat stress tolerance in reef coral symbionts. Coral Reefs 38(6):1241–1253

Samimi-Namin K, Riegl B (2012) Raised reefs on Larak suggest Acropora dominance on the Persian Gulf coral reefs since the Pleistocene. Coral Reefs 31:763

Samimi-Namin K, van Ofwegen L (2012) The octocoral fauna of the Gulf. In: Riegl B (ed) Coral reefs of the Gulf: adaptation to climatic extremes. Springer, Dordrecht, pp 225–252

Schoepf V, Baumann JH, Barshis DJ, Browne NK, Camp EF, Comeau S, Cornwall CE, Guzman H, Riegl B, Rodolfo-Metalpa R, Sommer B (2023) Corals at the edge of environmental limits: a new conceptional framework to re-define marginal and extreme coral communities. Sci Tot Environ 88(4):163688

Shahhosseiny MH, Mostafavi PG, Fatemi SMR, Karimi E (2011) Clade identification of symbiotic zooxanthellae of dominant sclerectinian coral species of intertidal pools in Hengam Island. Afr J Biotechnol 10(9):1502–1506

Sheppard CRC, Sheppard ALS (1991) Coral and coral communities of Arabia, Fauna of Saudi Arabia 12. National Commission for Wildlife Conservation and Development, Riyadh, p 170

Sheppard C, Price A, Roberts C (1992) Marine ecology of the Arabian region: patterns and processes in extreme tropical environments. Academic, London, p 359

Sheppard C, Al-Husiani M, Al-Jamali F, Al-Yamani F, Baldwin R, Bishop J et al (2010) The Gulf: a young sea in decline. Mar Poll Bull 60(1):13–38

Shinn EA (1976) Coral reef recovery in Florida and the Persian Gulf. Environ Geol 1:241–254

Shinn EA (2006) Submarine lithification of Holocene carbonate sediments in the Persian Gulf. Sedimentology 12:109–144

Shuail D, Wiedenmann J, D'Angelo C, Baird AH, Pratchett MS, Riegl B, Burt JA, Petrov P, Amos C (2016) Local bleaching thresholds established by remote sensing techniques vary among reefs with deviating bleaching patterns during the 2012 event in the Arabian/Persian Gulf. Mar Poll Bull 105(2):654–659

Smith EG, Hazzouri KM, Choi JY, Delaney P, Al-Kharafi M, Howells EJ, Aranda M, Burt JA (2022) Signatures of selection underpinning rapid coral adaptation to the world's warmest reefs. Science advances 8(2), p.eabl7287

Stat M, Gates RD (2011) Clade D *Symbiodinium* in scleractinian corals: a "nugget" of hope, a selfish opportunist, an ominous sign, or all of the above? J Mar Sci 2011:730715

Swift SA, Bower AS (2003) Formation and circulation of dense water in the Persian/Arabian Gulf. J Geophys Res Oceans 108(C1):4-1–4-21. https://doi.org/10.1029/2002JC001360

Teller JT, Glennie KW, Lancaster N, Singhvi AK (2000) Calcareous dunes of The United Arab Emirates and Noah's flood: the postglacial reflooding of the Persian (Arabian) Gulf. Quat Int 71:297–308

Thoppil PG, Hogan PJ (2009) On the mechanisms of episodic salinity outflow events in the strait of Hormuz. J Phys Oceanogr 39(6):1340–1360. https://doi.org/10.1175/2008jpo3941.1

Thoppil PG, Hogan PJ (2010) A modeling study of circulation and eddies in the Persian Gulf. J Phys Oceanogr 40(9):2122–2134. https://doi.org/10.1175/2010jpo4227.1

Toth LT, Precht WF, Modys AB, Stathakopoulos A, Robbart ML, Hudson JH, Oleinink AE, Riegl BM, Shinn EA, Aronson RB (2021) Climate and latitudinal limits of subtropical reef development. Sci Rep 11:13044. https://doi.org/10.1038/s41598-021-87883-8

Varasteh T, Shokri MR, Rajabi-Maham H, Behzadi S, Hume BC (2018) *Symbiodinium thermophilum* symbionts in *Porites harrisoni* and *Cyphastrea microphthalma* in the northern Persian Gulf, Iran. J Mar Biol Assoc UK 98(8):2067–2073

Voolstra CR, Buitrago-López C, Perna G, Cárdenas A, Hume BC, Rädecker N, Barshis DJ (2020) Standardized short-term acute heat stress assays resolve historical differences in coral thermotolerance across microhabitat reef sites. Glob Change Biol 26(8):4328–4343

Wiedenmann J, D'Angelo C, Smith EG, Hunt AN, Legiret FE, Postle AD, Achterberg EP (2013) Nutrient enrichment can increase the susceptibility of reef corals to bleaching. Nat Clim Chang 3(2):160–164

Wiedenmann J, D'Angelo C, Mardones ML, Moore S, Benkwitt CE, Graham NA, Hambach B, Wilson PA, Vanstone J, Eyal G, Ben-Zvi O (2023) Reef-building corals farm and feed on their photosynthetic symbionts. Nature 620(7976):1018–1024

Yao F, Johns WE (2010) A HYCOM modeling study of the Persian Gulf: 1. Model configurations and surface circulation. J Geophys Res Oceans 115:C11008

Where the Reef Meets the Land: Intertidal and Back-Reef Coral Environments

6

Verena Schoepf, Rene M. van der Zande, Courtney N. Klepac, and David Glassom

Abstract

The intertidal zone of tropical coral reefs has received relatively little attention compared to other reef habitats, despite the potential of intertidal and back-reef coral environments to serve as resilience hotspots and natural laboratories to study future ocean conditions. Here, we provide a classification of intertidal coral reefs based on spatial and temporal dynamics, and review the state of knowledge for five intertidal and back-reef coral habitats, focusing on both low- and high-latitude locations: the back-reef pools of Ofu Island, American Samoa; intertidal rock pools in South Africa; mesotidal reef flats in Phuket, Thailand, and on Heron Island, southern Great Barrier Reef; and the macrotidal Kimberley region in NW Australia. We show that intertidal and back-reef habitats are among the most extreme that tropical corals persist under, with tidal range, seawater residence time, and latitude being key drivers of environmental extremeness. However, this does not necessarily lead to ecological marginality, such as reduced coral cover or ecosystem functioning, though this is also site-specific. While information on physiological, symbiotic, molecular, genetic, and acclimatory traits continues to increase, determining whether common mechanisms facilitate coral survival in these extreme habitats remains challenging and represents but one of many future research priorities identified here.

Keywords

Intertidal · Environmental variability · Ofu Island · KwaZulu-Natal province · Phuket reef flats · Heron Island · Kimberley region

6.1 Introduction

Intertidal habitats, occupying the transition zone between terrestrial and marine realms, have long fascinated marine biologists because they are characterized by a range of extreme environmental conditions, marking them as some of the most dynamically challenging ecosystems on Earth (Garrison 2009). As the tides dictate dramatic fluctuations in abiotic conditions and regular emergence and submergence, intertidal organisms require highly specialized adaptive strategies for survival and display remarkable resistance to environmental extremes (Garrison 2009). Rather than tropical coastlines, intertidal habitats typically conjure up images of vast mudflats or the temperate rocky intertidal, such as the famous Mont Saint-Michel in France and Bay of Fundy in Canada. Yet, large parts of the tropics are also characterized by high tidal ranges ≫2 m (Fig. 6.1; Whitfield and Elliott 2011), and highly diverse coral reefs have developed under some of the most extreme tidal ranges found on this planet. For example, such coral reef systems exist off the Amazon coast at the Manuel Luís Reef (Cordeiro et al. 2021) or across the Australian North West Shelf where the tidal range is up to 12 m—the largest within the tropics (Wilson 2013; Richards et al. 2024). The tropical intertidal is therefore among the most extreme coral environments on Earth (*sensu* Schoepf et al. 2023) and can provide important insights into how corals and other reef-associated organisms adapt to environ-

V. Schoepf (✉)
Department of Freshwater and Marine Ecology, Institute for Biodiversity and Ecosystem Dynamics, University of Amsterdam, Amsterdam, The Netherlands

UWA Oceans Institute, University of Western Australia, Perth, WA, Australia
e-mail: v.schoepf@uva.nl

R. M. van der Zande
Department of Freshwater and Marine Ecology, Institute for Biodiversity and Ecosystem Dynamics, University of Amsterdam, Amsterdam, The Netherlands

C. N. Klepac
Hopkins Marine Station of Stanford University, Pacific Grove, CA, USA

D. Glassom
School of Life Sciences, University of KwaZulu-Natal, Durban, South Africa

M. O. Soares et al. (eds.), *Extreme and Marginal Coral Reefs and Communities*, Coral Reefs of the World 24,
https://doi.org/10.1007/978-3-032-28035-0_6

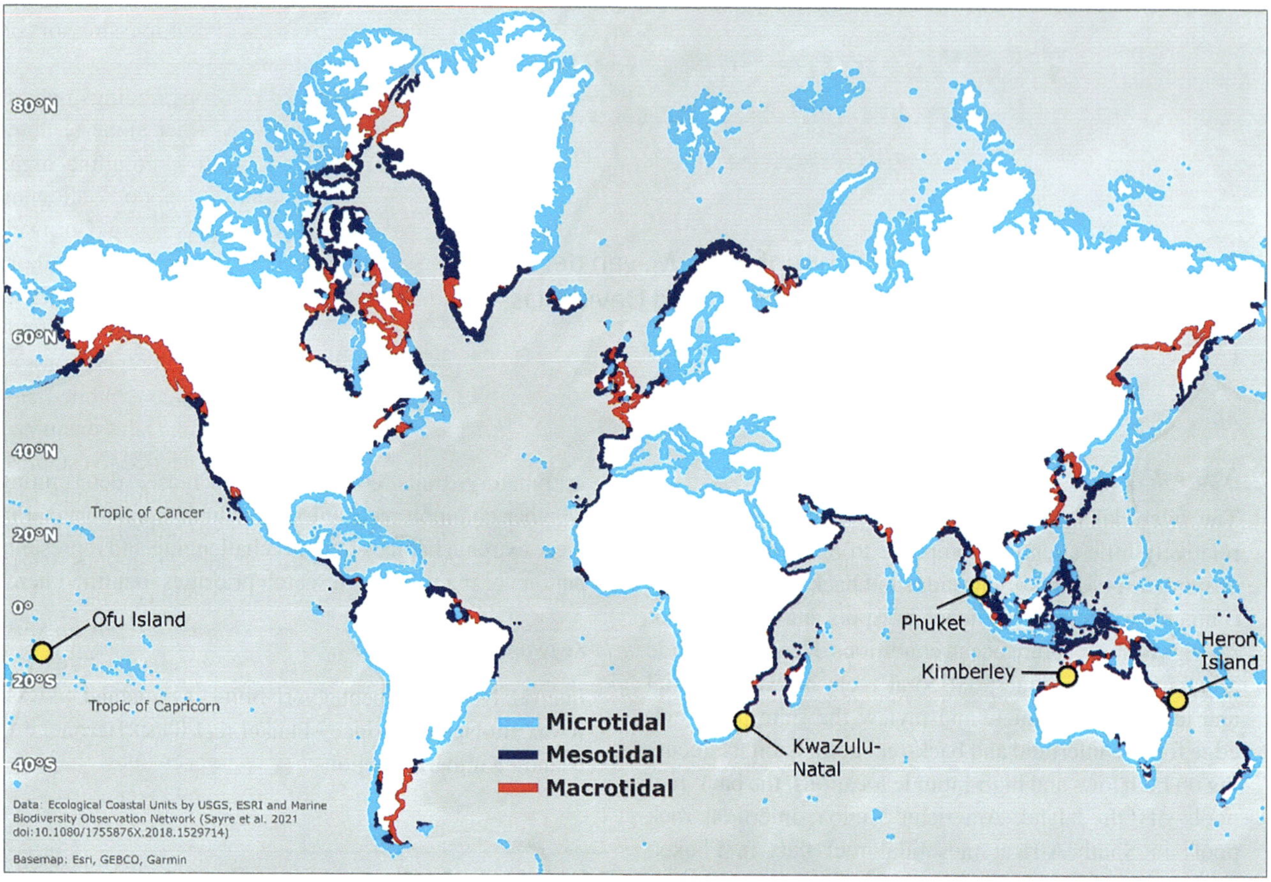

Fig. 6.1 Coastal regions of the world with microtidal, mesotidal, and macrotidal regimes, defined as a tidal range of <2, 2–4, and >4 m, respectively. The yellow circles indicate the five locations highlighted in this book chapter. (Map figure created by S. Shinneman using the ecological coastal units database from Sayre et al. (2019))

mental extremes, including those predicted to occur under climate change (e.g., Camp et al. 2018). By shifting attention to the intertidal zones of tropical reef environments, this chapter aims to expand our notion of the intertidal beyond temperate rocky shores and mudflats, and reviews its unique abiotic conditions as well as the adaptive mechanisms of scleractinian corals and other reef taxa that not only endure but seemingly thrive in intertidal and back-reef environments.

6.1.1 What Are Intertidal Coral Reefs? Defining Spatial Dynamics Using Temperate Rocky Shore Terminology

The intertidal is generally defined as “the marine zone between the highest high-tide point on a shoreline and the lowest low-tide point” (Garrison 2009) and can further be subdivided into various zones along the land-to-ocean continuum. These include the spray zone (or supratidal), the high intertidal, middle intertidal, and low intertidal, which are defined as follows (NOAA 2024a): the spray zone is dampened by ocean spray and high waves and is submerged only during very high tides or severe storms. The high intertidal floods during the peaks of daily high tides but remains dry for long stretches between high tides. In contrast, the tides ebb and flow over the middle intertidal twice a day whereas the low intertidal is virtually always underwater except during the lowest of spring tides. Less commonly, rocky shore ecosystems are also divided into supralittoral, eulittoral and sublittoral zones where the supralittoral corresponds to the spray or splash zone, the eulittoral to the high, middle and low intertidal, and the sublittoral to the subtidal zone (Garrison 2009; Whitfield and Elliott 2011).

Interestingly, the above classification of the intertidal zone has been widely used for the temperate rocky intertidal but is not commonly applied to tropical coral reefs. Although tidal ranges on many reefs are minimal (i.e., ≪2 m) (Hayes 1975; Whitfield and Elliott 2011), large stretches of tropical coasts, however, have meso- to macrotidal regimes with tidal ranges exceeding 2 or 4 m, respectively (Fig. 6.1). We therefore need to broaden our perceptions of what coral reefs traditionally ‘look like’—and under what environmental regimes, including tidal regimes, they occur (see also the discussion of the 'marginal reef' concept in Perry and Larcombe (2003) and Schoepf et al. (2023)). We attempt here to integrate the classification of the rocky intertidal with the classic zonation of tropical coral reefs to shed light on the spatial dynamics of intertidal coral communities and reefs.

Five standard reef zones are present on most swell- and trade-wind-dominated tropical coral reefs (Blanchon 2011): "a calm reef lagoon, with varying number of patch reefs; a largely barren, shallow reef flat strewn with detritus; a narrow, gently sloping reef front exposed to waves and swells; a steeper reef slope bearing corals to as deep as 100 m; and a less steeply inclined fore-reef apron, where coral and reef detritus accumulate". On reefs with microtidal regimes, only two of these five reef zones—the reef flat and reef front—can be characterized as intertidal and are restricted to the low intertidal given that it is "virtually always underwater except during the lowest of spring tides" (NOAA 2024a). This is also in agreement with the definition of reef flats as "intertidal platforms" and "the most recent expression of coral reef growth at sea level" (Blanchon 2011). We note, however, that subtidal habitats such as shallow lagoons or back-reef pools can nevertheless experience extreme fluctuations in abiotic conditions that rival those of the intertidal zone (e.g., Craig et al. 2001), even though they typically never get exposed to air.

In contrast, on coral reefs with meso- and macrotidal regimes, the intertidal zone expands substantially compared to microtidal reefs and includes both the low and middle intertidal. Specifically, the reef flat and front as well as parts of the shallow lagoon, lagoon-associated patch reefs and upper reef slope extend into the middle intertidal zone as they can experience extended periods of aerial exposure between low and high tides. On reefs with such large tidal ranges, some corals may even persist in small crevices or ponds at the transition from the middle to the high intertidal zone which floods only during the peaks of daily high tides but remains dry for long stretches between high tides. Therefore, depending on the tidal regime, up to four of the five standard reef zones—or at least parts thereof—can be characterized as intertidal, highlighting the importance of recognizing the pronounced intertidal nature of many shallow tropical coral reefs.

6.1.2 Temporal Dynamics of Intertidal Coral Environments

In addition to the strong spatial dynamics that determine the vertical zonation of intertidal coral environments, they are also characterized by temporal dynamics that can vary substantially across numerous time scales. These temporal dynamics not only represent their own unique set of challenges to resident organisms, but can also be used to further classify intertidal coral environments based on (sub-) daily to seasonal and inter-annual variations in sea level. On daily time scales, coral reefs—like any other coastal ecosystem—can be distinguished based on whether their tidal regimes are diurnal, semi-diurnal or mixed (Fig. 6.2). Diurnal tides have a tidal cycle of one high tide and one low tide per day, whereas semi-diurnal tides have two high tides and two low tides each lunar day, with the high tides being of nearly equal height (Garrison 2009). In contrast, mixed tidal regimes represent more complex tidal cycles that usually consist of two high tides and two low tides of unequal height per day (Garrison 2009).

The distinction between these tidal regimes is crucial because they determine the time period during which corals and other reef organisms become exposed to air and desiccation, as well as their frequency. While the degree of extremeness of an intertidal habitat depends considerably on the daily timing of low tide (Buckee et al. 2022), it is debatable whether diurnal or semi-diurnal tides represent more extreme environmental regimes: even though the duration of low tide is longer for diurnal regimes, the frequent switching between high and low tides under semi-diurnal regimes imposes its own challenges on intertidal organisms. Furthermore, nighttime low tides may provide relief from hot temperatures at low latitude but can induce cold stress at higher latitude (e.g., Hoegh-Guldberg et al. 2005). Even in the absence of aerial exposure, tidal dynamics strongly influence exposure to environmental extremes in shallow reef habitats with long seawater residence times, such as back-reef pools or reef flats (e.g., Koweek et al. 2015; Smit and Glassom 2017). Overall, these temporal dynamics not only have implications for the adaptations of marine organisms to these fluctuating conditions but in the case of coral reefs also determine the limit of upward reef growth (Dixon 2011).

Mean sea level can also vary significantly on seasonal and inter-annual time scales in certain regions. In many tropical regions affected by monsoons, seasonal variation in mean sea level is common, such as in the Red Sea and Gulf of Aqaba where sea level differs by 0.5 m between winter and summer monsoons (Dixon 2011). Regions of the western Pacific Ocean within latitudes of ±15° are an example for sea level variations on inter-annual time scales. For example, sea level changed by ~30 cm during the transition from the 1997/1998 El Niño to the ENSO-neutral year afterwards (Church et al. 2006). Given that El Niño years are often associated with coral mass bleaching events (Hughes et al. 2018), extreme low tides associated with sea level falls can augment heat-induced mortality on shallow coral reefs in such regions (e.g., Ampou et al. 2017). In the Indian Ocean, sea level also varies considerably and the eastern Indian Ocean has seen one of the largest rises globally from the 1950s to ~2000 (Bindoff et al. 2007), with significant influence on coral cover on shallow reef flats (Brown et al. 2011). These tempo-

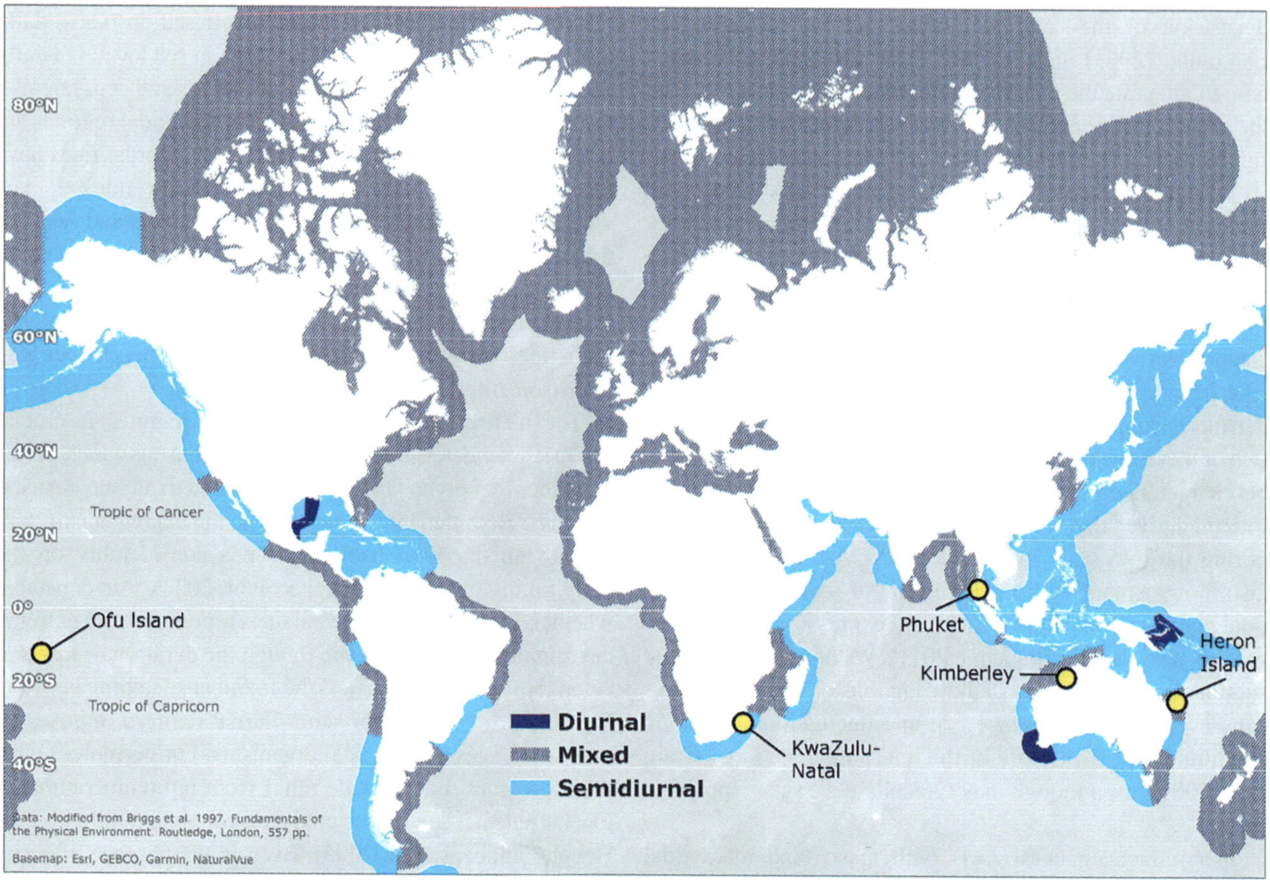

Fig. 6.2 Coastal regions of the world with diurnal, mixed, and semi-diurnal regimes. See text for definitions. The yellow circles indicate the five locations highlighted in this book chapter. (Map figure created by S. Shinneman using data from Briggs (1997))

ral dynamics in sea level variations, whether on daily or interannual time scales, strongly determine the environmental extremeness of these habitats and highlight that the "intertidal nature" of tropical coral reefs exists along a continuum, where some habitats are part of the intertidal zone on a daily basis while others turn into intertidal habitats only during certain seasons or years for limited periods of time.

6.1.3 Intertidal Zones as Naturally Extreme Coral Environments

The highly dynamic environmental conditions present in the tropical intertidal make it one of the most extreme coral environments found on this planet, enabling them to serve as natural laboratories and resilience hotspots to study how corals and other reef-associated organisms can acclimatize and/or adapt to sub-optimal environmental conditions *in situ* (Camp et al. 2018; Schoepf et al. 2023). The term 'extreme' is used here *sensu* Schoepf et al. (2023) who defined coral reef extremeness as the "combined suite of environmental conditions at a given location that represent large deviations from optimal conditions in terms of mean and/or variance". Although several parameters in the tropical intertidal can show large deviations in terms of mean (e.g., high solar and UV irradiance), it is the large variance, in particular, that makes it such an extreme coral environment. For example, corals in tide pools and reef flats in the macrotidal Kimberley region (NW Australia) experience some of the most extreme variance recorded for any type of coral community, with diel variations in temperature, pH and dissolved oxygen of up to 11 °C, 1.4 pH units and 440 µmol O_2, respectively (Gruber et al. 2017; Cornwall et al. 2018; Speelman et al. 2023). As there is now a renewed interest in these extreme coral communities, intertidal reef environments have provided valuable insights into the mechanisms underlying coral resistance to environmental extremes, including marine heatwaves and ocean acidification (Brown et al. 2014a, 2022b; Palumbi et al. 2014; Schoepf et al. 2015b). This highlights the high research and conservation value of coral communities at the edge of environmental limits and will continue to generate new knowledge to better understand how they can facilitate future coral reef persistence.

Several key factors determine the extremeness of intertidal coral environments. This includes, but is not limited to, the vertical position within the intertidal zone, the tidal range (with meso- and macrotidal areas having more extreme conditions than microtidal areas), the tidal regime (with diurnal tides being more extreme than semi-diurnal/mixed one, especially when combined with a large tidal range) (Fig. 6.1), and the daily timing of low tide (e.g., Buckee et al. 2022). In addition, the duration of the falling (ebb) tide can be greatly elongated relative to the duration of the rising (flood) tide on many tide-dominated coral reefs, which comprise about 30% of reefs worldwide (Lowe and Falter 2015; Lowe et al. 2015). Tidal currents can further significantly modulate environmental extremeness. On the one hand, they contribute to extremeness by creating strong currents and stirring up sediment, often leading to highly turbid waters (e.g., Wolanski and Spagnol 2003) (Fig. 6.3). Tidal currents also determine seawater residence times and water exchange, often leaving tide pools and other semi-enclosed bodies of water with stagnant water for extended periods of time during low tide (e.g., Koweek et al. 2015; Smit and Glassom 2017). On the other hand, tidal currents also mitigate extremeness as they determine exchange of water between the open ocean and semi-enclosed water bodies, such as atoll lagoons and semi-enclosed bays or tide pools.

Based on these factors, two types of intertidal and back-reef coral environments can be distinguished that differ in terms of tidal exposure and environmental extremeness: (1) habitats that are seldom or never exposed to air but are highly variable, and (2) 'true' intertidal habitats that are regularly exposed and highly variable. The first type is typified by tidal and back-reef pools where resident corals and other organisms rarely get exposed to air due to water being trapped during low tide but experience strong daily variability due to small water volumes and long seawater residence times. Examples include the well-studied back-reef pools in Ofu, American Samoa (e.g. Thomas et al. 2018) and high-latitude rock pools in South Africa (e.g. Smit and Glassom 2017). The second type mostly encompasses the intertidal zone on meso- and macrotidal coral reefs and includes reef flats, tide pools and shallow lagoons as well as the upper reef slope.

Fig. 6.3 Strong tidal currents in the inshore Kimberley region, NW Australia, creating whirlpools in the middle of the ocean. (Photo credit: Verena Schoepf)

Well-studied examples include shallow reef flats in the Andaman Sea (e.g., Brown et al. 2011), the Kimberley region in NW Australia (e.g., Schoepf et al. 2015b; Gruber et al. 2017), and high-latitude coral cays on the southern Great Barrier Reef (e.g., Shaw et al. 2015; Kekuewa et al. 2021). In the following sections, we review the state of knowledge for these two types of intertidal and back-reef coral environments, focusing on both low- and high-latitude locations. We discuss: (1) the environmental conditions that resident corals and other reef organisms persist under (i.e., coral reef extremeness *sensu* Schoepf et al. (2023)), (2) how the extreme conditions impact coral community composition and/or ecosystem functioning (i.e., coral reef marginality *sensu* Schoepf et al. (2023)), (3) which mechanisms support the survival and stress resilience of coral and other key taxa in these environments, and (4) the lessons that intertidal coral environments provide for the future of coral reefs under rapid climate change.

6.2 Back-Reef and Tide Pool Coral Environments

6.2.1 Ofu Island, American Samoa: Low-Latitude Back-Reef Pools

The back-reef pools of Ofu Island (14.2° S, Figs. 6.1 and 6.4), American Samoa, represent a well-studied natural laboratory for enhanced coral heat tolerance and resilience under rapid environmental change. The island is surrounded by a 80–180 m wide fringing reef (Craig et al. 2001) with shallow (1–3 m) back-reef pools (40—90 m wide) on the southeast coastline (Fig. 6.4). Although these pools experience a microtidal regime (1 m) (Craig et al. 2001), the shallow reef flat (<0.3 m) creates strong diurnal tidal variation, where offshore waters are exchanged at high tide and the pools become isolated at low tide. As a result of tidal modulation and wave flow, the thermal and biogeochemical regimes are exacerbated during austral summers, creating extreme environmental conditions (see below). Yet, coral diversity is high comparatively—85 scleractinian species with 25% coral cover—and incidences of coral bleaching are rare, suggesting adaptive resilience mechanisms for survival in these variable reef habitats (Craig et al. 2001; Birkeland et al. 2021).

The back-reef pools' coral composition, environmental parameters, and coral response metrics (e.g., growth, bleaching) have been characterized and compared to nearby exposed reefs (Craig et al. 2001; Smith et al. 2007; Barshis et al. 2010, 2018; Koweek et al. 2015; Morikawa and Palumbi 2019). Still, most coral thermal tolerance research has focused on two pools separated by <1 km, the Highly Variable (HV) and Moderately Variable (MV) pools (Fig. 6.4a) (Craig et al. 2001; Thomas et al. 2018; Klepac and Barshis 2020, 2022). Mechanisms underlying enhanced thermal tolerance in the genus *Acropora* have been reviewed by Thomas et al. (2018), therefore, this section focuses on the environmental conditions and coral community composition and broad mechanisms of resilience within the HV and MV pools.

Fig. 6.4 (**a**) Ofu and Olosega Island, American Samoa, with the back-reef pools on Ofu as seen from (**b**) shore and (**c**) the mountains above. (Photo credit: Courtney Klepac; Map credit: Thomas et al. (2018))

6.2.1.1 Abiotic Conditions

Tides, PAR, and Temperature Diurnal tides modulate the wave-driven flow into the pools and determine the depth at the reef crest and seawater residence time (Koweek et al. 2015). Austral winters produce greater wave energy striking the southern shore at high tide, which creates borelike, highly turbulent broken waves (Monismith et al. 2024) and facilitates net transfer alongshore (West to East) and offshore (Maticka et al. 2022; Monismith et al. 2024). Conversely, austral summers produce the iconic environmental variability, driving the biogeochemistry and ecology within the back-reef habitats. Wave-driven flow patterns alternate from 15 to 25 cm s^{-1} at high tide and 0–5 cm s^{-1} at low tide (Smith and Birkeland 2007; Koweek et al. 2015). Within the HV and MV pools specifically, the HV pool (4000 m^2) reaches a mean depth of 1.1 m at midday low tide whereas the MV pool (27,000 m^2) has a depth of 1.9 m (Craig et al. 2001). Low tide combined with low turbidity results in maximum irradiances of 1800–2300 μmol quanta m^{-2} s^{-1} of photosynthetically active radiation (PAR) (Smith and Birkeland 2007) in the upper meter, which is also the low tide depth within the HV pool.

Temperature variability has been the dominant focus when comparing Ofu's back-reef pools, and although most pools have similar mean summertime temperatures (29–30 °C) (Craig et al. 2001; Koweek et al. 2015; Klepac and Barshis 2020, 2022), the diel maxima and minima contribute to strong environmental and ecological gradients between the HV and MV pools, which are only separated by 0.5 km. The HV pool low-tide daily summertime temperature ranges from 24.5 to 35 °C (maximum range recorded 6.4 °C) with midday peaks of 34.5 °C for brief periods (Craig et al. 2001; Oliver and Palumbi 2011a; Klepac and Barshis 2020). In contrast, the MV pool summertime daily range is 25–32 °C with a maximum range of 4.5 °C. During the 2015 bleaching year, the HV pool had 125 days in which the daily maximum exceeded the bleaching threshold of 30.2 °C (NOAA CRW 5 km product) (Liu et al. 2018) versus 93 days for the MV pool. Satellite-derived NOAA CRW 5 km products indicated a DHW of 5 °C for both pools during 2015, but *in situ* calculated DHWs and Degree Heating Half-Hours (DHHH), or 'heatloading', revealed that the HV pool had a greater number of DHW and DHH than the MV pool (Klepac and Barshis 2022). Moreover, the number of DHWs has increased two-fold from 2000 to 2017 (Klepac and Barshis 2022). The small-scale variability, high magnitudes of temperature variation, and increased cumulative heatloading between the two pools likely contribute to bleaching tolerance, as mass bleaching has rarely been recorded yet annual non-lethal bleaching occurs regularly for some coral species (Fenner 2019).

pH and DO Compared to temperature, light and flow, other abiotic parameters are less well documented in the HV and MV back-reef pools, but pH and DO were characterized in a nearby (1km) pool, Pool 100 (Koweek et al. 2015). When they become virtually stagnant, pH, dissolved oxygen (DO), carbon flux, and temperature can be modified rapidly under physical and biogeochemical forcing and exceed nearby offshore parameters. Daily pH_T spends only 30% of the time within normal reef ranges of 8.0–8.2 (Koweek et al. 2015), where the diel range is 0.58 units (7.80–8.39). Thirty-four percent of the time, pH_T values are <8.0, and 36% of the time they are >8.2. DO saturation also deviates greatly from mean values, ranging from 32% to 178% (Koweek et al. 2015), and can get as low as 14.6% (i.e., hypoxia) during nighttime low tides (Craig et al. 2001), likely intensified by respiration.

6.2.1.2 Coral Diversity, Community Composition, and Ecosystem Functioning

Despite the extreme conditions, Ofu's back-reef pools boast a diversity of 85 scleractinian coral species (Craig et al. 2001). Mean coral cover is 25% (Brainard 2011) and is approximately 2.6-fold greater than the mean of 17 other inhabited U.S. Pacific islands and reefs (Smith et al. 2016). There are 59 scleractinian coral species within the HV pool while the MV pool contains 79 species; 88% of the species in the HV pool are also within the MV pool but only 58% of the MV species are found within the HV pool. Dominant species and life history strategies (Darling et al. 2012) also differ between the two pools. Generally, four life history strategies have been identified for scleractinian corals—competitive, weedy, stress-tolerant and generalist taxa—which are primarily separated by colony morphology, growth rate, and reproductive mode (Darling et al. 2012). Fifty-two percent of the total coral cover in the HV pool is attributed to *Porites lichen* (no strategy assigned), *Pocillopora damicornis* (weedy), *Pavona venosa* (no strategy assigned), and *Leptoria phrygia* (stress-tolerant). In contrast, *Porites cylindrica* (weedy), mounding *Porites* spp. (stress-tolerant), *Acropora muricata*, and *A. intermedia* (both competitive) account for 66% of coral cover within the MV pool (Craig et al. 2001). It appears weedy and stress-tolerant species dominate the HV pool while the MV pool is dominated by stress-tolerant, competitive, and a few weedy species.

In contrast to coral cover, mean fleshy algal cover is three-fold lower in Ofu than in other inhabited Pacific islands (Smith et al. 2016), and diver-towed estimates of crustose coralline algae (CCA) cover indicated estimates similar to coral cover at approximately 25% (Brainard 2011). The CCA species *Porolithon onkodes* dominates the reef crest, forereef slope, and other areas with higher wave action, rapidly cements rubble, and attracts successful coral recruitment (Birkeland et al. 2021). Coralline lethal orange disease is common and abundant in Ofu (Fenner 2019). However, coral disease prevalence is quite low at 0.34%, where disease states are commonly manifested as growth anomalies (Fenner 2019).

To the best of our knowledge, only one study to date has investigated community metabolism in Ofu's back-reef habitats. Koweek et al. (2015) quantified net community calcification to production ratios (NCC/NCP) and found that Ofu's back-reef represents a system dominated by photosynthesis and respiration, serving as a carbon sink during the day. This ratio is much lower than that reported for other shallow back-reefs with diel variability. Ofu pool carbon system variability is predicted to be greatest during spring low tides before sunrise and mid-afternoon. Such high production could result from accretion limitation in shallow depths and/or high production rates when waters warm rapidly (Koweek et al. 2015).

6.2.1.3 Mechanisms of Stress Tolerance

Thermal Tolerance and Growth The increased thermal tolerance of HV corals has previously been reviewed by Thomas et al. (2018), providing acclimatory and adaptive mechanisms underlying bleaching tolerance, but solely for *Acropora hyacinthus* and *A. nana.* For the remaining 83 coral species found within the backreef pools of Ofu Island, few studies have investigated tolerance capacities and potential mechanisms in massive *Porites* spp. (Smith et al. 2007; Barshis et al. 2010, 2018; Klepac and Barshis 2020, 2022), *Goniastrea retiformis* (Klepac and Barshis 2020), *Porites cylindrica* (Smith et al. 2007; Morikawa and Palumbi 2019), and *Pocillopora damicornis* (Morikawa and Palumbi 2019). Through transplantation between the HV and MV pools as well as between the HV pool and either a nearby forereef or less variable habitat, studies corroborated greater growth and bleaching resistance in back-reef corals, specifically from the HV pool environment (but see Klepac and Barshis 2020, 2022). HV corals consistently had greater growth rates and skeletal density (Smith et al. 2007; Barshis et al. 2018), except in massive *Porites* spp. and *P. cylindrica* where growth was greater in the MV pool (Morikawa and Palumbi 2019; Klepac and Barshis 2020, 2022). *Porites lobata* heat shock protein (HSP) 70 increases were more rapid and larger in back-reef or variable-thermal acclimated versus forereef or stable-thermal acclimated corals, suggesting the capacity to maintain homeostasis under thermal stress (Barshis et al. 2010, 2018). In transplantation studies, growth was strongly related to source pool (Morikawa and Palumbi 2019; Klepac and Barshis 2020). Fixed effects of natural or experimental bleaching responses were also observed, but which back-reef population was more thermally tolerant depended on the species and year of study. Morikawa and Palumbi (2019) confirmed greater natural bleaching tolerance in four branching coral species from the HV pool versus the MV pool. In contrast, Klepac and Barshis (2020, 2022) consistently documented reduced physiological and transcriptomic tolerance in experimental fragments of massive *Porites* spp. from the HV pool compared to the MV pool. It is tempting to speculate that corals from the MV pool have a broader tolerance to transplantation and/or acute heat stress, or that the HV environment has recently become chronic and native corals have not had sufficient recovery. Their work occurred during the warmest years in the last decade (2010–2017) compared to other thermal tolerance studies (but see Morikawa and Palumbi 2019) and could reveal the upper thermal limits for massive coral species in the chronically warm HV pool.

Coral Algal Symbionts and Microbiome Symbiotic associations within back-reef corals also contribute to elevated coral environmental tolerance but are environmentally and host-dependent. Oliver and Palumbi (2011b) identified the thermally tolerant genus *Durusdinium* in eight out of nine coral species sampled (cp23 and ITS genomic regions). *Durusdinium* proportions were also greater within five HV pool species in comparison to the MV pool, corroborating the higher prevalence of *Durusdinium* in hotter habitats (Ulstrup and Van Oppen 2003; Fabricius et al. 2004; Oliver and Palumbi 2009; LaJeunesse et al. 2010a, b). Conversely, tolerant massive *Porites* spp. and *Goniastrea retiformis* primarily associate with different variants of *Cladocopium* spp. in both HV and MV pools (Klepac and Barshis 2020) while *P. cylindrica* and *P. damicornis* host a combination of *Cladocopium* variants and *Durusdinium trenchii* (Morikawa and Palumbi 2019). Therefore, *Durusdinium* alone does not explain thermal tolerance differences between the two pools as transplantation experiments revealed no symbiont shuffling or switching (Palumbi et al. 2014; Klepac and Barshis 2020), and MV coral bleaching resistance was not explained by hosting *Durusdinium* compared to *Cladocopium* (Oliver and Palumbi 2011b). Although Symbiodiniaceae composition plays a role in the holobiont's thermal tolerance, the lack of shifting and subsequent increased thermal tolerance of corals moved into the HV pool suggest adaptive symbiont fidelity that does not change within short time frames ($\leq$12 months) (Palumbi et al. 2014; Klepac and Barshis 2020) or under non-bleaching years (Palumbi et al. 2014).

In contrast to Symbiodiniaceae, *Acropora hyacinthus's* microbial community shifts following coral transplantation from the MV to the HV pool to match that of the native HV community assemblages. Corals native and transplanted into the HV pool maintained their algal symbiont assemblages and exhibited greater bleaching resistance following acute heat stress experiments (Ziegler et al. 2017a). Whether this phenomenon exists in the remaining HV pool coral species is yet to be determined and a future directive to further understand the microbiome's role in stress tolerance.

6.2.1.4 Lessons and Future Outlook

The corals from the back-reef pools on Ofu Island thrive despite spatially varying environmental conditions that would result in bleaching and mortality elsewhere, even in nearby forereef habitats. Fixed and flexible responses to these extreme conditions have been demonstrated in as little as 5 km, therefore ecosystem-level outcomes under future warming depend on adapted genotypes and/or plastic responses that promote survival. Massive coral species were determined to be locally adapted to their native reef (Smith et al. 2007; Barshis et al. 2018; Klepac and Barshis 2020, 2022) while competitive fast-growing branching species were able to acclimate to novel conditions (Barshis et al. 2013; Palumbi et al. 2014; Bay and Palumbi 2015). Most coral species are broadcast spawners (except *P. damicornis*), yet high genetic differentiation exists among HV and MV coral populations (Bay and Palumbi 2014; Barshis et al. 2018). Spatial balancing selection under high gene flow (Levene 1953) has been posited as the mechanism whereby recruits from a mixed gene pool, under high tide wave-driven flows, become exposed to different selective forces in each pool, under low tide isolation, and recruits possessing locally-adapted alleles or epigenetic modifications from parents survive the environmental pressure within each pool. Provided moderate genetic diversity and gene flow, physiological variation in response to thermal stress, and evidence of acclimatization and local adaptation, Ofu's corals possess small-scale adaptation and the potential for rapid evolution over time (Bay et al. 2017; Thomas et al. 2018). However, increases in the frequency and severity of global bleaching events coupled with an absence of coral research on Ofu since 2018 posits the question of whether these adaptive mechanisms are sufficient to withstand future warming, especially if the conditions of the HV pool surpass thermal thresholds (see Klepac and Barshis 2020) under the recent extreme bleaching event in 2024. Further research adopting multi-species and multidisciplinary approaches investigating effective population sizes and rates of genetic adaptation would illuminate the capacity for survival under future emission scenarios. Moreover, assessing the susceptibility or tolerance to other environmental stressors, such hypoxia and acidification as observed by Koweek et al. (2015), could further enhance our understand of innate tolerance to these back-reef pools and the potential adaptability under future deoxygenation and acidification scenarios.

6.2.2 KwaZulu-Natal, South Africa: High-Latitude Intertidal Rock Pools

The northeastern coast of South Africa, constituted by the province of KwaZulu-Natal (colloquially KZN), runs from the southern border of Mozambique at 26.858° S to the border of the Eastern Cape province at 31.081° S (Fig. 6.1), a distance of approximately 580 km. The coast is high energy and swell dominated, and relatively linear with few bays and no islands or other offshore barriers (Palmer et al. 2011). About 80% of the coast is sandy beaches, with intermittent rocky outcrops making up the rest, increasing in proportion towards the south of the province. The area spans two 'bioregions', the Delagoa region and the Natal region, with the break at Cape Vidal (28° 7.596′ S, 32° 33.649′ E) (Sink et al. 2005). However, the boundary may not be clear for all faunal groups. For example, the region south of Cape Vidal has been described as an overlap region between the Indo-Pacific and Agulhas Province further south (Smit et al. 2013).

Hermatypic corals are found on subtidal reefs along most of the province's coastline. In the north, primarily within the Simangaliso Wetland Park, coral-dominated assemblages occur at depths of approximately 10–40 m, constituting the southernmost coral-dominated assemblages on the African continent and among the southernmost globally. These are not true reefs, however, as there is little or no accretion (Ramsay 1996). Rather they are assemblages of corals, growing on fossilized sand dunes (Ramsay 1996). In all, 96 species of hermatypic corals are recorded from the South African reefs (Riegl 1996 and references therein) and 39 species of soft corals (Schleyer and Celliers 2003). Despite the lower species richness, however, there is higher cover of soft corals on many of the reefs. South of the Simangaliso, corals occur on rocky reefs throughout the province, but coral cover and diversity are vastly less. Notably, the Aliwal Shoal, a rocky, algal dominated reef lying at a latitude of 30.25° S hosts 14 scleractinian coral species, especially in shallower areas of 12–15 m depth (Schleyer et al. 2006; Olbers 2017a).

On rocky shores along the coast, corals can be found in small tidal rock pools (Fig. 6.5). Corals have been formally recorded in tidal pools as far south as Munster Beach (31° 00′00.5″ S, 30° 16′05.4″ E, Fig. 6.5c) (Smit and Glassom 2017) but have been informally reported considerably further south, into the Eastern Cape province. Although there is a considerable body of work on the rocky intertidal shores of the region (e.g., Sink et al. 2005), this has focused almost entirely on the benthic assemblages of emergent shores, patterns of productivity, or resource use. As elsewhere, rock pools, where the gradients associated with rocky shores are not apparent, have been largely neglected and the abiotic conditions and coral fauna in the pools are thus poorly recorded. While much less is known about this intertidal

Fig. 6.5 The intertidal rock pools on South Africa's northeastern coast. (**a**) Treasure Beach, a wave-cut rocky platform. (**b**) Isipingo rock pool. (**c**) Deep rock pool at Munster. (**d**) *Anomastraea irregularis* competes with coralline algae. (Photo credit: (**a**, **d**) David Glassom, (**b**, **c**) Kaylee Smit)

coral habitat compared to the other locations highlighted in this chapter, we note that they have significant potential to serve as natural laboratories to study mechanisms of coral stress tolerance at higher latitudes and summarize here the little information that is available in the literature.

6.2.2.1 Abiotic Conditions

South Africa has a semi-diurnal, microtidal regime, with tidal ranges of 0.8 m (neap) to 1.8 m (spring). Most of the of KZN coastline is highly wave-exposed and rocky shores are classified either as wave-cut rocky platforms, with relatively shallow slopes (Fig. 6.5a), or as exposed rocky headlands (see Plate 1 in Jackson and Lipschitz 1984). Intertidal rock pools tend to range from 1 to 4 m in diameter and ≈0.1–0.5 m deep at spring low tide (Fig. 6.5) (see Plate 2 in Jackson and Lipschitz 1984). Possibly due to the high wave exposure, corals are seldom found on the exposed intertidal shore, but are common in rock pools (Fig. 6.5). Although sea surface temperature declines with latitude along the coast (Smit et al. 2013), mean temperature measured in rock pools was similar among sites covering the length of KZN (Smit and Glassom 2017). Summer mean temperatures in the pools are generally in the mid-20 s but temporarily reached 32.5 °C at spring low tides in some pools, with seasonal ranges of 13 °C (Smit and Glassom 2017). Maximum daily temperature ranges of up to 10 °C occurred in some locations at summer spring tides. However maximum temperatures in individual rock pools depend largely on the size and depth of the pool rather than on a latitudinal gradient. A similar pattern of higher variability of intertidal habitats, but similar mean temperatures, has been described for the Kimberley region of Australia (Speelman et al. 2023). Very little is known about other environmental parameters in the rock pools; however, mean annual irradiance generally decreases along the coast from 1768 kWh m^{-2} at Sodwana Bay to 1629 kWh m^{-2} at Munster (data from https://geosun.co.za/). Seasonal irradiance data are not available.

6.2.2.2 Coral Diversity, Community Composition, and Ecosystem Functioning

Coral fauna in South African rock pools is poorly documented, in part also because the transect/quadrat survey methods typically used on rocky shores easily bypass species that cover little of the substratum. Even recent monitoring programmes have largely overlooked corals in intertidal habitats, with one report recording only *Favia* sp. or 'hard coral' at 0.1% cover at three locations (Olbers 2017b) and another recording only the presence of *Pocillopora* along the coastline (Mvula 2020). To date, only one study has been devoted specifically to their distribution and abundance along the coast (Smit and Glassom 2017). Overall, they documented 16 species of scleractinian corals in rock pools at four sites along the coast, although corals recorded as *Pocillopora damicornis* may in fact be *P. verrucosa* (Chiazzari et al. 2019). Not all corals were identified to species level, largely because the constraints of shallow tidal pools resulted in unusual morphologies. However, those identified to species level represented all four life history strategies (Darling et al. 2012), though stress-tolerant and weedy species seem more common. In the northern and central regions of KZN province, 12 species were recorded at each site, although the species composition differed among sites. However, at the southernmost site, only six species were found. There was no obvious change in species composition with latitude. For example, *Favia speciosa* (now accepted as *Dipsastraea speciosa* (Dana, 1846)) occurred at the most northern and most southern sites, but not in between. It thus seems unlikely that coral species distributions in intertidal habitats are bound by the bioregions suggested above. Species richness in South African rock pools is similar to that found previously in Kenya, where *Anomastraea irregularis* is also common (Raker et al. 2014), further suggesting, albeit based on sparse data, that there is no steep latitudinal gradient in this region, with relatively few coral species thriving in intertidal pools. Species richness is also similar to that recorded on a large rocky reef, Aliwal Shoal (14 species), at a similar latitude to Isipingo (Olbers 2017a), although the species composition on the subtidal reef differed markedly from that in the intertidal pools. To date, no studies have investigated processes related to coral reef ecosystem functioning (Brandl et al. 2019) in the rock pools.

6.2.2.3 Mechanisms of Stress Tolerance

Two coral species, *Pocillopora verrucosa* (competitive life history strategy) and *Anomastrea irregularis* (unknown life history strategy), are the most common species in South African rock pools (Smit and Glassom 2017) and display considerably different traits—*P. verrucosa* being a fast-growing, widespread branching coral and *A. irregularis* a massive, relatively range and habitat-restricted species (Fig. 6.5d). Due to their abundance and contrasting traits, they have become the focus of ecological and physiological studies in the pools, providing most of our knowledge about coral adaptation to the conditions of these tidal pools. Since both species occur intertidally and subtidally in similar locations, comparisons between habitats have also been possible.

Ecophysiology Although the two species thrive in the tidal pools along the coast, they display considerably different physiological strategies in doing so (Smit 2014). For example, *P. verrucosa* relies heavily on stored lipids in winter, also reducing the number of algal symbionts, but increasing the chlorophyll content per cell. Stored lipids decline precipitously during winter and recover during spring, implying that low light in winter is more limiting than high temperatures in summer. Stored lipids declined more at more southerly sites, further supporting this notion. In contrast, *A. irregularis* displays little change in stored lipids, symbiont density or productivity throughout the year, implying that neither summer temperatures nor low winter light reach levels to cause stress. Isotope signatures suggest that *A. irregularis* also makes more use of heterotrophy than *Pocillopora* (Smit 2014).

Gene Expression and Species-Specific Heat Tolerance When subjected to a short heat stress test, host transcriptomes of each species responded differently (Onyango 2020; Onyango et al. 2021). However, the response was far greater in *P. verrucosa* than in *A. irregularis*, with 17,559 transcripts (33.59%) differing among treatments for *P. verrucosa* compared to only 1362 (1.23%) in *A. irregularis,* suggesting greater sensitivity to heat stress in *P. verrucosa*. The demands on energy reserves of mounting large transcriptomic responses to environmental stresses may partially explain *Pocillopora's* susceptibility to bleaching (Onyango et al. 2021). Responses in the two species differed in nature as well as magnitude. Overall, the heat stress response in *A. irregularis* comprised stress protective mechanisms, including up-regulation of apoptosis and a decrease in mitosis. There was no evidence of changes in metabolic genes or heat shock proteins, but transcripts coding for proteins involved in general cellular protection against heat damage differed between temperature treatments. In *P. verrucosa*, up-regulation of enzymes that activate the glyoxylate cycle (which allows the synthesis of carbohydrates from storage lipid breakdown) suggested that the host did not receive enough carbohydrates from its endosymbionts and thus relied on internal energy reserves during thermal stress.

Transcriptomic responses to heat stress also differed in symbionts from different host species (Onyango 2020; Onyango et al. 2021). Changes in the transcriptomes of heat stressed symbionts suggested reduced growth rates of thermally stressed *Symbiodinium*. *Symbiodinium* transcripts from *Pocillopora* also showed down-regulation of transcripts indicative of endocytosis, which is important in transferring nutrients between the host and symbiont. Its reduction may suggest a mechanism by which the coral-algal symbiosis may be disrupted (Onyango 2020; Onyango et al. 2021).

Photo-Physiology and Habitat-Specific Heat Tolerance Two physiological responses to heat stress, non-photochemical quenching (NPQ) and the ratio of variable to maximum fluorescence (F_v/F_m), were quantified in cultured symbionts extracted from *P. verrucosa* and *A. irregularis*. NPQ is a mechanism of dissipating excitation as heat to avoid damage by reactive oxygen species. Although increased NPQ is usually a reaction to high light levels on shallow reefs, it also increases in response to raised temperatures. A quicker and greater NPQ response may thus protect corals from bleaching. F_v/F_m is a measure of photochemical efficiency, with a higher ratio denoting greater efficiency. *Acropora irregularis* showed higher NPQ as temperatures increased than *P. verrucosa* and also maintained higher F_v/F_m. Algal symbionts from *A. irregularis* thus displayed greater thermotolerance than those of P. verrucosa. Interestingly, symbionts of both species from intertidal habitats maintained higher F_v/F_m under heat stress than those from *P. verrucosa from* subtidal habitats (D. Glassom/S. Babane, unpublished data). Similar habitat-specific differences in heat tolerance have also been observed in other study systems (e.g., Schoepf et al. 2015b; Thomas et al. 2022; Almeida et al. 2024), highlighting that temperature fluctuations characteristic of intertidal habitats enhance coral heat tolerance (Safaie et al. 2018).

6.2.2.4 Lessons and Future Outlook

Threats to the rock pools along the coast of KZN are largely due to the proximity of many of them to urban and industrial developments. Of the four sites discussed here, two of them, Treasure Beach and Isipingo, are close to Ethekwini, the largest metropolitan area of the province, and Isipingo, in particular is vulnerable to spills of industrial effluent from a nearby silt canal. Climate change also poses a threat to corals in pools, not only because of increasing temperatures, but also because of the risk of more frequent and intense storms (Guastella and Rossouw 2012) which could lead to sand inundation due to the shallow slope. At the southern limits of coral distribution, decreases in aragonite saturation due to increasing CO_2 levels is potentially limiting to coral growth. Last but not least, subsistence harvesting of mussels and other rocky shore invertebrates does not threaten corals directly, but does increase the risk of mechanical damage during harvesting. Although coral communities in intertidal pools along this coastline do not host large numbers of colonies or species, they may be important indicators of changing conditions in the intertidal environment, in addition to being useful for learning about coral adaptation to extreme and marginal conditions. Molecular studies on intertidal corals can show connectivity among populations and genetic changes at the edges of species ranges (Chiazzari et al. 2019). Research should continue to focus on this area.

6.3 'True' Intertidal Coral Environments with Regular Aerial Exposure

6.3.1 Phuket, Thailand: Low-Latitude Mesotidal Reef Flats

Coral reefs in the Andaman Sea represent unique ecosystems due to the meso- to macro-tidal regimes that expose intertidal reef areas to a wide range of environmental stresses not commonly experienced by their subtidal counterparts (see review by Brown 2007). The region predominantly experiences semi-diurnal tides with a range of 2–5 m, except for Myanmar's Mergui Archipelago, where tides exceed 5 m (Brown 2007). Extensive intertidal reef flats are common throughout the region which get exposed to air during spring low tides for up to 3 h (Fig. 6.6) (Brown 2007); however, the frequent occurrence of negative sea level anomalies in this region can modulate typical tidal dynamics and is an important driver of changes in coral cover over time (e.g., Brown et al. 2019; Dunne et al. 2021).

Many coral reefs in the Andaman Sea are characterized by significant turbidity because several large rivers discharge into the northern Andaman Sea and strong tidal currents stir up sediment. Nevertheless, extensive coral reefs exist in inshore turbid settings whereas offshore reefs are characterized by higher visibility (Brown 2007). On inshore reefs throughout the region, local drainage from mangrove forests also supplies waters enriched in particulate and dissolved organic matter which increases turbidity but may benefit coral health through enhanced heterotrophy (Brown 2007). However, sedimentation stress from strong tidal currents can also negatively affect coral health (Scoffin et al. 1997).

Much of our knowledge about coral resilience in extreme intertidal settings comes from long-term study sites on the SE tip of the island of Ko Phuket in Thailand (7.8° N, Figs. 6.1 and 6.6a). They are among the best-studied examples of intertidal reef flats and have yielded crucial insights into coral adaptations and mechanisms of heat tolerance (e.g., Brown 2007; Brown et al. 2019; Dunne et al. 2021). Notably, they have provided some of the strongest evidence of environmental stress memory in corals (Brown et al. 2002c, 2014a).

Fig. 6.6 (**a**) Intertidal reef flat on the SE tip of the Kho Phuket peninsula (Panwa), with (**b**, **c**) corals exposed at low tide. (Photo credit: Marlene Wall)

6.3.1.1 Abiotic Conditions

Tides, PAR, and Temperature Many coral reefs in the Andaman Sea belong to the roughly 30% of reef systems worldwide that are tide- instead of wave-dominated (Lowe and Falter 2015). The Phuket reef flats have a tidal range of 0.6 m (neap) to 3.1 m (spring) and get exposed to air only during spring low tides (Fig. 6.6). While the timing of low tides typically ensures that corals avoid mid-day aerial exposure (Brown et al. 2014b), afternoon exposure can nevertheless lead to bleaching and partial mortality caused by solar radiation, particularly on the western sides of coral colonies that face towards the sun (Brown et al. 1994; Scoffin et al. 1997; Sweet et al. 2017). Aerial exposure also varies seasonally, being greatest on afternoon tides in February–April and on early morning tides in January–February (Sweet et al. 2017). As a result of the mixed tides, the reef flat experiences strong fluctuations in abiotic conditions such as temperature, PAR and UV radiation (Brown et al. 1994, 2002c) but only few high-resolution datasets from *in situ* data loggers are available. They show that PAR can reach up to ~1800 μmol m^{-2} s^{-1} on a cloudless day (Sweet et al. 2017), while temperatures range from ~24 to >36 °C (Sweet et al. 2017; Conrad 2019). Reef flat organisms further experience daily fluctuations of up to 9 °C (Conrad 2019) or even up to 14 °C (Sweet et al. 2017), though the latter range includes measurements when the logger was exposed to air and solar radiation. These short-term maxima exceed the estimated local coral bleaching threshold of 30.1 °C (Brown 2007; Brown et al. 2019) by 5–6 °C, yet this does not typically lead to bleaching as exposure time is limited to a few hours (Sweet et al. 2017). Crucially, temperature and PAR maxima do not generally occur at the same time, as temperature maxima typically occur during the late afternoon/evening (Brown et al. 1994; Sweet et al. 2017), though exceptions are early afternoon low tides at certain times of the year (February—April) (Brown et al. 2002c; Sweet et al. 2017).

Turbidity and sedimentation Many inshore reefs in the Andaman Sea are characterized by high turbidity and sedimentation, with suspended terrigenous solids as high as 40 mg L^{-1} and visibility often <1 m (Scoffin et al. 1992). On the Phuket reef flats, terrigenous sediment supply in the form of clay minerals is greatest during the wet season (May to October) but reef flats facing the east are most affected by sedimentation from November to April when NE winds dominate (Scoffin et al. 1997). Sediment trap accumulation rates range from 13 to 16 kg m^{-2} $month^{-1}$ in the dry season and from 10 to 11 kg m^{-2} $month^{-1}$ in the wet season; however, times of highest turbidity and sedimentation do not necessarily coincide with times of greatest settlement on intertidal corals because this also depends on monsoon-related differences in wave agitation and currents (Scoffin et al. 1997). Accumulation of fine sediments is generally greater on the inner reef flat (Fig. 6.6) and often occurs on dead areas of the colony such as solar lesions, whereas bioerosion is more prevalent on the outer reef flat (Scoffin et al. 1997). These combined stresses lead to highly asymmetric growth patterns where massive colonies develop inclined, partially dead, western sides and actively growing, convex eastern sides (Scoffin et al. 1997).

pH and DO To the best of our knowledge, pH and DO dynamics on the Phuket reef flats have not yet been published but are likely highly variable as observed on other reef flats with large tidal ranges (e.g., Pedersen et al. 2016; Gruber et al. 2017).

6.3.1.2 Coral Diversity, Community Composition, and Ecosystem Functioning

Despite often sub-optimal environmental conditions, coral reefs in the Andaman Sea have high species richness (Brown 2007), with 353 coral species from 69 genera reported for the West coast of Thailand (Turak et al. 2005). Only a limited number of species can tolerate the extreme environmental conditions on intertidal reef flats, yet Phuket reef flats are nevertheless relatively diverse with over 30 scleractinian species (Brown et al. 2019; Dunne et al. 2021). Massive coral genera dominate (e.g., *Porites*, *Platygyra*, *Coelastrea*, *Goniastrea* and *Favites*), interspersed with some branching genera such as *Acropora*, *Pocillopora* and *Montipora* (Scoffin et al. 1997; Brown et al. 2019), whereas the more exposed offshore reefs are dominated by branching corals (Brown 2007). On the reef flat, the coral species *Goniastrea aspera* (weedy life history strategy), *Goniastrea retiformis* (stress-tolerant), *Platygyra sinensis* (stress-tolerant), *Platygyra daedalea* (stress-tolerant), and *Porites lutea* (stress-tolerant) are typically dominant (Scoffin et al. 1997), indicating that stress-tolerant and weedy life history strategies (Darling et al. 2012) facilitate survival and success in this extreme environment.

Long-term monitoring of permanent transects has revealed that coral cover on Phuket reef flats fluctuated dramatically between 1987 and 2020, ranging from as little as 4% to maxima of 67%, with key drivers being rising temperatures and sea level that are strongly modulated by (intra-) decadal climate processes associated with El Niño Southern Oscillation and the Indian Ocean Dipole (Dunne et al. 2021). Although rising sea level promotes coral cover on these shallow reef flats and may mitigate the impacts of marine heatwaves (Brown et al. 2019), severe reductions in coral cover were caused by negative sea level anomalies and bleaching events which sometimes coincide and are both expected to increase in frequency due to climate change (Dunne et al. 2021). In contrast to these well-documented trends in coral diversity, cover, and community composition, processes related to coral reef ecosystem functioning (Brandl et al. 2019) have not yet been studied on Phuket's reef flats.

6.3.1.3 Mechanisms of Stress Tolerance

Light, Heat, and Desiccation Stress Intertidal corals have evolved a range of mechanisms and adaptive strategies that facilitate survival and success in this extreme environment, with certain biochemical and behavioural adaptations facilitating resistance to heat, high light intensities and desiccation. For example, Phuket reef flat corals can rapidly retract their tissue to mitigate light stress for the algal symbionts (Brown et al. 2002b) and produce large amounts of mucus to avoid desiccation during aerial exposure (Jatkar et al. 2010). Furthermore, reef flat corals have high levels of antioxidant enzymes and heat-shock proteins as well as a large xanthophyll pool to dissipate excess light as heat via non-photochemical quenching (Brown et al. 2002a). These biochemical defences were found to be many-fold higher in intertidal reef flat corals compared to subtidal coral species from the Caribbean, and also in the western compared to the eastern colony sides which experience higher light levels and have enhanced heat tolerance (Brown et al. 2002a, c). Remarkably, the enhanced bleaching resistance of western compared to eastern colony sides of Phuket reef flat corals was maintained for at least 10 years after 180° rotation, providing the most robust evidence to date of long-term environmental stress memory in corals (Brown et al. 2014a).

Algal Symbionts and the Microbiome Reef flat corals frequently associate with heat-tolerant Symbiodiniaceae from the genus *Durusdinium* (Brown 2007), with various inshore corals such as *Coelastrea aspera* being dominated by *D. trenchii* (formerly D1a) (Pettay and LaJeunesse 2009; LaJeunesse et al. 2010a; Brown et al. 2014a). In fact, it has been suggested that the warm temperatures and variable light levels of the Andaman Sea have facilitated the ecological success and evolutionary radiation of *Durusdinium* (LaJeunesse et al. 2010a). Interestingly, however, the more bleaching-resistant western colony sides of reef flat *C. aspera* harbour the same symbiont type as the more susceptible, eastern sides (Brown et al. 2002c, 2014a), highlighting that the coral host also plays an important role in heat tolerance in this species (see above). Furthermore, coral-associated microbial communities can also influence coral health and stress tolerance (e.g., Reshef et al. 2006). For example, *C. aspera* rapidly shifts its microbial diversity over the tidal cycle on the reef flat (Sweet et al. 2017), which may help in maintaining a healthy microbiome under stressful conditions.

6.3.1.4 Lessons and Future Outlook

Given the remarkable stress tolerance and adaptations of intertidal corals in the Andaman Sea, it was long thought that coral reefs in this region have a higher resistance to marine heatwaves, particularly also because the turbid and nutrient-rich inshore waters may act as a refuge and benefit coral resilience (Brown 2007). While this may have been the case before bleaching events became increasingly common, Phuket reefs have now already experienced more than six bleaching events to date (1991, 1995, 1998, 2003, 2010, 2016, and 2024) (Phongsuwan and Chansang 2012; Brown et al. 2019; ThePhuketNews 2024), demonstrating that intensifying climate change poses a severe threat to even the most heat-resistant corals. Yet, the severe 2010 bleaching event also revealed differential resilience among reef flat types and coral genera: the shallowest reef flats experienced the greatest coral mortality and community structure disruption, and *Coelastrea* suffered extensive mortality while *Porites lutea* recovered fairly rapidly (Brown et al. 2014a, 2019). Negative sea-level anomalies and the frequency of marine heatwaves will therefore be key drivers of future shallow reef trajectories in the Andaman Sea, although rising sea level may mitigate these threats to some extent (Brown et al. 2019).

6.3.2 Heron Island, Southern Great Barrier Reef: A High-Latitude Mesotidal Reef Flat

The high-latitude Swain, Capricorn and Bunker groups mark the southernmost section of Australia's Great Barrier Reef (GBR). These reef systems are dominated by platform reefs and sandy coral cays, where a reef crest usually encloses an extensive shallow lagoon and/or reef flat. Southeastern trade winds prevail most of the year in this region, interspersed by calmer and more variable wind directions in summer (Connell et al. 1997). Heron Island (23.4° S, Figs. 6.1 and 6.7), a 0.16 km^2 coral cay surrounded by a large 27 km^2 platform reef lagoon, hosts the oldest and largest research station on the GBR. As such, the Heron Reef system is amongst the best studied reef habitats in the world. Located on the continental shelf but 65 km offshore, this region is primarily wind- and tide-dominated (Lowe and Falter 2015), with rela-

Fig. 6.7 Heron Island, southern Great Barrier Reef, Australia. (**a**) Reef flat with Heron Island in the background. (**b**, **c**) Coral communities of the reef flat. (**d**) Upward coral growth on the reef flat is limited by low tides. (Photo credit: Rene van der Zande)

tively little terrestrial influence from the Australian mainland, except during hurricanes (Devlin et al. 2001).

On platform reef systems such as Heron Reef, the reef flat often makes up the largest part of the area inside the reef crest (Fig. 6.7). During low tides, the lagoon is effectively isolated from the adjacent ocean, owing to a slightly elevated reef crest (Scopélitis et al. 2011). When this ponding occurs, the entire reef flat is subjected to reduced water movement for several hours while large sections also experience aerial exposure, leading to strong daily gradients in important environmental parameters such as PAR, temperature, pH and DO (see below). Despite the collective term, the reef flat is very heterogeneous in benthic composition, especially between areas closer and further from the crest (Phinn et al. 2012). Typical reef flat areas located more towards the center of the lagoon are characterized by a mostly sandy bottom with interspersed larger coral patches (Fig. 6.7b, c), whereas reef flat sections close to the crest are more dominated by pavement and a much higher coral cover (Phinn et al. 2012; Brown et al. 2020). Water depth is also slightly higher further away from the crest, and therefore this habitat generally is more stable in its environment.

6.3.2.1 Abiotic Conditions

Tides, Temperature, and PAR Tides on Heron Island and the rest of the Capricorn region follow a mixed tidal pattern. Tidal ranges differ considerably between neap and spring tide, varying between 0.8 and 2.8 m respectively (Connell et al. 1997). The timing of the tides continuously shifts throughout the tidal cycle, causing low tide and coral aerial exposure to regularly align with maximum daily temperatures and PAR. Due to its high latitude (23.4° S) and strong seasonality, Heron Island has a large yearly average sea surface temperature gradient compared to more tropical locations, with average monthly sea surface temperatures ranging from 20 to 22 °C in winter to 26–28 °C in summer (Connell et al. 1997). Likewise, both average and maximum daily PAR increase nearly two-fold between winter and summer (Kline et al. 2015), with the reef flat regularly experiencing daytime PAR levels exceeding 2000 μmol m^{-2} h^{-1}.

The interaction of seasonality with the shifting tidal cycle frequently produces large daily environmental variability as slack water over the reef flat heats or cools rapidly. Daily temperature fluctuations larger than 8 °C can be found locally (Jimenez et al. 2012; Safaie et al. 2018; Kekuewa et al. 2021; Brown et al. 2022a, 2024), although a 2–4 °C range is more common (Kline et al. 2015; Brown et al. 2024). However, excessive warming or cooling during low-tide exposure can reach local maximum and minimum temperatures of 36 and 13.3 °C in summer and winter (Hoegh-Guldberg et al. 2005; Brown et al. 2023), respectively, and Heron Reef has been one of the few places where both heat- and cold-stress related coral bleaching has been observed (Hoegh-Guldberg and Fine 2004; Hoegh-Guldberg et al. 2005; Jimenez et al. 2008; Fordyce et al. 2019).

pH, DO, and Nutrients During falling and incoming tides at calm weather, seawater has a long residence time over the reef flat (Mongin and Baird 2014) and interacts strongly with the underlying benthos (Anthony et al. 2013; Kline et al. 2015; McMahon et al. 2018; Perez et al. 2018). This leads to daily pH fluctuations between 0.1 and 0.8 units in reef flat habitats (Santos et al. 2011; Georgiou et al. 2015; Kline et al. 2015; Kekuewa et al. 2021), and fluctuations in seawater CO_2 saturation between 300 and 1650 μatm (Brown et al. 2022b), and in aragonite saturation state between 1.7 and 5.9 (Kline et al. 2015; McMahon et al. 2018). Similarly, DO fluctuates heavily, ranging between 0.45 mg L^{-1} and more than 7.41 mg L^{-1} during day and nighttime over a carbonate sand dominated reef flat section (Santos et al. 2011; Kekuewa et al. 2021), but is less variable close to the reef crest (Kline et al. 2012; Kekuewa et al. 2021). Despite its small size, the island hosts large seabird populations especially during brooding season across Austral summer. Transport of nutrients from guano can therefore locally and temporarily increase nutrient concentrations (Johnson 1995).

6.3.2.2 Coral Diversity, Community Composition, and Ecosystem Functioning

Heron Reef is home to approximately 72% of coral species found on the GBR (NOAA 2024b), in spite of its high latitude and low winter temperatures. However, the number of species inhabiting the highly variable crests and reef flat areas is considerably lower. Tanner (2017) identified a total of 63 reef-building species inhabiting a section of the exposed northern reef crest, whereas Connell et al. (1997) counted 72 species on a calmer section of the largely protected southern crest. A kilometer-long, north-south cross-section transect of the reef flat revealed at least 45 species (Bradbury and Young 1981). Similarly, another study found at least 28 massive coral species occupying a 15 × 7 m quadrat on the reef flat, though this list excluded *Porites* species (Endean et al. 1997). Surprisingly, more recent community assessments at species level are lacking or do not distinguish between reef slope and flat (Connell et al. 2004), making it difficult to determine whether corals with certain life history strategies dominate certain habitats. However, the 28 species reported in Endean et al. (1997) largely comprised corals with stress-tolerant life history strategies (Darling et al. 2012). In recent years, Heron Reef has been heavily surveyed (e.g., photo transects and remote sensing) but this data is so far only analyzed at community composition or functional group level (e.g., Scopélitis et al. 2011; Brown et al. 2020; Roelfsema et al. 2021; Tanner and Connell 2022).

The benthic composition of the reef flat is diverse, and communities can be locally dominated by corals or macroalgae (no seagrass), distributed across a mostly carbonate sands flat. Live coral cover on the reef flat is usually 10–20% (Brown et al. 2020, 2024; Fordyce et al. 2021; Kekuewa et al. 2021) and relatively stable throughout the year, but macroalgae cover varies strongly between seasons (Brown et al. 2018). Typical coral genera occurring on the reef flat are (branching) *Acropora*, *Pocillopora*, *Montipora*, branching and massive *Porites*, and many smaller colonies belonging to *Favites*, *Goniastrea* and *Platygyra* (Bradbury and Young 1981; Endean et al. 1997; Brown et al. 2018). Coral cover over the reef crest is also variable, but unlike on the flat it is largely dependent on cyclone activity, as the crest takes the brunt of the wave energy (Tanner and Connell 2022). Between cyclones as well as on the south crest which is sheltered by adjacent Wistari Reef, coral cover is high and dominated by branching and tabulate *Acropora*, locally reaching over 70% cover (Tanner 2017; Brown et al. 2020), but can be reset to zero after a major cyclone (Tanner 2017; Tanner and Connell 2022). As such, colonies over the reef crest are small, a feature that is aided by the lack of water covering the crest at low tide which further curtails vertical growth (Scopélitis et al. 2011).

Several studies have investigated processes related to reef ecosystem functioning (Brandl et al. 2019) on the Heron reef flat. Despite the extreme abiotic conditions, net community production and calcification rates are high (up to 200 and 40 mmol m^{-2} day^{-1}, respectively), and vary with benthic community composition, depth and time of day (McMahon et al. 2013, 2018; Albright et al. 2015; Perez et al. 2018; Lantz et al. 2022).

6.3.2.3 Mechanisms of Stress Tolerance

Coral Growth and Resistance to Ocean Acidification Reef flat corals on Heron Island are primarily limited in their vertical growth by the water height, and expansion rates closely follow sea-level variations. Development of the local harbor, and especially the construction of artificial bund walls raised seawater levels over the flat, which was met with the highest vertical and lateral growth rates over the last four decades (Scopélitis et al. 2011). However, calcification rates of several coral species from the variable reef flat are nevertheless lower than in conspecifics from the more stable reef slope (Brown et al. 2020; Clark et al. 2022), despite higher PAR on the reef flat, though this is not always the case (Brown et al. 2022b). This reduction in growth was traded though for elevated long-term survival rates and increased protein production (Clark et al. 2022), traits that have also been documented in other species. For example, reef flat *Isopora palifera* doubled tissue protein concentrations at the expense of growth compared to reef slope corals under a 1.5–2 °C and 150–200 ppm CO_2 increase, though there was no increase in survival (Rathbone et al. 2021). Similarly, branching *Porites cylindrica* realized lower growth but instead used greater lipid and protein concentrations to achieve higher survival (Mello-Athayde 2021), while *Galaxea fascicularis* had higher protein concentrations in colonies growing on the more extreme reef crest compared to the reef flat (Marshall and Clode 2004).

Reef flat corals are also able to exert firm control over the carbonate chemistry of their calcifying fluid, despite variable seawater pH in these habitats. The coral species *Porites cylindrica* maintained high and stable internal pH_T levels (8.4–8.6) in spite of strong external pH_T (7.7–8.3) fluctuations (Georgiou et al. 2015), indicating strong capacity for pH homeostasis. Similarly, reef flat *Pocillopora damicornis* displayed lower internal pH acidosis and faster recovery from acute pH_T stress (7.4) compared to conspecifics from the reef slope (Brown et al. 2022b), and additionally realized higher calcification rates. Kaniewska et al. (2012) found that *Acropora millepora* calcification was not reduced after 4 weeks under pH_{NBS} 7.6–7.7 despite 50% reductions in photosynthesis rates and symbiont densities, underscoring the ability to maintain calcification over longer timescales, though at the expense of fitness trade-offs.

Coral Microbiome and Heat Tolerance Direct comparisons of symbiont communities harbored by slope and reef flat corals are scarce on Heron Island (Ulstrup et al. 2008; Fisher et al. 2012). Habitat-specific shifts in the coral-associated microbiome have been documented on the Heron reef flat for at least one species. The coral *Pocillopora damicornis* hosts a distinctly different symbiont community on the reef flat (predominantly *Cladocopium* C42) compared to the reef slope (predominantly *Cladocopium* C33) and also had different bacterial communities (van Oppen et al. 2018; Marhoefer et al. 2021).

Reef flat-wide analysis of gene expression patterns in *Acropora millepora* revealed high inter-colony transcriptional variation, possibly reflecting local micro-environments (Granados-Cifuentes et al. 2013). In addition, genome-wide differentiation between reef flat and reef slope conspecifics was mostly centered around immune and stress response (van Oppen et al. 2018; Marhoefer et al. 2021), possibly indicating adaptations to the more variable and stressful reef flat conditions. Interestingly, thermal tolerance assessments with *Acropora millepora* did not reveal high resilience to heat stress in unacclimated (ambient conditions) fragments, but did demonstrate increased acclimation potential in fragments first subjected to a brief period of conditioning to intermediate temperatures (Bellantuono et al. 2012a, b). Short-term thermal stress assays revealed a higher thermal tolerance for reef flat populations of *Acropora aspera* and *Pocillopora damicornis* compared to the slope, but no difference in *Porites lobata* (Brown et al. 2024). Photo-protective mechanisms include the ability of reef flat corals to shift algal symbionts away from exposed surfaces to shaded areas during periods of high PAR during low tide (Ulstrup et al. 2006) and through high rates of dimethylsulfoniopropionate (DMSP) synthesis during low tide and high irradiance (Swan et al. 2017). Five-week exposure to moderate levels of warming and acidification (+2 °C and +220 pCO_2) did not lead to reduction in calcification, symbiont densities, and photosynthesis in reef flat *Acropora millepora*, although exposure to higher levels (+4 °C and +590 pCO_2) did (Kaniewska et al. 2015). This was also during springtime, before summer thermal maxima were reached. During peak summer in 2020, branching *Acropora* on the reef flat bleached significantly stronger compared to the reef slope, potentially due to higher temperature maxima (Ainsworth et al. 2021). Overall, variable responses between species and lack of standardization between studies make it unclear whether reef flat corals have higher heat tolerance than reef slope corals.

6.3.2.4 Lessons and Future Outlook

Its location within the Capricornia Cays National Park grants Heron Reef some protection from local anthropogenic pressures, such as (over)fishing, development, and extensive tourism, and its distance from Australia's mainland offers protection from land-based pollutants (Devlin et al. 2001). However, the proximity to Australia's main coal mining regions on the east coast remains a potential threat for the reefs in the region as future port development and shipping traffic intensifies. Additionally, the usual threats such as increased hurricane activity and intensity, sea level rise, and especially marine heat waves, are unabated under climate change. Heron Reef has experienced several bleaching events since 1998 (most recently in 2023–2024), which have significantly reduced coral cover (Wasim et al. 2024). There is no convincing evidence that reef flat corals were more tolerant to these events than corals on the more stable reef slope (Ainsworth et al. 2021), and long-term monitoring has demonstrated that coral cover from the flat and slope has decreased approximately the same over the last two decades (Wasim et al. 2024). The finding by Dove et al. (2013) that presently slope corals fared better under pre-industrial climate conditions compared to present day conditions, raises the question whether present-day corals are so far even adapted to the current situation, and therefore whether they can keep up with even more rapid future environmental changes.

6.3.3 Kimberley Region, NW Australia: Low-Latitude Macrotidal Tide Pools

The remote Kimberley region in NW Australia features some of the most unique and environmentally extreme coral reefs on this planet (Wilson 2013; Richards 2022). With a tidal range exceeding 11 m—the largest tropical tides in the world (Kowalik 2004; Wilson et al. 2011), coral reefs in this region are exposed to extreme fluctuations in environmental conditions, where periods of aerial exposure or slack water alternate with strong tidal currents (Fig. 6.3). Similarly, low tide exposes shallow reef organisms to temperature and light extremes that contrast sharply with the otherwise turbid waters in inshore areas. Despite these extreme conditions, abundant and highly diverse coral reefs exist throughout the region and remain relatively understudied compared to more accessible reef systems in Australia such as the Great Barrier Reef or Ningaloo Reef. However, the last two decades have seen a renewed research focus that has significantly improved our understanding of marine biodiversity in the Kimberley (Richards et al. 2015, 2024), the extent of environmental extremes that corals and other organisms can withstand (e.g., Dandan et al. 2015; Pedersen et al. 2016), and their resistance to various climate change stressors (e.g., Schoepf et al. 2015b; Cornwall et al. 2018).

Coral reefs of the Kimberley occur across two distinct bioregions. The offshore reef systems consist of large atolls and platform reefs characterized by clear, oligotrophic, oceanic waters whereas the inshore bioregion consists of fringing and submerged patch reefs exposed to high turbidity and sedimentation due to riverine input and tidal currents (Richards 2022). Notably, the inshore bioregion (13.8–19.5° S, Fig. 6.1) features several reef structures that are geomorphologically unique and rarely found elsewhere. For example, coralline algal and/or rodoliths banks can form extensive terraces that sit several meters above the reef during spring low tides and create impounded lagoons from which water drains in the form of waterfalls (Fig. 6.8) or 'tidal rivers' (Wilson et al. 2011; Richards and O'Leary 2015; Richards 2022). At low tide, emerged reef flats, intertidal pools and impounded lagoons expose corals and other reef organisms to some of the most extreme environmental conditions ever documented for tropical marine organisms. We will therefore primarily focus on these reef zones in the following review.

6.3.3.1 Abiotic Conditions

Tidal Regime Kimberley coral reefs are strongly tidally forced (Lowe and Falter 2015) but their complex hydro- and thermodynamics are relatively well-studied (Lowe and Falter 2015; Lowe et al. 2016; Green et al. 2019). The mixed tidal regime and large tidal range (8–12 m) lead to prolonged aerial exposure of shallow reef areas twice a day (Fig. 6.9), which can last more than 3 h (Purcell 2002; Richards et al.

Fig. 6.8 Tallon (Jalan) Island in the southern Kimberley region, with the water draining from the elevated reef flat and coralline algal banks in the form of "waterfalls". (Photo credit: Verena Schoepf)

Fig. 6.9 Coral communities exposed to air during spring low tide near Cygnet Bay, inshore Kimberley region. (Photo credit: Verena Schoepf)

2015). During the most extreme spring low tides of the year, around the autumn and spring equinoxes, even subtidal reef areas get aerially exposed for periods of up to 1 h for several days (Rosser and Veron 2011; Schoepf et al. 2020). However, low tides typically occur in the early morning and evening hours (Richards et al. 2015), thus preventing exposure to mid-day environmental extremes. Exceptions are (1) the most extreme tides of the year when low tide stretches into the mid to late morning (Dandan et al. 2015; Schoepf et al. 2020), and (2) impounded lagoons on elevated reef terraces where the duration of the falling (ebb) tide is greatly elongated relative to the duration of the rising (flood) tide (Lowe et al. 2015; Gruber et al. 2017). Importantly, the large tidal range and complex geomorphology of Australia's shallow NW shelf further result in strong, multi-directional tidal currents (Fig. 6.3) that can exceed 10 knots (Purcell 2002) and elevate turbidity due to sediment resuspension (Wilson 2013).

Turbidity, PAR and Temperature The generally high turbidity in the inshore Kimberley is due to substantial input of terrigenous sediment from various large rivers and monsoonal flooding, ranging from 0.8 to 13 mg L^{-1} but mostly being <2 mg L^{-1} (Wilson 2013). The diffuse down-welling attenuation coefficient K_d varied from 0.03 to 0.18 m^{-1} in one location (Richards et al. 2015). Direct light measurements from submersible data loggers are limited to short time periods but show that shallow intertidal pools and impounded lagoons can reach maximum daily PAR levels of 1800–2400 μmol m^{-2} s^{-1} (Dandan et al. 2015; Pedersen et al. 2016; Gruber et al. 2017; Schoepf et al. 2020), though average daily light levels are typically around 400–550 μmol m^{-2} s^{-1} (Schoepf et al. 2020). Much longer time series are available for temperature (Dandan et al. 2015; Schoepf et al. 2020; Speelman et al. 2023), with a 4-year dataset demonstrating very similar average daily temperatures of ~28.6 °C in an intertidal pool and shallow subtidal area, yet strong differences in daily maxima (38.1 °C vs 33.8 °C) and daily temperature range (average DTR: 2.6 °C vs 1.4 °C, maximum DTR: 8.5 °C vs 4.5 °C, DTR kurtosis: 3.5 vs 4.1, DTR skewness: 1.1 vs 0.8) (Speelman et al. 2023). Even more extreme daily fluctuations were recorded in a shallow impounded lagoon dominated by seagrass and macroalgae, with maximum temperatures of up to 38.2 °C and daily fluctuations of 11–12 °C (Pedersen et al. 2016; Gruber et al. 2017).

DO and pH High light and temperature levels, combined with low water volumes and long seawater residence times, can lead to extreme pH and dissolved oxygen (DO) fluctuations in shallow tide pools and impounded lagoons due to the metabolism of benthic communities, particularly when these are dominated by seagrass and macroalgae rather than corals and other calcifiers. Although logger-based time series are limited, the few available data sets have revealed some of the most extreme daily fluctuations for shallow tropical marine organisms. Diurnal pH_T varied less than 0.15 pH units in a coral-dominated tide pool at low tide (mean 7.96, range 7.8–8.2), but this was up to 1.4 pH units (mean 8.08, range 7.51–9.15) in a seagrass and macroalgae-dominated impounded lagoon at a high intertidal reef platform (Cornwall et al. 2018). Pedersen et al. (2016) reported similar pH_{NBS} fluctuations at this lagoon at low tide (7.64–8.84), but minimal fluctuations at high tide (8.12–8.18). The diel DO range was 450 μmol O_2, with 0.7–54.3 kPa during low tide and 9.5–22.6 kPa during high tide. Similarly, Gruber et al. (2017) documented diel DO ranges of up to 440 μM at this site, with saturation levels of up to 280% when low tide coincided with high light levels, and hypoxia (DO <63 μM or 2 mg L^{-1}) occurring nightly for 1–7.5 h each day.

6.3.3.2 Coral Diversity, Community Composition, and Ecosystem Functioning

Despite the extreme environmental conditions, the Kimberley region covers ~476,000 km^2 of reefs and represents a significant reservoir of tropical coral biodiversity in Australia (Richards 2022). Latest studies estimate that 438 species of zooxanthellate reef-building corals occur across the inshore and offshore Kimberley (Richards et al. 2024), supporting the most diverse intertidal coral communities in tropical Australia with >200 species from 44 genera (Richards et al. 2015, 2019). Twenty coral species from 13 genera were even documented in a very extreme reef zone, a seagrass- and algae-dominated lagoon on an intertidal reef platform (Richards and O'Leary 2015). Generally, coral species with stress-tolerant life history strategies (Darling et al. 2012) dominate the intertidal reef zones, such as *Goniastrea retiformis*, *Goniastrea favulus*, *Porites lutea*, *Dipsastraea pallida*, and *Coelastrea aspera* (unknown life history strategy), whilst *Porites lichen*, *Heliopora coerulea*, *Seriatopora hystrix*, *Goniastrea pectinata* and *Montipora aequituberculata* dominate the subtidal reefs (Richards et al. 2024). For the latter subtidal species, life history strategy is mostly unknown, except for weedy *Seriatopora hystrix*. Regional level of hard coral cover is 23%, but up to 76% in some locations (Richards et al. 2018).

Although ecosystem functioning remains severely understudied on most extreme and/or marginal reefs and coral communities, some information related to reef growth, community metabolism and coral population connectivity is available for the Kimberley. Compared to more traditional coral reefs, crustose coralline algae have been hypothesized to play a more significant role than corals in the construction of Kimberley coral reefs as they can, together with rhodoliths, form extensive terraces and intertidal reef platforms (Fig. 6.8) (Wilson 2013; Richards and O'Leary 2015) and dominate subtidal reef zones (Richards et al. 2013). Productivity of benthic communities within shallow impounded lagoons was further similar to other tropical seagrass and mixed reef flat communities despite experiencing much greater environmental extremes, with the production and consumption of fixed carbon being well-balanced on time scales greater than weeks (Gruber et al. 2017). Finally, it has been shown that there is limited connectivity between inshore and offshore Kimberley reefs as the extreme seascape drives local recruitment and genetic divergence in both brooding and spawning corals (e.g., Underwood et al. 2013, 2020). Interestingly, the strong tidal currents can act as both conduits for occasional larval dispersal between local populations as well as semi-permeable barriers (Underwood et al. 2020).

6.3.3.3 Mechanisms of Stress Tolerance

Metabolism and Calcification Given their unique stress tolerance, Kimberley marine animals and plants serve as model organisms to explore their ability to sustain physiological processes under extreme environmental conditions. These studies have also investigated the mechanisms behind their tolerance, potential fitness trade-offs, and resistance to various climate change stressors. While most of this work has focused on coral reef calcifiers, one study showed that two seagrass species were able to fully recover metabolic rates overnight after 4 h at 40 °C; however, exposure to 45 °C led to more permanent tissue damage (Pedersen et al. 2016). *In situ* investigations of coral calcification rates over 1.5 years demonstrated that corals generally calcified at rates comparable to similar species at Ningaloo Reef where environmental conditions are much more stable and benign (Dandan et al. 2015). While one branching species showed evidence of fitness trade-offs in summer at the most extreme (intertidal) site, two massive species calcified the most at that site (Dandan et al. 2015). A crustose coralline algae from an extreme intertidal reef platform also had significantly higher lab-based calcification and respiration rates than conspecifics from a less extreme setting, whereas a massive coral had higher gross photosynthesis rates (Cornwall et al. 2018). These studies demonstrate that adaptive strategies are highly taxon-dependent and that many species are able to cope with environmental extremes without apparent sublethal effects.

Resistance to Ocean Acidification (OA) Frequent exposure to environmental extremes could also increase the resistance of Kimberley organisms to OA and warming. Indeed, intertidal coral and crustose coralline algae were highly resistant to future OA levels (pH_T 7.65) in the lab, with resistance linked to the ability to actively modify the carbonate chemistry of the calcifying fluid and maintain high internal saturation state (Cornwall et al. 2018). These findings agree with long-term calcification rates (1919–2016) reconstructed from a subtidal *Porites* core, which remained remarkably stable over almost a century despite the coral registering a general warming trend and secular OA (Chen et al. 2019). However, intensifying climate change in recent years has impacted the typical seasonal variations in calcifying fluid chemistry and may be an early indicator of heat stress negatively affecting coral calcification in the long term (Chen et al. 2019; Schoepf et al. 2021).

Coral Heat Tolerance Most of the work investigating the heat tolerance of Kimberley corals has focused on an inshore site as a natural laboratory where corals from an intertidal pool are compared to conspecifics from a nearby subtidal habitat (see fig. 6.1 in Speelman et al. (2023) for photos). Despite the small spatial scale (only a few 100 m), strong differences in diurnal temperature variability but not mean temperatures (Speelman et al. 2023) drive significantly higher heat tolerance in intertidal compared to subtidal corals, as demonstrated in both aquarium experiments (Schoepf et al. 2015b) and during the 2016 global bleaching event (Le Nohaïc et al. 2017; Schoepf et al. 2020). Similarly, lagoon corals at an offshore atoll displayed higher heat tolerance than reef slope conspecifics (Thomas et al. 2022). Intertidal corals further demonstrated higher bleaching resilience as they recovered rapidly from the 2016 bleaching event and suffered much less mortality than subtidal corals (Schoepf et al. 2020; Speelman et al. 2023). Notably, intertidal corals were able to maintain this enhanced heat tolerance even when exposed to much cooler temperatures for ~1 year, pointing to long-term acclimatization and/or local adaptation (Schoepf et al. 2019).

Several potential mechanisms underlying the elevated heat tolerance of intertidal corals have been investigated but revealed a surprisingly small role for well-known drivers such as symbiont type and energy reserves. Kimberley corals generally host *Cladocopium* (Thomas et al. 2014; Schoepf et al. 2015b) and have stable symbiont associations across habitats at both inshore (*Cladocopium* C3; Jung et al. 2021) and offshore reef sites (*Cladocopium* C40; Thomas et al. 2022). However, small differences in symbiont community composition were nevertheless correlated with habitat and bleaching status during the 2016 bleaching event, indicating potential differences in symbiont heat tolerance (Jung et al. 2021). Cryptic genetic diversity is also present in a common intertidal *Acropora* species (Underwood et al. 2020) but did not influence divergent bleaching and recovery trajectories of intertidal and subtidal corals (Schoepf et al. 2020). Survival and high bleaching resilience were further largely decoupled from tissue energy reserves (Jung et al. 2021), even though they often promote recovery from bleaching (e.g., Schoepf et al. 2015a). At the molecular level, however, enhanced heat tolerance of offshore lagoon corals was linked to higher baseline expression of stress response genes as well as habitat-specific shifts in allele frequencies, indicating local adaptation despite high gene flow across habitats (Thomas et al. 2022).

6.3.3.4 Lessons and Future Outlook

The Kimberley region's isolation means that local stressors are relatively minimal, though oil and gas exploitation, fishing, tourism, and cyclones still pose threats. Climate change therefore remains the most significant risk to these coral reefs despite their remarkable stress tolerance. Offshore reefs have suffered repeatedly from severe bleaching in the past decades (Gilmour et al. 2019) but demonstrate high resilience to recurrent disturbances (Thomas et al. 2024). In contrast, the inshore reefs did not suffer from major bleaching until 2016 (Le Nohaïc et al. 2017), highlighting that even some of the most heat-tolerant corals are vulnerable to intensifying climate impacts (see also Klepac and Barshis 2020; Brown et al. 2022a). Experimental work has further confirmed the limited capacity of Kimberley corals to acclimate to rising temperatures (Schoepf et al. 2019). Yet, several factors suggest that Kimberley reefs may nevertheless cope somewhat better than others. The 2016 bleaching event did not affect the northern inshore reefs, indicating potential refuge areas within the region (Richards et al. 2019). Additionally, rising sea levels will reduce extreme temperature fluctuations in tide-dominated reef systems (Lowe et al. 2016), mitigating heat stress during future marine heatwaves. Inshore coral populations are also predicted to have a relatively low mismatch in genetic variation under climate change conditions, thus requiring less adaptive shift (Adam et al. 2022). Last but not least, a network of regional marine parks, jointly managed with Traditional Owners, has recently been established to protect the unique coral reefs of the Kimberley and improve monitoring efforts in this remote and understudied region.

6.4 Discussion

6.4.1 How Extreme Are Intertidal and Back-Reef Coral Environments, and How Does This Compare to Other Extreme Reef Systems?

Our review has summarized for the first time the abiotic conditions that exist in the transition zone where coral reefs meet the land, highlighting that these habitats are arguably among the most extreme that tropical corals persist under (e.g., Camp et al. 2018). For example, the temperature extremes reached in intertidal reef habitats (up to 38 °C; Speelman et al. 2023) can temporarily exceed those found in summer in the world's hottest sea, the Persian/Arabian Gulf (>35 °C; Riegl and Purkis 2012), or the southern Red Sea (>32 °C) (Fine et al. 2013). Even high-latitude intertidal habitats, such as the reef flat on Heron Island on the southern GBR (23.4° S) or rock pools in South Africa (29–31° S), reach maxima of 35 and >32 °C in summer, respectively (Smit and Glassom 2017; Brown et al. 2023). Similarly, daily pH fluctuations on intertidal reef flats can reach 0.8–1.4 pH units (Kline et al. 2015; Cornwall et al. 2018) and thus rival those found near CO_2 seeps (e.g., Comeau et al. 2022), while DO minima regularly exceed the nominal hypoxia threshold of 2 mg L^{-1} and may last for hours (e.g., Gruber et al. 2017; Kekuewa et al. 2021). Yet, several important characteristics also distinguish intertidal and back-reef environments from other extreme coral habitats. First and foremost, regular aerial exposure during (spring) low tides places unique stresses on corals and other reef organisms in 'true' intertidal environments and can last, for example, more than 3 h in Australia's Kimberley region (e.g., Purcell 2002). Second, due to the shallow depth of these habitats, UV and PAR levels by far exceed those typically experienced by corals in other reef habitats, regularly exceeding PAR levels of 2000 µmol m^{-2} s^{-1} (e.g., Brown et al. 1994; Kline et al. 2015; Schoepf et al. 2020), although this can be mitigated by otherwise turbid waters in certain locations (Scoffin et al. 1997; Wilson 2013).

Third, the highly dynamic nature of intertidal habitats exposes resident organisms to dramatic fluctuations in abiotic conditions on a (sub)daily basis, which are arguably stressful on their own but on the other hand provide periods of low stress intensity that may allow for relief and recovery (i.e., during high tide). This is in contrast to many other extreme coral environments where abiotic conditions are chronically extreme, such as the hot summer-time temperatures in the Persian/Arabian Gulf (e.g., Riegl and Purkis 2012) or persistent low light levels on turbid reefs (e.g., Browne et al. 2013). Finally, it should be noted that extreme coral environments outside of the intertidal zone can also be characterized by strong environmental variability, such as coral reefs growing near CO_2 seeps (e.g., Comeau et al. 2022) or in semi-enclosed bays and lagoons (e.g., Camp et al. 2019; Lucey et al. 2020; Maggioni et al. 2021).

While intertidal and back-reef environments clearly represent some of the most extreme coral habitats on this planet, it is difficult to directly compare their extremeness to other reef environments because (long-term) monitoring of abiotic conditions using high-resolution data loggers is often limited (e.g., Koweek et al. 2015) (Table 6.1). Nevertheless, the case studies reviewed here illustrate that three factors, in particular, determine the extremeness of these environments: tidal range, seawater residence time, and latitude. Reefs with meso- and macrotidal regimes (Fig. 6.1), such as those in the Andaman Sea and NW Australia, generally experience much more extreme environmental conditions than microtidal reefs, but the daily timing of low tide (Buckee et al. 2022)

Table 6.1 Future research priorities for studying intertidal and back-reef coral environments

Investigating intertidal and back-reef coral environments in understudied regions that are characterized by large tidal regimes, such as parts of Africa's eastern coast, northern and Central Brazil, and parts of the Eastern Tropical Pacific (Colombia to Guatemala)
Long-term monitoring of abiotic conditions using high-resolution data loggers
Impacts of sea level rise on intertidal and back-reef coral environments, its interaction with hydrodynamics, and consequences for environmental extremeness
Hydrodynamics of tide-dominated reefs and coral environments
Adaptive capacity and responses of intertidal and back-reef corals to novel environmental conditions which could become more extreme due to climate change or less extreme due to sea level rise
Molecular, cellular, physiological, and behavioral adaptations of intertidal and back-reef corals, particularly with respect to stressors such as aerial exposure and desiccation, intense light and UV stress, and fluctuating salinity
Population genetics and connectivity of intertidal coral populations
Exposure-dependent limits of stress tolerance of intertidal and back-reef corals, especially under multiple, co-varying stressors
Ecosystem functioning of intertidal and back-reef coral environments, such as net ecosystem productivity and calcification

These are complementary to those already defined more broadly for extreme coral environments in previous reviews (Camp et al. 2018; Schoepf et al. 2023)

and reef geomorphology (Lowe et al. 2015) are also crucial factors. For example, low tides typically occur in the early morning and evening in Australia's Kimberley region and on Phuket's reef flats in Thailand (e.g., Brown et al. 2014b; Richards et al. 2015), whereas coral aerial exposure is regularly aligned with maximum daily temperatures and PAR on the Heron Island reef flat (Kekuewa et al. 2021; Brown et al. 2023). However, Kimberley coral reefs are also prime examples of tide-dominated reefs where the duration of the falling (ebb) tide can be greatly elongated relative to the duration of the rising (flood) tide on impounded lagoons (Lowe et al. 2015), extending aerial exposure and low-tide extreme conditions far into mid-day (e.g., Pedersen et al. 2016; Gruber et al. 2017). The long seawater residence times on impounded reef flats and lagoons, combined with the small water volume of intertidal rock or back-reef pools, further allow for extreme temperature fluctuations (e.g., Oliver and Palumbi 2011a; Smit and Glassom 2017), and, depending on the benthic community composition, dramatic pH and DO fluctuations (e.g., Koweek et al. 2015; Gruber et al. 2017; Kekuewa et al. 2021). Importantly, on high-latitude intertidal and back-reef environments, seasonality further enhances the extremeness of daily abiotic fluctuations and can expose corals and other reef organisms to stressfully low temperatures during winter low tides. For example, temperatures can drop below 17 °C in South Africa's rock pools (27–31° S) (Smit and Glassom 2017) or even reach as low as 13.3 °C on the Heron Island reef flat (23.4° S), thus causing cold-stress induced bleaching (e.g., Hoegh-Guldberg et al. 2005). Overall, our findings highlight the crucial role of continuous, multi-physicochemical data sets in understanding coral reef extremeness and its implications for the stress tolerance and climate change resilience of reef organisms.

6.4.2 How Ecologically Marginal Are Intertidal and Back-Reef Coral Communities?

Coral reef marginality, as defined by Schoepf et al. (2023), refers to "coral communities with a community composition and/or ecosystem functioning that differs substantially from those on reefs with optimal conditions". They further distinguished two tightly linked components of ecological marginality, with structural marginality referring to ecosystem structure (e.g., coral species richness, coral cover and community composition) and functional marginality referring to ecosystem functioning (e.g., primary and carbonate production, Brandl et al. 2019). Since intertidal and back-reef coral environments are characterized by extreme environmental conditions, it might be expected that this also leads to ecological marginality. However, our review demonstrates that this is not necessarily the case and also highly site-specific, especially with respect to structural marginality for which data availability is much better than for functional marginality. Although both the inshore Andaman Sea and Kimberley region in NW Australia can be considered environmentally extreme on a regional level due to the large tidal range and high turbidity (Brown 2007; Wilson 2013), coral species diversity is nevertheless high, with up to 353 species reported for the West coast of Thailand (Turak et al. 2005) and 438 species for the Kimberley region (Richards et al. 2024). Similarly, high-latitude Heron Reef features more than 70% of coral species diversity found on the entire Great Barrier Reef (NOAA 2024b), while ~80 species are found in the back-reef pools of Ofu Island compared to ~200 species recorded for American Samoa (Craig et al. 2001). These locations should therefore be considered extreme but not marginal, even though in the case of Heron and Ofu Island site-specific diversity is lower than regional species estimates, yet coral cover is comparatively similar. On the other hand, other intertidal and back-reef locations are both extreme *and* marginal, such as the high-latitude rock pools in South Africa which feature similar coral species richness as a nearby subtidal reef but very different community composition (Schleyer et al. 2006; Olbers 2017a; Smit and Glassom 2017). In addition, specific sites within regions such as the Andaman Sea or Kimberley region can nevertheless be extreme and marginal, such as the reef flat in Phuket with >30 species (e.g., Brown et al. 2019) or Tallon Island (Jalan) with 20 species (Richards and O'Leary 2015).

The assessment of functional coral reef marginality on intertidal and back-reef coral environments is much more difficult due to a lack of data for the majority of sites. This is also reflected in our highlighted study locations, as to the best of our knowledge no such data exists for the rock pools in South Africa or the well-studied Phuket reef flats. Similarly, only one study has investigated benthic community metabolism on the well-studied back-reef of Ofu Island, showing lower net community calcification to net community productivity ratios than reported for other shallow back-reef environments (Koweek et al. 2015). In contrast, more information than one might expect is available for the remote Kimberley region, highlighting that productivity on elevated reef platforms is largely unimpaired by some of the most extreme conditions recorded for such communities (Gruber et al. 2017). In addition, crustose coralline algae likely play a much more prominent role in reef growth than on more traditional reefs (Fig. 6.8) (e.g., Wilson 2013; Birkeland et al. 2021), while strong tidal currents (Fig. 6.3) modulate coral connectivity across an extreme seascape (e.g., Underwood et al. 2020). Similarly, several studies have shown that community productivity and calcification are also high on Heron Island (McMahon et al. 2018), but here community productivity or factors associated with primary productivity (e.g., light) appear to be a better predictor of community calcification than aragonite saturation (Albright et al. 2015; McMahon et al. 2018; Perez et al. 2018), presumably because of the relatively low coral cover (Lantz et al. 2022). Overall, these findings suggest that, again, environmental extremes may not necessarily translate to reduced ecosystem functioning, though it may well be altered or feature unique aspects. However, much more data across a wide range of intertidal habitats and locations is needed for a robust assessment (Table 6.1). Furthermore, the scarcity of studies on ecosystem functioning not just on intertidal and back-reef environments, but coral reefs more generally (e.g., Brandl et al. 2019), represents a crucial knowledge gap.

6.4.3 Which Mechanisms Support Survival of Corals and Other Reef Taxa in Intertidal and Back-Reef Environments?

Naturally extreme reef systems can serve as natural laboratories to investigate how corals and other reef taxa will be able to cope with future climate change stressors (e.g., Camp et al. 2018; Burt et al. 2020). With research on these environments growing, increasing data availability across a wide range of locations will facilitate the assessment of the biological mechanisms that support the high stress tolerance and survival of organisms in such environments. For intertidal and back-reef environments, however, it remains challenging to determine whether commonalities exist or whether certain strategies may be species- or site-specific because few such mechanisms have been studied across different locations and species (Table 6.1). Nevertheless, assessment of corals' life history strategies (Darling et al. 2012) across the locations highlighted here indicates that stress-tolerant species clearly dominate in intertidal and back-reef habitats, followed by weedy species. Stress-tolerant species typically have traits that facilitate success and persistence in chronically harsh environments, such as slow growth, long generation times, and high fecundity (Darling et al. 2012). Notably though, corals with all four life history strategies can occur in such extreme environments, and dominant species may have competitive or other life history strategies (e.g. S African tide pools, Smit and Glassom 2017). Furthermore, a synthesis to identify commonalities across intertidal and back-reef environments can be attempted for certain genetic and physiological traits and mechanisms, namely association with certain Symbiodiniaceae, molecular and genetic responses, and calcification mechanisms.

The family Symbiodiniaceae is taxonomically and functionally diverse (LaJeunesse et al. 2018; Nitschke et al. 2022), with the genus *Durusdinium* often associated with higher heat tolerance (e.g., Berkelmans and van Oppen 2006; Swain et al. 2017) and prevalence in warmer environments or seasons (Ulstrup and Van Oppen 2003; Fabricius et al. 2004; Oliver and Palumbi 2009; LaJeunesse et al. 2010a, b). Yet, intertidal and back-reef corals do not exclusively associate with *Durusdinium*: while some coral species do have higher proportions of *Durusdinium* in the highly variable (HV) back-reef pool on Ofu Island and Phuket reef flats (e.g., Oliver and Palumbi 2009; Brown et al. 2014a; Morikawa and Palumbi 2019), both intertidal and subtidal/reef slope corals are dominated by *Cladocopium* in the Kimberley region and on the Heron Island reef flat (e.g., van Oppen et al. 2018; Jung et al. 2021; Thomas et al. 2022) and in massive species in the HV pool of Ofu (Barshis et al. 2018; Klepac and Barshis 2020). Corals in South African rock pools also primarily associate with *Cladocopium* C1 (D. Glassom/S. Ntuli, unpublished data). This highlights that heat tolerance is not always dictated by symbiont genus, that symbiont species can be locally adapted to environmental extremes (e.g., Howells et al. 2012) and that other members of the coral holobiont also play an important role in driving stress tolerance in these habitats (Brown et al. 2002b; Palumbi et al. 2014; Ziegler et al. 2017b).

Across intertidal and back-reef environments, emerging research shows that molecular and genetic mechanisms contribute to enhanced stress tolerance, most notably in gene expression patterns (e.g., Barshis et al. 2013; Seneca and Palumbi 2015; Thomas et al. 2022) and genetic differentiation (Bay and Palumbi 2014; Barshis et al. 2018; Thomas et al. 2022; Zhang et al. 2022; Fiesinger et al. 2023). At

Heron Island, Ofu Island, and offshore shallow atolls in the Kimberley, corals from habitats with elevated and variable temperatures display rapid acclimatization abilities in heat resilience by having a much lower magnitude of expression change before and after experimental heat stress (Bellantuono et al. 2012a; Barshis et al. 2013; Thomas et al. 2022; Klepac, unpubl.data). Additionally, baseline upregulation of heat stress-related genes prior to actual heat stress, called 'frontloading', has been identified as another mechanism of enhanced tolerance in corals from Ofu Island and the Kimberley (Barshis et al. 2013; Thomas et al. 2022; Klepac and Barshis, in prep). Work from the intertidal rock pools of South Africa has also shown that gene expression patterns, overall magnitude of the heat stress response, and RNA:DNA ratios differed significantly between heat-sensitive and heat-resistance coral species. Although the same symbiont type (C1) was found in both coral species, transcriptomic changes in the symbionts of each coral species also differed (Onyango 2020; Onyango et al. 2021).

Population genetic analyses have further revealed strong population differentiation between extreme environments and more traditional reef habitats. In spite of substantial regional gene flow, genetic structuring has been detected in as little as a few kilometers (Bay and Palumbi 2014; Barshis et al. 2018; Thomas et al. 2022; Zhang et al. 2022; Fiesinger et al. 2023). Local adaptation to extreme environments appears to occur within only a few genes that are responsive to habitat-specific selection, such as low dispersal and bottlenecks post-glacial colonization from offshore to inshore Kimberley reefs (Zhang et al. 2022), changing currents throughout each monsoonal season in the Andaman Sea (Fiesinger et al. 2023), and tidally-driven temperature profiles in the Kimberley region and Ofu Island (Barshis et al. 2018; Underwood et al. 2020; Thomas et al. 2022). Collectively, these studies demonstrate that extreme intertidal habitats are more genetically isolated and contain a molecular repertoire to withstand high levels of environmental variability.

Intertidal and back-reef environments can also provide important insights into calcification mechanisms under stressful environmental condition, including ocean acidification (OA). High pH variability, which is present in many intertidal and back-reef environments, has been hypothesized to enhance the resistance of coral reef calcifiers to ocean acidification (e.g., Rivest et al. 2017; Kapsenberg and Cyronak 2019). Indeed, intertidal coral and crustose coralline algae from the Kimberley region were highly resistant to future OA levels (pH_T 7.65), with resistance linked to the ability to actively modify the carbonate chemistry of the calcifying fluid and maintain high internal saturation state (Cornwall et al. 2018). Similar evidence comes from Heron Island, where reef flat corals were not affected by future OA (Kaniewska et al. 2012; Georgiou et al. 2015), maintained internal pH homeostasis despite strong external fluctuations (Georgiou et al. 2015), and displayed lower internal pH acidosis and faster recovery from acute pH stress compared to reef slope conspecifics (Brown et al. 2022b). While these findings suggest that some coral reef calcifiers will not be severely affected by OA in the future, fitness trade-offs in the form of reduced skeletal density (Brown et al. 2022b) or reduced photosynthesis and symbiont densities (Kaniewska et al. 2012) may occur and need to be better understood. Beyond resistance to OA, coral calcification also appears highly resistant to the extreme abiotic conditions of the intertidal zone, with studies from several locations reporting similar or higher calcification rates in intertidal compared to subtidal or forereef corals (Smith et al. 2007; Dandan et al. 2015; Barshis et al. 2018; Brown et al. 2022b), though this is not the case for all species (Brown et al. 2020; Mello-Athayde 2021; Rathbone et al. 2021; Clark et al. 2022).

6.4.4 Knowledge Gaps and Future Research Priorities for Intertidal and Back-Reef Coral Environments

Despite renewed scientific interest in extreme and marginal coral communities, they remain understudied compared to more 'traditional' coral reefs, and the research priorities outlined in previous reviews on this topic (Camp et al. 2018; Schoepf et al. 2023) therefore remain valid. Nevertheless, we want to highlight here research priorities that are particularly relevant and specific to intertidal and back-reef systems (Table 6.1), chiefly in their geographical identification, response to sea level rise, organismal adaptive capacity under rapid climate change, and genetic connectivity. One crucial knowledge gap is that very little is known about intertidal coral communities in many geographical areas, despite the fact that many areas of the tropics have meso- and macrotidal regimes (Fig. 6.1). This includes, for example, large parts of Africa's eastern coast, northern and central Brazil, and parts of the Eastern Tropical Pacific (Colombia to Guatemala). Some research in these locations is already leveraging the environmental gradient created by the tidal regimes to understand coral thermal resilience (Kenya: Almeida et al. 2024), while other work has focused on unique intertidal reef settings in Brazil (e.g., Barros et al. 2021) or the impacts of extreme low tide events on coral reefs in Colombia (e.g., Mejía-Rentería et al. 2020). Another major knowledge gap is how sea level rise will impact intertidal and back-reef environments. Sea level rise may mitigate environmental extremeness and the effects of negative sea-level anomalies on intertidal reefs to some extent (Lowe et al. 2016; Brown et al. 2019) but more research, particularly on the hydrodynamics of tide-dominated reefs, is needed across a wide range of reef settings. Moreover, it is imperative to understand whether corals could lose their environmental stress tolerance if environmental extremeness diminishes which ultimately affects long-term persistence under climate change in spite of sea level rise.

Other knowledge gaps concern the specific adaptations that intertidal corals require to cope with the unique stresses present in the intertidal zone (Table 6.1), such as aerial exposure and desiccation, intense light and UV stress, and fluctuating salinity. The effects of aerial exposure on coral physiology are generally severely understudied (but see Romaine et al. 1997; Castrillón-Cifuentes et al. 2017). Furthermore, while behavioral as well as thermal and oxidative stress responses are relatively well studied on reef flats in the Andaman Sea (Brown et al. 2002a, b, c; Jatkar et al. 2010), such research is lacking from many other locations where corals experience prolonged aerial exposure, including Australia's Kimberley region or the Heron Island reef flat. Another major knowledge gap is the population genetics and connectivity of intertidal coral populations (see above) which remain poorly understood in many, though not all, locations (Table 6.1). Although work on back-reef pools in American Samoa and an intertidal lagoon in the Kimberley region has demonstrated high gene flow between thermally divergent habitats (Bay and Palumbi 2014; Thomas et al. 2022), this may not be the case in more isolated intertidal habitats such as South Africa's rock pools where populations may depend on constant influx of recruits from elsewhere. Finally, more research is required to understand the limits of the remarkable stress tolerance of intertidal corals (Table 6.1), particularly in the context of climate change when future warming and marine heatwaves will be superimposed on already extreme thermal regimes. Work from several locations including the Kimberley, Heron Island and American Samoa has shown that prolonged heat stress can erode the otherwise elevated heat tolerance of corals in thermally variable reef environments (Le Nohaïc et al. 2017; Klepac and Barshis 2020; Brown et al. 2022b). Future research should therefore focus on exposure-dependent thresholds and potential tipping points in the thermal tolerance of intertidal corals (e.g., Schoepf et al. 2022; Brown and Barott 2022).

6.4.5 Lessons from Intertidal and Back-Reef Coral Environments

As is the case for many other extreme coral environments, intertidal and back-reef coral environments have high research and conservation value because they can play important roles as resilience hotspots and natural laboratories (Camp et al. 2018; Camp 2022; Schoepf et al. 2023). While the abiotic conditions in these environments do not perfectly simulate future ocean conditions, primarily due to the large fluctuations, they are nevertheless crucial to study how organisms respond to sub-optimal abiotic conditions under ecologically relevant conditions and time scales and which mechanisms enable their survival. Notably, temperature, pH and DO extremes often co-occur in these habitats and exceed those predicted under future climate scenarios, but an important limitation is their relatively short duration that is interspersed with periods of relief (Schoepf et al. 2023), which needs to be taken into account when using these habitats as natural laboratories and applying lessons to future ocean conditions.

Several studies from multiple locations have also highlighted the important role of intertidal and back-reef coral environments as resilience hotspots and "adaptive refugia" (*sensu* Kapsenberg and Cyronak 2019). Exposure to the extreme conditions in these environments often enhances the adaptive capacity and resistance of resident organisms to multiple climate change stressors (e.g., Palumbi et al. 2014; Cornwall et al. 2018; Brown et al. 2022b), with at least one study showing that such environmental "stress memory" can be maintained in the absence of the priming cue for at least a decade (Brown et al. 2014a). Protecting reefs with such extreme conditions and harnessing their natural evolutionary adaptation through inclusion in adaptive protective area networks will be a crucial conservation strategy in a warming world (Colton et al. 2022). Similarly, stress-resistant corals from these habitats could also be used to improve reef restoration in the Anthropocene (Caruso et al. 2021), although more research is required to understand maintenance of stress tolerance and potential fitness trade-offs in novel environments. However, we warn that the ability of intertidal and back-reef coral environments to serve as adaptive refugia may become increasingly constrained as climate change intensifies in the coming decades (Le Nohaïc et al. 2017; Klepac and Barshis 2020; Brown et al. 2022a). With record-breaking heat stress increasing in frequency (Frölicher et al. 2018), as seen at the time of writing during the unfolding fourth global bleaching event (Hoegh-Guldberg et al. 2023; Reimer et al. 2024), only time can tell whether intertidal and back-reef corals can maintain thermal resilience, perhaps buffered from increasing extremeness due to sea level rise (Lowe et al. 2016), or whether they may at some point be among the most vulnerable coral populations.

Acknowledgements We thank S. Shinneman of the University of Amsterdam Institute for Biodiversity and Ecosystem Dynamics for assistance with the map figures.

References

Adam AAS, Thomas L, Underwood J et al (2022) Population connectivity and genetic offset in the spawning coral *Acropora digitifera* in Western Australia. Mol Ecol 31:3533–3547. https://doi.org/10.1111/mec.16498

Ainsworth TD, Leggat W, Silliman BR et al (2021) Rebuilding relationships on coral reefs: coral bleaching knowledge-sharing to aid adaptation planning for reef users: bleaching emergence on reefs demonstrates the need to consider reef scale and accessibility when preparing for, and responding to, coral bleaching. BioEssays 43:e2100048. https://doi.org/10.1002/BIES.202100048

Albright R, Benthuysen J, Cantin N et al (2015) Coral reef metabolism and carbon chemistry dynamics of a coral reef flat. Geophys Res Lett 42:3980–3988. https://doi.org/10.1002/2015GL063488

Almeida L, Osinga R, Knoester EG (2024) Differential responses in recovery, growth and survival between intertidal and subtidal corals after acute thermal stress. Mar Biol 171:222

Ampou EE, Johan O, Menkes CE et al (2017) Coral mortality induced by the 2015–2016 El-Niño in Indonesia: the effect of rapid sea level fall. Biogeosciences 14:817–826. https://doi.org/10.5194/bg-14-817-2017

Anthony KRN, Diaz-Pulido G, Verlinden N et al (2013) Benthic buffers and boosters of ocean acidification on coral reefs. Biogeosciences 10:4897–4909. https://doi.org/10.5194/bg-10-4897-2013

Barros Y, Lucas CC, Soares MO (2021) An urban intertidal reef is dominated by fleshy macroalgae, sediment, and bleaching of a resilient coral (*Siderastrea stellata*). Mar Pollut Bull 173:112967. https://doi.org/10.1016/J.MARPOLBUL.2021.112967

Barshis DJ, Stillman JH, Gates RD et al (2010) Protein expression and genetic structure of the coral *Porites lobata* in an environmentally extreme Samoan back reef: does host genotype limit phenotypic plasticity? Mol Ecol 19:1705–1720. https://doi.org/10.1111/j.1365-294X.2010.04574.x

Barshis DJ, Ladner JT, Oliver TA et al (2013) Genomic basis for coral resilience to climate change. Proc Natl Acad Sci 110:1387–1392. https://doi.org/10.1073/pnas.1210224110

Barshis DJ, Birkeland C, Toonen RJ et al (2018) High-frequency temperature variability mirrors fixed differences in thermal limits of the massive coral *Porites lobata*. J Exp Biol 221:jeb188581. https://doi.org/10.1242/jeb.188581

Bay RA, Palumbi SR (2014) Multilocus adaptation associated with heat resistance in reef-building corals. Curr Biol 24:2952–2956. https://doi.org/10.1016/j.cub.2014.10.044

Bay RA, Palumbi SR (2015) Rapid acclimation ability mediated by transcriptome changes in reef-building corals. Genome Biol Evol 7:1602–1612. https://doi.org/10.1093/gbe/evv085

Bay RA, Rose NH, Logan CA, Palumbi SR (2017) Genomic models predict successful coral adaptation if future ocean warming rates are reduced. Sci Adv 3:e1701413. https://doi.org/10.1126/sciadv.1701413

Bellantuono AJ, Granados-Cifuentes C, Miller DJ et al (2012a) Coral thermal tolerance: tuning gene expression to resist thermal stress. PLoS One 7:e50685

Bellantuono AJ, Hoegh-Guldberg O, Rodriguez-Lanetty M (2012b) Resistance to thermal stress in corals without changes in symbiont composition. Proc R Soc Lond B Biol Sci 279:1100–1107. https://doi.org/10.1098/rspb.2011.1780

Berkelmans R, van Oppen MJH (2006) The role of zooxanthellae in the thermal tolerance of corals: a "nugget of hope" for coral reefs in an era of climate change. Proc R Soc B Biol Sci 273:2305–2312. https://doi.org/10.1098/rspb.2006.3567

Bindoff NL, Willebrand J, Artale V et al (2007) Observations: oceanic climate change and sea level. In: Climate change 2007: the physical science basis. Contribution of Working Group I to the fourth assessment report of the Intergovernmental Panel on Climate Change. Cambridge University Press, Cambridge, pp 384–428

Birkeland C, Green A, Lawrence A et al (2021) Different resiliencies in coral communities over ecological and geological time scales in American Samoa. Mar Ecol Prog Ser 673:55–68

Blanchon P (2011) In: Hopley D (ed) Geomorphic zonation—Encyclopedia of modern coral reefs: structure, form and process. Springer Netherlands, Dordrecht, pp 469–486

Bradbury RH, Young PC (1981) The effects of a major forcing function, wave energy, on a coral reef ecosystem. Mar Ecol Prog Ser 5:229–241

Brainard R (2011) Coral reef ecosystems of American Samoa: a 2002–2010 overview. US Department of Commerce, National Oceanic and Atmospheric Administration

Brandl SJ, Rasher DB, Côté IM et al (2019) Coral reef ecosystem functioning: eight core processes and the role of biodiversity. Front Ecol Environ 17:445–454. https://doi.org/10.1002/fee.2088

Briggs D (1997) Fundamentals of the physical environment. Routledge

Brown BE (2007) Coral reefs of the Andaman Sea—an integrated perspective. Oceanogr Mar Biol Annu Rev 45:173–194

Brown KT, Barott KL (2022) The costs and benefits of environmental memory for reef-building corals coping with recurring marine heatwaves. Integr Comp Biol 62:1748–1755. https://doi.org/10.1093/ICB/ICAC074

Brown BE, Dunne RP, Scoffin TP, Le Tissier MDA (1994) Solar damage in intertidal corals. Mar Ecol Prog Ser 105:219–230

Brown B, Downs C, Dunne R, Gibb S (2002a) Exploring the basis of thermotolerance in the reef coral *Goniastrea aspera*. Mar Ecol Prog Ser 242:119–129. https://doi.org/10.3354/meps242119

Brown BE, Downs CA, Dunne RP, Gibb SW (2002b) Preliminary evidence for tissue retraction as a factor in photoprotection of corals incapable of xanthophyll cycling. J Exp Mar Biol Ecol 277:129–144. https://doi.org/10.1016/s0022-0981(02)00305-2

Brown BE, Dunne RP, Goodson MS, Douglas AE (2002c) Experience shapes the susceptibility of a reef coral to bleaching. Coral Reefs 21:119–126

Brown BE, Dunne RP, Phongsuwan N, Somerfield PJ (2011) Increased sea level promotes coral cover on shallow reef flats in the Andaman Sea, eastern Indian Ocean. Coral Reefs 30:867–878. https://doi.org/10.1007/s00338-011-0804-9

Brown BE, Dunne RP, Edwards AJ et al (2014a) Decadal environmental 'memory' in a reef coral? Mar Biol 162:479–483. https://doi.org/10.1007/s00227-014-2596-2

Brown BE, Dunne RP, Phongsuwan N et al (2014b) The reef coral *Goniastrea aspera*: a 'winner' becomes a 'loser' during a severe bleaching event in Thailand. Coral Reefs 33:395–401. https://doi.org/10.1007/s00338-013-1120-3

Brown KT, Bender-Champ D, Kubicek A et al (2018) The dynamics of coral-algal interactions in space and time on the southern great barrier reef. Front Mar Sci 5:181

Brown BE, Dunne RP, Somerfield PJ et al (2019) Long-term impacts of rising sea temperature and sea level on shallow water coral communities over a ~40 year period. Sci Rep 9:8826. https://doi.org/10.1038/s41598-019-45188-x

Brown KT, Bender-Champ D, Hoegh-Guldberg O, Dove S (2020) Seasonal shifts in the competitive ability of macroalgae influence the outcomes of coral–algal competition. R Soc Open Sci 7:201797. https://doi.org/10.1098/rsos.201797

Brown KT, Eyal G, Dove SG, Barott KL (2022a) Fine-scale heterogeneity reveals disproportionate thermal stress and coral mortality in thermally variable reef habitats during a marine heatwave. Coral Reefs 42:131–142. https://doi.org/10.1007/s00338-022-02328-6

Brown KT, Mello-Athayde MA, Sampayo EM et al (2022b) Environmental memory gained from exposure to extreme pCO_2 variability promotes coral cellular acid–base homeostasis. Proc R Soc B 289:20220941. https://doi.org/10.1098/RSPB.2022.0941

Brown KT, Lenz EA, Glass BH et al (2023) Divergent bleaching and recovery trajectories in reef-building corals following a decade of successive marine heatwaves. Proc Natl Acad Sci 120:e2312104120. https://doi.org/10.1073/pnas.2312104120

Brown KT, Martynek MP, Barott KL (2024) Local habitat heterogeneity rivals regional differences in coral thermal tolerance. Coral Reefs 43:571–585. https://doi.org/10.1007/s00338-024-02484-x

Browne NK, Smithers SG, Perry CT (2013) Carbonate and terrigenous sediment budgets for two inshore turbid reefs on the central great barrier reef. Mar Geol 346:101–123. https://doi.org/10.1016/j.margeo.2013.08.011

Buckee J, Hetzel Y, Edge W et al (2022) Daily timing of low tide drives seasonality in intertidal emersion mortality risk. Front Mar Sci 9:904191

Burt JA, Camp EF, Enochs IC et al (2020) Insights from extreme coral reefs in a changing world. Coral Reefs 39:495–507

Camp EF (2022) Contingency planning for coral reefs in the Anthropocene; The potential of reef safe havens. Emerg Top Life Sci 6:107–124. https://doi.org/10.1042/ETLS20210232

Camp EF, Schoepf V, Mumby PJ et al (2018) The future of coral reefs subject to rapid climate change: lessons from natural extreme environments. Front Mar Sci 5:4. https://doi.org/10.3389/fmars.2018.00004

Camp EF, Edmondson J, Doheny A et al (2019) Mangrove lagoons of the great barrier reef support coral populations persisting under extreme environmental conditions. Mar Ecol Prog Ser 625:1–14

Caruso C, Hughes K, Drury C (2021) Selecting heat-tolerant corals for proactive reef restoration. Front Mar Sci 8:632027. https://doi.org/10.3389/fmars.2021.632027

Castrillón-Cifuentes AL, Lozano-Cortés DF, Zapata FA (2017) Effect of short-term subaerial exposure on the cauliflower coral, *Pocillopora damicornis*, during a simulated extreme low-tide event. Coral Reefs 36:401–414. https://doi.org/10.1007/s00338-017-1552-2

Chen X, D'Olivo Cordero JP, Wei G, McCulloch MT (2019) Anthropogenic Ocean warming and acidification recorded by Sr/Ca, Li/Mg, δ11B and B/Ca in Porites coral from the Kimberley region of northwestern Australia. Palaeogeogr Palaeoclimatol Palaeoecol 528:50–59. https://doi.org/10.1016/j.palaeo.2019.04.033

Chiazzari B, Magalon H, Gélin P, Macdonald A (2019) Living on the edge: assessing the diversity of South African Pocillopora on the margins of the Southwestern Indian Ocean. PLoS One 14:e0220477

Church JA, White NJ, Hunter JR (2006) Sea-level rise at tropical Pacific and Indian Ocean islands. Glob Planet Chang 53:155–168. https://doi.org/10.1016/j.gloplacha.2006.04.001

Clark V, Mello-Athayde MA, Dove S (2022) Colonies of *Acropora formosa* with greater survival potential have reduced calcification rates. PLoS One 17:e0269526

Colton MA, McManus LC, Schindler DE et al (2022) Coral conservation in a warming world must harness evolutionary adaptation. Nat Ecol Evol 6:1405–1407. https://doi.org/10.1038/s41559-022-01854-4

Comeau S, Cornwall CE, Shlesinger T et al (2022) pH variability at volcanic CO_2 seeps regulates coral calcifying fluid chemistry. Glob Chang Biol 28:2751–2763. https://doi.org/10.1111/GCB.16093

Connell JH, Hughes TP, Wallace CC (1997) A 30-year study of coral abundance, recruitment, and disturbance at several scales in space and time. Ecol Monogr 67:461–488

Connell JH, Hughes TP, Wallace CC et al (2004) A long-term study of competition and diversity of corals. Ecol Monogr 74:179–210. https://doi.org/10.1890/02-4043

Conrad V (2019) Effects of thermal stress on the calcification and physiological fitness of Porites lutea intertidal reef flat corals in Phuket, Thailand. University of Bremen

Cordeiro CAMM, Quimbayo JP, Nunes JACC et al (2021) Conservation status of the southernmost reef of the Amazon reef system: the parcel de Manuel Luís. Coral Reefs 40:165–185. https://doi.org/10.1007/s00338-020-02026-1

Cornwall CE, Comeau S, DeCarlo TM et al (2018) Resistance of corals and coralline algae to ocean acidification: physiological control of calcification under natural pH variability. Proc R Soc B Biol Sci 285:20181168. https://doi.org/10.1098/rspb.2018.1168

Craig P, Birkeland C, Belliveau S (2001) High temperatures tolerated by a diverse assemblage of shallow-water corals in American Samoa. Coral Reefs 20:185–189

Dandan SS, Falter JL, Lowe RJ, McCulloch MT (2015) Resilience of coral calcification to extreme temperature variations in the Kimberley region, Northwest Australia. Coral Reefs 34:1151–1163. https://doi.org/10.1007/s00338-015-1335-6

Darling ES, Alvarez-Filip L, Oliver TA et al (2012) Evaluating life-history strategies of reef corals from species traits. Ecol Lett 15:1378–1386. https://doi.org/10.1111/j.1461-0248.2012.01861.x

Devlin M, Waterhouse J, Taylor J, Brodie J (2001) Flood plumes in the great barrier reef: spatial and temporal patterns in composition and distribution. Great Barrier Reef Marine Park Authority

Dixon DJ (2011) Tidal effects on coral reefs. In: Hopley D (ed) Encyclopedia of modern coral reefs: structure, form and process. Springer Netherlands, Dordrecht, pp 1086–1091

Dove SG, Kline DI, Pantos O et al (2013) Future reef decalcification under a business-as-usual CO_2 emission scenario. Proc Natl Acad Sci 110:15342–15347. https://doi.org/10.1073/pnas.1302701110

Dunne RP, Brown BE, Phongsuwan N, Putchim L (2021) The Indian Ocean dipole and El Niño southern oscillation as major drivers of coral cover on shallow reefs in the Andaman Sea. Glob Chang Biol 27:3312–3323. https://doi.org/10.1111/gcb.15640

Endean R, Cameron AM, Fox HE et al (1997) Massive corals are regularly spaced: pattern in a complex assemblage of corals. Mar Ecol Prog Ser 152:119–130

Fabricius KE, Mieog JC, Colin PL et al (2004) Identity and diversity of coral endosymbionts (zooxanthellae) from three Palauan reefs with contrasting bleaching, temperature and shading histories. Mol Ecol 13:2445–2458. https://doi.org/10.1111/j.1365-294X.2004.02230.x

Fenner D (2019) Chapter 28: The Samoan archipelago. In: Sheppard C (ed) World seas: an environmental evaluation, 2nd edn. Academic Press, pp 619–644

Fiesinger A, Held C, Melzner F et al (2023) Population genetic differentiation of the ubiquitous brooding coral *Pocillopora acuta* along Phuket Island reefs in the Andaman Sea, Thailand. BMC Ecol Evol 23:42. https://doi.org/10.1186/s12862-023-02153-7

Fine M, Gildor H, Genin A (2013) A coral reef refuge in the Red Sea. Glob Chang Biol 19:3640–3647. https://doi.org/10.1111/gcb.12356

Fisher PL, Malme MK, Dove S (2012) The effect of temperature stress on coral–*Symbiodinium* associations containing distinct symbiont types. Coral Reefs 31:473–485. https://doi.org/10.1007/s00338-011-0853-0

Fordyce AJ, Ainsworth TD, Heron SF, Leggat W (2019) Marine heatwave hotspots in coral reef environments: physical drivers, ecophysiological outcomes, and impact upon structural complexity. Front Mar Sci 6:498. https://doi.org/10.3389/fmars.2019.00498

Fordyce AJ, Ainsworth TD, Page CE et al (2021) Microalgal blooms in the skeletons of bleached corals during the 2020 bleaching event on Heron Island, Australia. Mar Freshw Res 72:1689–1694

Frölicher TL, Fischer EM, Gruber N (2018) Marine heatwaves under global warming. Nature 560:360–364. https://doi.org/10.1038/s41586-018-0383-9

Garrison T (2009) Essentials of oceanography. Brooks/Cole, Cengage Learning

Georgiou L, Falter J, Trotter J et al (2015) pH homeostasis during coral calcification in a free ocean CO_2 enrichment (FOCE) experiment, Heron Island reef flat, great barrier reef. Proc Natl Acad Sci 112:13219–13224. https://doi.org/10.1073/pnas.1505586112

Gilmour JP, Cook KL, Ryan NM et al (2019) The state of Western Australia's coral reefs. Coral Reefs 38:651–667. https://doi.org/10.1007/s00338-019-01795-8

Granados-Cifuentes C, Bellantuono AJ, Ridgway T et al (2013) High natural gene expression variation in the reef-building coral Acropora millepora: potential for acclimative and adaptive plasticity. BMC Genomics 14:1–12

Green RH, Lowe RJ, Buckley ML et al (2019) Physical mechanisms influencing localized patterns of temperature variability and coral bleaching within a system of reef atolls. Coral Reefs 38:759–771. https://doi.org/10.1007/s00338-019-01771-2

Gruber RK, Lowe RJ, Falter JL (2017) Metabolism of a tide-dominated reef platform subject to extreme diel temperature and oxygen variations. Limnol Oceanogr 62:1701–1717. https://doi.org/10.1002/lno.10527

Guastella LA, Rossouw M (2012) Coastal vulnerability: what will be the impact of increasing frequency and intensity of coastal storms along the South African coast. Reef J 2:129–139

Hayes MO (1975) Morphology of sand accumulation in estuaries: an introduction to the symposium. In: Geology and engineering. Academic Press, pp 3–22

Hoegh-Guldberg O, Fine M (2004) Low temperatures cause coral bleaching. Coral Reefs 23:444

Hoegh-Guldberg O, Fine M, Skirving W et al (2005) Coral bleaching following wintry weather. Limnol Oceanogr 50:265–271. https://doi.org/10.4319/lo.2005.50.1.0265

Hoegh-Guldberg O, Skirving W, Dove SG et al (2023) Coral reefs in peril in a record-breaking year. Science 382:1238–1240. https://doi.org/10.1126/science.adk4532

Howells EJ, Beltran VH, Larsen NW et al (2012) Coral thermal tolerance shaped by local adaptation of photosymbionts. Nat Clim Chang 2:116–120. https://doi.org/10.1038/nclimate1330

Hughes TP, Anderson KD, Connolly SR et al (2018) Spatial and temporal patterns of mass bleaching of corals in the Anthropocene. Science 359:80–83. https://doi.org/10.1126/science.aan8048

Jackson LF, Lipschitz S (1984) Coastal sensitivity atlas of Southern Africa. Department of Transport of the Republic of South Africa

Jatkar AA, Brown BE, Bythell JC et al (2010) Coral mucus: the properties of its constituent mucins. Biomacromolecules 11:883–888. https://doi.org/10.1021/bm9012106

Jimenez IM, Kühl M, Larkum AWD, Ralph PJ (2008) Heat budget and thermal microenvironment of shallow-water corals: do massive corals get warmer than branching corals? Limnol Oceanogr 53:1548–1561. https://doi.org/10.4319/lo.2008.53.4.1548

Jimenez IM, Larkum AWD, Ralph PJ, Kühl M (2012) In situ thermal dynamics of shallow water corals is affected by tidal patterns and irradiance. Mar Biol 159:1773–1782

Johnson CR (1995) Nutrient inputs from seabirds and humans on a populated coral cay. Mar Ecol Prog Ser 124:189–200

Jung EMU, Stat M, Thomas L et al (2021) Coral host physiology and symbiont dynamics associated with differential recovery from mass bleaching in an extreme, macro-tidal reef environment in Northwest Australia. Coral Reefs 40:893–905. https://doi.org/10.1007/s00338-021-02094-x

Kaniewska P, Campbell PR, Kline DI et al (2012) Major cellular and physiological impacts of ocean acidification on a reef building coral. PLoS One 7:e34659

Kaniewska P, Chan C-KK, Kline D et al (2015) Transcriptomic changes in coral Holobionts provide insights into physiological challenges of future climate and ocean change. PLoS One 10:e0139223

Kapsenberg L, Cyronak T (2019) Ocean acidification refugia in variable environments. Glob Chang Biol 25:3201–3214. https://doi.org/10.1111/gcb.14730

Kekuewa SAH, Courtney TA, Cyronak T et al (2021) Temporal and spatial variabilities of chemical and physical parameters on the Heron Island coral reef platform. Aquat Geochem 27:241–268. https://doi.org/10.1007/s10498-021-09400-7

Klepac CN, Barshis DJ (2020) Reduced thermal tolerance of massive coral species in a highly variable environment. Proc R Soc B Biol Sci 287:20201379. https://doi.org/10.1098/rspb.2020.1379

Klepac CN, Barshis DJ (2022) High-resolution in situ thermal metrics coupled with acute heat stress experiments reveal differential coral bleaching susceptibility. Coral Reefs 41:1045–1057. https://doi.org/10.1007/s00338-022-02276-1

Kline DI, Teneva L, Schneider K et al (2012) A short-term in situ CO_2 enrichment experiment on Heron Island (GBR). Sci Rep 2:413

Kline DI, Teneva L, Hauri C et al (2015) Six month in situ high-resolution carbonate chemistry and temperature study on a coral reef flat reveals asynchronous pH and temperature anomalies. PLoS One 10:e0127648

Kowalik Z (2004) Tide distribution and tapping into tidal energy. Oceanologia 46:291–331

Koweek D, Dunbar R, Monismith S et al (2015) High-resolution physical and biogeochemical variability from a shallow back reef on Ofu, American Samoa: an end-member perspective. Coral Reefs 34:979–991. https://doi.org/10.1007/s00338-015-1308-9

LaJeunesse TC, Pettay DT, Sampayo EM et al (2010a) Long-standing environmental conditions, geographic isolation and host–symbiont specificity influence the relative ecological dominance and genetic diversification of coral endosymbionts in the genus Symbiodinium. J Biogeogr 37:785–800. https://doi.org/10.1111/j.1365-2699.2010.02273.x

LaJeunesse TC, Smith R, Walther M et al (2010b) Host-symbiont recombination versus natural selection in the response of coral-dinoflagellate symbioses to environmental disturbance. Proc R Soc B 277:2925–2934

LaJeunesse TC, Parkinson JE, Gabrielson PW et al (2018) Systematic revision of Symbiodiniaceae highlights the antiquity and diversity of coral endosymbionts. Curr Biol 28:2570–2580.e6. https://doi.org/10.1016/j.cub.2018.07.008

Lantz CA, Leggat W, Bergman JL et al (2022) Will daytime community calcification reflect reef accretion on future, degraded coral reefs? Biogeosciences 19:891–906. https://doi.org/10.5194/BG-19-891-2022

Le Nohaïc M, Ross CL, Cornwall CE et al (2017) Marine heatwave causes unprecedented regional mass bleaching of thermally resistant corals in northwestern Australia. Sci Rep 7:1–11. https://doi.org/10.1038/s41598-017-14794-y

Levene H (1953) Genetic equilibrium when more than one ecological niche is available. Am Nat 87:331–333. https://doi.org/10.1086/281792

Liu G, Mark Eakin C, Chen M et al (2018) Predicting heat stress to inform reef management: NOAA coral reef watch's 4-month coral bleaching outlook. Front Mar Sci 5:57. https://doi.org/10.3389/fmars.2018.00057

Lowe RJ, Falter J (2015) Oceanic forcing of coral reefs. Annu Rev Mar Sci 7:43–66. https://doi.org/10.1146/annurev-marine-010814-015834

Lowe RJ, Leon AS, Symonds G et al (2015) The intertidal hydraulics of tide-dominated reef platforms. J Geophys Res Ocean 120:4845–4868. https://doi.org/10.1002/2015jc010701

Lowe RJ, Pivan X, Falter J et al (2016) Rising sea levels will reduce extreme temperature variations in tide-dominated reef habitats. Sci Adv 2:e1600825. https://doi.org/10.1126/sciadv.1600825

Lucey NM, Collins M, Collin R (2020) Oxygen-mediated plasticity confers hypoxia tolerance in a corallivorous polychaete. Ecol Evol 10:1145–1157. https://doi.org/10.1002/ece3.5929

Maggioni F, Pujo-Pay M, Aucan J et al (2021) The Bouraké semi-enclosed lagoon (New Caledonia)—a natural laboratory to study the lifelong adaptation of a coral reef ecosystem to extreme environmental conditions. Biogeosciences 18:5117–5140. https://doi.org/10.5194/BG-18-5117-2021

Marhoefer SR, Zenger KR, Strugnell JM et al (2021) Signatures of adaptation and acclimatization to reef flat and slope habitats in the coral *Pocillopora damicornis*. Front Mar Sci 8:704709

Marshall AT, Clode P (2004) Calcification rate and the effect of temperature in a zooxanthellate and an azooxanthellate scleractinian reef coral. Coral Reefs 23:218–224. https://doi.org/10.1007/s00338-004-0369-y

Maticka SA, Rogers JS, Woodson CB et al (2022) Reef flat flow dynamics for a nearly closed fringing reef lagoon: Ofu, American Samoa. J Geophys Res Ocean 127:e2022JC018831. https://doi.org/10.1029/2022JC018831

McMahon A, Santos IR, Cyronak T, Eyre BD (2013) Hysteresis between coral reef calcification and the seawater aragonite saturation state. Geophys Res Lett 40:4675–4679

McMahon A, Santos IR, Schulz KG et al (2018) Determining coral reef calcification and primary production using automated alkalinity, pH and pCO_2 measurements at high temporal resolution. Estuar Coast Shelf Sci 209:80–88

Mejía-Rentería JC, Castellanos-Galindo GA, Osorio-Cano JD, Casella E (2020) Subaerial exposure of coral reefs during spring low tides in the eastern Pacific. Bull Mar Sci 96:219–220

Mello-Athayde MA (2021) Present and future coral physiology of the resilient coral Porites cylindrica (Dana, 1846). University of Queensland, Australia

Mongin M, Baird M (2014) The interacting effects of photosynthesis, calcification and water circulation on carbon chemistry variability on a coral reef flat: a modelling study. Ecol Model 284:19–34

Monismith SG, Maticka SA, Rogers JS et al (2024) Vertical structure of flows on a shallow reef flat: a coral reef surf zone. Coast Eng 190:104499. https://doi.org/10.1016/j.coastaleng.2024.104499

Morikawa MK, Palumbi SR (2019) Using naturally occurring climate resilient corals to construct bleaching-resistant nurseries. Proc Natl Acad Sci 116:10586–10591. https://doi.org/10.1073/pnas.1721415116

Mvula P (2020) Decadal changes in rocky shore communities in KwaZulu-Natal and survey methods for future monitoring. University of KwaZulu-Natal, South Africa

Nitschke MR, Rosset SL, Oakley CA et al (2022) The diversity and ecology of Symbiodiniaceae: a traits-based review. Adv Mar Biol 92:55–127. https://doi.org/10.1016/BS.AMB.2022.07.001

NOAA (2024a) What is the intertidal zone? National Ocean Service website. https://oceanservice.noaa.gov/facts/intertidal-zone.html. Accessed 16 Jun 2024

NOAA (2024b) Heron Island. https://www.pmelnoaagov/co2/story/Heron+Island. Accessed 19 Aug 2024

Olbers J (2017a) Rocky & coral reef biodiversity monitoring: Ezemvelo KZN wildlife research & monitoring of marine living resources programme for the Department of Environmental Affairs. Annual Report for 2016/2017. Pietermaritzburg, South Africa

Olbers J (2017b) Rocky shore biodiversity monitoring: Ezemvelo KZN wildlife research & monitoring of marine living resources programme for the Department of Environmental Affairs. Annual Report 2016/2107. Pietermaritzburg, South Africa

Oliver TA, Palumbi SR (2009) Distributions of stress-resistant coral symbionts match environmental patterns at local but not regional scales. Mar Ecol Prog Ser 378:103

Oliver TA, Palumbi SR (2011a) Do fluctuating temperature environments elevate coral thermal tolerance? Coral Reefs 30:429–440. https://doi.org/10.1007/s00338-011-0721-y

Oliver TA, Palumbi SR (2011b) Many corals host thermally resistant symbionts in high-temperature habitat. Coral Reefs 30:241–250. https://doi.org/10.1007/s00338-010-0696-0

Onyango CA (2020) Insights into the transcriptomic mechanisms underlying responses to heat stress and bleaching in two coral holobionts from Park Rynie, east coast South Africa. University of KwaZulu-Natal, South Africa

Onyango CA, Glassom D, MacDonald A (2021) De novo assembly of the transcriptome of scleractinian coral, Anomastraea irregularis and analyses of its response to thermal stress. Mol Biol Rep 483(48):2083–2092. https://doi.org/10.1007/S11033-021-06184-5

Palmer BJ, Van der Elst R, Mackay F et al (2011) Preliminary coastal vulnerability assessment for KwaZulu-Natal, South Africa. J Coast Res 64:1390–1395

Palumbi SR, Barshis DJ, Traylor-Knowles N, Bay RA (2014) Mechanisms of reef coral resistance to future climate change. Science 344:895–898. https://doi.org/10.1126/science.1251336

Pedersen O, Colmer TD, Borum J et al (2016) Heat stress of two tropical seagrass species during low tides—impact on underwater net photosynthesis, dark respiration and diel in situ internal aeration. New Phytol 210:1207–1218. https://doi.org/10.1111/nph.13900

Perez DI, Phinn SR, Roelfsema CM et al (2018) Primary production and calcification rates of algae-dominated reef flat and seagrass communities. J Geophys Res Biogeosci 123:2362–2375

Perry CT, Larcombe P (2003) Marginal and non-reef-building coral environments. Coral Reefs 22:427–432. https://doi.org/10.1007/s00338-003-0330-5

Pettay DT, LaJeunesse TC (2009) Microsatellite loci for assessing genetic diversity, dispersal and clonality of coral symbionts in 'stress-tolerant' clade D Symbiodinium. Mol Ecol Resour 9:1022–1025. https://doi.org/10.1111/j.1755-0998.2009.02561.x

Phinn SR, Roelfsema CM, Mumby PJ (2012) Multi-scale, object-based image analysis for mapping geomorphic and ecological zones on coral reefs. Int J Remote Sens 33:3768–3797

Phongsuwan N, Chansang H (2012) Repeated coral bleaching in the Andaman Sea, Thailand, during the last two decades. Phuket Mar Biol Cent Res Bull 71:19–41

Purcell S (2002) Intertidal reefs under extreme tidal flux in Buccaneer Archipelago, Western Australia. Coral Reefs 21:191–192. https://doi.org/10.1007/s00338-002-0223-z

Raker C, Cowburn B, Sindorf V et al (2014) Anomastraea irregularis, a vulnerable coral of the rocky intertidal zone of Watamu Marine National Park, Kenya

Ramsay PJ (1996) Quaternary marine geology of the Sodwana Bay continental shelf, northern KwaZulu-Natal. Oceanogr Lit Rev 11:1103

Rathbone M, Brown KT, Dove S (2021) Tolerance to a highly variable environment does not infer resilience to future ocean warming and acidification in a branching coral. Limnol Oceanogr 9999:1–13. https://doi.org/10.1002/LNO.11991

Reimer JD, Peixoto RS, Davies SW et al (2024) The fourth global coral bleaching event: where do we go from here? Coral Reefs 43:1121–1125. https://doi.org/10.1007/s00338-024-02504-w

Reshef L, Koren O, Loya Y et al (2006) The coral probiotic hypothesis. Environ Microbiol 8:2068–2073. https://doi.org/10.1111/j.1462-2920.2006.01148.x

Richards ZT (2022) Kimberley corals exposed. In: Hutchings P, Hamylton SM, Hoegh-Guldberg O (eds) Coral reefs of Australia: perspectives from beyond the water's edge, 1st edn. CSIRO Publishing, Melbourne, p 344

Richards ZT, O'Leary MJ (2015) The coralline algal cascades of Tallon Island (Jalan) fringing reef, NW Australia. Coral Reefs 34:595. https://doi.org/10.1007/s00338-015-1262-6

Richards ZT, Bryce M, Bryce C (2013) New records of atypical coral reef habitat in the Kimberley, Australia. J Mar Biol 2013:8. https://doi.org/10.1155/2013/363894

Richards ZT, Garcia RA, Wallace CC et al (2015) A diverse assemblage of reef corals thriving in a dynamic intertidal reef setting (Bonaparte archipelago, Kimberley, Australia). PLoS One 10:e0117791

Richards ZR, Bryce M, Bryce C (2018) The composition and structure of shallow benthic reef communities in the Kimberley, north-West Australia. Rec West Aust Museum Suppl 85:75. https://doi.org/10.18195/issn.0313-122x.85.2018.075-103

Richards ZT, Garcia R, Moore G et al (2019) A tropical Australian refuge for photosymbiotic benthic fauna. Coral Reefs 38:669–676. https://doi.org/10.1007/s00338-019-01809-5

Richards ZT, Pichon M, Wallace C (2024) Biodiversity and biogeography of zooxanthellate corals in Australasia revisited based on new data from the Kimberley. Rec West Aust Museum Suppl 85:201–276. https://doi.org/10.18195/issn.0313-122x.85.2024.201-276

Riegl B (1996) Hermatypic coral Fauna of subtropical Southeast Africa: a checklist. Pac Sci 50:404–414

Riegl B, Purkis SJ (2012) Coral reefs of the Gulf: adaptation to environmental extremes. Springer, Dordrecht

Rivest EB, Comeau S, Cornwall CE (2017) The role of natural variability in shaping the response of coral reef organisms to climate change. Curr Clim Chang Rep 3:271–281. https://doi.org/10.1007/s40641-017-0082-x

Roelfsema C, Kovacs EM, Markey K et al (2021) Benthic and coral reef community field data for Heron reef, southern great barrier reef, Australia, 2002–2018. Sci Data 8:84

Romaine S, Tambutté E, Allemand D, Gattuso J-P (1997) Photosynthesis, respiration and calcification of a zooxanthellate scleractinian coral under submerged and exposed conditions. Mar Biol 129:175–182. https://doi.org/10.1007/s002270050158

Rosser NL, Veron JEN (2011) Australian corals thriving out of water in an extreme environment. Coral Reefs 30:21. https://doi.org/10.1007/s00338-010-0689-z

Safaie A, Silbiger NJ, McClanahan TR et al (2018) High frequency temperature variability reduces the risk of coral bleaching. Nat Commun 9:1671. https://doi.org/10.1038/s41467-018-04074-2

Santos IR, Glud RN, Maher D et al (2011) Diel coral reef acidification driven by porewater advection in permeable carbonate sands, Heron Island, Great Barrier Reef. Geophys Res Lett 38:5

Sayre R, Suzanne N, Sharon H et al (2019) A new 30 meter resolution global shoreline vector and associated global islands database for the development of standardized ecological coastal units. J Oper Oceanogr 12:S47–S56. https://doi.org/10.1080/1755876X.2018.1529714

Schleyer MH, Celliers L (2003) Coral dominance at the reef–sediment interface in marginal coral communities at Sodwana Bay, South Africa. Mar Freshw Res 54:967–972

Schleyer MH, Heikoop JM, Risk MJ (2006) A benthic survey of Aliwal Shoal and assessment of the effects of a wood pulp effluent on the reef. Mar Pollut Bull 52:503–514. https://doi.org/10.1016/j.marpolbul.2005.09.038

Schoepf V, Grottoli AG, Levas SJ et al (2015a) Annual coral bleaching and the long-term recovery capacity of coral. Proc R Soc Lond B Biol Sci 282:20151887. https://doi.org/10.1098/rspb.2015.1887

Schoepf V, Stat M, Falter JL, McCulloch MT (2015b) Limits to the thermal tolerance of corals adapted to a highly fluctuating, naturally extreme temperature environment. Sci Rep 5:1–14. https://doi.org/10.1038/srep17639

Schoepf V, Carrion SA, Pfeifer SM et al (2019) Stress-resistant corals may not acclimatize to ocean warming but maintain heat tolerance under cooler temperatures. Nat Commun 10:4031. https://doi.org/10.1038/s41467-019-12065-0

Schoepf V, Jung MU, McCulloch MT et al (2020) Thermally variable, macrotidal reef habitats promote rapid recovery from mass coral bleaching. Front Mar Sci 7:245. https://doi.org/10.3389/FMARS.2020.00245

Schoepf V, D'Olivo JP, Rigal C et al (2021) Heat stress differentially impacts key calcification mechanisms in reef-building corals. Coral Reefs 40:459–471. https://doi.org/10.1007/s00338-020-02038-x

Schoepf V, Sanderson H, Larcombe E (2022) Coral heat tolerance under variable temperatures: effects of different variability regimes and past environmental history vs. current exposure. Limnol Oceanogr 67:404–418. https://doi.org/10.1002/lno.12000

Schoepf V, Baumann JH, Barshis DJ et al (2023) Corals at the edge of environmental limits: a new conceptual framework to re-define marginal and extreme coral communities. Sci Total Environ 884:163688. https://doi.org/10.1016/J.SCITOTENV.2023.163688

Scoffin TP, Tudhope AW, Brown BE et al (1992) Patterns and possible environmental controls of skeletogenesis of Porites lutea, South Thailand. Coral Reefs 11:1–11. https://doi.org/10.1007/BF00291929

Scoffin TP, Brown BE, Dunne RP, Tissier MDA (1997) The controls on growth form of intertidal massive corals, Phuket, South Thailand. PALAIOS 12:237–248. https://doi.org/10.2307/3515425

Scopélitis J, Andréfouët S, Phinn S et al (2011) Coral colonisation of a shallow reef flat in response to rising sea level: quantification from 35 years of remote sensing data at Heron Island, Australia. Coral Reefs 30:951–965

Seneca FO, Palumbi SR (2015) The role of transcriptome resilience in resistance of corals to bleaching. Mol Ecol 24:1467–1484. https://doi.org/10.1111/mec.13125

Shaw EC, Phinn SR, Tilbrook B, Steven A (2015) Natural in situ relationships suggest coral reef calcium carbonate production will decline with ocean acidification. Limnol Oceanogr 60:777–788. https://doi.org/10.1002/lno.10048

Sink KJ, Branch GM, Harris JM (2005) Biogeographic patterns in rocky intertidal communities in KwaZulu-Natal, South Africa. Afr J Mar Sci 27:81–96. https://doi.org/10.2989/18142320509504070

Smit K (2014) Ecophysiology of intertidal corals along the east coast of South Africa: coping in marginal habitats. University of KwaZulu-Natal

Smit K, Glassom D (2017) Large fluctuations but constant mean temperatures allow corals to persist in intertidal rock pools on the east coast of South Africa. Helgol Mar Res 711(71):1–9. https://doi.org/10.1186/S10152-017-0482-2

Smit AJ, Roberts M, Anderson RJ et al (2013) A coastal seawater temperature dataset for biogeographical studies: large biases between in situ and remotely-sensed data sets around the coast of South Africa. PLoS One 8:e81944

Smith LW, Birkeland C (2007) Effects of intermittent flow and irradiance level on back reef Porites corals at elevated seawater temperatures. J Exp Mar Biol Ecol 341:282–294. https://doi.org/10.1016/j.jembe.2006.10.053

Smith LW, Barshis D, Birkeland C (2007) Phenotypic plasticity for skeletal growth, density and calcification of *Porites lobata* in response to habitat type. Coral Reefs 26:559–567. https://doi.org/10.1007/s00338-007-0216-z

Smith JE, Brainard R, Carter A et al (2016) Re-evaluating the health of coral reef communities: baselines and evidence for human impacts across the Central Pacific. Proc R Soc B Biol Sci 283:20151985

Speelman PE, Parger M, Schoepf V (2023) Divergent recovery trajectories of intertidal and subtidal coral communities highlight habitat-specific recovery dynamics following bleaching in an extreme macrotidal reef environment. PeerJ 11:e15987. https://doi.org/10.7717/peerj.15987

Swain TD, Chandler J, Backman V, Marcelino L (2017) Consensus thermotolerance ranking for 110 Symbiodinium phylotypes: an exemplar utilization of a novel iterative partial-rank aggregation tool with broad application potential. Funct Ecol 31:172–183. https://doi.org/10.1111/1365-2435.12694

Swan HB, Jones GB, Deschaseaux ESM, Eyre BD (2017) Coral reef origins of atmospheric dimethylsulfide at Heron Island, southern great barrier reef, Australia. Biogeosciences 14:229–239

Sweet MJ, Brown BE, Dunne RP et al (2017) Evidence for rapid, tide-related shifts in the microbiome of the coral Coelastrea aspera. Coral Reefs 36:815–828. https://doi.org/10.1007/s00338-017-1572-y

Tanner JE (2017) Multi-decadal analysis reveals contrasting patterns of resilience and decline in coral assemblages. Coral Reefs 36:1225–1233

Tanner JE, Connell JH (2022) Coral community data Heron Island great barrier reef 1962–2016. Sci Data 9:617

ThePhuketNews (2024) Phuket coral reef health sparks concerns. ThePhuketNews. https://www.thephuketnews.com/phuket-coral-reef-health-sparks-concerns-92262.php. Accessed 19 Aug 2024

Thomas L, Kendrick GA, Kennington WJ et al (2014) Exploring Symbiodinium diversity and host specificity in Acropora corals from geographical extremes of Western Australia with 454 amplicon pyrosequencing. Mol Ecol 23:3113–3126. https://doi.org/10.1111/mec.12801

Thomas L, Rose NH, Bay RA et al (2018) Mechanisms of thermal tolerance in reef-building corals across a fine-grained environmental mosaic: lessons from Ofu, American Samoa. Front Mar Sci 4:434. https://doi.org/10.3389/fmars.2017.00434

Thomas L, Underwood JN, Noah RH et al (2022) Spatially varying selection between habitats drives physiological shifts and local adaptation in a broadcast spawning coral on a remote atoll in Western Australia. Sci Adv 8:eabl9185. https://doi.org/10.1126/sciadv.abl9185

Thomas L, Şahin D, Adam AS et al (2024) Resilience to periodic disturbances and the long-term genetic stability in Acropora coral. Commun Biol 7:410. https://doi.org/10.1038/s42003-024-06100-0

Turak E, Veron JEN, Sanpanich K (2005) Post-tsunami coral reef surveys of Thailand's north Andaman coast: coral reef status after 26 December 2004 tsunami, in Rapid Assessment Survey of Tsunami-affected Reefs of Thailand, Allen Gr and Stone GR (eds), New England Aquarium Technical Report.

Ulstrup KE, Van Oppen MJH (2003) Geographic and habitat partitioning of genetically distinct zooxanthellae (Symbiodinium) in Acropora corals on the great barrier reef. Mol Ecol 12:3477–3484. https://doi.org/10.1046/j.1365-294X.2003.01988.x

Ulstrup KE, Ralph PJ, Larkum AWD, Kühl M (2006) Intra-colonial variability in light acclimation of zooxanthellae in coral tissues of *Pocillopora damicornis*. Mar Biol 149:1325–1335. https://doi.org/10.1007/s00227-006-0286-4

Ulstrup KE, Hill R, Van Oppen MJH et al (2008) Seasonal variation in the photo-physiology of homogeneous and heterogeneous Symbiodinium consortia in two scleractinian corals. Mar Ecol Prog Ser 361:139–150

Underwood JN, Wilson SK, Ludgerus L, Evans RD (2013) Integrating connectivity science and spatial conservation management of coral reefs in north-West Australia. J Nat Conserv 21:163–172. https://doi.org/10.1016/j.jnc.2012.12.001

Underwood JN, Richards Z, Berry O et al (2020) Extreme seascape drives local recruitment and genetic divergence in brooding and spawning corals in remote north-West Australia. Evol Appl 13:2404–2421. https://doi.org/10.1111/eva.13033

van Oppen MJH, Bongaerts P, Frade P et al (2018) Adaptation to reef habitats through selection on the coral animal and its associated microbiome. Mol Ecol 27:2956–2971. https://doi.org/10.1111/mec.14763

Wasim M, Pandey AC, Basheer Ahammed KK, Dwivedi CS (2024) Two decades of coral bleaching in selected islands of Pacific Ocean: a holistic impact assessment. Reg Stud Mar Sci 77:103648. https://doi.org/10.1016/j.rsma.2024.103648

Whitfield A, Elliott M (2011) Ecosystem and biotic classifications of estuaries and coasts. In: Wolanski E, McLusky DS (eds) Treatise on estuarine and coastal science. Academic Press, Waltham, pp 99–124

Wilson B (2013) The biogeography of the Australian north west shelf: environmental change and life's response. Elsevier, Boston

Wilson B, Blake S, Ryan D, Hacker J (2011) Reconnaissance of species-rich coral reefs in a muddy, macro-tidal, enclosed embayment,—Talbot Bay, Kimberley, Western Australia. J R Soc West Aust 94:251–265

Wolanski E, Spagnol S (2003) Dynamics of the turbidity maximum in king sound, tropical Western Australia. Estuar Coast Shelf Sci 56:877–890. https://doi.org/10.1016/S0272-7714(02)00214-7

Zhang J, Richards ZT, Adam AAS et al (2022) Evolutionary responses of a reef-building coral to climate change at the end of the last glacial maximum. Mol Biol Evol 39:msac201. https://doi.org/10.1093/molbev/msac201

Ziegler M, Eguíluz VM, Duarte CM, Voolstra CR (2017a) Rare symbionts may contribute to the resilience of coral–algal assemblages. ISME J 12:161–172

Ziegler M, Seneca FO, Yum LK et al (2017b) Bacterial community dynamics are linked to patterns of coral heat tolerance. Nat Commun 8:14213

Long-Term Coral Responses to High CO_2: Insights from Volcanic Vents and Semi-Enclosed Bays

7

Sylvain Agostini and Riccardo Rodolfo-Metalpa

Abstract

This chapter provides a comprehensive review of natural laboratories used to study the effects of ocean acidification (OA) on coral reefs, emphasizing volcanic CO_2 vents and semi-enclosed bays. It examines key environmental conditions at these study sites and their implications for coral physiology, biodiversity, and reef resilience. Two prominent natural analogues are highlighted: the Shikine Island warm-temperate shallow CO_2 vent system in Japan and the semi-enclosed lagoon of Bouraké in New Caledonia.

Shikine Island represents a unique high-latitude coral environment influenced by the Kuroshio Current. The island's CO_2 vents provide a natural setting to investigate the community level effects of ocean acidification and how elevated CO_2 levels interact with ocean warming and influence coral tropicalization. Findings suggest that while some corals persist at high CO_2 levels, biodiversity declines significantly, with sensitive species unable to establish stable populations. Corals growing in these conditions exhibit reduced calcification, and the community structure shifts toward simpler, lower-complexity ecosystems dominated by turf algae. These results indicate that ocean acidification may act as a constraint on the poleward expansion of tropical corals, counteracting warming-induced range shifts.

In contrast, the semi-enclosed lagoon of Bouraké experiences extreme environmental fluctuations in pH, oxygen, and temperature, providing a window into the multifactorial stressors coral reefs may face in the future. Despite prolonged exposure to acidified, hypoxic, and thermally variable waters, corals in Bouraké demonstrate remarkable resilience. The system harbors diverse coral assemblages, with species exhibiting metabolic adjustments, increased heterotrophy, and shifts in symbiotic algae communities. This site suggests that coral ecosystems can develop physiological adaptations that enhance their survival in extreme conditions, offering valuable insights into the potential for long-term acclimatization.

By comparing these two contrasting natural analogues, this chapter highlights the complex interplay between acidification, warming, and other environmental stressors on coral reefs. While both sites provide critical insights into the future of coral ecosystems, they also underscore the urgent need for continued research and conservation efforts to mitigate the effects of climate change on marine biodiversity.

Keywords

Natural laboratories · Ocean acidification · Volcanic CO_2 vents · Semi-enclosed bays · Resilience

7.1 Introduction

7.1.1 Ocean Acidification Research

Anthropogenic CO_2 emissions have altered the carbonate equilibrium of the ocean, leading to increased CO_2 partial pressure, reduced carbonate ion concentration, and lower pH—a phenomenon known as ocean acidification (OA).

S. Agostini (✉)
Shimoda Marine Research Center, University of Tsukuba, Tsukuba, Japan

UMR ENTROPIE, IRD, Université de la Réunion, CNRS, IFREMER, Université de Nouvelle-Calédonie, Nouméa, New Caledonia

Labex ICONA, International CO2 Natural Analogues Network, Shimoda, Japan
e-mail: agostini.sylvain@shimoda.tsukuba.ac.jp; sylvain.agostini@ird.fr

R. Rodolfo-Metalpa
UMR ENTROPIE, IRD, Université de la Réunion, CNRS, IFREMER, Université de Nouvelle-Calédonie, Nouméa, New Caledonia

Labex ICONA, International CO2 Natural Analogues Network, Shimoda, Japan

M. O. Soares et al. (eds.), *Extreme and Marginal Coral Reefs and Communities*, Coral Reefs of the World 24,
https://doi.org/10.1007/978-3-032-28035-0_7

Laboratory experiments have demonstrated that OA can impair the growth and physiology of corals and other reef organisms, particularly calcifiers (Harvey et al. 2016). The impacts of OA extend across different life stages, from larvae to adults, affecting physiological processes and behaviour. While extensive laboratory studies have deepened our understanding of individual species' responses to OA, the broader community- and ecosystem-level effects remain poorly understood.

Many studies highlight the variability of responses to elevated CO_2, even within the most sensitive taxonomic groups, such as calcifiers, particularly corals. Furthermore, as most research is conducted in controlled laboratory conditions, there is a significant knowledge gap regarding the long-term effects of OA and its cascading impacts through biological interactions. Short-term, single-generation experiments primarily capture phenotypically plastic responses without considering potential multigenerational effects or genetic adaptation (Cornwall et al. 2020; Gattuso et al. 2015). Given the rapid pace of climate change and OA, the extent to which genetic adaptation can occur remains debated. However, studies have identified resistant species, populations, and even individuals, suggesting the potential for some degree of resilience (Torda et al. 2017).

Field-based CO_2 manipulation experiments have been used to assess community-level responses of coral reefs to increasing CO_2. These studies generally indicate that coral reef community calcification is highly sensitive to elevated CO_2 levels, with future projections suggesting a shift towards net dissolution. Both confined (e.g., Free Ocean Carbon Enrichment—FOCE, e.g., Doo et al. 2019) and unconfined (e.g. Albright et al. 2016) experiments confirm that rising CO_2 has already begun to affect reef metabolism and will continue to do so in the future. However, like laboratory experiments, these field studies are often limited in duration, typically exposing reef communities to elevated CO_2 for periods ranging from hours to days. As a result, they do not account for potential acclimatization, adaptation, or shifts in community structure over longer timescales.

7.1.2 Volcanic CO_2 Vents as Natural Analogues of Ocean Acidification

In 2008, Hall-Spencer et al. published a seminal study that, for the first time, used a volcanic vent system to assess the ecosystem-wide effects of OA (Hall-Spencer et al. 2008). This study focused on the volcanic CO_2 vents off the shore of Ischia, in Italy, which remains one of the most extensively studied sites. However, earlier works are worth noting, such as Tarasov et al. (1999), who described coral communities inhabiting the hydrothermal vent field of Matupi Bay, Papua New Guinea, and the early research conducted at the Milos CO_2 vents in Greece (Dando et al. 1999; Cocito et al. 2000; Morri et al. 2000). These earlier studies, however, did not frame their findings within the context of climate change and OA.

Volcanic vents are hydrothermal systems in volcanic regions where seawater or meteoric water infiltrates the seafloor, reacts with magmatic fluids and gases, and is then re-emitted into the water column as heated liquid fluids and/or streams of gas bubbles with diverse chemical compositions. Typically, vent fluids are acidic (though exceptions exist, such as the alkaline vent systems in Prony Bay, New Caledonia, and Eyjafjörður, northern Iceland). CO_2-rich gas plumes released at these sites often contain other magmatic volatiles, including sulfur dioxide (SO_2), hydrogen sulfide (H_2S), and methane (CH_4). Other geological systems also release CO_2-rich gases, such as cold seeps, mud volcanoes, and pockmarks, though these remain largely unexplored in the context of climate change and OA. Since the pioneering Ischia study, a handful of volcanic vent sites have been identified and used as "natural analogues" to study OA's long-term effects on marine ecosystems. Here, we reviewed studies using volcanic vents to research OA's physiological and ecological effects on corals. We describe the geochemical characteristics of each site and highlight the results and global trends gained from these systems on our understanding of the effects of OA on coral physiology and ecology.

7.2 Review on Coral Studies at Volcanic CO_2 Vents

Volcanic vents have been identified worldwide, but only a few—those with limited toxic fluid emissions and CO_2-rich gas releases—have been studied in the context of OA research. We reviewed studies investigating the effects of acidification exclusively on coral ecology and physiology in volcanic vent environments. However, many studies have investigated other taxa than corals using those systems. The following search query was used on Web of Science: AB = ((co2 OR acidif* OR warm*) AND (seep* OR vent* OR (natural NEAR/1 analogue))) AND AB = (coral*). This search yielded 153 papers, of which 47 were deemed relevant to the review. Additional searches on Google Scholar using the keywords "coral CO_2 seeps" and "coral CO_2 vents" resulted in 58 studies investigating coral responses to OA using volcanic vents.

The review identified 20 volcanic vent systems used for coral research (Fig. 7.1a, b). Upa-Upasina in Papua New Guinea (PNG) is the most extensively studied system, with 20 studies, followed by the nearby Dobu Island system (13 studies). Notably, the Dobu Island system has always been studied with Upa-Upasina. With the recent addition of two new sites (Grotta del Mago and Chiane del Lume), research around the historical Castello Aragonese site in Ischia, Italy,

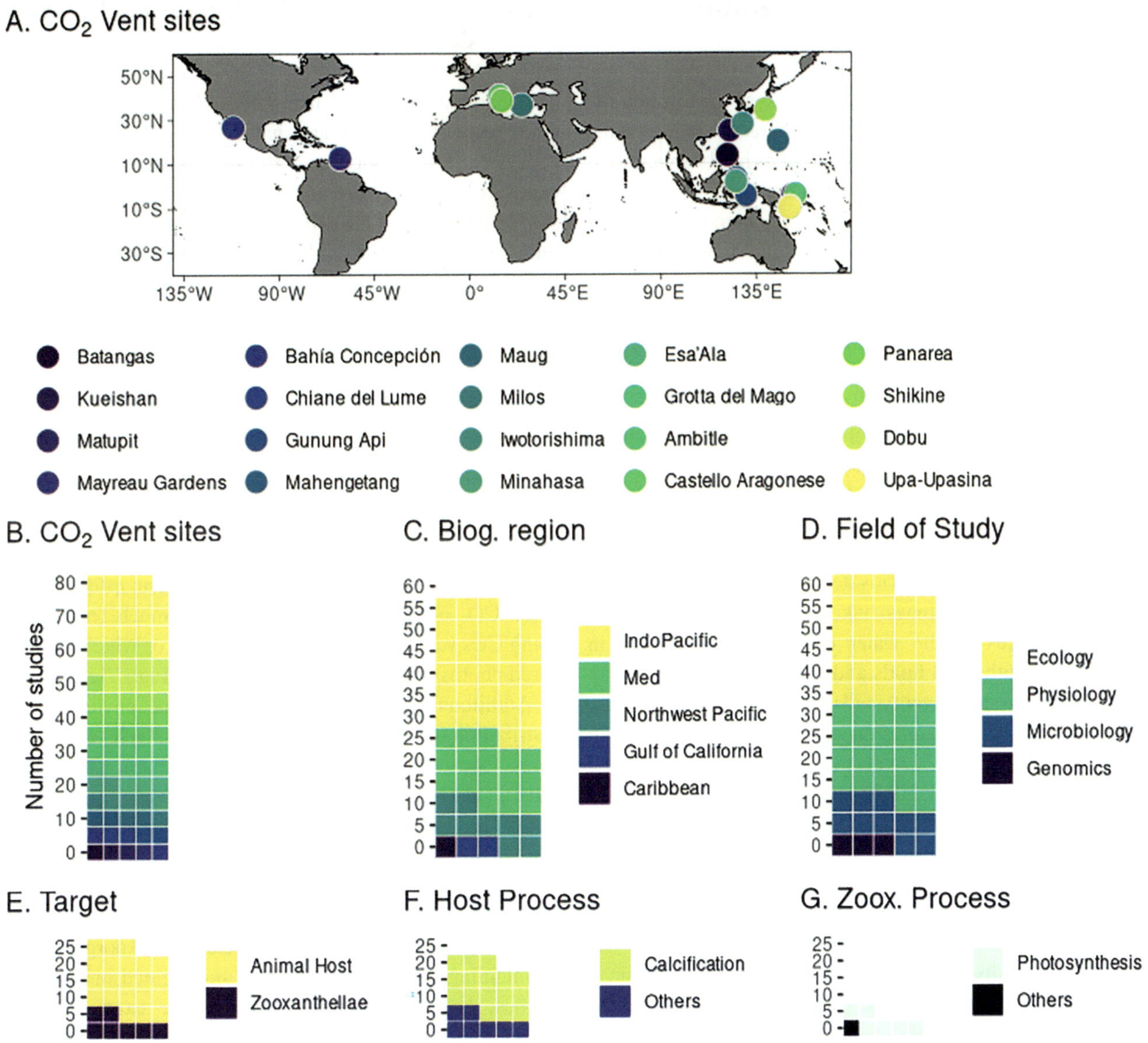

Fig. 7.1 (**a**) World map showing the different CO_2 vent systems used in studies on the long-term effects of ocean acidification on corals. The number of studies classified according to the CO_2 vent systems used (**b**, color for each site are according to the legend in **a**), biogeographic regions as defined by Costello et al. (2017) (**c**), field of study of the publications (**d**), the target of the study (host or zooxanthellae) (**e**). Number of studies investigating calcification among those targeting the animal host (**f**) and the number of studies investigating photosynthesis among those targeting the Symbiodiniaceae (**g**). Each square represents one study, in the case of one study using multiple sites, the number of squares is multiplied by the number of sites. Similarly, a study targeting both the host and Symbiodiniaceae would appear as two squares

has also expanded, bringing the total number of studies in this area to 11—making it the third most frequently studied region for corals under OA.

The majority of studies were conducted in the Indo--Pacific biogeographic region (Costello et al. 2017), with 30 studies, including those in the waters of Papua New Guinea, Indonesia (3 studies), and Taiwan (1 study) (Fig. 7.1c). The next most studied biogeographic region is the Mediterranean Sea, with 16 studies conducted at sites including Ischia and Panarea in the Aeolian Islands, Italy. Only one study spanned multiple biogeographic regions covering the Indo-Pacific and the Northwest Pacific (Agostini et al. 2021b). Despite the presence of both hydrothermal systems and coral reefs, the Caribbean region remains underrepresented, with only one study conducted there. The predominance of studies in the Indo-Pacific (particularly in Papua New Guinea) and the Mediterranean should be considered when extrapolating findings to other biogeographic regions.

7.2.1 Physico-Chemical Characteristics of the Acidified Sites

Since researchers began using CO_2 vent systems to study the effects of OA, the physico-chemical characterization of the water mass overlying benthic communities has been reported in most studies. Previously, research on these systems was primarily conducted by geologists, who understandably prioritized the characterization of hydrothermal fluids and nearby water properties. In most of the studies reviewed, surveys were conducted in shallow waters (<10 m), with researchers targeting zones where the mean pH and pCO_2 levels align with projected future values according to IPCC Representative Concentration Pathways (RCPs) (Table 7.1). All sites exhibit significant temporal variability in acidification, with increased variability at the lowest pH levels.

Long-term pH monitoring, standard deviation, minimum and maximum values, or percentiles, is often reported to highlight variability and facilitate comparisons across studies (Camp et al. 2018). This temporal variability—often unpredictable or sometimes linked to tidal effects—results from differences in the water circulation, patchiness in releasing hydrothermal fluids and gases, and fluctuations in hydrothermal system activity and their combination. Comeau et al. (2022) highlighted differences in pH variability between the two most studied CO2 vents in Papua New Guinea: Upa--Upasina (Fabricius et al. 2011) and Tutum Bay (Pichler et al. 2019). Upa-Upasina pH variability was significantly more prominent, with a lower pH value of 6.64. These factors make CO_2 vent systems inherently variable environments; therefore, they should be considered extreme sites for coral reefs rather than marginal sites (Schoepf et al. 2023). The extreme variability measured at specific CO_2 vents (Comeau et al. 2022) calls into question the validity of these sites as natural analogues for OA because the very variability and extreme values deviate from what might be the future reality.

At most sites—except for the most extreme ones where environmental conditions far exceed projected future levels—temperatures in acidified zones remain similar to those at reference sites (present-day CO_2 levels). However, OA occurs alongside ocean warming, and their combined effects may differ significantly from those observed under a single stressor. Thus, while CO_2 vents may not perfectly replicate future reef conditions, they serve as valuable "natural laboratories" for studying OA effects on marine organisms and ecosystems. Additionally, their limitations can be mitigated by selecting vent locations with lower variability or combining multiple sites to account for ocean warming and acidification (Agostini et al. 2021a). Finally, other types of natural laboratories, such as semi-enclosed lagoons—like the Bouraké lagoon in New Caledonia (Maggioni et al. 2021)—offer opportunities to study the impacts of multiple stressors.

7.2.2 Type of Studies Conducted at Volcanic CO_2 Vents

The main advantages of using hydrothermal vents to study coral responses to OA are that they provide insights into ecosystem-level responses and allow the study of life-long exposure. It is therefore not surprising that about half of the studies investigated the ecological response of coral communities in terms of abundance and diversity (28 studies; Fig. 7.1d). The following field of study is the physiological response to long-term exposure to elevated CO_2 (22 studies). Among these, 21 focused on the coral host (Fig. 7.1e), particularly on the effects of acidification on calcification (16 studies; Fig. 7.1f). Only seven studies investigated the response of both the animal host and its *Symbiodiniaceae*. Reciprocal transplantation of corals between control and high CO_2 areas is widely used, as it allows researchers to study effects under ecologically realistic conditions and over long-term exposure (months to years), taking full advantage of CO_2 vent systems. However, the long-term effects of elevated CO_2 on zooxanthellae have been much less studied using CO_2 vents, with only seven studies, of which six measured photosynthesis-related parameters (Fig. 7.1e, g).

Additionally, two studies examined the phylogeny and distribution of symbionts across different CO_2 levels, though these focused on non-hermatypic Zoantharia (Wee et al. 2019). One study investigated gene expression in symbionts under elevated CO_2 (Kenkel et al. 2018). This latter study is one of only three that have applied genomic or transcriptomic approaches (Kenkel et al. 2018; Leiva et al. 2023a; Teixidó et al. 2018), a field that requires further investigation. The importance of coral-associated microorganisms in maintaining coral health has been highlighted in numerous studies; however, only 10 studies have examined coral microbiomes at CO_2 vent sites.

7.2.3 Insights on the Effects of OA on the Ecology of Corals

Among the studies reviewed, 28 focus on the ecology of corals in CO_2 vents. A clear shift in the community structure was shown at most sites, with often, but not always, a decrease in species diversity and coral coverage. The most dramatic loss in coral diversity and abundance is observed for relatively high latitude reefs, such as a complete shift to soft corals at the Iwotorishima CO_2 vents in Japan (Inoue et al. 2013) and almost complete absence of corals at the CO_2 vents in Shikine Island, Japan (Agostini et al. 2018, 2021a). These drastic changes may be due to the combination with other stressors, such as cold temperatures that limit the growth of corals at high latitudes (Agostini et al. 2021a; Guinotte et al. 2003; Higuchi et al. 2018; Kleypas et al. 1999; Yara et al. 2012). Similarly, the decrease in abundance of the

Table 7.1 Characteristics of the different CO_2 vent systems regarding pH, temperature, pCO_2, gas composition and coral communities

CO_2 vent syst,	Lat.	Long.	Temp	pH	pCO2	Depth	Gas composition	Dominant biotope	References
Milos, Greece	36.7	24.5	NA	NA	NA	2–12 m	gas rich in CO2 (mean 80.5%) with 1.2% H2S (Dando et al. 1995)	Algal-dominated communities with some corals (*Caryophyllia inornata, Hoplangia durotrix, Leptopsammia pruvoti, Madracis pharensis, Phyllangia mouchezii, Polycyathus muellerae*) under over-hangs,	site SR in (Cocito et al. 2000)
Panarea, Italy	38.6	15.1	19.5 (19.3–19.7) 19.5 (19.3–19.7)	7.87 (7.83–7.90) 7.74 (7.68–7.81)	698 (638–759) 922 (764–1080)	10 m	Gas contains 98–99% CO2	Decreased in the population of the coral *Balanophyllia europaea,* which becomes absent at pH < 7.7.	sites S2 and S3 (Fantazzini et al. 2015)
Grotta del Mago, Ischia, Italy	40.7	14.0	14.7–26.5 (winter min to summer max)	7.65–7.88	2905 ± 1664	2 m	92%–95% CO2 with no detectable sulfur gas	Small colonies of *Astroides calycularis* with reduced population and increased signs of necrosis under elevated pCO2.	Carbonne et al. (2021), Teixidó et al. (2020)
Chiane del Lume, Ischia, Italy	40.7	14.0	16.8† ± 0.4 (25%–75%: 16.6, 17.1)	7.91 (25%–75%: 7.89, 7.99)	626 ± 311 (25%–75%:497, 643)	10 m	93.4 ± 1.0 CO2 with undetectable levels of hydrogen sulfide	Significant lower abundance of *C. caespitosa* with no large colonies under elevated pCO2.	Carbonne et al. (2021)
Castello Aragonese, Ischia, Italy	40.7	14.0	23.8 ± 0.7 17.5 ± 2.8	7.8 ± 0.2 7.8 ± 0.3	1075 ± 942 1581 ± 2711	<3 m	>98% CO_2 (Hall--Spencer et al. 2008)	*Caryophyllia smithii, Cladocora caespitosa* and *Balanophyllia europaea* are present at control sites but absent at pH < 7.8	geochemistry (Kroeker et al. 2013); ecology (Hall-Spencer et al. 2008)
Bahía Concepción, Mexico	26.7	−111.8	30	6.96 ± 0.19		<6 m	44% CO2	Monospecific reef patches of *Porites panamensis* in shallow waters surrounding the vents	Tortolero-Langarica et al. (2024)
Mayreau Gardens, St Vincent and the Grenadines	12.6	−61.4	28.5 ± 0.4	7.74 ± 0.41	2141 ± 4460	23 ± 0.5 m	NA	At least three species of corals (*Orbicella faveolata, Montastraea cavernosa, Porites astreoides*) living in areas that periodically experienced vent-associated acidification	mean and sd recalculated from (Enochs et al. 2020)
Maug, Northern Mariana Islands (USA)	20.0	145.2	30.3 ± 0.98 (min 27.4, max: 31.3)	7.94 ± 0.051 (min: 7.72, max: 8.07)		9 m	CO2: 61.1 ± 10.96% with 0.09 ± 0.067 ppb of total sulfur	Shift towards a macroalgae--dominated system, but persistence of corals such as *Leptastrea purpurea.*	Enochs et al. (2015)
Shikine, Japan	34.3	139.2	19.0 ± 0.7	7.90 ± 0.17	888.4 ± 471.3	5	NA	Turf algae with rare corals such as *Porites heronensis, Psammocora profundacella* and *Dispsatrea speciosa.*	Agostini et al. (2021a)
Iwotorishima, Japan	27.9	128.2	28.6 (25.8–31.3)	7.4	831 (at low tide)	<4	Volcanic gas does not include sulphur	Dense soft coral communities dominated by *S. elegans*. High abundance of *Nanipora* (Octocorallia: Helioporacea).	Inoue et al. (2013), Reimer et al. (2021)

(continued)

Table 7.1 (continued)

CO_2 vent syst,	Lat.	Long.	Temp	pH	pCO2	Depth	Gas composition	Dominant biotope	References
Kueishan, Taiwan	24.8	122.0	23.4 22.2	7.84 7.28	NA	NA	NA but vents fluids contain 8.95 and 15.2 mg/l of HS-	Absence of corals at the vent sites	bottom value of "yellow" and "white" vent sites (Chan et al. 2012)
Batangas, Philippines	13.8	120.9	NA	7.98 ± 0.04	460.23 ± 51.88	10–25 m	NA	14% coverage of hard corals (7.68 colonies per m2) with "stress--tolerant" species (including *Porites* spp.) accounting for 40.5% of adult colonies.	Maningas et al. (2025)
Gunung Api, Indonesia	−4.5	129.9	28.80 ± 0.80	7.85 ± 0.05	659.72 ± 71.99	4–6 m	NA	Increase in Acroporidae and decrease in Poritidae at the moderate acidification sites, showing the highest coral cover.	moderate acidification sites from (Januar et al. 2017)
Minahasa, Indonesia	1.7	125.1	28.80 ± 0.30	7.91 ± 0.04	545.26 ± 61.56	4–6 m	NA	Increase in Acroporidae and decrease in Poritidae at the moderate acidification sites, which show the lowest coral cover.	moderate acidification sites from (Januar et al. 2017)
Mahengetang, Indonesia	3.1	125.4	28.80 ± 0.60	7.84 ± 0.04	659.72 ± 71.99	4–6 m	NA	Higher cover of *Helipora* spp. at the moderate acidification site.	moderate acidification sites from (Januar et al. 2017)
Dobu, Papua New Guinea	−9.8	150.9	28.6 28.2 29.1	7.73 7.09 7.89	9,536,065,213	<5 m	>99% CO2, 163 ppm H2S	The coral community at the three sites are not detailed for each. Generally, a high coral cover is observed at pH ≥ 7.8; however, the coral communities have become dominated by massive *Porites*.	Fabricius et al. (2011)
Upa-Upasina, Papua New Guinea	−9.8	150.8	27.6 (5%–95%: 27.4–28.6)	7.77 7.02 7.99	8,464,545,737	<5 m	>99% CO2, 0 ppm H2S		
Esa'Ala, Papua New Guinea	−9.7	150.8	27.7 27.6 28.0	8.00 7.99 8.02	444,423,460	<5 m	>98% CO2, 0 ppm H2S		
Matupit, Papua New Guinea	−4.2	152.2	32	7.5	1030	<27 m	86.5% CO2, 0% H2S	Rich corals near the vents have lots of *Porites*, *Pectinia*, and *Tubinaria*, and they have a 90–100% coral cover. *Pachyseris* was observed at 0.5–1 m at temperatures exceeding 60 °C.	Area III in (Tarasov et al. 1999). The community is described for shallow areas (3–4 m), and geochemistry is taken for water mass at 27 m.
Ambitle, Papua New Guinea	−4.1	153.6	30.52 ± 0.12 (min: 30.34, max: 31.0	7.637 ± 0.10 (min: 7.346, max: 7.933	1140 ± 306 (min: 515, max: 2369)	4 m	>95% CO2, absence of H2S (Pichler et al. 1999)	High diversity and coverage of corals in the elevated CO_2 areas. Species found in these areas include *Acropora nana, Dipsastraea pallida, Heliopora coerulea, Pocillopora damicornis, Pocillopora verrucosa, Porites cylindrica* (Biscéré et al. 2019).	Station 4 (Pichler et al. 2019).

Mediterranean corals *Astroides calycularis*, *Caryophyllia smithii*, *Cladocora caespitosa* and *Balanophyllia europaea* (Carbonne et al. 2021; Fantazzini et al. 2015; Hall-Spencer et al. 2008; Teixidó et al. 2020), while these observations may help in understanding the effects of OA on hermatypic corals, they are not entirely relevant to understand the future community structure of coral reefs. However, Mediterranean and high-latitude systems highlight the limit of high latitudes as a coral refuge and tropicalization under combined ocean warming and acidification.

Rather than focusing solely on the loss of coral diversity and abundance, it is also important to highlight species and reefs that show high resilience and resistance to elevated CO_2 (Enochs et al. 2020; Maningas et al. 2025; Tortolero-Langarica et al. 2024). Early work at the CO_2 vent systems in Matupit, Rabaul (Papua New Guinea), and Milos (Greece) already highlighted the presence of corals in even the most extreme conditions, where hydrothermal fluids strongly influence temperature and CO_2 levels (Bianchi et al. 2011; Cocito et al. 2000; Tarasov et al. 1999). Some taxonomic groups, such as Poritidae, have been consistently found to resist and persist under elevated CO_2 (Agostini et al. 2021a; Enochs et al. 2020; Fabricius et al. 2011; Januar et al. 2017; Maningas et al. 2025; Oporto-Guerrero et al. 2018; Tarasov et al. 1999). Poritidae are known to withstand multiple abiotic stresses, including high temperatures (van Woesik et al. 2011), cold temperatures (Chen et al. 2016), and high turbidity (Dikou and van Woesik 2006). Due to their diverse tolerance traits, Poritidae are often found in marginal and degraded reefs, including urban reefs (Heery et al. 2018).

However, the simple morphologies of resistant massive species inhabiting high-CO_2 areas lead to reduced habitat complexity, with cascading effects on reef biodiversity (Cattano et al. 2020; Fabricius et al. 2011, 2014; Sunday et al. 2017). Observations at multiple sites in Indonesia have shown high coral coverage and diversity near CO_2 vents, suggesting that other abiotic factors, rather than pH alone, can drive differences in community structure across sites (Januar et al. 2017). Additionally, studies at the CO_2 vents in Tumtum Bay, Ambitle Island (Papua New Guinea), revealed a coral community not dominated by *Porites* but instead featuring a significant presence of branching corals, highlighting potential resistance mechanisms even within coral families (such as Acroporidae) generally considered sensitive to environmental stressors (Biscéré et al. 2019; Pichler et al. 2019).

This review emphasizes the need to replicate and compare coral community structures exposed to elevated CO_2 under different abiotic conditions and across biogeographic regions. Ultimately, community structure is determined not only by the ecological and physiological effects of CO_2 but also by the interplay of other abiotic factors on coral populations and individuals.

7.2.4 Insights on the Effects of OA on the Physiology of Corals

Early laboratory experiments on the effects of OA on corals highlighted the sensitivity of hermatypic corals, particularly a decrease in calcification rates. However, later studies have shown that the response of corals to OA is highly variable depending on the species. Studies at CO_2 vents also support the view that responses to OA are species-specific. Among the 21 studies conducted at CO_2 vents on coral host physiology, 16 have examined the calcification process and its effects on skeletal properties. An early study at the Papua New Guinean sites demonstrated the resistance of massive *Porites*, showing only minor declines in calcification rates and skeletal density (Fabricius et al. 2011). However, this study also reported an increase in bioeroder density, a finding that has been replicated in other systems (Enochs et al. 2016). The physiological resistance of the genus *Porites*, which can maintain calcification rates despite elevated CO_2, has been observed in different species and biogeographical contexts (Agostini et al. 2021a; Tortolero-Langarica et al. 2024).

The lack of adverse effects on calcification rates or growth has also been observed in other coral species, both in tropical coral reefs and temperate regions, including *Caryophyllia smithii, Dendrophyllia cornigera, Desmophyllum dianthus, Cladocora caespitosa, Balanophyllia europaea*, and *Astroides calycularis* (Hulver et al. 2024; Rodolfo-Metalpa et al. 2011, 2015). The resistance of calcification to OA has been attributed to various physiological traits, including a high capacity to maintain elevated pH at the calcification site (Comeau et al. 2022; Wall et al. 2016) —particularly in species adapted to variable pH conditions, such as those found in some CO_2 vent systems— a higher heterotrophic capacity (Houlbrèque et al. 2003; Hulver et al. 2024), and alternative energy allocation strategies (Agostini et al. 2021b).

Despite the resistance of the calcification process to elevated CO_2, the skeletons of corals exposed to high CO_2 for their entire lifetime have been shown to degrade, with lower density and higher porosity (Fabricius et al. 2011; Fantazzini et al. 2015; Prada et al. 2021; Teixidó et al. 2020; Tortolero-Langarica et al. 2024), though not in all species (Strahl et al. 2015). This discrepancy between the resistance of the calcification process and sclerochronology may be due to the ability of corals to control the chemistry at the calcification site biologically. In contrast, the dissolution of the formed skeleton is primarily regulated by environmental chemistry. Studies examining coral skeletons not covered by living tissue have shown increased dissolution and greater effects of micro- and macro-bioeroders in high CO_2 areas surrounding vents (Enochs et al. 2016; Rodolfo-Metalpa et al. 2011). Degraded skeletal properties are particularly concerning, as warming-induced bleaching and mortality events will expose more dead coral skeletons to environmental conditions. OA could, therefore, accelerate reef flattening, impacting habitat provision and triggering cascading effects throughout the ecosystem, ultimately leading to decreased biodiversity and a lower abundance of other taxa (Alvarez-Filip et al. 2009; Bozec et al. 2015; Hughes et al. 2017; Kochan et al. 2023; Perry and Alvarez-Filip 2019).

7.2.5 Adaptations of Corals to High CO_2 Concentrations

The potential adaptation of corals to future environmental conditions must be considered when predicting the composition of future coral reef assemblages (Kleypas et al. 2021). The adaptive capacity of corals to future increased temperature will certainly largely determine the structure of future coral reefs (Logan et al. 2014), but the adverse effects of OA should also be considered along with corals' potential to resist elevated CO_2. Resistance to OA can be acquired within an individual's lifetime through phenotypic plasticity or genetic adaptation, the latter requiring the selection and reproduction of resistant genotypes. In this context, volcanic CO_2 vents offer a unique opportunity to study coral adaptation, as corals found in acidified areas have already been naturally selected for their resistance to OA. However, an inherent characteristic of volcanic CO_2 vents is their open nature, allowing corals from non-acidified areas to recruit near the vents, which may limit our conclusions about the potential for genetic adaptation given the uncertainty that the corals present at the CO_2 vents have reproduced and adapted over several generations. Nevertheless, rapid acclimatization can facilitate the quick acquisition of resistance to stressors such as warming and OA (Torda et al. 2017; Leiva et al. 2025). This process can occur at the host, symbiont, or microbiome level but requires the presence of adaptive alleles within the population.

While most studies have examined the physiological traits of corals exposed to elevated CO_2 for their entire lifetime—highlighting traits linked to OA resistance at both the host (Agostini et al. 2021b; Strahl et al. 2015, 2016) and symbiont (Biscéré et al. 2019; Noonan et al. 2013; Strahl et al. 2015) levels—few have explored these traits at the genomic level. Notably, studies by Kenkel et al. (2018), Leiva et al. (2023b), and Teixidó et al. (2020) have identified alleles under selective pressure in coral populations near volcanic vents. Among these, several alleles were found in genes associated with calcification (Leiva et al. 2023a; Teixidó et al. 2020). However, genomic analyses of symbionts have been less consistent, and specific adaptations to elevated CO_2 in natural populations around CO_2 vents remain to be demonstrated (Kenkel et al. 2018; Noonan et al. 2013).

The coral microbiome is essential for maintaining coral health, including resistance to stressors and diseases, as is the case for other organisms (Bang et al. 2018). Moreover, microbiome restructuring has been highlighted as a mechanism for rapid acclimatization (Torda et al. 2017). However, studies on coral microbiomes at volcanic CO_2 vents remain limited (10 studies in this review), and their findings are inconsistent. Despite this, some patterns have emerged. Resistant coral species, such as massive *Porites*, exhibit a stable microbiome even under high CO_2 conditions, suggesting that microbial stability may contribute to *Porites*' resistance to OA (Marcelino et al. 2017; O'Brien et al. 2018; Shore et al. 2021). The potential role of prokaryotes in conferring OA resistance is further supported by observations of increased nitrogen-fixing bacteria under high CO_2 (Biagi et al. 2020; Meunier et al. 2021) and the selection of bacteria resistant to stress (Palladino et al. 2022). However, some studies have also reported a decline in *Endozoicomonadaceae* abundance (Morrow et al. 2014; Shore et al. 2021), a widely distributed prokaryotic symbiont considered essential for coral health (Hochart et al. 2023). Overall, further research is needed to clarify the functional roles of these microbes in response to OA, and this within the different coral compartments (mucus, tissue, gastric cavity, and skeleton) (Agostini et al. 2012; Biagi et al. 2020; Palladino et al. 2022; Sweet et al. 2010) and using metagenomics and metatranscriptomics approaches.

Rather than volcanic CO_2 vents, other naturally acidified environments, such as semi-enclosed bays and lagoons, may be better suited for studying coral acclimatization and adaptation to future ocean conditions. Unlike CO_2 vents, these sites offer limited connectivity with surrounding coral populations not exposed to elevated CO_2, allowing for the study of long-term adaptation and interactions among multiple stressors. Two extensively studied sites include the semi-enclosed lagoon of Bouraké in New Caledonia and the semi-enclosed bay of Nikko in Palau. The former is discussed later in this chapter, with insights on coral responses to chronic exposure to multiple and variable stressors. Despite experiencing high CO_2 levels, low total alkalinity, and elevated temperatures, Nikko Bay supports high coral cover and biodiversity (Barkley et al. 2015; Kurihara et al. 2021). This unexpectedly rich reef ecosystem may result from local adaptation, facilitated by the community's isolation due to the bay's high water residence time (Golbuu et al. 2016). Additionally, recent studies have shown that thermally tolerant coral populations from the bay can spread to neighbouring areas, underscoring the potential role of natural analogues as genetic reservoirs of pre-acclimatized populations (Rivera et al. 2022) (Fig. 7.2).

Volcanic CO_2 vents have provided valuable insights into the effects of OA on coral reefs and their potential for rapid adaptation (Fig 7.2). Although their use as "natural

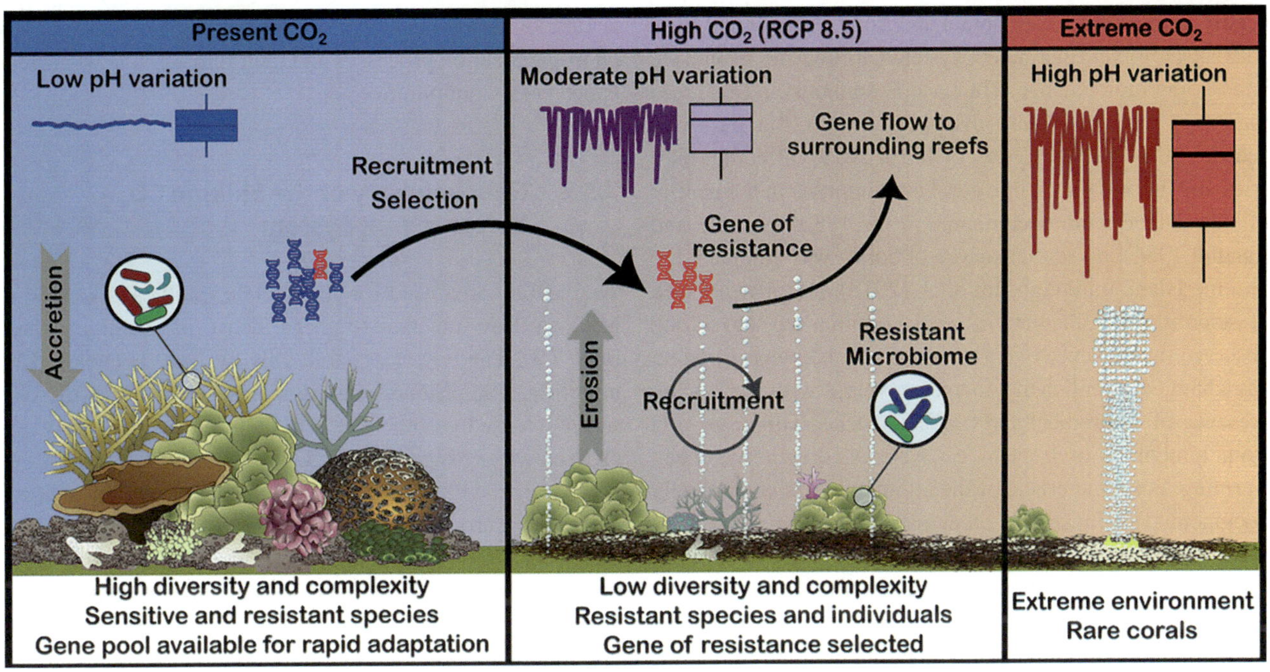

Fig. 7.2 Synthesis figure illustrating the shift in coral assemblages along CO_2 gradients created by volcanic vents. Near the vents, extreme CO_2 levels and large environmental fluctuations limit the growth of most corals. Farther from the vents, coral communities consist of species exhibiting resistance traits at the host, symbiont, and microbiome levels and individuals from sensitive species that have undergone selection for alleles conferring resistance. In contrast, nearby reefs unaffected by the vents display high diversity and structural complexity, with low bioerosion and high accretion, serving as a genetic reservoir for rapidly adapting resistant individuals within high-CO_2 areas. Coral communities surrounding CO_2 vents may act as a source of resistance genes for adjacent reefs. Symbols courtesy of Integration and Application Network (ian.umces.edu/media-library)

analogues" can be limited by inherent characteristics such as unpredictable CO_2 level variations, the open nature of these systems allows recruitment of non-adapted individuals, and the lack of multiple stressors—particularly warming—which may obscure interactions between OA and other factors, volcanic CO_2 vents remain one of the most direct approaches to predict and anticipate the effects of OA on coral ecosystems. However, their number remains limited due to their unique nature. In addition, the logistics to access and conduct research at these systems are often complicated and expensive, making multiple sites and long-term surveys rare. To foster research at volcanic CO_2 vents and other natural analogues, we created the ICONA network (International CO_2 Natural Analogues Network https://www.shimoda.tsukuba.ac.jp/~icona/index.html). Using this network, we hope to foster collaboration, harmonization and capacity building to support active and reproducible research at natural analogues.

Here, using two among the natural analogues actively used within the ICONA network, namely the Shikine CO_2 vents in Japan and the semi-enclosed lagoon of Bourake in New Caledonia, we illustrate the insights that can be gained using the specificity of each site. Specifically, we highlight how the marginality of the latitude of the Shikine CO_2 vents allowed for a better understanding of future limitations of tropicalization of marine ecosystems and how semi-enclosed bays, such as the Bourake lagoon, can allow research on the effects of multiple abiotic factors: high temperature, reduced oxygen in addition to elevated CO_2.

7.3 The Shikine Island Warm-Temperate Shallow CO_2 Vent System

7.3.1 Biogeographical Context

Shikine Island (34°19.5′N, 139°13′E) is part of the Izu Islands archipelago, located in the northernmost section of the Izu-Bonin-Mariana Arc, south of Tokyo, Japan. Covering an area of 3.67 km^2, Shikine Island has a population of 495 people (as of 2022). Its economy is primarily driven by fisheries and tourism, with more than 17,000 visitors annually. The island is a popular destination for recreational fishing and marine activities such as snorkelling and diving. It is renowned for its natural hot springs, several of which are found along the shoreline (Planning Coordination Office n.d.).

Strongly influenced by the Kuroshio Current, Shikine Island experiences a humid subtropical climate despite being in a temperate zone. The mean seawater temperature is 21.1 °C (2011–2020), with an interquartile range of 17.4–24.2 °C (Agostini et al. 2021a). The region is a biodiversity hotspot where temperate and tropical fauna and flora coexist. However, marine ecosystems around the island are undergoing rapid change. The once-abundant kelp (*Ecklonia radicosa*) has completely disappeared over the past two decades (Biodiversity Center of Japan 2019), affecting fisheries and marine life. While coral communities in the nearby Miyake Island were documented in the 1980s (Tribble and Randall 1986), no formal survey of corals was conducted at Shikine Island until Agostini et al. (2018). Despite this lack of scientific documentation, local communities have long observed the presence of corals and noted that the ecosystem has changed significantly in recent years, with a growing presence of subtropical and tropical species. This transition from temperate to tropical ecosystems, driven by ocean warming, is characteristic of the region and has been widely documented in Japan and other parts of the world (Vergés et al. 2019).

Japan, which is located along the Pacific Ring of Fire, is home to 111 active volcanoes, making it one of Earth's most volcanically active regions (Fujita et al. 2020). Local fishermen and hot spring visitors have long been aware of the site. While these geological features pose risks to local populations, they also provide unique scientific opportunities. Although Japan's deep-sea hydrothermal vents have been extensively studied (Miyazaki et al. 2017), much less is known about the biological communities inhabiting its shallow volcanic vents. The suitability of Shikine Island's CO_2 vents for OA research was first described by Agostini et al. (2015). The bay where the most active venting occurs is known as Mikawa Bay, which translates to "pot" or "volcanic crater." Shikine Island's volcanic CO_2 vents are the second site in Japan to be used for acidification research, following Iwotori Island (Inoue et al. 2013). More than 20 scientific publications have utilized this system, not only to study the effects of OA on marine communities but also for the calibration of paleontological CO_2 proxies (Witkowski et al. 2019) and as a natural analogue for CO_2 leakage in Carbon Dioxide Capture and Storage (Tanaka et al. 2019).

The coral communities around Shikine Island are best described as high-latitude marginal coral communities (Kleypas et al. 1999; Schoepf et al. 2023), which are considered critical for coral conservation in Japan (Veron 1992). These high-latitude sites may serve as refuges for corals (Nakabayashi et al. 2019), particularly as warming threatens coral populations in lower latitudes (Nakamura 2017; Reimer et al. 2024). While the warm Kuroshio Current and ocean warming facilitate the rapid poleward expansion of coral distributions (Kumagai et al. 2018; Yamano et al. 2011), models suggest that OA may constrain this expansion (Yara et al. 2012, 2016). In this context, Shikine Island's volcanic CO_2 vents provide a unique opportunity to study the effects of OA on tropicalization processes and their impact on newly established coral communities.

7.3.2 Geochemistry of the Shikine CO_2 Volcanic Vent System

The venting area used for coral-related studies is located in Mikawa Bay on the southern coast of Shikine Island (Fig. 7.3). The bay, accessible only by boat, is enclosed by tall cliffs. The primary venting site is northwest of the bay, where intense hydrothermal activity occurs. A hot spring discharges water exceeding 80 °C in the intertidal zone in this area. Large white microbial mats, most likely *Beggiatoa* mats, thrive around the vents, fueled by the hydrogen sulfide emissions (Fig. 7.4). This venting site is also home to the only known macrofaunal species endemic to shallow CO_2 vents, the crab *Xenograpsus testudinatus* (Miyake et al. 2019). In the shallowest sections (1–2 m depth), pH levels can drop below 7.0, with CO_2 concentrations exceeding 10,000 µatm and a negative redox potential (−45 mV). Gas released from these vents has been shown to contain up to 90 ppm of hydrogen sulfide (Agostini et al. 2015).

For OA research, a far less extreme area is used, located in the northern section of the bay. Here, gas venting is limited to sparse bubble streams in the northernmost parts of the cove. Long-term pH monitoring using ISFET sensors (Seafet, USA) has recorded an average pH_T of 7.75 ± 0.10 (mean ± SD, interquartile: 7.71~7.82), corresponding to CO_2 levels of 874 ± 256 µatm (mean ± SD, interquartile: 713~951 µatm; recalculated from Agostini et al. 2018). CO_2 fluctuations can still be significant, with short-term spikes reaching up to 2500 µatm over several minutes (Fig. 7.3). However, these variations are much lower than those observed near the primary vents. The redox potential in this area is comparable to regions unaffected by venting (+80 mV), and no detectable sulfides indicate the rapid oxidation of reduced sulfur released by the vents. The substrate in this region consists mainly of volcanic rocks and large boulders, with depths ranging from the intertidal zone to approximately 10 m. Further south, the depth increases, and the substrate transitions to sand. On the eastern side of the bay, areas with an average CO_2 concentration of 450 ppm are found, where no visible bubbling occurs. These findings suggest that elevated CO_2 levels result from water acidified near the vents flowing into the area. Additional bubble streams and venting sites are at greater depths (15–20 m), with sonar detections indicating active venting at 40–50 m outside the bay. However, these deeper venting areas remain largely unexplored due to strong tidal currents.

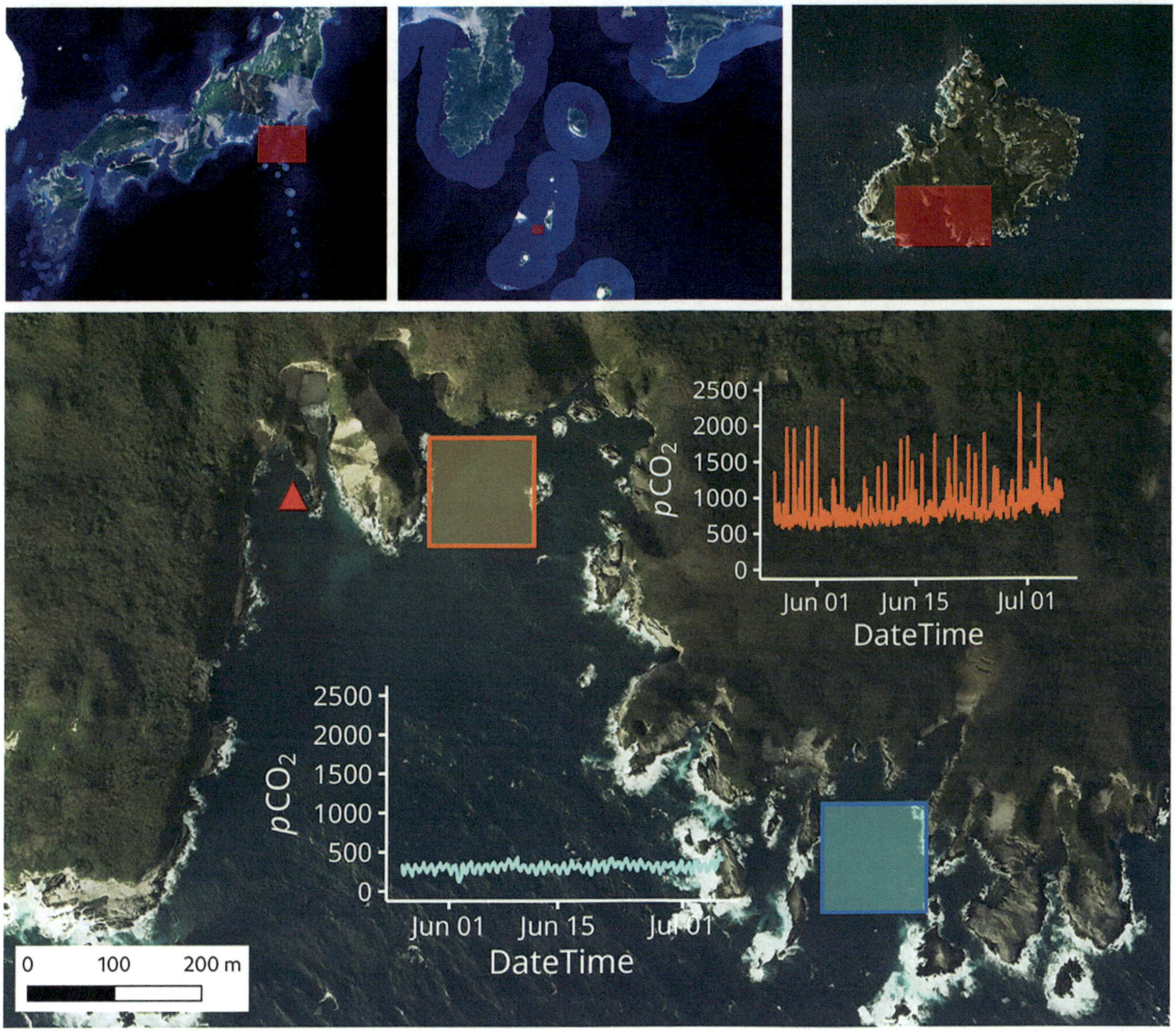

Fig. 7.3 Maps of Shikine (basemaps data obtained from Geospatial Information Authority of Japan (GSI; https://www.gsi.go.jp)). The main vent is shown as a red triangle. The areas used as a natural analogue of OA (orange) and present condition (cyan) are approximately delimited by a square. Levels of pCO_2 recorded in 2016 (Agostini et al. 2018) are shown for the respective areas following the same colors

7.3.3 Effects of OA on Warm-Temperate Coral Communities

Coral abundance and diversity drastically decrease with increasing CO_2 until corals are absent at around 1000 ppm CO_2 at the Shikine volcanic CO_2 vent (Fig. 7.4) (Agostini et al. 2021a; Harvey et al. 2021). Corals are rare even at the CO_2 value of ca. 450 ppm—expected in the region as early as 2050— with only a few colonies of *Porites heronensis, Psammocora profundacella,* and *Dipsastrea speciosa* observed. This drastic decline, which is not commonly seen in other volcanic CO_2 vents, may be due to the marginality of Shikine Island as a high-latitude environment, where winter seawater temperatures drop as low as 15 °C. The presence of corals is likely determined by their physiological resistance to elevated CO_2 and ecological resilience, particularly their ability to compete with macroalgae and algal turf.

Coral species found in elevated CO_2 areas exhibit higher mitochondrial activity per biomass (Agostini et al. 2021b) and maintain calcification rates regardless of CO_2 levels and seasonal variations (Agostini et al. 2021a). These eco-physiological traits are common to other "winning" species, i.e., those found in both control and high-CO_2 areas at the volcanic CO2 vents in Papua New Guinea (Agostini et al. 2021b). Another major challenge for corals in these high-CO_2 zones is the intense competition with turf algae, which rapidly occupy ecological space (Wada et al. 2021). Although typhoons naturally create open spaces each year by dislodging algae attached to the rocks, turf algae—whose growth is favoured under elevated CO_2—trap sediment, forming an

Fig. 7.4 (**a**) Main shallow volcanic vents at Shikine Island, showing large amounts of gas released and abundant microbial mats around. (**b**) Seascape at the areas typically used for OA research where CO_2 levels are around 900 ppm and some rare corals and soft corals are found among turf algae. (**c**) Seascape at control areas in the bay next to the venting sites (CO_2 300 ppm). Abundant and diverse corals are observed. Locations of the photos are shown in Fig. 7.3 (main vent, OA natural analogue and present conditions for (**a–c**), respectively). (Photos courtesy of Nicolas Floc'h)

anoxic, low-pH layer that can prevent coral recruits from settling and surviving (Harvey et al. 2021).

Despite the highly unfavourable conditions at the Shikine CO_2 vents, some coral species appear resistant, and a few individuals of species typically classified as sensitive—such as table *Acropora* (most likely *Acropora solitaryensis*)—persist, albeit in very low numbers (~5 known colonies across the entire area). This suggests that extreme environmental conditions act as a filter, allowing only the most resistant species and individuals to survive. However, the few individuals found severely limit the use of a genomic approach to understanding selection mechanisms. *Acropora solitaryensis* has rapidly expanded its range poleward in Japan (Yamano et al. 2011). With ocean warming driving kelp loss in warm-temperate regions and OA restricting the poleward expansion of corals while favoring low-complexity morphologies, marine ecosystems in these latitudes risk becoming highly simplified. Such shifts toward lower-complexity benthic communities could cascade through the ecosystem, reducing fish diversity and abundance (Cattano et al. 2020).

The Shikine CO_2 vents have been instrumental in understanding the limitations of tropicalization under OA conditions. While warming-driven range expansion has been observed in many marine taxa, including corals, the findings at Shikine suggest that elevated CO_2 levels may be a significant barrier to this process. The lack of coral recruitment and survival at CO_2 levels projected for the near future highlights that OA can counteract the potential benefits of ocean warming for tropical species moving into higher latitudes. The persistence of only a few resistant individuals and the near-complete loss of key reef-building species suggest that coral communities in warm-temperate regions may not transition into functional coral reefs under future climate scenarios. Instead, these ecosystems risk becoming dominated by turf algae, leading to fundamental shifts in habitat structure and ecosystem function.

Findings at the Shikine CO_2 vents warn that warm-temperate latitudes—often considered potential refuges for reef-building corals—may not provide suitable environments for coral growth under the combined pressures of ocean warming and acidification. This region, which currently provides essential ecosystem services to local populations, may face severe ecological and economic limitations if CO_2 emissions are not curbed shortly (Hall-Spencer et al. 2022; Hall-Spencer and Harvey 2019). Located within the Fuji-Hakone-Izu National Park and under relatively low anthropogenic pressure, the Shikine CO_2 vent system benefits from some protection. However, given the rarity of resistant populations of high-complexity coral species—such as those observed around the volcanic CO_2 vents—and the limited scope of existing conservation measures (e.g., fishing is still permitted), we strongly recommend that these systems, along with other shallow CO_2 vents, be specifically targeted for conservation.

7.4 The Semi-Enclosed Lagoon of Bouraké in New Caledonia

Bouraké is a unique site that combines features typically found in semi-enclosed bays with features of mangrove natural analogues. It exhibits alterations in seawater caused by intense oxidative reactions in sediments and mangrove roofs while possessing characteristics of a semi-enclosed bay with limited seawater exchange (Fig. 7.5). Combining these factors creates an environment that is even more extreme than what we anticipate for the future, particularly concerning specific environmental parameters such as seawater pH and temperature. In addition, the absence of freshwater inputs from the surrounding semi-arid region makes it a valuable natural laboratory where extreme conditions have persisted over ecologically realistic timescales, incorporating environmental complexity across natural daily and seasonal fluctuations (Schoepf et al. 2023).

Since 2016, Bouraké has been utilized as a natural analogue for studying the effects of future extreme conditions on coral reefs and their associated organisms (e.g. Camp et al. 2017; Maggioni et al. 2021). Dubuc et al. (2019) investigated how changes in dissolved oxygen levels affect the distribution and metabolism of reef fish at the boundary between the mangrove forest and the bay. In November 2015, an initial visit to the site by a team of scientists, including David Suggett and Emma Camp from the University of Technology Sydney, highlighted the unique nature of this location. Starting in 2020, Bouraké has become one of the natural analogues studied by the ICONA consortium, including several CO_2 vents in the Pacific, the Mediterranean Sea, and the semi-enclosed Nikko Bay in Palau.

Located 150 km from Nouméa, the central city of New Caledonia, this site experiences lagoon waters that flow into the system during flood current, circulate within, and then exit as the tide falls, following a semi-diurnal cycle. The system's depth varies from just a few centimetres near the mangroves to over 6 m in the channel. More than 80 m wide, this channel penetrates the mangrove area and creates large pools covering a total area of over 60,000 m^2, a value not including the extensive area occupied by the mangrove and the inner reservoir. Corals form stable fringing reefs in several zones of the system, particularly at the ends of the terraces that extend from the mangrove toward the channel, where the densest and most compact coral colonies thrive. The terraces are exposed to the air during spring tides, especially near the mangrove roots (Fig. 7.6).

With each rising tide, new lagoon water enters the vast inner basin of the mangroves through the channel. During this transit, the water's chemistry changes due to metabolic reactions occurring in the sediments and mangrove roots (as referenced in Maggioni et al. 2021) and interactions with coral reefs and mangrove habitats. This water also mixes with the existing volume that has resided in the semi-enclosed bay, which has become more acidic and deoxygenated (Fig. 7.7). A 3D water circulation model (Rodolfo-Metalpa, unpublished data) indicated that the current can vary significantly, ranging from very strong

Fig. 7.5 Aerial picture of the semi-enclosed lagoon of Bouraké, surrounded by a rich mangrove forest

Fig. 7.6 Aerial picture of the entrance of the Bouraké semi-enclosed system during low tide

(up to 1.5 knots) to almost absent, depending on the area in question and the tide phase. Initial simulations demonstrated that the water mass moves in and out of the system up to seven times before effectively mixing with water outside Bouraké. This data is invaluable for understanding larval dispersion during spawning events and measuring water flow's effects on the diffusion boundary layer (DBL) and coral physiology. Furthermore, because of the limited depth in the extensive mangrove area beside the semi-enclosed bay, seawater temperatures become exceptionally high during the summer and, conversely, cool in the winter.

The first daily fluctuations in pH levels were measured in 2016 during a 24-h cycle (Camp et al. 2017). This unique site has proven valuable for studying the ability of corals to acclimatize and adapt to extreme conditions. Even at high tide, the water in this system does not return to normal, reef-like pH levels. For instance, the maximum pH values (using total scale pH_T) were around 7.9, while the current normal ocean pH is approximately 8.05, which is predicted to decrease to 7.7–7.8 by the year 2100.

As the tide falls, seawater becomes more acidic and oxygen-depleted. The system drains, and extremely hot, acidic, and deoxygenated water flows through the mangrove forest and is eventually discharged into the lagoon. At low tide, the pH in the entire system reaches the extreme value of 7.3, while the oxygen concentration drops to 2 mg L^{-1}. The expected oxygen concentration at the coast is typically between 4 and 8 mg L^{-1}.

The changes in this area are influenced by both the tides and the bottom topography, as well as the geomorphology of the bay. This relationship means that environmental fluctuations occur regularly and can be predicted. For specific parameters, these fluctuations can be effectively modelled. They remain consistent with the tides and seasons, suggesting that the topography of the system has likely remained unchanged for the past 60–100 years (according to the witnesses collected) or likely much longer than a century. As a result, the organisms inhabiting this area have been exposed to these conditions throughout their lifetimes and have likely adapted to them.

Consequently, Bouraké presents a unique combination of the three main factors that will impact coral reefs in the future—referred to as the "deadly trio" —warming, acidification, and deoxygenation. This site is particularly valuable compared to other natural laboratories, where the duration and intensity of environmental stress can vary significantly over time and space.

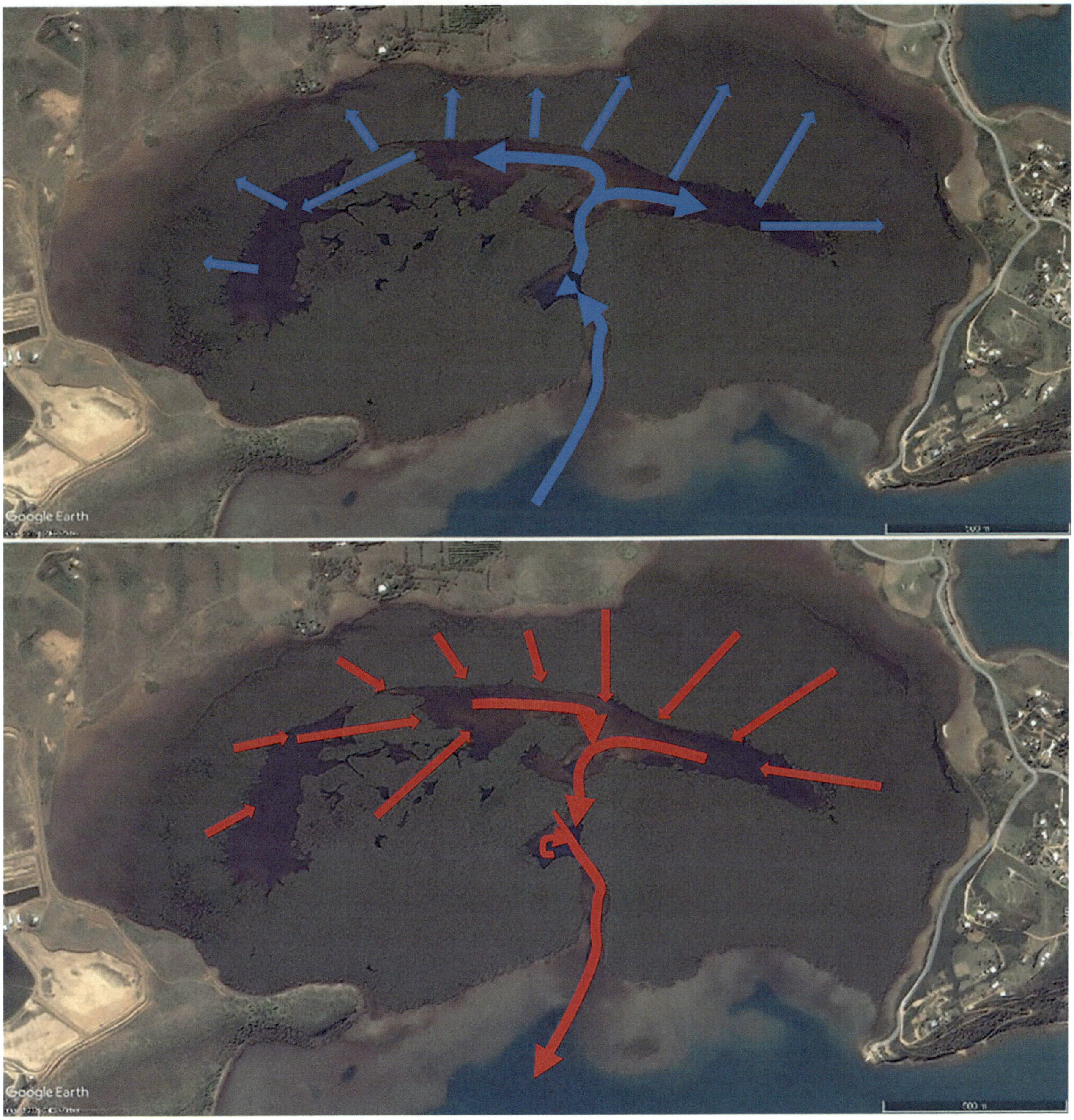

Fig. 7.7 Satellite picture of the mangrove area of Bouraké. The blue arrow suggests the direction of the rising tide that brings new seawater from the external lagoon, through the channel and the back reservoir, and finally, the mangrove forest. The red arrow suggests the opposite displacement of the hot, acidic, deoxygenated seawater from the mangrove toward the external lagoon. (Source Google Earth)

7.4.1 Environmental Variability and Community Composition

In 2016, a research team composed of Emma Camp, David Suggett, Matthew Nitschke, and Riccardo Rodolfo-Metalpa collected preliminary data to describe this site and assess the corals' ability to withstand extreme conditions. They described the seawater environmental conditions, produced a list of coral species, and provided the first analyses of the bacterial community that lives in symbiosis with the corals (Camp et al. 2017).

First, we were surprised by the diverse and abundant species that comprised the reef, which spent 44% of the time at a pH of 7.7–7.8 or lower while also experiencing temperatures predicted for the end of the century under the IPCC scenario RCP4.5. We observed high respiration rates and increased heterotrophy in the corals from Bouraké, which we interpreted as an adaptive response to the harsh conditions. This adaptability was potentially enhanced by the autotrophic conditions of the seawater, such as organic input from the mangrove.

However, it's important to note that during these first measurements in 2016, seawater temperatures were abnormally high, even for Bouraké, stressing the corals and altering their physiological responses. These preliminary studies set the stage for further research focused on enhancing environmental measurements with long-term, high-frequency data and diel cycles in the chemical parameters of seawater. All collected data culminated in a compelling assessment of this unique natural laboratory (Maggioni et al. 2021).

In Bouraké, various abiotic parameters fluctuate due to factors such as tides, time of day, season, and specific events like La Niña. These parameters include pH, temperature, dissolved oxygen, salinity, nutrients, and dissolved and suspended organic carbon. All of these factors can lead to slight or dramatic changes in the metabolism of reef organisms.

Seawater pH is the most significant physical environmental parameter characterizing the Bouraké site, making it one of the most extreme coral habitats currently known. The extensive literature, particularly studies examining the effects of variable and low seawater pH on coral calcification and skeleton dissolution, indicates that corals cannot effectively form reef-building structures below a specific seawater pH value (Fine and Tchernov 2007; Kleypas et al. 1999). In contrast, we observed highly productive reefs in Bouraké with a mean pH of 7.67 ± 0.23 pH_T units. The variability of pH is closely correlated with the tidal cycle, showing lower values at low tide. Throughout each tidal phase, pH levels fluctuated by an average of about 0.6 units, reaching a minimum of 7.23 at low tide and a maximum of 8.06 at high tide near the system's entrance, with values of 7.96 in the inner lagoon.

Let us consider arbitrary threshold values for seawater pH required for net reef accretion to be around pH_T 7.7–7.6, which is approximately at the undersaturated threshold of carbonate chemistry (i.e., $\Omega_{aragonite} = 1$). It becomes evident that corals in Bouraké are exposed to unfavourable seawater pH conditions for calcification for about 12 h a day and at least 5 h in conditions undersaturated for $\Omega_{aragonite}$.

Dissolved oxygen (DO) in seawater can significantly impact coral metabolism. In marine ecosystems, the typical concentration of DO ranges from approximately 4–8 mg O_2 mg L^{-1}. Hypoxia, a level of oxygen insufficient to support healthy physiological processes, varies among species and life stages (Limburg et al. 2020). Generally, hypoxia is considered to occur when DO levels fall below 2 mg O_2 mg L^{-1} (Díaz and Rosenberg 1995). The effects of hypoxia on marine benthic organisms are species-specific; some species can tolerate considerable reductions in DO. This low oxygen availability can compromise behaviours, metabolism, growth, and other biological processes, including reproduction (e.g. Vaquer-Sunyer and Duarte 2008). On a day in Bouraké, we recorded fluctuations in DO levels of up to 4.91 mg L^{-1}, with minimum DO values dropping below 2 mg L^{-1} during low tide and a maximum DO value of 7.24 mg L^{-1} observed during high tide.

Seawater temperatures in mangrove forest areas vary significantly from those in the surrounding environment due to stagnation. During winter, this can expose corals to cold atmospheric temperatures, dropping to 13 °C in the region. Conversely, summer temperatures can rise dramatically due to solar insolation, reaching as high as 36 °C in the air. Consequently, over 1 year, the water in these areas tends to be, on average, 1 °C colder in winter and between 0.5 °C and 3 °C warmer in summer compared to adjacent reefs.

These temperature shifts can subject corals to marine cold and heat waves, with daily temperature fluctuations of up to 6 °C, particularly in summer. Observations have shown that SST variations of up to 3.4 °C occur at an adjacent reef and up to 6.5 °C in Bouraké. During the summer of 2021, a four-month study (Alessi et al. 2024a) revealed that the daily seawater maximum temperature in Bouraké exceeded 30 °C for 75 days, with a peak temperature of 32.89 °C. In contrast, the maximum temperature reached 30.95 °C at the reference fringing reef over only 12 days.

Seawater salinity can increase due to insolation and tidal changes, rising by 1–3 units during the summer, especially at low tide. Additionally, during rare instances of heavy rainfall, typically associated with La Niña phenomena, Bouraké collects rainwater from the surrounding land, experiencing significant drops in salinity. While it is speculated that corals can withstand short-term osmotic changes (Mayfield and Gates 2007), low salinity levels can induce osmotic shock, potentially leading to the death of entire reefs. This scenario unfolded in Bouraké during the 2021–2022 period, which marked two consecutive years of La Niña, possibly the most intense event since 1981 (Alessi et al. 2024a). During 18 days of heavy rainfall in Bouraké, precipitation levels were 5.3 times higher than the average recorded from 1981–2000 for the same period. This excessive rainfall triggered mass coral mortality. During these episodes, salinity dropped to 19 and remained low for several days within Bouraké. By the end of the 2 years of La Niña, following multiple intense rain events, up to 60% of the corals in Bouraké had died (R. Rodolfo-Metalpa, unpublished data) due to the repeated occurrences of low salinity and high temperatures, both negatively acting individually and in combination.

While pH, temperature, and dissolved oxygen (DO) are certainly the most significant abiotic factors influencing marine life in the future due to their direct connection to climate change, other parameters may also change—especially for fringing reefs—primarily because of increased land discharge from heavier rainfall events.

In general, most of the chemical parameters we measured were: (i) up to five times more concentrated in Bouraké compared to reference reefs; (ii) their concentrations increased from the outer to the inner part of the semi-enclosed bay; and (iii) they were strongly influenced by tidal movements, with concentrations fluctuating accordingly (see Maggioni et al. 2021 for a comprehensive data set).

Mangrove habitats are known to be highly productive ecosystems with significant carbon processing (e.g. Leopold et al. 2017). In Bouraké, various heterotrophic inputs, such as nutrients, carbon, and nitrogen sources, are released from the mangrove and made available to corals. We recorded high concentrations of dissolved organic carbon (DOC), silicates, and phosphorus in the Bouraké lagoon. These likely provide sustainable energy to help corals endure extreme environmental conditions and maintain energy levels during stress (Anthony and Fabricius 2000).

Environmental variability, increased water flow, and food availability may interact in ways that enhance or offset an organism's metabolism. This interaction could help explain why, in Bouraké, 66 species of corals (representing 33 genera), 28 species of macroalgae, and 11 species of sponges flourish despite extreme seawater conditions, including low pH, low dissolved oxygen, and high temperatures.

7.4.2 Coral Sexual Reproduction and Recruitment: A Bottleneck for Population Replenishment and Persistence in Bouraké?

Coral sexual reproduction and the ability of coral recruits to settle, establish, and grow into new colonies represent a critical bottleneck for coral populations exposed to hostile conditions. Factors such as ocean warming, acidification, and other abiotic stressors will likely alter corals' capacity to maintain healthy and diverse populations in the future. While laboratory experiments have predominantly demonstrated the negative impact of ocean acidification (OA) on coral recruitment and larval growth, there are few examples of its effects in natural environments, particularly concerning the combined influence of multiple environmental stressors. The consequences of climate change on coral sexual reproduction are even less understood due to the following reasons: (i) coral reproduction is a rare event, occurring only once a year over a few days; (ii) gametogenesis is highly sensitive to environmental changes, even within its normal range; and (iii) reproductive success varies from year to year, among species, and from colony to colony (Harrison and Wallace 1990).

Although Bouraké is not an isolated system—since larvae can freely enter from external reefs—it may serve as a natural laboratory for studying the effects of multiple stressors on coral recruitment and reproduction.

Tanvet et al. (2022) confirmed previous findings regarding the negative impact of low pH, whether alone or combined with other parameters, on crustose coralline algae (CCA). CCA provides essential biotic support, allowing coral larvae to settle and establish themselves (e.g. Cornwall et al. 2021). However, after 2 years of incubation, the authors observed that coral recruitment in Bouraké (Fig. 7.8) was three times higher than at adjacent fringing reefs. This finding was not corroborated by Alessi et al. (in preparation), as the artificial substrates used to collect larvae exhibited significantly lower recruitment compared to the sites examined in the previous study. Further research is needed to establish whether coral recruitment is the bottleneck for future reef health. During the study period (2020–2022), corals in Bouraké experienced extraordinary and anomalous warm and rainy periods during a three-year La Niña event.

In contrast, at least for species such as *Acropora tenuis*, *Montipora digitata*, and *M. stellata,* coral spawning and fecundity seem not adversely affected in Bouraké (Alessi et al. 2024b). Although the corals in Bouraké spawned at different times compared to southern reference sites, their egg count (though smaller in size) and overall fecundity were similar to those of adjacent and southern reefs.

These preliminary studies suggest that extreme conditions in Bouraké do not fundamentally hinder both coral reproduction and do not influence the early larval stages following planula fixation. These results may help explain the abundance and diversity of corals in Bouraké, where 66 species cover up to 90% of the surface (Maggioni et al. 2021).

7.4.3 Physiological Adjustments in Corals: From Acclimatization to Adaptation

Bouraké corals exhibit unique metabolic adaptations, indicating a significant degree of physiological plasticity in response to fluctuating environmental conditions. A notable observation is the species-specific changes in photosynthetic performance and metabolic respiration rates compared to conspecifics from more stable reef environments. Despite these variations, corals from Bouraké maintain comparable, or in some instances even higher, calcification rates, suggesting the presence of compensatory mechanisms (Jacquemont et al. 2022; Tanvet et al. 2023). Typically, a change in photo-

Fig. 7.8 A reef dominated by *Acropora* spp. in Bouraké with two terracotta tiles positioned on the reef to assess coral recruits, one having CCA and one deprived

synthetic rates is the first response corals exhibit when facing stressful environmental conditions. Camp et al. (2017) initially reported no effect of contrasting ecological conditions but noted increased respiration rates in two Bouraké coral species compared to corals from reference reefs. Enhanced respiration rates are believed to result from increased heterotrophy in corals living in Bouraké, where abundant dissolved organic carbon and nutrients, particularly during ebb current and low tide conditions, have been measured (Maggioni et al. 2021).

Furthermore, Jacquemont et al. (2022) incubated seven coral species from Bouraké and an adjacent reference reef under varying seawater conditions (with water samples collected at high and low tide in Bouraké). Their findings confirmed that corals from Bouraké exhibited altered metabolic rates compared to conspecifics from reference reefs, and these differences persisted regardless of water conditions (i.e., similar responses at both high and low tide). This suggests that the metabolic rates of corals inhabiting the semi-enclosed lagoon of Bouraké may be more indicative of adaptive mechanisms rather than solely physiological plasticity.

The roles of OA on coral growth and the nutrient supply in enhancing coral resilience to harsh conditions in Bouraké were investigated by Tanvet et al. (2023), who conducted long-term aquarium incubations of corals from both reference and Bouraké reefs. Fragments of *Acropora tenuis, Montipora digitata*, and a massive *Porites* sp. were maintained under different pH conditions (8.1, 7.7, 7.5, and fluctuating between 7.5 and 8.1). They found that low pH decreased overall calcification rates by 16% for corals from both origins. However, corals from Bouraké consistently exhibited higher growth rates than those from reference sites. This difference could not be attributed to an increased heterotrophic ability or better energy expenditure, as all corals received the same oligotrophic seawater during the incubation. Additionally, it could not be linked to the depletion of lipid reserves in Bouraké corals since the incubation lasted 4 months, while lipid turnover occurs much more quickly.

Tanvet et al. (2023), along with Alessi et al. (2024a), demonstrated that the enhanced performance of Bouraké corals was associated with divergent and more homogeneous symbiont communities, all belonging to the genus *Cladocopium*. The authors found that the significant *Symbiodiniaceae* ITS2 type sequences in *Acropora muricata* from Bouraké were *Cladocopium proliferum* type C1 (formerly *C. goreaui*), while corals from reference sites exhibited type C50b, which was highly variable among specimens. *Cladocopium* type C1 has been suggested to be efficient in carbon fixation and enhancing nitrogen acquisition (Baker et al. 2013), potentially boosting the energy budget of corals by improving their photo-physiological traits. Whether specific *Symbiodiniaceae* could enhance overall coral growth rates or if other mechanisms are at play in Bouraké corals remains to be elucidated.

Alessi et al. (2024a) found that *Cladocopium* C1 was predominant in unbleached *A. muricata* in Bouraké but absent in bleached corals, which showed ITS2 sequences similar to those of unbleached conspecifics from reference reefs. Interestingly, the few bleached colonies that survived and fully recovered their metabolic and calcification rates all acquired *Cladocopium* type C1, suggesting that a specific Symbiodiniaceae community can support physiological traits in corals exposed to extreme conditions.

Recently, Leiva et al. (2025) not only showed the rapid adaptation strategy of corals from Bouraké, confirming the corals' flexibility to associate with different algae species that are capable of tolerating extreme conditions, but almost provided the first high-quality whole genomes for a coral from an extreme natural laboratory, *Acropora tenuis*. The researchers discovered that these corals' survival is linked to specific genes involved in stress response, including TRPA1, a gene that helps organisms sense changes in temperature and oxygen. Mutations in TRPA1, along with other genes related to cellular stress, showed signals of genomic adaptation to Bouraké and appear to play a crucial role in the corals' ability to cope with the harsh environment. This research emphasizes how corals may adapt to future climate change, stressing that their adaptive potential depends on the current genetic diversity of their populations, as it is from this diversity that natural selection acts upon.

7.5 Conclusion

Since the early studies, research on coral communities surrounding volcanic CO_2 vents has significantly expanded our understanding of the effects of OA on coral reef ecosystems, from the single organisms to the ecosystem level. Despite their limitations as "perfect natural analogues" of OA, volcanic CO_2 vents remain one of the few tools available to study the chronic effects of long-term high-CO_2 exposure on reef communities. They are likely the best to assess potential species adaptation to OA. These studies confirm that OA will contribute to coral reef degradation, with future communities comprising only the most resistant species and individuals, reducing diversity and structural complexity. Although we have focused on the effects of OA on coral physiology and ecology, research at volcanic CO_2 vents has also revealed cascading effects driven by shifts in benthic communities, as well as the direct impacts of high CO_2 exposure across nearly all levels of reef ecosystems, from microbes and plankton to fish.

Semi-enclosed bays, like volcanic CO_2 vents, provide valuable opportunities to study the chronic effects of long-term CO_2 exposure. Additionally, they allow researchers to investigate the interactions between OA and other environmental stressors, such as high temperatures and low oxygen levels. Together with enclosed lagoons, urban reefs, and acidic underwater springs, these marginal and extreme environments are great opportunities for the study of long-term ecosystem broad effects of stressors and opportunities for developing mitigation and adaptation strategies. In particular, studies from these natural analogues have shed light on the remarkable adaptive capacity of corals, providing hope that they may be able to survive in increasingly harsh ocean conditions. While there are various approaches to coral conservation, including human-driven methods like artificial selection, the first full-genome sequencing suggests that corals already have the genetic tools to adapt naturally. The findings emphasize the importance of preserving genetic diversity in coral populations, as this diversity fuels natural selection and potentially allows corals to adjust to the challenges posed by climate change. These insights could inform conservation strategies, helping us protect coral reefs, which are essential to marine biodiversity and the well-being of millions of people worldwide.

Despite the often harmful effects of OA observed at CO_2 vents and semi-enclosed bays, these environments also highlight the remarkable capacity of corals to acclimatize rapidly to high-CO_2 conditions. There is much to learn from the corals inhabiting these areas, making it a research priority. However, with only 17 volcanic CO_2 vent systems studied in coral reef regions—and even fewer semi-enclosed bays—it is crucial to coordinate, facilitate, and standardize research at these sites, a goal set forth by the ICONA network and recognized by the UN Ocean Decade. Certain regions, such as the South Pacific, the East Pacific, and the Caribbean, remain underexplored and should be further investigated. Scientists should also strive to replicate studies across different systems and biogeographical regions to better understand coral responses to OA.

Located within the Fuji-Hakone-Izu National Park and under relatively low anthropogenic pressure, the Shikine CO_2 vent system benefits from some protection. However, given the rarity of resistant populations of high-complexity coral species—such as those observed around the volcanic CO_2 vents—and the limited scope of existing conservation measures (e.g., commercial fishing is still permitted), it is critical to establish targeted conservation strategies. These systems and other shallow CO_2 vents should be systematically protected as unique natural laboratories that provide crucial insights into coral resilience under future ocean conditions. While these environments are often remote and presumed relatively free from local anthropogenic pressures, volcanic hydrothermal systems are rich in rare earth elements and metals, making them increasingly attractive for mining and other extractive activities. Given their potential as reservoirs of genetic resistance that could support assisted gene flow and adaptation, it is urgent to consider their systematic protection, similar to the conservation efforts directed at deep-sea hydrothermal vent ecosystems. The case for protecting semi-enclosed volcanic hydrothermal systems is even more complex, as these areas have often been overlooked or deemed "degraded" and of low conservation value. Nevertheless, the ecological processes occurring in these natural analogues of future ocean conditions are invaluable for understanding the long-term effects of OA. Unlike acute stressors, OA exerts slow, pervasive impacts on marine ecosystems that may not yet be readily visible. However, volcanic CO_2 vents and semi-enclosed bays serve as powerful windows into the future, offering tangible, real-world examples of how acidification reshapes marine biodiversity. This predictive capacity makes them not only essential for scientific research but also for public engagement and awareness.

The use of natural analogues extends beyond science into media and artistic fields, where they serve as compelling visual and narrative tools to communicate the urgency of climate change. These environments have been featured in various media outlets, documentaries, and educational initiatives, helping to bridge the gap between scientific research and public perception. The artistic work of Prof. Nicolas Floc'h, for instance, highlights the potential of these natural laboratories for raising awareness about OA (Floc'h 2018, 2022). His systematic documentation of ecosystems around volcanic vents in Italy and Japan provides a striking visual representation of how acidification transforms marine landscapes, appealing to scientific and artistic audiences. Such interdisciplinary approaches are instrumental in conveying the potentially drastic effects of OA, making the case for stronger conservation measures and climate action more accessible and impactful to a broader audience.

References

Agostini S, Suzuki Y, Higuchi T, Casareto BE, Yoshinaga K, Nakano Y, Fujimura H (2012) Biological and chemical characteristics of the coral gastric cavity. Coral Reefs 31:147–156. https://doi.org/10.1007/s00338-011-0831-6

Agostini S, Wada S, Kon K, Omori A, Kohtsuka H, Fujimura H, Tsuchiya Y, Sato T, Shinagawa H, Yamada Y, Inaba K (2015) Geochemistry of two shallow CO_2 seeps in Shikine Island (Japan) and their potential for ocean acidification research. Reg Stud Mar Sci 2:45–53. https://doi.org/10.1016/j.rsma.2015.07.004

Agostini S, Harvey BP, Wada S, Kon K, Milazzo M, Inaba K, Hall-Spencer JM (2018) Ocean acidification drives community shifts towards simplified non-calcified habitats in a subtropical−temperate transition zone. Sci Rep 8:11354. https://doi.org/10.1038/s41598-018-29251-7

Agostini S, Harvey BP, Milazzo M, Wada S, Kon K, Floc'h N, Komatsu K, Kuroyama M, Hall-Spencer JM (2021a) Simplification, not "tropicalization", of temperate marine ecosystems under ocean warming and acidification. Glob Change Biol gcb.15749. https://doi.org/10.1111/gcb.15749

Agostini S, Houlbrèque F, Biscéré T, Harvey BP, Heitzman JM, Takimoto R, Yamazaki W, Milazzo M, Rodolfo-Metalpa R (2021b) Greater mitochondrial energy production provides resistance to ocean acidification in "Winning" Hermatypic Corals. Front Mar Sci 7:600836. https://doi.org/10.3389/fmars.2020.600836

Albright R, Caldeira L, Hosfelt J, Kwiatkowski L, Maclaren JK, Mason BM, Nebuchina Y, Ninokawa A, Pongratz J, Ricke KL, Rivlin T, Schneider K, Sesboüé M, Shamberger K, Silverman J, Wolfe K, Zhu K, Caldeira K (2016) Reversal of ocean acidification enhances net coral reef calcification. Nature 531:362–365. https://doi.org/10.1038/nature17155

Alessi C, Lemonnier H, Camp EF, Wabete N, Payri C, Rodolfo Metalpa R (2024a) Algal symbiont diversity in Acropora muricata from the extreme reef of Bouraké associated with resistance to coral bleaching. PLoS One 19:e0296902

Alessi C, Randall CJ, Dumas M, Lemonnier H, Rodolfo-Metalpa R (2024b) Spawning window and fecundity in three Acroporid corals from the environmentally variable semi-enclosed lagoon of Bouraké. Coral Reefs 43:1469–1482. https://doi.org/10.1007/s00338-024-02551-3

Alvarez-Filip L, Dulvy NK, Gill JA, Côté IM, Watkinson AR (2009) Flattening of Caribbean coral reefs: region-wide declines in architectural complexity. Proc R Soc Lond B Biol Sci 276:3019–3025. https://doi.org/10.1098/rspb.2009.0339

Anthony KRN, Fabricius KE (2000) Shifting roles of heterotrophy and autotrophy in coral energetics under varying turbidity. J Exp Mar Biol Ecol 252:221–253. https://doi.org/10.1016/S0022-0981(00)00237-9

Baker DM, Andras JP, Jordán-Garza AG, Fogel ML (2013) Nitrate competition in a coral symbiosis varies with temperature among Symbiodinium clades. ISME J 7:1248–1251. https://doi.org/10.1038/ismej.2013.12

Bang C, Dagan T, Deines P, Dubilier N, Duschl WJ, Fraune S, Hentschel U, Hirt H, Hülter N, Lachnit T, Picazo D, Pita L, Pogoreutz C, Rädecker N, Saad MM, Schmitz RA, Schulenburg H, Voolstra CR, Weiland-Bräuer N, Ziegler M, Bosch TCG (2018) Metaorganisms in extreme environments: do microbes play a role in organismal adaptation? Zoology 127:1–19. https://doi.org/10.1016/j.zool.2018.02.004

Barkley HC, Cohen AL, Golbuu Y, Starczak VR, DeCarlo TM, Shamberger KEF (2015) Changes in coral reef communities across a natural gradient in seawater pH. Sci Adv 1:e1500328. https://doi.org/10.1126/sciadv.1500328

Biagi E, Caroselli E, Barone M, Pezzimenti M, Teixido N, Soverini M, Rampelli S, Turroni S, Gambi M, Brigidi P, Goffredo S, Candela M (2020) Patterns in microbiome composition differ with ocean acidification in anatomic compartments of the Mediterranean coral Astroides calycularis living at CO2 vents. Sci Total Environ 724. https://doi.org/10.1016/j.scitotenv.2020.138048

Bianchi CN, Dando PR, Morri C (2011) Increased diversity of sessile epibenthos at subtidal hydrothermal vents: seven hypotheses based on observations at Milos Island, Aegean Sea. Adv Oceanogr Limnol 2:1–31. https://doi.org/10.1080/19475721.2011.565804

Biodiversity Center of Japan (2019) Report of Seagrass and algal beds Survey 2019 (Survey Report No. 2019/03), Monitoring Sites1000 Project. Ministry of the Environment

Biscéré T, Zampighi M, Lorrain A, Jurriaans S, Foggo A, Houlbrèque F, Rodolfo-Metalpa R (2019) High pCO2 promotes coral primary production. Biol Lett 15:20180777. https://doi.org/10.1098/rsbl.2018.0777

Bozec Y-M, Alvarez-Filip L, Mumby PJ (2015) The dynamics of architectural complexity on coral reefs under climate change. Glob Change Biol 21:223–235. https://doi.org/10.1111/gcb.12698

Camp EF, Nitschke MR, Rodolfo-Metalpa R, Houlbreque F, Gardner SG, Smith DJ, Zampighi M, Suggett DJ (2017) Reef-building corals thrive within hot-acidified and deoxygenated waters. Sci Rep 7. https://doi.org/10.1038/s41598-017-02383-y

Camp EF, Schoepf V, Mumby PJ, Hardtke LA, Rodolfo-Metalpa R, Smith DJ, Suggett DJ (2018) The future of coral reefs subject to rapid climate change: lessons from natural extreme environments. Front Mar Sci 5. https://doi.org/10.3389/fmars.2018.00004

Carbonne C, Teixidó N, Moore B, Mirasole A, Guttierez T, Gattuso J, Comeau S (2021) Two temperate corals are tolerant to low pH regardless of previous exposure to natural CO_2 vents. Limnol Oceanogr 66:4046–4061. https://doi.org/10.1002/lno.11942

Cattano C, Agostini S, Harvey BP, Wada S, Quattrocchi F, Turco G, Inaba K, Hall-Spencer JM, Milazzo M (2020) Changes in fish communities due to benthic habitat shifts under ocean acidification conditions. Sci Total Environ 725:138501. https://doi.org/10.1016/j.scitotenv.2020.138501

Chan I, Peng S, Chang C, Hung J, Hwang J (2012) Effects of acidified seawater on the skeletal structure of a Scleractinian Coral from evidence identified by SEM. Zool Stud 51:1319–1331

Chen T, Li S, Shi Q, Chen T (2016) Cold tolerance of subtropical Porites lutea from the northern South China Sea. Acta Oceanol Sin 35:58–64. https://doi.org/10.1007/s13131-016-0940-8

Cocito S, Bianchi C, Morri C, Peirano A (2000) First survey of sessile communities on subtidal rocks in an area with hydrothermal vents: Miles Island, Aegean Sea. Hydrobiologia 426:113–121. https://doi.org/10.1023/A:1003991117108

Comeau S, Cornwall CE, Shlesinger T, Hoogenboom M, Mana R, McCulloch MT, Rodolfo-Metalpa R (2022) pH variability at volcanic CO2 seeps regulates coral calcifying fluid chemistry. Glob Change Biol 28:2751–2763. https://doi.org/10.1111/gcb.16093

Cornwall CE, Comeau S, DeCarlo TM, Larcombe E, Moore B, Giltrow K, Puerzer F, D'Alexis Q, McCulloch MT (2020) A coralline alga gains tolerance to ocean acidification over multiple generations of exposure. Nat Clim Chang 10:143–146. https://doi.org/10.1038/s41558-019-0681-8

Cornwall CE, Comeau S, Kornder NA, Perry CT, van Hooidonk R, DeCarlo TM, Pratchett MS, Anderson KD, Browne N, Carpenter R, Diaz-Pulido G, D'Olivo JP, Doo SS, Figueiredo J, Fortunato SAV, Kennedy E, Lantz CA, McCulloch MT, González-Rivero M, Schoepf V, Smithers SG, Lowe RJ (2021) Global declines in coral reef calcium carbonate production under ocean acidification and warming. Proc Natl Acad Sci 118:e2015265118. https://doi.org/10.1073/pnas.2015265118

Costello MJ, Tsai P, Wong PS, Cheung AKL, Basher Z, Chaudhary C (2017) Marine biogeographic realms and species endemicity. Nat Commun 8:1057. https://doi.org/10.1038/s41467-017-01121-2

Dando PR, Hughes JA, Leahy Y, Niven SJ, Taylor LJ, Smith C (1995) Gas venting rates from submarine hydrothermal areas around the island of Milos, Hellenic Volcanic Arc. Cont Shelf Res 15:913–929. https://doi.org/10.1016/0278-4343(95)80002-U

Dando PR, Stüben D, Varnavas SP (1999) Hydrothermalism in the Mediterranean Sea. Prog Oceanogr 44(1–3):333–367

Díaz R, Rosenberg R (1995) Marine benthic hypoxia: a review of its ecological effects and the behavioural responses of benthic macrofauna. Oceanogr Mar Biol 33:245

Dikou A, van Woesik R (2006) Survival under chronic stress from sediment load: spatial patterns of hard coral communities in the southern islands of Singapore. Mar Pollut Bull 52:1340–1354. https://doi.org/10.1016/j.marpolbul.2006.02.011

Doo SS, Edmunds PJ, Carpenter RC (2019) Ocean acidification effects on in situ coral reef metabolism. Sci Rep 9:12067. https://doi.org/10.1038/s41598-019-48407-7

Dubuc A, Baker R, Marchand C, Waltham NJ, Sheaves M (2019) Hypoxia in mangroves: occurrence and impact on valuable tropical fish habitat. Biogeosciences 16:3959–3976. https://doi.org/10.5194/bg-16-3959-2019

Enochs IC, Manzello DP, Donham EM, Kolodziej G, Okano R, Johnston L, Young C, Iguel J, Edwards CB, Fox MD, Valentino L, Johnson S, Benavente D, Clark SJ, Carlton R, Burton T, Eynaud Y, Price NN (2015) Shift from coral to macroalgae dominance on a volcanically acidified reef. Nat Clim Chang 5:1083–1088. https://doi.org/10.1038/nclimate2758

Enochs IC, Manzello DP, Kolodziej G, Noonan SHC, Valentino L, Fabricius KE (2016) Enhanced macroboring and depressed calcification drive net dissolution at high-CO2 coral reefs. Proc R Soc B 283:20161742. https://doi.org/10.1098/rspb.2016.1742

Enochs IC, Formel N, Manzello D, Morris J, Mayfield AB, Boyd A, Kolodziej G, Adams G, Hendee J (2020) Coral persistence despite extreme periodic pH fluctuations at a volcanically acidified Caribbean reef. Coral Reefs 39:523–528. https://doi.org/10.1007/s00338-020-01927-5

Fabricius KE, Langdon C, Uthicke S, Humphrey C, Noonan S, De'ath G, Okazaki R, Muehllehner N, Glas MS, Lough JM (2011) Losers and winners in coral reefs acclimatized to elevated carbon dioxide concentrations. Nat Clim Chang 1:165–169. https://doi.org/10.1038/nclimate1122

Fabricius KE, De'ath G, Noonan S, Uthicke S (2014) Ecological effects of ocean acidification and habitat complexity on reef-associated macroinvertebrate communities. Proc R Soc B Biol Sci 281:20132479. https://doi.org/10.1098/rspb.2013.2479

Fantazzini P, Mengoli S, Pasquini L, Bortolotti V, Brizi L, Mariani M, Di Giosia M, Fermani S, Capaccioni B, Caroselli E, Prada F, Zaccanti F, Levy O, Dubinsky Z, Kaandorp JA, Konglerd P, Hammel JU, Dauphin Y, Cuif J-P, Weaver JC, Fabricius KE, Wagermaier W, Fratzl P, Falini G, Goffredo S (2015) Gains and losses of coral skeletal porosity changes with ocean acidification acclimation. Nat Commun 6. https://doi.org/10.1038/ncomms8785

Fine M, Tchernov D (2007) Scleractinian coral species survive and recover from decalcification. Science 315:1811–1811. https://doi.org/10.1126/science.1137094

Floc'h N (2018) Nicolas Floc'h - Glaz. Roma Publication/Stichting Roma Publications, Amsterdam

Floc'h N (2022) Initium maris: carnet de bord 2020/21, II/III. Éditions GwinZegal, Guingamp

Fujita E, Ueda H, Nakada S (2020) A new Japan volcanological database. Front Earth Sci 8. https://doi.org/10.3389/feart.2020.00205

Gattuso J-P, Magnan A, Billé R, Cheung WWL, Howes EL, Joos F, Allemand D, Bopp L, Cooley SR, Eakin CM, Hoegh-Guldberg O, Kelly RP, Pörtner H-O, Rogers AD, Baxter JM, Laffoley D, Osborn D, Rankovic A, Rochette J, Sumaila UR, Treyer S, Turley C (2015) Contrasting futures for ocean and society from different anthropogenic CO2 emissions scenarios. Science 349:aac4722. https://doi.org/10.1126/science.aac4722

Golbuu Y, Gouezo M, Kurihara H, Rehm L, Wolanski E (2016) Long-term isolation and local adaptation in Palau's Nikko Bay help corals thrive in acidic waters. Coral Reefs 35:909–918. https://doi.org/10.1007/s00338-016-1457-5

Guinotte JM, Buddemeier RW, Kleypas JA (2003) Future coral reef habitat marginality: temporal and spatial effects of climate change in the Pacific basin. Coral Reefs 22:551–558. https://doi.org/10.1007/s00338-003-0331-4

Hall-Spencer JM, Harvey BP (2019) Ocean acidification impacts on coastal ecosystem services due to habitat degradation. Emerg Top Life Sci 3:197–206. https://doi.org/10.1042/ETLS20180117

Hall-Spencer JM, Rodolfo-Metalpa R, Martin S, Ransome E, Fine M, Turner SM, Rowley SJ, Tedesco D, Buia M-C (2008) Volcanic carbon dioxide vents show ecosystem effects of ocean acidification. Nature 454:96–99. https://doi.org/10.1038/nature07051

Hall-Spencer JM, Belfiore G, Tomatsuri M, Porzio L, Harvey BP, Agostini S, Kon K (2022) Decreased diversity and abundance of marine invertebrates at CO2 seeps in warm-temperate Japan. Zool Sci 39:41–51. https://doi.org/10.2108/zs210061

Harrison P, Wallace C (1990) Reproduction, dispersal and recruitment of scleractinian corals. Ecosyst World

Harvey BP, McKeown NJ, Rastrick SPS, Bertolini C, Foggo A, Graham H, Hall-Spencer JM, Milazzo M, Shaw PW, Small DP, Moore PJ (2016) Individual and population-level responses to ocean acidification. Sci Rep 6:20194. https://doi.org/10.1038/srep20194

Harvey BP, Allen R, Agostini S, Hoffmann LJ, Kon K, Summerfield TC, Wada S, Hall-Spencer JM (2021) Feedback mechanisms stabilise degraded turf algal systems at a CO2 seep site. Commun Biol 4:219. https://doi.org/10.1038/s42003-021-01712-2

Heery EC, Hoeksema BW, Browne NK, Reimer JD, Ang PO, Huang D, Friess DA, Chou LM, Loke LHL, Saksena-Taylor P, Alsagoff N, Yeemin T, Sutthacheep M, Vo ST, Bos AR, Gumanao GS, Syed

Hussein MA, Waheed Z, Lane DJW, Johan O, Kunzmann A, Jompa J, Suharsono T, Bauman AG, Todd PA (2018) Urban coral reefs: degradation and resilience of hard coral assemblages in coastal cities of East and Southeast Asia. Mar Pollut Bull 135:654–681. https://doi.org/10.1016/j.marpolbul.2018.07.041

Higuchi T, Yuyama I, Agostini S (2018) Physiology of winter coral bleaching in temperate zone. In: Proceedings of the COAST symposium. Presented at the COAST Bordeaux 2018, France-Japan Society of Oceanography, Bordeaux, France

Hochart C, Paoli L, Ruscheweyh H-J, Salazar G, Boissin E, Romac S, Poulain J, Bourdin G, Iwankow G, Moulin C, Ziegler M, Porro B, Armstrong EJ, Hume BCC, Aury J-M, Pogoreutz C, Paz-García DA, Nugues MM, Agostini S, Banaigs B, Boss E, Bowler C, de Vargas C, Douville E, Flores M, Forcioli D, Furla P, Gilson E, Lombard F, Pesant S, Reynaud S, Thomas OP, Troublé R, Wincker P, Zoccola D, Allemand D, Planes S, Thurber RV, Voolstra CR, Sunagawa S, Galand PE (2023) Ecology of Endozoicomonadaceae in three coral genera across the Pacific Ocean. Nat Commun 14:3037. https://doi.org/10.1038/s41467-023-38502-9

Houlbrèque F, Tambutté E, Ferrier-Pagès C (2003) Effect of zooplankton availability on the rates of photosynthesis, and tissue and skeletal growth in the scleractinian coral Stylophora pistillata. J Exp Mar Biol Ecol 296:145–166. https://doi.org/10.1016/S0022-0981(03)00259-4

Hughes TP, Kerry JT, Alvarez-Noriega M, Alvarez-Romero JG, Anderson KD, Baird AH, Babcock RC, Beger M, Bellwood DR, Berkelmans R, Bridge TC, Butler IR, Byrne M, Cantin NE, Comeau S, Connolly SR, Cumming GS, Dalton SJ, Diaz-Pulido G, Eakin CM, Figueira WF, Gilmour JP, Harrison HB, Heron SF, Hoey AS, Hobbs J-PA, Hoogenboom MO, Kennedy EV, Kuo C, Lough JM, Lowe RJ, Liu G, Cculloch MTM, Malcolm HA, Mcwilliam MJ, Pandolfi JM, Pears RJ, Pratchett MS, Schoepf V, Simpson T, Skirving WJ, Sommer B, Torda G, Wachenfeld DR, Willis BL, Wilson SK (2017) Global warming and recurrent mass bleaching of corals. Nature 543:373. https://doi.org/10.1038/nature21707

Hulver AM, Carbonne C, Teixidó N, Comeau S, Kemp DW, Keister EF, Gattuso J-P, Grottoli AG (2024) Elevated heterotrophic capacity as a strategy for Mediterranean corals to cope with low pH at CO2 vents. PLoS One 19:e0306725. https://doi.org/10.1371/journal.pone.0306725

Inoue S, Kayanne H, Yamamoto S, Kurihara H (2013) Spatial community shift from hard to soft corals in acidified water. Nat Clim Chang 3:683–687. https://doi.org/10.1038/nclimate1855

Jacquemont J, Houlbrèque F, Tanvet C, Rodolfo-Metalpa R (2022) Long-term exposure to an extreme environment induces species--specific responses in corals' photosynthesis and respiration rates. Mar Biol 169:82. https://doi.org/10.1007/s00227-022-04063-6

Januar H, Zamani N, Soedarma D, Chasanah E, Wright A (2017) Tropical coral reef coral patterns in Indonesian shallow water areas close to underwater volcanic vents at Minahasa Seashore, and Mahengetang and Gunung Api Islands. Mar Ecol- Evol Perspect 38. https://doi.org/10.1111/maec.12415

Kenkel CD, Moya A, Strahl J, Humphrey C, Bay LK (2018) Functional genomic analysis of corals from natural CO2-seeps reveals core molecular responses involved in acclimatization to ocean acidification. Glob Change Biol 24:158–171. https://doi.org/10.1111/gcb.13833

Kleypas JA, McManus JW, Menez LAB (1999) Environmental limits to coral reef development: where do we draw the line? Integr Comp Biol 39:146–159. https://doi.org/10.1093/icb/39.1.146

Kleypas J, Allemand D, Anthony K, Baker AC, Beck MW, Hale LZ, Hilmi N, Hoegh-Guldberg O, Hughes T, Kaufman L, Kayanne H, Magnan AK, Mcleod E, Mumby P, Palumbi S, Richmond RH, Rinkevich B, Steneck RS, Voolstra CR, Wachenfeld D, Gattuso J-P (2021) Designing a blueprint for coral reef survival. Biol Conserv 257:109107. https://doi.org/10.1016/j.biocon.2021.109107

Kochan DP, Mitchell MD, Zuercher R, Harborne AR (2023) Winners and losers of reef flattening: an assessment of coral reef fish species and traits. Oikos 2023:e10011. https://doi.org/10.1111/oik.10011

Kroeker KJ, Gambi MC, Micheli F (2013) Community dynamics and ecosystem simplification in a high-CO2 ocean. Proc Natl Acad Sci 110:12721–12726. https://doi.org/10.1073/pnas.1216464110

Kumagai NH, Molinos JG, Yamano H, Takao S, Fujii M, Yamanaka Y (2018) Ocean currents and herbivory drive macroalgae-to-coral community shift under climate warming. Proc Natl Acad Sci 115:8990–8995. https://doi.org/10.1073/pnas.1716826115

Kurihara H, Watanabe A, Tsugi A, Mimura I, Hongo C, Kawai T, Reimer JD, Kimoto K, Gouezo M, Golbuu Y (2021) Potential local adaptation of corals at acidified and warmed Nikko Bay. Palau Sci Rep 11:11192. https://doi.org/10.1038/s41598-021-90614-8

Leiva C, Pérez-Portela R, Lemer S (2023a) Genomic signatures suggesting adaptation to ocean acidification in a coral holobiont from volcanic CO2 seeps. Commun Biol. 6:1–14. https://doi.org/10.1038/s42003-023-05103-7

Leiva C, Pérez-Portela R, Lemer S (2023b) Synergistic genomic mechanisms of adaptation to ocean acidification in a coral holobiont. https://doi.org/10.1101/2023.06.18.545179

Leiva C, Torda G, Zhou C, Pan Y, Harris J, Xiang X, Tan S, Tian W, Hume B, Miller DJ, Li Q, Zhang G, Cooke I, Rodolfo-Metalpa R (2025) Rapid evolution in action: environmental filtering supports coral adaptation to a hot, acidic, and deoxygenated extreme habitat. Glob Chang Biol 31(3):e70103

Leopold A, Marchand C, Deborde J, Allenbach M (2017) Water biogeochemistry of a mangrove-dominated estuary under a semi-arid climate (New Caledonia). Estuar Coasts 40:773–791. https://doi.org/10.1007/s12237-016-0179-9

Limburg KE, Breitburg D, Swaney DP, Jacinto G (2020) Ocean deoxygenation: a primer. One Earth 2:24–29. https://doi.org/10.1016/j.oneear.2020.01.001

Logan CA, Dunne JP, Eakin CM, Donner SD (2014) Incorporating adaptive responses into future projections of coral bleaching. Glob Change Biol 20:125–139. https://doi.org/10.1111/gcb.12390

Maggioni F, Pujo-Pay M, Aucan J, Cerrano C, Calcinai B, Payri C, Benzoni F, Letourneur Y, Rodolfo-Metalpa R (2021) The Bouraké semi-enclosed lagoon (New Caledonia)—a natural laboratory to study the lifelong adaptation of a coral reef ecosystem to extreme environmental conditions. Biogeosciences 18:5117–5140. https://doi.org/10.5194/bg-18-5117-2021

Maningas JAC, Ticzon VS, Rodriguez MVB (2025) Characterization of coral communities in the shallow hydrothermal vents of Mabini, Batangas, Philippines. Reg Stud Mar Sci 81:103943. https://doi.org/10.1016/j.rsma.2024.103943

Marcelino VR, Morrow KM, van Oppen MJH, Bourne DG, Verbruggen H (2017) Diversity and stability of coral endolithic microbial communities at a naturally high pCO2 reef. Mol Ecol 26:5344–5357. https://doi.org/10.1111/mec.14268

Mayfield AB, Gates RD (2007) Osmoregulation in anthozoan–dinoflagellate symbiosis. Comp Biochem Physiol A Mol Integr Physiol 147:1–10

Meunier V, Geissler L, Bonnet S, Rädecker N, Perna G, Grosso O, Lambert C, Rodolfo-Metalpa R, Voolstra CR, Houlbrèque F (2021) Microbes support enhanced nitrogen requirements of coral holobionts in a high CO2 environment. Mol Ecol 30:5888–5899. https://doi.org/10.1111/mec.16163

Miyake H, Oda A, Wada S, Kodaka T, Kurosawa S (2019) First record of a shallow hydrothermal vent crab, Xenograpsus testudinatus, from Shikine-jima Island in the Izu archipelago. Biogeography 21:31–36

Miyazaki J, Kawagucci S, Makabe A, Takahashi A, Kitada K, Torimoto J, Matsui Y, Tasumi E, Shibuya T, Nakamura K, Horai S, Sato S, Ishibashi J, Kanzaki H, Nakagawa S, Hirai M, Takaki

Y, Okino K, Watanabe HK, Kumagai H, Chen C (2017) Deepest and hottest hydrothermal activity in the Okinawa Trough: the Yokosuka site at Yaeyama Knoll. R Soc Open Sci 4:171570. https://doi.org/10.1098/rsos.171570

Morri C, Vafidis D, Peirano A, Chintiroglou CC, Bianchi CN (2000) Anthozoa from a subtidal hydrothermal area of Milos Island (Aegean Sea), with notes on the construction potential of the scleractinian coralMadracis pharensis. Ital J Zool 67:319–325. https://doi.org/10.1080/11250000009356331

Morrow KM, Bourne DG, Humphrey C, Botté ES, Laffy P, Zaneveld J, Uthicke S, Fabricius KE, Webster NS (2014) Natural volcanic CO2 seeps reveal future trajectories for host–microbial associations in corals and sponges. ISME J. https://doi.org/10.1038/ismej.2014.188

Nakabayashi A, Yamakita T, Nakamura T, Aizawa H, Kitano YF, Iguchi A, Yamano H, Nagai S, Agostini S, Teshima KM, Yasuda N (2019) The potential role of temperate Japanese regions as refugia for the coral Acropora hyacinthus in the face of climate change. Sci Rep 9:1892. https://doi.org/10.1038/s41598-018-38333-5

Nakamura T (2017) Mass coral bleaching event in Sekisei lagoon observed in the summer of 2016. J Jpn Coral Reef Soc 19:29–40. https://doi.org/10.3755/jcrs.19.29

Noonan SHC, Fabricius KE, Humphrey C (2013) Symbiodinium community composition in scleractinian corals is not affected by life--long exposure to elevated carbon dioxide. PLoS One 8:e63985. https://doi.org/10.1371/journal.pone.0063985

O'Brien P, Smith H, Fallon S, Fabricius K, Willis B, Morrow K, Bourne D (2018) Elevated CO2 has little influence on the bacterial communities associated with the pH-tolerant coral, massive Porites spp. Front Microbiol 9. https://doi.org/10.3389/fmicb.2018.02621

Oporto-Guerrero T, Reyes-Bonilla H, Ladah LB (2018) Presence of the reef-building coral, Porites panamensis, in a shallow hydrothermal field in the Gulf of California. Mar Biodivers 48:703–708. https://doi.org/10.1007/s12526-016-0546-6

Palladino G, Caroselli E, Tavella T, D'Amico F, Prada F, Mancuso A, Franzellitti S, Rampelli S, Candela M, Goffredo S, Biagi E (2022) Metagenomic shifts in mucus, tissue and skeleton of the coral Balanophyllia europaea living along a natural CO2 gradient. ISME Commun 2:65. https://doi.org/10.1038/s43705-022-00152-1

Perry CT, Alvarez-Filip L (2019) Changing geo-ecological functions of coral reefs in the Anthropocene. Funct Ecol 33:976–988. https://doi.org/10.1111/1365-2435.13247

Pichler T, Veizer J, Hall G (1999) The chemical composition of shallow-water hydrothermal fluids in Tutum Bay, Ambitle Island, Papua New Guinea and their effect on ambient seawater. Mar Chem 64:229–252. https://doi.org/10.1016/S0304-4203(98)00076-0

Pichler T, Biscéré T, Kinch J, Zampighi M, Houlbrèque F, Rodolfo-Metalpa R (2019) Suitability of the shallow water hydrothermal system at Ambitle Island (Papua New Guinea) to study the effect of high pCO2 on coral reefs. Mar Pollut Bull 138:148–158. https://doi.org/10.1016/j.marpolbul.2018.11.003

Planning Coordination Office (n.d.) Data Niijima 2022. Niijima Village Planning and Finance Division, Tokyo

Prada F, Brizi L, Franzellitti S, Mengoli S, Fermani S, Polishchuk I, Baraldi N, Ricci F, Palazzo Q, Caroselli E, Pokroy B, Giorgini L, Dubinsky Z, Fantazzini P, Falini G, Goffredo S, Fabricius K (2021) Coral micro- and macro-morphological skeletal properties in response to life-long acclimatization at CO2 vents in Papua New Guinea. Sci Rep 11. https://doi.org/10.1038/s41598-021-98976-9

Reimer JD, Kurihara H, Ravasi T, Ide Y, Izumiyama M, Kayanne H (2021) Unexpected high abundance of aragonite-forming Nanipora (Octocorallia: Helioporacea) at an acidified volcanic reef in southern Japan. Mar Biodivers 51:1–5. https://doi.org/10.1007/s12526-021-01165-0

Reimer JD, Peixoto RS, Davies SW, Traylor-Knowles N, Short ML, Cabral-Tena RA, Burt JA, Pessoa I, Banaszak AT, Winters RS, Moore T, Schoepf V, Kaullysing D, Calderon-Aguilera LE, Wörheide G, Harding S, Munbodhe V, Mayfield A, Ainsworth T, Vardi T, Eakin CM, Pratchett MS, Voolstra CR (2024) The fourth global coral bleaching event: where do we go from here? Coral Reefs 43:1121–1125. https://doi.org/10.1007/s00338-024-02504-w

Rivera HE, Cohen AL, Thompson JR, Baums IB, Fox MD, Meyer-Kaiser KS (2022) Palau's warmest reefs harbor thermally tolerant corals that thrive across different habitats. Commun Biol 5:1–12. https://doi.org/10.1038/s42003-022-04315-7

Rodolfo-Metalpa R, Houlbrèque F, Tambutté É, Boisson F, Baggini C, Patti FP, Jeffree R, Fine M, Foggo A, Gattuso J-P, Hall-Spencer JM (2011) Coral and mollusc resistance to ocean acidification adversely affected by warming. Nat Clim Chang 1:308–312. https://doi.org/10.1038/nclimate1200

Rodolfo-Metalpa R, Montagna P, Aliani S, Borghini M, Canese S, Hall-Spencer JM, Foggo A, Milazzo M, Taviani M, Houlbrèque F (2015) Calcification is not the Achilles' heel of cold-water corals in an acidifying ocean. Glob Change Biol 21:2238–2248. https://doi.org/10.1111/gcb.12867

Schoepf V, Baumann JH, Barshis DJ, Browne NK, Camp EF, Comeau S, Cornwall CE, Guzmán HM, Riegl B, Rodolfo-Metalpa R, Sommer B (2023) Corals at the edge of environmental limits: a new conceptual framework to re-define marginal and extreme coral communities. Sci Total Environ 884:163688. https://doi.org/10.1016/j.scitotenv.2023.163688

Shore A, Day R, Stewart J, Burge C (2021) Dichotomy between regulation of coral bacterial communities and calcification physiology under ocean acidification conditions. Appl Environ Microbiol 87. https://doi.org/10.1128/AEM.02189-20

Strahl J, Stolz I, Uthicke S, Vogel N, Noonan SHC, Fabricius KE (2015) Physiological and ecological performance differs in four coral taxa at a volcanic carbon dioxide seep. Comp Biochem Physiol A Mol Integr Physiol 184:179–186. https://doi.org/10.1016/j.cbpa.2015.02.018

Strahl J, Francis DS, Doyle J, Humphrey C, Fabricius KE (2016) Biochemical responses to ocean acidification contrast between tropical corals with high and low abundances at volcanic carbon dioxide seeps. ICES J Mar Sci 73:897–909. https://doi.org/10.1093/icesjms/fsv194

Sunday JM, Fabricius KE, Kroeker KJ, Anderson KM, Brown NE, Barry JP, Connell SD, Dupont S, Gaylord B, Hall-Spencer JM, Klinger T, Milazzo M, Munday PL, Russell BD, Sanford E, Thiyagarajan V, Vaughan MLH, Widdicombe S, Harley CDG (2017) Ocean acidification can mediate biodiversity shifts by changing biogenic habitat. Nat Clim Chang 7:81–85. https://doi.org/10.1038/nclimate3161

Sweet MJ, Croquer A, Bythell JC (2010) Bacterial assemblages differ between compartments within the coral holobiont. Coral Reefs 30:39–52. https://doi.org/10.1007/s00338-010-0695-1

Tanaka S, Suenaga H, Kubota K (2019) Investigation of underground and seabed gas discharge cases in Japan as natural analog of CO2 leakages in carbon dioxide capture and storage

Tanvet C, Benzoni F, Peignon C, Thouzeau G, Rodolfo-Metalpa R (2022) High coral recruitment despite coralline algal loss under extreme environmental conditions. Front Mar Sci:9, 837877

Tanvet C, Camp EF, Sutton J, Houlbrèque F, Thouzeau G, Rodolfo-Metalpa R (2023) Corals adapted to extreme and fluctuating seawater pH increase calcification rates and have unique symbiont communities. Ecol Evol 13:e10099. https://doi.org/10.1002/ece3.10099

Tarasov VG, Gebruk AV, Shulkin VM, Kamenev GM, Fadeev VI, Kosmynin VN, Malakhov VV, Starynin DA, Obzhirov AI (1999) Effect of shallow-water hydrothermal venting on the biota of Matupi Harbour (Rabaul Caldera, New Britain Island, Papua New Guinea). Cont Shelf Res 19:79–116. https://doi.org/10.1016/S0278-4343(98)00073-9

Teixidó N, Gambi MC, Parravacini V, Kroeker K, Micheli F, Villéger S, Ballesteros E (2018) Functional biodiversity loss along natural CO 2 gradients. Nat Commun 9:5149. https://doi.org/10.1038/s41467-018-07592-1

Teixidó N, Caroselli E, Alliouane S, Ceccarelli C, Comeau S, Gattuso J-P, Fici P, Micheli F, Mirasole A, Monismith SG, Munari M, Palumbi SR, Sheets E, Urbini L, De Vittor C, Goffredo S, Gambi MC (2020) Ocean acidification causes variable trait-shifts in a coral species. Glob Change Biol 26:6813–6830. https://doi.org/10.1111/gcb.15372

Torda G, Donelson JM, Aranda M, Barshis DJ, Bay L, Berumen ML, Bourne DG, Cantin N, Foret S, Matz M, Miller DJ, Moya A, Putnam HM, Ravasi T, van Oppen MJH, Thurber RV, Vidal-Dupiol J, Voolstra CR, Watson S-A, Whitelaw E, Willis BL, Munday PL (2017) Rapid adaptive responses to climate change in corals. Nat Clim Chang 7:627–636. https://doi.org/10.1038/nclimate3374

Tortolero-Langarica J, García-Malo C, Calderon-Aguilera L, Norzagaray-Lopez C, Balart E, Paz-García D, Cabral-Tena R, Carricart-Ganivet J (2024) Calcification and growth of the reef coral Porites panamensis in a shallow hydrothermal field in the Gulf of California. Mexico Reg Environ CHANGE 24. https://doi.org/10.1007/s10113-024-02311-7

Tribble G, Randall R (1986) A description of the high-latitude shallow water coral communities of Miyake-jima, Japan. Coral Reefs 4:151–159

van Woesik R, Sakai K, Ganase A, Loya Y (2011) Revisiting the winners and the losers a decade after coral bleaching. Mar Ecol Prog Ser 434:67–76. https://doi.org/10.3354/meps09203

Vaquer-Sunyer R, Duarte CM (2008) Thresholds of hypoxia for marine biodiversity. Proc Natl Acad Sci 105:15452–15457. https://doi.org/10.1073/pnas.0803833105

Vergés A, McCosker E, Mayer-Pinto M, Coleman MA, Wernberg T, Ainsworth T, Steinberg PD (2019) Tropicalisation of temperate reefs: implications for ecosystem functions and management actions. Funct Ecol 33:1000–1013. https://doi.org/10.1111/1365-2435.13310

Veron JEN (1992) Conservation of biodiversity: a critical time for the hermatypic corals of Japan. Coral Reefs 11:13–21. https://doi.org/10.1007/BF00291930

Wada S, Agostini S, Harvey BP, Omori Y, Hall-Spencer JM (2021) Ocean acidification increases phytobenthic carbon fixation and export in a warm-temperate system. Estuar Coast Shelf Sci 250:107113. https://doi.org/10.1016/j.ecss.2020.107113

Wall M, Fietzke J, Schmidt GM, Fink A, Hofmann LC, de Beer D, Fabricius KE (2016) Internal pH regulation facilitates in situ long--term acclimation of massive corals to end-of-century carbon dioxide conditions. Sci Rep 6:30688. https://doi.org/10.1038/srep30688

Wee HB, Kurihara H, Reimer JD (2019) Reduced Symbiodiniaceae diversity in Palythoa tuberculosa at a heavily acidified coral reef. Coral Reefs 38:311–319

Witkowski CR, Agostini S, Harvey BP, van der Meer MTJ, Sinninghe Damste JS, Schouten S (2019) Validation of carbon isotope fractionation in algal lipids as a PCO_2 proxy using a natural CO_2 seep (Shikine Island, Japan). Biogeosci Discuss 1–18. https://doi.org/10.5194/bg-2019-158

Yamano H, Sugihara K, Nomura K (2011) Rapid poleward range expansion of tropical reef corals in response to rising sea surface temperatures. Geophys Res Lett 38:L04601. https://doi.org/10.1029/2010GL046474

Yara Y, Vogt M, Fujii M, Yamano H, Hauri C, Steinacher M, Gruber N, Yamanaka Y (2012) Ocean acidification limits temperature-induced poleward expansion of coral habitats around Japan. Biogeosci Discuss 9:7165–7196. https://doi.org/10.5194/bgd-9-7165-2012

Yara Y, Yamano H, Steinacher M, Fujii M, Vogt M, Gruber N, Yamanaka Y (2016) Potential future coral habitats around Japan depend strongly on anthropogenic CO_2 emissions. In: Nakano S, Yahara T, Nakashizuka T (eds) Aquatic biodiversity conservation and ecosystem services, ecological research monographs. Springer, Singapore, pp 41–56

Genomic Responses of Fishes to Ocean Acidification: Insights from Volcanic CO_2 Seeps

8

Michael Izumiyama, Roger Huerlimann, Erina Kawai, Jeffrey Jolly, Ben P. Harvey, Shigeki Wada, Sylvain Agostini, and Timothy Ravasi

Abstract

Extreme and marginal reef environments provide valuable opportunities to examine the effects of altered seawater parameters that resemble predicted end-of-century ocean conditions under climate change. Volcanic CO_2 seeps, in particular, have proven useful for studying the effects of ocean acidification (OA) in natural settings. In this chapter, we present a case study from a volcanic CO_2 seep at Shikine Island, Japan, investigating the genomic response of the neon damselfish (*Pomacentrus coelestis*) to OA conditions. We examined transcriptomic profiles from brain and gill tissues collected across two seasons (July and November) and compared them to samples from multiple control sites to assess consistent patterns of gene expression under elevated CO_2. Our results found that genes involved in blood oxygen regulation, immune response, and circadian rhythm were differentially expressed across seasons and tissues, suggesting their key role in the response to OA in *P. coelestis*. We also detected seasonal shifts in gene expression patterns, highlighting the complexity of the physiological responses to OA in natural environments. Many of the differentially expressed genes align with findings from previous laboratory and field studies, reinforcing the value of volcanic CO_2 seeps for OA research. Finally, we discuss the challenges of working in natural seep environments, including variability across control sites and seasons, and emphasize the future potential of CO_2 seeps for advancing our understanding of fish responses to ocean acidification.

Keywords

Transcriptomics · Shikine Island · Differential gene expression · *Pomacentrus coelestis* · Adaptation · Climate change

8.1 Introduction

As established in previous chapters, unique sites such as volcanic CO_2 seeps provide valuable opportunities to study the effects of elevated partial pressure of carbon dioxide (pCO_2) on marine organisms and allow us to examine the potential impacts of ocean acidification (OA) on marine life in their natural environment. This chapter focuses on molecular responses to OA in fishes inhabiting these environments. We

M. Izumiyama · E. Kawai · J. Jolly · T. Ravasi (✉)
Marine Climate Change Unit, Okinawa Institute of Science and Technology (OIST), Onna-son, Okinawa, Japan

ICONA International CO2 Natural Analogues Network, Shimoda, Japan
e-mail: timothy.ravasi@oist.jp

R. Huerlimann
Marine Climate Change Unit, Okinawa Institute of Science and Technology (OIST), Onna-son, Okinawa, Japan

B. P. Harvey
ICONA International CO2 Natural Analogues Network, Shimoda, Japan

Shimoda Marine Research Center, University of Tsukuba, Shizuoka, Japan

S. Wada
ICONA International CO2 Natural Analogues Network, Shimoda, Japan

Shimoda Marine Research Center, University of Tsukuba, Shizuoka, Japan

Seto Inland Sea Carbon-neutral Research Center, Hiroshima University, Higashi Hiroshima, Japan

S. Agostini
ICONA International CO2 Natural Analogues Network, Shimoda, Japan

Shimoda Marine Research Center, University of Tsukuba, Shizuoka, Japan

ENTROPIE, IRD, Université de la Réunion, IFREMER, Nouméa, New Caledonia

M. O. Soares et al. (eds.), *Extreme and Marginal Coral Reefs and Communities*, Coral Reefs of the World 24,
https://doi.org/10.1007/978-3-032-28035-0_8

also present a case study from a volcanic CO_2 seep in Japan, highlighting both the insights and challenges of using such sites to investigate the molecular effects of OA on fishes.

Fishes tightly regulate acid-base balance to maintain physiological function and can tolerate high pCO_2, as seen in aquaculture (Ellis et al. 2017). However, OA has adverse effects on fishes, including changes in the energy budget and food availability (see Moreira et al. 2022; Yoon et al. 2024), and various physiological effects have been documented (see Esbaugh 2018).

Various studies have examined the effects of OA on various organisms in laboratory experiments (Clements and Hunt 2015; Kroeker et al. 2010, 2013; Nagelkerken and Munday 2016). Fishes raised under high CO_2 have exhibited multiple effects, including an inability to distinguish predator olfactory cues (Dixson et al. 2010; Munday et al. 2013; Sundin and Jutfelt 2015), a disruption in the homing ability of larval fishes caused by the inability to distinguish olfactory cues (Munday et al. 2009), a reduction in visual response to predators (Ferrari et al. 2012), and impaired hearing ability (Radford et al. 2021; Simpson et al. 2011). Additionally, the cognitive abilities of fishes under OA have been altered, including their inability to identify predators (Chivers et al. 2014) and behavioral lateralization (preference for right or left) (Domenici et al. 2012, 2014). However, numerous other studies have shown minimal or no effect of increased CO_2 on fish behavior (Clark et al. 2020; Jutfelt and Hedgärde 2015; Sundin et al. 2017, 2019). The varied responses of fishes to elevated CO_2 may be affected by factors such as life stages, taxonomic groups, habitat, or experimental design (Clements and Hunt 2015; Munday et al. 2020). Molecular studies can complement previous behavior studies, enriching our understanding of how fishes respond to OA.

As genes are transcribed, they can provide a snapshot of which cellular processes are active. The transcribed genes can provide insight into the organism's physiological function (Huang et al. 2002). By exposing fishes to a stressor such as OA and examining the transcriptomics, we can understand how the organism responds to the conditions and potential pathways that may be affected, which can inform conservation (Connon et al. 2018). Numerous studies investigating the molecular responses of fishes to OA have been carried out in laboratory settings and have varied in terms of species studied and duration of exposure. Studies exploring the effects of OA on fishes have spanned from short-term investigations lasting several days or weeks (Wang et al. 2021) to long-term, including developmental or transgenerational experiments (Schunter et al. 2016, 2018). These studies have played a pivotal role in expanding our knowledge of OA effects on fishes, bridging behavioral and physiological responses with potential genetic mechanisms. However, many initial laboratory studies employed consistent pCO_2 levels, representing predicted average pCO_2 under future OA scenarios, without considering daily variations in the wild (Hofmann et al. 2011; Shaw et al. 2012). Indeed, studies have indicated that the effects of OA on fishes may be less substantial when accounting for daily variations in pCO_2 levels, which are similar to natural conditions expected on coral reefs (Schunter et al. 2021).

While numerous studies have shown the adverse effects of OA, some fish may be more resilient than others, and fishes may also be capable of adapting to OA conditions (Schunter et al. 2016, 2018). Research on the impact of OA on marine organisms has been crucial in emphasizing concerns about anthropogenic CO_2 emissions and the future of the ocean.

Indeed, laboratory studies have been essential for understanding the molecular responses of fishes to OA in a controlled setting, replicating the complexities and dynamics of natural ecosystems remains challenging. Most experiments are short-term and do not replicate daily fluctuations in ocean chemistry. Unaccounted variables, such as seasonal upwelling in intake water, could affect laboratory experiments (Galloway et al. 2020). Furthermore, laboratory experiments cannot capture the dynamic complexity of ecosystems and species interactions, nor the long-term adaptive responses of species to OA conditions across generations.

This limitation makes it difficult to fully understand how fishes and their habitats will respond to OA. Sites, such as CO_2 seeps, provide a valuable opportunity to examine how fishes may adapt to OA conditions within an environment where intricate interactions between species and ecosystems are present. However, research on seeps exploring the molecular responses of fishes to OA has been limited (Kang et al. 2022; Petit-Marty et al. 2021; Suresh et al. 2023). These studies have shown diverse responses to OA in fishes, with some similarities to findings from laboratory experiments, which will be discussed later in this chapter.

Here we present a case study using the volcanic CO_2 seep of Shikine Island, Japan, and examine the genomic response of the neon damselfish (*Pomacentrus coelestis*) to conditions of OA. Our study examined how *P. coelestis* in the seep responded to elevated CO_2 compared to surrounding reefs across seasons and demonstrates how natural analogues can be used to understand the molecular mechanisms for adaptation to OA. We also highlight the difficulties associated with natural systems and how interpreting molecular signals can be challenging. Our study highlights how natural analogues can be valuable in understanding molecular response to OA and provide insight into the adaptive potential of fishes to the future ocean.

8.2 Methods

8.2.1 Study Site and Sample Collection

For our study, we collected neon damselfish (*Pomacentrus coelestis*) from the volcanic island of Shikine, situated east of the Izu Peninsula in Japan (34°19′9″ N, 139°12′18″ E). Shikine island hosts numerous shallow water CO_2 seeps, many of which have been well-characterized (Agostini et al. 2015, 2018). Specifically, we focused on the 900 μatm pCO_2 seep at Mikama Bay (from here on referred to as seep) and the nearby 300 μatm pCO_2 reef as control (from here on referred to as control) described by Agostini et al. (2018) (Fig. 8.1). Additionally, we collected *P. coelestis* from two sites as alternative controls located north of the island, Nakano Ura and Ohura (Fig. 8.1).

We conducted two sampling surveys in 2021, from July sixth to 12th and November second to third, during which we collected a total of 68 *P. coelestis* for transcriptomic analysis. In July, we collected 15 fish from the seep site, 14 from the control site, 12 from Nakano Ura, and nine from Ohura. In November, we collected nine fish each from the seep and control sites. All samples were collected in accordance with the Animal Care and Use Program at the Okinawa Institute of Science and Technology Graduate University (OIST). Fishes were collected by SCUBA diving with hand nets and kept alive in holding nets underwater during sampling. Once sampling was completed, fishes were transferred into 20 L buckets filled with water from the sampling site, with approximately five fish per bucket, containing around 10~15 L of water per bucket to reduce the change in water parameters from respiration of *P. coelestis* during transportation. Fishes were taken to the Shikine Marine Station (University of Tsukuba) on the island, where they were immediately dissected. Euthanasia was performed via cervical dislocation, and the brain and gills were removed and transferred to a 1.5 mL tube and snap-frozen in liquid nitrogen. All samples were kept in liquid nitrogen and transferred to a −80 °C freezer until further analysis.

8.2.2 RNA Extraction and Sequencing

Total RNA was extracted from the brain and gills of 67 *P. coelestis* using the Maxwell® RSC simply RNA Tissue Kit and Maxwell® RSC Instrument (Promega, USA) following the manufacturer's protocol. Total RNA concentration was quantified using the **Qubit flex benchtop fluorometer**, and extraction quality was verified using the Agilent TapeStation 4200. The sequencing library was prepared using the NEBNext Ultra II Directional RNA Library Prep Kit for Illumina (New England Biolabs) and was sequenced on a shared lane of Illumina NovaSeq6000 S4 flowcells with 150 bp PE reads. Library preparation and sequencing were conducted by the Sequencing Section (SQC) at OIST.

8.2.3 Differential Gene Expression Analysis

The RNA-sequencing reads were quality checked using FastQC (Andrews 2010) and MultiQC (Ewels et al. 2016), and trimmed using Trimmomatic v0.39 (Bolger et al. 2014) with the parameters: "TruSeq3-PE.fa:2:30:10:8:keep BothReads LEADING:3 TRAILING:3 MINLEN:36.". Quality trimmed reads were mapped to the *P. coelestis genome* (*Izumiyama, unpublished dataset*) using HISAT2 v2.2.1 (Kim et al. 2019), the resulting SAM files were converted to BAM files using SAMtools v1.10 (Li et al. 2009). BAM files and the *P. coelestis* gene annotation (*Izumiyama, unpublished dataset*) were used as input for StringTie v2.1.4 (Pertea et al. 2016) to quantify expression levels. Gene-specific raw read counts were generated from the StringTie

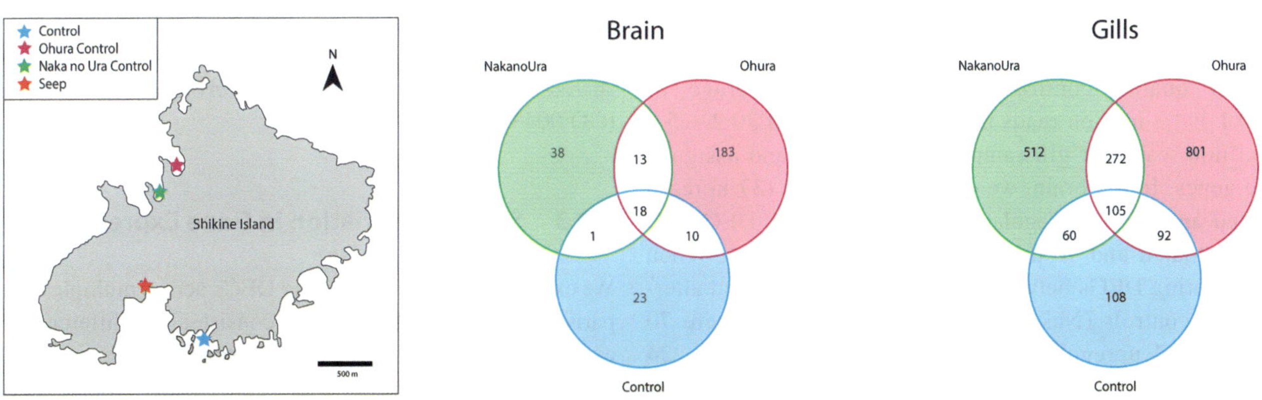

Fig. 8.1 Map of Shikine Island showing the sample collection sites (left panel), Venn diagram of differentially expressed genes (DEGs) in the brain (center panel), and Venn diagram of DEGs in the gills (right panel) of *Pomacentrus coelestis* collected in July. The DEGs are shown in comparison between the seep site and the control sites (Control, Nakano Ura, and Ohura)

output using the Python script prepDE.py3 (Pertea et al. 2016). One gill sample was excluded from downstream analysis due to failure to pass quality control. The remaining analysis was conducted in R v4.4.2 (R Core Team 2023). Pairwise differential gene expression analysis was conducted using DESeq2 (Love et al. 2014) to identify differentially expressed genes (DEGs). Functional enrichment analysis was performed with the clusterProfiler R package (Yu 2024) to identify significantly over-represented Gene Ontology (GO) terms. The Benjamini-Hochberg false discovery rate (FDR) method was used to adjust *p*-values and GO terms with an adjusted *p*-value <0.05 were considered significantly enriched.

Each differential gene expression analysis was conducted by comparing the seep samples to either control samples (July and November) or alternative control sites (Nakano Ura and Ohura to the July seep samples). Raw gene count tables were filtered prior to downstream analysis to reduce noise from lowly expressed and uninformative genes. For each comparison, count matrices were first filtered to remove genes with fewer than 10 reads in the smallest sample group (ranging from nine to 14 samples). Second, genes with low variability (variance less than 1) across samples were removed. Finally, genes with zero counts in more than 50% of samples were excluded. We identified the final number of differentially expressed genes (DEGs) for each tissue with an FDR < 0.05. Genes were categorized as upregulated in the seep with a log2 fold change >1 and downregulated with a log2 fold change <-1.

8.3 Results

8.3.1 Tissues Specific Differentially Expressed Genes of *Pomacentrus coelestis* at the CO_2 Seep

For the transcriptomic analysis of *Pomacentrus coelestis*, we sequenced an average of 42.1 ± 5.7 million reads per brain sample and 41.7 ± 4.6 million reads per gill sample. After quality control and mapping, an average of 30.1 ± 4.3 million reads per brain sample and 39.2 ± 5.0 million reads per gill sample were retained and assigned to genes. In the brain, we detected 52 DEGs (37 upregulated and 15 downregulated) in July and 171 DEGs (75 upregulated and 96 downregulated) in November. When comparing DEGs between seep samples in July and alternative controls (Nakano Ura and Ohura), there were 70 DEGs (44 upregulated and 26 downregulated) and 224 DEGs (208 upregulated and 16 downregulated), respectively. In the gills, we detected 365 DEGs (255 upregulated and 110 downregulated) in July and 730 DEGs (376 upregulated and 354 downregulated). When comparing gill DEGs of seep and alternative controls in July, we found 949 DEGs (368 upregulated and 581 downregulated) against the Nakano Ura control site, and 1270 DEGs (558 upregulated and 712 downregulated) against the Ohura control site.

8.3.2 Differentially Expressed Pathways

The GO enrichment analysis for each tissue-specific and site comparison revealed the following number of enriched GO categories. In the brain, comparing seep to control conditions by season, 24 GO terms were enriched in July and 27 in November. Two GO categories were shared across both seasons: ATP-dependent protein folding chaperone (GO:0140662) and blood microparticle (GO:0072562). When comparing DEGs between alternative control sites in July (i.e., seep vs. Nakano Ura and seep vs. Ohura), 24 GO terms were enriched relative to Nakano Ura and 59 relative to Ohura. Nine GO categories were shared across both comparisons, suggesting these functions are consistently enriched in July at the seep site. These GO categories include: peptidyl-proline hydroxylation to 4-hydroxy-L-proline (GO:0018401), blood microparticle (GO:0072562), L-ascorbic acid binding (GO:0031418), regulation of neuron apoptotic process (GO:0043523), ferrous iron binding (GO:0008198), cellular response to leukemia inhibitory factor (GO:1990830), caspase activation (GO:0006919, obsolete), regulation of cell population proliferation (GO:0042127), and cellular response to hypoxia (GO:0071456).

The number of enriched GO categories in the gills was seven in July and 32 in November, with no overlapping GO terms shared across both seasons. When comparing DEGs between alternative control sites in July, 40 and 60 GO terms were enriched in Nakano Ura and Ohura, respectively. None of the GO terms were common across gill tissue comparisons. However, the control and Nakano Ura comparison shared two GO terms: ferrous iron binding (GO:0008198), and peptidyl-proline hydroxylation to 4-hydroxy-L-proline (GO:0018401), and the control and Ohura comparison shared one GO term: regulation of neuron apoptotic process (GO:0043523).

8.3.3 Seasonal Variation in Gene Expression

We examined the overlap among DEGs across multiple comparisons to identify genes consistently differentially expressed in *P. coelestis* at the seep site. Notably, one gene, endothelin 2 (*edn2*), was consistently upregulated ($log_2FC > 1$) in both the brain and gill samples across all control comparisons (Control, Nakano Ura, and Ohura), and comparison across both seasons (July and November) between the seep and control site (Fig. 8.2).

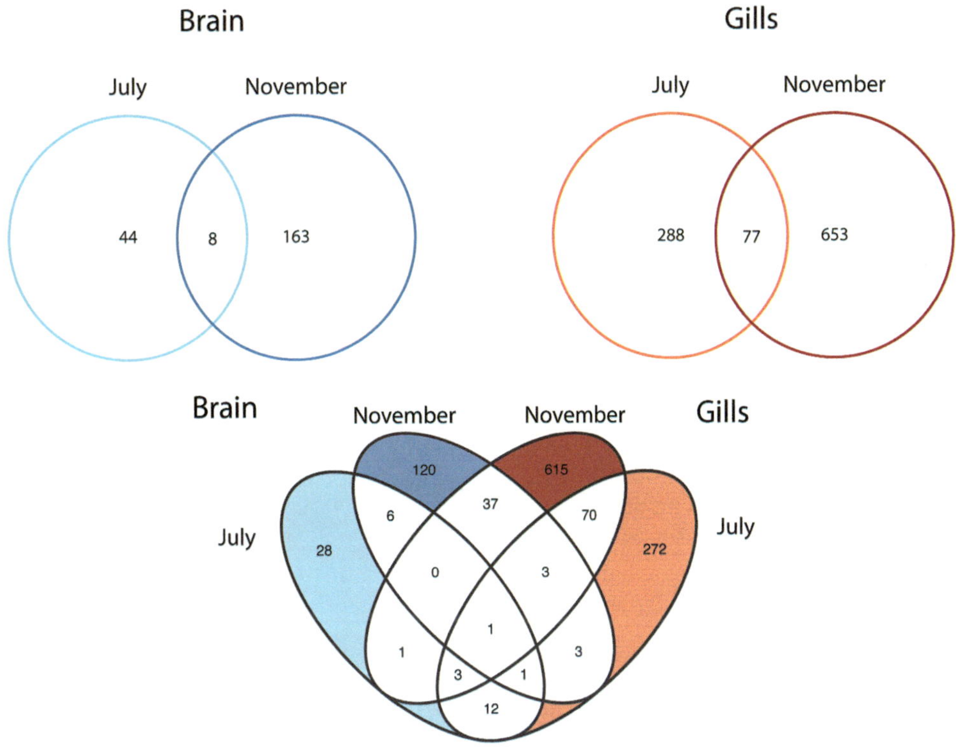

Fig. 8.2 Venn diagrams of differentially expressed genes (DEGs) in *Pomacentrus coelestis* across seasons. The top panels show the number of DEGs in July and November between the seep and control sites in the brain (top left) and gills (top right). The bottom panel compares all DEGs across tissues and seasons

When comparing the DEGs across seasons (July and November) of *P. coelestis* in the seep compared to the control site, we found eight overlapping genes across both seasons in the brain and 77 in the gills (Fig. 8.2). In the brain, four of the eight overlapping genes were isoforms of the heat shock 70 kDa protein 1 (*hspa1a*) and heat shock 70 kDa protein 1-like, all of which were consistently downregulated ($log_2FC < -1.6$) across seasons (also in the July gill samples, $log_2FC < -1$), and two of the *hspa1a* were differentially expressed in the alternative controls but were upregulated ($log_2FC > 1.2$). The hypermethylated in cancer one protein isoform (*hic1*) was upregulated in all brain samples (including alternative controls $log_2FC > 1.2$). Two uncharacterized protein-coding genes were also differentially expressed in the brain across seasons; however, they exhibited opposite expression patterns between July and November (i.e., one was upregulated and the other downregulated in July and vice versa in November).

In the gills of the 77 genes that overlapped across the seasons, 27 were also differentially expressed in the alternative controls. Twenty-five of the 27 genes were upregulated across all samples, and two were upregulated in July but downregulated in November. These genes were associated with processes such as the formation of vascular tissue, such as *vegfa* and *dusp1* involved in blood oxygen regulation and hypoxia response, and various genes related to the immune system, such as interleukin and its receptor *il1r-like* and *il8* or chemokine receptors *cxcr-7*.

When examining DEGs common between tissues within the same season (i.e., between the brain and gills in July and November), 17 DEGs overlapped in July and 41 overlapped in November. In July, many isoforms of the egl-9 family hypoxia inducible factor 3 (*egln3*) and hemoglobin subunit alpha-D like (*hba-d*) overlapped. In November, several CYP genes and genes involved in circadian rhythm (*arntl2, cip*) overlapped.

When examining all differentially expressed genes (DEGs) across the samples, many of the previously mentioned gene groups also appeared as differentially expressed, though not always involving the same specific genes. These groups included various heat shock proteins (HSPs), cytochrome P450 (CYP) enzymes, immune-related genes such as chemokines, interleukins, and their receptors, as well as genes linked to vascular function and blood oxygen regulation, such as endothelin, egl nine homolog, and dual specificity phosphatases (DUSPs). Additionally, circadian rhythm genes, including CLOCK-interacting pacemaker (CIP) and period circadian protein (PER), were differentially expressed (Fig. 8.3). These genes were identified as differentially expressed in at least one of the comparisons, and their expression profiles were visualized using mean z-scores. In differential gene expression analysis, the z-score represents the standardized difference from the mean expression level, accounting for variability across samples. This provides a useful means of visualizing expression patterns (Fig. 8.3).

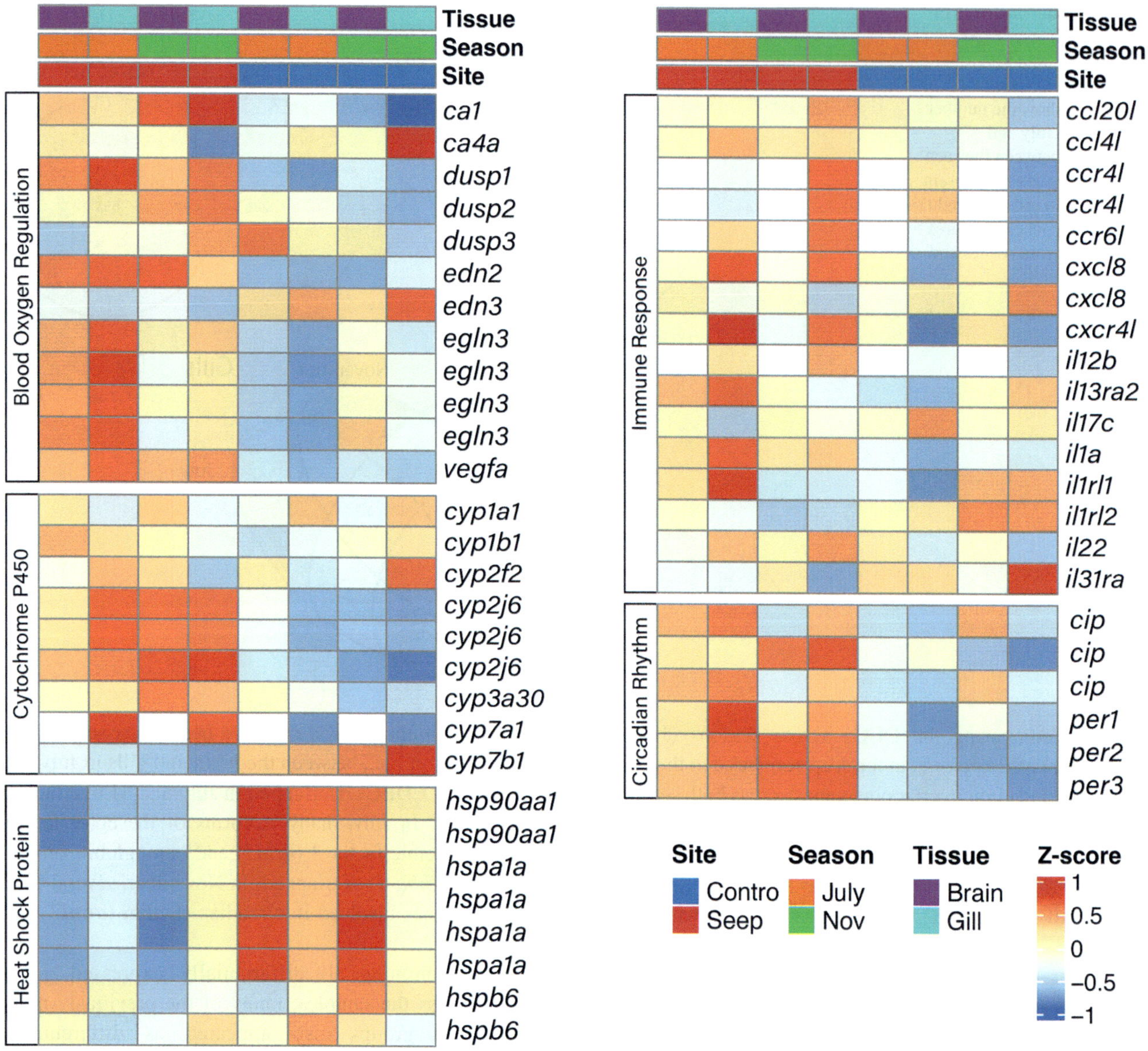

Fig. 8.3 Heatmap of the mean *z*-scores of selected genes of interest. Rows represent individual genes, grouped by function or category. Columns correspond to tissue groups arranged by site (Control, Seep), season (July, November), and tissue type (brain, gill)

In addition to these gene groups, we identified a subset of differentially expressed genes that do not fit neatly into groups but are noteworthy, such as period circadian protein homolog (*per 1*, *per 2*, and *per 3*) and carbonic anhydrase (*ca1* and *ca4a*). These genes are of particular interest because previous studies have implicated them in responses to ocean acidification, suggesting they may play a role in the adaptation of *P. coelestis* to elevated CO_2 conditions at the seep site. The accessibility of the seep site, combined with the availability of nearby facilities, enabled repeated sampling and the inclusion of multiple surrounding control sites. This approach allowed us to assess consistent patterns of gene expression in individuals inhabiting the seep environment, enabling the identification of unique transcriptional signatures that may reflect potential adaptive mechanisms in *P. coelestis*.

8.4 Discussion

Our study leveraged the unique opportunity volcanic CO_2 seeps provide to examine the molecular response of fishes to ocean acidification (OA) conditions predicted for the future.

8.4.1 The Role of Endothelin

One of the major challenges facing fishes in an elevated CO_2 environment is maintaining acid-base balance in the blood and tissue to prevent acidosis. Acid-base balance is a tightly controlled process in fish and is crucial for survival (Brauner et al. 2019). Endothelin peptides and their receptors are closely involved in regulating blood pressure and Na homeostasis (Kohan et al. 2011), and endothelin 1 (*end1*) and its receptors have been shown to be a key component of acid-base regulation in zebrafish (Guh et al. 2014). In our study, endothelin 2 (*end2*) was the only gene consistently differentially expressed across seasons, tissues, and comparisons across different controls in July. Additionally, in the gills, we found endothelin 3 (*end3*) downregulated across the season, indicating that endothelin likely plays a major role in acid-base regulation in *P. coelestis*.

To further support the role of endothelin in fishes inhabiting CO_2 seeps, a study by Kang et al. (2022) found that endothelin converting enzyme 1 (*ece1*) was differentially expressed in the spiny chromis (*Acanthochromis polyacanthus*) living within a CO_2 seep in Papua New Guinea compared to individuals from nearby control sites. Interestingly, this study took advantage of a storm during their expedition, which temporarily flushed out the acidified water inside the seep and returned it to near-control conditions. They collected *A. polyacanthus* from the seep during the storm and found that *ece1* was also differentially expressed between fish during and before the storm (Kang et al. 2022). This evidence aligns with our findings and supports our proposition that endothelin likely plays a role in adaptation to conditions of OA. Further studies are needed to validate the role of endothelin in the physiological response of fishes to OA.

8.4.2 Gills Specific Gene Expression Under OA

As mentioned above, regulating acid-base balance is a major challenge for fishes under OA. The gills serve multiple critical functions, including gas exchange, regulation of osmotic and ion balance, and excretion of waste (Evans et al. 2005; Kovac and Goss 2024). In our study, many of the genes differentially expressed in the gills involved ion regulation and oxygen transport, both of which are essential processes for maintaining homeostasis under OA. The expression of these genes in the gills is consistent with the role of this organ as the primary site of interaction with the external environment.

Genes associated with vascular tissue formation in the gills, such as vascular endothelial growth factor A (*vegfa*), which can enhance oxygen transport (Domingues and Fantin 2021), were differentially expressed across all seasons. Similarly, a study on the Japanese flounder (*Paralichthys olivaceus*) found that under hypoxia stress, *vegfa* signaling pathways were upregulated (Liu et al. 2024), suggesting a conserved role of *vegfa* in oxygen transport under environmental stress.

Dual specificity protein phosphatase 1 (*dusp1*) was another gene constantly expressed in the gills. Studies have shown that DUSP is involved in thermal tolerance and maintaining mitochondrial integrity in zebrafish gills (Wang et al. 2023), in response to hypoxia in pufferfish gills (Zhang et al. 2023), and is involved in response to temperature and immune stress in flatfish (Han et al. 2022).

Some interesting genes differentially expressed in the gills include carbonic anhydrase (*ca4a*), which catalyzes the conversion of CO_2 to H^+ and HCO_3^- and the reverse, and is important to CO_2 transportation (Brauner et al. 2019). Carbonic anhydrase (CA) has been shown to be differentially expressed in experiments of elevated CO_2 and acid-base regulation (Georgalis et al. 2006) and found to be differentially expressed in two species of damselfish from a natural analogue in Papua New Guinea (Kang et al. 2022) and gobies found at a CO_2 seep in Italy (Suresh et al. 2023). Studies have demonstrated that the presence of CA in the tissue and its absence from the gills may enhance O_2 delivery through the Bohr-Root effect (Rummer and Brauner 2011, 2015). In our study, *ca4a* was down-regulated at the gills in November, which aligns with the proposed mechanism for increasing O_2 delivery by the Bohr-Root effect in *P. coelestis* under the assumption that CA is present at some level in the tissue, but a downregulation at the gills. Similar expression patterns of ca4a were seen in Atlantic cod larvae in laboratory experiments (Mittermayer et al. 2019).

Many genes involved in the immune response, including interleukins, chemokines, and their receptors, which play key roles in immune defense (Zou et al. 2015), were differentially expressed in the gills. These genes were commonly expressed across the seasons in the gills, and many were also differentially expressed in the brain. Several studies have reported immunological responses to OA in fishes in laboratory studies (Bresolin de Souza et al. 2014; Cohen-Rengifo et al. 2022; Huth and Place 2016; Machado et al. 2020), as well as in wild populations inside CO_2 seeps in Papua New Guinea, where five of the six species examined exhibited differential expression of immune response genes (Kang et al. 2022).

A potential explanation for the upregulation of immune genes in fishes at the Shikine CO_2 seep may lie in ecosystem-level changes driven by OA. The shift from coral to turf algae-dominated habitat has been reported at the seep of Shikine (Agostini et al. 2018), and a recent study found that under warming conditions, the senescence of turf algae has been shown to promote the spread of coral diseases (Heitzman et al. 2023). Similar processes may be occurring at Shikine, where the proliferation and decay of turf algae could increase pathogen loads, potentially contributing to the upregulation of immune genes in resident fishes. Future studies could examine whether fishes from CO_2 seeps at

Shikine are more susceptible to infections or disease, further validating our findings.

8.4.3 Heat Shock Protein (HSP)

In the brain, among the commonly expressed genes across seasons, four of the eight overlapping genes were isoforms of heat shock 70 kDa protein 1 (*hspa1a*). These *hsp1a* and *hsp1a-like* genes were down-regulated in the brain, and *hsp1a* was down-regulated in the gills in July. Heat shock protein (HSP) beta-9 (*hspb6*) and HSP 90-alpha 1 (*hsp90aa1*) were also down-regulated across both brain and gill tissues. Hsp70 and Hsp90 have been utilized as biomarkers for response to immune and stress responses (Taylor et al. 2013), and HSP has been shown to be affected by OA in limpets and coral (Moya et al. 2015; Wang et al. 2018). In fishes, HSP has been shown to be differentially expressed in several tissues under ocean warming (Moore et al. 2024). A review by Yusof et al. (2022) proposed HSP70 as a biomarker for ecological risk assessments in the Antarctic to detect OA warming and salinity stress. Interestingly, most studies have shown up-regulation of HSPs in response to stressors, yet in our study, HSP was downregulated in all our samples. This could be associated with HSP's high-energy and amino acid requirements (Krebs and Feder 1997). In Antarctic plunderfish (*Harpagifer antarcticus*), qPCR from liver, muscle, and digestive gland tissues found downregulation of the Hsp70 during heat shock experiments (Clark et al. 2008). Indeed, HSP seems to be a good marker for detecting OA stress in *P. coelestis*, as it was differentially expressed in both the brain and gills of fishes exposed to OA.

8.4.4 Seasonal Patterns of Gene Expression

In the *P. coelestis* collected in July, several genes belonging to the **egl-9 family hypoxia-inducible factor 3** (*egln3*), were found to be differentially expressed in the brain and gills, which are involved in hypoxia signaling and act as oxygen sensors (Berra et al. 2006). Fishes collected in November had several genes involved in the circadian rhythm (CR) pathway that were differentially expressed. Circadian rhythm genes are involved in many processes across eukaryotes (Dunlap and Loros 2016). Many osmoregulation processes are governed by the CR (Dmitriev and Mangel 2000; Peterson and Gilmore 1988). Elevated CO_2 experiments in laboratories have shown expression changes in the circadian rhythm (CR) pathways (Schunter et al. 2016, 2021), as well as in fishes found at the natural analogue in Papua New Guinea (Kang et al. 2022). Our study found many CR genes differentially expressed in the gills, including period circadian protein homologs (*per1, 2*), CLOCK-interacting pacemaker (*cipc*). These genes mentioned above align with gene expression patterns from previous studies in both laboratories and natural CO_2 seeps and illustrate how many of these genes may be involved in responding to environmental stressors, including OA. Shikine is near the northern distribution of *P. coelestis* and has larger seasonal temperature differences than a tropical habitat. The fishes found in the seep may be using different pathways to cope with the conditions found inside the seep, which could explain seasonal differences in expression patterns.

Another group of genes that exhibited seasonal expression patterns belonged to the Cytochrome P450 (CYP) superfamily. In November, three copies of *cyp2J6* and one copy of *cyp3A30* were upregulated in the brain and gills of *P. coelestis*. Two of the *cyp2J6* copies were also upregulated in the gills in July. In addition to these CYP genes, several other members of the CYP family were differentially expressed across tissues and seasons, including *cyp1a1*, *cyp1b1*, *cyp2f2*, *cyp7a1*, and *cyp7b1*. The CYP family consists of many genes that encode P450 enzymes that catalyze monooxygenase reactions involving oxygen and play important roles in helping organisms adjust to environmental changes and stress (Uno et al. 2012).

Cytochrome P450 likely plays a role in responding to the seep's environmental conditions, helping protect cells from damage. An important question is whether this response reflects an immediate, plastic reaction to the stressor or a longer-term adaptive mechanism that has been selected for. A pioneering study on transgenerational acclimation to ocean warming in fishes found that a small number of genes were significantly differentially expressed, including cytochrome P450 2j2 (*cyp2j2*), but only in offspring whose parents had also been exposed to elevated temperatures (Veilleux et al. 2015). In this study, *cyp2j2* expression may provide a cytoprotective role only to transgenerational fishes, potentially contributing to physiological adaptation to warming conditions (Veilleux et al. 2015).

Similarly, fishes found in the Shikine seeps have likely experienced multiple generations of exposure to these conditions, and their molecular responses may reflect long-term adaptation to the unique environment inside the seep. Additional reciprocal transplant experiments, introducing control fish into the seep environment, could help determine whether these responses are immediate and plastic or the result of adaptive processes. Combined with population genetic studies within the island, could further clarify whether there is evidence of selection for traits that enhance tolerance to ocean acidification within the seep population, an area that remains intriguing and warrants further investigation.

The ability of fish to tolerate acidic conditions is not surprising, as seen in dense aquaculture settings (Ellis et al.

2017). However, there are elevated energetic costs associated with acid-base regulation in fishes (Lefevre 2019). Similar findings have been reported in the common triplefin (*Forsterygion lapillum*) found in the CO_2 seeps in New Zealand. Many of the genes differentially expressed were associated with maintaining pH homeostasis (Petit-Marty et al. 2021). Petit-Marty et al. (2021) also found genes involved in protein and energy production indicating an increase in metabolic cost associated with maintaining pH homeostasis. Previous work from the seep in New Zealand has found that bottom-up CO_2 enrichment causes higher food availability for *F. lapillum* (Doubleday et al. 2019), and *F. lapillum* behavior inside the seep allows them to gain better access to food compared to other fishes inside the seep (Nagelkerken et al. 2017; Petit-Marty et al. 2021), potentially offsetting the energetic cost associated with acid-base regulation.

Examining the energetic costs associated with adaptation to seep conditions is interesting. In our study, *P. coelestis* was one of the most abundant species inside the seep during the July sampling. Most individuals were reproductively active, as indicated by mature eggs and sperm, suggesting that *P. coelestis* inside the seep were likely obtaining sufficient calories to support reproduction despite the energetic challenges associated with OA.

8.5 Future Directions

Extreme and marginal reefs provide valuable natural settings for examining the effects of climate change, and CO_2 seeps have proven especially useful for studying how fishes respond to ocean acidification (OA). Our results on the molecular signature of acclimation to OA in *Pomacentrus coelestis* support previous findings from laboratory and natural analogue studies (Kang et al. 2022; Petit-Marty et al. 2021; Suresh et al. 2023).

One key advantage of using natural CO_2 seeps is their ability to capture the complexity and dynamics of natural ecosystems when studying OA effects. However, it is important to recognize that factors other than OA, such as seasonal temperature fluctuations and biological cycles like reproduction, also influence gene expression patterns inside the seep, as evidenced by our observed seasonal differences. Choosing appropriate controls is another challenge; in our study, the number of differentially expressed genes varied between control sites on the island. Local environmental conditions and population structure may have influenced these gene expression patterns and should be carefully considered in future studies investigating molecular adaptation mechanisms.

Finally, a critical question remains: are the molecular signals we detected the result of phenotypic plasticity in response to acidification, as seen in *Acanthochromis polyacanthus* from Papua New Guinea (Kang et al. 2022), or do they reflect transgenerational acclimation demonstrated in controlled tank experiments (Veilleux et al. 2015)? These are important factors when designing and interpreting experiments conducted at these natural sites.

Our study benefits from a large sample size, sampling across multiple seasons, and comparison across numerous control sites to account for natural variability, enabling us to detect generalized molecular responses to OA in *P. coelestis*. Gene expression responses to climate change vary across tissue types (Moore et al. 2024), and we identified several gene groups consistently expressed in both brain and gills, including heat shock proteins, genes involved in oxygen regulation, and immune-related genes.

Many opportunities remain to validate and build upon our findings to better understand the mechanisms underlying ocean acidification adaptation using CO_2 seeps. Our work highlights the value of CO_2 seeps in advancing knowledge of fish responses to OA. Seasonal variability will continue to influence gene expression patterns, especially at Shikine Island, which lies near the northern range limit of tropical species. With the expected increase in extreme weather events, such as marine heatwaves, Shikine offers a valuable natural laboratory to examine the combined effects of ocean acidification and temperature stress. This approach can enhance our understanding of multiple impacts of environmental change, creating significant opportunities for future research.

Acknowledgements We are grateful for the assistance provided by the Shikine-jima Fisherman's Union, the staff at the Shimoda Marine Research Center, University of Tsukuba, Sequencing Section and Scientific Computing and Data Analysis section of the Core Facilities at Okinawa Institute of Science and Technology Graduate University (OIST) and the continued support of members of the Marine Climate Change Unit (OIST). Research reported in this chapter was supported by funding from Japan Society for the Promotion of Science (JSPS) DC1 fellowship, OIST, and the International CO2 Natural Analogues (ICONA) network by the JSPS as core-to-core funding (Advanced Research Network, Grant Number: JPJSCCA20210006).

Data Accessibility Statement All transcriptomic sequencing reads generated in this study have been deposited in the DDBJ BioProject database under accession number PRJDB35466. Supporting data, including gene count matrices, differentially expressed genes, and gene ontology (GO) analyses, are openly accessible in the Figshare repository at https://doi.org/10.6084/m9.figshare.29253056.

References

Agostini S, Wada S, Kon K, Omori A, Kohtsuka H, Fujimura H, Tsuchiya Y, Sato T, Shinagawa H, Yamada Y, Inaba K (2015) Geochemistry of two shallow CO_2 seeps in Shikine Island (Japan) and their potential for ocean acidification research. Reg Stud Mar Sci 2:45–53. https://doi.org/10.1016/j.rsma.2015.07.004

Agostini S, Harvey BP, Wada S, Kon K, Milazzo M, Inaba K, Hall-Spencer JM (2018) Ocean acidification drives community shifts towards simplified non-calcified habitats in a subtropical-temperate transition zone. Sci Rep 8(1):11354. https://doi.org/10.1038/s41598-018-29251-7

Andrews S (2010) FastQC: a quality control tool for high throughput sequence data. http://www.bioinformatics.babraham.ac.uk/projects/fastqc/

Berra E, Ginouvès A, Pouysségur J (2006) The hypoxia-inducible-factor hydroxylases bring fresh air into hypoxia signalling. EMBO Rep 7(1):41–45. https://doi.org/10.1038/sj.embor.7400598

Bolger AM, Lohse M, Usadel B (2014) Trimmomatic: a flexible trimmer for illumina sequence data. Bioinformatics 30(15):2114–2120. https://doi.org/10.1093/bioinformatics/btu170

Brauner CJ, Shartau RB, Damsgaard C, Esbaugh AJ, Wilson RW, Grosell M (2019) 3—acid-base physiology and CO_2 homeostasis: regulation and compensation in response to elevated environmental CO_2. In: Grosell M, Munday PL, Farrell AP, Brauner CJ (eds) Fish physiology, vol 37. Academic Press, pp 69–132. https://doi.org/10.1016/bs.fp.2019.08.003

Bresolin de Souza K, Jutfelt F, Kling P, Förlin L, Sturve J (2014) Effects of increased CO_2 on fish gill and plasma proteome. PLoS One 9(7):e102901. https://doi.org/10.1371/journal.pone.0102901

Chivers DP, McCormick MI, Nilsson GE, Munday PL, Watson S-A, Meekan MG, Mitchell MD, Corkill KC, Ferrari MCO (2014) Impaired learning of predators and lower prey survival under elevated CO_2: a consequence of neurotransmitter interference. Glob Chang Biol 20(2):515–522. https://doi.org/10.1111/gcb.12291

Clark MS, Fraser KPP, Burns G, Peck LS (2008) The HSP70 heat shock response in the Antarctic fish *Harpagifer antarcticus*. Polar Biol 31(2):171–180. https://doi.org/10.1007/s00300-007-0344-5

Clark TD, Raby GD, Roche DG, Binning SA, Speers-Roesch B, Jutfelt F, Sundin J (2020) Ocean acidification does not impair the behaviour of coral reef fishes. Nature 577:370–375. https://doi.org/10.1038/s41586-019-1903-y

Clements JC, Hunt HL (2015) Marine animal behaviour in a high CO_2 ocean. Mar Ecol Prog Ser 536:259–279. https://doi.org/10.3354/meps11426

Cohen-Rengifo M, Danion M, Gonzalez A-A, Bégout M-L, Cormier A, Noël C, Cabon J, Vitré T, Mark FC, Mazurais D (2022) The extensive transgenerational transcriptomic effects of ocean acidification on the olfactory epithelium of a marine fish are associated with a better viral resistance. BMC Genomics 23(1):448. https://doi.org/10.1186/s12864-022-08647-w

Connon RE, Jeffries KM, Komoroske LM, Todgham AE, Fangue NA (2018) The utility of transcriptomics in fish conservation. J Exp Biol 221(2):jeb148833. https://doi.org/10.1242/jeb.148833

Dixson DL, Munday PL, Jones GP (2010) Ocean acidification disrupts the innate ability of fish to detect predator olfactory cues. Ecol Lett 13(1):68–75. https://doi.org/10.1111/j.1461-0248.2009.01400.x

Dmitriev AV, Mangel SC (2000) A circadian clock regulates the pH of the fish retina. J Physiol 522(1):77–82. https://doi.org/10.1111/j.1469-7793.2000.0077m.x

Domenici P, Allan B, McCormick MI, Munday PL (2012) Elevated carbon dioxide affects behavioural lateralization in a coral reef fish. Biol Lett 8(1):78–81. https://doi.org/10.1098/rsbl.2011.0591

Domenici P, Allan BJM, Watson S-A, McCormick MI, Munday PL (2014) Shifting from right to left: the combined effect of elevated CO_2 and temperature on behavioural lateralization in a coral reef fish. PLoS One 9(1):e87969. https://doi.org/10.1371/journal.pone.0087969

Domingues A, Fantin A (2021) Neuropilin 1 regulation of vascular permeability signaling. Biomolecules 11(5):666. https://www.mdpi.com/2218-273X/11/5/666

Doubleday ZA, Nagelkerken I, Coutts MD, Goldenberg SU, Connell SD (2019) A triple trophic boost: how carbon emissions indirectly change a marine food chain. Glob Chang Biol 25(3):978–984. https://doi.org/10.1111/gcb.14536

Dunlap JC, Loros JJ (2016) Yes, circadian rhythms actually do affect almost everything. Cell Res 26(7):759–760. https://doi.org/10.1038/cr.2016.65

Ellis RP, Urbina MA, Wilson RW (2017) Lessons from two high CO_2 worlds—future oceans and intensive aquaculture. Glob Chang Biol 23(6):2141–2148. https://doi.org/10.1111/gcb.13515

Esbaugh AJ (2018) Physiological implications of ocean acidification for marine fish: emerging patterns and new insights. J Comp Physiol B 188(1):1–13. https://doi.org/10.1007/s00360-017-1105-6

Evans DH, Piermarini PM, Choe KP (2005) The multifunctional fish gill: dominant site of gas exchange, osmoregulation, acid-base regulation, and excretion of nitrogenous waste. Physiol Rev 85(1):97–177. https://doi.org/10.1152/physrev.00050.2003

Ewels P, Magnusson M, Lundin S, Kaller M (2016) MultiQC: summarize analysis results for multiple tools and samples in a single report. Bioinformatics 32(19):3047–3048. https://doi.org/10.1093/bioinformatics/btw354

Ferrari MCO, McCormick MI, Munday PL, Meekan MG, Dixson DL, Lönnstedt O, Chivers DP (2012) Effects of ocean acidification on visual risk assessment in coral reef fishes. Funct Ecol 26(3):553–558. https://doi.org/10.1111/j.1365-2435.2011.01951.x

Galloway AWE, Dassow G, Schram JB, Klinger T, Hill TM, Lowe AT, Chan F, Yoshioka RM, Kroeker KJ (2020) Ghost factors of laboratory carbonate chemistry are haunting our experiments. Biol Bull 239(3):183–188. https://doi.org/10.1086/711242

Georgalis T, Perry SF, Gilmour KM (2006) The role of branchial carbonic anhydrase in acid-base regulation in rainbow trout (*Oncorhynchus mykiss*). J Exp Biol 209(3):518–530. https://doi.org/10.1242/jeb.02018

Guh Y-J, Tseng Y-C, Yang C-Y, Hwang P-P (2014) Endothelin-1 regulates H+-ATPase-dependent transepithelial H+ secretion in zebrafish. Endocrinology 155(5):1728–1737. https://doi.org/10.1210/en.2013-1775

Han P, Qiao Y, He J, Men Y, Liu Y, Liu X, Wang X (2022) Identification and functional analysis of dual-specificity phosphatases (DUSP) genes in Japanese flounder (*Paralichthys olivaceus*) against temperature and Edwardsiella tarda stress. Fish Shellfish Immunol 130:453–461. https://doi.org/10.1016/j.fsi.2022.09.051

Heitzman JM, Mitushasi G, Spatafora D, Agostini S (2023) Seasonal coral-algae interactions drive white mat syndrome coral disease outbreaks. Sci Total Environ 900:166379. https://doi.org/10.1016/j.scitotenv.2023.166379

Hofmann GE, Smith JE, Johnson KS, Send U, Levin LA, Micheli F, Paytan A, Price NN, Peterson B, Takeshita Y, Matson PG, Crook ED, Kroeker KJ, Gambi MC, Rivest EB, Frieder CA, Yu PC, Martz TR (2011) High-frequency dynamics of ocean pH: a multi-ecosystem comparison. PLoS One 6(12):e28983. https://doi.org/10.1371/journal.pone.0028983

Huang W, Sher Y-P, Peck K, Fung YCB (2002) Matching gene activity with physiological functions. Proc Natl Acad Sci 99(5):2603–2608. https://doi.org/10.1073/pnas.042684399

Huth TJ, Place SP (2016) Transcriptome wide analyses reveal a sustained cellular stress response in the gill tissue of *Trematomus bernacchii* after acclimation to multiple stressors. BMC Genomics 17(1):127. https://doi.org/10.1186/s12864-016-2454-3

Jutfelt F, Hedgärde M (2015) Juvenile Atlantic cod behavior appears robust to near-future CO_2 levels. Front Zool 12(1):11. https://doi.org/10.1186/s12983-015-0104-2

Kang J, Nagelkerken I, Rummer JL, Rodolfo-Metalpa R, Munday PL, Ravasi T, Schunter C (2022) Rapid evolution fuels transcriptional plasticity to ocean acidification. Glob Chang Biol 28(9):3007–3022. https://doi.org/10.1111/gcb.16119

Kim D, Paggi JM, Park C, Bennett C, Salzberg SL (2019) Graph-based genome alignment and genotyping with HISAT2

and HISAT-genotype. Nat Biotechnol 37(8):907–915. https://doi.org/10.1038/s41587-019-0201-4
Kohan DE, Rossi NF, Inscho EW, Pollock DM (2011) Regulation of blood pressure and salt homeostasis by endothelin. Physiol Rev 91(1):1–77. https://doi.org/10.1152/physrev.00060.2009
Kovac A, Goss GG (2024) Cellular mechanisms of ion and acid-base regulation in teleost gill ionocytes. J Comp Physiol B 194(5):645–662. https://doi.org/10.1007/s00360-024-01560-6
Krebs RA, Feder me (1997) Natural variation in the expression of the heat-shock protein hsp70 in a population of drosophila melanogaster and its correlation with tolerance of ecologically relevant thermal stress. Evolution. 51(1):173–179. https://doi.org/10.1111/j.1558-5646.1997.tb02398.x. PMID: 28568793.
Kroeker KJ, Kordas RL, Crim RN, Singh GG (2010) Meta-analysis reveals negative yet variable effects of ocean acidification on marine organisms. Ecol Lett 13(11):1419–1434. https://doi.org/10.1111/j.1461-0248.2010.01518.x
Kroeker KJ, Kordas RL, Crim R, Hendriks IE, Ramajo L, Singh GS, Duarte CM, Gattuso JP (2013) Impacts of ocean acidification on marine organisms: quantifying sensitivities and interaction with warming. Glob Chang Biol 19(6):1884–1896. https://doi.org/10.1111/gcb.12179
Lefevre S (2019) Effects of high CO_2 on oxygen consumption rates, aerobic scope and swimming performance. In: Grosell M, Munday PL, Farrell AP, Brauner CJ (eds) Fish physiology, vol 37. Academic Press, pp 195–244. https://doi.org/10.1016/bs.fp.2019.08.001
Li H, Handsaker B, Wysoker A, Fennell T, Ruan J, Homer N, Marth G, Abecasis G, Durbin R (2009) The sequence alignment/map format and SAMtools. Bioinformatics 25(16):2078–2079. https://doi.org/10.1093/bioinformatics/btp352
Liu B, Li G, Li X, Wang H, Yang J, Wen H, He F (2024) The responsive mechanisms of DNA methylation and transcriptional regulation to acute hypoxia stress in HIF-1/VEGFA signal pathway of Japanese flounder (*Paralichthys olivaceus*). Aquaculture 578:740021. https://doi.org/10.1016/j.aquaculture.2023.740021
Love MI, Huber W, Anders S (2014) Moderated estimation of fold change and dispersion for RNA-seq data with DESeq2. Genome Biol 15(12):550. https://doi.org/10.1186/s13059-014-0550-8
Machado M, Arenas F, Svendsen JC, Azeredo R, Pfeifer LJ, Wilson JM, Costas B (2020) Effects of water acidification on Senegalese sole *Solea senegalensis* health status and metabolic rate: implications for immune responses and energy use. Front Physiol 11:26. https://doi.org/10.3389/fphys.2020.00026
Mittermayer FH, Stiasny MH, Clemmesen C, Bayer T, Puvanendran V, Chierici M, Jentoft S, Reusch TBH (2019) Transcriptome profiling reveals exposure to predicted end-of-century ocean acidification as a stealth stressor for Atlantic cod larvae. Sci Rep 9(1):16908. https://doi.org/10.1038/s41598-019-52628-1
Moore B, Jolly J, Izumiyama M, Kawai E, Ravasi T, Ryu T (2024) Tissue-specific transcriptional response of post-larval clownfish to ocean warming. Sci Total Environ 908:168221. https://doi.org/10.1016/j.scitotenv.2023.168221
Moreira JM, Mendes AC, Maulvault AL, Marques A, Rosa R, Pousão-Ferreira P, Sousa T, Anacleto P, Marques GM (2022) Impacts of ocean warming and acidification on the energy budget of three commercially important fish species. Conserv Physiol 10(1):coac048. https://doi.org/10.1093/conphys/coac048
Moya A, Huisman L, Forêt S, Gattuso J-P, Hayward DC, Ball EE, Miller DJ (2015) Rapid acclimation of juvenile corals to CO_2-mediated acidification by upregulation of heat shock protein and Bcl-2 genes. Mol Ecol 24(2):438–452. https://doi.org/10.1111/mec.13021
Munday PL, Dixson DL, Donelson JM, Jones GP, Pratchett MS, Devitsina GV, Døving KB (2009) Ocean acidification impairs olfactory discrimination and homing ability of a marine fish. Proc Natl Acad Sci 106(6):1848–1852. https://doi.org/10.1073/pnas.0809996106
Munday PL, Pratchett MS, Dixson DL, Donelson JM, Endo GGK, Reynolds AD, Knuckey R (2013) Elevated CO_2 affects the behavior of an ecologically and economically important coral reef fish. Mar Biol 160:2137–2144. https://doi.org/10.1594/PANGAEA.833440
Munday PL, Dixson DL, Welch MJ, Chivers DP, Domenici P, Grosell M, Heuer RM, Jones GP, McCormick MI, Meekan M, Nilsson GE, Ravasi T, Watson S-A (2020) Methods matter in repeating ocean acidification studies. Nature 586(7830):E20–E24. https://doi.org/10.1038/s41586-020-2803-x
Nagelkerken I, Munday PL (2016) Animal behaviour shapes the ecological effects of ocean acidification and warming: moving from individual to community-level responses. Glob Chang Biol 22(3):974–989. https://doi.org/10.1111/gcb.13167
Nagelkerken I, Goldenberg SU, Ferreira CM, Russell BD, Connell SD (2017) Species interactions drive fish biodiversity loss in a high-CO(2) world. Curr Biol 27(14):2177–2184. https://doi.org/10.1016/j.cub.2017.06.023
Pertea M, Kim D, Pertea GM, Leek JT, Salzberg SL (2016) Transcript-level expression analysis of RNA-seq experiments with HISAT, StringTie and Ballgown. Nat Protoc 11(9):1650–1667. https://doi.org/10.1038/nprot.2016.095
Peterson MS, Gilmore RG (1988) Hematocrit, osmolality, and ion concentration in fishes: consideration of circadian patterns in the experimental design. J Exp Mar Biol Ecol 121(1):73–78. https://doi.org/10.1016/0022-0981(88)90024-X
Petit-Marty N, Nagelkerken I, Connell SD, Schunter C (2021) Natural CO_2 seeps reveal adaptive potential to ocean acidification in fish. Evol Appl 14(7):1794–1806. https://doi.org/10.1111/eva.13239
R Core Team (2023) R: a language and environment for statistical computing. R Foundation for Statistical Computing, Vienna. https://www.R-project.org/
Radford CA, Collins SP, Munday PL, Parsons D (2021) Ocean acidification effects on fish hearing. Proc R Soc B Biol Sci 288(1946):20202754. https://doi.org/10.1098/rspb.2020.2754
Rummer JL, Brauner CJ (2011) Plasma-accessible carbonic anhydrase at the tissue of a teleost fish may greatly enhance oxygen delivery: in vitro evidence in rainbow trout, Oncorhynchus mykiss. J Exp Biol 214(14):2319–2328. https://doi.org/10.1242/jeb.054049
Rummer JL, Brauner CJ (2015) Root effect haemoglobins in fish may greatly enhance general oxygen delivery relative to other vertebrates. PLoS One 10(10):e0139477. https://doi.org/10.1371/journal.pone.0139477
Schunter C, Welch MJ, Ryu T, Zhang H, Berumen ML, Nilsson GE, Munday PL, Ravasi T (2016) Molecular signatures of transgenerational response to ocean acidification in a species of reef fish. Nat Clim Chang 6(11):1014–1018. https://doi.org/10.1038/nclimate3087
Schunter C, Welch MJ, Nilsson GE, Rummer JL, Munday PL, Ravasi T (2018) An interplay between plasticity and parental phenotype determines impacts of ocean acidification on a reef fish. Nat Ecol Evol 2(2):334–342. https://doi.org/10.1038/s41559-017-0428-8
Schunter C, Jarrold MD, Munday PL, Ravasi T (2021) Diel pCO_2 fluctuations alter the molecular response of coral reef fishes to ocean acidification conditions. Mol Ecol 30(20):5105–5118. https://doi.org/10.1111/mec.16124
Shaw EC, McNeil BI, Tilbrook B (2012) Impacts of ocean acidification in naturally variable coral reef flat ecosystems. J Geophys Res Oceans 117(C3):C03038. https://doi.org/10.1029/2011JC007655
Simpson SD, Munday PL, Wittenrich ML, Manassa R, Dixson DL, Gagliano M, Yan HY (2011) Ocean acidification erodes crucial auditory behaviour in a marine fish. Biol Lett 7(6):917–920. https://doi.org/10.1098/rsbl.2011.0293
Sundin J, Jutfelt F (2015) 9–28 d of exposure to elevated pCO_2 reduces avoidance of predator odour but had no effect on behavioural lateralization or swimming activity in a temperate wrasse (*Ctenolabrus rupestris*). ICES J Mar Sci 73(3):620–632. https://doi.org/10.1093/icesjms/fsv101

Sundin J, Amcoff M, Mateos-González F, Raby GD, Jutfelt F, Clark TD (2017) Long-term exposure to elevated carbon dioxide does not alter activity levels of a coral reef fish in response to predator chemical cues. Behav Ecol Sociobiol 71(8):108. https://doi.org/10.1007/s00265-017-2337-x

Sundin J, Amcoff M, Mateos-González F, Raby GD, Clark TD (2019) Long-term acclimation to near-future ocean acidification has negligible effects on energetic attributes in a juvenile coral reef fish. Oecologia 190(3):689–702. https://doi.org/10.1007/s00442-019-04430-z

Suresh S, Mirasole A, Ravasi T, Vizzini S, Schunter C (2023) Brain transcriptome of gobies inhabiting natural CO(2) seeps reveal acclimation strategies to long-term acidification. Evol Appl 16(7):1345–1358. https://doi.org/10.1111/eva.13574

Taylor DA, Thompson EL, Nair SV, Raftos DA (2013) Differential effects of metal contamination on the transcript expression of immune- and stress-response genes in the Sydney rock oyster, Saccostrea glomerata. Environ Pollut 178:65–71. https://doi.org/10.1016/j.envpol.2013.02.027

Uno T, Ishizuka M, Itakura T (2012) Cytochrome P450 (CYP) in fish. Environ Toxicol Pharmacol 34(1):1–13. https://doi.org/10.1016/j.etap.2012.02.004

Veilleux HD, Ryu T, Donelson JM, van Herwerden L, Seridi L, Ghosheh Y, Berumen ML, Leggat W, Ravasi T, Munday PL (2015) Molecular processes of transgenerational acclimation to a warming ocean. Nat Clim Chang 5(12):1074–1078. https://doi.org/10.1038/nclimate2724

Wang J, Russell BD, Ding MW, Dong YW (2018) Ocean acidification increases the sensitivity of and variability in physiological responses of an intertidal limpet to thermal stress. Biogeosciences 15(9):2803–2817. https://doi.org/10.5194/bg-15-2803-2018

Wang L, Sun F, Wen Y, Yue GH (2021) Effects of ocean acidification on transcriptomes in Asian seabass juveniles. Mar Biotechnol 23(3):445–455. https://doi.org/10.1007/s10126-021-10036-5

Wang Y, Wang H-M, Zhou Y, Hu L-H, Wan J-M, Yang J-H, Niu H-B, Hong X-P, Hu P, Chen L-B (2023) *Dusp1* regulates thermal tolerance limits in zebrafish by maintaining mitochondrial integrity. Zool Res 44(1):126–141. https://doi.org/10.24272/j.issn.2095-8137.2022.397

Yoon GR, Bozai A, Porteus CS (2024) Could future ocean acidification be affecting the energy budgets of marine fish? Conserv Physiol 12(1):coae069. https://doi.org/10.1093/conphys/coae069

Yu G (2024) Thirteen years of clusterProfiler. Innovation 5(6):100722. https://doi.org/10.1016/j.xinn.2024.100722

Yusof NA, Masnoddin M, Charles J, Thien YQ, Nasib FN, Wong CMVL, Abdul Murad AM, Mahadi NM, Bharudin I (2022) Can heat shock protein 70 (HSP70) serve as biomarkers in Antarctica for future ocean acidification, warming and salinity stress? Polar Biol 45(3):371–394. https://doi.org/10.1007/s00300-022-03006-7

Zhang H, Li R, Wang Y, Zhou J, Xu H, Gou M, Ye J, Qiu X, Wang X (2023) Transcriptomic analysis of Takifugu obscurus gills under acute hypoxic stress. Animals 13(10):1572. https://doi.org/10.3390/ani13101572

Zou J, Redmond AK, Qi Z, Dooley H, Secombes CJ (2015) The CXC chemokine receptors of fish: insights into CXCR evolution in the vertebrates. Gen Comp Endocrinol 215:117–131. https://doi.org/10.1016/j.ygcen.2015.01.004

Turbid Reefs

9

Shannon Dee and Adi Zweifler (Zvifler)

Abstract

Turbid coral reefs are increasingly recognized as ecologically distinct components of the global reef mosaic. Once regarded primarily as poor or degraded reef systems, these reefs are now understood to host diverse coral communities uniquely adapted to dynamic, sediment-rich environments. This chapter explores the ecological significance, adaptive strategies, and conservation potential of turbid coral reefs in the context of global environmental change. Turbid reef corals exhibit specialized traits that enable them to withstand chronic low light, elevated sedimentation, and fluctuating temperature regimes, often supporting stress-tolerant symbiont assemblages and enhanced heterotrophic capabilities. These features may facilitate resistance to climate-induced coral bleaching, positioning turbid reefs as potential climate refugia in an increasingly warming ocean. However, ongoing coastal development and nutrient enrichment pose significant risks to these systems, threatening to undermine their resilience. A nuanced understanding of the differences between naturally turbid and anthropogenically impacted reefs is essential for effective conservation and management. This chapter highlights the importance of addressing knowledge gaps surrounding the cumulative impacts of local and global stressors on turbid reef ecosystems and calls for targeted research to clarify their role in the future of coral reef conservation.

S. Dee (✉)
Centre for Marine Ecosystem Research, School of Science, Edith Cowan University, Joondalup, WA, Australia
e-mail: s.dee@ecu.edu.au

A. Zweifler (Zvifler)
School of Life and Environmental Sciences, Faculty of Science, The University of Sydney, Camperdown, NSW, Australia

Keywords

Turbidity · Low light · Nearshore reefs · Sedimentation · Adaptation

9.1 What Are Turbid Reefs?

Turbid reefs are coral reef systems that develop in environments with high levels of suspended particulate matter (SPM). Turbidity is the scattering of light in the water column due to SPM, which significantly affects water clarity (Van Duin et al. 2001; Flores et al. 2012). SPM includes both inorganic components, such as terrestrial sediments transported by rivers and wind or resuspended from the seafloor, and organic matter including phytoplankton, zooplankton, and microbial organisms (Rogers 1990; Anthony 1999). The organic fraction of SPM is often quantified through chlorophyll-*a* levels, which serve as an indicator of phytoplankton biomass (Flores et al. 2012). Approaches to quantifying turbidity vary across studies. Direct measurements are often reported in nephelometric turbidity units (NTU) or formazin turbidity units (FTU) using turbidity loggers (Anthony and Larcombe 2000; Morgan et al. 2020), whereas other studies employ light attenuation as a proxy (e.g., Kd490, PAR), derived from satellite remote sensing or measured with light loggers (Loiola et al. 2019; Zweifler et al. 2024; Gaspar et al. 2025).

Unlike clear water reefs, which are typically found offshore in nutrient-poor waters, turbid reefs are predominantly found in shallow coastal waters, usually within 20 km of the shoreline and at depths of less than 10 m (Morgan et al. 2017). While they occur in very shallow waters, they are characterized by their light-limited environment (Fig. 9.1), similar to deeper mesophotic reefs (>30 m). These ecosystems represent a significant portion of global reef systems, comprising 12% of the world's reefs and 30% of those within the Coral Triangle (Sully and van Woesik 2020). However, scientific understanding of turbid reef ecosystems remains limited, primarily due to the challenges of conducting

M. O. Soares et al. (eds.), *Extreme and Marginal Coral Reefs and Communities*, Coral Reefs of the World 24,
https://doi.org/10.1007/978-3-032-28035-0_9

Fig. 9.1 Scenes from turbid inshore reefs of Western Australia

research in low-visibility environments, whether through direct observation or remote sensing techniques (Morgan et al. 2017).

Historically, scientists emphasized the negative impacts of turbid environments on coral health, including reduced photosynthetic capacity, physical damage to coral tissues, and decreased biodiversity (Larcombe et al. 2001; Sanders and Baron-Szabo 2005; Weber et al. 2012). However, recent research has challenged these assumptions, revealing that many turbid reef systems support high coral coverage (Anthony and Larcombe 2000; Browne et al. 2013; Browne et al. 2019; Zweifler et al. 2024) and can demonstrate enhanced resistance to thermal stress compared to their clear-water counterparts (Perry et al. 2008; Wagner et al. 2010; Morgan et al. 2017). Furthermore, studies of ancient reef environments indicate that coral reefs successfully established and grew under turbid conditions throughout the mid- to late-Holocene period (Perry and Smithers 2010; Santodomingo et al. 2015; Solihuddin et al. 2015; Johnson et al. 2017; Potts and Jacobs 2000). These findings suggest that turbid reef environments may serve as important refuge for coral conservation in the face of global climate change.

Turbidity in coral reef environments can occur from both natural and anthropogenic (human-induced) sources (Fig. 9.2). Naturally turbid reef systems have evolved under consistent exposure to suspended particles generated by wave action, tidal flows, or river discharge (Potts and Jacobs 2000). These reefs have adapted to fluctuating sediment loads over long timescales, allowing corals and associated organisms to develop physiological and ecological strategies to cope with reduced light availability. Notable examples include the inshore regions of the Great Barrier Reef (GBR; Larcombe et al. 2001; Palmer et al. 2010), the Kimberley coast of Western Australia (Richards et al. 2018), and Brazil's Abrolhos reefs (Segal and Castro 2011), where turbidity is a natural occurring environmental feature.

In contrast, anthropogenic turbid reefs have experienced relatively recent (<100 years) increases in turbidity due to human activities including coastal development, dredging, deforestation, and agricultural expansion (Jones et al. 2016; Smith et al. 2007; Foster et al. 2011; Dsikowitzky et al. 2016). These human activities accelerate sediment resuspension, and when combined with climate change-induced shifts in precipitation patterns, they contribute to increased sediment loads in coastal waters, further altering reef conditions. One well-documented example is the reefs surrounding Singapore (Tan et al. 2016), where rapid urbanization and land reclamation have led to sustained high turbidity levels.

The distinction between natural and anthropogenic turbid reefs is particularly relevant for conservation management efforts. While naturally turbid reefs have evolved under stable yet dynamic sediment regimes, anthropogenic turbid systems often exhibit signs of stress and compromised ecological function (Perry et al. 2018a; Januchowski-Hartley et al. 2020) including reduced coral cover and community composition as well as restricted vertical distribution of coral communities (Morgan et al. 2020a). These differences suggest that anthropogenic turbid reefs may represent a distinct ecological category requiring specialized management approaches compared to naturally turbid systems.

9.1.1 Sedimentary Settings and Turbidity Regimes

Turbid reefs experience extreme environmental conditions including high turbidity and sedimentation (Table 9.1). Under Schoepf et al. (2023) framework, such conditions might predict marginal ecological outcomes (i.e. low coral cover and diversity). However, turbid reefs can be highly productive, functional, and resilient, demonstrating that extreme environmental conditions do not necessarily yield marginal ecological states. The characterization of turbid reefs as extreme and/or marginal depends on local conditions and turbidity source. Naturally turbid reefs may exhibit extreme conditions while maintaining high ecological function, whereas anthropogenically impacted reefs often show both extreme conditions and marginal ecology. This decoupling of environmental extremes from ecological marginality highlights coral adaptability to suboptimal conditions, offer-

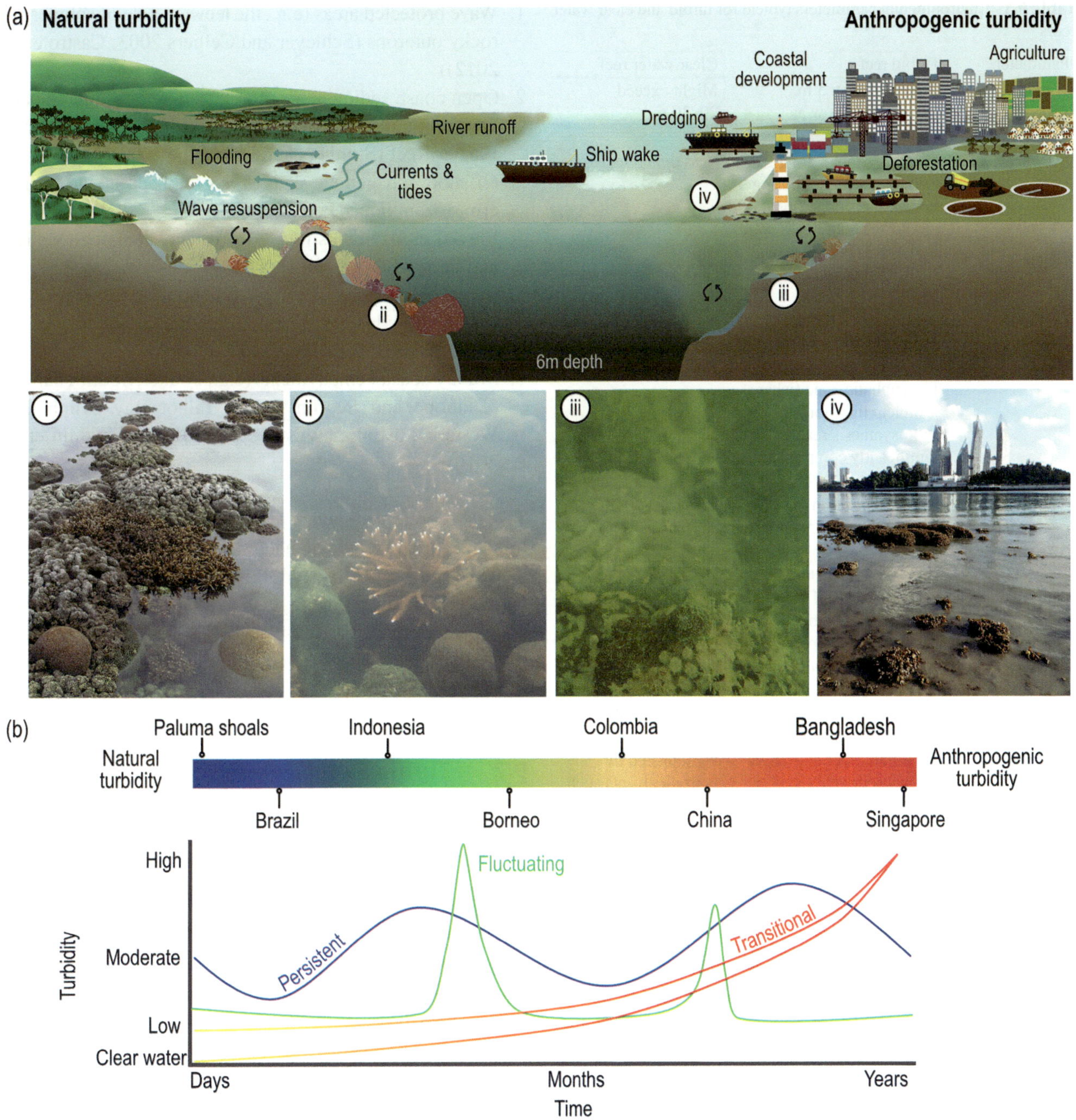

Fig. 9.2 (**a**) Natural turbid reef (on the left, two photos of PSRC by Nicola Browne) versus anthropogenic turbid reef (on the right, two photos of Singapore reefs by Kyle Morgan). (**b**) Types of turbidity regimes on a temporal scale: persistent (blue), fluctuating (green) and transitional (yellow-red). The y-axis turbidity scale shows: low (<5 mg L^{-1}/<15 NTU), moderate (5 mg L^{-1}/15 NTU) and high (>50 mg L^{-1}/>150 NTU). Reproduced with permission from Zweifler et al. Diversity; published by MDPI, 2021

ing valuable insights into how corals might respond to climate-driven ocean changes.

It is important to note that not all turbid reefs exist under the same sediment regimes. Zweifler et al. (2021) defined three conceptual sediment exposure regimes: (1) persistent, (2) fluctuating, and (3) transitional.

Reefs with persistent turbidity experience daily suspended sediment concentrations exceeding 5 mg L^{-1} or 15 NTU, as observed in persistent sediment-rich coral reefs. These environments typically support stress-tolerant corals, and their reef structures contain substantial terrigenous sediment components.

In contrast, fluctuating turbidity reefs undergo episodic severe suspended sediment events (exceeding 50 mg L^{-1} or 150 NTU) alternating with periods of relative clarity (below 5 mg L^{-1} or 15 NTU). Examples include Marino Ballena in

Table 9.1 Environmental parameters typical for turbid and clear water reefs

Parameter	Turbid reef	Clear water reef
Water visibility	Often less than 5 m, sometimes as low as <1–2 m	Might exceed 20–30 m
Sediment loads	Typically, above 5 mg L^{-1} (sometimes higher during storm events or seasonal runoff, see below) (Browne et al. 2013; Fabricius et al. 2014; Cartwright et al. 2023)	Less than 5 mg/L
Light attenuation	Decrease rapidly with depth due to scattering and absorption by particles. The euphotic zone (where sufficient light for photosynthesis reaches) is significantly compressed (Morgan et al. 2020a)	Can be 40 m deep or more (Jones et al. 2021)
Sedimentation rates	Active deposition of 10–50 mg/cm^2/day on reef surfaces (Browne et al. 2012; Morgan et al. 2020a)	Less than 10 mg/cm^2/day
Nutrient levels	Typical dissolved inorganic nitrogen (DIN) above 1.0 μM and phosphate above 0.2 μM (Fabricius 2005)	DIN <1.0 μM and phosphate 0.05–0.2 μM (Fabricius et al. 2012)
Temporal variability	Seasonal or episodic changes in turbidity related to rainfall, river discharge, and wind-driven resuspension events (Browne et al. 2013; Fabricius et al. 2013; Zweifler et al. 2021)	Less temporal variability but can be impacted by episodic events such as storms or ENSO driven temperature anomalies

Pacific Costa Rica, the Pilbara region in Western Australia, and areas of Borneo. These reefs also feature predominantly stress-tolerant coral communities adapted to variable energy conditions, though their reef matrices contain lower terrigenous sediment proportions compared to persistent turbidity reefs.

Transitional reefs demonstrate progressive or stepped increases in turbidity over annual to decadal timeframes, often evolving from initially clear conditions toward higher turbidity states, as seen in Singapore. Having experienced relatively recent increases in sediment exposure (within the past century), these reefs typically exhibit signs of depth compression, declining coral cover, and reduced biodiversity. The severity of these impacts depends on baseline conditions and the rate at which sediment exposure has intensified.

Globally, reefs that have initiated and developed within natural turbid conditions are typically found in one of six environmental settings. These include:

1. Wave protected areas (e.g., the leeward side of submerged rocky outcrops (Schleyer and Celliers 2003; Castro et al. 2012))
2. Open coast, sedimentary shorelines (e.g., Paluma Shoals, situated on intertidal terrigenous sand/mud, central Great Barrier Reef (Smithers and Larcombe 2003))
3. Offshore terrigenous shelves (e.g., Inhaca Island, southern Mozambique (Perry 2005))
4. Fluvial embayments (e.g., Rio Bueno, Jamaica (Mallela and Perry 2007))
5. River deltas (e.g., Bay of Baten, Indonesia (Hoitink 2004); Magdalena River, Colombia (Restrepo et al. 2006); Pearl River, Hong Kong (Goodkin et al. 2011))
6. Muddy coastal embayments (e.g., Phuket, South Thailand (Tudhope and Scoffin 1994); Talbot Bay, Kimberley, Western Australia (Wilson et al. 2011; Solihuddin et al. 2016))

These different sedimentary and geomorphic settings highlight the broad range of natural environmental conditions where turbid reefs have developed. Understanding these variations could help distinguish natural turbid reefs (persistent, fluctuating) from those reefs that have transitioned to turbid (or to more turbid) conditions during the Anthropocene (i.e. Fig. 9.3).

9.2 Turbid Reefs Through Time

Palaeoecological research provides valuable insights into the development and long-term persistence of turbid coral reef ecosystems through time. Identifying evidence of community shifts throughout the Holocene establishes a natural baseline against which recent and future changes can be compared, helping to predict how reefs might respond to ongoing environmental pressures (Pandolfi et al. 2003). Research into the historical record and growth history of turbid reefs has primarily concentrated in Australia, particularly the GBR, with additional research locations including Western Australia, Singapore, China, Japan, Brazil, Costa Rica, and Thailand (Chan et al. 2025).

Researchers typically recover vertical reef framework samples using percussion or rotary drilling, most commonly from intertidal or shallow subtidal reef flat environments accessible during low tide. These cores can reach up to 6.5 m in length, capturing approximately 7000 years of reef growth history. While submerged reef slopes have also been cored using manual percussion methods with SCUBA equipment, these cores are shorter (up to 4.5 m) and represent narrower time windows of reef development. Studies utilizing reef core data have identified intervals throughout the Holocene marked by alternating phases of active reef growth and periods of interrupted growth, often triggered by significant fluctuations in sea level.

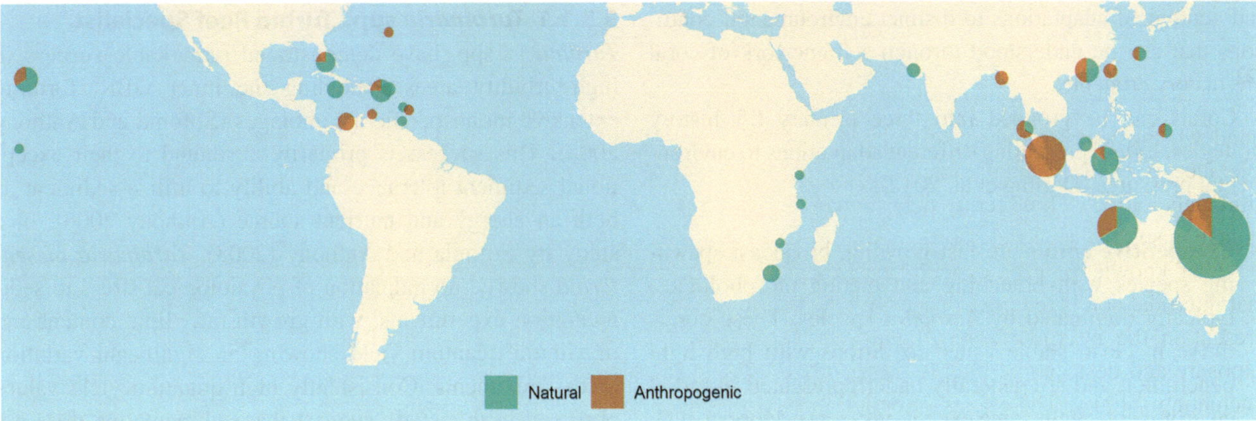

Fig. 9.3 Global distribution of naturally turbid, and anthropogenic turbid coral reefs adapted with permission from Zweifler et al. Diversity; published by MDPI, 2021

Despite spatial limitations inherent in core sampling (typically 75–100 mm in diameter), researchers can analyse multiple indicators within cores to reconstruct past environmental conditions:

Sediment Composition: The ratio of land-derived versus carbonate sediments provides insights into historical turbidity conditions. For example, cores from the Kimberley region (NW Australia) contain consistently high proportions of siliciclastic sediments, indicating that these reefs developed under long-term turbidity regimes.

Coral Assemblages: Changes in coral community structure through time can indicate shifts in environmental conditions. For instance, a transition from *Acropora*-dominated to *Pavona*-dominated communities on some GBR sites coincided with European settlement, suggesting increased sediment and nutrient inputs.

Benthic Foraminifera: These microscopic organisms preserved in reef sediments can inform interpretations about past water depth, turbidity, and hydrodynamic conditions.

Extensive reef coring campaigns reveal that many present-day turbid reef communities have developed under persistent turbidity conditions. Studies of inshore and fringing reef systems further reveal that coastal reefs have long accreted in the turbid environments typical of continental shelf settings (Bonesso et al. 2023; Morgan et al. 2020b; Smithers and Larcombe 2003). Core records often show a historically greater abundance of *Acropora* compared to modern assemblages (Yamano et al. 2019), although *Acropora* has been consistently present throughout various stages of reef development (Zhao et al. 2022).

In Singapore, Chan et al. (2025) identified evidence of a naturally turbid environment existing prior to urbanisation within a relatively sheltered inshore reef system. Their findings also suggest a potential mid-development growth disruption, similar to patterns observed in other regional reef environments. It is likely that Singapore's reefs have remained immersed in turbid conditions throughout much of the Holocene, with community structures dominated by stress-tolerant coral species. However, increased turbidity linked to recent urbanisation may have intensified these conditions, driving further shifts toward communities increasingly adapted to sediment-rich environments. Together, these findings suggest that turbid reefs are naturally adapted ecosystems rather than degraded remnants of formerly clear-water reefs.

9.3 Coral Adaptations

Corals persisting in turbid environments exhibit specialized adaptations that allow them to thrive despite challenging conditions. These include enhanced heterotrophic feeding, morphological plasticity, efficient sediment-shedding mechanisms, increased tolerance to low light conditions, resistance to burial events, and physiological adaptations to fluctuating salinity and temperature regimes. These adaptations represent sophisticated evolutionary responses to persistent environmental stressors.

9.3.1 Coral Cover, Species Composition and Morphology

Researchers have long observed that coral community structure on offshore reefs differs significantly from those found on coastal turbid reefs. These differences reflect evolutionary and ecological adaptations to distinct environmental conditions that can be understood through a framework of coral life-history strategies.

Corals can be grouped into three primary life-history strategies, each representing different adaptations to environmental pressures (Darling et al. 2012):

1. **Competitive corals** are fast-growing, broadcast spawning species with branching and plating morphologies, typically dominated by *Acropora* species. These corals thrive in clear, stable water conditions with high light penetration and are generally underrepresented in turbid environments, with some exceptions (sensu Morgan et al. 2017; Zweifler et al. 2024).
2. **Weedy corals** reproduce primarily through brooding and form smaller colonies that opportunistically colonize recently disturbed habitats. This group includes branching *Porites*, some *Pocilloporidea*, *Leptastrea* and specific *Merulinidae* species such as *Cyphastrea and Goniastrea.* Their reproductive strategy and growth patterns make them well-adapted to withstand variable environmental conditions.
3. **Stress-tolerant corals** are slow-growing, broadcast spawning species that primarily display domed morphologies with large corallites. These species maintain high fecundity and substantial energy reserves. In the Atlantic, they include *Montastraea annularis*, *M. cavernosa*, *Colpophyllia natans*, and *Diploria* species, while in the Indo-Pacific, massive *Poritidea*, Favidea and *Merulinidae* species represent this strategy. These stress-tolerant corals are particularly successful in turbid reef environments where their physiological adaptations provide competitive advantages.

Turbid reef environments predominantly favour stress-tolerant and weedy coral species rather than competitive species. This selection pressure creates distinctive community compositions with surprising ecological characteristics. Morgan et al. (2017) documented large areas of topographically complex reef in shallow, muddy nearshore zones of the GBR with surprisingly high coral cover (exceeding 30%) and relatively diverse communities comprising 21 genera. This finding stands in contrast to many clear water reefs of the GBR, where average coral cover of only 12% is common. Further expanding our understanding of coral adaptability, Richards et al. (2015) documented diverse assemblages (60 genera) of stress-tolerant corals in the Bonaparte Archipelago, Western Australia. These remarkable communities survive extreme environmental conditions, including large fluctuations in sea surface temperature, variable exposure conditions, high turbidity, and extreme tidal ranges. Such discoveries have significantly broadened our understanding of coral tolerance thresholds in nearshore marginal settings.

9.3.1.1 *Turbinaria* spp.: Turbid Reef Specialist

Turbinaria spp. have demonstrated remarkable success in high-turbidity areas, including the inner GBR, forming extensive monospecific assemblages (Sofonia and Anthony 2008). This success is primarily attributed to their exceptional sediment tolerance and ability to utilize sediment as both an energy and nutrient source (Anthony 2006). In a study by Sofonia and Anthony (2008), *Turbinaria mesenterina* showed no indication of physiological stress to sedimentation experiments, with growth rate, lipid content and maximum quantum yield showing no significant variation across treatments. Consistently high quantum yield values observed in this study suggest that sedimentation does not significantly damage the coral's photosystem unless accumulated material cannot be cleared, which can lead to tissue necrosis. Remarkably, *T. mesenterina* colonies were able to clear sediment loads exceeding 100 mg cm^{-2} within just 4–5 h, representing some of the highest clearance rates reported for corals (Fig. 9.4; Sofonia and Anthony 2008; Stafford-Smith 1993).

This rapid sediment removal occurs through both active mechanisms, such as cilia-mediated transport of sediment-laden mucus across the surface, and passive mechanisms related to their distinctive morphology. Additionally, raised corallites (1–2 mm above the coenosarc) facilitate gravity or flow-induced sediment removal from polyps (Veron 2000), reducing the energy costs associated with sediment clearing (Lasker 1980).

Further adaptations of this genera are shown by Juhi et al. (2021) where *T. mesenterina* colonies from nearshore turbid environments were shown to have the ability to readily acclimate to rapid fluctuations in light intensity within 2–5 days. In addition, *Turbinaria* species have long shown enhanced resilience to thermal stress, which is common in inshore reef systems (Marshall and Baird 2000; Dias et al. 2018).

However, it is important to note that combinations of stressors typical of nearshore environments—such as high nutrient loads, low salinity, and high temperatures—can negatively impact *Turbinaria* species when present at high levels (Fig. 9.5; Faxneld et al. 2010). This highlights that despite their adaptive capacity, their resilience may still be limited under future climate change scenarios.

9.3.2 Growth Rates and Energy Stores

Physiological responses, such as variations in coral growth and calcification rates (Highsmith 1979; Cruz-Pinon et al. 2003), and coral lipid content (Harriott 1993) provide fine-scale spatial and temporal data that help determine how environmental variables influence coral health and productivity (Browne 2012).

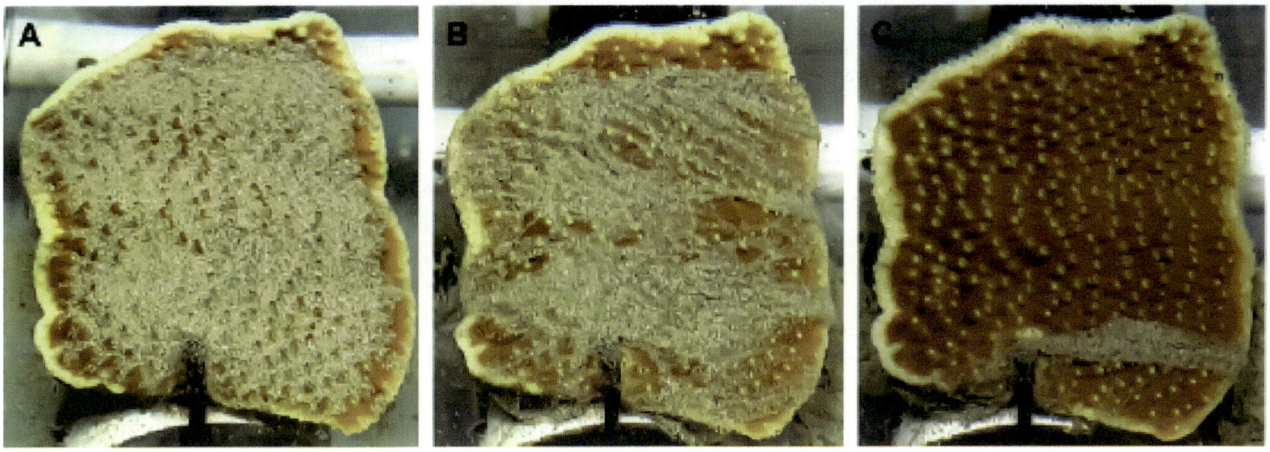

Fig. 9.4 *Turbinaria mesenterina* shedding sediment loads. Time sequence: (**a**) t = 0 h, (**b**) t = 1.5 h, and (**c**) t = 4.5 h. Image reproduced from Sofonia and Anthony (2008) Estuarine, Coastal and Shelf Science; published by Elsevier, 2008, with permission from KRN Anthony

Fig. 9.5 (**a**) large *Turbinaria* colony existing in naturally turbid conditions. (**b**) *Turbinaria reniformis* withstanding high sedimentation. (**c**) variation in thermal tolerance between neighbouring colonies of *Turbinaria reniformis* with the left colony being bleached, and the right colony retaining healthy conditions

While turbid reefs were traditionally believed to support slower coral growth rates compared to clear-water reefs due to the associated negative effects of high sediment loads and reduced water quality, over the past 15 years, research has shown that the growth and calcification rates of some genera are comparable. This is particularly evident in fast growing species (e.g., *Acropora*) and those well adapted to environmental fluctuations (e.g., *Turbinaria, Montipora*; Table 9.2).

Further, Anthony and Fabricius (2000) found that tissue growth rates in *Porites cylindrica* and *Goniastrea retiformis* nearly doubled under turbidity loads of 4 to 8 and 16 mg l^{-1}, respectively, whereas calcification rates showed only minor variation across turbidity regimes. At Middle Reef on the inshore GBR, Browne et al. (2012) recorded mean linear extension in *Acropora* (6.3 cm/year) comparable to those measured at similar depths on offshore reefs on the GBR. Similarly, *Montipora* exhibited linear extension of 2.9 cm/year, exceeding estimates available from both clear-water and turbid reefs (Table 9.2). Perry et al. (2012) investigated historic reef accretion at Middle Reef and recorded an average accretion rate of 8.3 mm $year^{-1}$, with the most rapid phases of accretion coinciding with periods dominated by fine-grained terrigenoclastic sediment accumulation.

However, many studies continue to report slow growth and calcification rates at turbid sites globally. These findings support the notion that while many coral genera can exist in turbid conditions, not all have the physiological adaptations necessary for high performance under these extreme conditions. Moreover, not all turbidity regimes are equally favourable for supporting coral growth and energy stores.

A commonly used proxy for assessing coral physiological health and energy reserves is lipid content. Anthony (2006) showed that corals within the inshore turbid reefs of GBR had higher lipid concentrations compared to those from offshore clearwater reefs, with particularly elevated levels observed in *Turbinaria mesenterina*. It has been hypothesised that the high concentrations of particulate matter on the nearshore reefs of the GBR lagoon represent a significant food source for these corals, contributing to their high lipid stores.

Jones et al. (2020) found that a multitude of coral species experienced a decline in lipid content over a seven-week

Table 9.2 Calcification rates (g cm^2 year^{-1}) of various genera across turbid and non-turbid sites within the Indo Pacific

Morphology	Genus	Calcification Rate (g cm^{-2} year^{-1})	Location	Turbid reef?	Source
Branching	*Acropora spp.*	0.25	Pilbara	Yes	Dee et al. (2024)
		0.44	Pilbara and Kimberley	Yes	Dandan et al. (2015)
		0.42	Coral Bay	No	Foster et al. (2014)
	Pocillopora damicornis	0.21	Pilbara	Yes	Dee et al. (2024)
		0.34	Coral Bay	No	Foster et al. (2014)
Foliose/plate	*Montipora aequituberculata*	1.52	Inshore GBR	Yes	Browne et al. (2012)
		0.30	Average of multiple sites withing Indo-Pacific		ReefBudget Indo-Pacific; Perry et al. (2018b)
	Turbinaria	1.37	Inshore GBR	Yes	Browne et al. (2012)
		0.25	Average of multiple sites withing Indo-Pacific		ReefBudget Indo-Pacific; Perry et al. (2018b)
Massive	*Porites*	1.62	Pilbara	Yes	Lough et al. (2016)
		1.48	Average of multiple sites withing Indo-Pacific		ReefBudget Indo-Pacific; Perry et al. (2018b)
	Faviidae	0.37	Pilbara	Yes	Dandan et al. (2015)
		0.69	Average of multiple sites withing Indo-Pacific		ReefBudget Indo-Pacific; Perry et al. (2018b)
	Lobophyllia	0.23	Pilbara	Yes	Dandan et al. (2015)
		1.155	Average of multiple sites withing Indo-Pacific		ReefBudget Indo-Pacific; Perry et al. (2018b)

period of increased turbidity, representing impacts of dredging on coral health. This displays the variation in corals being exposed to rapid onset of human induced turbidity, rather than those that have adapted over time to naturally occurring turbidity regimes. Such rapid anthropogenic disturbances can further affect coral populations by reducing energy available for vital processes such as reproduction.

Coral reproductive success is closely tied to the composition and size of eggs, which are composed of proteins, carbohydrates, and lipids. These components not only aid in embryo floatation but also provide crucial energy source for the developing larvae (Harii et al. 2007, 2010). Generally, egg size acts as an indicator of reproductive health and success, as smaller eggs hold fewer lipids, reducing fertilisation success and limits dispersal potential (Levitan 2006). Further, Harii et al. (2010) found that lipids may be the major energy source for larvae for some coral species under light limited conditions, with more than 85% of larval metabolic demand was found to be derived from lipids under light limited conditions. Alternatively, under light conditions, coral larvae hosting symbiotic algae are relatively less dependent on stored lipids (Harii et al. 2010).

9.3.3 Heterotrophy in Corals: An Adaptive Feeding Strategy

Corals exist across the earth under variable environmental conditions that influence their physiology and adaptive feeding strategies. Most scleractinian corals maintain symbiotic relationship with photosynthetic dinoflagellates (*Symbiodiniaceae*) within their tissue, which support the coral host's nutritional requirements. Under optimal conditions of light and temperature, these symbiotic dinoflagellates provide the majority (>90%) of the coral's metabolic needs via autotrophic nutrient uptake (Houlbrèque and Ferrier-Pagès 2009; Goldberg 2018). However, reef-building corals can also obtain their energy and nutrient intake thorough heterotrophy, capturing organic particles from the water column, including plankton, detritus and dissolved organic matter (Houlbrèque and Ferrier-Pagès 2009; Denis et al. 2024). In turbid reef environments, where light availability is reduced, many corals exhibit elevated ratios of mixotrophy and heterotrophy to compensate for decreased photosynthetic efficiency (Anthony et al. 2009; Houlbrèque and Ferrier-Pagès 2009; Wooldridge 2014; Travaglione et al. 2013). These flexible feeding strategies help offset energy deficits and support coral growth and survival under suboptimal light conditions (Grottoli et al. 2006; Hughes and Grottoli 2013).

External nutrient sources such as riverine discharge, submarine groundwater inputs, lateral nutrient fluxes and airborne nutrient deposition (e.g., desert dust) can enhance heterotrophic feeding opportunities by increasing the concentration of organic material in the water column. Additionally, oceanographic processes, including internal waves, eddies and upwelling- facilitate vertical nutrient transport, influencing the distribution of suspended organic particles and dissolved nutrients, further shaping coral feeding dynamics.

Research has demonstrated that many coral species can shift between autotrophic and heterotrophic feeding strategies across various environmental gradients, including water depth (Palardy et al. 2005), turbidity levels (Anthony and Fabricius 2000), and nutrient availability (Fox et al. 2019). The degree of heterotrophic reliance varies among coral species, with some genera, such as *Platygyra* and *Turbinaria*, exhibiting enhanced particle-capture capabilities (Saper et al. 2023). This is likely due to larger polyp sizes, lower surface area to volume ratios, or the ability to produce particle-trapping mucus (Porter 1976; Conti-Jerpe et al. 2020). Nevertheless, Houlbrèque and Ferrier-Pagès (2009) suggest that feeding success is more closely related to increased feeding effort rather than morphological traits alone.

For corals less efficient at heterotrophic feeding, one key acclimation strategy involves increasing symbiont densities and chlorophyll concentrations, allowing corals to maximize light absorption under low-light conditions (Titlyanov and Titlyanova 2002; Cooper et al. 2008). These adaptations have been observed in turbid reef systems in Singapore (Browne et al. 2015a, 2015b) and the nearshore GBR (Rocker et al. 2017), where corals maintain photosynthetic productivity despite chronically high turbidity levels.

For more details on heterotrophy, see Rossi (2026) in this volume.

9.3.4 Symbiont Relationships

The symbiotic relationship between corals and *Symbiodiniaceae* algae plays a fundamental role in coral reef ecosystems. Through photosynthesis, symbionts convert sunlight and carbon dioxide into organic carbon and oxygen, providing corals with a significant proportion of their metabolic energy (Muscatine and Porter 1977; Page et al. 2019). However, as stated above, in turbid reef environments, where light penetration is reduced, corals must modify their symbiont communities and physiological traits to optimize photosynthetic efficiency.

Different *Symbiodiniaceae* clades exhibit varying degrees of tolerance to environmental stressors and shifts in symbiont community composition can enhance coral resilience (Berkelmans and Van Oppen 2006; Howells et al. 2012). While higher diversity in *Symbiodiniaceae* may facilitate more adaptive resilience to harsher environmental conditions, studies from turbid environments have found mixed results. Smith et al. (2020) found that corals in Singapore hosted low diversity of common *Cladocopium* species, with comparatively low presence of the more stress tolerant *Durusdinium* species. In contrast, on the naturally turbid reefs of Palk Bay, India, Hussain et al. (2025) reported high diversity of *Symbiodiniaceae,* with elevated concentrations of *Durusdinium* in most coral samples. This shift is hypothesised to be an adaptive response following previous bleaching events, highlighting the ability of corals to dynamically regulate and shift their symbiont populations. Such strategies allow corals to better respond to changing environmental conditions, improving their long-term survival prospects. These strategies highlight the complex interactions between coral hosts and their symbionts, underscoring the ecological flexibility that enables corals to thrive in dynamic and variable reef habitats.

9.3.5 Case Study

9.3.5.1 Natural vs. Anthropogenic Turbid Reefs

Paluma Shoals Reef Complex (PSRC) and Singapore represent two different types of turbid reefs, natural and anthropogenic respectively. The contrasting turbidity regimes in Singapore reefs and PSRC may, in part, explain important differences in coral coverage and carbonate production rates in various types of turbid reefs (Table 9.3). In Singapore, turbidity levels are lower than in PSRC, but reefs experience higher sedimentation rates (Browne et al. 2012, 2015a, b). Frequent exposure to wind-driven waves at PSRC leads to elevated sediment resuspension and near persistent turbidity. Whereas in Singapore, the sheltered tidal-controlled system is characterized by low energy and therefore, higher levels of sedimentation (Morgan et al. 2020a).

Generally, sedimentation is considered to be more detrimental to coral settlement, growth and survival than lower light levels, potentially resulting in lower coral coverage (Birrell et al. 2005; Perez et al. 2014; Ricardo et al. 2016). As such, the balance between sediment settling and resuspension is as important as the volume of sediments entering the nearshore environment. In the coastal waters of Southeast Asia, including Singapore, the high presence of terrestrial derived dissolved organic matter contributes to low light availability that further compounds suspended sediment impacts (Morgan et al. 2020a). In contrast, PSRC, although located in the wet Australian tropics, exhibits lower nutrient levels due to its remoteness from an urban centre along with effective regulations and management of water catchments in the area (Macdonald et al. 2013). Thus, the dominant source of turbidity is suspended sediment.

Despite differences in the turbidity regime (length of exposure and source), both reef systems are largely composed of sediment-tolerant coral species that have also demonstrated resilience to warm water temperatures. In Singapore, it has been suggested that the historical record of warmer waters has resulted in a coral community adapted and more tolerant to elevated sea surface temperatures (Guest et al. 2016a, b). PSRC does not have the same history of exposure to warm waters, which suggests that resilience may partially be explained by a turbidity-driven reduction in UV, which acts as a synergistic stressor decreasing rates of thermal stress (Anthony et al. 2007; Welle et al. 2017).

Table 9.3 Comparison of environmental, physical and ecological parameters at PSRC and Singapore

		Paluma shoals reef complex	Singapore
Nearest urban development		Townsville ~30 km, 201,313 people (in 2024)	Singapore <6 km, 6.04 million (in 2024)
Reef initiation period		1700–1000 YBP	No data available
Stressors	Global	Cyclones, heat waves, crown-of-thorns starfish	Heat waves
	Local	N/A	Dredging, coastal development, ship traffic
Sea surface temperature (°C)		25–28	27–31
Turbidity regime		Natural-persistent (wind-waves, tidal currents, river plums)	Anthropogenic-transitional (dredging, coastal development)
Turbidity (NTU)		15–50	4.8–6.6
Sedimentation rate (average)[a]		60.5 g m^2 d^{-1}	176 g m^2 d^{-1}
Coral genera[b]		*Montipora* (50%), *Acropora* (15%), *Turbinaria* (12%), *Porites* (1.5%), *Lobophyllia, Stylophora, Seriatopora, Pavona, Goniastrea, Favia, Favites Platygyra, Goniopora, Galaxea, Psammocora, Cyphastrea, Hydnophora, Symphyllia, Echinopora, Pachyseris, Alveopora, Fungia, Euphyllia*	*Pectinia* (11–19%), *Pachyseris* (7–14%), *Merulina* (6–12%), *Montipora* (7%), *Porites* (6%), *Echinopora* (4%), *Platygyra* (4%), *Acropora, Pocillopora, Pavona, Goniastrea, Favia, Favites, Lobophyllia, Goniopora, Galaxea, Montastraea, Diploastrea, Cyphastrea, Hydnophora, Symphyllia, Echinophyllia, Oxypora, Leptoseris, Leptastrea, Fungia*
Coral cover (average)		38%	31%
Reef geomorphology		Fringing (inner-shelf, coastal reefs) and offshore patch reefs	Fringing or patch reefs near the southern islands
Coral growth depth range		<6 m	<6 m
Reef area		~16 km^2	~9.5 km^2
Carbonate budget (CaCO3)		~6.9 kg m^2 $year^{-1}$	~3.7 kg m^2 $year^{-1}$
Reef accretion potential (average, based on carbonate budget values)		2.97 mm $year^{-1}$	1.55 mm $year^{-1}$

Reproduced with permission from Zweifler et al. Diversity; published by MDPI, 2021

[a]In PSRC measured as net sedimentation using sediment trays and in Singapore measured as gross sediment accumulation using sediment traps

[b]Coral genera (%) is percentage cover at that site. Species without (%) are <1% of coral cover

The differences in reef setting and turbidity regime together with differences in reef ecology, functionality (e.g., carbonate production) and potential resilience support the acknowledgment that natural and anthropogenic turbid reefs are two different reef types. This distinction is particularly important when assessing their value for future reef conservation plans.

9.3.6 Bioerosion and Reef Stability

Bioerosion on coral reefs is the process by which various organisms break down coral structures. Bioerosion can be caused by external grazing from taxa such as parrotfish and urchins, as well as endolithic boring and chemical dissolution from "microborers" (fungus or algae), and larger "macroborers" (sponges, worms, or bivalves). While bioerosion plays a natural role in reef ecosystem dynamics by recycling calcium carbonate, excessive bioerosion—often linked to environmental stressors like ocean acidification, pollution, and overfishing—can weaken reef structures and contribute to reef degradation.

Rates of bioerosion were previously expected to be higher on inshore turbid reefs due to warmer waters and increased nutrients from terrestrial sources. However, recent studies that have looked at external and endolithic bioerosion in more detail have recorded relatively low rates of bioerosion among these reefs. It is expected that lower light levels and higher sedimentation are not favourable conditions for key grazing taxa such as parrot fish and urchins. Further, sedimentation may smother and reduce substrate access for endolithic boring taxa.

These environmental impacts are evident when comparing two studies investigating bioerosion across two reef systems of Western Australia—The clear waters of Ningaloo Reef (Thomson et al. 2024), and the adjacent reefs of Exmouth Gulf (Dee et al. 2023) which are exposed to fluctuating levels of turbidity (Table 9.4). Both studies deployed blocks of coral skeleton (Porites lutea) for 12 months and measured the area of bioeroded substrate via micro-CT scans. While macrobioerosion was greater among the turbid reef sites due to rapid colonisation by boring bivalves, grazing was far greater on Ningaloo due to high densities of parrotfish and urchins, which drove total bioerosion rates to be almost 16 times higher than that recorded within Exmouth Gulf (Table 9.4).

These unfavourable conditions for grazing taxa, may contribute to positive carbonate budgets for turbid reef systems, facilitating sustained reef accretion even in the case of lower coral cover and calcification rates (Dee et al. 2024).

Table 9.4 Bioerosion rates on clear waters of Ningaloo and turbid reefs of nearby Exmouth Gulf recorded by Thomson et al. (2024), and Dee et al. (2023), respectively

	Ningaloo	Exmouth Gulf
Micro-bioerosion (kg m^{-2} $year^{-1}$)	0.12 ± 0.03	0.05 ± 0.01
Macro-bioerosion (kg m^{-2} $year^{-1}$)	0.02 ± 0.01	0.09 ± 0.01
External grazing (kg m^{-2} $year^{-1}$)	2.4 ± 0.20	0.03 ± 0.01
Total (kg m^{-2} $year^{-1}$)	~2.54	~0.16

9.4 The Future of Turbid Coral Reefs

9.4.1 Environmental Interactions and the Refugia Hypothesis

The interaction between temperature, sedimentation, and coral physiology creates a highly complex adaptive landscape. Turbid reefs, by virtue of their physical and biological characteristics often exhibited lower bleaching indices compared to clear-water environments, challenging traditional perceptions of reef vulnerability (Zweifler et al. 2024). On a broader evolutionary scale, Dimitrijević et al. (2024) provided a broad historical perspective, demonstrating that turbid and mesophotic environments have consistently served as critical refugia, with coral occurrences in these zones increasing by approximately 8% following hyperthermal events. Their research suggests that turbid environments have long served as important ecological refugia for reef-building corals. Their analysis, however, spans geological time scales of millions of years, during which adaptive processes could occur. In contrasts, contemporary anthropogenic climate change is unfolding on the order of centuries. This rapid pace raises uncertainty as to whether turbid reefs can function as refugia under present-day conditions (Gaspar et al. 2025).

9.4.2 Coral Bleaching in Turbid Reef Environments: Complexity and Adaptive Strategies

The twenty-first century has witnessed unprecedented marine heatwave events that have devastated coral reefs worldwide. However, turbid reef ecosystems present a complex and nuanced picture of coral resilience under thermal stress. Contrary to expectations, these environments often demonstrate remarkable adaptive capacities during marine heatwave events, challenging the traditional narrative of coral as universally vulnerable to warming oceans (Cardoso et al. 2025; Morgan et al. 2017). A growing body of research—from the GBR to the Philippines and Western Australia—highlights the multiple, often complementary, mechanisms that enable corals in turbid environments to survive increasingly extreme marine conditions.

9.4.3 Physiological Mechanisms of Protection

The complex interplay of environmental, physiological, and ecological mechanisms that shape bleaching responses in turbid reefs holds important implications for the future of these systems. Turbid reef environments can moderate the effects of marine heatwaves through light attenuation, enhanced heterotrophy, and symbiont flexibility—mechanisms that are increasingly recognized as providing these reefs with an edge in the face of climate stress.

9.4.3.1 Turbidity, Light Attenuation and Stress Mitigation

Recent research has highlighted the protective role of turbid waters in mitigating coral bleaching during periods of thermal stress, and the critical role turbidity plays in protecting corals from bleaching through several sophisticated mechanisms (Hock et al. 2017; Browne et al. 2019; Page et al. 2019; Sully and van Woesik 2020). High turbidity levels significantly attenuate harmful UV radiation and excessive light exposure, with studies showing that up to 90% of light can be lost within two meters of depth in turbid waters (Morgan et al. 2017). This natural "umbrella" effectively reduces radiative stress on coral colonies, creating a protective environment that mitigates some of the most damaging effects of marine heatwaves thereby, minimizing or even preventing coral bleaching responses (Warner et al. 1999; Anthony and Connolly 2004; Morgan et al. 2017). The environmental buffering provided by turbid waters extends beyond light attenuation. Suspended particles help moderate local water temperatures, offering a potential buffer against extreme heat (Keighan et al. 2023). As ocean temperatures continue to rise, this protective effect may become increasingly significant allowing turbid reef communities to persist while their clear-water counterparts experience substantial coral mortality (van Woesik and McCaffrey 2017; Camp et al. 2018; Burt et al. 2020).

9.4.3.2 Enhanced Heterotrophy

As discussed above, in addition to shielding corals from light and heat stress, turbid environments may promote heterotrophic feeding strategies in corals. Keighan et al. (2023) revealed that moderate chlorophyll-a concentrations can provide alternative energy sources, enhancing coral resilience during thermal stress. This heterotrophic capability allows corals to maintain their energy reserves, potentially offsetting the metabolic challenges posed by bleaching events (Guest et al. 2016a, b; Banha et al. 2019).

The resilience of these systems is also underpinned by the dominance of stress-tolerant coral taxa and the frequent establishment of thermally resilient symbiont partnerships, both of which are common in turbid habitats that experience naturally fluctuating temperature and light regimes (Van Woesik et al. 2012; Morgan et al. 2017; Teixeira et al. 2019).

9.4.3.3 Variability in Coral Bleaching Responses

Observations from turbid reef systems show remarkably low bleaching rates, with Morgan et al. (2017) observing only 1.5% of coral colonies showing bleaching signs in turbid GBR environments. Conversely, other studies like Valino et al. (2021) in the Philippines found more moderate bleaching levels, with 13–15% of colonies experiencing some degree of stress. This variability is potentially influenced by coral taxa and growth forms.

Coral genera exhibit notably different levels of resilience in these environments. Typically, genera such as *Porites* and *Cyphastrea* demonstrate higher thermal tolerance, while *Acropora* and *Platygyra* tend to be more susceptible to bleaching (Keighan et al. 2023). However, this is not always the case as Zweifler et al. (2024) observed massive and encrusting genera being more susceptible to bleaching at increasingly turbid sites than branching *Acropora*, potentially due to the combined stress of heat and sedimentation on those flatter morphologies (Fig. 9.6). This highlights the complexity of coral responses to thermal stress in turbid environments, suggesting that morphological and species-specific adaptations play a crucial role in survival.

9.4.3.4 Reproductive Impacts and Long-Term Consequences

Bleaching events in turbid environments have subtle yet profound implications for coral reproduction. Research by Briggs et al. (2024) has uncovered subtle yet significant impacts of bleaching events among turbid corals. Even when mortality remains low, these events can substantially reduce reproductive output, with some studies documenting up to a 21% decrease in egg production. Moreover, symbiont densities can often remain supressed for extended periods, indicating potential long-term physiological challenges that extend beyond immediate visual signs of stress.

9.4.4 Emerging Pressures and Shifting Baselines

The expansion of coastal development and increasing human population pressures are projected to elevate sediment discharge into tropical marine environments (Ogston and Field 2010; Hoegh-Guldberg et al. 2017; Morgan et al. 2020b). This trend may result in both the geographical expansion of turbid reef habitats (Guinotte et al. 2003; Oppenheimer et al. 2019) and increased turbidity levels within existing turbid reef systems. Compounding this is nutrient enrichment through coastal water eutrophication, which frequently accompanies increased terrestrial sediment input. This phenomenon poses a significant threat to coastal reef ecosystems since it can erode the ecological stability of turbid reefs over time and reduce their capacity to withstand other global change stressors. The impact is especially pronounced near urban centres, where nutrient concentrations reach their highest levels (Wittenberg and Hunte 1992; Duprey et al. 2016).

9.4.4.1 Limitations and Future Challenges

Despite their remarkable adaptive potential, turbid reefs are not impervious to environmental change. Cardoso et al. (2025) highlighted that prolonged or intense thermal events could overwhelm protective mechanisms, leading to significant coral mortality. Excessive sedimentation poses additional challenges, potentially compromising long-term ecosystem health.

9.4.5 Knowledge Gaps and Research Priorities

The capacity of reef systems to withstand global climate-related challenges, including ocean warming, increased cyclone activity, and sea level rise, will depend significantly

Fig. 9.6 Coral colonies during a marine heatwave event in Exmouth Gulf, Western Australia in 2021. (**a**) bleached encrusting coral among healthy foliose colonies. (**b**) a bleached *Astreopora* with a competing healthy *Acropora*. (**c**) bleached encrusting coral neighboured by healthy foliose and branching corals

on their ability to cope with these localized environmental pressures. However, our current understanding of how these various stressors interact with water quality to affect reef functionality remains inadequate. Filling this knowledge gap is essential for our ability to accurately assess whether turbid reef systems possess inherent resilience to future environmental challenges and for identifying the thresholds at which their adaptive advantages might fail. A comprehensive multidisciplinary examination is needed to evaluate the potential responses of turbid reefs to both local and global stressors, identify key resilience indicators, and establish priorities for future research directions in turbid reef ecosystems.

9.5 Conclusions

Turbid reefs represent a significant and often overlooked component of the global coral reef ecosystem. Far from being merely degraded versions of clear-water reefs, these ecosystems demonstrate remarkable adaptations to challenging environmental conditions.

Key insights from the research presented in this chapter include:

1. Turbid reef environments support diverse and resilient coral communities that have evolved specialized adaptations to cope with reduced light, elevated sedimentation, and variable environmental conditions.
2. The distinction between natural and anthropogenic turbid reefs is crucial, as these systems differ fundamentally in their ecological function, resilience capabilities, and conservation value.
3. Turbid reef environments may serve as important refugia for corals facing global climate change pressures, as they demonstrate enhanced resistance to thermal stress and bleaching events.
4. The predominance of stress-tolerant coral species on turbid reefs, combined with their nutritional flexibility and energy reserves, provides these ecosystems with inherent resilience capabilities.
5. Future research should focus on better understanding the complex interactions between local and global stressors on turbid reef ecosystems, particularly as coastal development continues to expand.

As global climate change continues to threaten coral reef ecosystems worldwide, turbid reefs may play an increasingly important role in coral conservation efforts. Their ecological characteristics and demonstrated resilience make them valuable natural laboratories for understanding coral adaptation mechanisms and potential evolutionary pathways. Conservation strategies that recognize the unique properties and conservation value of turbid reef ecosystems will be essential for preserving coral biodiversity into the future.

References

Anthony KRN (1999) Coral suspension feeding on fine particulate matter. J Exp Mar Biol Ecol 232:85–106

Anthony KRN (2006) Enhanced energy status of corals on coastal, high-turbidity reefs. Mar Ecol Prog Ser 319:111–116

Anthony KRN, Connolly SR (2004) Environmental limits to growth: physiological niche boundaries of corals along turbidity-light gradients. Oecologia 141:373–384

Anthony KRN, Fabricius KE (2000) Shifting roles of heterotrophy and autotrophy in coral energetics under varying turbidity. J Exp Mar Biol Ecol 252:221–253

Anthony K, Larcombe P (2000) Coral reefs in turbid waters: sediment-induced stresses in corals and likely mechanisms of adaptation. In: Proceedings of the 9th international coral reef symposium I, pp 239–244

Anthony KRN, Connolly SR, Hoegh-Guldberg O (2007) Bleaching, energetics, and coral mortality risk: effects of temperature, light, and sediment regime. Limnol Oceanogr 52:716–726

Anthony KRN, Hoogenboom MO, Maynard JA, Grottoli AG, Middlebrook R (2009) Energetics approach to predicting mortality risk from environmental stress: a case study of coral bleaching. Funct Ecol 23:539–550

Banha TNS, Capel KCC, Kitahara MV, Francini-Filho RB, Francini CLB, Sumida PYG, Mies M (2019) Low coral mortality during the most intense bleaching event ever recorded in subtropical Southwestern Atlantic reefs. Coral Reefs:1–7

Berkelmans R, Van Oppen MJH (2006) The role of zooxanthellae in the thermal tolerance of corals: a "nugget of hope" for coral reefs in an era of climate change. Proc R Soc B Biol Sci 273:2305–2312

Birrell CL, McCook LJ, Willis BL (2005) Effects of algal turfs and sediment on coral settlement. Mar Pollut Bull 51:408–414

Bonesso JL, Cuttler MVW, Browne NK, Mather CC, Paumard V, Hiscock W et al (2023) Reef island evolution in a turbid-water coral reef province of the Indo-Pacific. Depositional Rec 9:921–934. Available from:. https://doi.org/10.1002/dep2.242

Briggs ND, Page CA, Giuliano C et al (2024) Dissecting coral recovery: bleaching reduces reproductive output in *Acropora millepora*. Coral Reefs **43**:557–569. https://doi.org/10.1007/s00338-024-02483-y

Browne NK (2012) Spatial and temporal variations in coral growth on an inshore turbid reef subjected to multiple disturbances. Mar Environ Res 77:71–83

Browne NK, Smithers SG, Perry CT (2012) Coral reefs of the turbid inner-shelf of the Great Barrier Reef, Australia: an environmental and geomorphic perspective on their occurrence, composition and growth. Earth Sci Rev 115:1–20

Browne NK, Smithers SG, Perry CT (2013) Spatial and temporal variations in turbidity on two inshore turbid reefs on the Great Barrier Reef, Australia. Coral Reefs 32:195–210

Browne NK, Tay J, Todd PA (2015a) Recreating pulsed turbidity events to determine coral-sediment thresholds for active management. J Exp Mar Biol Ecol 466:98–109

Browne NK, Tay JKL, Low J, Larson O, Todd PA (2015b) Fluctuations in coral health of four common inshore reef corals in response to seasonal and anthropogenic changes in water quality. Mar Environ Res 105:39–52

Browne N, Braoun C, McIlwain J, Nagarajan R, Zinke J (2019) Borneo coral reefs subject to high sediment loads show evidence of resilience to various environmental stressors. PeerJ 7:e7382

Burt JA, Camp EF, Enochs IC, Johansen JL, Morgan KM, Riegl B, Hoey AS (2020) Insights from extreme coral reefs in a changing world. Coral Reefs

Camp EF, Schoepf V, Mumby PJ, Hardtke LA, Rodolfo-Metalpa R, Smith DJ, Suggett DJ (2018) The future of coral reefs subject to rapid climate change: lessons from natural extreme environments. Front Mar Sci 5:4

Cardoso FC, Carneiro IM, Castro GM et al (2025) Nowhere to hide: massive corals' mortality after cumulative heat stress in a putative climate change refugium. Coral Reefs 44:321–337. https://doi.org/10.1007/s00338-024-02609-2

Cartwright PJ, Browne NK, Belton D et al (2023) Long-term spatial variations in turbidity and temperature provide new insights into coral-algal states on extreme/marginal reefs. Coral Reefs 42:859–872. https://doi.org/10.1007/s00338-023-02393-5

Castro CB, Segal B, Negrão F, Calderon EN (2012) Four-year monthly sediment deposition on turbid southwestern Atlantic Coral Reefs, with a comparison of benthic assemblages. Braz J Oceanogr 60:49–63

Chan YKS, Ang ACF, Choo M, Oh RM, Morgan KM, Todd PA, O'Leary MJ, Huang D (2025) Holocene coral assemblages reveal similarities to living communities in Singapore's urban reef environment. Palaeogeogr Palaeoclimatol Palaeoecol 669:112930. https://doi.org/10.1016/j.palaeo.2025.112930

Conti-Jerpe IE, Thompson PD, Wong CWM, Oliveira NL, Duprey NN, Moynihan MA, Baker DM (2020) Trophic strategy and bleaching resistance in reef-building corals. Sci Adv 6:2–4

Cooper TF, Ridd PV, Ulstrup KE, Humphrey C, Slivkoff M, Fabricius KE (2008) Temporal dynamics in coral bioindicators for water quality on coastal coral reefs of the Great Barrier Reef. Mar Freshw Res 59:703–716

Cruz-Pinon G, Carricart-Ganivet JP, Espinoza-Avalos J (2003) Monthly skeletal extension rates of the hermatypic corals Montastrea annularis and Montastrea faveolata: biological and environmental controls. Mar Biol 143:491e500

Dandan SS, Falter JL, Lowe RJ, McCulloch MT (2015) Resilience of coral calcification to extreme temperature variations in the Kimberley region, northwest Australia. Coral Reefs 34:1151–1163. https://doi.org/10.1007/s00338-015-1335-6

Darling ES, Alvarez-Filip L, Oliver TA, McClanahan TR, Côté IM (2012) Evaluating life-history strategies of reef corals from species traits. Ecol Lett 15:1378–1386. https://doi.org/10.1111/j.1461-0248.2012.01861.x

Dee S, DeCarlo T, Lozić I, Nilsen J, Browne (2023) Low bioerosion rates on inshore turbid reefs of Western Australia. Diversity, doi:https://doi.org/10.3390/d15010062

Dee S, Zweifler A, Cuttler M, Nilsen J, Bonesso J, O'Leary M, Browne NK (2024) The application of carbonate and sediment budgets to assess the stability of marginal reef systems. Mar Geol 473:107324

Denis V, Ferrier-Pagès C, Schubert N, Coppari M, Baker DM, Camp EF, Gori A, Grottoli AG, Houlbrèque F, Maier SR, Mancinelli G, Martinez S, Yalçın Özdilek Ş, Radice VZ, Ribes M, Richter C, Viladrich N, Rossi S (2024) Heterotrophy in marine animal forests in an era of climate change. Biol Rev Camb Philos Soc 99(3):965–978. https://doi.org/10.1111/brv.13053

Dias M, Ferreira A, Gouveia R, Cereja R, Vinagre C (2018) Mortality, growth and regeneration following fragmentation of reef-forming corals under thermal stress. J Sea Res 141:71–82. https://doi.org/10.1016/j.seares.2018.08.008

Dimitrijević D, Santodomingo N, Kiessling W (2024) Reef refugia in the aftermath of past episodes of global warming. Coral Reefs 43:1431–1442. https://doi.org/10.1007/s00338-024-02548-y

Dsikowitzky L, Ferse S, Schwarzbauer J, Vogt TS, Irianto HE (2016) Impacts of megacities on tropical coastal ecosystems – the case of Jakarta, Indonesia. Mar Pollut Bull 110:621–623

Duprey NN, Yasuhara M, Baker DM (2016) Reefs of tomorrow: eutrophication reduces coral biodiversity in an urbanized seascape. Glob Chang Biol 22:3550–3565

Fabricius K (2005) Effects of terrestrial runoff on the ecology of corals and coral reefs: review and synthesis. Mar Pollut Bull 50:125–146. https://doi.org/10.1016/j.marpolbul.2004.11.028

Fabricius KE, Cooper TF, Humphrey C, Uthicke S, De'ath G, Davidson J, LeGrand H, Thompson A, Schaffelke B (2012) A bioindicator system for water quality on inshore coral reefs of the Great Barrier Reef. *Mar Pollut Bull 65*(4–9):320–332. https://doi.org/10.1016/j.marpolbul.2011.09.004

Fabricius KE, De'ath G, Humphrey C, Zagorskis I, Schaffelke B (2013) Intra-annual variation in turbidity in response to terrestrial runoff on near-shore coral reefs of the Great Barrier Reef. *Estuar Coast Shelf Sci 116*:57–65. https://doi.org/10.1016/j.ecss.2012.03.010

Fabricius KE, Logan M, Weeks S, Brodie J (2014) The effects of river run-off on water clarity across the central Great Barrier Reef. *Mar Pollut Bull 84*(1–2):191–200. https://doi.org/10.1016/j.marpolbul.2014.05.012

Faxneld S, Jörgensen TL, Tedengren M (2010) Effects of elevated water temperature, reduced salinity and nutrient enrichment on the metabolism of the coral Turbinaria mesenterina. Estuar Coast Shelf Sci 88(4):482–487. https://doi.org/10.1016/j.ecss.2010.05.008

Flores F, Hoogenboom MO, Smith LD, Cooper TF, Abrego D, Negri AP (2012) Chronic exposure of corals to fine sediments: lethal and sub-lethal impacts. PLoS One 7:1–12

Foster T, Smith A, Jury M, Driscoll A (2011) Overview of PIANC report 108 – dredging and port construction around coral reefs. In: 20th Australasian Coastal and Ocean Engineering Conference 2011 and the 13th Australasian Port and Harbour Conference 2011, COASTS and PORTS 2011, pp 573–578

Foster T, Short JA, Falter JL, Ross C, Mcculloch MT (2014) Reduced calcification in Western Australian corals during anomalously high summer water temperatures. J Exp Mar Biol Ecol 461:133–143. https://doi.org/10.1016/j.jembe.2014.07.014

Fox MD, Smith EAE, Smith JE, Newsome SD (2019) Trophic plasticity in a common reef-building coral: Insights from δ13C analysis of essential amino acids. Funct Ecol 33:2203–2214

Gaspar TL, Francini-Filho R, Mies M, Rodrigues RR, Neto AG, da Silva Junior CA, Banha T, Longo GO, Segal B (2025) Partial attenuation of coral bleaching by turbidity along the Brazilian coast. Coral Reefs. https://doi.org/10.1007/s00338-025-02745-3K

Goldberg WM (2018) Coral food, feeding, nutrition, and secretion: a review. In: Kloc M, Kubiak JZ (eds) Marine organisms as model systems in biology and medicine. Springer, pp 377–421

Goodkin N, Switzer A, McCorry D, DeVantier L, True J, Hughen K, Angeline N, Yang T (2011) Coral communities of Hong Kong: long-lived corals in a marginal reef environment. Mar Ecol Prog Ser 426:185–196

Grottoli AG, Rodrigues LJ, Palardy JE (2006) Heterotrophic plasticity and resilience in bleached corals. Nature 440:1186–1189

Guest JR, Low J, Tun K, Wilson B, Ng C, Raingeard D, Ulstrup KE, Tanzil JTI, Todd PA, Toh TC, McDougald D, Chou LM, Steinberg PD (2016a) Coral community response to bleaching on a highly disturbed reef. Sci Rep:6

Guest JR, Tun K, Low J, Vergés A, Marzinelli EM, Campbell AH, Bauman AG, Feary DA, Chou LM, Steinberg PD (2016b) 27 years of benthic and coral community dynamics on turbid, highly urbanised reefs off Singapore. Sci Rep 6:1–10

Guinotte JM, Buddemeier ARW, Kleypas AJA (2003) Future coral reef habitat marginality: temporal and spatial effects of climate change in the Pacific basin. Coral Reefs 22:551–558

Harii S, Nadaoka K, Yamamoto M, Iwao K (2007) Temporal changes insettlement, lipid content and lipid composition of larvae of the spawning hermatypic coral Acropora tenuis. Mar Ecol Prog Ser 346:89–96

Harii S, Yamamoto M, Hoegh-Guldberg O (2010) The relative contribution of dinoflagellate photosynthesis and stored lipids to the survivorship of symbiotic larvae of the reef-building corals. Mar Biol 157:1215–1224

Harriott VJ (1993) Coral lipids and environmental stress. Environ Monit Assess 25:131e139

Highsmith RC (1979) Coral growth rates and environmental control of density banding. J Exp Mar Biol Ecol 37:105e125

Hock K, Wolff NH, Ortiz JC, Condie SA, Anthony KRN, Blackwell PG, Mumby PJ (2017) Connectivity and systemic resilience of the Great Barrier Reef. PLoS Biol 15:e2003355

Hoegh-Guldberg O, Poloczanska ES, Skirving W, Dove S (2017) Coral reef ecosystems under climate change and ocean acidification. Front Mar Sci 4:158

Hoitink AJF (2004) Tidally-induced clouds of suspended sediment connected to shallow-water coral reefs. Mar Geol 208:13–31

Houlbrèque F, Ferrier-Pagès C (2009) Heterotrophy in tropical scleractinian corals. Biol Rev 84:1–17

Howells EJ, Beltran VH, Larsen NW, Bay LK, Willis BL, Van Oppen MJH (2012) Coral thermal tolerance shaped by local adaptation of photosymbionts. Nat Clim Chang 2(2):116–120

Hughes AD, Grottoli AG (2013) Heterotrophic compensation: a possible mechanism for resilience of coral reefs to global warming or a sign of prolonged stress? PLoS One 8:e81172

Hussain A, Kumar SHK, Kumar AA, Prathiviraj R, Renjith K, Kiran GS, Selvin J (2025) Delineating the emergence of thermally tolerant Symbiodiniaceae genotypes across the dominant coral species of a turbid reef: adaptation and resilience strategy. Sci Total Environ 963:178255. https://doi.org/10.1016/j.scitotenv.2024.178255. RetryClaude can make mistakes. Please double-check responses. 3.7 Sonnet

Januchowski-Hartley FA, Bauman AG, Morgan KM, Seah JCL, Huang D, Todd PA (2020) Accreting coral reefs in a highly urbanized environment. Coral Reefs:1–23

Johnson JA, Perry CT, Smithers SG, Morgan KM, Santodomingo N, Johnson KG (2017) Palaeoecological records of coral community development on a turbid, nearshore reef complex: baselines for assessing ecological change. Coral Reefs 36:685–700

Jones R, Bessell-Browne P, Fisher R, Klonowski W, Slivkoff M (2016) Assessing the impacts of sediments from dredging on corals. Mar Pollut Bull 102:9–29

Jones R, Giofre N, Luter HM, Neoh TL, Fisher R, Duckworth A (2020) Responses of corals to chronic turbidity. Sci Rep 10

Jones R, Pineda M-C, Luter HM, Fisher R, Francis D, Klonowski W, Slivkoff M (2021) Underwater light characteristics of turbid coral reefs of the inner Central Great Barrier Reef. Front Mar Sci 8:727206. https://doi.org/10.3389/fmars.2021.727206

Juhi ZS, Abdul Mubin NAA, Jonik MGG, Salleh S, Mohammad M (2021) Impact of short-term light variability on the photobiology of turbid water corals. J Sea Res 175:102088. https://doi.org/10.1016/j.seares.2021.102088

Keighan R, van Woesik R, Yalon A et al (2023) Moderate chlorophyll-a environments reduce coral bleaching during thermal stress in Yap, Micronesia. Sci Rep 13:9338. https://doi.org/10.1038/s41598-023-36355-2

Larcombe P, Costen A, Woolfe KJ (2001) The hydrodynamic and sedimentary setting of nearshore coral reefs, Central Great Barrier Reef Shelf, Australia: Paluma Shoals, a case study. Sedimentology 48:811–835

Lasker HR (1980) Sediment rejection by reef corals: the roles of behavior and morphology in Montastrea covernosa (Linnaeus). J Exp Mar Biol Ecol 47:77–87

Levitan DR (2006) Relationship between egg size and fertilization success in broadcast spawning marine invertebrates. Integrative and Comparative Biology 46:298–311

Loiola M, Cruz IC, Lisboa DS, Mariano-Neto E, Leão ZM, Oliveira MD, Kikuchi RK (2019) Structure of marginal coral reef assemblages under different turbidity regimes. Mar Environ Res 147:138–148

Lough JM, Cantin NE, Benthuysen JA, Cooper TF (2016) Environmental drivers of growth in massive Porites corals over 16 degrees of latitude along Australia's northwest shelf. Limnol Oceanogr 61:684–700. https://doi.org/10.1002/lno.10244

Macdonald RK, Ridd PV, Whinney JC, Larcombe P, Neil DT (2013) Towards environmental management of water turbidity within open coastal waters of the Great Barrier Reef. Mar Pollut Bull 74:82–94

Mallela J, Perry CT (2007) Calcium carbonate budgets for two coral reefs affected by different terrestrial runoff regimes, Rio Bueno, Jamaica. Coral Reefs 26:129–145

Marshall PA, Baird AH (2000) Bleaching of corals on the Great Barrier Reef: differential susceptibilities among taxa. Coral Reefs 19:155–163

Morgan KM, Perry CT, Johnson JA, Smithers SG (2017) Nearshore turbid-zone corals exhibit high bleaching tolerance on the Great Barrier Reef following the 2016 ocean warming event. Front Mar Sci 4:1–13

Morgan KM, Moynihan MA, Sanwlani N, Switzer AD (2020a) Light limitation and depth-variable sedimentation drives vertical reef compression on turbid coral reefs. Front Mar Sci 7

Morgan KM, Perry CT, Arthur R, Williams HTP, Smithers SG (2020b) Projections of coral cover and habitat change on turbid reefs under future sea-level rise. Proc R Soc B Biol Sci 287:20200541

Muscatine L, Porter JW (1977) Reef corals: mutualistic symbioses adapted to nutrient-poor environments. Bioscience 27:454–460

Ogston AS, Field ME (2010) Predictions of turbidity due to enhanced sediment resuspension resulting from sea-level rise on a fringing coral reef: evidence from Molokai, Hawaii. J Coast Res 26:1027–1037

Oppenheimer M, Glavovic BC, Hinkel J, Wal R van de, Magnan AK, Abd-Elgawad A, Cai R, Cifuentes-Jara M, DeConto RM, Ghosh T, Hay J, Isla F, Marzeion B, Meyssignac B, Sebesvari Z (2019) Sea level rise and implications for low lying islands, coasts and communities

Page CE, Leggat W, Heron SF, Choukroun SM, Lloyd J, Ainsworth TD (2019) Seeking resistance in coral reef ecosystems: the interplay of biophysical factors and bleaching resistance under a changing climate: the interplay of a reef's biophysical factors can mitigate the coral bleaching response. BioEssays 41:1–12

Palardy JE, Grottoli AG, Matthews KA (2005) Effects of upwelling, depth, morphology and polyp size on feeding in three species of Panamanian corals. Mar Ecol Prog Ser 300:79–89

Palmer SE, Perry CT, Smithers SG, Gulliver P (2010) Internal structure and accretionary history of a nearshore, turbid-zone coral reef: Paluma Shoals, Central Great Barrier Reef, Australia. Mar Geol 276:14–29

Pandolfi JM, Bradbury RH, Sala E, Hughes TP, Bjorndal KA, Cooke RG et al (2003) Global trajectories of the long-term decline of coral reef ecosystems. Science 301:955–958

Perez K, Rodgers KS, Jokiel PL, Lager CV, Lager DJ (2014) Effects of terrigenous sediment on settlement and survival of the reef coral Pocillopora damicornis. PeerJ 2014

Perry CT (2005) Structure and development of detrital reef deposits in turbid nearshore environments, Inhaca Island, Mozambique. Mar Geol 214:143–161

Perry CT, Smithers SG (2010) Evidence for the episodic "turn on" and "turn off" of turbid-zone coral reefs during the late Holocene sea-level highstand. Geology 38:119–122

Perry CT, Smithers SG, Palmer SE, Larcombe P, Johnson KG (2008) 1200 year paleoecological record of coral community development from the terrigenous inner shelf of the Great barrier reef. Geology 36:691–694

Perry CT, Smithers SG, Gulliver P, Browne NK (2012) Evidence of very rapid reef accretion and reef growth under high turbidity and terrigenous sedimentation. Geology 40:719–722

Perry CT, Alvarez-Filip L, Graham NAJ, Mumby PJ, Wilson SK, Kench PS, Manzello DP, Morgan KM, Slangen ABA, Thomson DP, Januchowski-Hartley F, Smithers SG, John S, Labrador C (2018a) Loss of coral reef growth capacity to track future increases in sea level. Nature 558:396–400

Perry CT, Lange I, Januchowski-Hartley FA (2018b) ReefBudget Indo Pacific: online resource and methodology. Retrieved from http://geography.exeter.ac.uk/reefbudget/

Porter JW (1976) Autotrophy, heterotrophy, and resource partitioning in Caribbean reef-building corals, vol 110. The University of Chicago Press for The American Society of Naturalists Stable, pp 731–742

Potts D, Jacobs J (2000) Evolution of reef-building scleractinian corals in turbid environments: a paleo-ecological hypothesis. Proc 9th Int Coral Reef Symp Bali 1:249–254

Restrepo JD, Zapata P, Díaz JM, Garzón-Ferreira J, García CB (2006) Fluvial fluxes into the Caribbean Sea and their impact on coastal ecosystems: the Magdalena River, Colombia. Glob Planet Chang 50:33–49

Ricardo GF, Jones RJ, Negri AP, Stocker R (2016) That sinking feeling: Suspended sediments can prevent the ascent of coral egg bundles. Sci Rep 6

Richards ZT, Garcia RA, Wallace CC, Rosser NL, Muir PR (2015) A diverse assemblage of reef corals thriving in a dynamic intertidal reef setting (Bonaparte archipelago, Kimberley, Australia). PLoS One 10

Richards Z, Bryce M, Bryce C (2018) The composition and structure of shallow benthic reef communities in the Kimberley, North-West Australia. Record West Aus Mus Suppl 85:75

Rocker MM, Francis DS, Fabricius KE, Willis BL, Bay LK (2017) Variation in the health and biochemical condition of the coral Acropora tenuis along two water quality gradients on the Great Barrier Reef, Australia. Mar Pollut Bull 119:106–119

Rogers CS (1990) Responses of coral reefs and reef organisms to sedimentation. Mar Ecol Prog Ser 62:185–202

Sanders D, Baron-Szabo RC (2005) Scleractinian assemblages under sediment input: their characteristics and relation to the nutrient input concept. Palaeogeogr Palaeoclimatol Palaeoecol 216:139–181

Santodomingo N, Novak V, Pretkovic V, Marshall N, Di Martino E, Capelli ELG, Rosler A, Rcich S, Braga JC, Renema W, Johnson KG (2015) A divers patch reef from turbid habitats in the middle miocene (East Kalimantan, Indonesia). Palaios 30:128–149

Saper J, Høj L, Humphrey C et al (2023) Quantifying capture and ingestion of live feeds across three coral species. Coral Reefs 42:931–943. https://doi.org/10.1007/s00338-023-02397-1

Schleyer MH, Celliers L (2003) Coral dominance at the reef-sediment interface in marginal coral communities at Sodwana Bay, South Africa. Mar Freshw Res 54:967–972

Schoepf V, Baumann JH, Barshis DJ, Browne NK, Camp EF, Comeau S, Cornwall CE, Guzmán HM, Riegl B, Rodolfo-Metalpa R, Sommer B (2023) Corals at the edge of environmental limits: a new conceptual framework to re-define marginal and extreme coral communities. Sci Total Environ 884:163688

Segal B, Castro CB (2011) Coral community structure and sedimentation at different distances from the coast of the Abrolhos Bank, Brazil. Braz J Oceanogr 59:119–129

Smith A, Foster T, Corcoran E, Monkivitch J (2007) Dredging and material relocation in sensitive coral environments. In: Proceedings 18th world dredging congress (WODCON XVIII), pp 945–955

Smith EG, Gurskaya A, Hume BCC, Voolstra CR, Todd PA, Bauman AG, Burt JA (2020) Low Symbiodiniaceae diversity in a turbid marginal reef environment. Coral Reefs 39:545–553

Smithers S, Larcombe P (2003) Late Holocene initiation and growth of a nearshore turbid-zone coral reef: Paluma Shoals, Central Great Barrier Reef, Australia. Coral Reefs 22:499–505

Sofonia JJ, Anthony KRN (2008) High-sediment tolerance in the reef coral Turbinaria mesenterina from the inner Great Barrier Reef lagoon (Australia). Estuar Coast Shelf Sci 78:748–752

Solihuddin T, Collins LB, Blakeway D, O'Leary MJ (2015) Holocene coral reef growth and sea level in a macrotidal, high turbidity setting: Cockatoo Island, Kimberley Bioregion, Northwest Australia. Mar Geol 359:50–60

Solihuddin T, O'Leary MJ, Blakeway D, Parnum I, Kordi M, Collins LB (2016) Holocene reef evolution in a macrotidal setting: Buccaneer Archipelago, Kimberley Bioregion, Northwest Australia. Coral Reefs 35:783–794

Stafford-Smith MG (1993) Sediment-rejection efficiency of 22 species of Australian scleractinian corals. Mar Biol 115:229–243

Sully S, vanWoesik R (2020) Turbid reefs moderate coral bleaching under climate-related temperature stress. Glob Change Biol.;26:1367–1373. https://doi.org/10.1111/gcb.14948

Tan KS, Acerbi E, Lauro FM, Siang Tan K, Acerbi E, Lauro FM, Tan KS, Acerbi E, Lauro FM (2016) Marine habitats and biodiversity of Singapore's coastal waters: a review. Reg Stud Mar Sci 8:340–352

Teixeira CD, Leitão RLL, Ribeiro FV, Moraes FC, Neves LM, Bastos AC, Pereira-Filho GH, Kampel M, Salomon PS, Sá JA, Falsarella LN, Amario M, Abieri ML, Pereira RC, Amado-Filho GM, Moura RL (2019) Sustained mass coral bleaching (2016–2017) in Brazilian turbid-zone reefs: taxonomic, cross-shelf and habitat-related trends. Coral Reefs 38:801–813

Thomson DP, Dee S, Doropoulos C, Orr M, Wilson SK, Hoey AS (2024) High rates oferosion on a wave-exposed fringing coral reef. Limnol Oceanogr. https://doi.org/10.1002/lno.12586

Titlyanov EA, Titlyanova TV (2002) Reef-building corals—symbiotic autotrophic organisms: 1. General structure, feeding pattern, and light-dependent distribution in the shelf. Russ J Mar Biol 28(1):S1–S15

Travaglione N, Evans R, Moustaka M et al (2013) Scleractinian corals rely on heterotrophy in highly turbid environments. Coral Reefs 42:997–1010. https://doi.org/10.1007/s00338-023-02407-2

Tudhope AW, Scoffin TP (1994) Growth and structure of fringing reefs in a muddy environment, South Thailand. J Sediment Res A Sediment Petrol Process:752–764

Valino DAM, Baria-Rodriguez MV, Dizon RM, Aliño PM (2021) Responses of Buluan Island turbid fringing reefs, Southern Philippines to the 2016 thermal anomaly. Reg Stud Mar Sci 43:101704. https://doi.org/10.1016/j.rsma.2021.101704

Van Duin EHS, Blom G, Los FJ, Maffione R, Zimmerman R, Cerco CF, Dortch M, Best EPH (2001) Modeling underwater light climate in relation to sedimentation, resuspension, water quality and autotrophic growth. Hydrobiologia 444:25–42

van Woesik R, McCaffrey KR (2017) Repeated thermal stress, shading, and directional selection in the Florida Reef Tract. Front Mar Sci 4:182

Van Woesik R, Houk P, Isechal AL, Idechong JW, Victor S, Golbuu Y (2012) Climate-change refugia in the sheltered bays of Palau: analogs of future reefs. Ecol Evol 2:2474–2484

Veron JEN (2000) Corals of the world, vol 1–3. Australian Institute of Marine Science and CRR, Townsville

Wagner DE, Kramer P, Van Woesik R (2010) Species composition, habitat, and water quality influence coral bleaching in Southern Florida. Mar Ecol Prog Ser 408:65–78

Warner ME, Fitt WK, Schmidt GW (1999) Damage to photosystem II in symbiotic dinoflagellates: a determinant of coral bleaching. Proc Natl Acad Sci 96:8007–8012

Weber M, De Beer D, Lott C, Polerecky L, Kohls K, Abed RMM, Ferdelman TG, Fabricius KE (2012) Mechanisms of damage to corals exposed to sedimentation. Proc Natl Acad Sci USA 109

Welle PD, Small MJ, Doney SC, Azevedo IL (2017) Estimating the effect of multiple environmental stressors on coral bleaching and mortality. PLoS One 12:e0175018

Wilson B, Blake S, Ryan D, Hacker J (2011) Reconnaissance of species-rich coral reefs in a muddy, macro-tidal, enclosed embayment, Talbot Bay, Kimberley, Western Australia. J R Soc West Aust 94:251–265

Wittenberg M, Hunte W (1992) Effects of eutrophication and sedimentation on juvenile corals. Mar Biol 138:131–138

Wooldridge SA (2014) Formalising a mechanistic linkage between heterotrophic feeding and thermal bleaching resistance. Coral Reefs 33:1131–1136

Yamano H, Inoue T, Adachi H, Tsukaya K, Adachi R, Baba S (2019) Holocene sea-level change and evolution of a mixed coral reef and mangrove system at Iriomote Island, Southwest Japan. Estuar Coast Shelf Sci 220:166–175

Zhao M, Zhong Y, Zhang S, Zhang H, Yang H, Shi Q, Yan H, Tong H, Xu W, Yan W (2022) Coral community of Holocene coral reef in the southern South China Sea and its significance for reef growth. Palaeogeogr Palaeoclimatol Palaeoecol 602:111152. https://doi.org/10.1016/j.palaeo.2022.111152

Zweifler A, O'Leary M, Morgan K, Browne NK (2021) Turbid coral reefs: past, present and future – a review. Diversity (Basel) 13:251

Zweifler A, Dee S, Browne NK (2024) Resilience of turbid coral communities to marine heatwave. Coral Reefs 43:1303–1315. https://doi.org/10.1007/s00338-024-02538-0

Upwelling: Curse or Blessing for Coral Reefs?

10

Marlene Wall and Claudio Richter

Abstract

Coral reef and upwelling systems occupy opposite ends of the marine ecological spectrum. While coral reefs are rich benthic ecosystems thriving in warm, nutrient-depleted and aragonite-supersaturated waters, upwelling systems are just the contrary—rich pelagic systems flourishing in cold, nutrient-rich and aragonite-depleted water. In the light of these contrasts, coral reefs and upwelling systems appear mutually exclusive. But not only do corals cope with upwelling waters, they may also benefit from upwelling under thermal stress. In this chapter, we explore the different mechanisms and spatio-temporal scales of upwelling, and how this 'extremeness' in the corals' environment affects their composition, growth, and survival vis-a-vis potential space competitors in so-called 'marginal' reefs. Finally, we review the role of upwelling and, more generally, of ocean variability and dynamics, on the future of coral reefs under anthropogenic threats, particularly warming, acidification and hypoxia. The answer to the question in the chapter's title appears context-dependent and needs to be assessed against other prevailing drivers both in the past and present. Identifying thresholds and tipping points across reefs affected by upwelling can contribute to better understanding the evolution of reefs in a changing ocean.

Keywords

Upwelling · Internal waves · Coral reefs · Thermal refugia · Resource subsidies · Community shifts · Environmental limits · Tipping points

M. Wall (✉)
The GEOMAR—Helmholtz Centre for Ocean Research Kiel, University of Kiel, Kiel, Germany

Department of Functional and Evolutionary Ecology, University of Vienna, Vienna, Austria
e-mail: mwall@geomar.de

C. Richter
Alfred Wegener Institute (AWI), Bremerhaven, Germany

10.1 Introduction

The first global systematic treatise on coral reefs was the monograph of Charles Darwin from 1842 on 'The Structure and Distribution of Coral Reefs', where he related the conspicuous absence of coral reefs in some low latitude regions he visited to anomalous cold water. He stated in Chap. 4—On the Growth of Coral Reefs—and Sect. 4.1—*On the distribution of Coral-Reefs, and on the Conditions favourable to their Increase*—on page 61: *'The entire absence of coral reefs in certain large areas within the tropical seas, is a remarkable fact. Thus no coral-reefs were observed, during the surveying voyages of the Beagle and her Tender on the west coast of South America south of the equator, or round the Galápagos Islands. It appears, also, that there are none* north of the Equator; Mr. Lloyd, who surveyed the isthmus of Panamá, remarked to me, that although he had seen corals living in the Bay of Panamá, yet he had never observed any reefs formed by them. I at first attributed this absence of reefs on the coasts of Peru and of the Galápagos Islands*[†]*, to the coldness of the currents from the south, but the Gulf of Panamá is one of the hottest pelagic districts in the world.*[‡] *In the central parts of the Pacific there are islands, entirely free from reefs; in some few of these cases I have thought, that this was owing to recent volcanic action; but the existence of reefs round the greater part of Hawaii, one of the Sandwich Islands, shows that recent volcanic action does not necessarily prevent their growth.'*[1]

[1]Note added by Charles Darwin: * I have been informed that this is the case, by Lieut. Ryder, R.N. and others who have had ample opportunities for observation.

[†] The mean temperature of the surface sea from observations made by the direction of Capt. FitzRoy on the shores of the Galápagos Islands, between the 16th of September and the 20th of October, 1835, was 68°

M. O. Soares et al. (eds.), *Extreme and Marginal Coral Reefs and Communities*, Coral Reefs of the World 24,
https://doi.org/10.1007/978-3-032-28035-0_10

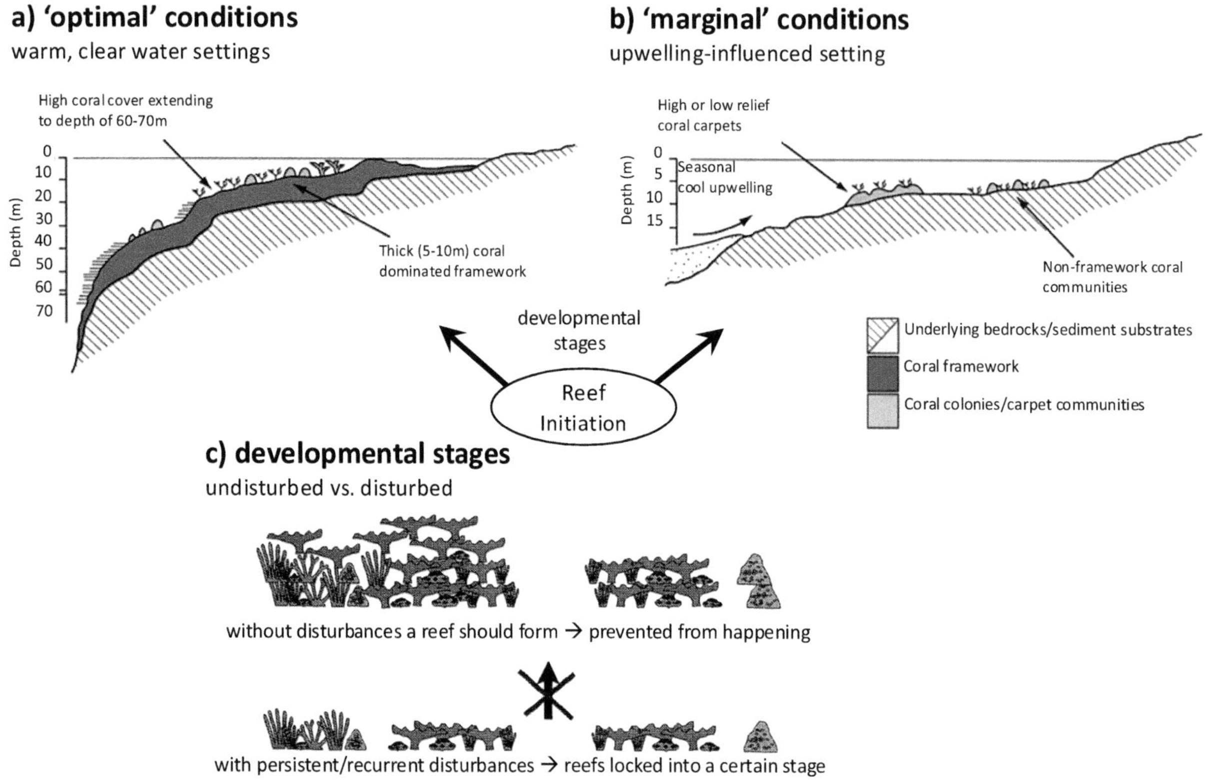

Fig. 10.1 Some schematic illustrations of normal and marginal coral reefs in the literature. Coral reefs thriving in (**a**) 'optimal' environmental conditions develop a three-dimensional reef framework, while limiting or 'marginal' conditions in upwelling-influenced habitats (**b**) prevent reef framework formation. This has also be described as alternative developmental stages of reef formation (adapted from Perry and Larcombe 2003). (**c**) Model for framework destruction as driven by repeated 'disturbance' or stunting of a dominant frame building genus (adapted from Riegl 2001) that does not allow return to normal reef framework developmental stage

This description includes areas covered in the present chapter on 'extreme' and 'marginal' coral reefs. In this chapter, we focus on areas where benthic communities are subjected to upwelling of varying intensity and exposure to deep-water that is usually cooler, elevated in nutrients, reduced in oxygen and pH compared to surface water (e.g. Manzello et al. 2008; Schmidt et al. 2012). Darwin's early account of the adverse effects of upwelling on shallow-water corals adapted to warm, oligotrophic, oxygen-replete and alkaline waters was followed by numerous discoveries in other areas where reef formation was altered through the influence of cool, nutrient-rich upwelling waters. They eventually became summarized and conceptualized by a flurry of terms like 'marginal reef settings' or 'alternative states of [reef] development' (Kleypas et al. 1999; Perry and Larcombe 2003)—as opposed to 'normal' reefs under 'optimal' environmental conditions (Fig. 10.1a, b). The habitats in these 'extreme' environments were characterized as 'oligo- or monotypic coral carpets' (sensu Riegl and Piller 2003; Benzoni et al. 2003), 'thermally marginal areas' (Glynn 1977) subjected to a 'pseudo high-latitude effect' (Sheppard and Salm 1988), in particular, when reef framework development was restricted or lacking (Fig. 10.1b).

A common feature in 'marginal' (e.g. upwelling-influenced) areas are frequent 'disturbances' that may lock ecosystem into a specific developmental state (e.g. Riegl 2001, Fig. 10.1c). However, not always prominent shifts in reef state (e.g. community composition) occur and changes become effective at the organismal and ecophysiological level. Examples are stress-hardening of corals in high-variability habitats providing new opportunities to foster resilience (Hackerott et al. 2021) that would not arise under 'optimal' conditions.

Fahr. The lowest temperature observed was 58°.5 at the S.W. end of Albemarle Island; and on the west coast of this island, it was several times 62° and 63°. The mean temperature of the sea in the Low Archipelago of atolls, and near Tahiti, from similar observations made on board the *Beagle*, was (although further from the equator) 77°.5, the lowest any day being 76°.5. Therefore we have here a difference of 9°.5 in mean temperature, and 18° in extremes; a difference doubtless quite sufficient to affect the distribution of organic beings in the two areas.

‡ Humboldt's Personal Narrative, vol. vii. p. 434.

The term 'upwelling-influenced environment' summarizes all possible oceanographic settings of cold, nutrient-rich and aragonite-depleted deep water intrusions into the shallows (see Sect. 10.2.1) affecting the structure and function of the coral communities (e.g. Manzello et al. 2008; Schmidt et al. 2012). To account for the myriad of possible community states along the *continuum* from 'scattered coral' to 'coral reef', and to standardize the definition and terms used, Schoepf et al. (2023) introduced a novel conceptual framework for corals at the edge of environmental boundaries that integrates variation in both abiotic conditions and ecological states. The framework was motivated by the bleak future projected for reefs and the search for reef resilience hotspots (e.g. Riegl et al. 2011; Palumbi et al. 2014) and refugia (e.g. Kavousi and Keppel 2018; Kapsenberg and Cyronak 2019) that allow corals to survive in a changing climate. These refugia could also be used as natural laboratories to gain a better understanding of possible future reef states (Camp et al. 2018a; Schoepf et al. 2023). In this context, upwelling-influenced environments are of interest as they can serve as refugia from increasing thermal stress (Wall et al. 2015; Wyatt et al. 2020), as 'training grounds' preparing organisms to cope with future climate change stressors (Buerger et al. 2015; Hackerott et al. 2021; Wall et al. 2023) as well as natural analogs to future more acidic ocean (Manzello et al. 2014). The novel framework provides a nuanced and well differentiated view on upwelling-influenced areas that helps to identify, characterize and compare conservation or management priority areas. The characterization should be based on a sound evaluation of both environmental and ecological criteria, with the two levels 'optimal/extreme' and 'normal/marginal', respectively. Here, significant deviations from average abiotic conditions define 'extremeness' and significant differences in at least two ecological traits (e.g. cover, diversity, accretion) define 'marginality'. One would expect that environmental 'extremeness' invariably leads to 'marginal' reefs. However, both ecophysiological and ecological responses may yield unexpected pairings of environmental 'extremeness' with reef ecological 'normality', i.e. coral and reef resilience to notorious environmental disturbance. This makes Schoepf et al.'s (2023) distinction so important, despite the fact that 'optimal' environmental and 'normal' ecological conditions are defined in relative (not: absolute) terms and despite the fact that 'optimal' conditions today may deviate strongly from 'optimal' conditions in pristine reef settings, i.e. before human intervention (Jackson 1997). While an unexpected pairing of 'extremeness' and ecological 'normality' may occur, the opposite of a 'marginal' reef under 'optimal' environmental conditions is unlikely, i.e. 'marginality' is invariably linked to some sort of 'extremeness', but not all 'extreme' environments lead to 'marginal' reefs. This suggests that a suite of interacting environmental and biological factors govern the state of the benthic community in upwelling areas. The source of 'extremeness' may not necessarily be related to the most obvious disturbance or the disturbance expectedly responsible for 'marginality', as we will outline in this chapter and further discuss in more detail in a later section of the chapter. Regardless, a common framework is useful to identify a) commonalities and differences in responses to disturbances and limiting conditions coral reef ecosystem experience to day and b) reef trajectories that allow for improved future projections.

We start this chapter with a first categorization of 'extremeness' and 'marginality' based on available data for a few selected upwelling-influenced settings (Fig. 10.2) showing the full range of reef states observed under upwelling-influence from 'optimal' to 'marginal' conditions (Fig. 10.1a–c). We focus on locations where the upwelling influence of shallow coral habitats is directly and explicitly assessed along with the environmental conditions. However, often only one abiotic parameter (commonly temperature) is assessed, while concomitant data on oxygen, aragonite saturation, etc., are lacking. Similarly, at some sites, a range of ecological criteria are available, while in others such data are either lacking at all or very limited. This underscores the need to fill these knowledge gaps. While this is the first attempt at applying this framework in this context, still other drivers need to be factored in, like past 'extreme' events or additional local pressures, which may contribute to reef 'extremeness' and 'marginality' besides the influence of upwelling of deep-water. As opposed to laboratory studies where controlling factors can be vigorously tested by experiments, assessing the effect of upwelling in empirical field studies can only be done by inference. To this effect, we chose to minimize spatio-temporal differences between upwelling-affected and upwelling protected sites, to minimize the influence of the many other factors that may have biased the comparison. Examples are the comparison between reef faces, which are exposed to or protected from upwelling on Jarvis in the Pacific (e.g. Vargas-Ángel et al. 2019) or Dongsha in the South China Sea (e.g. Tkachenko and Soong 2017). Another is the comparison between the Gulf of Chiriquí and the Gulf of Panamá (e.g. Randall et al. 2020) or along the coast of Oman (e.g. Glynn 1993; Benzoni et al. 2003). However, 'normal' reefs under 'optimal' environmental conditions *sensu* Schoepf et al. (2023) in such close-by locations may not represent ideal counterparts as they may be located at different depth ranges (eg. Dongsha 'normal' reef image from a lagoonal setting versus Dongsha 'extreme' reef on the internal wave exposed fore-reef slope, Tkachenko and Soong 2017) or the non-upwelling Gulf of Chiriquí reefs which does not compare well with other 'normal' reefs given the high dominance of pioneer species following past 'extreme' events—classical features of 'marginality' (e.g. Smith et al. 2014).

Fig. 10.2 A new classification framework was provided to differentiate 'extreme' versus 'marginal' coral communities using environmental and ecological criteria (Schoepf et al. 2023) and used to characterize upwelling-influenced areas spanning all oceans and include different upwelling drivers. At some sites, a full gradient from 'extremeness' to 'marginality' occurs, while in others reefs do not differ much ecologically (specifically in structural reef formation capacity) from 'normal' reefs. Commonly such sites are compared to close-by non-upwelling-influenced reef counterparts, yet they may not fulfill all traits and characteristics of 'normal' reefs under 'optimal' environmental conditions (*sensu* Schoepf et al. 2023). Here we provide a non-exhaustive selection of sites that are upwelling-influenced and where both 'extremeness' and 'marginality' have been assessed at some point in time. The photograph from the Line Islands (top row) shows a western facing reef in Jarvis

(continued)

Here, we aim to compile the available environmental and ecological criteria that have been assessed in the context of upwelling-influenced areas to better understand the range of 'extremeness' and 'marginality' in such habitats. We outline the different drivers responsible for shoreward transport and mixing of sub-thermocline water into nearshore regions, how they affect corals and framework development, and explore potential shifts in ecological reef states, considering changes in structural and functional states *sensu* Schoepf et al. (2023). However, these changes must be placed in the context of additional factors prevalent in shallow coral reef habitats that also have the potential to influence reef corals and their ability to form reefs. Therefore, their contribution to the transition from one reef condition to another must also be considered in order to understand the relative role of upwelling on reef conditions. Only then can we more accurately assess how reefs will change in the future. The focus here is on shallow water coral reef habitats (down to a depth of approx. 30 m) within the tropics. Deeper mesophotic and high-latitude reef areas are not the focus of this chapter and will be discussed in more detail in Chaps. 12 and 13. However, mesophotic habitats will be briefly treated at the end of the chapter, as upwelling intensifies with depth with important ramifications for deeper reefs and their refugia potential (Foreman et al. 2024).

10.2 Upwelling-Influenced Areas—'Extremeness'

10.2.1 Upwelling—Main Routes of Cold Deep Water into Shallow Reef Habitats

In a seminal paper, Margalef (1978) pointed out that, because of the pull of gravity on particles but not on solutes and the microbial remineralization of organic particles to nutrients, the sinking of particles through a thermally stratified water column leads to the vertical segregation of light (above) and nutrients (below) so that in a stagnant ocean where no external (or 'auxiliary') kinetic energy returns vital nutrients to the surface, primary production eventually ceases.

To sustain primary production, it is therefore essential to replenish the euphotic layer with new nutrients: either from above through atmospheric deposition, from the side through land run-off, or from below through mixing (bi-directional, up and down) and upwelling (unidirectional, up). While atmospheric deposition and run-off follow gravity and are accompanied by a loss of potential energy, upwelling and mixing take place against gravity and hence require substantial inputs of external energy. One of these external inputs is wind energy, which drives ocean circulation, and the other is tidal energy.

Several physical mechanisms have been invoked in the upwelling of dense deep water to the surface (Fig. 10.3).

Longshore winds along with Coriolis forces associated with the earth's rotation may cause horizontal transport of surface waters away from the coast (so-called Ekman transport), drawing a compensatory upwelling of deep-water near the coast (Ekman 1905; Price et al. 1987, Fig. 10.3b,c). Examples for this coastal upwelling are the Arabian and Oman coastal upwelling regions (Glynn 1993) causing sea-surface temperature contrasts of up to 15 °C with the adjacent Persian/Arabian Gulf (Green 1986).

Trade winds blowing across the equator, along with Coriolis forces acting in opposite directions on either side of the equator, cause oceanic upwelling along the equator (Fig. 10.3d). This *equatorial upwelling* affects e.g. the Galápagos archipelago in the Eastern Tropical Pacific (Wyrtki 1981) or the Line Islands chain in the central Pacific Ocean (Johnson et al. 2020).

Another mechanism is advection interacting with topography, both convex (e.g. seamounts, promontories) and concave (e.g. submarine canyons). In the former case, strong zonal currents become deflected upwards over seamounts in their path. Prominent examples of *topographic upwelling* are, again, the Galápagos Islands, where the archipelago obstructs the core of the Equatorial Undercurrent (formerly: Cromwell Current) causing topographic upwelling (Jakoboski et al. 2020), as well as Jarvis Island and Gilbert Islands in the central equatorial Pacific (Gove et al. 2006; Karnauskas and Cohen 2012). Variations in the strength and path of alongshore currents interacting with cross-shore fissures, grooves and canyons may cause pulsed upwelling that enhances the supply of larvae and nutrients upslope along these channels (Andrews and Gentien 1982; Rautenbach et al. 2023). Wind-driven submesoscale currents interacting

Fig. 10.2 (continued) exposed to internal wave upwelling (Vargas-Ángel et al. 2019) similar to the Andaman Sea (second row, Schmidt et al. 2012; Wall et al. 2012) and Dongsha atoll (third row, Tkachenko and Soong 2017) where internal waves of exceptional amplitude affect the reefs. While the Andaman Sea reefs influenced by internal waves are located on the continental shelf (<200 m water depth), Dongsha represents an Atoll in the center of the northern part of the South China Sea surrounded by deep ocean (>500 m water depth). Reefs under the influence of upwelling (sometimes referred to as the 'high-latitude effect') are often defined as 'marginal' reefs. Examples of reefs along an upwelling gradient are found on the coast of Oman and Yemen (Benzoni et al. 2003), at different locations in the Galápagos archipelago: from Darwin with its perceived 'normal' conditions that allow reef framework development towards Florena island and Devils Point with restricted reef formation (Manzello et al. 2014; Glynn et al. 2018). Similarly, the coast of Panamá is divided into non-upwelling and upwelling reefs found in the Gulf of Chiriquí and the Gulf of Panamá, respectively (Glynn et al. 2017; Fernández-García et al. 2021)

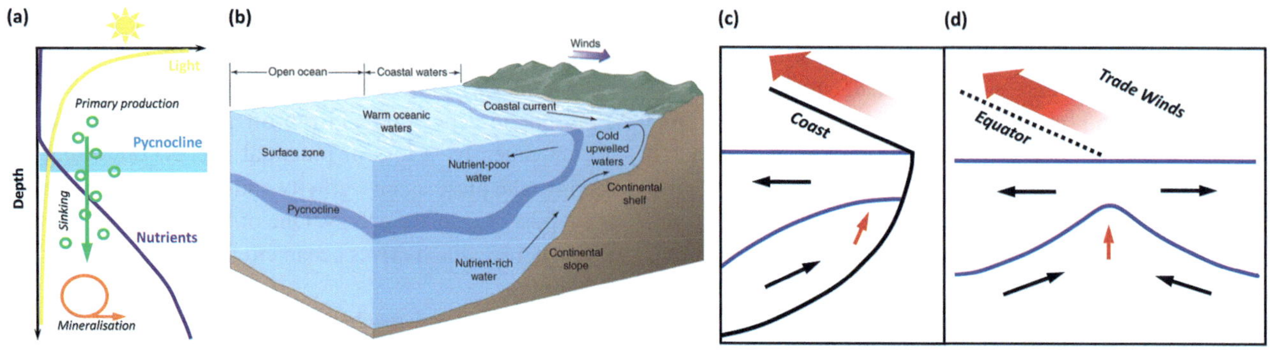

Fig. 10.3 Upwelling reverses the gravitational segregation between light and nutrients. (**a**) In the stratified ocean, the sinking and subsequent microbial mineralization of photosynthetically produced organic matter across the pycnocline leads to a surface layer with light but no nutrients and to a deep nutrient-rich ocean without light. (**b**, **c**) Longshore winds induce Ekman transport of nutrient-poor surface waters away from the coast, causing a compensatory coastal upwelling of cold and nutrient-rich deep waters fertilizing the surface layer. (**d**) Trade winds blowing across the equator induce opposing Ekman transport causing equatorial upwelling of nutrient rich deep waters. (**a**) Modified after (Tréguer and Jacques 1986) (**b**) adapted from Abel and McConnell 2009; (**c**, **d**) by Kämpf and Chapman (2016)

with topography may drive local upwelling in oceanic settings (Forryan et al. 2021).

Large-scale oceanic gyres are associated with deflections of the pycnoclines: where anticyclonic circulation causes a depression of the pycnocline and cyclonic circulation causes a doming of the pycnocline in the center of the gyre, bringing nutrients nearer to the surface. One example for such an *upwelling dome* is the the Costa Rica dome (Wyrtki 1964; Hofmann et al. 1981) driven by the Equatorial Countercurrent in the south, its northward deflection along the Costa Rican coast (Costa Rican Coastal Current) and the North Equatorial Current turning west (e.g. Fiedler 2002).

While winds and geostrophic effects cause directional currents with net transport, tides generated by the pull of the moon and sun cause oscillatory flows with no significant local net transport. Tidal forces cause sea-level changes in the order of a few meters. Because of the three orders of magnitude density contrast at the ocean-atmosphere interface, between seawater (1027 kg/m^3) and air (1.204 kg/m^3), and the much lower density contrast at the pycnocline separating surface and deep waters (<55 kg/m^3), internal waves are ~20-fold larger than the surface waves (Fig. 10.4a). On the continental slope, internal tides interact with the sloping bottom and cause *tidal pumping* (Kämpf and Chapman 2016) associated with pulses of nutrients and plankton pushed upslope (Leichter et al. 1998, 2003, 2006, 2012). Tidal pumping may be modulated both, by topography (e.g. funneling by submarine canyons, Walter and Phelan 2016) and oceanography, such as mesoscale and submesoscale eddies associated with strong along-slope boundary currents (such as the Florida Current, cf. Sponaugle et al. 2005).

In stratified waters (thermal and/or saline, e.g. through river runoff), the interaction of strong tidal currents with topography may cause nonlinear amplification of internal waves. For example, as stratified water is forced by strong tidal forces across a sill between two islands, the narrowing of the cross-sectional area above the sill causes a local acceleration of flow over the sill (Apel 2002). As the water cascading over the shallow sill decelerates into the deeper water downstream, a stationary wave forms in the lee of the sill, characterized by a marked local depression of the pycnocline (Fig. 10.4b, left panel). This lee-wave is the result of a dynamic equilibrium between kinetic and buoyant forces and prevails as long as the tide flows. However, when the tide turns, the balance is overturned and buoyant forces cause the lee wave to propagate in the opposite direction. As it moves across the sill, non-linear effects cause the solitary wave, or *soliton*, to break up into secondary wave packages (Fig. 10.4b, middle panel). The resulting wave-train is conspicuous, showing a large lead soliton and 5–7

Fig. 10.4 (continued) dynamical (internal wave) state, respectively). Like surface waves, internal waves may break and generate bores that travel upslope into shallow coastal areas. Shelf and slope bathymetric features like submarine canyons often channel this upward transport (**a**, adapted from Shanmugam 2013). (**b**) In a stratified ocean, currents interacting with underwater features (like seamounts or sills) may generate stationary internal waves (IWs) in the lee of the obstacles (blue depression of pycnocline, left panel). Changing currents cause the lee waves to move and generate secondary waves that travel in packages (middle and right panel, adapted from Jackson et al. 2012). The surface manifestations of these IWs are rip bands caused by alternating zones of convergence and divergence. They can be seen by synthetic aperture radar images or optical images of the ocean surface (**c**, adapted from Jackson 2004). When IWs interact with the continental shelf/shelf slope they may break and generate solibores, which travel upslope, resuspending bedload material that can be transported into shallow reef habitats (**d**, coral communities at 15 m water depth on Miang Island, Thailand exposed to internal waves, image courtesy J. Laudien)

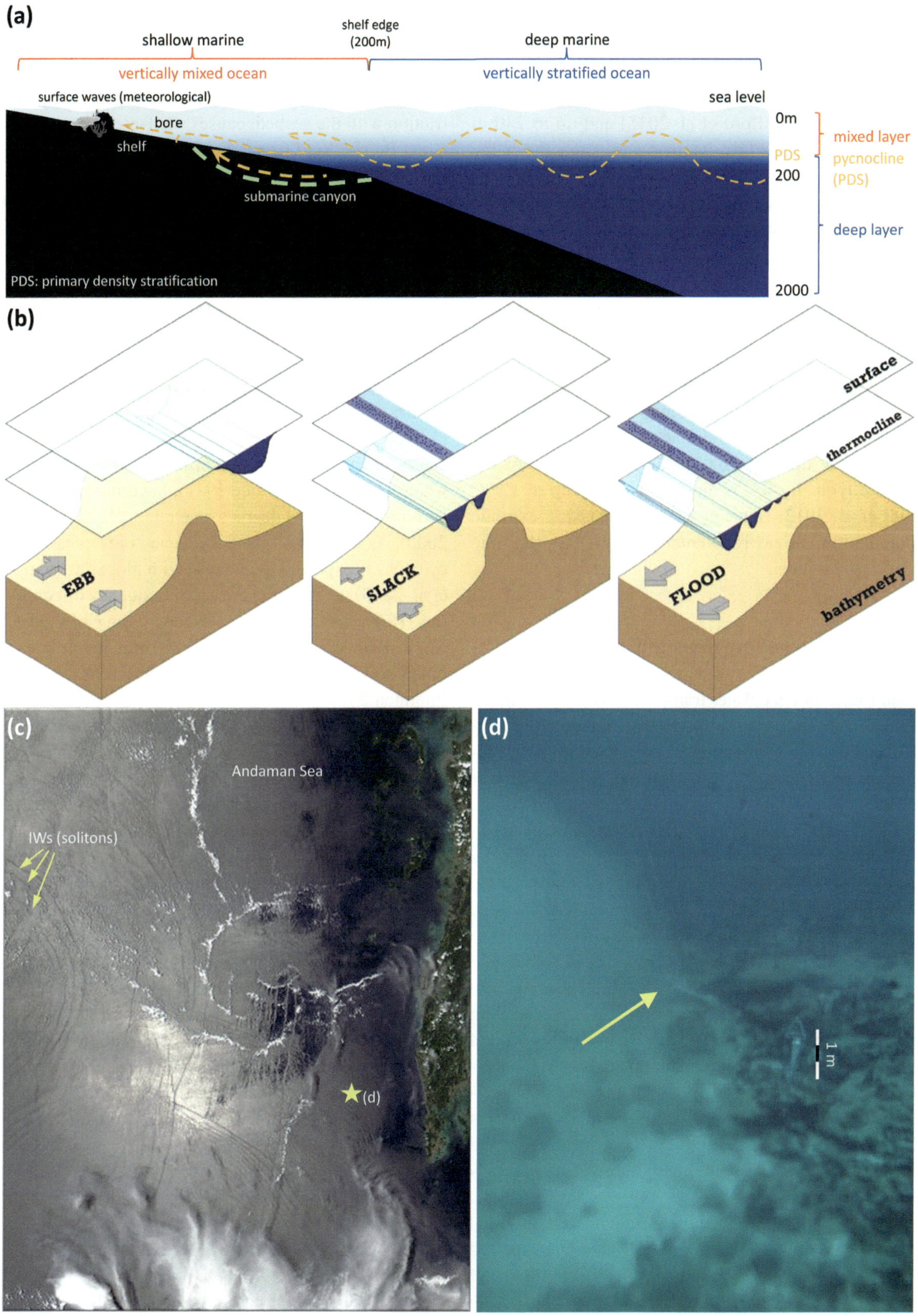

Fig. 10.4 Internal waves are generated along strong density gradients within the water column between the mixed surface layer and the colder deep layer (pycnocline (or primary density stratification PDS): both lines (solid, dashed) represent the pycnocline, in a stationary and

consecutively smaller trailing waves (Fig. 10.4b, right panel) which can propagate long distances over deep water at 2–3 m/s with little dissipation of energy. Solitons are steep, non-linear and can have amplitudes of >150 m (Klymak et al. 2006; Alford et al. 2015) and up to 240 m (Huang et al. 2016)! While internal waves, as the name implies, are subsurface features, the passage of solitons leads to convection in the water column and the generation of rip bands at the ocean surface which can be monitored from space (Fig. 10.4b, c). The magnitude of the gargantuan internal solitons and their arrival in packages or wave trains, make them an important, yet poorly investigated, source of intense upwelling at subtidal frequencies. When they approach shallow waters, friction with the seabed causes steepening and breaking of the waves and the formation of bores that can travel upslope into shallow water (Fig. 10.4a, d). On the fore-reef slope, temperature differences across the thermal front at the leading edge of the *solibore* are typically >2 °C and occasionally up to 10 °C (Schmidt et al. 2012; Davis et al. 2020), causing extreme local fluctuations in environmental conditions in the time frame of only minutes.

10.2.2 Other Upwelling Mechanisms

The main routes of deep water transport towards the surface are well known (wind, tides and currents), a number of additional mechanisms can also cause upwelling of sub-thermocline water that are less generally known (Walker et al. 2005; Kämpf and Chapman 2016). Away from the coast, differences in wind stress cause concomitant variations in Ekman transport, resulting in zones of divergence and convergence. In contrast to coastal up- and downwelling, which have vertical but also horizontal components, these compensatory flows are exclusively vertical and referred to as Ekman suction or *Ekman pumping*. It was found that Ekman pumping fuels one quarter of new (i.e. nitrate-based) phytoplankton production in the major coastal upwelling systems off Benguela, NW Africa, Peru and California (Messié et al. 2009). Ekman-pumping-enhanced phytoplankton production was invoked as a possible mechanism triggering Crown-of-Thorns (*Acanthaster* spp.) starfish outbreaks in the Pacific Ocean (Houk et al. 2007), expanding on the hypothesis that the survival and recruitment of *A. planci* is driven by nutrients (Brodie 1992).

Dynamic lift or *shelf break upwelling* has been observed in regions where strong western boundary currents, such as the Gulf Stream or Kuroshio, move swiftly along the shelf break (Kämpf and Chapman 2016). The geostrophic nature and high velocities of these currents are associated with steep pycnoclines that slope upwards from the ocean towards the shelf break, thus bringing sub-thermocline waters closer to the surface. As boundary currents are part of large-scale anticyclonic gyres, they are largely independent of local wind conditions. Meandering of the East Australian Current, in response to planetary wave perturbations, along with local friction with the seabed, causes shelf-break upwelling in the Great Barrier Reef from a bottom Ekman layer on the continental slope (Andrews and Gentien 1982).

Flow disturbances in the wake of islands have long been known to generate turbulence and vertical flows that bring nutrients to the surface and enhance biological productivity. This so-called *island mass effect* (Doty and Oguri 1956) has been found in the vicinity of reefs (Hamner and Hauri 1981) and basin-scale investigations have shown that island mass effect are ubiquitous among more than 90% of the coral islands and atolls of the tropical western Pacific (Gove et al. 2006).

Tropical storms and cyclones (hurricanes, typhoons) cause intense mixing and upwelling of sub-thermocline water to the surface (Price 1981), lowering sea-surface temperatures by several degrees on a regional scale (Wentz et al. 2000). As the intensity and frequency of tropical storms and cyclones is expected to increase with global warming (Sobel et al. 2016), as is the severity and frequency of coral bleaching (Hughes et al. 2017), surface cooling in the wakes of tropical storms and cyclones has been invoked to play a potential role in mitigating coral bleaching (Carrigan and Puotinen 2014).

Finally, the passage of *coastal trapped waves* can induce transient down- and upwelling along their path (Gill and Clarke 1974). They are generated when meteorological or oceanographic disturbances such as typhoons or coastal upwelling events cause perturbations in sea-level and density in the distance. Coastal trapped waves depend crucially upon the earth's rotation and, for a given hemisphere, can travel only in one direction (Mysak 1980). At low latitudes, the equator acts as a wave guide, which is interrupted at the coasts. The incoming equatorial wave energy can either be reflected or transmitted poleward along the coast as coastal trapped waves—in a cyclonic direction along basins, and around islands in an anticyclonic direction (Brink 1991). Coastal trapped waves around Palau have been invoked to bring cold sub-thermocline water onto reefs that may either mitigate heat stress or can even exert cold stress (Schramek et al. 2019).

10.2.3 Global Distribution of Upwelling-Influenced Coral Habitats

Upwelling is an important feature, not only within the eastern boundary upwelling systems, but throughout the tropics in all ocean basins (Fig. 10.5). It is strongest where thermal (or on occasion haline) stratification coincides with a pycno-

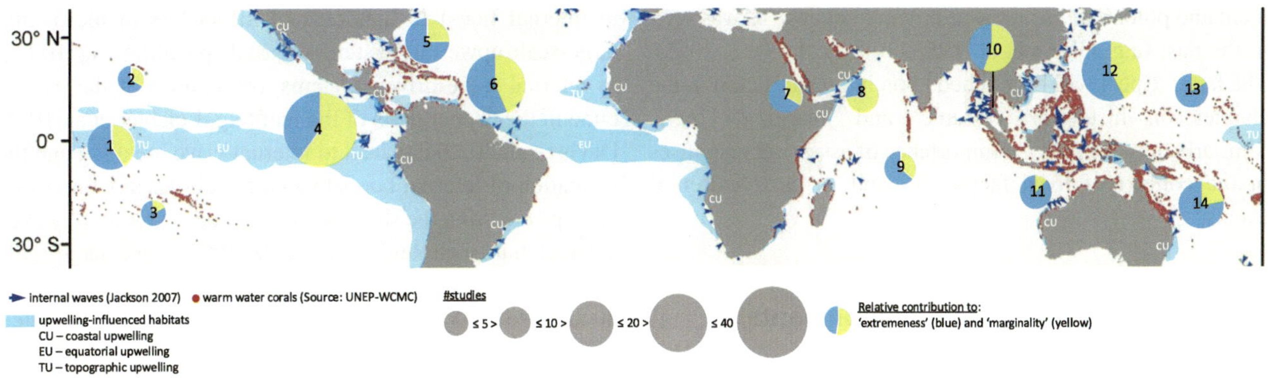

Fig. 10.5 Global map of the tropical ocean showing the occurrence of warm water corals (red dots) and upwelling areas (blue shading and arrows). Upwelling driven by internal waves/tides/large amplitude internal waves can be documented by SAR imagery (Jackson 2007) and are denoted by blue arrows. Coastal and equatorial upwelling areas (including also topographic upwelling, upwelling domes, etc.) are highlighted by light blue areas. Pie charts show the number and nature of upwelling-related coral reef studies addressed to date (>200 references compiled for this chapter and included here) in the respective reef provinces, specifically with regard to reef 'extremeness' and 'marginality' (see definition in Sect. 10.1). The pie chart represents the relative contribution to 'extremeness' (blue) and 'marginality' (yellow) in the different areas. The exact reef location summarized for specific areas (1–14) are the following: 1: reefs along the Line Islands (e.g. Jarvis, Palmyra, Howe, Baker, Gilbert), 2: Hawaii reefs, 3: French Polynesia (e.g. Moorea, Tahiti), 4: reefs within the Eastern Tropical Pacific (e.g. Galápagos, Gulf of Panamá, Golf of Chiriquí, Gulf of Papagayo), 5: Northern Caribbean reefs (e.g. Florida reef tract, Bahamas), 6: Southern Caribbean reefs (e.g. Venezuela, Colombia), 7: Red Sea (e.g. Farsan islands), 8: Oman (e.g.), 9: Central Indian Ocean (e.g. Chagos, Madagascar, South Africa), 10: Andaman Sea (e.g. Similan islands, Racha island), 11: Western Australia (Scott reef), 12: South China Sea (e.g. Dongsha atoll, Taiwan, Vietnam), 13: Western Pacific (e.g. Palau, Guam), 14: Great Barrier Reef/Coral Sea (e.g. Myridon reef, Ribbon reef, New Caledonia). Note the upwelling-influenced areas highlighted in blue are approximations but not exact representations of upwelling habitats due to their heterogeneous, often small-scale extension at the coasts and dynamic nature. The distribution of upwelling is approximated from the map in (Froyland et al. 2014). The locations of internal waves are derived from (Jackson 2007) and the coral distribution based on (UNEP-WCMC et al. 2010)

cline near the ocean surface (Kämpf and Chapman 2016). Trade winds blowing to the west pile up water in the western parts of ocean basins (Forryan et al. 2021), resulting in a compensatory depression of the pycnocline in the west and a shoaling of the pycnocline in the east. The effect is stronger the larger the basin: this is why we have the strongest upwelling in the largest ocean basin: the strongest coastal and equatorial upwelling is found in the tropical Pacific, followed by the Atlantic, while there is only moderate upwelling in the Indian Ocean (Fig. 10.5).

Upwelling has been investigated in shallow reef habitats worldwide including coastal areas (e.g. Florida keys, Leichter et al. 2006), off-shore islands (e.g. Andaman Sea islands, Schmidt et al. 2012) and atolls (e.g. Williams et al. 2013; Ford et al. 2020; Fox et al. 2023) with a few focal areas that received more attention (Fig. 10.5). This compilation of upwelling-influenced coral reefs (in all its ecological states) focused explicitly on studies that addressed aspects of either 'extremeness' or 'marginality' or the combination of both. In this context, reefs in the Eastern Tropical Pacific received the most attention. They are situated between the Humboldt and California Current systems both are known to be strong upwelling systems (Bograd et al. 2023). Over 60 studies were compiled, addressing the 'extremeness' and/or 'marginality' of extant shallow coral reefs within the Eastern Tropical Pacific. Other focal areas are located in the South China Sea (especially the Dongsha atoll has been extensively studied; e.g. Fu et al. 2012; Reid et al. 2019), along the Line Islands in the Central Pacific, and within the Caribbean as well as the Andaman Sea. The ability to modulate environmental conditions and likely alter the local shallow reef ecological structures is less intensively studied in the remaining locations (Fig. 10.5). The lower numbers of studies from these regions to some extent can be attributed to the fact that research in this context started more recently during the past decades. However, it also indicates that further research is needed in these areas to understand the extent and consequences of upwelling for these specific sites, but also more generally for reef development under such likely 'extreme' settings.

Upwelling has historically been a focus of reef studies within the Eastern Tropical Pacific. However, the severe El Niño Southern Oscillation (ENSO) of 1982/83 and its far-reaching ecological consequences (Glynn 1985, 1988) have led to localized reef shutdown, resulting in reefs being reduced to a 'marginal' state (Fig. 10.1c) and severely hindered from recovery (e.g. Glynn 1988, 1994), particularly along the upwelling gradient within the central Galápagos (e.g. Glynn 1994; Manzello et al. 2014). On the other hand, such locations may provide a glimpse into the future and insights into how upwelling in combination with other strong drivers like heat waves may further modulate reef development and potentially lead towards a reef collapse as was seen in the past (e.g. Glynn et al. 1983; Toth et al. 2012, 2015). The latter aspect will be touched upon in greater detail in the chapter on historical 'extreme' and 'marginal' reefs. Similarly, it highlights the importance of assessing responses in the context of local factors as well as past 'extreme' events.

10.2.4 Living at the Edge—Environmental Conditions in Upwelling Areas

The majority of studies on upwelling-influenced reefs assessed the abiotic conditions. However, only a few studies considered more than one factor at high temporal resolution (e.g. Schmidt et al. 2012; Enochs et al. 2021) revealing that 'extreme' habitats are characterized by a wealth of facets in abiotic changes (as also outlined in Camp et al. 2018a). We find that upwelling exposes corals to the opposite end of environmental conditions that commonly favour vigorous reef growth. Temperature was the factor assessed in most of these studies and thus, represents the environmental criteria that can be more readily compared across and within locations. Temperature is commonly monitored at a high temporal (less often: high spatial) resolution. It allows for extracting information on the frequency, and duration of changes, in addition to the magnitude or mean changes during upwelling periods (e.g. Leichter et al. 1996; Guillaume-Castel et al. 2021). Overall, our analyses shows that oceanographic and environmental studies provide the most thorough insight into the 'extremeness' and the extent to which conditions deviate from 'optimal' abiotic conditions. They also indicate so far not considered conditions that could represent environmental limits for coral performance and thus, reef development and functioning. The dynamics of inorganic and organic nutrients as well as carbon chemistry are still not described in great detail as outlined below, however, they are essential for a comprehensive understanding of the underlying ecological changes (see Sects. 10.3 and 10.4). Studies that included more than one parameter revealed coupled changes, e.g. that temperature co-varied with inorganic nutrients, the carbonate system, oxygen and salinity (e.g. Leichter et al. 2003; Schmidt et al. 2012; Reid et al. 2019; Enochs et al. 2021). However, we often lack environmental data, especially at high temporal resolution for instance in nutrient dynamics, as such parameters are generally sampled discretely and are rarely assessed continuously—a common limitation in 'extreme' and 'marginal' environments (see review Camp et al. 2018a).

The type of abiotic changes reflects the different drivers influencing reef areas with their marked variation both at spatial and temporal scales. They can affect reef habitats ranging in spatial extent from hundreds of meters (e.g. internal waves to internal tides) to hundreds of kilometers in meso- and large-scale upwelling systems (coastal upwelling, Fig. 10.6a). In internal wave driven systems, oscillations of the pycnocline in the deeper parts of the reef (e.g. Leichter et al. 1996; Davis et al. 2020) can lead to internal wave breaking and the formation of internal bores that carry sub-thermocline water upslope affecting the shallow fore-reef and occasionally even the reef flat (e.g. Reid et al. 2019). Meso- and large-scale upwelling systems can affect the shallowest parts of reefs as documented in lowered sea-surface temperatures in the satellite records (Gulf of Panamá, e.g. Glynn et al. 1983, 2001). Similarly, these drivers differ in the frequency and duration of deviations from expectedly 'optimal' environmental conditions. Obviously, the meso- and large-scale features show lower frequency changes (weeks to seasons, Fig. 10.6d) than the high frequency oscillations associated with internal waves (hours to weeks, Fig. 10.6b, c). While internal waves can induce step changes between 'normal', 'extreme' and again 'normal' conditions, within only minutes (e.g. Schmidt et al. 2012; Reid et al. 2019), more gradual and prolonged changes occur in internal tide and wind-driven or topographic upwelling settings. Especially, larger scales upwelling conditions can affect entire regions and the deviations from 'optimal' can persist for days or even weeks (e.g. Glynn et al. 2001; Randall et al. 2020). However, temporally large-scale features do govern the gating of IW onto the shelf. An example is the Indian Ocean Dipole that determines the intensity of IW impinging on the Andaman Sea continental shelf, explaining the large seasonal but also interannual variations in IW upwelling forcing in shallow reef habitats (Yadidya and Rao 2022). Likewise, small-scale and mesoscale features such as internal wave activity in the Northern South China Sea at Dongsha atoll and the intrusions of the Kuroshio Current into Luzon strait (DeCarlo et al. 2015) are influenced by basin-scale and planetary disturbances, e.g. ENSO. These deviations from 'optimal' also vary inter annually, leading to stronger or weakerforcing by deep-water delivery (e.g. (e.g. Glynn et al. 2001; Leichter et al. 2006, 2012; Roder et al. 2010; Watanabe et al. 2017; illustrated in Fig. 10.10). This variability is driven by aforementioned changes in large scale features that affect water column stratification or ocean currents strength (e.g. Gove et al. 2006; Fox et al. 2023) and contribute to the dynamic nature of such sites.

Overall, deep-water intrusions reduce the reef temperature and these temperature reductions can be substantial. The measured changes were found to range from a few degrees to more than 10 °C. For instance, the weaker upwelling strength at Hainan Island, China changed the reef temperature by 1.7 °C (e.g. Li et al. 2015) and internal waves caused changes in the range of 1–2 °C at tidal frequency in 20–30 m in Northern Caribbean (e.g. Belize, St. Croix and Bahamas, Leichter et al. 2006). Larger and often more sudden temperature drops were observed by large

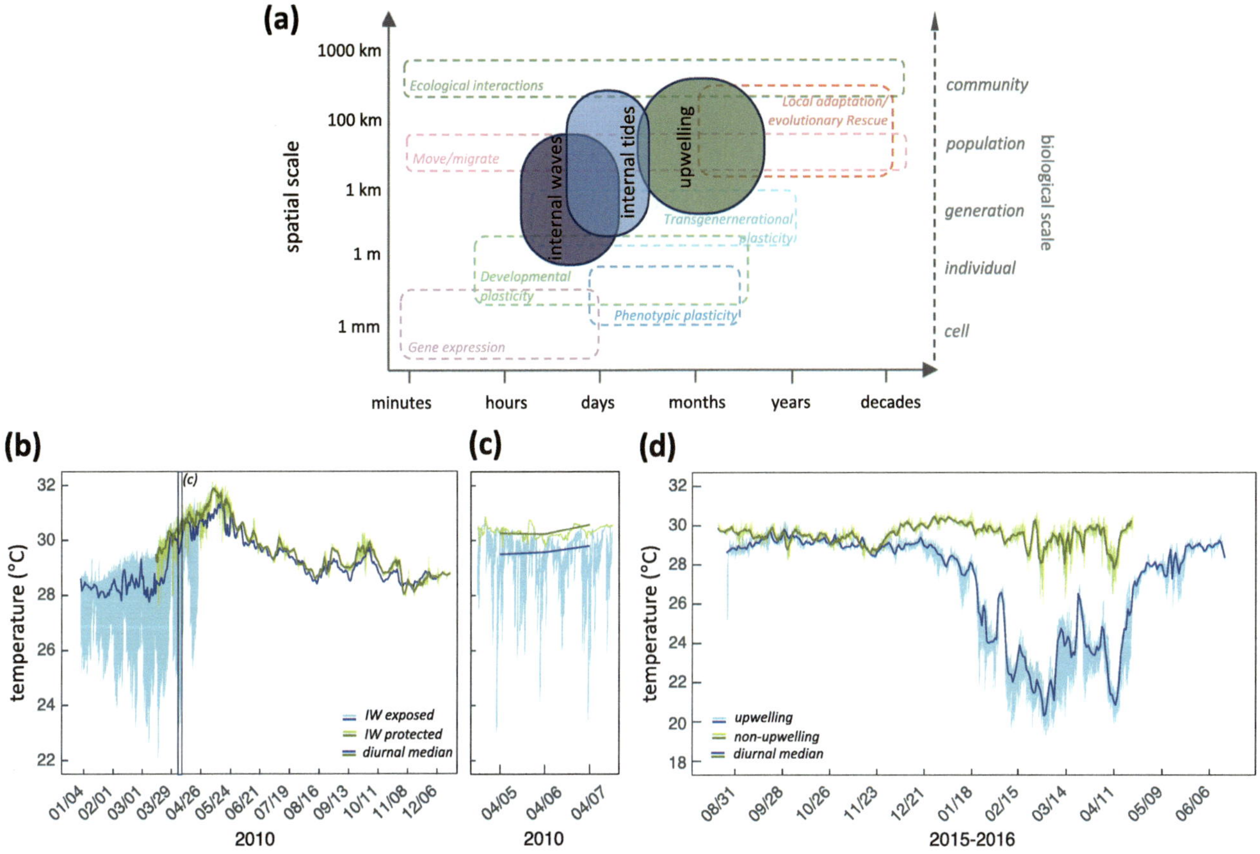

Fig. 10.6 Upwelling drivers differ in both spatial and temporal scale and thus, affect different biological processes that can shape ecosystems states (**a**, adapted from Kroeker et al. 2020). Here the different temporal scales are illustrated for (**b**, **c**) internal wave high-frequency deep-water intrusions (data from Wall et al. 2015) and (**d**) seasonal upwelling in the Gulf of Panamá compared to the non-upwelling reefs in Gulf of Chiriquí (data from Enochs et al. 2021). Internal wave exposure has little effect on diurnal median temperature conditions in internal wave exposed and protected reefs compared to marked differences between upwelling and non-upwelling reef settings

amplitude internal waves. For example, temperature drops in the Andaman Sea Island reefs range up to 10 °C reaching temperatures as low as 19 °C at 25 m (Schmidt et al. 2012). Similar temperature changes of up to 12 °C at 25 m were measured in Diego Garcia, Chagos Archipelago (Sheppard 2009) or 10 °C temperature plunges in 18 m depth at Dongsha Atoll, South China Sea (e.g. Davis et al. 2020). Internal tides at Conch Reef, Florida slope, however, can also induce temperature plunges of up to 9 °C at 30 m water depth (Leichter et al. 2003). Given the upslope transport of cold sub-thermocline water into the shallows, the magnitude of change varies with depth, i.e. less frequent and less severe temperature swings are expected in shallow water. In this context, a 10.1 °C temperature drop (from 28 °C to 17.9 °C) at only 3 m water depth on reefs in the Gulf of Panamá must be considered particularly 'extreme' (Enochs et al. 2021). As mentioned before, deep-water intrusions can happen instantaneously within a few minutes (e.g. Sheppard 2009; Reid et al. 2019) or occur more gradually in the course of a few hours (e.g. Leichter et al. 2005) to a few days (e.g. Glynn et al. 2001). The strength of these changes commonly decreases towards shallower reef habitats and can even be exempt from upwelling influence (e.g. Leichter and Miller 1999; Leichter et al. 2012; Schmidt et al. 2012; Reid et al. 2019).

Concomitantly to temperature changes, other parameters deviate from 'optimal' conditions (Fig. 10.7a, b). For instance, increased salinity and nutrients, reduced oxygen and changed carbonate chemistry parameters are characteristic of the deep-water with below pycnocline origin (e.g. Nielsen et al. 2004; Baum 2011; Noowong 2011; Fig. 10.7a). Oxygen values as low as 1.9 mg/l (2.0 mgl/l Andaman Sea Islands, Thailand, Schmidt et al. 2012; 4.2 - 1.9 mg/l Andaman Sea Islands, Thailand, Wall et al. 2012; 3.7–3.0 mg/l on Gorgona Island, Colombia, Castrillón-Cifuentes et al. 2023a) have been recorded when deep water reaches into in shallow reef habitats, compared to oxygen values ranging from 6.4–4.75 mg/l on respective reference sites or prior to deep water upwelling (e.g. Schmidt et al. 2012; Castrillón-Cifuentes et al. 2023a; Fig. 10.7). Similarly, car-

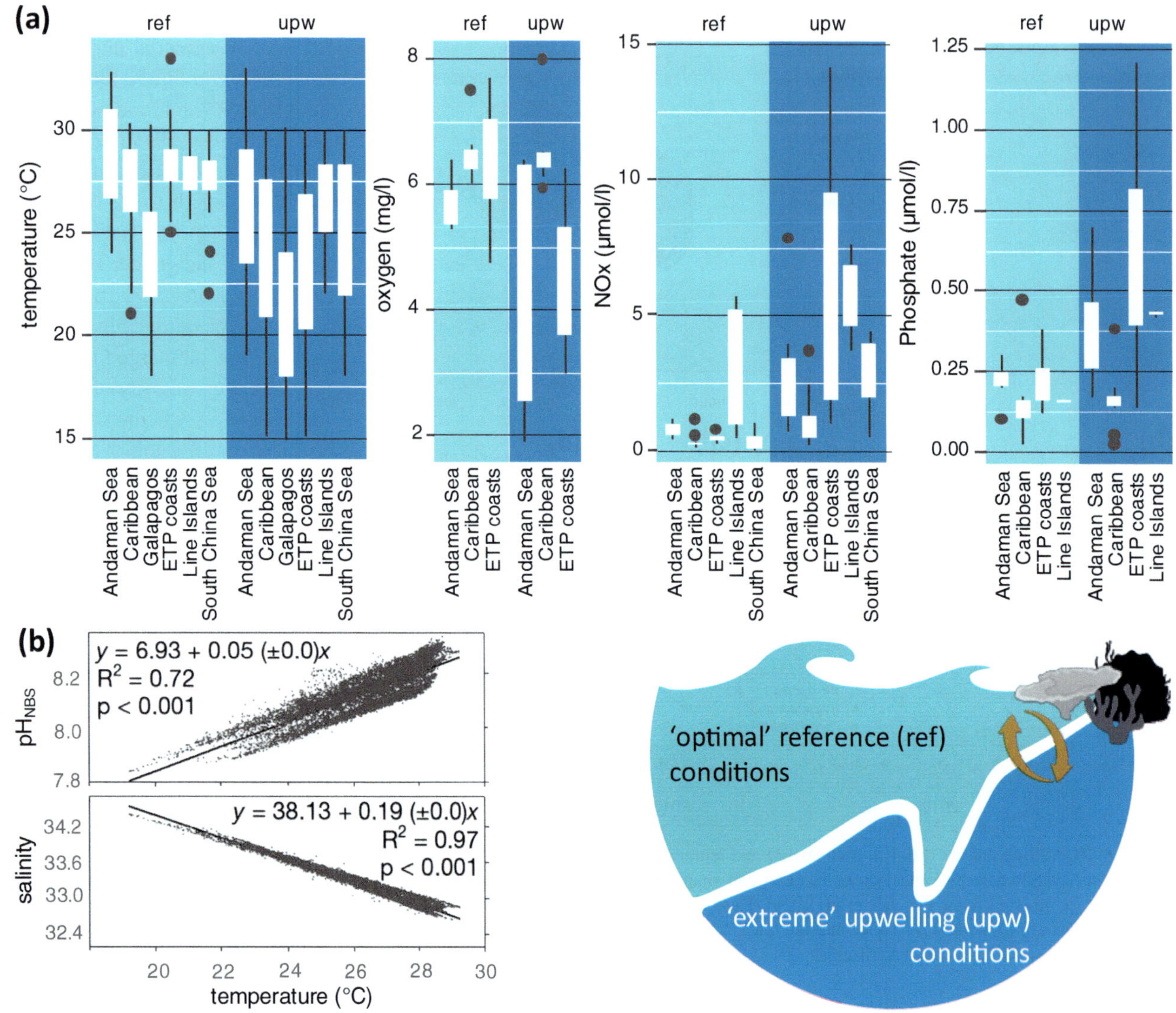

Fig. 10.7 Compilation of recorded ranges of 'optimal' reference (ref, light blue background) and 'extreme' environmental conditions experienced in nearby non-upwelling sites/prior and during upwelling (upw, dark blue background), respectively (**a**). The data were summarized from a number of regions (cf. Fig. 10.5) and indicate how environmental changes in temperature correlate with other parameters. Changes in temperature result in concomitant changes in other parameters like salinity and pH (here exemplified from the study Schmidt et al. 2012 for Similan island large amplitude internal wave diapycnal mixing in shallow reef environments at 20 m). Environmental ranges are retrieved from a number of studies and based on the observed ranges and/or mean values if provided: Andaman Sea (Baum 2011; Jantzen et al. 2013; Noowong 2011; Roder et al. 2011; Schmidt et al. 2012, 2016; Schmidt and Richter 2013; Wall et al. 2012, 2015, 2023; Wall et al. unpublished data), Caribbean (Antonius 1980; Arroyave Gómez et al. 2021; Astor et al. 2003; Bayraktarov et al. 2014, 2016; Chollett et al. 2010; Davis and Monismith 2011; Fox et al. 2018; Leichter et al. 1996, 2003, 2005, 2014; Leichter and Miller 1999Gómez et al. 2023), Eastern Tropical Pacific (ETP) coasts separated for Galápagos (Glynn and Stewart 1973; Dana 1975; Abele 1976; Glynn 1977, 1988; Glynn et al. 1983, 1991Jiménez 2001; Jiménez et al. 2001; Podestá and Glynn 2001; Castrillón-Cifuentes et al. 2023a, b; Chollett et al. 2010; Combillet et al. 2022; D'Croz and O'Dea 2007; Enochs et al. 2021; Giraldo et al. 2008; Krutwa 2014; Manzello 2010; Manzello et al. 2008, 2014; Randall et al. 2020; Riegl et al. 2019; Rixen et al. 2015; Sánchez-Noguera et al. 2018; Smith 2006; Stuhldreier et al., 2015a, b, c; Witman and Smith 2003; Wyatt et al. 2020), Line Islands (Aston et al. 2019; Fox et al. 2018, 2023; Gove et al. 2006, 2015, 2016; Guillaume-Castel et al. 2021; Johnson et al. 2020; Karnauskas and Cohen 2012; Miller et al. 2008; Williams et al. 2018,*Note at Jarvis relatively high nutrient concentrations occur even at depth below pycnocline values up to 20 µmol/l were recorded) and the South China Sea (Wang et al. 2007; DeCarlo et al. 2015; Tkachenko and Soong 2017; Reid et al. 2019; Wyatt et al. 2020; Davis et al. 2020; Hung et al. 2021). The data are not exhaustive, reflect the observed ranges across depth and provide an overview of 'optimal' reference versus 'extreme' upwelling conditions. Note the strong bias towards temperature, while data on other important parameters (oxygen, phosphate, pH, Chl *a)* are limited. However, changes in temperature strongly correlate with changes in other parameters like pH and salinity (b) as illustrated from the Andaman Sea environmental monitoring data (Schmidt et al. 2012)

bonate chemistry is changing with deep water intrusions into shallow reef habitats (e.g. Schmidt et al. 2012; Manzello et al. 2014; Stuhldreier et al. 2015b; Enochs et al. 2021; Fig. 10.6); a deep water that is commonly characterized by reduced pH, increased inorganic carbon concentration and lower aragonite saturation state (e.g. Feely et al. 2008). With measured seawater pH values as low as 7.8 in 20 m water depth in off-shore reefs in the Andaman Sea (Schmidt et al. 2012) or in 3–8 m water depth across the Eastern Tropical Pacific (e.g. Manzello et al. 2008; Rixen et al. 2015; Sánchez-Noguera et al. 2018). Water samples collected in the water column above the reef revealed an increase in nutrient concentrations (up to 20-fold increase, e.g. Leichter et al. 2003; Bayraktarov et al. 2014; Stuhldreier et al. 2015b) that correlate with an enhancement of nutrient flux (by 1–2 orders of magnitude Leichter et al. 2003; or 5-fold increase Reid et al. 2019) due to internal waves relative to non-internal wave periods. At the same time, the breaking internal waves can resuspend bedload material that is transported upslope at high velocities (up to 0.6 m/s; e.g. Leichter et al. 1996; Witman and Smith 2003; Roder et al. 2010; Davis and Monismith 2011) as cold and turbulent bores.

Temperature is the most commonly recorded parameter that allows quantifying the upwelling strength at high temporal resolution. Mean temperature data are sensitive to 'extreme' temperature excursions but may still not do full justice to the degree of thermal variability to which coral reefs are exposed to in internal wave exposed environments. For example, internal wave exposed and protected sites at 15 m water depth experienced a mean temperature of 29.05 and 29.69 °C, respectively during a more intense internal wave upwelling period (Wall et al. 2023). Similarly, at Conch reef, Florida mean reef temperatures changed from 28.7 °C at 7–10 m depth to 28.1 °C at 21 m during summer with more intense internal wave activity that increases with depth (Leichter and Miller 1999). Different metrics are needed to represent more accurately the degree of variability. In contrast, yearly mean temperature differ markedly between the non-upwelling Gulf of Chiriquí and Gulf of Panamá (for instance 28.41 °C ± 0.12 (sem) compared to 26.52 °C ± 0.47, Glynn et al. 1988; or 28.8 °C ± 0.7 (sd) and 25.1 °C ± 2.5, Enochs et al. 2021) with its more continuous shift in pycnocline. To assess the degree of variability other measures have been introduced and in particular, applied for locations exposed to internal wave activity. The raw time-series are often converted to summary parameters, such as degree days cooling (e.g. Leichter and Miller 1999; Wall et al. 2012), %-time below a certain temperature threshold (e.g. Leichter and Miller 1999; Sevadjian et al. 2012), and their derivatives, to obtain measures of temperature stress (or relief) for corals (and other biota). These measures foremost focus on the accumulated upwelling strength/exposure, but less on individual cold-pulse characteristics as in more recent approaches (e.g. Sevadjian et al. 2012; Williams et al. 2018; Guillaume-Castel et al. 2021). In respect to ecological 'marginality' the strength of individual deep-water pulses, the temporal dimension as well as the accumulated exposure matters and altogether will define physiological responses and allow a more refined comparison across upwelling-influenced areas (Fig. 10.6a).

The outlined changes in abiotic conditions certainly fall within what commonly is perceived as 'non-optimal' or 'extreme' (Fig. 10.7) and irrespective of its temporal extent (from hours to weeks) affect the corals' metabolism and their performance (Fig. 10.6a). In the long run, it may translate into ecological consequences affecting coral reef functioning and state (Fig. 10.2). Whether detectable ecological changes occur also depends on the different physiological thresholds or tolerance limits of the corals to the prevailing abiotic changes. As upwelling is often seasonal, it is also related to the potential of corals to endure 'extreme' periods and recover in times of 'optimal' conditions without upwelling. The first characterization within the novel framework on reef 'extremeness' and 'marginality' (Fig. 10.2) outlines the very distinct reef responses suggesting that some abiotic changes may lie well within the corals' tolerance limits, even being beneficial fueling their metabolism and well-being while in other areas they exceed tolerance limits negatively impacting coral reefs. The suite of co-varying abiotic conditions may also entail synergistic or antagonistic responses, together dictating the overall response. In the next section, we will focus on what is known in respect to reef 'marginality' and its deviations from 'normal'. We compile studies addressing the individual corals' responses towards potential shifts at the community level. The aim is to shed light on how the changes in the condition of the reefs are related to the extent of upwelling and whether they depend on the region (Sect. 10.3). Upwelling does not occur in isolation and responses may also be driven by other prevailing factors (both current and past) that need to be taken into account as they can represent 'non-optimal' conditions (Sect. 10.4). We will touch upon underlying physiological and ecological mechanisms observed and described for upwelling settings.

10.3 Upwelling and 'Ecological Marginality'

Upwelling can cause 'extreme' local fluctuations in environmental conditions. Mobile organisms (e.g. fish, crabs, stomatopods, shrimp and squid) are able to migrate away from low-oxygen areas associated with seasonal upwelling (Levin et al. 2009). On much shorter time-scales, reef fish have been observed to literally escape from advancing fronts of turbulent solibore discharging cold hypoxic sub-thermocline water into the shallow reef (Richter, pers. obs., see also

Fig. 10.4d for the leading edge of a solibore). Sessile organisms, however, obviously have no means to escape and thus have to cope with adverse environmental conditions—or perish. As outlined above, this 'extremeness' can be very different, depending on the nature of upwelling: It can persist for long, in the case of seasonal upwelling, and involve gradual change. On the other end of the spectrum, it can be acute, and entail sharp changes in the physico-chemical parameters, in the case of internal tidal bores. Relevant environmental factors for reef development and growth are temperature, aragonite saturation, nutrients and light, but also oxygen levels, sediment load (turbidity) and food supply (see detailed discussion of factors below). Under 'optimal' environmental conditions, coral reefs thrive at tropical to subtropical temperatures in shallow aragonite-supersaturated water, in high light but low nutrient conditions. All of these parameters are altered under upwelling conditions and they act at the organismal and the ecosystem level, however, changes in these parameters may not always act in the same direction.

10.3.1 Upwelling Effects on Reef Corals

On the organismal (holobiont) level, corals flourish where the average winter minimum temperature does not drop below the climatological value of 18 °C (Wells 1957). Numerous studies have shown, however, that corals can physiologically acclimate to a given temperature range of their respective geographic areas and may show remarkable acclimatisation to both, upper and lower temperature 'extremes' with the most striking examples of thermal adaptations to summer maxima of 36–38 °C and periodic cold winter temperature below 10 °C (e.g. Coles and Fadlallah 1991). Thermal stress, both heat and cold stress, has been shown to affect coral feeding (Palardy et al. 2005; Ferrier-Pagès et al. 2010), growth (Glynn 1977; Cantin et al. 2010), health and survival (Glynn and D'Croz 1990; Guzmán et al. 1990; Hoegh-Guldberg 1999; Hoegh-Guldberg and Fine 2004). Coral growth is directly related to temperature (Lough and Barnes 2000; Álvarez-Noriega et al. 2023) so that growth rates may be 50% lower during the upwelling period (Glynn 1977). Like heat stress, cold stress appears to entail an acute stress response (Roth et al. 2012) including oxidative stress in the Symbiodiniaceae photosymbionts that can lead to paling (Marangoni et al. 2021). While most cases of cold stress relate to meteorological events, i.e. cold spells affecting shallow water corals (see review by González-Espinosa and Donner 2020), there are reports of bleaching in deep and cool waters following 'cold shock' events—up to 12 °C declines over a week to several months—attributed to unusually strong upwelling in the Galápagos (Glynn et al. 2009), the Gulf of Panamá (Zapata et al. 2011) and the Caribbean (Kobluk and Lysenko 1994; Laboy-Nieves et al. 2001).

Upwelling also entails nutrient enrichment. While it is generally assumed that coral calcification is favoured in oligotrophic waters, Wellington and Glynn (1983) proposed that the high growth rates of corals (*Pavona clavus*) in the upwelling areas of the Eastern Tropical Pacific are related to a higher nutrient availability during the upwelling season. High nutrients, in particular phosphates, have been shown to reduce skeletal density (Dunn et al. 2012). This may explain why corals in some upwelling areas (e.g. Gulf of Papagayo, Gulf of Panamá, Galápagos Islands, Arabian Sea) are able to maintain similar or even higher calcification rates than corals from non-upwelling regions (e.g. Gulf de Chiriquí, Cano, Galápagos non-upwelling locations, Gulf of Oman) because of high linear extension rates, albeit at the expense of a reduced skeletal density (Jiménez and Cortés 2003; Manzello et al. 2014; Spreter et al. 2022). Increased exposure to nitrogen levels has also been documented to enhance bleaching susceptibility under heat and light stress (Wiedenmann et al. 2013) and could add to the potential bleaching susceptibility of corals at upwelling sites (D'Croz and Maté 2004). Changes in nutrient availability and composition affect the fine balance between coral and its photosymbionts that is not always beneficial for coral growth (Wiedenmann et al. 2013; Rädecker et al. 2015). It also alters coral associated bacterial taxa favouring taxa involved in nutrient cycling (Kelly et al. 2014). Enhanced nutrient delivery, however, can also favour plankton growth (Wyatt et al. 2010), and plankton enrichment in combination with upwelling currents enhance food supply to corals (Leichter et al. 1998; Roder et al. 2010). These heterotrophic subsidies are considered increasingly important for coral reefs and other marine animal forests in an era of climate change (Denis et al. 2024). Mixotrophic corals benefit from enhanced plankton productivity (Fox et al. 2018) by switching to increased heterotrophic feeding (Roder et al. 2010; Radice et al. 2019; Fox et al. 2023) that may contribute to increased resilience during stress like heat waves (Grottoli et al. 2006; Fox et al. 2023Conti-Jerpe et al. 2020).

The effects of temperature, nutrients and food supply are, however, difficult to disentangle from other parameters co-varying with upwelling, such as light or aragonite saturation. While warm shallow waters in the tropics are supersaturated with regard to aragonite, the solubility of the coral aragonite skeleton increases with the partial pressure of CO_2 so that the aragonite saturation of seawater decreases with increasing depth and CO_2 concentration. This 'acidified' aragonite-undersaturated water is considered 'corrosive' for organisms with aragonite skeletons (Feely et al. 2008), depressing reef growth (Kleypas et al. 1999) and enhancing bioerosion, often in concert with elevated nutrients (DeCarlo et al. 2015; Alvarado et al. 2017). While there is overall agreement on negative effects of ocean acidification on marine organisms in general and calcifying organisms including corals in particular

(Kroeker et al. 2010; Leung et al. 2022) upwelling areas often feature variable effects likely linked to co-varying factors.

Overall, upwelling appears to provide a 'trick or treat' to corals which depends on both, the intensity or 'extremeness', and the specific circumstance or context (e.g. presence or absence of a heat wave). The upwelling-related environmental changes can provide a 'treat' to corals, e.g. material and/or energy subsidies in an oligotrophic environment, but also a 'trick' when exposing corals to limiting conditions. The latter becomes evident when 'extreme' upwelling conditions restrict reef formation and promote other benthic organisms resulting in 'marginal' reefs which are visibly different from 'normal' reefs. While we do find reefs in an upwelling-influenced setting that are prospering and thriving, showing no to limited expression of 'marginality' traits (Benzoni et al. 2003; Aston et al. 2019; Vargas-Ángel et al. 2019), we also find hampered reef formation (Fig. 10.2) and a loss of hard coral cover under more intense upwelling conditions (Fig. 10.8a; Spring and Williams 2023). Most striking is the restriction of reef accretion and thus net calcification rate that decreases with increasing strength of upwelling intensity (Fig. 10.8b; Benzoni et al. 2003; Schmidt et al. 2012; Wall et al. 2012; Glynn et al. 2015, 2017).

10.3.2 Ecological Interactions and Processes Affected by Upwelling

Above we have outlined the effects of upwelling on the physico-chemical characteristics of the environment and their repercussions on the reef corals. Here, we briefly address the ecological interactions and processes we believe are most affected by upwelling.

Recurrent or prolonged exposure to upwelling pressure appears to have selected for 'marginal' reefs characterized by a biodiversity as low as only one to three reef-building coral species, as found in the Eastern Tropical Pacific (Guzmán and Cortés 1993; Glynn et al. 2017), a feature prompting the term 'minimum reefs' (Cortés 1997). Similar mono- and oligotypic coral communities with limited or no framework development or so-called '*coral carpets*' (Riegl and Piller 2003) have also been reported from upwelling

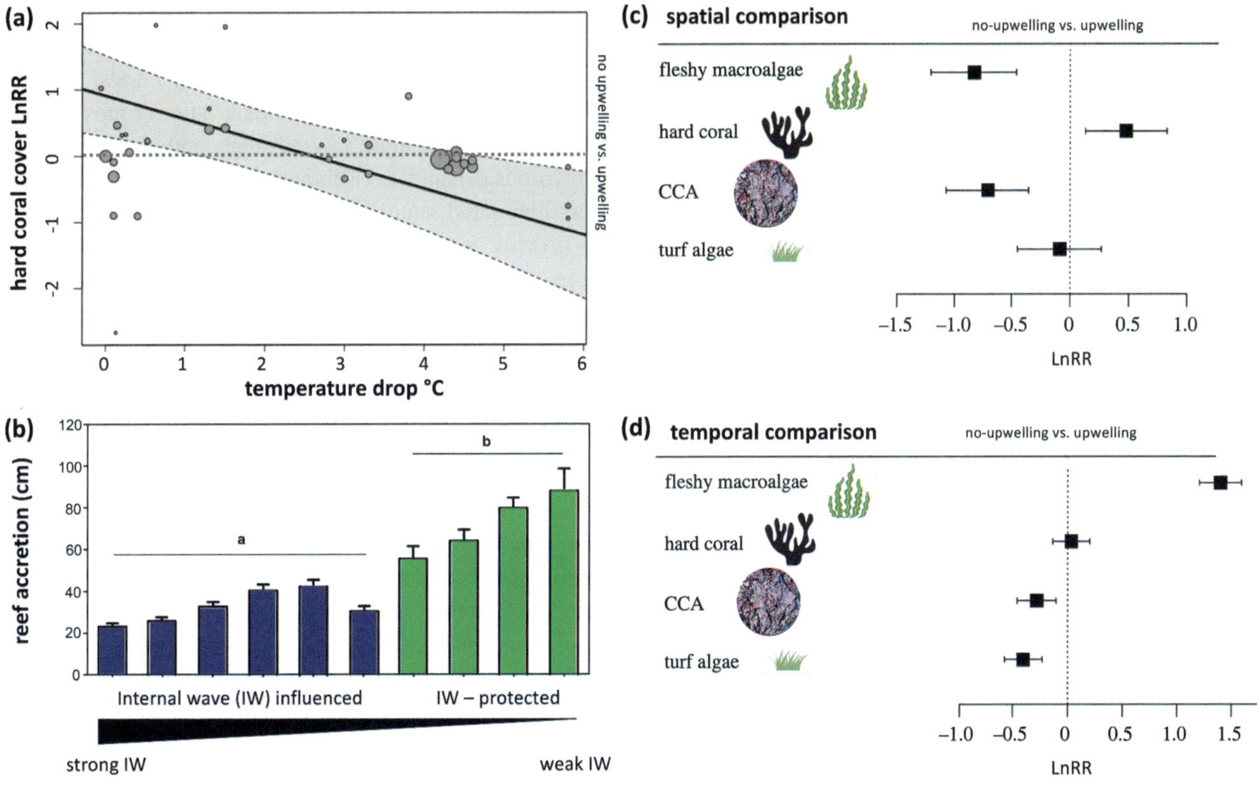

Fig. 10.8 Upwelling can entail changes in the composition and structure of the benthos and contribute to reef 'marginality'. Most commonly recorded changes as a function of upwelling intensity (i.e. temperature drops or internal wave (IW) intensity) are a decrease in hard coral cover (**a**) and reef accretion height (**b**) with increasing intensity of upwelling or internal waves. Effect size (LnRR—natural logarithm of response ratio) of upwelling influence on the percentage cover of key benthic groups (fleshy macroalgae, hard corals, crustose coralline algae (CCA) and turf algae) with enhanced cover due to upwelling when effect sizes are positive and vice versa when negative (**c**, **d**). Upwelling effects were assessed by comparing adjacent reefs exposed to non-upwelling and upwelling conditions, respectively (**c**) and by comparing within upwelling reefs during upwelling and non-upwelling season (**d**). Panels (**a**), (**c**) and (**d**) are modified from Spring and Williams (2023) and (**b**) is modified from Wall et al. (2012)

areas in other reef provinces (Glynn 1993; Benzoni et al. 2003). Yet these coral communities or carpets may be of similar hard coral cover or even exceed hard coral cover of reference locations (Fig. 10.8c: summarized in Spring and Williams 2023). Coral reproduction in the Eastern Tropical Pacific was higher (for instance in number of reproductive cycles per year, percentage of reproductively active corals and fecundity per coral surface area) in the thermally stable environments (Costa Rica, Gulf of Chiriquí) than in the upwelling areas (Gulf of Panamá, Galápagos), where it was also higher in the non-upwelling than in the upwelling season (Glynn et al. 1994, 1996). While upwelling did not appear to affect coral spawning behaviour in the Caribbean Cubagua Island, Venezuela, where spawning occurred at dates and times similar to those reported for well-developed reefs in the Caribbean, larval settlement and coral recruitment appeared to be suppressed and dominated by one tax (a brooder) at the upwelling sites (Rodríguez et al. 2009). This underscores the role of both pre- and post-settlement processes (e.g. space competition with other benthic taxa, predation of early recruits) to contribute to a restricted repopulation in upwelling areas (Birkeland 1977; Wellington 1982). However, asexual reproduction via fragmentation (known to occur predominantly in the region for some taxa of *Porites* and *Pocillopora*) may facilitate population maintenance in some areas (Glynn et al. 1994, 1996). All processes together can contribute to the development of monospecific carpets.

Another important factor for reef growth is the consolidation of the reef structure by crustose coralline algae (CCA), a group of calcifying algae negatively affected by upwelling (Fig. 10.8c, d, cf. also Spring and Williams 2023). As CCA are important in cementation, binding of framework components and porosity occlusion (Manzello et al. 2008), but also in coral settlement (Tebben et al. 2015), upwelling-related CCA depression may play a role in preventing reef framework formation, in spite of positive upwelling effect sizes on coral carpet cover (Fig. 10.8c). On the contrary, other reef framework consolidator or builder could make up for such process like bryozoan framework builder in 'marginal' conditions that where shown to benefit from heterotrophic food supply (Bastos et al. 2018).

Internal waves may cause profound benthic community differences at small spatial scales: opposite sides of small islands. In the Similan Islands of the Andaman Sea, the ocean-facing western sides of the islands are exposed to 'extreme' conditions of frequent and intense solibores. The shore-facing eastern sides, by contrast, are largely sheltered from the solitons and thus feature only weak deviations from 'optimal' environmental conditions. The coral communities exposed to 'extreme' and 'optimal' conditions, respectively, on opposite sides of the islands, are barely a kilometer apart, yet they are strikingly different (see Fig. 10.2, row Andaman Sea, left and right panels). While 'optimal' conditions allow the growth of erect colonies and the formation of a 3D framework, the 'marginal' reefs show stunted growth and predominance of encrusting 2D growth forms (Fig. 10.8b). Closer inspection, however, revealed that species richness was higher in the solibore-exposed 'marginal' reefs than in the 'normal' reef (Schmidt et al. 2012). This surprising finding is the first and only report, to the best of our knowledge, that upwelling may also promote coral biodiversity. We believe this may have to do with the intermittent nature of upwelling due to internal waves. As opposed to prolonged 'extremeness', as in the case of coastal upwelling, intermittent upwelling due to internal waves allows for alternating periods of stress and relief exerting competition but not competitive exclusion among the coral community. Also there are prolonged periods (during the wet southeast monsoon) where the pycnocline is depressed so that solibores do not reach the reef. Intermittent upwelling may thus provide intermediate levels of disturbance, promoting biodiversity, in compliance with Connell's (1978) intermediate disturbance hypothesis.

Upwelling can provide material and energy subsidies to corals and other reef biota, affecting not only the ecophysiology of the organisms but also the ecological interactions across taxa. Competition can act among corals, but also between corals, macroalgae and other major groups. In 'normal' settings, high light, low nutrients and high grazing by herbivorous fish and invertebrates (e.g. sea-urchins), favour a coral-dominated state, where seaweeds are suppressed by low growth (low nutrient concentrations) and high grazing losses. Here, the fast-calcifying branching corals (e.g. *Acropora* spp.) become competitively dominant. In 'extreme' settings, cold shocks, nutrient subsidies, turbidity and other factors previously described may shift and narrow the spectrum to oligo- and monotypic coral communities dominated by hardier genera like *Porites* or *Pocillopora* (Guzmán and Cortés 1993). However, it may also foster macroalgae that can directly metabolize the excess inorganic nutrients (Diaz-Pulido and Garzón-Ferreira 2002). Fleshy macroalgae can benefit from enhanced nutrient supplies during the upwelling period, promoting the waxing and waning of macroalgal dominance (Diaz-Pulido and Garzón-Ferreira 2002) in sync with the seasonal upwelling (Fig. 10.8d, Diaz-Pulido and Garzón-Ferreira 2002; Eidens et al. 2015; Stuhldreier et al. 2015c). In addition, grazing pressure can be reduced in mobile fishes that either graze less (Smith 2008) or avoid the upwelling waters in advancing solibores (Richter, pers. obs.). However, other grazers like sea urchins may be less sensitive to upwelling and maintain grazing rates sufficient to keep macroalgae at check, thus preventing the above seasonal dominance of fleshy macroalgae (Stuhldreier et al. 2015c). Such macroalgal boost was not always observed, especially when compared with close-by reference sites under 'optimal' environmental conditions (Fig. 10.8c, d). However, shifts in herbivore pressure and climatic perturbations like ENSO events may tip the balance from coral to macroalgae dominated state (Fong et al. 2017).

Upwelling may favour also other functional groups competing with corals. For example, heterotrophic benthic planktivores like mussels or barnacles are favoured in plankton-enriched upwelling waters, to the point of outgrowing the corals (Wizemann et al. 2018; Enochs et al. 2021). A similar benefit was found for bryozoan that can actually contribute to framework builder in 'marginal' conditions (Bastos et al. 2018).

Nutrient enrichment is also known to enhance erosion rates (Reaka-Kudla et al. 1996; Alvarado et al. 2017; Enochs et al. 2021) affecting the delicate balance between calcium carbonate accretion and erosion, thus governing the establishment or disappearance of a reef framework (Glynn 1988; Smith 2008; Schmidt and Richter 2013). Bioerosion is an important ecological process (Alvarado et al. 2017), where the boring, etching and scraping organisms involved in the internal and external erosion of the reefal carbonate are differentially affected by the various facets of upwelling. In particular, macroborer communities where favoured in upwelling regions (Enochs et al. 2021). The boring bivalves (*Lithophaga*, *Gastrichaena*) and etching sponges (*Cliona*) are filter feeders benefiting from nutrient-enhanced phytoplankton productivity in upwelling areas, and areas with high phytoplankton productivity feature corals with weakened skeletons riddled by internal bioeroders (Scott and Risk 1988). Scraping and excavating herbivores, by contrast, appeared to favour areas with less intense upwelling in the Galápagos Islands (Kingsford et al. 2023), in spite of putatively higher macroalgal growth. This somewhat counterintuitive finding may be related to the temperature effect of upwelling on the fish (see above).

Furthermore, not only the supply but also the nature of plankton that matters: While neutrally buoyant and passive particles drift passively with the currents, buoyant particles (e.g. eggs) and/or vertically swimming plankton are able to decouple from fluid motion and aggregate in areas of convergence and divergence associated with down- and upwelling (Genin et al. 2005) and propagating internal waves (Garwood et al. 2020). The concentration mechanism resulting from opposing vertical motion of physical (flow, buoyancy) and biological drivers (vertical migration) is a vital prerequisite for the existence of 'walls of mouths' of planktivores (Hamner et al. 1988) ranging from coral polyps to whale sharks gorging on an ample supply of food in an oligotrophic environment. It is also crucial in the concentration and transport of larval invertebrates and fish (Shanks 1983; Lee et al. 1992; Pineda 1994). Its role in the functionality of coral reef ecosystems, however, has not yet been studied in detail, although its likely importance to sustain fish and invertebrate populations has been emphasized (Leichter et al. 1998; Sponaugle et al. 2005; Skinner et al. 2021) or supply macro borer larvae (Wizemann et al. 2018). Within this plankton pump at work other less beneficial larvae may concentrate, like the coral predator *Acanthaster* (Houk et al. 2007; Mendonca et al. 2010), and contribute to observed outbreaks in upwelling regions (Glynn 1993). However, such ecological interactions warrant further investigation as they can be critical for coral reef functionality and recovery in a future changing ocean.

10.4 Local Conditions and Implications for a Future Ocean

10.4.1 Upwelling—Local and Regional Effects

A plethora of environmental and biological factors govern the structure and function of reefs at a given upwelling location. A recent review on upwelling influence on coral reefs introduced the factor 'remoteness' to account for gradients in human activity as an important additional factor influencing coral reef habitats (Spring and Williams 2023). This differentiation acknowledges that the status of a reef is context-specific, i.e. upwelling-induced differences need to be assessed against nearby references with little or no upwelling influence. In other words: there is no global 'gold standard' for a 'normal' reef. This lack of an absolute benchmark, and the use of context-specific regional benchmarks take into account the regional differences in the long-term decline of coral reef ecosystems (Pandolfi et al. 2003), as also described above (Sect. 10.1) and discussed in Schoepf et al. (2023). Even though reference reefs do not represent pristine conditions (Jackson 1997) and 'optimal' environmental conditions may vary between regions, they provide adequate regional benchmarks to assess the effect of upwelling. Proximity to dense human populations can lead to an increase in nutrients in reef habitats (Pastorok and Bilyard 1985; Aguiar et al. 2023) with the underlying mechanisms ranging from direct run-off, river input to groundwater discharge (e.g. Fabricius 2005; Arroyave Gómez et al. 2021) and influence from coastal development (Jimenez 1997). Fishing pressure may also be increased in such areas (Weijerman et al. 2013; Holbrook et al. 2022), affecting the balance between predators and prey and leading to shifts in trophic interactions (Wilson et al. 2010; Graham et al. 2017). Altogether can have an impact on the ecosystem structure of coral reefs (Fig. 10.9a, e.g. Nyström et al. 2000; Norström et al. 2009; Ford et al. 2020) and thus, modulate the interaction with other drivers.

The comparison between populated and remote reef sites in upwelling areas shows that upwelling alone allows for 'normal' coral-dominated and macroalgae-suppressed con-

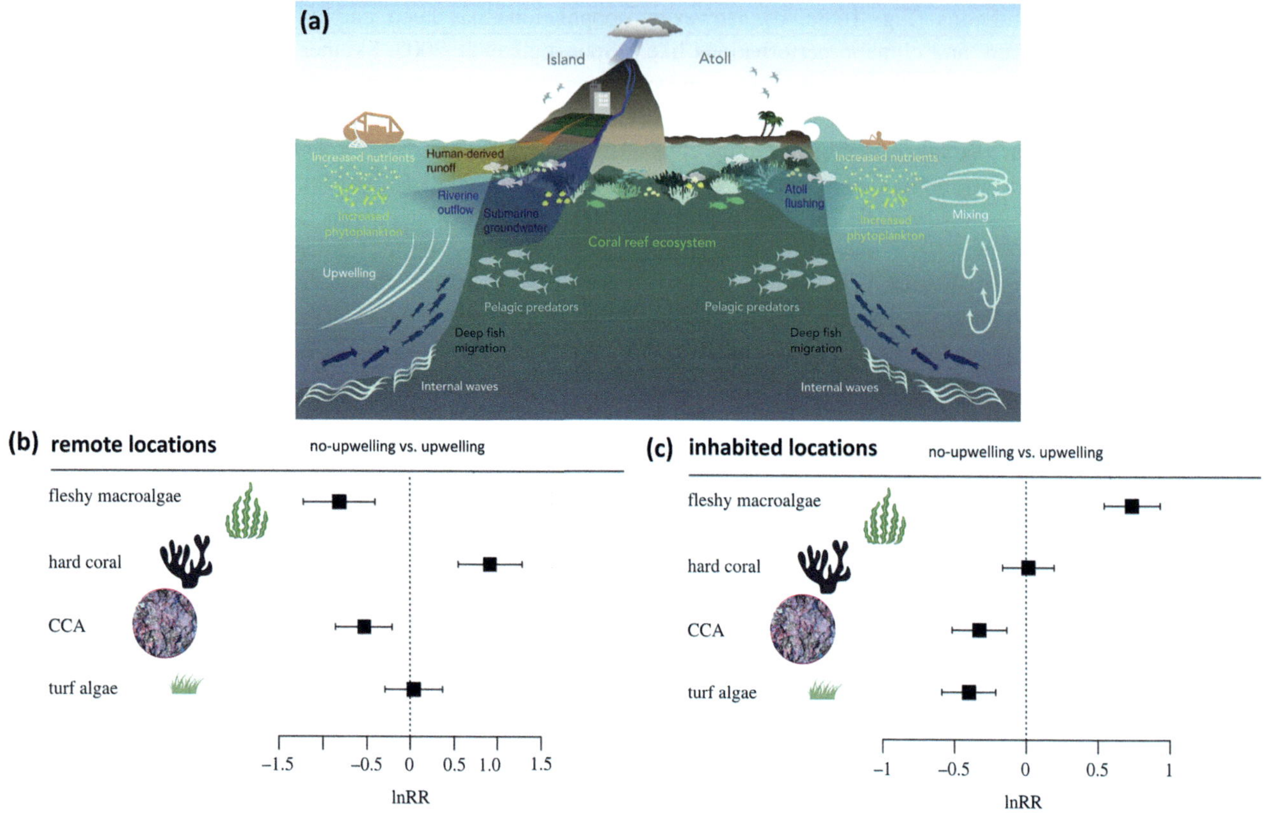

Fig. 10.9 Upwelling occurs on the backdrop of other prevalent drivers in different reef habitats (**a**, adapted from Gove et al. 2016). Besides local natural conditions like wave energy, mixing, river runoff, increased nutrient and sediment loads due to anthropogenic changes in the watershed, coastal development and fishery activity can influence or even change the reef state from 'normal' to 'marginal'. This can contribute to shifts in benthic composition as demonstrated by Spring and Williams (2023) showing differential upwelling effects between remote (**b**) and inhabited locations (**c**). Positive effect sizes (lnRR—natural logarithm of response ratio) indicate an enhancement of the percentage of benthic coverage of the respective benthic group group (fleshy macroalgae, hard coral, crustose coralline algae (CCA) and turf algae) by upwelling, while negative effect size indicate that the group is favoured by non-upwelling conditions. (Figures b, c modified from Spring and Williams (2023))

ditions, whereas upwelling, in combination with anthropogenic impact (nutrient runoff, fishing), may tip the balance towards a fleshy macroalgae dominated state (Fig. 10.9b, c). Note, however, that different sample sizes may have biased the results: the remote reefs were limited to Jarvis Island, while the inhabited locations encompassed 12 locations across all ocean basins. But in spite of the resulting larger variability in the results of the inhabited locations, the observed differences in effect sizes were significant. Reduced hard coral cover in upwelling-influenced reefs were observed in average across the Andaman Sea Islands in 20 m water depth with highest internal wave activity compared to the reference site (Schmidt et al. 2012) likewise in Colombian reefs within the Caribbean (Eidens et al. 2015), in contrast to studies on the Pacific coast of Panamá that revealed an enhanced hard coral cover in upwelling influenced reefs of the Gulf of Panamá compared to Gulf of Chiriquí (Smith 2005; Randall et al. 2020, but also see the role of global change in this context in the next section).

Overall, changes in benthic cover is dynamic at various spatial and temporal scales and underlying drivers may include differences in the type and strength of anthropogenic and upwelling pressure. Changes related to human impact can manifest at all scales, from the ecosystem to the microbial level. Ecosystem-wide changes include e.g. shifts in relative composition of key functional groups (Bellwood et al. 2003). Microscopic changes are e.g. shuffling of symbionts in response to environmental stressors (LaJeunesse et al. 2010; Palacio-Castro et al. 2023) or shifts in coral-associated microbes in response to anthropogenic stressors (Rosales et al. 2019; Zhu et al. 2022; Rutsch et al. 2025). The remote location with its enhanced hard coral cover on upwelling-influenced reefs (Aston et al. 2019; Vargas-Ángel et al. 2019; Huntington et al. 2022), may stem from moderately elevated nutrients that help overcome nutrient limitation in corals (Falkowski et al. 1993), fuel reef metabolism, growth and energetic status (D'Angelo and Wiedenmann 2014; Fox et al. 2023). At the inhabited locations, the enhanced fleshy algal cover is probably due to nutrients exceeding local critical thresholds (e.g. Gil 2013; Fabricius et al. 2013) and/or reduction or absence of herbivores (Littler and Littler 1985). For reasons of comparability and to derive response effect sizes only comparable sites under upwelling and non-upwelling conditions were considered and this limited the assessment (Spring and Williams 2023). This highlights the lack of comprehensive data that include information on both 'extremeness' and 'marginality' in upwelling-influenced areas (Fig. 10.5). Comparability is further compromised by sometimes pronounced differences in other local factors that are generally not reported, like physical impact by surface waves, chlorophyll levels, intensity and timing of rainfall and run-off between reference reef regions (e.g. Glynn 1977; Jiménez et al. 2010; Gove et al. 2013; Eidens et al. 2015; Ford et al. 2020). While the compiled data on upwelling influence emphasized the potential for modulation of ecological coral reef traits, other prevailing factors need to be considered (e.g. Arroyave Gómez et al. 2021). Integrating local factors will allow a more accurate assessment of reef states and conditions and help to clarify the relative importance of specific factors responsible for changes in coral reef conditions. Understanding thresholds and tipping points across natural and anthropogenic gradients is crucial and requires more holistic and concerted work (e.g. Gove et al. 2015).

10.4.2 Implications for a Future Ocean

Climate change has a profound effect on the ocean (Bijma et al. 2013; Visbeck and Keiser 2021): the absorption of heat and atmospheric CO_2 makes the ocean warmer and more acidic. The stronger thermal stratification implies less mixing and less oxygenated deep waters. Especially the change in temperature results in cascading effects of ice melt, marine heat waves, changing circulation and stratification patterns that have profound effects on marine ecosystems and biodiversity (Doney et al. 2012). 'Extreme' events are projected to increase in frequency and intensity as underscored by the recurrence of massive global bleaching events (Hughes et al. 2017, 2020). Upwelling takes place against the backdrop of such global changes, and this can lead to synergistic but also antagonistic effects. The upwelling of cold water has the potential to mitigate the severity of heat waves (Wall et al. 2015; Wyatt et al. 2020) and strengthen coral reef resilience (Wall et al. 2023). Mass bleaching and mortality of corals is considered the most severe threat to coral reefs today (Hughes et al. 2018) as warming continues to increase to unprecedented levels (Hoegh-Guldberg et al. 2023), prompting the search for naturally resistant populations or sites that provide refugia from adverse conditions projected for the future (Camp et al. 2018b; Schoepf et al. 2023).

In an earlier paper on 'Coral reef bleaching: facts, hypotheses and implications', Glynn (1996) raised the hypothesis that possible refuge habitats, which may be protected from rapid temperature rise, may be found in upwelling areas, where cold deep water reaching the surface would abate heating and thus, mitigate coral bleaching in a warming ocean. This nugget of hope prompted the reef conservation community to prioritize thermal refuges for protection and conservation (e.g. McClanahan et al. 2024) and triggered intense research on thermal refuges and 'marginal' reefs over the past decades. First support for this concept arrived from Costa Rica (Jiménez et al. 2001), Panamá (e.g. Fong and Glynn 2001; Podestá and Glynn 2001) and the Bahamas (Riegl and Piller 2003), where upwelling was shown to provide thermal protection to corals during the first global mass bleaching event in 1997/1998 (Fig. 10.10a shows the thermal trajectories leading to differential bleaching in the Gulf of Panamá and Gulf of Chiriquí, Glynn et al. 2001). The spatial scale of the refugia can range from reef provinces (e.g. Glynn et al. 2001; Riegl et al. 2019) to individual reefs (e.g. Wall et al. 2015; DeCarlo et al. 2017), down to specific depth ranges (e.g. Bridge et al. 2013; Smith et al. 2014) and reef features like grooves and channels (Andrews and Gentien 1982).

In the Andaman Sea, continuous *in situ* temperature records together with monitoring of the coral reef community and their health state during the 2010 severe bleaching event allowed to assess the bleaching severity in response to localized upwelling by internal waves (Wall et al. 2015). This was done by ranking the reefs investigated according to their exposure to internal waves, calculating the associated *in situ* cooling (degree days cooling) counteracting the heat stress (degree days heating) and monitoring the fate of bleaching-sensitive and bleaching-resistant coral taxa in the aftermath of the event (Schmidt et al. 2016, Fig. 10.10c). The breaking of internal waves and subsequent propagation of cold bores carrying sub-thermocline waters upslope, caused pulsed cooling at the height of the heat-wave mitigating coral bleaching and mortality in the offshore-facing western sides of islands. While modal *in situ* temperatures from the onset to the peak of the heat wave differed only marginally (internal wave exposed vs protected: 30.3 vs 30.6 °C), the accumulated net heat stress (heating minus cooling) deviated markedly (degree heating weeks of 2.3 compared to 7.8, Wall et al. 2015), explaining to a large extent the observed differences in bleaching and mortality. The findings underscore the importance of internal wave induced upwelling, characterized by frequent step changes in temperature, in reducing heat stress to corals (Fig. 10.6b, c). The global significance of internal waves providing thermal refugia has subsequently been confirmed with quantitative support across numerous reef regions (e.g. Sabasaki, Irimoto, Japan, Iglesias, Gulf of Chiriquí, Panamá, Haapati, Moorea, French

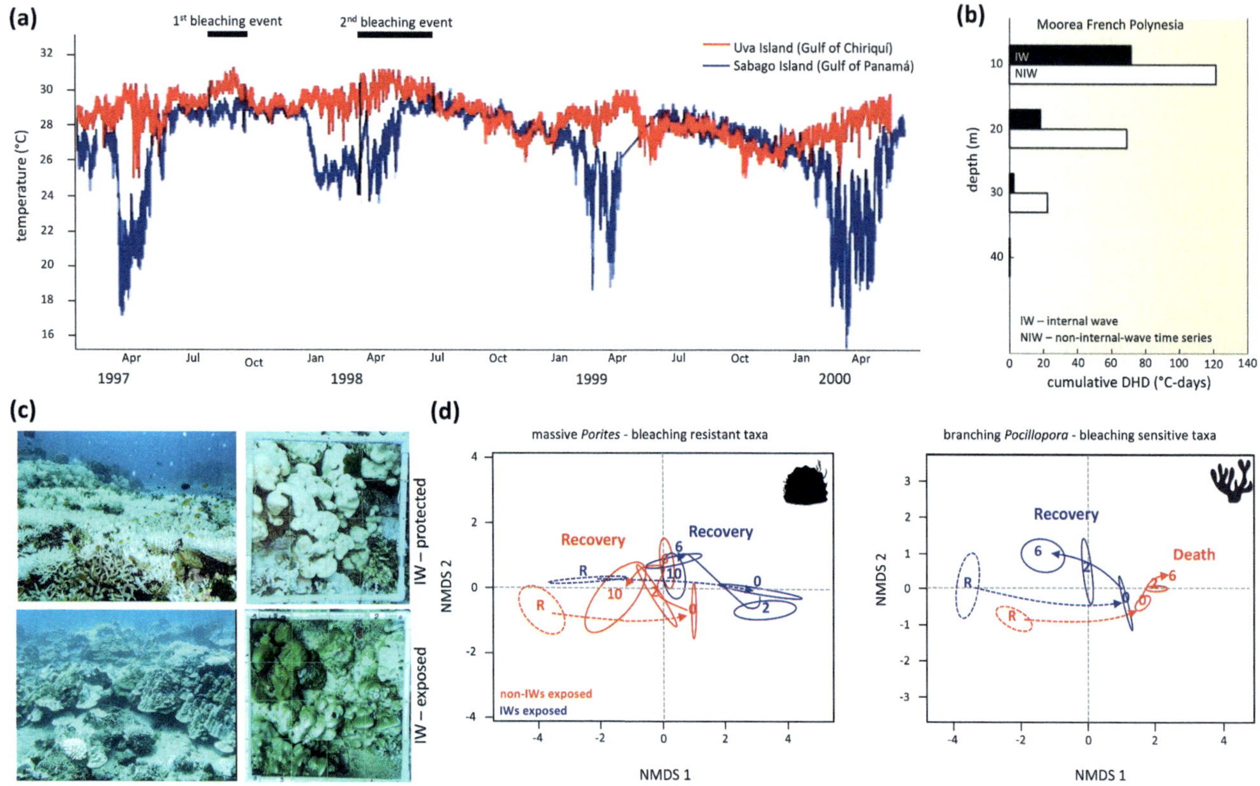

Fig. 10.10 Upwelling during anomalously elevated temperatures, the effect of environmental 'extremeness' and the ecological consequences when both thermal stress and upwelling coincide, thus creating thermal refugia. (**a**) During the 1997/1998 first global massive bleaching event, the reefs in the Gulf of Chiriquí experienced severe heat stress while reefs in the Gulf of Panamá remained protected by upwelling (adapted from Glynn et al. 2001). (**b**) Low-pass filtering of *in situ* temperature time series of coral reefs in French Polynesia at different water depths (simulating non-internal-wave (NIW) conditions) allowed to quantify the extent by which internal wave upwelling can decrease the thermal stress (calculated as degree heating days DHD) compared to when internal waves (IWs) are present (adapted from Wyatt et al. 2020). (**c**, **d**) During the second global massive bleaching event in 2010, with the most severe thermal stress in the Andaman on record, IWs significantly mitigated thermal stress. (**c**) IW-exposed reefs showed a lower bleaching response and fully pigmented corals during the peak of the heatwave (bottom row), compared to heavy bleaching on reefs without IWs (top row). Even though the IWs could not completely protect the corals from bleaching, the extent of bleaching was significantly reduced (adapted from Wall et al. 2015). (**d**) Bleaching-sensitive (*Pocillopora*) and robust (*Porites*) corals from IW-exposed (blue) and IW-protected (red) reefs show differences in the trajectory of recovery from bleaching in the months before (R = −24 month), during (0 month) and after this bleaching event (2–10 months) (adapted from Schmidt et al. 2016)

Polynesia; Wyatt et al. 2020, Fig. 10.10b) and numerical models (Bachman et al. 2020, 2022). Although bleaching in the Andaman Sea could not be completely prevented, the short periods of thermal relief reduced bleaching and gave the corals a chance to recover from the heat stress (Fig. 10.10c), while in the absence of thermal relief in the reference reefs, the heat-sensitive taxa almost completely vanished (Wall et al. 2015; Schmidt et al. 2016, Fig. 10.10c). Along with experimental work on the effect of simulated internal waves on coral stress response and survival (Buerger et al. 2015), these independent lines of evidence highlight the importance of temporary relief during periods of chronic thermal stress.

Although thermal refugia have been reported in some locations and years, upwelling does not appear to provide a full coverage collision insurance for corals against heat stress. One reason is that planetary disturbances like ENSO events are associated with basin-wide adjustments in sea levels and thermocline depths (Firing et al. 1983), so that even when upwelling conditions prevail (e.g. wind-driven water mass movements), a depression in the thermocline may only allow for upwelling of warm above-thermocline water (Huyer et al. 1987). For example, a lowered thermocline exacerbated heat waves during ENSO 2019 in Moorea, French Polynesia, as less cooling from internal waves was available to counteract heat stress throughout depth (surface to 40 m). This resulted in pronounced bleaching at a depth of 10 m, compared to the ENSO in 2016 where internal wave upwelling was detected and more pronounced at this depth (Wyatt et al. 2023). Another reason is that upwelling may cease when it is most needed to mitigate heat stress. In the Pacific, the Equatorial Undercurrent is the main source of equatorial upwelling of cold nutrient-rich water. However, strong ENSOs appear to be associated with the shut-down of the Equatorial Undercurrent and cessation of upwelling (Firing et al. 1983; Stramma et al. 2016; Fox et al. 2023). This has led to catastrophic coral mortality (>95%) in the Galápagos during the 1982–83 ENSO (Glynn 1990) and severe bleaching and mortality in the central equatorial Pacific reefs in the 2015–16 event (Brainard et al. 2018; Claar et al. 2020). Furthermore, upwelling and heat waves do not always coincide in time (Chollett et al. 2010; Chollett and Mumby 2013; Vargas-Ángel et al. 2019): For example, a heat wave that caused massive bleaching in the Gulf of Panamá in 1982/1983 had a delayed onset compared to the heat wave of 1997/1998, which started earlier and did not result in extensive bleaching (Glynn et al. 2001). Curiously, shelf-break upwelling does not seem to play an important role in mitigating coral bleaching within the Great Barrier Reef: strongest periods of shelf-break upwelling coincided with Great Barrier Reef-wide bleaching events in 1997–1998 and 2001–2002 (Berkelmans et al. 2010). Conversely, an anomalously sluggish East Australian Current with weak shelf-break upwelling coincided with the absence of bleaching in 2004—the hottest year on the 1982–2017 satellite record (DeCarlo and Harrison 2019). In contrast, the shutdown of the Northern Queensland Coastal current that would supply cooler what caused severe bleaching during 2015/2016 ENSO and upwelling is suggested to protect from heat waves otherwise in the Northern Great Barrier Reef (Wolanski et al. 2017; Huang et al. 2024).

While the suppression of heat stress by upwelling is a key factor, (additional) bleaching protection may be imparted by other co-occurring environmental or biological factors. Upwelling increases the planktonic food supply to corals (Roder et al. 2010) improving coral energetics (Gove et al. 2006; Elliott et al. 2012) and providing energy subsidies against stress (Grottoli et al. 2006; Riegl et al. 2019). Upwelling also enhances turbidity, thus reducing radiative stress (Sully and van Woesik 2020). It has been found that habitats characterized by high environmental variability can increase corals thermal tolerance through stress-hardening, particularly thermal fluctuations near the corals' upper thermal threshold (Oliver and Palumbi 2011; Palumbi et al. 2014). Similar effects have been observed with environmental fluctuations near the corals' lower thermal limit, characteristic for upwelling areas. For the Andaman Sea, it was shown that step changes in temperature (up to 10 °C) associated with the waxing and waning of internal wave generated bores across the fore-reef slope rendered corals more resistant to heat stress (Buerger et al. 2015; Doering et al. 2021; Wall et al. 2023). First attempts to elucidate the underlying mechanisms emphasize the role of coral associated bacterial taxa (Doering et al. 2021; Rutsch et al. 2025) as well as energy and material subsidies or physiological trade-offs that enhance lipid reserves (Weichler et al. 2024) that altogehter may contribute to enhanced stress resistance. However, if temperature fluctuations exceed the thermal limits of an organism, stress-hardening may give way to collapse when the energetic resources become exhausted (Hackerott et al. 2021). In Panamá, thermal stress sensitivity was higher for corals from upwelling sites compared to non-upwelling sites and it was suggested that (chronic) upwelling may be stressful, draining energy stores and thus rendering corals more susceptible to heat stress (Glynn et al. 1988; D'Croz et al. 2001). We believe that the apparent conflict between the Andaman Sea and Panamá findings reflects the nature and frequency of upwelling. In the former, upwelling is wave-driven and occurs at high frequency (minutes to hour pulses). In the latter it is wind-driven and happens at low frequency (weeks to months). While intermittent cooling conveys hardening, notorious cooling causes weakening of the corals. This may explain the hampered recovery of reefs following massive bleaching events in 1982/83 in Galápagos (Glynn and Colgan 1992) that was exacerbated by cascading effects from enhanced bioerosion (Glynn et al. 1988) and a *Acanthaster* outbreak (Glynn 1985) without recovery up to 30 years later (Glynn et al. 2015). Similarly, excess nutrients through upwelling exacerbated stressful conditions during a heat wave in the Red Sea (DeCarlo et al. 2020).

In addition to seasonal thermal stress, global change inflicts a chronic pressure on coral reefs by warming and ocean acidification, which is expected to become exacerbated in the future. The CO_2 tipping points in terms of ocean acidification identified so far today suggest that coral reefs under 'optimal' conditions in Papua New Guinea cease to exist and express 'marginal' traits at a pH of 7.7 (or aragonite saturation state (Ω_{arag}) of 2.3, Fabricius et al. 2011, 2014). However, such an inflection point was reached earlier in upwelling-influenced regions along a natural acidification gradient within the Eastern Tropical Pacific around Galápagos (Manzello et al. 2014) at a pH of 8.0 (or Ω_{arag} of 3) and no growth in different coral taxa within in the Eastern Tropical Pacific between Ω_{arag} 2.2 and 2.5 (Rixen et al. 2015; Sánchez-Noguera et al. 2018). The cessation of reef development was driven not necessarily by a reduction in accretion (both from coral calcification (Randall et al. 2020) nor other benthic calcifiers (Enochs et al. 2021)) but due to an enhancement of bioerosion, specifically macroboring activity (e.g. Wizemann et al. 2018; Enochs et al. 2021) and/or reduced skeletal density (Manzello et al. 2014) that may further facilitate bioerosion. In the past, coral calcification has been reduced in upwelling reefs within the Eastern Tropical Pacific (Glynn 1977) but recently a reversal was observed, probably as a result from global warming that restricts growth near the corals' upper thermal limit in non-upwelling reefs (Randall et al. 2020; Rodriguez-Ruano et al. 2023). Warming is expected to foster growth and contribute to a higher coral cover in upwelling reefs of the Eastern Tropical Pacific where thermal optima are not yet exceeded (Randall et al. 2020). Climate change is expected to contribute to a strengthening of the thermal stratification of the water column, leading to a reduction in tropical ocean productivity and more oligotrophic conditions (see Sect. 2.1) However, climate change is also predicted to contribute to stronger winds and stronger wind-driven upwelling of cold and nutrient-rich waters (Sydeman et al. 2014; Wang et al. 2015). Upwelling has thus been invoked to reduce warming rates (up to 0.7 °C per century, Karnauskas and Cohen 2012; Randall et al. 2020) and replenish the depleted surface layer with vital nutrients, thus sustaining marine productivity in an otherwise desolate tropical ocean. Such an effect of enhanced material and energy subsidy is expected for the equatorial upwelling areas and their coral reefs (Karnauskas and Cohen 2012).

Climate change can both ameliorate or exacerbate stressful conditions. Therefore, it is important to understand the underlying upwelling mechanisms in order to be able to assess possible changes in the future. Changes in temperature, stratification and regional atmospheric forces together with underwater bathymetry will decide whether upwelling will be enhanced or diminished in the future, and how the above outlined consequences for the shallow reef ecosystems will change in the future. More intense upwelling may bring organisms closer to their lower physiological limit, reverse beneficial effects, surpass ecological tipping points (Palmer et al. 2022) and lead to collapse, as evidenced for coral reefs in the Eastern Tropical Pacific following the 1982/83 ENSO or during the little ice age and its coral reef growth hiatus (Glynn et al. 1983; Toth et al. 2012). For other regions, an intensification may turn out to be a blessing rather than a curse, and safeguard their persistence into a warmer future. Yet, to be able to predict the regionally varying changes in upwelling intensity and timing along with their consequences on the reef biota, in-depth regional assessments of physiological thresholds and ecological tipping points are required.

Ocean warming is unfolding as a series of consecutive intensifying 'extreme' events causing coral mass bleaching and mortality around the globe (Hughes et al. 2017). It is not clear, if and to what extent the other 'extreme', i.e. upwelling, may play a role in reducing or even reversing this trend (Bograd et al. 2023). As mentioned earlier, upwelling is expected to intensify in response to increased wind speeds, with most notable effects in the coastal and equatorial upwelling regions (Bakun 1990; Bakun et al. 2010), albeit with regional deviations from this trend (Sydeman et al. 2014). Amplification is expected to be largest in the largest basins (e.g. the Pacific), while in small basins like the Caribbean, long-term sea-surface temperature data analyses from 1985 to 2009, failed to detect clear trends in upwelling strength (Chollett et al. 2012). It cannot be ruled out that temperature trends emerge from the analysis of longer time periods (e.g. coral proxy records) exceeding the 40 year satellite record. Based on coral skeletal and instrumental temperature records of the twentieth century, the warming trends could be confirmed in many parts of the Eastern Tropical Pacific (Clipperton, Palmyra, Wolf). However, other areas (Puerto Ayora, Galápagos; Puerto Chicama, Peru) revealed no trend (Jimenez et al. 2018) or pointed towards an intensification of the Galápagos cool pool through a strengthening of the equatorial and topographic upwelling (Karnauskas et al. 2015; Jimenez et al. 2018). Increased thermal stratification is expected to intensify internal waves in the future, as internal waves propagate along the pycnocline. A strengthening of internal waves is expected to enhance the associated upwelling and affect the aerial extent and location of corresponding thermal refuges (DeCarlo et al. 2015; Storlazzi et al. 2020; Yadidya and Rao 2022). While many of the changes may be beneficial for corals in a warming world, changes in timing and intensity may also have negative effects, as discussed above. They include for instance losses in the reliability of internal wave refugia (Liu et al. 2022; Wyatt et al. 2023), due to shifts in monsoon onset (Liu et al. 2022). Overall, projections on upwelling occurrence and strength require a deeper understanding of fine-scale baroclinic processes (Storlazzi et al. 2020), interaction with other oceanographic features and climatic phenomena (DeCarlo et al. 2015, Camelia et al. accepted) and their incorporation into global models (Storlazzi et al. 2020). Such improved projections and understanding are key to understand the fate of 'marginal' reef systems in the future.

10.5 Implications for Mesophotic Coral Reefs

Intense upwelling in the Eastern Tropical Pacific has restricted reefs to shallower waters (Cortés 1997). Yet, the role of internal waves at deeper depth and its contribution to structuring or modulating these mesophotic habitats is less well known compared to what is known from its influence on shallow reef habitats. While the environmental assessment more commonly extends towards mesophotic depth, the biological assessment is often limited to depths shallower than 30 m. The 'extremeness' (i.e. deviation from 'optimal' conditions) increases with depth and can be quite substantial in mesophotic habitats compared to their shallow reef counterparts (Fig. 10.11). All aspects discussed in terms of physiological and ecological consequences of internal wave activity, therefore, are even more profound within the mesophotic environment and coupled with shifts in other environmental conditions like reduced light availability and decreased mean temperature conditions. It has been discussed that upwelling may play a role in many ocean regions, like the Bahamas and Bonaire within the Caribbean (Frade et al. 2019; Goodbody-

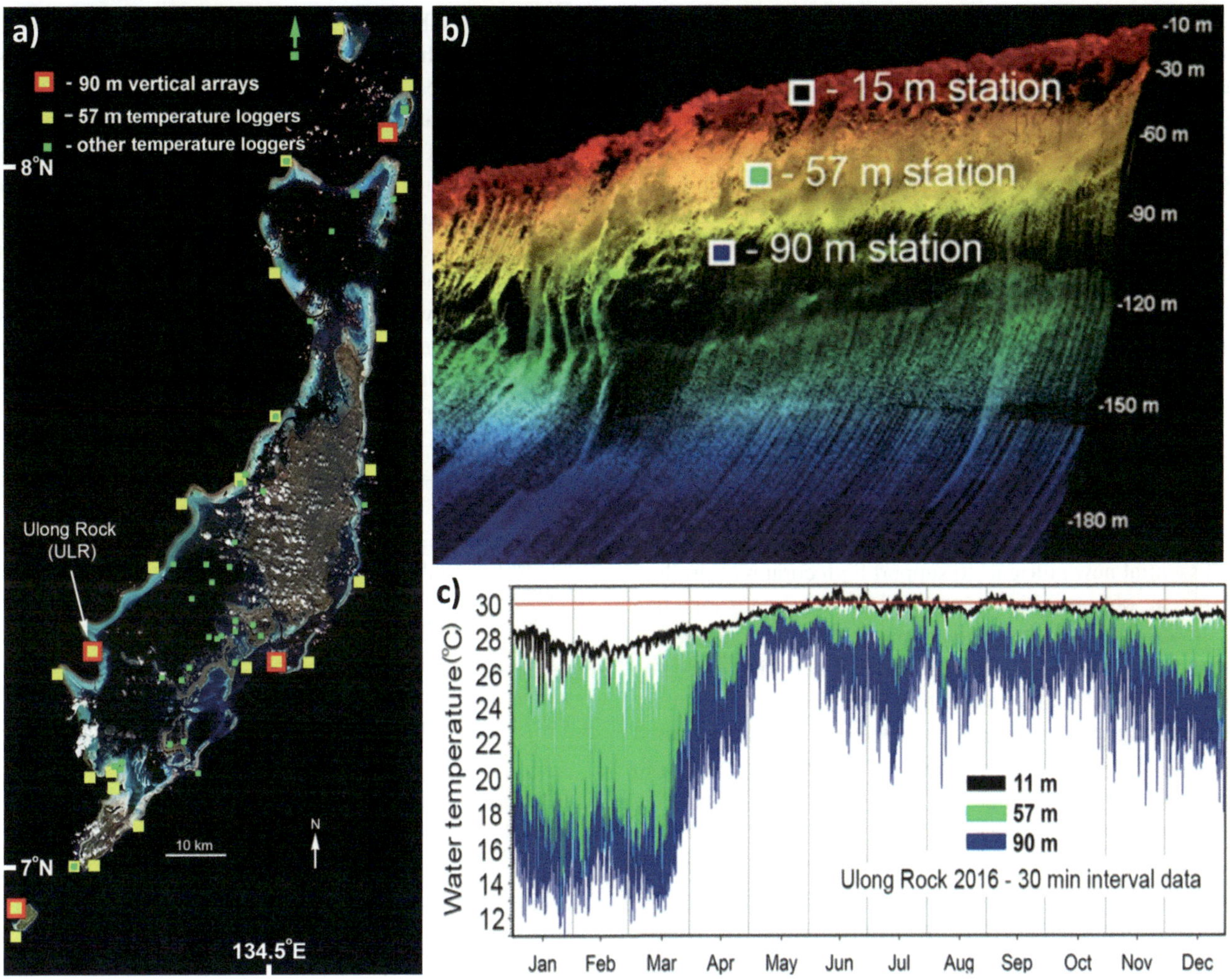

Fig. 10.11 'Extremeness' of environmental deviations from 'optimal' in mesophotic environments. Upwelling as sourced from below pycnocline increases in strength the closer the location is situated to the pycnocline. Therefore, the strength of environmental variations increases with depth and has been recorded for instance in Palau (**a**) at the outer slope (**b**) in both shallow (11 m) as well as mesophotic coral reef habitats (57 m and 90 m, **c**). (Adapted from Colin and Johnston (2020))

Gringley et al. 2019), Palau (Colin and Lindfield 2019) and Micronesia (Rowley et al. 2019) in the Central Pacific and the Eastern Tropical Pacific (Cortés 2019). Across these sites, upwelling is expected to represent a key mechanism for the delivery of nutrients where it can increase plankton productivity (Goodbody-Gringley et al. 2019) and promote heterotrophic energy acquisition (Lesser et al. 2010). Upwelling has also been hypothesized to contribute to coral health along Florida as long as temperatures do not exceed the coral's lower critical level that can cause bleaching (Reed et al. 2019). Generally, its role in these habitats is commonly regarded as beneficial but it may also induce stress or limiting conditions leading to depauperated benthic assemblages (Wolanski et al. 2004). While upwelling was shown to enhance growth in the upper mesophotic environment (Leichter and Genovese 2006) and protect from bleaching at some mesophotic locations (Pérez-Rosales et al. 2021; Diaz et al. 2023), formation of an *Orbicella*-dominated basin in the Hind Bank Marine Conservation District, U.S Virgin Islands was only found in internal wave sheltered habitats and not observed in benthos exposed to internal waves (Smith et al. 2010). Recently, cold-water bleaching was reported for mesophotic coral reefs due to exceptional pycnocline shoaling in the Clipperton Atoll underscoring the potential for limiting conditions acting from below (Foreman et al. 2024). While shallow reef environments benefit from internal wave upwelling of cold deep waters during heat stress (Wall et al. 2015), mesophotic reefs may actually suffer when the backslash of the bores that mixed with hot surface water advects surface heat to mesophotic depths, exacerbating bleaching. Such internal wave enhanced mesophotic bleaching has been observed at the western island side at Ile des Rats on Etmont Atoll, Chagos (Diaz et al. 2023). These unexpected findings highlight that mesophotic habitats, which are yet poorly explored, still harbor many hidden surprises. Unearthing this information may help us better understand the role of upwelling across the full depth range of reef habitats. We believe it is imperative to understand the coupling of oceanographic processes like currents and waves with coral reef structure and functioning to define the limits of occurrence of corals in such 'extreme' environments and identify thresholds and tipping points of coral reef existence.

10.6 Conclusions

Upwelling is a mixed blessing: its negative effect on reef development has already been recognized by Darwin (1842) in his monograph 'Structure and Distribution of Coral Reefs'. Positive effects of upwelling are e.g. the provision of essential new nutrients to support high coral reef productivity in oligotrophic waters. While coral-zooxanthellae symbiosis is central to efficient nutrient recycling, external inputs of nutrients can provide material subsidies and contribute to coral health and performance. Whether upwelling is a blessing or a curse depends on the nature and strength of the upwelling, and thus transition from net reef accretion to erosion. Increasing frequency and intensity of change in environmental conditions can push corals and associated reef organisms to their ecological limits and beyond, and govern the balance of antagonistic processes like carbonate accretion and erosion. This can further change the competitive relationships between organisms and favour non-reef-building communities at the expense of reef-building processes. On the other hand, changes within the corals' optimal range or environmental limits can fuel their growth and render them more stress tolerant. In a warming ocean, the upwelling of cold deep water can provide a thermal refuge and contribute to the continued existence of reefs. Future warmer temperatures may bring corals closer to their optimal range during upwelling periods and promote growth in regions where growth was previously restricted as the lower temperature limits were reached. Furthermore, upwelling of cold water has been shown to protect against severe marine heat waves and contribute to the survival of reefs from catastrophic mass bleaching and mortality. Yet an intensification of upwelling in the future can further restrict reef development and may even lead to a reef shutdown. Conservation of coral reef ecosystems has become a major focus under the current rate of ocean change. The necessary endeavors and protection measures differ, however, between 'normal' and 'marginal' reef habitats (Bates et al. 2019). Additionally, conservation may also take advantage of coral reefs under 'extreme' conditions as they may contribute to reef persistence either due to thermal refugia, energy subsidies, stress hardening or the combination of these. It is essential to establish thresholds and tipping points at a regional scale when coral framework establishment transitions into coral community formation and ultimately leads to conditions that prevent coral occurrence. This requires the assessment of regional reference conditions on the backdrop of which environmental 'extremes' occur. It is clear that 'extremes' are, by definition, context-dependent and need to be assessed against the prevailing drivers in the past and present. We also need a deeper understanding of climate feedback on upwelling and consider the dual nature of upwelling in future projections for reef persistence: as a friend or foe—providing coral reef refugia or preventing coral reef recovery.

References

Abel D, McConnell R (2009) Environmental oceanography: topics and analysis. Jones and Bartlett Publishers, Sudbury, p 311. Modified by https://slideplayer.com/slide/4279467/14/images/49/Coastal+upwelling+in+the+Northern+Hemisphere.jpg

Abele LG (1976) Comparative species richness in fluctuating and constant environments: coral-associated decapod crustaceans. Science 192:461–463. https://doi.org/10.1126/science.192.4238.461

Aguiar DK, Wiegner TN, Colbert SL, Burns J, Abaya L, Beets J, Couch C, Stewart J, Panelo J, Remple K, Nelson C (2023) Detection and impact of sewage pollution on South Kohala's coral reefs, Hawai'i. Mar Pollut Bull 188:114662. https://doi.org/10.1016/j.marpolbul.2023.114662

Alford MH, Peacock T, MacKinnon JA, Nash JD, Buijsman MC, Centurioni LR, Chao S-Y, Chang M-H, Farmer DM, Fringer OB, Fu K-H, Gallacher PC, Graber HC, Helfrich KR, Jachec SM, Jackson CR, Klymak JM, Ko DS, Jan S, Johnston TMS, Legg S, Lee I-H, Lien R-C, Mercier MJ, Moum JN, Musgrave R, Park J-H, Pickering AI, Pinkel R, Rainville L, Ramp SR, Rudnick DL, Sarkar S, Scotti A, Simmons HL, St Laurent LC, Venayagamoorthy SK, Wang Y-H, Wang J, Yang YJ, Paluszkiewicz T, Tang T-Y (2015) The formation and fate of internal waves in the South China Sea. Nature 521:65–69. https://doi.org/10.1038/nature14399

Alvarado JJ, Grassian B, Cantera-Kintz JR, Carballo JL, Londoño-Cruz E (2017) Coral reef bioerosion in the eastern tropical Pacific. In: Glynn PW, Manzello DP, Enochs IC (eds) Coral reefs of the Eastern Tropical Pacific: persistence and loss in a dynamic environment. Springer, Dordrecht, pp 369–403

Álvarez-Noriega M, Marrable I, Noonan SHC, Barneche DR, Ortiz JC (2023) Highly conserved thermal performance strategies may limit adaptive potential in corals. Proc R Soc B Biol Sci 290:20221703. https://doi.org/10.1098/rspb.2022.1703

Andrews JC, Gentien P (1982) Upwelling as a source of nutrients for the great barrier Reef ecosystems: a solution to Darwin's question? Mar Ecol Prog Ser 8:257–269

Antonius A (1980) Occurrence and distribution of stony corals in the Gulf of Cariaco, Venezuela. Int Rev Gesamten Hydrobiol Hydrogr 65:321–338. https://doi.org/10.1002/iroh.19800650302

Apel JR (2002) Oceanic internal waves and solitons. An atlas of oceanic internal solitary waves. Global Ocean Associates, pp 1–40

Arroyave Gómez DM, Bartoli M, Bresciani M, Luciani G, Toro-Botero M (2021) Biogeochemical modelling of a tropical coastal area undergoing seasonal upwelling and impacted by untreated submarine outfall. Mar Pollut Bull 172:112771. https://doi.org/10.1016/j.marpolbul.2021.112771

Aston EA, Williams GJ, Green JAM, Davies AJ, Wedding LM, Gove JM, Jouffray J-B, Jones TT, Clark J (2019) Scale-dependent spatial patterns in benthic communities around a tropical Island seascape. Ecography 42:578–590. https://doi.org/10.1111/ecog.04097

Astor Y, Muller-Karger F, Scranton MI (2003) Seasonal and interannual variation in the hydrography of the Cariaco Basin: implications for basin ventilation. Cont Shelf Res 23:125–144. https://doi.org/10.1016/S0278-4343(02)00130-9

Bachman SD, Shakespeare CJ, Kleypas J, Castruccio FS, Curchitser E (2020) Particle-based Lagrangian filtering for locating wave-generated thermal refugia for coral reefs. J Geophys Res Oceans 125:e2020JC016106. https://doi.org/10.1029/2020JC016106

Bachman SD, Kleypas JA, Erdmann M, Setyawan E (2022) A global atlas of potential thermal refugia for coral reefs generated by internal gravity waves. Front Mar Sci 9:921879. https://doi.org/10.3389/fmars.2022.921879

Bakun A (1990) Global climate change and intensification of Coastal Ocean upwelling. Science 247:198–201. https://doi.org/10.1126/science.247.4939.198

Bakun A, Field DB, Redondo-Rodriguez A, Weeks SJ (2010) Greenhouse gas, upwelling-favorable winds, and the future of coastal ocean upwelling ecosystems. Glob Change Biol 16:1213–1228. https://doi.org/10.1111/j.1365-2486.2009.02094.x

Bastos AC, Moura RL, Moraes FC, Vieira LS, Braga JC, Ramalho LV, Amado-Filho GM, Magdalena UR, Webster JM (2018) Bryozoans are major modern builders of South Atlantic oddly shaped reefs. Sci Rep 8(1):9638. https://doi.org/10.1038/s41598-018-27961-6

Bates AE, Cooke RS, Duncan MI, Edgar GJ, Bruno JF, Benedetti-Cecchi L, Côté IM, Lefcheck JS, Costello MJ, Barrett N, Bird TJ, Fenberg PB, Stuart-Smith RD (2019) Climate resilience in marine protected areas and the 'Protection Paradox'. Biol Conserv 236305–314. https://doi.org/10.1016/j.biocon.2019.05.005

Baum G (2011) Calcification and photosynthesis of *Porites lutea* in response to Large Amplitude Internal Waves (LAIW) in the Andaman Sea. MSc thesis, University of Bremen

Bayraktarov E, Pizarro V, Wild C (2014) Spatial and temporal variability of water quality in the coral reefs of Tayrona National Natural Park, Colombian Caribbean. Environ Monit Assess 186:3641–3659. https://doi.org/10.1007/s10661-014-3647-3

Bayraktarov E, Bastidas Salamanca ML, Wild C (2016) The physical environment in coral reefs of the Tayrona National Park (Colombia Caribbean) in response to seasonal upwelling. Bull Mar Coast Res 43:137–157. https://doi.org/10.25268/bimc.invemar.2014.43.1.38

Bellwood DR, Hoey AS, Choat JH (2003) Limited functional redundancy in high diversity systems: resilience and ecosystem function on coral reefs. Ecol Lett 6:281–285. https://doi.org/10.1046/j.1461-0248.2003.00432.x

Benzoni F, Bianchi CN, Morri C (2003) Coral communities of the northwestern Gulf of Aden (Yemen): variation in framework building related to environmental factors and biotic conditions. Coral Reefs 22:475–484. https://doi.org/10.1007/s00338-003-0342-1

Berkelmans R, Weeks SJ, Steinberga CR (2010) Upwelling linked to warm summers and bleaching on the Great Barrier Reef. Limnol Oceanogr 55:2634–2644. https://doi.org/10.4319/lo.2010.55.6.2634

Bijma J, Pörtner H-O, Yesson C, Rogers AD (2013) Climate change and the oceans—what does the future hold? Mar Pollut Bull 74:495–505. https://doi.org/10.1016/j.marpolbul.2013.07.022

Birkeland C (1977) The importance of rate of biomass accumulation in early successional stages of benthic communities to the survival of coral recruits. In: Proc 3rd Int Coral Reef Sym. Proc Third Int Coral Reef Symp Miami 1

Bograd SJ, Jacox MG, Hazen EL, Lovecchio E, Montes I, Pozo Buil M, Shannon LJ, Sydeman WJ, Rykaczewski RR (2023) Climate change impacts on eastern boundary upwelling systems. Annu Rev Mar Sci 15:303–328. https://doi.org/10.1146/annurev-marine-032122-021945

Brainard RE, Oliver T, McPhaden MJ, Cohen A, Venegas R, Heenan A, Vargas-Ángel B, Rotjan R, Mangubhai S, Flint E, Hunter SA (2018) Ecological impacts of the 2015/16 El Niño in the central equatorial Pacific. American Meteorological Society (AMS). https://doi.org/10.1175/BAMS-D-17-0128.1

Bridge TCL, Hoey AS, Campbell SJ, Muttaqin E, Rudi E, Fadli N, Baird AH (2013) Total hard coral cover and Acropora cover for each transect pre-bleaching (2009) and post-bleaching (2011). F1000Research 2:4762 Bytes. https://doi.org/10.6084/M9.FIGSHARE.777781

Brink KH (1991) Coastal-trapped waves and wind-driven currents over the continental shelf. Annu Rev Fluid Mech 23:389–412. https://doi.org/10.1146/annurev.fl.23.010191.002133

Brodie JE (1992) Enhancement of larval and juvenile survival and recruitment in *Acanthatser planci* from the effects of terrestrial runoff: a review. Mar Freshw Res 43:539–553. https://doi.org/10.1071/mf9920539

Buerger P, Schmidt GM, Wall M, Held C, Richter C (2015) Temperature tolerance of the coral Porites lutea exposed to simulated large amplitude internal waves (LAIW). J Exp Mar Biol Ecol 471:232–239. https://doi.org/10.1016/j.jembe.2015.06.014

Camp EF, Schoepf V, Mumby PJ, Hardtke LA, Rodolfo-Metalpa R, Smith DJ, Suggett DJ (2018a) The future of coral reefs subject to rapid climate change: lessons from natural extreme environments. Front Mar Sci 5:4. https://doi.org/10.3389/fmars.2018.00004

Camelia H, Felis T, Hargreaves JA, Koelling KM, Scheffers S, Chavanich S, Sangmanee C, Wall M (accepted) Tropical climate modes control strength and distribution of thermal stress mitigation in a coral reef refugia. Sci Rep

Camp EF, Schoepf V, Mumby PJ, Suggett DJ, Hardtke LA, Rodolfo-Metalpa R, Smith DJ, Suggett DJ (2018b) Editorial: the future of coral reefs subject to rapid climate change: lessons from natural extreme environments. Front Mar Sci 5:1–21. https://doi.org/10.3389/fmars.2018.00433

Cantin NE, Cohen AL, Karnauskas KB, Tarrant AM, McCorkle DC (2010) Ocean warming slows coral growth in the central Red Sea. Science 329:322–325. https://doi.org/10.1126/science.1190182

Carrigan AD, Puotinen M (2014) Tropical cyclone cooling combats region-wide coral bleaching. Glob Change Biol 20:1604–1613. https://doi.org/10.1111/gcb.12541

Castrill6n-Cifuentes AL, Zapata FA, Giraldo A, Wild C (2023a) Spatiotemporal variability of oxygen concentration in coral reefs of Gorgona Island (Eastern Tropical Pacific) and its effect on the coral *Pocillopora capitata*. PeerJ 11:e14586. https://doi.org/10.7717/peerj.14586

Castrill6n-Cifuentes AL, Zapata FA, Wild C (2023b) Physiological responses of *Pocillopora* corals to upwelling events in the Eastern Tropical Pacific. Front Mar Sci 10:1212717. https://doi.org/10.3389/fmars.2023.1212717

Chollett I, Mumby PJ (2013) Reefs of last resort: locating and assessing thermal refugia in the wider Caribbean. Biol Conserv 167:179–186. https://doi.org/10.1016/j.biocon.2013.08.010

Chollett I, Mumby P, Cortés J (2010) Upwelling areas do not guarantee refuge for coral reefs in a warming ocean. Mar Ecol Prog Ser 416:47–56. https://doi.org/10.3354/meps08775

Chollett I, Müller-Karger FE, Heron SF, Skirving W, Mumby PJ (2012) Seasonal and spatial heterogeneity of recent sea surface temperature trends in the Caribbean Sea and Southeast Gulf of Mexico. Mar Pollut Bull 64:956–965. https://doi.org/10.1016/j.marpolbul.2012.02.016

Claar DC, Starko S, Tietjen KL, Epstein HE, Cunning R, Cobb KM, Baker AC, Gates RD, Baum JK (2020) Dynamic symbioses reveal pathways to coral survival through prolonged heatwaves. Nat Commun 11:6097. https://doi.org/10.1038/s41467-020-19169-y

Coles SL, Fadlallah YH (1991) Reef coral survival and mortality at low temperatures in the Arabian Gulf: new species-specific lower

temperature limits. Coral Reefs 9:231–237. https://doi.org/10.1007/BF00290427
Colin PL, Johnston TMS (2020) Measuring temperature in Coral Reef environments: experience, lessons, and results from Palau. J Mar Sci Eng 8:680. https://doi.org/10.3390/jmse8090680
Colin PL, Lindfield SJ (2019) Palau. In: Loya Y, Puglise KA, Bridge TCL (eds) Mesophotic coral ecosystems. Springer International Publishing, Cham, pp 285–299
Combillet L, Fabregat-Malé S, Mena S, Marín-Moraga JA, Gutierrez M, Alvarado JJ (2022) *Pocillopora* spp. growth analysis on restoration structures in an Eastern Tropical Pacific upwelling area. PeerJ 10:e13248. https://doi.org/10.7717/peerj.13248
Connell JH (1978) Diversity in tropical rain forests and coral reefs: high diversity of trees and corals is maintained only in a nonequilibrium state. Science 199:1302–1310. https://doi.org/10.1126/science.199.4335.1302
Conti-Jerpe IE, Thompson PD, Wong CWM, Oliveira NL, Duprey NN, Moynihan MA, Baker DM (2020) Trophic strategy and bleaching resistance in reef-building corals. Sci Adv 6(15):eaaz5443. https://doi.org/10.1126/sciadv.aaz5443
Cortés J (1997) Biology and geology of eastern Pacific coral reefs. Coral Reefs 16:S39–S46. https://doi.org/10.1007/s003380050240
Cortés J (2019) Isla del Coco, Costa Rica, Eastern Tropical Pacific. In: Loya Y, Puglise KA, Bridge TCL (eds) Mesophotic coral ecosystems. Springer International Publishing, Cham, pp 465–475
D'Angelo C, Wiedenmann J (2014) Impacts of nutrient enrichment on coral reefs: new perspectives and implications for coastal management and reef survival. Curr Opin Environ Sustain 7:82–93. https://doi.org/10.1016/j.cosust.2013.11.029
D'Croz L, Maté JL (2004) Experimental responses to elevated water temperature in genotypes of the reef coral *Pocillopora damicornis* from upwelling and non-upwelling environments in Panama. Coral Reefs 23:473–483. https://doi.org/10.1007/s00338-004-0397-7
D'Croz L, O'Dea A (2007) Variability in upwelling along the Pacific shelf of Panama and implications for the distribution of nutrients and chlorophyll. Estuar Coast Shelf Sci 73:325–340. https://doi.org/10.1016/j.ecss.2007.01.013
D'Croz L, Maté JL, Oke JE (2001) Responses to elevated sea water temperature and UV radiation in the coral Porites lobata from upwelling and non-upwelling environments on the Pacific Coast of Panama. Bull Mar Sci 69:203–214
Dana TF (1975) Development of contemporary eastern Pacific coral reefs. Mar Biol 33:355–374. https://doi.org/10.1007/BF00390574
Darwin C (1842) The structure and distribution of coral reefs: being the first part of the geology of the voyage of the beagle, under the command of Capt Fitzroy RN During the Years 1832 to 1836. Smith, Elder and Co p, London
Davis KA, Monismith SG (2011) The modification of bottom boundary layer turbulence and mixing by internal waves shoaling on a Barrier Reef. J Phys Oceanogr 41:2223–2241. https://doi.org/10.1175/2011JPO4344.1
Davis KA, Arthur RS, Reid EC, Rogers JS, Fringer OB, DeCarlo TM, Cohen AL (2020) Fate of internal waves on a shallow shelf. J Geophys Res Oceans 125. https://doi.org/10.1029/2019JC015377
DeCarlo TM, Harrison HB (2019) An enigmatic decoupling between heat stress and coral bleaching on the Great Barrier Reef. PeerJ 7:e7473. https://doi.org/10.7717/peerj.7473
DeCarlo TM, Karnauskas KB, Davis KA, Wong GTF (2015) Climate modulates internal wave activity in the northern South China Sea: climate modulates internal waves. Geophys Res Lett 42:831–838. https://doi.org/10.1002/2014GL062522
DeCarlo TM, Cohen AL, Wong GTF, Davis KA, Lohmann P, Soong K (2017) Mass coral mortality under local amplification of 2 °C ocean warming. Sci Rep 7:44586. https://doi.org/10.1038/srep44586
DeCarlo TM, Gajdzik L, Ellis J, Coker DJ, Roberts MB, Hammerman NM, Pandolfi JM, Monroe AA, Berumen ML (2020) Nutrient-supplying ocean currents modulate coral bleaching susceptibility. Sci Adv 6:eabc5493. https://doi.org/10.1126/sciadv.abc5493
Denis V, Ferrier-Pagès C, Schubert N, Coppari M, Baker DM, Camp EF, Gori A, Grottoli AG, Houlbrèque F, Maier SR, Mancinelli G, Martinez S, Özdilek ŞY, Radice VZ, Ribes M, Richter C, Viladrich N, Rossi S (2024) Heterotrophy in marine animal forests in an era of climate change. Biol Rev 99:965–978. https://doi.org/10.1111/brv.13053
Diaz C, Foster NL, Attrill MJ, Bolton A, Ganderton P, Howell KL, Robinson E, Hosegood P (2023) Mesophotic coral bleaching associated with changes in thermocline depth. Nat Commun 14:6528. https://doi.org/10.1038/s41467-023-42279-2
Diaz-Pulido G, Garzón-Ferreira J (2002) Seasonality in algal assemblages on upwelling-influenced coral reefs in the Colombian Caribbean. Bot Mar 45:284–292. https://doi.org/10.1515/BOT.2002.028
Doering T, Wall M, Putchim L, Rattanawongwan T, Schroeder R, Hentschel U, Roik A (2021) Towards enhancing coral heat tolerance: a "microbiome transplantation" treatment using inoculations of homogenized coral tissues. Microbiome 9:102. https://doi.org/10.1186/s40168-021-01053-6
Doney SC, Ruckelshaus M, Duffy JE, Barry JP, Chan F, English CA, Galindo HM, Grebmeier JM, Hollowed AB, Knowlton N, Polovina J, Rabalais NN, Sydeman WJ, Talley LD (2012) Climate change impacts on marine ecosystems. Annu Rev Mar Sci 4:11–37. https://doi.org/10.1146/annurev-marine-041911-111611
Doty MS, Oguri M (1956) The Island mass effect. ICES J Mar Sci 22:33–37. https://doi.org/10.1093/icesjms/22.1.33
Dunn JG, Sammarco PW, LaFleur G (2012) Effects of phosphate on growth and skeletal density in the scleractinian coral *Acropora muricata*: a controlled experimental approach. J Exp Mar Biol Ecol 411:34–44. https://doi.org/10.1016/j.jembe.2011.10.013
Eidens C, Hauffe T, Bayraktarov E, Wild C, Wilke T (2015) Multiscale processes drive benthic community structure in upwelling-affected coral reefs. Front Mar Sci 2. https://doi.org/10.3389/fmars.2015.00002
Ekman VW (1905) On the influence of the earth's rotation on ocean-currents
Elliott J, Patterson M, Gleiber M (2012) Detecting 'Island Mass Effect' through remote sensing. In: Proc 12th Int Coral Reef Symp 5A Remote sensing of reef environments
Enochs IC, Toth LT, Kirkland A, Manzello DP, Kolodziej G, Morris JT, Holstein DM, Schlenz A, Randall CJ, Maté JL, Leichter JJ, Aronson RB (2021) Upwelling and the persistence of coral-reef frameworks in the eastern tropical Pacific. Ecol Monogr 91:e01482. https://doi.org/10.1002/ecm.1482
Fabricius KE (2005) Effects of terrestrial runoff on the ecology of corals and coral reefs: review and synthesis. Mar Pollut Bull 50:125–146. https://doi.org/10.1016/j.marpolbul.2004.11.028
Fabricius KE, Langdon C, Uthicke S, Humphrey C, Noonan S, De'ath G, Okazaki R, Muehllehner N, Glas MS, Lough JM (2011) Losers and winners in coral reefs acclimatized to elevated carbon dioxide concentrations. Nat Clim Chang 1:165–169. https://doi.org/10.1038/nclimate1122
Fabricius KE, Cséke S, Humphrey C, De'ath G (2013) Does trophic status enhance or reduce the thermal tolerance of Scleractinian corals? A review, experiment and conceptual framework. PLoS One 8:e54399. https://doi.org/10.1371/journal.pone.0054399
Fabricius KE, De'ath G, Noonan S, Uthicke S (2014) Ecological effects of ocean acidification and habitat complexity on reef-associated macroinvertebrate communities. Proc R Soc B Biol Sci 281:20132479. https://doi.org/10.1098/rspb.2013.2479
Falkowski PG, Dubinsky Z, Muscatine L, McCloskey L (1993) Population control in symbiotic corals: ammonium ions and organic materials maintain the density of zooxanthellae. Bioscience 43:606–611. https://doi.org/10.2307/1312147

Feely RA, Sabine CL, Hernandez-Ayon JM, Ianson D, Hales B (2008) Evidence for upwelling of corrosive "acidified" water onto the continental shelf. Science 320:1490–1492. https://doi.org/10.1126/science.1155676

Fernández-García C, Salas-Moya C, Mena S, Azofeifa-Solano JC, Alvarado JJ (2021) Diversidad de los hábitats submareales de la Península de Santa Elena e Islas Murciélago, Pacífico Norte, Costa Rica. Rev Biol Trop 69:S160–S179. https://doi.org/10.15517/rbt.v69iSuppl.2.48774

Ferrier-Pagès C, Rottier C, Beraud E, Levy O (2010) Experimental assessment of the feeding effort of three scleractinian coral species during a thermal stress: effect on the rates of photosynthesis. J Exp Mar Biol Ecol 390:118–124. https://doi.org/10.1016/j.jembe.2010.05.007

Fiedler PC (2002) Environmental change in the eastern tropical Pacific Ocean: review of ENSO and decadal variability. Mar Ecol Prog Ser 244:265–283. https://doi.org/10.3354/meps244265

Firing E, Lukas R, Sadler J, Wyrtki K (1983) Equatorial undercurrent disappears during 1982-1983 El Niño. Science 222:1121–1123. https://doi.org/10.1126/science.222.4628.1121

Fong P, Glynn PW (2001) Population abundance and size-structure of an eastern tropical Pacific reef coral after the 1997-98 ENSO: a simulation model predicts field measures. Bull Mar Sci 69:187–202

Fong P, Smith TB, Muthukrishnan R (2017) Algal dynamics: alternate stable states of reefs in the eastern tropical Pacific. In: Glynn PW, Manzello DP, Enochs IC (eds) Coral reefs of the eastern tropical Pacific: persistence and loss in a dynamic environment. Springer, Dordrecht, pp 339–367

Ford AK, Jouffray J-B, Norström AV, Moore BR, Nugues MM, Williams GJ, Bejarano S, Magron F, Wild C, Ferse SCA (2020) Local human impacts disrupt relationships between benthic Reef assemblages and environmental predictors. Front Mar Sci 7. https://doi.org/10.3389/fmars.2020.571115

Foreman AD, Duprey NN, Yuval M, Dumestre M, Leichliter JN, Rohr MC, Dodwell RCA, Dodwell GAS, Clua EEG, Treibitz T, Martínez-García A (2024) Severe cold-water bleaching of a deep-water reef underscores future challenges for mesophotic coral ecosystems. Sci Total Environ 951:175210. https://doi.org/10.1016/j.scitotenv.2024.175210

Forryan A, Naveira Garabato AC, Vic C, Nurser AJG, Hearn AR (2021) Galápagos upwelling driven by localized wind–front interactions. Sci Rep 11:1277. https://doi.org/10.1038/s41598-020-80609-2

Fox MD, Williams GJ, Johnson MD, Radice VZ, Zgliczynski BJ, Kelly ELA, Rohwer FL, Sandin SA, Smith JE (2018) Gradients in primary production predict trophic strategies of mixotrophic corals across spatial scales. Curr Biol 28:3355–3363.e4. https://doi.org/10.1016/j.cub.2018.08.057

Fox MD, Guillaume-Castel R, Edwards CB, Glanz J, Gove JM, Green JAM, Juhlin E, Smith JE, Williams GJ (2023) Ocean currents magnify upwelling and deliver nutritional subsidies to reef-building corals during El Niño heatwaves. Sci Adv 9:eadd5032. https://doi.org/10.1126/sciadv.add5032

Frade PR, Bongaerts P, Baldwin CC, Trembanis AC, Bak RPM, Vermeij MJA (2019) Bonaire and Curaçao. In: Loya Y, Puglise KA, Bridge TCL (eds) Mesophotic coral ecosystems. Springer International Publishing, Cham, pp 149–162

Froyland G, Stuart RM, van Sebille E (2014) How well-connected is the surface of the global ocean? Chaos Interdiscip J Nonlinear Sci 24:033126. https://doi.org/10.1063/1.4892530

Fu K-H, Wang Y-H, St. Laurent L, Simmons H, Wang D-P (2012) Shoaling of large-amplitude nonlinear internal waves at Dongsha Atoll in the northern South China Sea. Cont Shelf Res 37:1–7. https://doi.org/10.1016/j.csr.2012.01.010

Garwood JC, Musgrave RC, Lucas AJ (2020) Life in internal waves. Oceanography 33:38–49. https://doi.org/10.5670/oceanog.2020.313

Genin A, Jaffe JS, Reef R, Richter C, Franks PJS (2005) Swimming against the flow: a mechanism of zooplankton aggregation. Science 308:860–862. https://doi.org/10.1126/science.1107834

Gil MA (2013) Unity through nonlinearity: a unimodal coral–nutrient interaction. Ecology 94:1871–1877. https://doi.org/10.1890/12-1697.1

Gill AE, Clarke AJ (1974) Wind-induced upwelling, coastal currents and sea-level changes. Deep Sea Res Oceanogr Abstr 21:325–345. https://doi.org/10.1016/0011-7471(74)90038-2

Giraldo A, Rodríguez-Rubio E, Zapata F (2008) Condiciones oceanográficas en isla Gorgona, Pacífico oriental tropical de Colombia. Lat Am J Aquat Res 36:121–128. https://doi.org/10.3856/vol36-issue1-fulltext-12

Glynn PW (1977) Coral growth in upwelling and nonupwelling areas off the Pacific coast of Panama. J Mar Res 35:567–585

Glynn P (1985) El Nino-associated disturbance to coral reefs and post disturbance mortality by *Acanthaster planci*. Mar Ecol Prog Ser 26:295–300. https://doi.org/10.3354/meps026295

Glynn PW (1988) El Nino-southern oscillation 1982-1983: nearshore population, community, and ecosystem responses. Ann Rev Ecol Syst 19:309–345

Glynn PW (1990) Coral mortality and disturbances to coral reefs in the tropical eastern Pacific. In: Elsevier oceanography series, vol 52. Elsevier, pp 55–126. https://doi.org/10.1016/S0422-9894(08)70033-3

Glynn PW (1993) Monsoonal upwelling and episodic *Acanthaster* predation as probable controls of coral reef distribution and community structure in Oman, Indian Ocean. Atoll Res Bull 379:1–66

Glynn PW (1994) State of coral reefs in the gal pagos islands: natural vs anthropogenic impacts. Mar Pollut Bull 29:131–140

Glynn PW (1996) Coral reef bleaching: facts, hypotheses and implications. Glob Change Biol 2:495–509. https://doi.org/10.1111/j.1365-2486.1996.tb00063.x

Glynn PW, Colgan MW (1992) Sporadic disturbances in fluctuating coral reef environments: El Niño and coral reef development in the eastern Pacific. Am Zool 32:707–718

Glynn PW, D'Croz L (1990) Experimental evidence for high temperature stress as the cause of El Nifio-coincident coral mortality. Coral Reefs 8:181–191

Glynn PW, Stewart RH (1973) Distribution of coral reefs in the Pearl Islands (Gulf of Panama) in relation to thermal conditions. Limnol Oceanogr 18:367–379. https://doi.org/10.4319/lo.1973.18.3.0367

Glynn PW, Druffel EM, Dunbar RB (1983) A dead Central American coral reef tract: possible link with the Little Ice Age. J Mar Res 41:605–637. https://doi.org/10.1357/002224083788519740

Glynn PW, Guzman HM, Cortes J, Richmond RH (1988) El Niño (1982-83) associated coral mortality and relationship to sea surface temperature deviations in the tropical eastern Pacific. In: Proc 6th Int Coral Reef Symp. Springer

Glynn PW, Gassman NJ, Eakin CM, Cortes J, Smith DB, Guzman HM (1991) Reef coral reproduction in the eastern Pacific: Costa Rica, Panama, and Galapagos Islands (Ecuador): I. Pocilloporidae. Mar Biol 109:355–368. https://doi.org/10.1007/BF01313501

Glynn PW, Colley SB, Eakin CM, Smith DB, Cortés J, Gassman NJ, Guzmán HM, Del Rosario JB, Feingold JS (1994) Reef coral reproduction in the eastern Pacific: Costa Rica, Panamá, and Galápagos Islands (Ecuador). II Poritidae. Mar Biol 118:191–208. https://doi.org/10.1007/BF00349785

Glynn PW, Colley SB, Gassman NJ, Black K, Cortés J, Maté JL (1996) Reef coral reproduction in the eastern Pacific: Costa Rica, Panamá, and Galápagos Islands (Ecuador). III. Agariciidae (Pavona gigantea and Gardineroseris planulata). Mar Biol 125:579–601. https://doi.org/10.1007/BF00353270

Glynn PW, Maté JL, Baker AC, Calderón MO (2001) Coral bleaching and mortality in Panama and Ecuador during the 1997–1998

El Niño-Southern Oscillation event: Spatial/temporal patterns and comparison with the 1982–1983 event. Bull Mar Sci 69:79–109

Glynn PW, Riegl B, Correa AMS, Baums I (2009) Rapid recovery of a coral reef at Darwin Island, Galapagos Islands. Galapagos Res 66:6–13

Glynn PW, Riegl B, Purkis S, Kerr JM, Smith TB (2015) Coral reef recovery in the Galápagos Islands: the northernmost islands (Darwin and Wenman). Coral Reefs 34:421–436. https://doi.org/10.1007/s00338-015-1280-4

Glynn PW, Alvarado JJ, Banks S, Cortés J, Feingold JS, Jiménez C, Maragos JE, Martínez P, Maté JL, Moanga DA, Navarrete S, Reyes-Bonilla H, Riegl B, Rivera F, Vargas-Ángel B, Wieters EA, Zapata FA (2017) Eastern Pacific Coral Reef Provinces, Coral Community structure and composition: an overview. In: Glynn PW, Manzello DP, Enochs IC (eds) Coral reefs of the eastern tropical Pacific: persistence and loss in a dynamic environment. Springer, Dordrecht, pp 107–176

Glynn PW, Feingold JS, Baker A, Banks S, Baums IB, Cole J, Colgan MW, Fong P, Glynn PJ, Keith I, Manzello D, Riegl B, Ruttenberg BI, Smith TB, Vera-Zambrano M (2018) State of corals and coral reefs of the Galápagos Islands (Ecuador): past, present and future. Mar Pollut Bull 133:717–733. https://doi.org/10.1016/j.marpolbul.2018.06.002

Gómez C, Acosta-Chaparro A, Bernal C, Gómez-López D, Navas-Camacho R, Alonso D (2023) Seasonal upwelling conditions modulate the calcification response of a tropical Scleractinian Coral. Oceans 4:170–184. https://doi.org/10.3390/oceans4020012

González-Espinosa PC, Donner SD (2020) Predicting cold-water bleaching in corals: role of temperature, and potential integration of light exposure. Mar Ecol Prog Ser 642:133–146. https://doi.org/10.3354/meps13336

Goodbody-Gringley G, Noyes T, Smith SR (2019) Bermuda. In: Loya Y, Puglise KA, Bridge TCL (eds) Mesophotic coral ecosystems. Springer International Publishing, Cham, pp 31–45

Gove JM, Merrifield MA, Brainard RE (2006) Temporal variability of current-driven upwelling at Jarvis Island. J Geophys Res 111:C12011. https://doi.org/10.1029/2005JC003161

Gove JM, Williams GJ, McManus MA, Heron SF, Sandin SA, Vetter OJ, Foley DG (2013) Quantifying climatological ranges and anomalies for Pacific Coral Reef ecosystems. PLoS One 8:e61974. https://doi.org/10.1371/journal.pone.0061974

Gove J, Williams G, McManus M, Clark S, Ehses J, Wedding L (2015) Coral reef benthic regimes exhibit non-linear threshold responses to natural physical drivers. Mar Ecol Prog Ser 522:33–48. https://doi.org/10.3354/meps11118

Gove JM, McManus MA, Neuheimer AB, Polovina JJ, Drazen JC, Smith CR, Merrifield MA, Friedlander AM, Ehses JS, Young CW, Dillon AK, Williams GJ (2016) Near-island biological hotspots in barren ocean basins. Nat Commun 7:10581. https://doi.org/10.1038/ncomms10581

Graham NAJ, McClanahan TR, MacNeil MA, Wilson SK, Cinner JE, Huchery C, Holmes TH (2017) Human disruption of coral reef trophic structure. Curr Biol 27:231–236. https://doi.org/10.1016/j.cub.2016.10.062

Green F (1986) The coral seas of Muscat. In: Middle East economic digest. GlobalData Plc, London

Grottoli AG, Rodrigues LJ, Palardy JE (2006) Heterotrophic plasticity and resilience in bleached corals. Nature 440:1186–1189. https://doi.org/10.1038/nature04565

Guillaume-Castel R, Williams GJ, Rogers JS, Gove JM, Green JAM (2021) Quantifying upwelling in tropical shallow waters: a novel method using a temperature stratification index. Limnol Oceanogr Methods 19:566–577. https://doi.org/10.1002/lom3.10449

Guzmán HM, Cortés J (1993) Arrecifes coralinos del Pacífico Oriental Tropical: Revisión y perspectivas. Rev Biol Trop 41:535–557

Guzmán HM, Cortés J, Glynn PW, Richmond RH (1990) Coral mortality associated with dinoflagellate blooms in the eastern Pacific (Costa Rica and Panama). Mar Ecol Prog Ser 60:299–303

Hackerott S, Martell HA, Eirin-Lopez JM (2021) Coral environmental memory: causes, mechanisms, and consequences for future reefs. Trends Ecol Evol 36:1011–1023. https://doi.org/10.1016/j.tree.2021.06.014

Hamner WM, Hauri IR (1981) Effects of Island mass: water flow and plankton pattern around a reef in the Great Barrier Reef lagoon, Australia1. Limnol Oceanogr 26:1084–1102. https://doi.org/10.4319/lo.1981.26.6.1084

Hamner WM, Jones MS, Carleton JH, Hauri IR, Williams DMB (1988) Zooplankton, planktivorous fish, and water currents on a windward reef face: Great Barrier Reef, Australia. Bull Mar Sci 42:459–479

Hoegh-Guldberg O (1999) Climate change, coral bleaching and the future of the world's coral reefs. Mar Freshw Res 50:839–866. https://doi.org/10.1071/MF99078

Hoegh-Guldberg O, Fine M (2004) Low temperatures cause coral bleaching. Coral Reefs 23:444–444. https://doi.org/10.1007/s00338-004-0401-2

Hoegh-Guldberg O, Skirving W, Dove SG, Spady BL, Norrie A, Geiger EF, Liu G, De La Cour JL, Manzello DP (2023) Coral reefs in peril in a record-breaking year. Science 382:1238–1240. https://doi.org/10.1126/science.adk4532

Hofmann EE, Busalacchi AJ, O'Brien JJ (1981) Wind generation of the Costa Rica Dome. Science 214:552–554. https://doi.org/10.1126/science.214.4520.552

Holbrook SJ, Wencélius J, Dubel AK, Adam TC, Cook DC, Hunter CE, Lauer M, Lester SE, Miller SD, Rassweiler A, Schmitt RJ (2022) Spatial covariation in nutrient enrichment and fishing of herbivores in an oceanic coral reef ecosystem. Ecol Appl 32:e2515. https://doi.org/10.1002/eap.2515

Houk P, Bograd S, van Woesik R (2007) The transition zone chlorophyll front can trigger *Acanthaster planci* outbreaks in the Pacific Ocean: historical confirmation. J Oceanogr 63:149–154. https://doi.org/10.1007/s10872-007-0013-x

Huang X, Chen Z, Zhao W, Zhang Z, Zhou C, Yang Q, Tian J (2016) An extreme internal solitary wave event observed in the northern South China Sea. Sci Rep 6:30041. https://doi.org/10.1038/srep30041

Huang Z, Feng M, Dalton SJ, Carroll AG (2024) Marine heatwaves in the great barrier Reef and Coral Sea: their mechanisms and impacts on shallow and mesophotic coral ecosystems. Sci Total Environ 908:168063. https://doi.org/10.1016/j.scitotenv.2023.168063

Hughes TP, Kerry JT, Álvarez-Noriega M, Álvarez-Romero JG, Anderson KD, Baird AH, Babcock RC, Beger M, Bellwood DR, Berkelmans R, Bridge TC, Butler IR, Byrne M, Cantin NE, Comeau S, Connolly SR, Cumming GS, Dalton SJ, Diaz-Pulido G, Eakin CM, Figueira WF, Gilmour JP, Harrison HB, Heron SF, Hoey AS, Hobbs JPA, Hoogenboom MO, Kennedy EV, Kuo CY, Lough JM, Lowe RJ, Liu G, McCulloch MT, Malcolm HA, McWilliam MJ, Pandolfi JM, Pears RJ, Pratchett MS, Schoepf V, Simpson T, Skirving WJ, Sommer B, Torda G, Wachenfeld DR, Willis BL, Wilson SK (2017) Global warming and recurrent mass bleaching of corals. Nature 543:373–377. https://doi.org/10.1038/nature21707

Hughes TP, Kerry JT, Baird AH, Connolly SR, Dietzel A, Eakin CM, Heron SF, Hoey AS, Hoogenboom MO, Liu G, Mcwilliam MJ, Pears RJ, Pratchett MS, Skirving WJ, Stella JS, Torda G (2018) Global warming transforms coral reef assemblages. Nature 556:492–496

Hughes DJ, Alderdice R, Cooney C, Kühl M, Pernice M, Voolstra CR, Suggett DJ (2020) Coral reef survival under accelerating ocean deoxygenation. Nat Clim Chang 10:296–307. https://doi.org/10.1038/s41558-020-0737-9

Hung J-J, Wang Y-H, Fu K-H, Lee I-H, Tsai S-S, Lee C-Y, Lu W-T, Shen Y-J, Lin Y-H (2021) Biogeochemical responses to internal-wave impacts in the continental margin off Dongsha Atoll in the northern South China Sea. Prog Oceanogr 199:102689. https://doi.org/10.1016/j.pocean.2021.102689

Huntington B, Weible R, Halperin A, Winston M, McCoy K, Amir C, Asher J, Vargas-Angel B (2022) Early successional trajectory of benthic community in an uninhabited reef system three years after mass coral bleaching. Coral Reefs 41:1087–1096. https://doi.org/10.1007/s00338-022-02246-7

Huyer A, Smith RL, Paluszkiewicz T (1987) Coastal upwelling off Peru during normal and El Niño times, 1981–1984. J Geophys Res Oceans 92:14297–14307. https://doi.org/10.1029/JC092iC13p14297

Jackson JBC (1997) Reefs since Columbus. Coral Reefs 16:S23–S32. https://doi.org/10.1007/s003380050238

Jackson C (2004) The Andaman Sea. Atlas Ocean Intern Solitary Waves. Global Ocean Associates, pp 485–500

Jackson C (2007) Internal wave detection using the moderate resolution imaging spectroradiometer (MODIS). J Geophys Res Oceans 112. https://doi.org/10.1029/2007JC004220

Jackson C, Da Silva J, Jeans G (2012) The generation of nonlinear internal waves. Oceanography 25:108–123. https://doi.org/10.5670/oceanog.2012.46

Jakoboski J, Todd RE, Owens WB, Karnauskas KB, Rudnick DL (2020) Bifurcation and upwelling of the Equatorial Undercurrent West of the Galápagos Archipelago. American Meteorological Society. https://doi.org/10.1175/JPO-D-19-0110.1

Jantzen C, Schmidt GM, Wild C, Roder C, Khokiattiwong S, Richter C (2013) Benthic reef primary production in response to large amplitude internal waves at the Similan Islands (Andaman Sea, Thailand). PLoS One 8:e81834. https://doi.org/10.1371/journal.pone.0081834

Jimenez C (1997) Corals and coral reefs of Culebra Bay, Pacific Coast of Costa Rica: anarchy in the reef, vol 1. STRI, pp 329–334

Jiménez C (2001) Seawater temperature measured at the surface and at two depths (7 and 12 m) in one coral reef at Culebra Bay, Gulf of Papagayo, Costa Rica. Rev Biol Trop 49:153–161

Jiménez C, Cortés J (2003) Growth of seven species of scleractinian corals in an upwelling environment of the eastern Pacific (Golfo de Papagayo, Costa Rica). Bull Mar Sci 72:187–198

Jiménez C, Cortés J, León A, Ruíz E (2001) Coral bleaching and mortality associated with the 1997-98 El Niño in an upwelling environment in the eastern Pacific (Gulf of Papagayo, Costa Rica). Bull Mar Sci 69:151–169

Jiménez C, Bassey G, Segura Á, Cortés J (2010) Characterization of the coral communities and reefs of two previously undescribed locations in the upwelling region of papagayo (Costa Rica). Rev Cienc Mar Costeras 2:95. https://doi.org/10.15359/revmar.2.8

Jimenez G, Cole JE, Thompson DM, Tudhope AW (2018) Northern Galápagos corals reveal twentieth century warming in the Eastern Tropical Pacific. Geophys Res Lett 45:1981–1988. https://doi.org/10.1002/2017GL075323

Johnson MD, Fox MD, Kelly ELA, Zgliczynski BJ, Sandin SA, Smith JE (2020) Ecophysiology of coral reef primary producers across an upwelling gradient in the tropical Central Pacific. PLoS One 15:e0228448. https://doi.org/10.1371/journal.pone.0228448

Kämpf J, Chapman P (2016) Upwelling systems of the world. Springer International Publishing, Cham

Kapsenberg L, Cyronak T (2019) Ocean acidification refugia in variable environments. Glob Change Biol 25:3201–3214. https://doi.org/10.1111/gcb.14730

Karnauskas KB, Cohen AL (2012) Equatorial refuge amid tropical warming. Nat Clim Chang 2:530–534. https://doi.org/10.1038/nclimate1499

Karnauskas KB, Jenouvrier S, Brown CW, Murtugudde R (2015) Strong sea surface cooling in the eastern equatorial Pacific and implications for Galápagos penguin conservation. Geophys Res Lett 42:6432–6437. https://doi.org/10.1002/2015GL064456

Kavousi J, Keppel G (2018) Clarifying the concept of climate change refugia for coral reefs. ICES J Mar Sci 75:43–49. https://doi.org/10.1093/icesjms/fsx124

Kelly LW, Williams GJ, Barott KL, Carlson CA, Dinsdale EA, Edwards RA, Haas AF, Haynes M, Lim YW, McDole T, Nelson CE, Sala E, Sandin SA, Smith JE, Vermeij MJA, Youle M, Rohwer F (2014) Local genomic adaptation of coral reef-associated microbiomes to gradients of natural variability and anthropogenic stressors. Proc Natl Acad Sci 111:10227–10232. https://doi.org/10.1073/pnas.1403319111

Kingsford MJ, Brandt M, Alava-Jurado JM (2023) Levels of upwelling are important to consider for conservation. In: Walsh SJ, Mena CF, Stewart JR, Muñoz Pérez JP (eds) Island ecosystems: challenges to sustainability. Springer International Publishing, Cham, pp 271–289

Kleypas JA, Mcmanus JW, Meñez LAB (1999) Environmental limits to coral reef development: where do we draw the line? Am Zool 39:146–159. https://doi.org/10.1093/icb/39.1.146

Klymak JM, Pinkel R, Liu C-T, Liu AK, David L (2006) Prototypical solitons in the South China Sea. Geophys Res Lett 33. https://doi.org/10.1029/2006GL025932

Kobluk DR, Lysenko MA (1994) "Ring" bleaching in southern Caribbean Agaricia Agaricites during rapid water cooling. Bull Mar Sci 54:142–150

Kroeker KJ, Kordas RL, Crim RN, Singh GG (2010) Meta-analysis reveals negative yet variable effects of ocean acidification on marine organisms. Ecol Lett 13:1419–1434. https://doi.org/10.1111/j.1461-0248.2010.01518.x

Kroeker KJ, Bell LE, Donham EM, Hoshijima U, Lummis S, Toy JA, Willis-Norton E (2020) Ecological change in dynamic environments: accounting for temporal environmental variability in studies of ocean change biology. Glob Change Biol 26:54–67. https://doi.org/10.1111/gcb.14868

Krutwa A (2014) Small-scale differences in tropical subtidal rocky Reef communities of Floreana Island, Galápagos. Doctoral thesis

Laboy-Nieves EN, Klein E, Conde JE, Losada F, Cruz JJ, Bone D (2001) Mass mortality of tropical marine communities in morrocoy, Venezuela. Bull Mar Sci 68:163–179

LaJeunesse TC, Smith R, Walther M, Pinzón J, Pettay DT, McGinley M, Aschaffenburg M, Medina-Rosas P, Cupul-Magaña AL, Pérez AL, Reyes-Bonilla H, Warner ME (2010) Host–symbiont recombination versus natural selection in the response of coral–dinoflagellate symbioses to environmental disturbance. Proc R Soc B Biol Sci 277:2925–2934. https://doi.org/10.1098/rspb.2010.0385

Lee TN, Rooth C, Williams E, McGowan M, Szmant AF, Clarke ME (1992) Influence of Florida current, gyres and wind-driven circulation on transport of larvae and recruitment in the Florida Keys coral reefs. Cont Shelf Res 12:971–1002. https://doi.org/10.1016/0278-4343(92)90055-O

Leichter J, Genovese S (2006) Intermittent upwelling and subsidized growth of the scleractinian coral *Madracis mirabilis* on the deep fore-reef slope of Discovery Bay, Jamaica. Mar Ecol Prog Ser 316:95–103. https://doi.org/10.3354/meps316095

Leichter JJ, Miller SL (1999) Predicting high-frequency upwelling: spatial and temporal patterns of temperature anomalies on a Florida coral reef. Cont Shelf Res 19:911–928. https://doi.org/10.1016/S0278-4343(99)00004-7

Leichter JJ, Wing SR, Miller SL, Denny MW (1996) Pulsed delivery of subthermocline water to Conch Reef (Florida Keys) by internal tidal bores. Limnol Oceanogr 41:1490–1501. https://doi.org/10.4319/lo.1996.41.7.1490

Leichter J, Shellenbarger G, Genovese S, Wing S (1998) Breaking internal waves on a Florida (USA) coral reef: a plankton pump at work? Mar Ecol Prog Ser 166:83–97. https://doi.org/10.3354/meps166083

Leichter JJ, Stewart HL, Miller SL (2003) Episodic nutrient transport to Florida coral reefs. Limnol Oceanogr 48:1394–1407. https://doi.org/10.4319/lo.2003.48.4.1394

Leichter JJ, Deane GB, Stokes MD (2005) Spatial and temporal variability of internal wave forcing on a coral Reef. J Phys Oceanogr 35:1945–1962. https://doi.org/10.1175/JPO2808.1

Leichter JJ, Helmuth B, Fischer AM (2006) Variation beneath the surface: quantifying complex thermal environments on coral reefs in the Caribbean, Bahamas and Florida. J Mar Res 64:563–588. https://doi.org/10.1357/002224006778715711

Leichter JJ, Stokes MD, Hench JL, Witting J, Washburn L (2012) The Island-scale internal wave climate of Moorea, French Polynesia: internal waves on Moorea. J Geophys Res Oceans 117:1–16. https://doi.org/10.1029/2012JC007949

Leichter JJ, Stokes MD, Vilchis LI, Fiechter J (2014) Regional synchrony of temperature variation and internal wave forcing along the Florida Keys reef tract: regional synchrony of internal waves. J Geophys Res Oceans 119:548–558. https://doi.org/10.1002/2013JC009371

Lesser MP, Slattery M, Stat M, Ojimi M, Gates RD, Grottoli A (2010) Photoacclimatization by the coral *Montastraea cavernosa* in the mesophotic zone: light, food, and genetics. Ecology 91:990–1003. https://doi.org/10.1890/09-0313.1

Leung JYS, Zhang S, Connell SD (2022) Is ocean acidification really a threat to marine Calcifiers? A systematic review and meta-analysis of 980+ studies spanning two decades. Small 18:2107407. https://doi.org/10.1002/smll.202107407

Levin LA, Ekau W, Gooday AJ, Jorissen F, Middelburg JJ, Naqvi SWA, Neira C, Rabalais NN, Zhang J (2009) Effects of natural and human-induced hypoxia on coastal benthos. Biogeosciences 6:2063–2098. https://doi.org/10.5194/bg-6-2063-2009

Li X, Wang D, Huang H, Zhang J, Lian J, Yuan X, Yang J, Zhang G (2015) Linking benthic community structure to terrestrial runoff and upwelling in the coral reefs of northeastern Hainan Island. Estuar Coast Shelf Sci 156:92–102. https://doi.org/10.1016/j.ecss.2014.09.021

Littler MM, Littler DS (1985) Factors controlling relative dominance of primary producers on biotic reefs. In: Proc 5th Int Coral Reef Symp, vol 4, pp 35–39

Liu Y, Putchim L, Li K, Gao H, Sun L, Khokiattiwong S, Yu W (2022) Late monsoon threatens coral refugia in the Andaman Sea. Environ Res Lett 17:034038. https://doi.org/10.1088/1748-9326/ac4a30

Lough JM, Barnes DJ (2000) Environmental controls on growth of the massive coral *Porites*. J Exp Mar Biol Ecol 245:225–243. https://doi.org/10.1016/S0022-0981(99)00168-9

Manzello DP (2010) Ocean acidification hotspots: spatiotemporal dynamics of the seawater CO2 system of eastern Pacific coral reefs. Limnol Oceanogr 55:239–248. https://doi.org/10.4319/lo.2010.55.1.0239

Manzello DP, Kleypas JA, Budd DA, Eakin CM, Glynn PW, Langdon C (2008) Poorly cemented coral reefs of the eastern tropical Pacific: Possible insights into reef development in a high-CO_2 world. Proc Natl Acad Sci 105:10450–10455. https://doi.org/10.1073/pnas.0712167105

Manzello DP, Enochs IC, Bruckner A, Renaud PG, Kolodziej G, Budd DA, Carlton R, Glynn PW (2014) Galápagos coral reef persistence after ENSO warming across an acidification gradient. Geophys Res Lett 41:9001–9008. https://doi.org/10.1002/2014GL062501

Marangoni LFDB, Rottier C, Ferrier-Pagès C (2021) Symbiont regulation in *Stylophora pistillata* during cold stress: an acclimation mechanism against oxidative stress and severe bleaching. J Exp Biol 224:jeb235275. https://doi.org/10.1242/jeb.235275

Margalef R (1978) Life-forms of phytoplankton as survival alternatives in an unstable environment. Oceanol Acta 1:493–509

McClanahan TR, Darling ES, Beger M, Fox HE, Grantham HS, Jupiter SD, Logan CA, Mcleod E, McManus LC, Oddenyo RM, Surya GS, AmeliaS W, Zinke J, Maina JM (2024) Diversification of refugia types needed to secure the future of coral reefs subject to climate change. Conserv Biol 38:e14108. https://doi.org/10.1111/cobi.14108

Mendonca V, Jabri MM, Ajmi I, Muharrami M, Areimi M, Aghbari HA (2010) Persistent and expanding population outbreaks of the corallivorous starfish *Acanthaster planci* in the Northwestern Indian Ocean: are they really a consequence of unsustainable starfish predator removal through overfishing in coral reefs, or a response to a changing environment? Zool Stud 49:108–123

Messié M, Ledesma J, Kolber DD, Michisaki RP, Foley DG, Chavez FP (2009) Potential new production estimates in four eastern boundary upwelling ecosystems. Prog Oceanogr 83:151–158. https://doi.org/10.1016/j.pocean.2009.07.018

Miller J, Maragos J, Brainard R, Asher J, Vargas-Ángel B, Kenyon J, Schroeder R (2008) The state of coral reef ecosystems of the Pacific Remote Island Areas. NOAA

Mysak LA (1980) Topographically trapped waves. Annu Rev Fluid Mech 12:45–76. https://doi.org/10.1146/annurev.fl.12.010180.000401

Nielsen T, Bjørnsen P, Boonruang P, Fryd M, Hansen P, Janekarn V, Limtrakulvong V, Munk P, Hansen O, Satapoomin S, Sawangarreruks S, Thomsen H, Østergaard J (2004) Hydrography, bacteria and protist communities across the continental shelf and shelf slope of the Andaman Sea (NE Indian Ocean). Mar Ecol Prog Ser 274:69–86. https://doi.org/10.3354/meps274069

Noowong A (2011) Input of various nitrogen species by large amplitude internal waves (LAIW) and intake by *Pocillopora verrucosa*. MSc thesis, University of Bremen

Norström AV, Nyström M, Lokrantz J, Folke C (2009) Alternative states on coral reefs: beyond coral–macroalgal phase shifts. Mar Ecol Prog Ser 376:295–306. https://doi.org/10.3354/meps07815

Nyström M, Folke C, Moberg F, Nyström M, Folke C, Moberg F, Nyström M, Folke C, Moberg F (2000) Coral reef disturbance and resilience in a human-dominated environment. Trends Ecol Evol 15:413–417. https://doi.org/10.1016/S0169-5347(00)01948-0

Oliver TA, Palumbi SR (2011) Do fluctuating temperature environments elevate coral thermal tolerance? Coral Reefs 30:429–440. https://doi.org/10.1007/s00338-011-0721-y

Palacio-Castro AM, Smith TB, Brandtneris V, Snyder GA, van Hooidonk R, Maté JL, Manzello D, Glynn PW, Fong P, Baker AC (2023) Increased dominance of heat-tolerant symbionts creates resilient coral reefs in near-term ocean warming. Proc Natl Acad Sci 120:e2202388120. https://doi.org/10.1073/pnas.2202388120

Palardy J, Grottoli A, Matthews K (2005) Effects of upwelling, depth, morphology and polyp size on feeding in three species of Panamanian corals. Mar Ecol Prog Ser 300:79–89. https://doi.org/10.3354/meps300079

Palmer C, Jimenez C, Bassey G, Ruiz E, Villalobos Cubero T, Chavarria Diaz MM, Harrison XA, Puschendorf R (2022) Cold water and harmful algal blooms linked to coral reef collapse in the eastern tropical Pacific. PeerJ 10:e14081. https://doi.org/10.7717/peerj.14081

Palumbi SR, Barshis DJ, Traylor-Knowles N, Bay RA (2014) Mechanisms of reef coral resistance to future climate change. Science 344:895–898. https://doi.org/10.1126/science.1251336

Pandolfi JM, Bradbury RH, Sala E, Hughes TP, Bjorndal KA, Cooke RG, McArdle D, McClenachan L, Newman MJH, Paredes G, Warner RR, Jackson JBC (2003) Global trajectories of the long-term decline of coral reef ecosystems. Science 301:955–958. https://doi.org/10.1126/science.1085706

Pastorok R, Bilyard G (1985) Effects of sewage pollution on coral-reef communities. Mar Ecol Prog Ser 21:175–189. https://doi.org/10.3354/meps021175

Pérez-Rosales G, Rouzé H, Torda G, Bongaerts P, Pichon M, Under the Pole Consortium, Parravicini V, Hédouin L (2021) Mesophotic coral communities escape thermal coral bleaching in French Polynesia. R Soc Open Sci 8:210139. https://doi.org/10.1098/rsos.210139

Perry CT, Larcombe P (2003) Marginal and non-reef-building coral environments. Coral Reefs 22:427–432. https://doi.org/10.1007/s00338-003-0330-5

Pineda J (1994) Internal tidal bores in the nearshore: warm-water fronts, seaward gravity currents and the onshore transport of neustonic larvae. J Mar Res 52:427–458. https://doi.org/10.1357/0022240943077046

Podestá GP, Glynn PW (2001) The 1997-98 El Niño event in Panama and Galápagos; an update of thermal stress indices relative to coral bleaching. Bull Mar Sci 69:43–59

Price JF (1981) Upper Ocean response to a hurricane. J Phys Oceanogr 11:153–175

Price JF, Weller RA, Schudlich RR (1987) Wind-driven ocean currents and Ekman transport. Science 238:1534–1538. https://doi.org/10.1126/science.238.4833.1534

Rädecker N, Pogoreutz C, Voolstra CR, Wiedenmann J, Wild C (2015) Nitrogen cycling in corals: the key to understanding holobiont functioning? Trends Microbiol 23:490–497. https://doi.org/10.1016/j.tim.2015.03.008

Radice VZ, Brett MT, Fry B, Fox MD, Hoegh-Guldberg O, Dove SG (2019) Evaluating coral trophic strategies using fatty acid composition and indices. PLoS One 14:e0222327. https://doi.org/10.1371/journal.pone.0222327

Randall CJ, Toth LT, Leichter JJ, Maté JL, Aronson RB (2020) Upwelling buffers climate change impacts on coral reefs of the eastern tropical Pacific. Ecology 101:e02918. https://doi.org/10.1002/ecy.2918

Rautenbach G, Hermes J, Halo I, Morris T, Veitch J (2023) Wind- and eddy-driven upwelling over submarine canyons inshore of the northern Agulhas current. Afr J Mar Sci 45:1–14. https://doi.org/10.2989/1814232X.2022.2163288

Reaka-Kudla ML, Feingold JS, Glynn W (1996) Experimental studies of rapid bioerosion of coral reefs in the Galfipagos Islands. Coral Reefs 15:101–107

Reed JK, Farrington S, David A, Harter S, Pomponi SA, Cristina Diaz M, Voss JD, Spring KD, Hine AC, Kourafalou VH, Smith RH, Vaz AC, Paris CB, Dennis Hanisak M (2019) Pulley Ridge, Gulf of Mexico, USA. In: Loya Y, Puglise KA, Bridge TCL (eds) Mesophotic coral ecosystems. Springer International Publishing, Cham, pp 57–69

Reid EC, DeCarlo TM, Cohen AL, Wong GTF, Lentz SJ, Safaie A, Hall A, Davis KA (2019) Internal waves influence the thermal and nutrient environment on a shallow coral reef. Limnol Oceanogr 64:1949–1965. https://doi.org/10.1002/lno.11162

Riegl B (2001) Inhibition of reef framework by frequent disturbance: examples from the Arabian Gulf, South Africa, and the Cayman Islands. Palaeogeogr Palaeoclimatol Palaeoecol 175:79–101. https://doi.org/10.1016/S0031-0182(01)00387-X

Riegl B, Piller WE (2003) Possible refugia for reefs in times of environmental stress. Int J Earth Sci 92:520–531. https://doi.org/10.1007/s00531-003-0328-9

Riegl BM, Purkis SJ, Al-Cibahy AS, Abdel-Moati MA, Hoegh-Guldberg O (2011) Present limits to heat-adaptability in corals and population-level responses to climate extremes. PLoS One 6:e24802. https://doi.org/10.1371/journal.pone.0024802

Riegl B, Johnston M, Glynn PW, Keith I, Rivera F, Vera-Zambrano M, Banks S, Feingold J, Glynn PJ (2019) Some environmental and biological determinants of coral richness, resilience and reef building in Galápagos (Ecuador). Sci Rep 9:10322. https://doi.org/10.1038/s41598-019-46607-9

Rixen T, Jiménez C, Cortés J (2015) Impact of upwelling events on the sea water carbonate chemistry and dissolved oxygen concentration in the Gulf of Papagayo (Culebra Bay), Costa Rica: implications for coral reefs. Rev Biol Trop 60:187. https://doi.org/10.15517/rbt.v60i2.20004

Roder C, Fillinger L, Jantzen C, Schmidt G, Khokiattiwong S, Richter C (2010) Trophic response of corals to large amplitude internal waves. Mar Ecol Prog Ser 412:113–128. https://doi.org/10.3354/meps08707

Roder C, Jantzen C, Schmidt GM, Kattner G, Phongsuwan N, Richter C (2011) Metabolic plasticity of the corals *Porites lutea* and *Diploastrea heliopora* exposed to large amplitude internal waves. Coral Reefs 30:57–69. https://doi.org/10.1007/s00338-011-0722-x

Rodríguez S, Alvizu A, Tagliafico A, Bastidas C (2009) Low natural repopulation of marginal coral communities under the influence of upwelling. Hydrobiologia 624:1–11. https://doi.org/10.1007/s10750-008-9662-z

Rodriguez-Ruano V, Toth LT, Enochs IC, Randall CJ, Aronson RB (2023) Upwelling, climate change, and the shifting geography of coral reef development. Sci Rep 13:1770. https://doi.org/10.1038/s41598-023-28489-0

Rosales SM, Sinigalliano C, Gidley M, Jones PR, Gramer LJ (2019) Oceanographic habitat and the coral microbiomes of urban-impacted reefs. PeerJ 7:e7552. https://doi.org/10.7717/peerj.7552

Roth MS, Goericke R, Deheyn DD (2012) Cold induces acute stress but heat is ultimately more deleterious for the reef-building coral *Acropora yongei*. Sci Rep 2:240. https://doi.org/10.1038/srep00240

Rowley SJ, Roberts TE, Coleman RR, Spalding HL, Joseph E, Dorricott MKL (2019) Pohnpei, Federated States of Micronesia. In: Loya Y, Puglise KA, Bridge TCL (eds) Mesophotic coral ecosystems. Springer International Publishing, Cham, pp 301–320

Rutsch M, Bayer T, Wall M, Putchim L, Ratanawongwan T, Roik A (2025) Towards coral probiotics: linking microbiome traits with thermal resistance of corals from the Andaman Sea, Thailand. Rosenstiel Sch Mar Atmos Sci

Sánchez-Noguera C, Stuhldreier I, Cortés J, Jiménez C, Morales Á, Wild C, Rixen T (2018) Natural Ocean acidification at papagayo upwelling system (North Pacific Costa Rica): implications for reef development. Biogeosciences 15:2349–2360. https://doi.org/10.5194/bg-15-2349-2018

Schmidt GM, Richter C (2013) Coral growth and bioerosion of Porites lutea in response to large amplitude internal waves. PLoS One 8:e73236. https://doi.org/10.1371/journal.pone.0073236

Schmidt G, Phongsuwan N, Jantzen C, Roder C, Khokiattiwong S, Richter C (2012) Coral community composition and reef development at the Similan Islands, Andaman Sea, in response to strong environmental variations. Mar Ecol Prog Ser 456:113–126. https://doi.org/10.3354/meps09682

Schmidt GM, Wall M, Taylor M, Jantzen C, Richter C (2016) Large-amplitude internal waves sustain coral health during thermal stress. Coral Reefs 35:869–881. https://doi.org/10.1007/s00338-016-1450-z

Schoepf V, Baumann JH, Barshis DJ, Browne NK, Camp EF, Comeau S, Cornwall CE, Guzmán HM, Riegl B, Rodolfo-Metalpa R, Sommer B (2023) Corals at the edge of environmental limits: a new conceptual framework to re-define marginal and extreme coral communities. Sci Total Environ 884:163688. https://doi.org/10.1016/j.scitotenv.2023.163688

Schramek TA, Terrill EJ, Colin PL, Cornuelle BD (2019) Coastally trapped waves around Palau. Cont Shelf Res 191:104002. https://doi.org/10.1016/j.csr.2019.104002

Scott PJB, Risk MJ (1988) The effect of Lithophaga (Bivalvia: Mytilidae) boreholes on the strength of the coral *Porites lobata*. Coral Reefs 7:145–151. https://doi.org/10.1007/BF00300974

Sevadjian JC, McManus MA, Benoit-Bird KJ, Selph KE (2012) Shoreward advection of phytoplankton and vertical re-distribution of zooplankton by episodic near-bottom water pulses on an insular shelf Oahu, Hawaii. Cont Shelf Res 50–51:1–15

Shanks A (1983) Surface slicks associated with tidally forced internal waves may transport pelagic larvae of benthic invertebrates and fishes shoreward. Mar Ecol Prog Ser 13:311–315. https://doi.org/10.3354/meps013311

Shanmugam G (2013) Modern internal waves and internal tides along oceanic pycnoclines: challenges and implications for ancient

deep-marine baroclinic sands. AAPG Bull 97:799–843. https://doi.org/10.1306/10171212101
Sheppard C (2009) Large temperature plunges recorded by data loggers at different depths on an Indian Ocean atoll: comparison with satellite data and relevance to coral refuges. Coral Reefs 28:399–403. https://doi.org/10.1007/s00338-009-0476-x
Sheppard CRC, Salm RV (1988) Reef and coral communities of Oman, with a description of a new coral species (Order Scleractinia, genus *Acanthastrea*). J Nat Hist 22:263–279. https://doi.org/10.1080/00222938800770201
Skinner C, Mill AC, Fox MD, Newman SP, Zhu Y, Kuhl A, Polunin NVC (2021) Offshore pelagic subsidies dominate carbon inputs to coral reef predators. Sci Adv 7:eabf3792. https://doi.org/10.1126/sciadv.abf3792
Smith TB (2005) The dynamics of coral reef algae in an upwelling system. PhD thesis
Smith TB (2006) Benthic algal production in an upwelling coral reef environment (Gulf of Panamá): evidence for control by fish herbivores. In: Proc 10th Int Coral Reef Symp
Smith TB (2008) Temperature effects on herbivory for an Indo-Pacific parrotfish in Panamá: implications for coral–algal competition. Coral Reefs 27:397–405. https://doi.org/10.1007/s00338-007-0343-6
Smith TB, Blondeau J, Nemeth RS, Pittman SJ, Calnan JM, Kadison E, Gass J (2010) Benthic structure and cryptic mortality in a Caribbean mesophotic coral reef bank system, the Hind Bank marine Conservation District, U.S. Virgin Islands. Coral Reefs 29:289–308. https://doi.org/10.1007/s00338-009-0575-8
Smith TB, Glynn PW, Maté JL, Toth LT, Gyory J (2014) A depth refugium from catastrophic coral bleaching prevents regional extinction. Ecology 95:1663–1673. https://doi.org/10.1890/13-0468.1
Sobel AH, Camargo SJ, Hall TM, Lee C-Y, Tippett MK, Wing AA (2016) Human influence on tropical cyclone intensity. Science 353:242–246. https://doi.org/10.1126/science.aaf6574
Sponaugle S, Lee T, Kourafalou V, Pinkard D (2005) Florida current frontal eddies and the settlement of coral reef fishes. Limnol Oceanogr 50:1033–1048. https://doi.org/10.4319/lo.2005.50.4.1033
Spreter PM, Reuter M, Mertz-Kraus R, Taylor O, Brachert TC (2022) Calcification response of reef corals to seasonal upwelling in the northern Arabian Sea (Masirah Island, Oman). Biogeosciences 19:3559–3573. https://doi.org/10.5194/bg-19-3559-2022
Spring DL, Williams GJ (2023) Influence of upwelling on coral reef benthic communities: a systematic review and meta-analysis. Proc R Soc B Biol Sci 290:20230023. https://doi.org/10.1098/rspb.2023.0023
Storlazzi CD, Cheriton OM, van Hooidonk R, Zhao Z, Brainard R (2020) Internal tides can provide thermal refugia that will buffer some coral reefs from future global warming. Sci Rep 10:13435. https://doi.org/10.1038/s41598-020-70372-9
Stramma L, Fischer T, Grundle DS, Krahmann G, Bange HW, Marandino CA (2016) Observed El Niño conditions in the eastern tropical Pacific in October 2015. Ocean Sci 12:861–873. https://doi.org/10.5194/os-12-861-2016
Stuhldreier I, Sánchez-Noguera C, Rixen T, Cortés J, Morales A, Wild C (2015a) Effects of seasonal upwelling on inorganic and organic matter dynamics in the water column of Eastern Pacific coral reefs. PLoS One 10:e0142681. https://doi.org/10.1371/journal.pone.0142681
Stuhldreier I, Sánchez-Noguera C, Roth F, Cortés J, Rixen T, Wild C (2015b) Upwelling increases net primary production of corals and reef-wide gross primary production along the Pacific Coast of Costa Rica. Front Mar Sci 2. https://doi.org/10.3389/fmars.2015.00113
Stuhldreier I, Sánchez-Noguera C, Roth F, Jiménez C, Rixen T, Cortés J, Wild C (2015c) Dynamics in benthic community composition and influencing factors in an upwelling-exposed coral reef on the Pacific coast of Costa Rica. PeerJ 3:e1434. https://doi.org/10.7717/peerj.1434
Sully S, van Woesik R (2020) Turbid reefs moderate coral bleaching under climate-related temperature stress. Glob Change Biol 26:1367–1373. https://doi.org/10.1111/gcb.14948
Sydeman WJ, García-Reyes M, Schoeman DS, Rykaczewski RR, Thompson SA, Black BA, Bograd SJ (2014) Climate change and wind intensification in coastal upwelling ecosystems. Science 345:77–80. https://doi.org/10.1126/science.1251635
Tebben J, Motti CA, Siboni N, Tapiolas DM, Negri AP, Schupp PJ, Kitamura M, Hatta M, Steinberg PD, Harder T (2015) Chemical mediation of coral larval settlement by crustose coralline algae. Sci Rep 5:10803. https://doi.org/10.1038/srep10803
Tkachenko KS, Soong K (2017) Dongsha Atoll: a potential thermal refuge for reef-building corals in the South China Sea. Mar Environ Res 127:112–125. https://doi.org/10.1016/j.marenvres.2017.04.003
Toth LT, Aronson RB, Vollmer SV, Hobbs JW, Urrego DH, Cheng H, Enochs IC, Combosch DJ, Van Woesik R, Macintyre IG (2012) ENSO drove 2500-year collapse of Eastern Pacific coral reefs. Science 337:81–84. https://doi.org/10.1126/science.1221168
Toth LT, Aronson RB, Cobb KM, Cheng H, Edwards RL, Grothe PR, Sayani HR (2015) Climatic and biotic thresholds of coral-reef shutdown. Nat Clim Chang 5:369–374. https://doi.org/10.1038/nclimate2541
Tréguer P, Jacques G (1986) Ecosystèmes pélagiques marins. Masson, Paris
UNEP-WCMC, WorldFish, World Resources Institute, The Nature Conservancy (2010) Global distribution of coral reefs. UNEP-WCMC. 1.33 GB
Vargas-Ángel B, Huntington B, Brainard RE, Venegas R, Oliver T, Barkley H, Cohen A (2019) El Niño-associated catastrophic coral mortality at Jarvis Island, central Equatorial Pacific. Coral Reefs 38:731–741. https://doi.org/10.1007/s00338-019-01838-0
Visbeck M, Keiser S (2021) Climate change and its impact on the ocean. In: Ekau W, Hornidge A-K (eds) Transitioning to sustainable life below water. MDPI
Walker ND, Leben RR, Balasubramanian S (2005) Hurricane-forced upwelling and chlorophyll a enhancement within cold-core cyclones in the Gulf of Mexico. Geophys Res Lett 32:1–5. https://doi.org/10.1029/2005GL023716
Wall M, Schmidt GM, Janjang P, Khokiattiwong S, Richter C (2012) Differential impact of monsoon and large amplitude internal waves on coral reef development in the Andaman Sea. PLoS One 7:e50207. https://doi.org/10.1371/journal.pone.0050207
Wall M, Putchim L, Schmidt GM, Jantzen C, Khokiattiwong S, Richter C (2015) Large-amplitude internal waves benefit corals during thermal stress. Proc R Soc B Biol Sci 282:20140650. https://doi.org/10.1098/rspb.2014.0650
Wall M, Doering T, Pohl N, Putchim L, Ratanawongwan T, Roik A (2023) Natural thermal stress-hardening of corals through cold temperature pulses in the Thai Andaman Sea. Limnol Oceanogr 69:2023.06.12.544549
Walter RK, Phelan PJ (2016) Internal bore seasonality and tidal pumping of subthermocline waters at the head of the Monterey submarine canyon. Cont Shelf Res 116:42–53. https://doi.org/10.1016/j.csr.2016.01.015
Wang Y, Dai C, Chen Y (2007) Physical and ecological processes of internal waves on an isolated reef ecosystem in the South China Sea. Geophys Res Lett 34:2007GL030658. https://doi.org/10.1029/2007GL030658
Wang D, Gouhier TC, Menge BA, Ganguly AR (2015) Intensification and spatial homogenization of coastal upwelling under climate change. Nature 518:390–394. https://doi.org/10.1038/nature14235
Watanabe TK, Watanabe T, Yamazaki A, Pfeiffer M, Garbe-Schönberg D, Claereboudt MR (2017) Past summer upwelling events in the Gulf of Oman derived from a coral geochemical record. Sci Rep 7:4568. https://doi.org/10.1038/s41598-017-04865-5

Weichler M, Roik A, Putchim L, Ratanawongwan T, Wall M (2024) Does shift in trophic strategy contribute to an enhanced bleaching resilience of corals regularly exposed to internal wave induced upwelling? Christian-Albrechts-Universität zu Kiel

Weijerman M, Fulton EA, Parrish FA (2013) Comparison of coral reef ecosystems along a fishing pressure gradient. PLoS One 8:e63797. https://doi.org/10.1371/journal.pone.0063797

Wellington GM (1982) Depth zonation of corals in the Gulf of Panama: control and facilitation by resident Reef fishes. Ecol Monogr 52:223–241. https://doi.org/10.2307/2937329

Wellington GM, Glynn PW (1983) Environmental influences on skeletal banding in eastern Pacific (Panama) corals. Coral Reefs 1:215–222. https://doi.org/10.1007/BF00304418

Wells JW (1957) Chapter 20: Coral reefs. In: Hedgpeth JW (ed) Treatise on marine ecology and paleoecology. Geological Society of America

Wentz FJ, Gentemann C, Smith D, Chelton D (2000) Satellite measurements of sea surface temperature through clouds. Science 288:847–850. https://doi.org/10.1126/science.288.5467.847

Wiedenmann J, D'Angelo C, Smith EG, Hunt AN, Legiret F-E, Postle AD, Achterberg EP (2013) Nutrient enrichment can increase the susceptibility of reef corals to bleaching. Nat Clim Chang 3:160–164. https://doi.org/10.1038/nclimate1661

Williams GJ, Smith JE, Conklin EJ, Gove JM, Sala E, Sandin SA (2013) Benthic communities at two remote Pacific coral reefs: effects of reef habitat, depth, and wave energy gradients on spatial patterns. PeerJ 1:e81. https://doi.org/10.7717/peerj.81

Williams GJ, Sandin SA, Zgliczynski BJ, Fox MD, Gove JM, Rogers JS, Furby KA, Hartmann AC, Caldwell ZR, Price NN, Smith JE (2018) Biophysical drivers of coral trophic depth zonation. Mar Biol 165:60. https://doi.org/10.1007/s00227-018-3314-2

Wilson SK, Fisher R, Pratchett MS, Graham NAJ, Dulvy NK, Turner RA, Cakacaka A, Polunin NVC (2010) Habitat degradation and fishing effects on the size structure of coral reef fish communities. Ecol Appl 20:442–451. https://doi.org/10.1890/08-2205.1

Witman JD, Smith F (2003) Rapid community change at a tropical upwelling site in the Galápagos Marine Reserve. Biodivers Conserv 12:25–45. https://doi.org/10.1023/A:1021200831770

Wizemann A, Nandini SD, Stuhldreier I, Sánchez-Noguera C, Wisshak M, Westphal H, Rixen T, Wild C, Reymond CE (2018) Rapid bioerosion in a tropical upwelling coral reef. PLoS One 13:e0202887. https://doi.org/10.1371/journal.pone.0202887

Wolanski E, Colin P, Naithani J, Deleersnijder E, Golbuu Y (2004) Large amplitude, leaky, Island-generated, internal waves around Palau, Micronesia. Estuar Coast Shelf Sci 60:705–716. https://doi.org/10.1016/j.ecss.2004.03.009

Wolanski E, Andutta F, Deleersnijder E, Li Y, Thomas CJ (2017) The Gulf of Carpentaria heated Torres Strait and the Northern Great Barrier Reef during the 2016 mass coral bleaching event. Estuar Coast Shelf Sci 194:172–181. https://doi.org/10.1016/j.ecss.2017.06.018

Wyatt ASJ, Lowe RJ, Humphries S, Waite AM (2010) Particulate nutrient fluxes over a fringing coral reef: relevant scales of phytoplankton production and mechanisms of supply. Mar Ecol Prog Ser 405:113–130. https://doi.org/10.3354/meps08508

Wyatt ASJ, Leichter JJ, Toth LT, Miyajima T, Aronson RB, Nagata T (2020) Heat accumulation on coral reefs mitigated by internal waves. Nat Geosci 13:28–34. https://doi.org/10.1038/s41561-019-0486-4

Wyatt ASJ, Leichter JJ, Washburn L, Kui L, Edmunds PJ, Burgess SC (2023) Hidden heatwaves and severe coral bleaching linked to mesoscale eddies and thermocline dynamics. Nat Commun 14:25. https://doi.org/10.1038/s41467-022-35550-5

Wyrtki K (1964) Upwelling in the Costa Rica dome. Fish Bull 63:355–372

Wyrtki K (1981) An estimate of equatorial upwelling in the Pacific. J Phys Oceanogr 11:1205–1214

Yadidya B, Rao AD (2022) Interannual variability of internal tides in the Andaman Sea: an effect of Indian Ocean dipole. Sci Rep 12:11104. https://doi.org/10.1038/s41598-022-15301-8

Zapata FA, Jaramillo-González J, Navas-Camacho R (2011) Extensive bleaching of the coral *Porites lobata* at Malpelo Island, Colombia, during a cold water episode in 2009. Bol Investig Mar Costeras - INVEMAR 40:185–193

Zhu W, Ren Y, Liu X, Huang D, Xia J, Zhu M, Yin H, Chen R, Li X (2022) The impact of coastal upwelling on coral reef ecosystem under anthropogenic influence: coral reef community and its response to environmental factors. Front Mar Sci 9:888888. https://doi.org/10.3389/fmars.2022.888888

Mangroves and Seagrasses as Habitat for Corals

11

Heather A. Stewart, Rafael J. Araújo, Karina Scavo Lord, and Jennifer L. Au

Abstract

This chapter explores the intricate relationships and ecological interdependencies among corals, mangroves, and seagrasses, emphasizing their significance in maintaining biodiversity and providing essential ecosystem services. Tropical coastal regions host unique habitats where corals coexist within mangrove forests and seagrass meadows, forming distinct ecosystems known as coexisting mangrove-coral (CMC) and coexisting seagrass-coral (CSC) habitats. These systems, often considered marginal or extreme, serve as refugia, supporting coral persistence in the face of climate change, ocean acidification, and other anthropogenic threats. CMC and CSC habitats exhibit distinct environmental conditions, including diel temperature and pH fluctuations, which influence coral physiology, morphology, and community structure. Corals in these habitats display remarkable adaptability, such as flattened growth forms for low light conditions and sediment-clearing mechanisms in soft substrates. Furthermore, mangroves and seagrasses mitigate environmental stressors through carbon sequestration, nutrient cycling, and water quality improvement, providing essential support for coral survival. These habitats also serve as nurseries for reef-associated fish and invertebrates, enhancing fisheries and contributing to coastal resilience. By highlighting case studies and current research, this chapter underscores the urgent need for integrated conservation strategies. Furthermore, we provide a baseline of 177 coral species observed to occupy CMC habitats and 39 coral species in CSC habitats. Protecting and restoring these interdependent ecosystems will enhance their role in mitigating climate change impacts, sustaining biodiversity, and supporting coastal livelihoods globally.

Keywords

Coexisting mangrove-coral · Seagrass-coral · Nested assemblage · Refugia · Reef-mangrove

11.1 Introduction

Most coral reefs in the tropics are nearshore and thus, corals are associated with mangrove and seagrass habitats even if the coral reef is not in a back-reef system (McMahon et al. 2012). These mangrove, seagrass, and coral seascapes are vital for the health of our oceans, coastal defense, and fisheries (Laegdsgaard and Johnson 1995; Guannel et al. 2016; Nordlund et al. 2018; Rogers and Mumby 2019). Strong ecological linkages bind these ecosystems to one another. For example, these three ecosystems are joined by movement of reef fishes which utilize one ecosystem as nurseries and/or shelters while foraging in another. Mangroves can alleviate nutrient pollution for adjacent coral reefs and seagrasses through rapid nutrient uptake, long-term burial, or denitrification (Mack et al. 2024). Both mangroves and seagrass can provide shade for light-stressed corals. Mangroves also maintain coastal stability, reducing sedimentation and safeguarding biodiversity (Laegdsgaard and Johnson 1995; Nagelkerken et al. 2001; Dubuc et al. 2019; Rogers and Mumby 2019; Sánchez-Núñez et al. 2019), while seagrasses can locally counteract ocean acidification (Manzello et al. 2012). In many parts of the world, corals are not only found adjacent to mangroves and seagrass beds (i.e., coral reefs), but coexisting in nested assemblages where coral colonies utilize mangrove forests and seagrass beds as habitat. Nested

H. A. Stewart (✉)
U.S. Fish and Wildlife Service, Florida Keys National Wildlife Refuge Complex, Big Pine Key, FL, USA

R. J. Araújo · J. L. Au
Department of Marine Biology and Ecology, Rosenstiel School of Marine, Atmospheric & Earth Science, University of Miami, Miami, FL, USA

K. Scavo Lord
Department of Biology, Boston University, Boston, MA, USA

M. O. Soares et al. (eds.), *Extreme and Marginal Coral Reefs and Communities*, Coral Reefs of the World 24,
https://doi.org/10.1007/978-3-032-28035-0_11

mangrove-coral assemblages, known as coexisting mangrove-coral (CMC) habitats (Stewart et al. 2022), have been observed for decades and differ notably from adjacent coral reefs and mangrove forests, where these ecosystems form distinct, nonoverlapping zones. Similarly, nested seagrass-coral assemblages have been documented in the literature, though they have not yet been designated as a distinct habitat type (e.g., coexisting seagrass-coral, CSC). This chapter examines the structure and ecological dynamics of both coexisting mangrove-coral and seagrass-coral systems. We present case studies from around the world to demonstrate the prominence of these ecosystems' structure and explore the conditions of these habitats, which may be marginal in some regions of the world and serve as coral refugia in others. We discuss conservation and management strategies for these nested ecosystems and provide a baseline of coral species observed to occupy these habitats.

11.2 Coral Habitats in Extreme-Marginal Environments

Globally, coral reefs are experiencing precipitous declines, and many species face potential extirpation (IUCN 2024a). While reef-building corals have existed over 500 million years, in the Anthropocene, reefs have been faced with a combination of local-scale anthropogenic factors (e.g., coastal development, nutrient runoff, eutrophication, pollution, and overfishing) and large-scale climate change impacts (e.g., warming ocean temperatures, ocean acidification, hurricanes) (Hughes et al. 2007; Bellwood et al. 2019). Shallow-water coral reefs and back-reef habitats are particularly threatened due to their proximity to coasts and more direct negative impacts from humans (Sealey et al. 2024). Furthermore, shallow corals experience more thermal extremes than deeper corals (Tavakoli-Kolour et al. 2024). Stressed corals are more likely to succumb to disease, which has been increasing in prevalence over the past couple of decades (Burke et al. 2023). Although reefs have the capacity to recover from disturbances, the magnitude and rate of multiple stressors and disturbance events outpace the natural recovery rates. The result is threatened coral persistence (ability to survive over time) and resilience (capacity to withstand and recover from disturbances) and weakened ecological services and function of the reef (Harborne et al. 2017). As species on the reef are lost, biodiversity decreases, and the structural complexity of the reef erodes. Habitat selection is complex and can have significant repercussions on post-settlement survival and growth. Factors include substrate orientation, microtopography, biochemical cues, and sound (Davies et al. 2014; Foster and Gilmour 2016; Aoki et al. 2024). With global conditions unlikely to return to historic levels, these novel ecosystems provide vital strongholds and warrant investigation to understand how these environments function and how to appropriately manage them. However, records of corals living in mangrove and seagrass habitats for nearly 100 years provide some legacy for corals persisting in nonreef habitats.

Mangroves are resilient carbon-rich forests that occupy intertidal areas in tropical and subtropical coastal regions and regulate nutrient cycles within ecosystems. To overcome the challenges posed by high temperatures, salinity, and anoxia, mangroves have developed physiological and morphological adaptations, most notably the presence of aerial roots and vivipary. While having aerial roots and being viviparous are not mandatory characteristics of mangroves, *strict* or *true* mangrove species must exhibit fidelity to the mangrove habitat, a distinct ecosystem from other terrestrial communities (Tomlinson 1986). Mangrove adaptations allow them to alter their environment, having cascading impacts on secondary species.

In CMC habitats, corals have bound to the submersed roots and increased habitat complexity (Wright et al. 2023). The tides are a dominating factor in CMC habitats, creating an upper limit to coral growth by exposure intolerance around the low tide mark and tidal current flushing excessive nutrients. Tidal amplitude is a limiting parameter because prolonged air exposure is stressful to corals and permanent inundation is stressful to mangroves. When conditions align to allow for overlap, the habitat is frequently considered marginal for both the corals and the mangroves. For example, CMC habitats have currents that regularly flush the system. This limits nutrient loads and fine sediment, which benefits nutrient-sensitive corals (Nalley et al. 2023) but is detrimental to mangrove growth and results in dwarf trees. The absence of regular freshwater input further stresses mangroves at the benefit of corals. These conditions are essential for corals to persist in the habitat.

Low light levels and shallow water of the mangrove habitat require corals in CMC habitats to adapt through the utilization of a variety of flattened, plate-like and encrusting growth forms (Stewart et al. 2021) (Fig. 11.1). This morphological adaptation maximizes incoming light per unit area of coral tissue surface and reduces self-shading (Stambler and Dubinsky 2005). By increasing light-harvesting pigments, corals further enhance light capture (Kahng et al. 2012, 2019). Additionally, this growth form may be an adaptation to avoid air exposure in shallow water conditions within the mangrove (~0.5 m depth on average). Growing laterally has a secondary advantage of allowing coral colonies to stabilize roots, which may otherwise move and rub coral polyps causing physical damage. A trade-off of plate-like growth forms is that the fragile structure is more prone to breaking. Corals such as *Manicina areolata* (Linnaeus, 1758) can attach to mangrove roots and then detach to diffuse out from the mangrove canopy. While these corals would risk being smothered in sediment or covered by algae in the process of diffusing out of the man-

Fig. 11.1 Mangrove coral growth forms and examples of coral species found within the mangrove canopy. Coral species from top left to bottom right: *Mycetophyllia aliciae*, *Diploria labyrinthiformis*, *Pseudodiploria clivosa*, *Colpophyllia natans*, *Isophyllia rigida*, *Siderastrea siderea*, *Montastraea cavernosa*, *Porites furcata*, and *Porites astreoides*. (Photographs taken by H.A. Stewart)

groves in a typical mangrove forest, CMC habitats have far fewer risks. Among types of CMC habitats, water quality parameters (e.g., dissolved oxygen, pH, chlorophyll, dissolved organic matter) range from stressful (i.e., marginal habitats) to superior relative to the adjacent reef (i.e., refugia). In many cases, CMC habitats exist on a spectrum with individual habitats having some conditions better suited for corals than on the reef, while other conditions are suboptimal (Stewart et al. 2021).

Corals living in mangrove habitats of Indonesia and the Caribbean have been reported having tentacles extended from corallites during the day (Polapa et al. 2021) (Fig. 11.2). These observations support the theory that corals living in CMC habitats may be heterotrophic or use heterotrophy as an adaptation for survival in these habitats. In environments with high levels of primary productivity, some coral species demonstrate a preference for heterotrophic nutrition. Heterotrophy allows corals to be adaptable to turbid waters and more resilient to bleaching events (Conti-Jerpe et al. 2020; Travaglione et al. 2023).

Corals living in seagrass habitats experience greater diel temperature fluctuations compared to adjacent reef sites, although no significant difference of mean temperature between seagrass-coral and adjacent reef sites have been reported (Camp et al. 2016a; Lohr et al. 2017). Additionally, seagrass-coral habitats have greater pH variability compared to adjacent reef environments, primarily driven by seagrass photosynthesis and respiration, which in turn affects coral calcification rates (Semesi et al. 2009; Saderne et al. 2013; Challener et al. 2016; Lohr et al. 2017). Seagrass density and shading may impact coral distribution and presence as seen by high mortality in low-density beds from predation and siltation and high stress in corals transplanted to high-density beds, likely due to low light conditions (Ruiz-Zárate et al. 2000; Landau et al. 2013). These coral distribution patterns with seagrass density are species specific and more research is needed to understand the impacts of seagrass density on coral community structure. It is suspected that soft-bottomed substrate of seagrass beds may result in coral smothering or burial, limiting habitats to coral species effective at clearing sediment and able to withstand short-term burial periods with minor physiological consequences (Lirman and Manzello 2009).

11.3 Mangroves as Coral Habitats

Mangrove structure and ecology

Mangrove structure is diverse. Forests are the result of many forces that either contribute to mangrove development (e.g., nutrients, tides, hydrology) or act as a strain to it (e.g., drought, storms, excess salinity, frost). The result is a wide

Fig. 11.2 *Montastraea cavernosa* within the mangrove canopy with tentacles extended from corallites during the day. (Photographs taken by H.A. Stewart)

array of forest types that capture the combination of the magnitudes and periodicities of these forces. Since mangroves occupy vast regions of the tropics and subtropics, regional and local variations are common. Mangrove structure is also affected by speciation. There are 82 species in 32 genera and 18 families, including commonly known genera such as *Rhizophora*, *Avicennia*, *Bruguiera*, and *Sonneratia*. Most species (63 including hybrids) occur in the Indo-West Pacific region, the remaining (19 including hybrids) are inhabitants of the Atlantic-East Pacific region (Leal and Spalding 2024).

Recent estimates (Leal and Spalding 2024) report global mangrove extent at 147,256 km^2, with most of the distribution in Southeast Asia (33.6%), West and Central Africa (15.5%), North and Central America & Caribbean (14.4%), and South America (13.2%). The top ten mangrove holding countries are Indonesia (21% of total), Brazil (7.4%), Australia (6.9%), Mexico (5.8%), Nigeria (5.8%), Malaysia (3.9%), Myanmar (3.8%), India (3.1%), Bangladesh (2.9%), and Papua New Guinea (2.9%).

Mangroves have a series of singular adaptations that allows them to colonize and flourish in tropical and subtropical regions characterized by high temperatures, salinity variations, and unstable and anaerobic substrates (Tomlinson 2016). Some of these adaptations include complex mechanisms to avoid salt, a root system that allows mangroves to breathe in waterlogged, oxygen-poor soils, and viviparous propagation.

Living in a saline environment is only possible due to specialized morphological and physiological adaptations that result in complex mechanisms for salt exclusion or secretion (salt tolerance or accumulation has also been documented, but it is less common). Salt exclusion is done at the root level and is a typical mechanism in *Rhizophora, Avicennia, Ceriops* and *Sonneratia* (Hogarth 2015). Salt secretion has been observed in *Avicennia* and *Laguncularia* and is facilitated by specialized glands that secrete excess salt, which crystallizes and is then washed away by rain or wind (Hogarth 2015; Lonard et al. 2021).

For many species of mangroves, roots extend aboveground, providing access to air and stabilizing the plants in sometimes shifting coastal sediments. Moreover, the intricate root systems trap sediments and filter pollutants, improving water quality and serving as a physical barrier that breaks wave and wind energy in stormy conditions. Additionally, mangrove roots (especially the submerged roots of mangroves within the family Rhizophoraceae) are known for housing a variety of organisms that include corals (Camp et al. 2016a, b; Stewart et al. 2021). Aerial mangrove roots are striking to behold and come in a variety of forms that include stilt, looping roots (*Rhizophora*), pneumatophores (*Avicennia*, *Sonneratia*), root knees (*Ceriops*, *Bruguiera*), and planks and buttresses (*Xylocarpus*, *Pelliciera*) (Tomlinson 2016, Fig. 11.3). There are also mangrove species without any aerial roots (*Osbornia*, *Excoecaria*) and the occasional mangrove species that normally lacks aerial roots

Fig. 11.3 Root systems in mangroves. (**a**) Stilt root of *Rhizophora,* (**b**) pneumatophores of *Avicennia,* (**c**) planks and/or buttresses of *Pelliciera,* and (**d**) root knees of *Bruguiera*. (Photographs taken by H.A. Stewart and R.J. Araújo)

but can develop them in challenging conditions (*Laguncularia*, *Kandelia*) (Tomlinson 2016).

Many (but not all) mangroves exhibit viviparous propagation (as is the case in *Rhizophora, Kandelia, Bruguiera, Ceriops*). Vivipary is an evolutionary reproductive adaptation where embryos develop and germinate while still attached to the parent. In the case of plants, vivipary results in a seedling that emerges directly from the fruit while it is still attached to the parent tree, giving the new plant a head start in growth. Vivipary is relatively rare in the plant kingdom and is primarily observed in environments where early development increases the chances of seed survival. This adaptation is particularly beneficial in challenging environments like saline coastal zones, where the success of the seedling depends on its ability to quickly establish itself before being carried away by tides or waves, or in the case that the seedling is transported away, give it a chance to survive during longer periods of time (Hogarth 2015). Nonviviparous mangroves (*Laguncularia, Sonneratia, Xylocarpus, Lumnitzera*, etc.) produce seeds like most angiosperms. For some mangrove genera, there exists an intermediate state called cryptovivipary (*Avicennia, Pelliciera, Aegialitis, Aegicieras*) in which the embryos start to develop before the seed is shed, but do not break through the outside of the fruit capsule (Tomlinson 2016).

There are various ways to categorize mangrove forests. One common approach is to divide mangrove communities based on coastal environmental settings (e.g., large rivers, small deltas, tidal systems, lagoons, carbonate settings, arheic; *see* Rovai et al. 2021). An older and popular approach, especially for categorizing mangroves of the Western Hemisphere, was devised by Lugo and Snedaker (1974). In this straightforward and intuitive classification, mangroves are grouped as overwash, fringe, riverine, basin, hammock, or dwarf forests. This classification relies on the physiognomy of the forest resulting from different geological and hydrological processes. It also allows for useful comparisons of forests between sites. The Lugo and Snedaker (1974) classification system was the inspiration used for the coexisting mangrove-coral habitat scheme described later in the chapter.

As unique as mangroves and their ecology are, these ecosystems have a profound and interconnected relationship with human communities, particularly those living in tropical coastal areas. This relationship can be understood through the role mangroves play in the livelihoods and economies of nations, especially for their role in coastal fisheries (Bimrah et al. 2022). Mangroves serve as nurseries for many fish and crustacean species, including commercially valuable ones like shrimp, crabs, and fish (Huxham et al. 2017; Bimrah et al. 2022). This supports local and regional fisheries, providing a critical source of income and food for coastal communities. The mangrove fish connection goes from ecosystem to individual mangrove trees, as these solitary mangroves have been noted to provide alternative habitat on backreefs (von Hammerstein et al. 2024). Mangroves also provide forestry products (timber, fuelwood) and nontimber products like honey and tannins. These resources are often harvested by local people for both subsistence and trade (Bhowmik et al. 2022).

Also acknowledged is the role that mangroves play in coastal protection (Spalding et al. 2014). As stated before, mangrove roots stabilize shorelines by trapping sediments, reducing coastal erosion, and protecting against storm surges. Moreover, mangroves act as natural barriers against hurricanes, tsunamis, and cyclones, reducing the impact on coastal communities and infrastructure. In other words, they strengthen our coastlines and make them resilient. This protective function can save lives and reduce property damage (Spalding et al. 2014).

Added to these ecosystem functions is the role that mangroves play in carbon sequestration and climate regulation (Alongi 2018). Mangrove forests are highly efficient at capturing and storing carbon dioxide, which helps mitigate climate change. This storage is part of what is known as *blue carbon*—carbon captured by coastal and marine ecosystems, including mangroves, seagrasses, and salt marshes (Macreadie et al. 2021). This makes mangroves vital for global carbon management, especially at the local scale, and particularly in countries that contain important tracts of mangrove forest that might provide opportunities for carbon credits and conservation financing (Kumari and Rathore 2021). Conserving and restoring these forests is now a priority at many levels and is often encouraged by governments, NGOs, citizen groups, and other organizations (Leal and Spalding 2024).

Recognition of the importance of mangroves as ecosystems is now nearly universal, and many countries and local communities are actively working to conserve and restore them (Leal and Spalding 2024). Mangroves are widely regarded as one of the most effective nature-based solutions to address some of the most severe impacts of climate change (Ashrafuzzaman 2023). Their ecological significance is not only well-established but also celebrated. For many coastal communities, mangroves hold deep cultural, spiritual, and historical value, serving as sites for traditional practices, rituals, and folklore, and forming a crucial part of local identity (Scott et al. 2024). Additionally, mangroves are popular destinations for tourism and recreation, activities that promote environmental awareness and generate income for these communities (Kumar 2023).

Mangrove and coral biogeography

Mangrove and corals have similar wide distribution in the tropics and subtropics. Mangrove biogeography is characterized by two biodiversity hot spots, the Indo-West Pacific and the Atlantic–East Pacific and these two regions have few shared species (Tomlinson 2016; Duke 2017). Coastal mangroves are recognized for having long-distance positive interactions with coral reefs (e.g., retaining nutrients, reducing sediment flux), and coral reefs allow mangroves to thrive by dissipating wave energy and creating calm conditions for propagules to settle (Harborne et al. 2006; Gillis et al. 2014; van de Koppel et al. 2015). Coral reefs and mangrove forests can coexist mere kilometers apart, while being essentially isolated from one another through hydrodynamic processes, frequently with seagrass beds in the middle acting as a buffer zone (Kitheka 1996). The driver of such linkages being water. With geological shifts in sea-level rise and fall and movements of tectonic plates, coral reefs and mangroves have repeated their dance of expanding and waning (Alongi et al. 2008), going through periods of direct overlap while transitioning. However, in some regions, these mangrove-coral systems are not merely transitional but represent stable ecosystems where diverse coral communities are permanently integrated within mangrove habitats.

In the Pacific, where tidal range is large, mangrove roots and corals may be air-exposed, limiting which coral species may survive and the stressors they are exposed to. In the Caribbean, where there is a limited tidal range (typically ranging between 0.3 and 0.6 m), and the topographic relief is not too steep, mangrove roots and corals might remain submerged for longer periods, which allows a wider variety of coral species to thrive but also exposes them to different stressors. Moreover, tidal amplitude combined with mangrove biogeography results in zonation patterns that are common in some geographical regions. For instance, Caribbean mangroves present a characteristic zonation that progressing inland from the coast situates *Rhizophora mangle* (L.) shoreward, followed by *Avicennia germinans* (L.) L., and *Laguncularia racemosa* (L.) C.F.Gaertn. (Stehlé

1945). In Indo-Pacific locations, higher mangrove diversity results in different zonation pattern combinations to those observed in the Caribbean. In these regions, shoreline species might be *Sonneratia*, *Avicennia*, or *Rhizophora* (Chapman 1976). Since mangrove species show notable variations in root morphology, the seaward positioning of one species might be a critical factor in determining which species of coral might be suitable for coexisting with the mangroves.

Coral biogeography is shaped by historical, environmental and ecological factors and the presence of coral diversity hot spots. As with mangroves, the Indo-Pacific is the most biodiverse coral region (over 650 species of reef-building corals; DeVantier and Turak 2017) including genera such as *Acropora*, *Porites*, *Montipora*, and *Pocillopora*. In the Western Indian Ocean, there are approximately 369 species of reef-building corals (Obura 2012). The main genera include *Acropora, Porites, Favia, Favites, Goniastrea, Platygyra,* and *Cyphastrea.* In the Red Sea, around 300 species have been documented, and the area is known for its high levels of endemism, many coral species found there are unique to the region (Berumen et al. 2019). The Atlantic (Caribbean) contains approximately 75 species of corals and these constitute the backbone of coral ecosystems for the region. Lastly, the Eastern Pacific contains about 47 species of reef-building corals, with eight being endemic to this area (Souter et al. 2021).

Key aspects of coral biogeography are driven by the high level of endemism observed in certain areas due to geographical isolation and unique environmental conditions. As with mangroves, connectivity is limited by physical barriers, temperature gradients, and ocean currents leading to distinct coral assemblages in each biogeographical region. Coral reefs in all regions face significant threats to their existence from climate change, ocean acidification, overfishing, pollution, disease, and coastal development making conservation and restoration efforts crucial for maintaining diversity (Frieler et al. 2013).

Coexisting mangrove-coral habitats

Just as abiotic conditions (e.g., light, water quality, tidal inundation) determine types of mangrove formations (e.g., overwash islands, basin, coastal fringe, lagoon, riverine) and mangrove species zonation patterns, CMC habitat type can be characterized by the physical setting and location of coral relative to the mangrove forest. CMC habitat types fall amongst intersecting continuums of light and water flow (Stewart et al. 2022).

Coexisting mangrove-coral habitat types and descriptions sensu Stewart et al. (2022):

- *Lagoon CMC habitat:* Corals grow within lagoons that are semienclosed by mangroves and have restricted water flow. In these habitats, corals are positioned outside the shade of the mangrove canopy during high noon (Fig. 11.4b). Observed in the Caribbean, South Pacific (New Caledonia), Indonesia (Sulawesi), Indian Ocean (Seychelles), the Great Barrier Reef, and the Northern Red Sea.
- *Inlet CMC habitat:* Corals grow along mangrove channels or creeks (e.g., between mangrove islands or cays, natural tidal channels or man-made canals) partially to fully under the shade of the mangrove canopy (Fig. 11.4c). Observed in the Caribbean region, in the Northern Territory of Australia and the Great Barrier Reef.
- *Edge CMC habitat:* Corals grow on and around the mangrove roots along the fringe of the mangroves partially or completely under the shade of their canopy. Edge CMC habitats are typically fringe mangrove or overwash islands exposed to open ocean and have unrestricted water flow unlike Lagoon or Inlet CMC habitats. This category may include portions of fringe reefs where corals are growing under the shade of mangrove canopies (Fig. 11.4d). Observed throughout the Caribbean region and the Great Barrier Reef.
- *Canopy CMC habitat:* This habitat encompasses the most interior mangrove-coral setting and is connected to Edge CMC habitats through a transition zone. Corals grow on and around mangrove roots, completely under the shade of the mangrove canopy. Prop roots from the mangrove canopy edge restrict water flow to this habitat, cutting off direct connection with the open ocean (Fig. 11.4d). Only reported in Caribbean Panama.
- *Non-CMC habitat:* Adjacent assemblage of coral reefs and mangrove forests where these ecosystems create discrete zones and there is no influence of canopy shade on the reef (Fig. 11.4a). A non-CMC habitat corresponds to the classic zonation pattern in which mangroves are near, but not overlapping with, a coral reef, oftentimes with seagrasses in between them. Common in the tropics, especially in areas with large tides, such as the Tropical Eastern Pacific, as well as in the Caribbean Sea.

Requirements of CMC habitats

CMC habitat type is largely driven by tides, which leads to geographic differences in the mangrove-coral interactions. However, there are similarities between all CMC habitats which are believed to be requirements for mangroves and corals to coexist. These conditions include a connection to the open ocean or constant flushing by tidal currents, high salinity, low turbidity, limited subaerial exposure, proximity to coral reefs (~5 m), and light attenuation (Ferns 1994;

Fig. 11.4 Coexisting mangrove-coral (CMC) assemblage types. (**a**) Non-CMC: discrete adjacent assemblage of coral reefs and mangrove forests. (**b**) Lagoon CMC: corals along semienclosed mangrove lagoons outside the shade of the mangrove canopy, but close enough for biogeochemical influence. (**c**) Inlet CMC: corals along mangrove channels partially under shade. (**d**) Edge CMC: partially or completely shaded corals grow on and around mangrove roots along the fringe of the mangrove. Canopy CMC: corals grow completely under the mangrove canopy with the lowest light among CMC habitats. (**e**) CMC habitats relative to each other. Graphics created by H.A. Stewart and J.L. Au with vector files from Tracey Saxby, Dieter Tracey, and Joanna Woerner Integration and Application Network, University of Maryland Center for Environmental Science (ian.umces.edu/imagelibrary/). (Modified from Stewart et al. 2022)

Stewart et al. 2022). The connection between the ocean and mangrove forest and the proximity to coral reefs is important as a source of coral planulae. Water movement, in the form of tidal current, regularly flushes the system to remove excessive nutrient loads and fine sediment, as well as replenishes depleted oxygen, and maintains consistent temperature and salinity levels. CMC habitats predominantly have sandy and coarse calcareous sediment surrounding the system which is low in silt levels (Ferns 1994). Moreover, the absence of large rivers significantly reduces the levels of sedimentation that would typically occur in a mangrove habitat. One of the most striking features of CMC habitats is the clarity of the water, which is likely a result of the regular flushing. Low turbidity allows corals to persist in CMC habitats with drastically lower light levels than the reef. Within a CMC habitat, the depth of standing water sets the upper limit to colony size and structures the coral community based on the tolerance of different species to subaerial exposure (Ferns 1994). In the Caribbean, most corals in CMC habitats remain below the tidal range; however, in the Pacific, where coral more commonly is exposed to air, the increased humidity from the forest and shade from the mangrove canopy may mitigate some of the stress caused by subaerial exposure. Proximity to land development, pollution, sedimentation, and freshwater inflow negatively correlate with the number of coral species present in CMC habitats and can completely limit coral presence or survival (Stewart et al. 2022). Reports of CMC habitats in the scientific literature describe areas without freshwater influence other than rain and subsoil water or with creeks or rivers being >1.2 km away (Bengtsson et al. 2019; Stewart et al. 2022).

Abiotic parameters that appear to correspond with CMC habitat type distinctions include, but are not limited to, (a) current speed/tidal velocity, (b) temperature, (c) salinity, (d) light, (e) dissolved oxygen and (f) pH (Stewart et al. 2022). The difference in these parameters is likely the reason why chlorophyll and dissolved organic matter levels are distinct between CMC types (Stewart et al. 2022). Lagoon CMC habitats have been described in the scientific literature as being hotter and more acidic with lower oxygen levels compared to adjacent reef (Camp et al. 2017). Although Lagoon CMC habitats have the lowest pH range (7.31–8.10) observed of the CMC habitat types, abiotic conditions (e.g., temperature, pH, and dissolved oxygen) are strongly correlated with the tidal cycle (Camp et al. 2017; Maggioni et al. 2021). Canopy CMC habitats have lower reported dissolved oxygen levels (averaging 3.74 mg/L), compared to Lagoon (Stewart et al. 2022). Light attenuation is a major influence on delineating CMC habitat type as well as coral community structure. Within the mangrove habitat, corals show great plasticity in growth forms and photophysiology in response to the complexity of mangrove architecture (e.g., sunflecks). Flattened and encrusting morphologies allow coral colonies to expand laterally, avoiding subaerial exposure. It appears that coral distribution with CMC habitats is closely linked to light availability (Stewart et al. 2022). Some studies have compared nutrient levels of CMC habitat types and found them to be relatively uniform and not a determinant of zonation or habitat type distinction (Ferns 1994).

Global distribution of CMC habitats

Globally, seven species of mangroves and 177 species of corals have been identified as part of CMC habitats (Table 11.1). More coral species may occupy mangrove habitat but were unable to be extracted from the scientific literature due to the lack of species or abundance data mentioned (Stewart et al. 2022). Observations have been made in locations spanning the Caribbean Sea, Red Sea, Indian Ocean, Arabian Sea, and South Pacific (Table 11.2). The most frequently reported coral genera in decreasing order are *Porites* (n = 77), *Acropora* (n = 51), *Agaricia* (n = 27), *Siderastrea* (n = 24), *Millepora* (n = 23), *Orbicella* (n = 18), *Favia* (n = 16), *Montipora* (n = 15), *Pseudodiploria* (n = 14), *Manicina* (n = 13), and *Pocillopora* (n = 12). In the Western Hemisphere, CMC habitats have been exclusively described in association with *R. mangle*. However, white mangrove *L. racemosa* has been observed in CMC habitats following hurricane disturbances which reshape mangrove community structure. In the Eastern Hemisphere, a greater variety of mangrove species are reported making up CMC habitats. These species are within the genera *Rhizophora*, *Avicennia*, *Bruguiera*, *Lumnitzera*, and *Ceriops*.

Using 10-m resolution Copernicus Sentinel-2 satellite imagery, Global Mangrove Watch measured the 2020 global mangrove distribution to be 147,256 km^2 (Bunting et al. 2024). Mangrove forests and coral reefs were previously forecasted to overlap in a global area of 805 km^2, using global mangrove and coral habitat extents from 1999 to 2003 (Stewart et al. 2022). Since air exposure of corals can be a limiting factor for the existence of CMC habitats, if areas with tidal regimes greater than 2 m were removed, the potential global CMC habitat extent was reduced to 710 km^2. Since the publication of the work by Stewart et al. (2022), updated global extents for both mangroves (Bunting et al. 2024) and coral reefs (Lyons et al. 2022) have been released. Repeated global mapping of CMC habitats based on these updated mangrove and coral reef extents resulted in an increased overlapping global area of 963 km^2 (Table 11.2 and Fig. 11.5). Understanding the distribution and extent of coexisting mangrove-coral habitats is crucial for several reasons. Even if we take conservative estimates of their potential overlap, these areas represent biodiversity hotspots that contribute to the health of the oceans (Guannel et al. 2016). The precise mapping and monitoring of CMC habitats using advanced satellite imagery allows for better conservation and management strategies. Given the threats posed by climate change, sea level rise, and human activities, protecting CMC habitats becomes even more urgent as their preservation not only safeguards marine biodiversity but also

Table 11.1 Coral species observed growing within coexisting mangrove-coral (CMC) habitats. Summary of coral species observed by CMC type, country, and reported mangrove species

Coral species	CMC type	Country	Mangrove species
Acropora aculeus	I	NCL	Am, Rs
Acropora aspera	I, L	NCL	Am, Rs
Acropora austera	L	SYC, IDN	Am, Bg, Lr, Rmu, Rs
Acropora carduus	I	NCL	Am, Rs
Acropora cerealis	I	NCL	Am, Rs
Acropora cervicornis	E, L	BLZ	Rma
Acropora cytherea	L	AUS	Rmu
Acropora digitifera	I	AUS	Rs
Acropora divaricata	I	NCL	Am, Rs
Acropora echinata	I	NCL	Am, Rs
Acropora gemmifera	L	SYC	Am, Bg, Lr, Rmu
Acropora horrida	I	AUS, NCL	Am, Rs
Acropora humilis	I, L	AUS, NCL	Am, Rmu, Rs
Acropora intermedia	L	AUS	Rmu
Acropora kirstyae	I, L	NCL	Am, Rs
Acropora microphthalma	I, L	NCL	Am, Rs
Acropora millepora	L	AUS	Rmu
Acropora muricata	I, L	AUS, NCL, SYC, IDN	Am, Bg, Lr, Rmu, Rs
Acropora nana	I	NCL	Am, Rs
Acropora palmata	E	VIR	Rma
Acropora prolifera	I	BHS	Rma
Acropora pulchra	I, L	AUS, NCL	Am, Rmu, Rs
Acropora robusta	L	AUS	Rmu
Acropora rudis	L	IDN	
Acropora samoensis	I	NCL	Am, Rs
Acropora tenuis	I, L	AUS, NCL	Am, Rmu, Rs
Acropora valenciennesi	I	AUS	Rs
Acropora valida	I, L	AUS, NCL	Am, Rmu, Rs
Acropora vaughani	I, L	NCL	Am, Rs
Acropora verweyi	I	AUS	Rs
Agaricia agaricites	C, E	CUB, PAN, PRI, VIR	Rma
Agaricia fragilis	C, E	PAN	Rma
Agaricia grahamae	E	VIR	Rma
Agaricia humilis	C	PAN	Rma
Agaricia lamarcki	C, E	PAN, VIR	Rma
Agaricia tenuifolia	C, E, L	BLZ, PAN	Rma
Anacropora forbesi	I	NCL	Am, Rs
Anacropora matthaii	I	NCL	Am, Rs
Astrea curta	I	AUS	Rs
Astreopora gracilis	L	AUS	Rmu
Cantharellus noumeae	I	NCL	Am, Rs
Cladocora arbuscula	E	CUB, VIR	Rma
Coelastrea aspera	I, L	AUS, NCL	Am, Rmu, Rs
Coeloseris mayeri	L	IDN	
Colpophyllia natans	C, E	PAN, VIR	Rma

(continued)

Table 11.1 (continued)

Coral species	CMC type	Country	Mangrove species
Ctenactis echinata	I	NCL	Am, Rs
Cyphastrea chalcidicum	I	AUS	Rs
Cyphastrea microphthalma	I	AUS	Rs
Cyphastrea serailia	I	AUS	Rs
Cyphastrea spp.	I, L	AUS, NCL, IDN	Am, Rmu, Rs
Danafungia horrida	L	IDN	Rs
Danafungia scruposa	L	IDN	
Dendrogyra cylindrus	E	VIR	Rma
Dichocoenia stokesii	C, E, I	BLZ, CUB, PAN, VIR	Rma
Diploria labyrinthiformis	C, E	PAN, VIR	Rma
Dipsastraea lizardensis	I	NCL	Am, Rs
Dipsastraea pallida	I	AUS, NCL	Am, Rs
Dipsastraea rotumana	I, L	AUS, IDN	Rs
Dipsastraea speciosa	I, L	AUS, IDN	Rs
Dipsastraea spp.	L	AUS, IDN	Rmu
Echinophyllia aspera	I	AUS, NCL	Am, Rs
Echinopora lamellosa	I	NCL	Am, Rs
Echinopora spp.	L	IDN	
Euphyllia cristata	L	NCL	
Euphyllia glabrescens	L	IDN	
Eusmilia fastigiata	C, E	PAN, VIR	Rma
Favia fragum	C, E, I	BLZ, CUB, USA, PAN, PRI, VIR	
Favia spp.		IND	
Favites abdita	I	AUS, NCL	Am, Rs
Favites melicerum	I	NCL	Am, Rs
Favites spp.	L	AUS, IDN	Rmu
Fimbriaphyllia divisa	I	AUS	
Fungia fungites	I, L	AUS, NCL, IDN	Am, Rmu, Rs
Galaxea astreata	I, L	AUS, IDN	Rs
Galaxea fascicularis	I, L	NCL	Am, Rs
Gardineroseris planulata	L	IDN	
Goniastrea edwardsi	I, L	AUS, IDN	Rs
Goniastrea favulus	I, L	AUS, NCL	Am, Rs
Goniastrea pectinata	I	AUS	
Goniastrea retiformis	I	AUS	Rs
Goniastrea spp.	L	IDN	
Goniopora palmensis	I	AUS	Rs
Goniopora pedunculata	I	NCL	Am, Rs
Halomitra pileus	I	NCL	Am, Rs
Heliofungia actiniformis	I, L	AUS, NCL, IDN	Am, Rs
Heliopora spp.	L	AUS	Rmu
Helioseris cucullata	C	PAN	Rma
Herpolitha limax	I	AUS	Rs
Hydnophora rigida	I, L	NCL, IDN	Am, Rs
Isophyllia rigida	C	PAN	Rma
Isophyllia sinuosa	C	PAN	Rma

(continued)

Table 11.1 (continued)

Coral species	CMC type	Country	Mangrove species
Isopora brueggemanni	I	AUS	Rs
Leptastrea purpurea	I	NCL	Am, Rs
Leptastrea transversa	I	AUS	Rmu, Rs
Leptastrea spp.	L	IDN	
Leptoria phrygia	L	IDN	
Leptoseris scabra	I	AUS	Rs
Lithophyllon edwardsi	I	AUS	Rs
Lithophyllon repanda	I	AUS	Rs
Lobophyllia corymbosa	I, L	AUS	Rmu, Rs
Lobophyllia hemprichii	I, L	AUS, NCL	Am, Rmu, Rs
Lobophyllia radians	I	AUS	Rs
Lobophyllia recta	I	AUS	Rs
Lobophyllia spp.	L	IDN	
Madracis auretenra	I	BLZ	Rma
Manicina areolata	E, I, L	BLZ, CUB, PAN, VIR	Rma
Meandrina meandrites	E	VIR	Rma
Merulina ampliata	I	NCL	Am, Rs
Merulina scabricula	I	NCL	Am, Rs
Millepora alcicornis	C, E, L	BLZ, CUB, PAN, VIR	Rma
Millepora complanata	C, E, I, L	BLZ, PAN	Rma
Millepora tenera	L	IDN	
Millepora spp.	E, L	AUS, BLZ, PRI, VIR	Rma, Rmu
Montastraea cavernosa	C, E, I	CUB, PAN, VIR	Rma
Montipora aequituberculata	I	AUS	Rs
Montipora cactus	I	NCL	Am, Rs
Montipora crassituberculata	I	AUS	Rs
Montipora digitata	I, L	AUS, NCL	Am, Rmu, Rs
Montipora hispida	I	NCL	Am, Rs
Montipora spongodes	I	AUS	Rs
Montipora stellata	I	NCL	Am, Rs
Montipora turgescens	I	AUS	Rs
Montipora turtlensis	L	AUS	Rmu
Montipora spp.	E, L	AUS, IDN	
Mussa angulosa	C, E	PAN	Rma
Mycetophyllia aliciae	C, E	PAN, VIR	Rma
Mycetophyllia spp.	I	BLZ	Rma
Oculina diffusa	E	USA, VIR	Rma
Orbicella annularis	E, L	BLZ, PAN, VIR	Rma
Orbicella faveolata	C, E, I	BLZ, PAN, VIR	Rma
Orbicella franksi	E	PAN, VIR	Rma
Oulophyllia crispa	L	IDN	
Oulophyllia spp.	L	AUS	Rmu
Pachyseris rugosa	I	NCL	Am, Rs

(continued)

Table 11.1 (continued)

Coral species	CMC type	Country	Mangrove species
Pachyseris speciosa	I, L	NCL, IDN	Am, Rs
Pavona cactus	I, L	NCL, IDN	Am, Rs
Pavona decussata	I, L	NCL	Am, Rs
Pavona duerdeni	L	AUS	Rmu
Pavona varians	L	IDN	Rs
Pectinia paeonia	I, L	NCL, IDN	Am, Rs
Phyllangia americana	C, E	CUB, PAN, VIR	Rma
Platygyra daedalea	I	AUS, NCL	Am, Rs
Platygyra lamellina	I	AUS	Rs
Platygyra pini	I	AUS	Rs
Platygyra sinensis	I, L	AUS, NCL	Am, Rmu, Rs
Platygyra verweyi	I	AUS	Rs
Plerogyra sinuosa	L	IDN	
Pocillopora acuta	E, L	AUS	Rmu
Pocillopora damicornis	I, L	AUS, NCL, IDN, SYC, IDN	Am, Bg, Lr, Rmu, Rs
Pocillopora verrucosa	L	SYC	Am, Bg, Lr, Rmu
Polycyathus fulvus	I	NCL	Am, Rs
Porites annae	I	AUS, NCL	Am, Rs
Porites astreoides	C, E, I	BLZ, CUB, PAN, VIR	Rma
Porites attenuata	L	SYC, IDN	Am, Bg, Lr, Rmu, Rs
Porites colonensis	I	BLZ	Rma
Porites cylindrica	I, L	AUS, NCL, IDN	Am, Rmu, Rs
Porites divaricata	C, E, I, L	BLZ, CUB, PAN, VIR	Rma
Porites furcata	C, E, I	BLZ, CUB, PAN, VIR	Rma
Porites lobata	I, L	AUS, NCL, SYC, IDN	Am, Bg, Lr, Rmu, Rs
Porites lutea	I, L	AUS, NCL, SYC, IDN	Am, Bg, Lr, Rmu, Rs
Porites nigrescens	I	AUS	Rs
Porites porites	E, I	BLZ, COL, CUB, USA, PAN, VIR	Rma
Porites rus	L	IDN	
Psammocora contigua	I	NCL	Am, Rs
Pseudodiploria clivosa	C, E	PAN, PRI, VIR	Rma
Pseudodiploria strigosa	C, E, I	BLZ, PAN, VIR	Rma
Pseudosiderastrea tayamai	I	NCL	Am, Rs
Sandalolitha dentata	I	NCL	Am, Rs
Scolymia cubensis	C, E	PAN, VIR	Rma
Scolymia lacera	C, E	PAN, VIR	Rma
Seriatopora hystrix	L	IDN	
Seriatopora spp.	L	AUS	Rmu
Siderastrea radians	C, E, I	CUB, USA, PAN, VIR	Rma

(continued)

Table 11.1 (continued)

Coral species	CMC type	Country	Mangrove species
Siderastrea siderea	C, E, I, L	BLZ, CUB, USA, PAN, VIR	Rma
Solenastrea bournoni	E, I	USA, VIR	Rma
Stephanocoenia intersepta	C, E, I	USA, PAN, VIR	Rma
Stephanocoenia spp.	I	BLZ	Rma
Stylocoeniella guentheri	I	AUS	Rs
Stylophora pistillata	I, L	AUS, EGY, IDN	Am, Ct, Rmu
Trachyphyllia geoffroyi	I	NCL	Am, Rs
Tubastraea aurea	E	VIR	Rma
Tubastraea coccinea	E, I	NCL, VIR	Am, Rma, Rs
Tubastraea micranthus	I	NCL	Am, Rs
Turbinaria mesenterina	I, L	AUS, NCL	Am, Rmu, Rs
Turbinaria reniformis	L	AUS	Rmu

When only the coral genus was given, spp. was used. Coral names were standardized using approved nomenclature according to the World Register of Marine Species

CMC type codes are as follows: C = Canopy, E = Edge, I = Inlet, L = Lagoon. Country codes as assigned by the International Organization for Standardization. AUS = Australia, BHS = Bahamas, BLZ = Belize, COL = Colombia, CUB = Cuba, EGY = Egypt, IND = India, IDN = Indonesia, NCL = New Caledonia, PAN = Panama, PRI = Puerto Rico, SYC = Seychelles, USA = United State of America, VIR = U.S. Virgin Islands. Mangrove species codes as follows: Am = *Avicennia marina,* Bg = *Bruguiera gymnorhiza,* Ct = *Ceriops tagal,* Lr = *Lumnitzera racemosa,* Rma = *Rhizophora mangle,* Rmu = *Rhizophora mucronata,* Rs = *Rhizophora stylosa*

Table 11.2 Potential global extent of coexisting mangrove-coral and mangrove-seagrass habitats

Region	Global coral mangrove overlap (km^2)	Global coral mangrove overlap (%)	Global coral seagrass overlap (km^2)	Global coral seagrass overlap (%)
Red Sea	17.93	1.86%	3462.51	8.38%
Persian Gulf	1.77	0.18%	1558.49	3.77%
Indian Ocean	167.48	17.39%	3816.94	9.23%
South China Sea	30.77	3.20%	743.73	1.80%
North Pacific	129.97	13.50%	3222.16	7.79%
South Pacific	293.22	30.45%	6847.29	16.57%
Caribbean	321.77	33.42%	21,059.70	50.95%
North Atlantic	0.04	0.00%	626.05	1.51%
South Atlantic	0.01	0.00%	0.00	0.00%
Total	962.96	100%	41,336.87	100%

enhances the resilience of coastal regions against environmental changes and extreme weather events.

11.4 Seagrasses as Coral Habitats

Seagrass structure and ecology

Seagrasses are a group of near globally distributed marine flowering plants that contain approximately 60 different species across 11 genera and 4 families (Green and Short 2003; den Hartog and Kuo 2006) living in both tropical and temperate ecosystems along all continents except Antarctica (Kennedy and Björk 2009). Short et al. (2001) provide a comprehensive list of species distribution across 10 temperate and tropical global regions. Global spatial distribution of seagrass has been estimated in past studies ranging from 177,000 to 600,000 km^2 (Duarte et al. 2010; Green and Short 2003) in 191 countries across six global bioregions including temperate and tropical seas (Short et al. 2007). However, more recent studies indicate global seagrass area ranges between 160,387 and 266,562 km^2 based on moderate to high and lower confidence estimates, respectively, across 103–136 countries and territories (McKenzie et al. 2020). Based on moderate to high confidence estimates of seagrass distribution, McKenzie et al. (2020) indicate that tropical bioregions account for approximately 91% of the total global seagrass area, with the Tropical Indo-Pacific bioregion accounting for 87,791 km^2 (55%), followed by the Tropical Atlantic with 44,222 km^2 (27%), and the Mediterranean with 14,167 km^2 (9%). Comparatively, the global distribution of coral reefs covers approximately 284,803 km^2 (McKenzie et al. 2020).

The seagrass plants contain roots, known as rhizomes, which grow downward and anchor into the muddy or sandy substrate, stabilizing it, as well as long, narrow leaves that extend into the water column. Living primarily in shallow coastal waters (e.g., estuaries, lagoons, along open coasts) across both temperate and tropical ecosystems, seagrasses build extensive meadows composed of one or multiple species (Kennedy and Björk 2009; Jayathilake and Costello 2018). These meadows provide essential habitat for a diversity of marine invertebrate and vertebrate species including commercially and recreationally important fish species (Nordlund et al. 2018). In addition to their roles in supporting marine biodiversity, seagrass meadows provide a host of ecosystem services. The seagrass canopy reduces wave attenuation while the dense roots promote sediment deposition and mitigate erosion (Lima et al. 2023). Seagrass meadows also improve water quality by filtering pollutants and trapping and cycling nutrients from the water column (de Boer 2007). Seagrasses also mitigate climate change through the removal and sequestration of carbon dioxide from the atmosphere (Fourqurean et al. 2012).

Seagrasses rely on specific environmental conditions to support their growth and survival. These include ample light necessary for photosynthesis, adequate nutrient levels, and calm waters to facilitate settlement and establishment. Seagrasses grow in marine, estuarine, and brackish waters where salinity is typically above 5 ppt and underwater irradiance is more than 10% of that at the water's surface (Duarte 1991; Hemminga and Duarte 2000). Their depth distribution is limited by irradiance needed for photosynthesis and their requirement for immersion in seawater (Hemminga and Duarte 2000; Duarte 2002). Thus, seagrass meadows are typically found in subtidal and intertidal environments between depths of 0–30 m along all the continents except Antarctica (den Hartog 1970). Other physical and environmental variables that affect seagrass distribution include temperature, waves, currents, seafloor substrate, day length, nutrient availability, and either natural or anthropogenic alteration of these variables (Short et al. 2001). Variations in these physical and environmental conditions at local and regional scales regulate physiological activity of seagrasses, including photosynthetic activity, and/or availability of resources and seafloor substrate for growth (Short et al. 2001).

Seagrass meadows are among the most productive ecosystems globally and act as significant carbon sinks (Duarte et al. 2010). Despite supporting abundant heterotrophic organisms (Hemminga and Duarte 2000), they are typically net autotrophic due to slow tissue decomposition and root preservation in sediments (Duarte et al. 2005, 2010). Duarte et al. (2010) analyzed 403 seagrass metabolism estimates from 155 sites, finding gross primary production (GPP) exceeded respiration, resulting in positive net community production (NCP) and P/R ratios above 1, with 63% of sites being net autotrophic. They estimated global NCP at 20.73–1010.39 Tg C yr-1 based on seagrass extent of 300,000–600,000 km^2. However, updated global seagrass distribution estimates of 160,387–266,562 km^2 (McKenzie et al. 2020) suggest that global NCP may be significantly lower.

Coexisting seagrass-coral habitats

Coral communities have been reported living in seagrass meadows in the Atlantic, including the Florida Keys (Manzello et al. 2012; Landau et al. 2013), the Caribbean, including the Cayman Islands (Camp et al. 2016a, b; Lohr et al. 2017), Mexico (Ruiz-Zárate et al. 2000), Belize (Rützler and Macintyre 1982; Wulff 2008), Puerto Rico (Uhrin et al. 2005), Panama (Johnson 1992), South China Sea, including Taiwan (Pezner et al. 2024), as well as the Indo-Pacific, including Japan (Nanami and Yamada 2009; Higuchi et al. 2014; Meekaew et al. 2014), the Seychelles Archipelago (Camp et al. 2016a), and Indonesia (Wakatobi, Southeast Sulawesi; Camp et al. 2016a). These

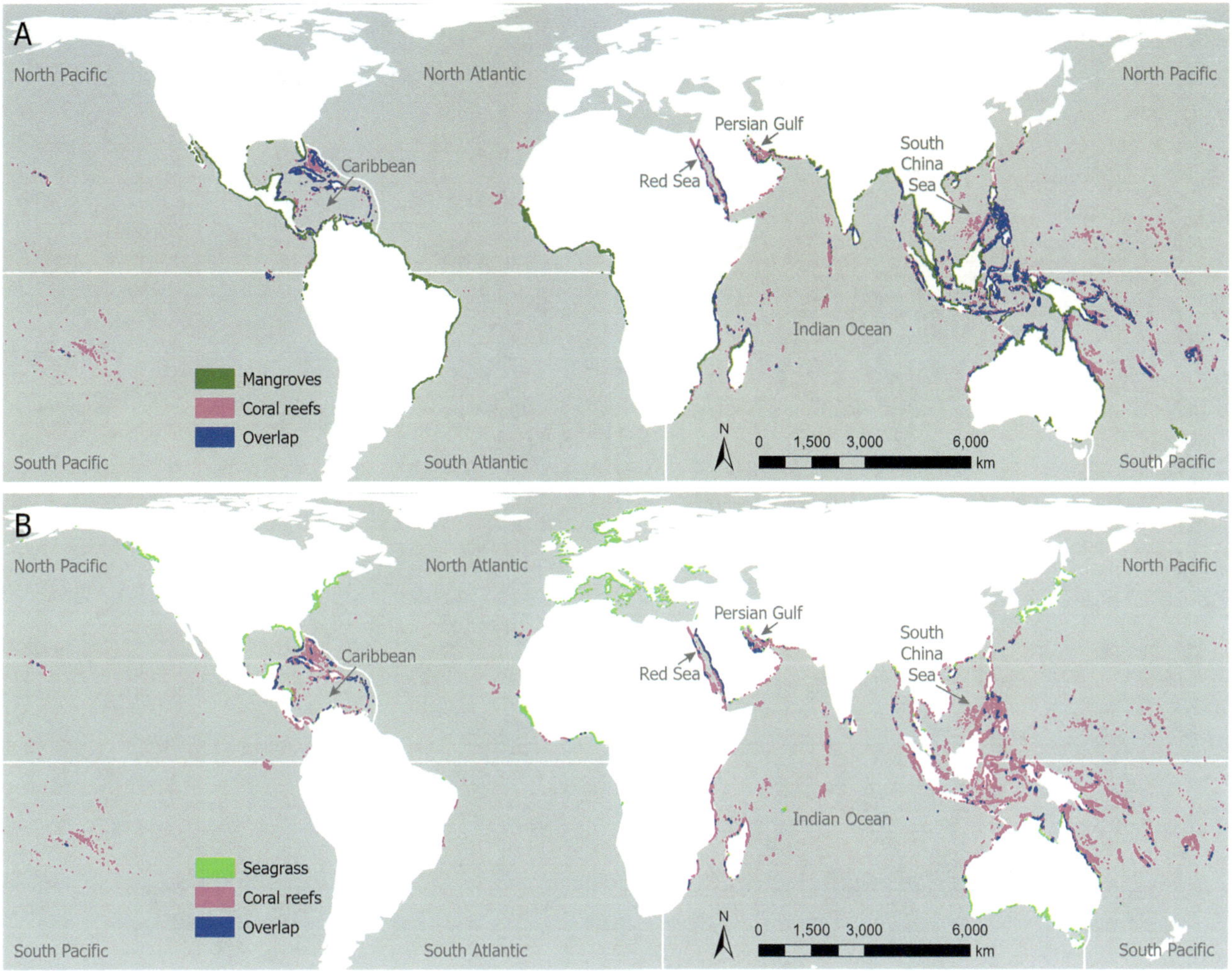

Fig. 11.5 Global distribution of coral reefs (pink) and (**a**) mangroves (dark green) and (**b**) seagrass (light green), showing overlap of their extents (blue)

communities have been described as dominated by one or multiple species of seagrasses, and may include other species of photoautotrophs, including both calcifying and noncalcifying algae (Lohr et al. 2017; Ruiz-Zárate et al. 2000) as well as other seagrass associated marine invertebrates (e.g., sponges, echinoderms, ascidians, etc.) (Wulff 2008; Landau et al. 2013; Manikandan et al. 2011) (Fig. 11.6).

Studies characterizing seagrass-coral habitats have described sites based on their position (e.g., inshore vs lagoon, mangrove vs. reef adjacent), depth, environmental conditions, as well as benthic composition. All these characteristics may help shape coral community structure within these habitats. At least 39 species of corals are known to live in seagrass meadows comprised of *Oceana serrulata* (R.Br.) Byng & Christenh., *Halodule wrightii* Asch., *Syringodium filiforme* Kütz., *Thalassia hemprichii* (Ehrenb. ex Solms) Asch., and/or *Thalassia testudinum* Banks & Sol. ex K.D.Koenig (Table 11.3). The most described coral genera in decreasing order are *Porites* (n = 19), *Acropora* (n = 9), *Siderastrea* (n = 8), and *Manicina* (n = 6), which are also among the most frequently reported coral genera to grow in mangrove habitats.

Seagrass-coral habitats have been reported in leeward inshore locations (Manzello et al. 2012; Thanopoulou et al. 2022) and reef lagoons (Ruiz-Zárate et al. 2000; Rützler and Macintyre 1982). Some studies have also described these communities as mangrove-adjacent (Johnson 1992) or reef-adjacent (Camp et al. 2016a; Lohr et al. 2017). The extent to which the seagrass habitat's position impacts the corresponding coral community and population structure within the site is not well known and merits future work. Presumably, such placement may drive differences in substrate type, as well as environmental parameters including average and diel temperature, pH, salinity, etc. The location of seagrass-coral habitats near reefs or mangroves can affect coral diversity, as these habitats support distinct coral communities. Dispersal between habitat types may influence the species present,

Fig. 11.6 Various coral species growing in seagrass beds on Turneffe Atoll, Belize, including (**a**) *Dichocoenia stokesii*, (**b**) a mixed assemblage of *Siderastrea siderea*, *Porites astreoides*, and branching *Porites* species, (**c**) *Orbicella annularis* and *S. siderea* and (**d**) *S. siderea*. (Photographs taken by K. Novelo and J.R. Finnerty)

driving differences between mangrove- and reef-adjacent seagrass-coral habitats. In addition, mangroves and reefs interact with seagrass beds in different ways and can provide different biogeochemical services, which may drive differences in coral community composition and structure.

The depths of these habitats have all been reported to be relatively shallow (~2–4 m) (Johnson 1992; Ruiz-Zárate et al. 2000; Camp et al. 2016a; Lohr et al. 2017), with the shallowest seagrass-coral habitat reported as 0.5–1.5 m deep (at low tide) (Landau et al. 2013), and the deepest at 4–5 m (Manzello et al. 2012). Diel tidal cycles within sites are not often documented (except for Camp et al. 2016a who reported a diel range of 1.8 ± 0.3 m within their study sites) but may be important in explaining daily fluctuations of other environmental parameters (Camp et al. 2016a).

Studies examining coral communities within seagrass beds have extensively analyzed associated environmental parameters (Camp et al. 2016a, b, Lohr et al. 2017). These include temperature, light, photosynthetic active radiation, nutrient levels (e.g., NO_3^-), salinity, pH, total alkalinity, dissolved oxygen (DO), and various measures of seawater carbonate chemistry (e.g., $p\mathrm{CO_2}$, TCO_2, Ω_{arg} - aragonite saturation state). These parameters often differ from those in typical reef environments, with seagrass beds exhibiting greater variability. Studies comparing temperatures between seagrass sites and reefs have yielded mixed results; some show higher mean temperatures in seagrass beds, while others find no significant difference, although temperature variability is generally higher in seagrass sites. For instance, Lohr et al. (2017) reported higher mean temperatures in seagrass beds compared to outer reefs, but no difference com-

Table 11.3 Coral species observed growing within coexisting seagrass-coral habitats. Summary of coral species observed by country and seagrass species when reported

Coral species	Country	Seagrass species
Acropora austera	SYC, IDN	Th
Acropora cervicornis	BLZ	Sf, Tt
Acropora gemmifera	SYC, IDN	
Acropora muricata	SYC, IDN	
Acropora palmata	CYM	Sf, Tt
Acropora spp.	TWN	Cr, Cs
Agaricia agaricites	CYM	
Agaricia humilis	CYM	
Cladocora arbuscula	USA	
Danafungia horrida	IDN	
Dichocoenia stokesii	CYM, USA	Sf, Tt
Diploria labrinthiformis	BLZ	
Faviidae spp.	JPN	
Favites abdita	IDN	
Galaxea cryptoramosa	IDN	
Goniastrea edwardsi	IDN	
Goniastrea pectinata	SYC	
Isophyllia rigida	CYM	
Lobophyllia hataii	IDN	
Manicina areolata	CYM, USA, MEX, PAN, PRI	Hw, Sf, Tt
Meandrina meandrites	USA	
Millepora alcicornis	BLZ, CYM	
Millepora spp.	SYC	
Monitopora digitata	JPN	Th
Orbicella annularis	BLZ, CYM	
Pavona varians	IDN	
Pocillopora damicornis	SYC, IDN	
Porites astreoides	CYM	
Porites attenuata	SYC, IDN	
Porites divaricata	CYM, USA	Tt
Porites furcata	BLZ, CYM	Tt
Porites lobata	IDN	
Porites lutea	SYC, IDN	
Porites porites	CYM	
Porites spp.	JPN, TWN	Cr, Cs, Os, Th, Tt
Pseudodiploria clivosa	CYM	
Pseudodiploria strigosa	BLZ, CYM	
Scolymia lacera	CYM	
Siderastrea radians	BLZ, CYM, USA	
Siderastrea siderea	BLZ, CYM	Tt
Solenastrea bournoni	CYM, USA	Tt
Solenastrea hyades	USA	
Stephanocoenia intersepta	CYM	

When only the coral family or genus was given, spp. is used. Coral names standardized using approved nomenclature according to the World Register of Marine Species. Country codes as assigned by the International Organization for Standardization. BLZ = Belize, CYM = Cayman Islands, IDN = Indonesia, JPN = Japan, MEX = Mexico, PAN = Panama, PRI = Puerto Rico, SYC = Seychelles, TWN = Taiwan, USA = United State of America. Seagrass species codes as follows: Cr = *Cymodocea rotundata*, Cs = *Cymodocea serrulata*, Hw = *Halodule wrightii*, Os = *Oceana serrulata*, Sf = *Syringodium filiforme*, Th = *Thalassia hemprichii*, Tt = *Thalassia testudinum*

pared to back-reef sites. Similarly, Camp et al. (2016a) found no significant mean temperature difference between seagrass-coral and outer-reef sites, yet observed significantly higher temperature variability in seagrass-coral sites (Atlantic Ocean: 29.1 ± 0.11, Indian Ocean: 30.5 ± 0.11, Pacific Ocean: 27.4 ± 0.05). Both studies reported diel temperature fluctuations, reaching up to 6.7 °C in seagrass-coral sites (Lohr et al. 2017).

Seagrass beds also exhibit significantly greater pH variability compared to adjacent reef environments (Camp et al. 2016a; Lohr et al. 2017). Lohr et al. (2017) and Barry et al. (2013) reported diel pH fluctuations of 0.51–0.62 units and 0.47 units, respectively, in seagrass beds in the Cayman Islands. Furthermore, seagrass beds may maintain an overall elevated pH relative to offshore reef environments (Camp et al. 2016a; Lohr et al. 2017). This variability in pH is primarily driven by seagrass photosynthesis and respiration, influencing changes in pCO_2 levels within the bed (Semesi et al. 2009; Saderne et al. 2013; Challener et al. 2016; Lohr et al. 2017). These daily pCO_2 fluctuations impact Ω_{arg} (aragonite saturation state), which in turn affects coral calcification rates. The benefits of these natural processes within seagrass beds to co-occurring corals are discussed in subsequent sections below.

Surveys of benthic composition within these habitats have included measures of the percent cover of abiotic hard- and soft-bottom substrate, seagrasses, and noncalcifying and calcifying algae (Camp et al. 2016a; Lohr et al. 2017). Abiotic substrate type has been identified as an important characteristic of seagrass-coral habitats given that the availability of hard substrate is critical for structuring benthic communities, particularly corals which require hard-bottomed substrate for successful recruitment and subsequent survival (Jackson 1977; Taylor and Wilson 2003). Soft-bottomed substrate is considered a poor habitat for corals (Manzello et al. 2012; Jackson 1986), as it may result in the smothering and/or burial of colonies, leading to mortality. Hard-bottomed substrate is typically limited in seagrass-coral sites; however, some seagrass sites have more hard-bottomed substrate than others (Ruiz-Zárate et al. 2000; Lohr et al. 2017; Camp et al. 2016a). Hard-bottomed substrate that may occur in seagrass beds include carbonate reefs (Johnson 1992), while soft-bottom substrate includes sand (different grain sizes) and silt.

Benthic surveys of seagrass-coral habitats may also contain estimates of seagrass density or biomass within the site. Coral communities have been reported in seagrass beds with high (Camp et al. 2016a), medium (Ruiz-Zárate et al. 2000), and low (Camp et al. 2016a) overall seagrass density and/or biomass despite definitions of these densities not being universally established. Ruiz-Zárate et al. (2000) defined seagrass beds with 542 shoots m^{-2} as high density, 239 shoots m^{-2} as medium density, and 35 shoots m^{-2} as low density. Within coexisting seagrass-coral habitats, seagrass density may impact the distribution and/or presence or absence of coral colonies (Ruiz-Zárate et al. 2000; Landau et al. 2013). For example, Ruiz-Zárate et al. (2000) observed that colonies of the stony coral, *Manicina areolata* (Linnaeus, 1758), were only naturally distributed within medium-density seagrass beds in the Mexican Caribbean. Upon transplantation to high-density beds, these colonies exhibited high stress likely due to low light conditions, as indicated by lowered zooxanthellae density (Ruiz-Zárate et al. 2000). Transplantation to low-density beds resulted in high mortality from predation and siltation (Ruiz-Zárate et al. 2000). However, seagrass density may not impact all coral species. Landau et al. (2013) observed no significant relationship between seagrass density and *Cladocora arbuscula* (Le Sueur, 1820) and *Porites* spp. More research is needed to understand the impacts of seagrass density on coral community structure. Research on coexisting seagrass-coral habitats has concentrated on the abundance, diversity, population structure, and distribution of coral communities within seagrass beds. Overall low mean coral cover has been reported in seagrass beds across geographic regions. Reported estimates of percent coral cover range from 2.7 ± 0.4% to 3.1% in seagrass meadows in the Atlantic (Lohr et al. 2017, Camp et al. 2016a), 4.2% in the Indian Ocean (Camp et al. 2016a), and 5.4% in the Pacific Ocean (Camp et al. 2016a). Adjacent back-reef environments had twice the mean cover compared to seagrass beds (6.5 ± 1.2% vs. 2.7 ± 0.4%, respectively) at Little Cayman, Cayman Islands (Lohr et al. 2017). Lower overall coral diversity has also been reported in seagrass beds compared to adjacent reefs across geographic regions (Camp et al. 2016a; Lohr et al. 2017). For example, Lohr et al. (2017) observed 15 species of coral in the seagrass beds at Little Cayman, Cayman Islands, while 19 species were found in the adjacent back-reef.

Seagrass beds host coral species from three (out of four) life history strategies: competitive, weedy, and stress-tolerant (Darling et al. 2012). In the Caribbean, common species include *M. areolata*, *Siderastrea radians* (Pallas, 1766), *Porites* spp., *Orbicella* spp., *Pseudodiploria* spp., and *S. siderea* with many belonging to the weedy or stress-tolerant guilds (Lohr et al. 2017). Seagrass-coral communities in Little Cayman consist largely of Poritidae (49%) and Siderastreidae (50%) (Lohr et al. 2017). Indo-Pacific seagrass beds also host competitive (*Acropora* spp.), stress-tolerant (*Pavona varians* (Verrill, 1864)), and weedy (*Pocillopora damicornis* (Linnaeus, 1758)) corals, along with unclassified species (Darling et al. 2012; Camp et al. 2016a). In the Atlantic, seagrass corals represent a larger proportion of regional species (20–30%) compared to the Indo-Pacific (1–2%), possibly due to differences in species pools, environmental histories, or evolutionary factors (Camp et al. 2016a).

Percent coral cover and species diversity in seagrass beds are likely a reflection of coral species' tolerance to limited hard-bottomed substrates and highly variable environmental physiochemical conditions (e.g., temperature, pH) (Camp et al. 2016a; Lohr et al. 2017). Some predominant species in seagrass beds may be tolerant to sedimentation through behavioral or morphological adaptations. Predominant species, such as free-living *M. areolata* possess several behavioral strategies to avoid burial by sedimentation, including the ability to right itself if overturned (Johnson 1992). The Caribbean coral *S. radians* is effective at clearing sediment and able to withstand short-term burial periods with minor physiological consequences (Lirman and Manzello 2009). Similarly, some members of the genus *Porites* in both the Caribbean and Pacific are also considered tolerant to sediment stress (Rogers 1990; Yeemin et al. 2013). Mechanisms of sediment clearing may include behaviors such as tissue swelling, mucus production, tentacular and ciliary action, and/or be controlled by colony shape (Hubbard and Pocock 1972; Stafford-Smith and Ormond 1992; Lirman and Manzello 2009). Corals may capitalize on existing hard substrate, as Ruiz-Zárate et al. (2000) observed in Mexican Caribbean seagrass sites where 92% of *M. areolata* recruits were found attached to the calcareous algae, *Neogoniolithon* spp. Recruitment on *Neogoniolithon* spp. was thought to prevent the smothering of recruits by sediment and may only be important at early life stages after which the colony becomes separated from the alga after reaching a certain size and weight (Ruiz-Zárate et al. 2000). Successful recruitment by weedy species is attributed to their brooding reproductive mode, which is presumed to be predominantly local (Harrison and Wallace 1990) and could thereby increase local population abundance. Weedy corals may also capitalize on other asexual reproductive methods to colonize habitats, such as fragmentation and parthenogenesis, which may allow for rapid population expansion at low population densities (Darling et al. 2012). Successful coral recruitment in seagrass beds may not entirely depend on substrate types, as coral recruitment has also been linked to many other factors, including spatial and temporal trends in coral fecundity, distance to nearby conspecific populations, and presence of other species that facilitate recruitment (e.g., crustose coralline alga, microbial films) (Lohr et al. 2017).

Both suboptimal and highly variable physiochemical conditions are thought to contribute to the lower coral cover and diversity observed within seagrass beds. Due to seagrass respiration and photosynthesis, seagrass beds experience diel pH cycles with low pH levels at night. These low pH levels can create unfavorable conditions for some coral species (Camp et al. 2016a; Lohr et al. 2017). For example, for *A. palmata*, low pH may inhibit larval settlement processes (Albright et al. 2010), which can exclude this species from some seagrass meadows (Lohr et al. 2017). Other limitations could include modifications to coral growth and calcification. Additionally, significant physiochemical fluctuations increase the energetic demands required for physiological maintenance (Hofmann and Todgham 2010). Weedy and stress-tolerant coral species that thrive in seagrass beds are known to colonize and survive in disturbed or harsh environments. These species exhibit traits that enable them to withstand such conditions. For instance, stress-tolerant species typically have slow growth rates, long generation times, and large corallites, which facilitate energy storage (Darling et al. 2012). These energy reserves become critical under stressful conditions. Weedy species are generally smaller in size and show the greatest variation in species traits, which may allow them to quickly acclimatize or adapt to fluctuating environments (Darling et al. 2012).

Benefits of seagrass habitat for corals

Understanding the benefits of seagrass habitats for co-occurring corals has become an essential area of study amid unprecedented environmental changes. Research has shown that seagrass habitats may offer several critical advantages for corals including, but not limited to, counteracting ocean acidification locally, subsidizing coral metabolic needs by supplying nutrients, and providing shade during periods of high light stress. These advantages may be crucial for coral resistance and survival and could play a significant role in enhancing coral evolutionary potential.

One of the most prominent fields of study in this domain focuses on understanding the capacity of seagrasses to buffer co-occurring corals from decreasing pH levels due to ocean acidification. Ocean acidification is caused by the addition of excess anthropogenic carbon dioxide in seawater, which shifts the balance of dissolved inorganic carbon (DIC) and leads to lower levels of carbonate ions (CO_3^{2-}) (Unsworth et al. 2012). This reduced concentration of carbonate ions (or carbonate saturation) can negatively impact Scleractinian corals, which rely on these minerals for skeletal construction. Through the process of photosynthesis, seagrasses may help mitigate ocean acidification by absorbing DIC, and thereby maintaining or elevating carbonate saturation levels in the surrounding environment. The elevation of carbonate saturation and corresponding pH during daylight (active photosynthetic periods) is reversed at night due to seagrass respiration, which increases DIC and reduces pH and carbonate saturation (Camp et al. 2016a). The interplay of these biological processes explains the observed diel fluctuations in pH within seagrass beds. Seagrasses may also counteract the effects of ocean acidification by initiating the dissolution of calcareous sediments in the benthos (Manzello et al. 2012). In this process, oxygen produced during photosynthesis is transported to their roots, where it drives aerobic respiration. This activity releases CO_2 into the sediment pore water, where it reacts with water to dissolve calcium carbonate minerals, thereby increasing the total alkalinity (TA) of the surrounding seawater. TA can help reduce the partial pressure of CO_2 (pCO_2) in seawater, potentially mitigating ocean acidification. Higher seagrass density and photosynthetic productivity further enhance this process (Manzello et al. 2012). Thus, the anticipated increase in seagrass productivity due to ocean acidification could create a positive feedback loop, enhancing their role as refugia (Palacios and Zimmerman 2007).

Seagrass habitats have been found to have elevated local mean pH compared to adjacent habitats in the Caribbean (Manzello et al. 2012; Camp et al. 2016a), Mediterranean (Hendriks et al. 2014), and Indo-Pacific (Anthony et al. 2013; Camp et al. 2016a). However, low mean pH levels that are considered stressful for corals have been observed in seagrass beds at night (Camp et al. 2016a; Lohr et al. 2017) and may lead to reduced coral calcification. Comparisons of coral calcification rates within seagrass beds and adjacent outer reefs revealed mixed results (Camp et al. 2016a). Some species maintained their calcification rates within seagrass beds, while more opportunistic species known to frequent these habitats, such as *P. astreoides*, *S. radians*, and *P. attenuata,* exhibited an increase in calcification rates (1–3%) compared to their outer reef counterparts (Camp et al. 2016a). Conversely, calcification rates in *Acropora* spp. were substantially reduced in seagrass beds (68%) (Camp et al. 2016b). The variation in these responses suggests that the capacity of seagrass beds to serve as refugia from ocean acidification may be species-specific (Camp et al. 2016a); however, more research is needed to understand these responses across species and sites. It is also important to note that the extent to which seagrasses alter seawater carbonate chemistry may be species-specific and influenced by additional factors such as water residence time and depth (Unsworth et al. 2012; Lohr et al. 2017).

Apart from serving as a refugia for corals, the highly variable physiochemical conditions within seagrass beds may enhance tolerance to thermal and pH stress (Camp et al. 2016b). Coral exposure to intermittent periods of stress can act as a form of preconditioning, potentially enhancing the corals' physiological mechanisms to cope with and adapt to extreme thermal and pH variations (Guadayol et al. 2014). This adaptive response might make co-occurring seagrass corals more resilient to the chronic stressors associated with climate change and ocean acidification and improve their chances of survival in increasingly harsh marine environments. The potential for enhanced tolerance to thermal and chemical stress due to exposure to physiochemical fluctuations in seagrass beds was investigated in seagrass colonies of *P. astreoides* and *A. palmata* (Camp et al. 2016b). However, no enhanced resistance was observed when exposed to pH and temperature conditions predicted for the year 2100 (Camp et al. 2016b). The capacity for seagrass beds to enhance coral resistance and/or resilience merits much further investigation.

Regarding nutrient dynamics, investigations into the benefits of these co-occurring communities have revealed that seagrasses may derive more benefits from co-inhabitation with corals than corals themselves (Higuchi et al. 2014). In chamber experiments, the NO_x uptake rate was higher in coral-seagrass habitats than in coral reef or seagrass beds alone (Higuchi et al. 2014). Because nutrients are not limited in this system, this suggests that seagrasses may benefit more from co-existing with corals because they can uptake more nitrogen than in seagrass-only habitats (Higuchi et al. 2014). Similarly, seagrass exhibited greater uptake of ammonium in coral-seagrass habitats compared to seagrass beds alone (Higuchi et al. 2014). Within these habitats, corals release ammonium due to sufficient nitrate levels, thereby enhancing its availability for seagrasses (Higuchi et al. 2014).

Co-occurring corals in seagrass beds may also benefit from the organic matter and nutrients provided by seagrasses, which can potentially enhance their growth and reproduction (Lai et al. 2013). Corals inhabiting coral reef habitats are known to incorporate the nutrients provided by adjacent habitats, like seagrass beds and mangroves (Granek et al. 2009). Lai et al. (2013) observed that the coral *Oulastrea crispata* (Lamarck, 1816) had ingested particles of the seagrass *Halophila ovalis* (R.Br.) Hook.f., providing evidence that corals may assimilate nitrogen-based nutrients by feeding directly on seagrass particulate matter and/or absorbing leached dissolved organic matter. The extent to which this occurs among different coral species is unknown and requires further investigation.

Additional benefits of the synergistic relationship between seagrass and corals have been documented. Higuchi et al. (2014) suggested that reduced carbonate saturation at seagrass roots and rhizomes may facilitate coral fragmentation after observing the fragility of *Montipora digitata* (Dana, 1846) bases compared to other parts of the colony. Propagated branch fragments may be trapped by the surrounding seagrasses (Landau et al. 2013) increasing the local population density of the coral and further coral habitat development. Dense seagrass canopies may also provide shade for small or fragmented coral colonies, which may protect them from photooxidative stress leading to bleaching (Higuchi et al. 2014). Similar trends were observed in understory corals living beneath canopy forming macrophyte (Jompa and McCook 2003). Ninomiya et al. (2006) suggested a physically mutualistic relationship between co-occurring corals and seagrass whereby seagrass bed stability is enhanced by the interlacing of seagrass stems and coral branches both on and under the surface of the benthos.

11.5 Coral-Mangrove-Seagrass Interactions

Ecosystem interdependencies of coral, mangrove, and seagrass

Coral reefs, mangroves, and seagrasses coexist and interact within tropical regions worldwide (Carlson et al. 2021). Regions with particularly strong intersections of these three ecosystems include Central America (notably Belize), the Caribbean, the Red Sea, the Coral Triangle (especially Malaysia), Madagascar, and the Great Barrier reef (Carlson et al. 2021). These ecosystems inhabit a dynamic land-sea transition zone, facilitating biological, physical, and biogeochemical mutualistic relationships (Carlson et al. 2021). The nature and strength of these interactions have been widely explored and may be spatially heterogeneous particularly across geographic regions.

Among these interactions, mangroves and seagrasses serve as vital nurseries and shelters for juvenile coral reef fish species (Laegdsgaard and Johnson 1995; Whitfield 2017; Rogers and Mumby 2019). Research has shown that these habitats support higher proportions of juvenile fish, highlighting their critical role in the early life stages of many fish species (Nagelkerken et al. 2001). The higher proportions of juvenile fish species in these habitats has been attributed to factors including greater abundance of food (Odum and Heald 1972; Nagelkerken et al. 2000), shelter provision due to high structural complexity (Parrish 1989; Robertson and Blaber 1992; Nagelkerken et al. 2000), lower predator abundance due to distance from offshore reefs (Shulman 1985; Parrish 1989), decreased predator foraging efficiency as a result of turbid waters in these habitats (Blaber and Blaber 1980; Robertson and Blaber 1992), and the increased likelihood that these habitats are intercepted by planktonic fish larvae (Parrish 1989). While this role has been recognized globally, it may not always be the case in some regions of the Indo-Pacific (Quinn and Kojis 1985; Blaber and Milton 1990).

Mangroves and seagrasses also play a role in improving water quality in coral reef ecosystems (Carlson et al. 2021). These habitats mitigate sediment flow from land or riverine sources (Victor et al. 2004; de Boer 2007; Golbuu et al. 2008), which is crucial for protecting coral reefs from sedimentation. Sedimentation can smother corals and/or reduce the light needed for photosynthesis (Rogers 1990) resulting in a decline in coral growth rates, species diversity, abundance, and successful recruitment in reef habitats (Rogers 1990). By extending into the water column with their blades, roots, or peat banks, seagrasses and mangroves effectively slow down water currents, allowing suspended sediment particles to settle in the substrate (Victor et al. 2004; Chen et al. 2018). Seagrass blades or mangrove roots may also provide a substrate for the settlement of passing sediment particles (Du et al. 2020). Mangrove forests and seagrass meadows also act as sinks for heavy metals, such as copper, lead, cobalt, and zinc (Attri and Kerkar 2011; Ahmad et al. 2014; Li et al. 2023). High concentrations of heavy metals are toxic to marine organisms and negatively affect biological processes including photosynthesis, development, growth, and gametogenesis (Jin et al. 2021). For example, exposure to elevated levels of cobalt resulted in the decreased growth rates of two major reef-building coral species in the Pacific, *Stylophora pistillata* (Esper, 1792) and *Acropora muricata* (Linnaeus, 1758) (Biscéré et al. 2015). Mangroves and seagrasses can accumulate heavy metals in their leaves, roots, or rhizomes (MacFarlane et al. 2007; Li et al. 2023), or cause these metals to lose bioavailability through processes such as precipitation as sulfides or complexation with organic matter (MacFarlane et al. 2003). Mangroves provide additional water quality improvements for both seagrass and coral reef ecosystems (Carlson et al. 2021). For example, mangrove forests reduce freshwater flow to these through transpiration, which helps maintain stable salinity levels in seagrass beds and coral reefs mitigating the effects of heavy rainfall events (Carlson et al. 2021).

Coral reefs and seagrasses also act as natural protective barriers, which protect mangroves from erosion due to wave action (Ogden 1988). Coral reefs and seagrasses may also facilitate mangrove propagule recruitment and settlement by reducing nearshore water currents (Guannel et al. 2016). Together these habitats are essential to mitigating wave forces leading to the coastal erosion or destruction of coastal communities (Guannel et al. 2016).

Furthermore, nutrient exchange is essential for sustaining these interconnected ecosystems. This exchange creates a lifeline that supports their mutual health and productivity and promotes the growth of the inhabiting species. Nutrients derived from the high net primary productivity of mangroves and seagrasses are exported to adjacent nutrient-poor coral reef habitats through several processes. Nutrients may be exported through transportation via mobile organisms including zooplankton, fish, and invertebrates (Odum and Heald 1972, 1975). For example, juvenile reef *Haemulon* spp. transport seagrass derived nutrients to corals on the reefs via solid waste deposition (Meyer et al. 1983). Coral colonies with these resident fish schools were found to have faster growth rates than those without them (Meyer et al. 1983).

Nutrients may also be exchanged via outwelling of particulate and dissolved organic matter, such as during tides (Lee 1995; Granek et al. 2009). Marine species may incorporate these nutrients in varying quantities (Granek et al. 2009). For example, Granek et al. (2009) found that in the Caribbean, mangrove-derived organic matter contributed up to 40% in some coral species, while other species, such as the tunicate *Phallusia nigra* Savigny, 1816, derived almost no organic matter from mangroves. This study also showed that mangrove forests up to 10 km away could contribute to nutrient levels in marine species (Granek et al. 2009).

Importance of preserving and restoring coral-mangrove-seagrass ecosystems

Preserving and restoring coral reef, mangrove, and seagrass ecosystems is crucial for sustaining the vitality of marine life and the well-being of adjacent coastal communities by providing several critical ecosystem services. Together, these habitats act as natural barriers and stabilize shorelines; they play a significant role in reducing incoming wave energy, lowering inundation levels, and preventing the loss of mud sediment (Guannel et al. 2016). Live corals, seagrasses, and mangroves collectively offer greater protection services than any single habitat or any pairing of two habitats (Guannel et al. 2016).

These habitats achieve their protective functions through various mechanisms. Coral reefs create physical barriers that break waves before they reach the shore, while mangroves' dense root systems trap sediments and reduce water flow, which helps prevent erosion. Seagrasses anchor sediments with their roots and absorb wave energy. The capacity of these habitats to offer coastal protection services is influenced by local geomorphology and environmental conditions (Guannel et al. 2016). For example, a study conducted in Belize found that under normal weather conditions, habitats associated with long and deep barrier reefs dissipate more wave energy than those in short and shallow fringing reefs (Guannel et al. 2016). However, during hurricanes, the barrier reef profile experiences higher surge and peak wave heights than the fringing reef profile (Guannel et al. 2016). This variability underscores the importance of maintaining a diversity of habitats to enhance coastal resilience.

In the context of climate change, these natural defenses are becoming increasingly vital. Rising sea levels and more frequent and intense storms pose significant risks to coastal areas. Effective conservation and restoration of these ecosystems can provide a cost-effective strategy for climate adaptation and disaster risk reduction, ensuring long-term protection and sustainability for coastal communities.

Coral reefs, seagrass meadows, and mangrove forests play a vital role in sustaining commercial fisheries, which support economic growth and stability for coastal communities. These ecosystems provide essential habitats for various marine species, many of which are commercially valuable (Moberg and Rönnbäck 2003). Healthy fisheries contribute significantly to local and national economies through both direct harvesting and the associated industries that rely on these resources. For instance, they support the livelihoods of fishers, processors, and distributors by creating jobs and generating income. In Florida, USA, it is estimated that 80% of all marine species of commercial or recreational significance rely on mangrove estuarine areas during some stage of their life cycles (Hamilton and Snedaker 1984). Furthermore, coral reef fisheries are estimated to comprise up to 10% of the world's fisheries catch (Smith 1978), and up to 25% in some developing countries (McManus 1988).

In addition to their economic value, reliable fisheries catch are crucial for enhancing food security for coastal populations by ensuring a stable supply of protein (Hughes et al. 2012). However, the sustainability of these benefits is threatened by overfishing, habitat destruction, and climate change. To safeguard these ecosystems and their services for future generations, it is essential to implement effective management practices, including sustainable fishing methods, habitat restoration efforts, and climate adaptation strategies. By prioritizing these actions, we can promote resilience and sustainability in coastal communities, securing both their ecological and economic well-being.

The socio-economic benefits of these three ecosystems extend beyond their direct impact on fisheries. They play a crucial role in driving ecotourism by attracting visitors from around the world, which generates substantial revenue for local communities. The rich marine biodiversity supported by coral reefs, seagrass meadows, and mangrove forests draws tourists interested in recreational activities such as scuba diving and snorkeling. This influx of visitors boosts local economies through spending on tours, accommodation, and related services. Furthermore, the prominence of these ecosystems in ecotourism often fosters greater awareness and appreciation for marine conservation, encouraging both visitors and local communities to support and engage in sustainable practices.

Beyond these economic contributions, coral reefs, seagrass meadows, and mangrove forests provide additional critical ecological services. They enhance water quality by filtering out pollutants such as heavy metals, excess nitrogen, oil spills, and marine debris like microplastics (Carlson et al. 2021). Importantly, these ecosystems also play a significant role in climate mitigation through carbon sequestration. They capture and store substantial amounts of carbon dioxide, with mangroves, seagrasses, and corals each contributing uniquely to this process. Mangrove forests, for instance, are among the most effective carbon sinks, sequestering carbon both above (leaves, branches) and below (sediments, roots) the ground (Hilmi et al. 2023) at rates up to five times higher than tropical forests (Donato et al. 2011). Seagrass meadows also sequester carbon efficiently by trapping organic matter in their sediments, contributing to long-term carbon storage (Fourqurean et al. 2012). Although coral reefs are traditionally considered to play a smaller role in carbon sequestration, they can act as carbon sinks under specific conditions by facilitating the growth of calcium carbonate structures that store carbon (James et al. 2024). Additionally, by offering protection to adjacent seagrass meadows and allowing for greater accumulation of organic sediment, coral reefs can enhance the carbon storage capacity of seagrass ecosystems (Guerra-Vargas et al. 2020). Collectively, these ecosystems help mitigate climate change by reducing greenhouse gas concentrations in the atmosphere, underscoring the need for their conservation and restoration to maintain their vital role in carbon sequestration (Carlson et al. 2021).

Given their strong interdependencies, threats to one habitat can lead to cascading effects on the others. Protecting coral reefs, seagrass meadows, and mangrove forests in tandem can yield synergistic benefits (Carlson et al. 2021). First, such integrated protection enhances ecosystem resilience by increasing resistance to disturbances that impact multiple habitat types (Carlson et al. 2021). Second, it supports high biodiversity by offering a variety of resources suitable for organisms at different stages of their life cycles (Carlson et al. 2021). Third, it creates portfolio effects, where habitats with overlapping ecological assets can allow for a rearrangement of species where displaced species can be accommodated in an alternative habitat, thereby stabilizing each ecosystem (Carlson et al. 2021). Lastly, this comprehensive approach can also reduce the incentive for harmful human activities by highlighting the value of maintaining these interconnected systems.

11.6 Case Studies and Research Findings

Corals have been documented thriving in mangrove habitats going back to 1928. Although these assemblages were not the object of research in most studies, observations of corals growing on or between mangrove roots are found in biodiversity surveys, technical reports, and through personal communication (Stephenson et al. 1931; Almy and Carrion-Torres 1963; Fishelson 1971; Loya 1976). Likewise, corals within seagrasses have been documented worldwide since at least 1982 (Rützler and Macintyre 1982) with their unique habitat characteristics and environmental parameters. Coral species composition and community structure information within the mangrove and seagrass habitat is limited prior to the 1990s as are empirical studies (Stewart et al. 2022). Here we highlight some notable seagrass-coral and CMC research findings by geographic region, based on major water bodies. Scientific names have been updated, as verified by the World Register of Marine Species (WoRMS) (https://www.marinespecies.org), and the Global Biodiversity Information Facility (https://gbif.org) databases and thus may appear different from those in the original reports.

Caribbean Sea

In the Caribbean Sea CMC habitats have been observed in Florida, Cuba, The Bahamas, Puerto Rico, the U.S. Virgin Islands, Belize, Panama, and Colombia. The chronology of CMC studies in the Caribbean is relatively contemporaneous, with the sporadic derivative observation of corals seen in mangroves in a few reports in the second half of the twentieth century. Notably, within the Caribbean, CMC studies have established a global classification system (Stewart et al. 2022) while also enhancing our understanding of coral-mangrove interactions and environmental influences (Yates et al. 2014; Rogers 2017; Scavo Lord et al. 2020, 2023; Stewart et al. 2021).

Likewise, coral-seagrass habitats have been identified in several regions, including Florida (Manzello et al. 2012; Landau et al. 2013), the Cayman Islands (Camp et al. 2016a), Cuba (González-Ferrer et al. 2006), the U.S. Virgin Islands (Yates et al. 2014), and Mexico (Ruiz-Zárate et al. 2000), etc., with potential for additional locations not yet documented or reported in this chapter. These studies highlight the interdependence between coral and seagrass communities, illustrating how seagrasses can buffer environmental stressors for corals while also contributing to enhanced biodiversity and ecosystem resilience.

Florida

The role of seagrass beds in sheltering corals from the effects of ocean acidification was recognized by Manzello et al. (2012) while trying to shed light on the spatial and temporal variability of carbonate chemistry within reefs. Ocean acidification is expected to reduce calcification in marine organisms, but its effects vary across systems. Since 2009, data from the Florida Reef Tract show that inshore reefs experience higher aragonite saturation due to CO_2 uptake by seagrasses and macroalgae, potentially creating refugia from ocean acidification. This pattern contrasts with offshore reefs, which reflect typical oceanic conditions, suggesting that inshore reefs may provide temporal and spatial protection from acidification.

Landau et al. (2013) described the ecology of corals and sponges in seagrass beds dominated by *T. testudinum* at Long Key State Park in the Florida Keys. This survey provided baseline estimates of densities and distributions of common sponges and corals (*M. areolata*, *Porites* sp., *C. arbuscula*). Of all organisms examined, only *Manicina* showed a significant (but negative) correlation with seagrass, while weak positive relationships were observed between certain sponge and coral species. All species exhibited aggregated distributions, with clumping quantified using statistical models.

Acknowledging that seagrass beds are habitats for corals, Thanopoulou et al. (2022) set out to examine the patterns of occurrence of stony corals with respect to nearshore water quality measurements along the Florida Keys reef tract. Small corals characterized the nearshore areas, with the most common coral species being *S. radians* and *P. divaricata*, which are also tolerant to eutrophic conditions. Seagrass was the dominant benthic type and occurred in over half of the area surveyed. Stony coral occurrence was limited across most study sites. Two of the 13 sites had no coral species. Significant differences in water quality parameters, such as Chlorophyll-*a*, pH, dissolved oxygen, temperature, NO_x, N:P ratio, total nitrogen, and total phosphorus, were observed between sites with and without corals. Multivariate analysis revealed distinct clustering between these two groups.

Increasing numbers of global CMC reports and the lack of information in Florida waters motivated Kellogg et al. (2020) to survey the mangrove shoreline of the Upper and Lower Keys. Survey sites were determined through personal communication with local community members and researchers and satellite imagery. The surveys revealed patterns in the distribution of corals among the mangroves. *P. porites*, *P. furcata*, *P. divaricata*, and *Favia fragum* (Esper, 1793) grew directly on the mangrove prop-roots while *S. siderea*, *S. bournoni* and *S. intersepta* grew in the tidal mangrove channels. *S. radians* was observed growing in both patterns. The coral colonies in the tidal mangrove channels were shaded by the mangrove canopy, at least part of the day.

Similarly, Flowers et al. (2023) explored the Miami metropolitan area. This was the first CMC survey in a heavily urbanized coastal environment and resulted in five coral species being observed growing under the mangrove roots: *S. siderea*, *S. radians*, *P. porites*, *Oculina diffusa* Lamarck, 1816, and *S. bournoni*. The mean temperature at these sites ranged from 26.7 to 30.91 °C and light was recorded as 303.9–505.3 lux (July–August). DNA extraction revealed *Breviolum* (>99% abundance) as the dominant algal symbiont in all five species, except for one colony of *P. porites* and two colonies of *S. radians*, which were dominated by *Symbiodinium* (>99% abundance). The thermotolerant symbiont *Durusdinium* was not detected in any of the corals sampled, which is noteworthy given its occurrence at other nearby nearshore sites (Rubin et al. 2021), which may suggest corals under the mangrove roots are experiencing less stressful conditions than their neighbors.

Cuba

In Cuba, CMCs have been observed in both inlets (de la Guardia et al. 2004) and edge habitats (Hernández-Fernández 2015). Working in the diving zone of Punta Francés, located in the southwest corner of the Isle of Youth, de la Guardia et al. (2004) observed corals in mangrove channels, particularly in the canal connecting La Siguanea with the southern coast of the peninsula. These channels had high water turbidity, sediment covering seagrass blades, and abundant mangrove litterfall along the canal edges. Small colonies of

Montastraea cavernosa (Linnaeus, 1767) and *S. siderea* were found sporadically in these channels attached to *R. mangle* roots.

Another study by González-Ferrer et al. (2006) explored the interactions between mangroves, seagrasses, and coral reefs in Cuba. Authors noted the presence of corals such as *P. furcata* and *P. astreoides*, *S. radians*, and *C. arbuscula* within red mangroves, though the specific CMC environment was not clearly identified. These corals established themselves in mangrove environments, often on the roots of red mangroves, where they found conditions for growth despite the sediment-rich waters.

Hernández-Fernández (2015) conducted a study at Jardines de la Reina National Park to determine the composition of stony corals on submerged roots of *R. mangle* in five sites between 2010 and 2012. A total of 11 stony coral species from seven genera were identified. Although the predominant species found were *P. astreoides* and *P. divaricata*, the study also included *Millepora alcicornis* Linnaeus, 1758, *S. radians, M. areolata,* and *P. furcata*. All coral colonies were found in association with crustose coralline algae, mainly from the genus *Neogoniolithon*. Based on these findings, Hernández-Fernández (2015) concluded that the presence of crustose coralline algae appears to be crucial for the successful settlement of stony coral colonies on mangrove roots.

The Bahamas

A rapid ecological assessment (REA) was conducted from July 22 to 27, 2013, to evaluate the condition of marine habitats and resources in eastern Grand Bahama. The study aimed to assess coral reefs, mangroves, and associated fish communities to support the potential establishment of a national park (Dahlgren et al. 2014). Authors found low coral diversity but adequate coral cover. Reef fish communities were diverse and mangroves and adjacent creeks provided habitat for fish, as well as seagrass and macroalgae, all serving as important nurseries. The mangroves in East Grand Bahama are expansive and unique, consisting of creek systems with naturally occurring mangrove islands and channels bordered by red mangroves (*R. mangle*). These systems contain numerous blue holes and mini reef habitats. The report identified several coral species within the mangrove habitats of East Grand Bahama. Notably, the mangrove channels and islands contained coral patch reefs containing *Acropora prolifera* (Lamarck, 1816).

Puerto Rico

Among the earliest reports of CMCs in the Caribbean, we have the investigations of Almy and Carrion-Torres (1963) on shallow-water stony corals. The report provides an illustrated checklist of 38 species of nearshore corals with a focus on La Parguera, on the southwest coast of the island, although the report also includes other locations with significant reef development, such as the northeastern end of Puerto Rico. Corals were observed growing on mangroves. *R. mangle* was noted as the primary mangrove species supporting coral growth. Among the corals, species such as *Agaricia agarictes* (Linnaeus, 1758), *F. fragum*, and *P. clivosa* were seen attached to red mangrove roots.

Uhrin et al. (2005), working at La Parguera, reported free-living colonies of *M. areolata*, which inhabit shallow, high-energy environments, are often overturned by waves and currents but have adaptations that allow them to right themselves. The team tested how coral morphology, such as area, meander density, and shape, affected righting time. On average, colonies took 116 h to right themselves, with 86% succeeding within 23 days. Colony area was the only significant factor influencing righting ability, with larger colonies being less efficient. Meander density and shape had minimal effects, and while higher meander density aids sediment rejection, the associated increase in area hindered righting ability.

Also working at La Parguera, Jeffrey et al. (2010) reported "hard and soft corals" growing within mangrove habitats. Although the article provides a detailed list of organisms found in different sediment types, it is impossible to parse which species of corals were specifically attached to mangrove prop roots.

Cayman Islands

The north coast of Little Cayman (Bloody Bay Marine Park) served as one of the sites for the global study of Camp et al. (2016a) to study the biochemical services that mangroves and seagrasses provide to corals. Seagrass beds show elevated mean pH values (8.15 ± 0.01) when compared to adjacent outer reefs, yet they also experience periods of high and low pH. As a result, corals in the seagrasses sustained their calcification or experienced a reduction of calcification relative to the outer reef. The dominant seagrass species were *T. testudinum* and *S. filiforme*.

Once established that corals in seagrasses are regularly exposed to broad fluctuations in temperature, pH, in conditions like those predicted for coral reefs (Camp et al. 2016a), Camp et al. (2016b) used the same location in the Cayman Islands to determine whether corals experiencing these changing conditions were able to enhance their tolerance from low-variance habitats to high-variance conditions. Experimentally, no changes were detected in growth or metabolic rates between corals from seagrass (high variance) to outer reef (low variance) in present-day conditions. Moreover, simulation of future scenarios (exposing corals to elevated pH and temperature) resulted in no enhanced tolerance either for seagrass or outer-reef corals, with calcification rates decreasing in all situations. The results suggest that seagrass habitats may not offer refugia from climate change if future temperature and pH changes match those predicted for reefs.

If seagrass meadows are not an ideal site for corals, then why are coral communities increasingly found in nonreef habitats? This is the question at the center of research by Lohr et al. (2017), who aimed to understand how corals persist in seagrass meadows by surveying established and newly recruited corals in Little Cayman and comparing them to back-reef sites. Despite once again finding higher variability in physicochemical conditions in seagrasses, especially pH, when compared to back-reefs (*see* Camp et al. 2016b), the authors found 14 taxa, with multiple families being able to recruit to these habitats. These numbers, however, represent lower coral cover, species diversity, richness, and evenness when compared to coral populations on back reefs suggesting that, although suboptimal, seagrass meadows can still function as habitats for corals, providing important insights into coral adaptability to variable environments.

U.S. Virgin Islands

At least 33 scleractinian coral species have been observed growing in mangroves of Hurricane Hole, St. John, U.S. Virgin Islands prior to hurricanes Irma and Maria (Rogers 2017; Yates et al. 2014). Hurricane Hole is made up of four mangrove-lined bays with no permanent source of freshwater inflow. Water clarity is high, nutrient concentrations are low, the tidal range is <0.25 m and the area is protected from high winds. The CMC habitat has seagrass *T. testudinum* intermixed, as well as calcareous algae *Halimeda* spp. and *Penicillus* spp. (Yates et al. 2014). Coral colonies grow directly on *R. mangle* prop roots or on hard substrate between roots. The most abundant corals in the mangroves are *Orbicella* spp., *Colpophyllia natans* (Houttuyn, 1772), *Diploria labyrinthiformis* (Linnaeus, 1758), *Agaricia* spp., and *Porites* spp. (Rogers and Herlan 2012). Both *C. natans* and *D. labyrinthiformis* have greater abundance in the mangroves than on the adjacent reef. *C. natans* also were in better condition, with fewer dead portions, in the mangroves than on the reefs (Rogers 2017). Many of the colonies of *C. natans*, *D. labyrinthiformis*, *Siderastrea*, *Montastraea*, and *Orbicella faveolata* (Ellis & Solander, 1786) reach over 1 m in diameter (Rogers 2009; Rogers and Herlan 2012). Corals like *Mycetophyllia* spp., *Scolymia* spp., *Eusmilia fastigiata* (Pallas, 1766) and *D. stokesii*, which are rare on coral reefs, are found in the protective shade of the mangrove canopy. Not all corals that settle in the mangrove habitat survive long enough to grow into large colonies. While multiple small colonies of *A. palmata* have been found in Hurricane Hole, those colonies did not persist (Rogers 2017). Outside of Hurricane Hole, *Agaricia grahamae* Wells, 1973, *Agaricia lamarcki* Milne-Edwards & Haime, 1851, and *M. areolata* were observed growing on mangrove roots in Great Lameshur Bay, St. John (Buob 2019). The sponge community of CMC habitats consists of more reef-associated species than mangrove species.

Variations in the benthic community structure and physical environment create chemical microclimates in these coastal habitats (Yates et al. 2014). The chemical conditions of CMC habitats more closely resemble those of rock outcrop coral sites than those of mangrove habitats without coral.

Belize

In Belize, mangrove-coral communities have been documented on offshore mangrove cays both leeward of the Belize Barrier Reef complex (e.g., Farnsworth and Ellison 1996; Rützler et al. 2004; Macintyre et al. 2000), as well as within offshore Belize's atolls, specifically Turneffe Atoll (Bengtsson et al. 2019; Scavo Lord et al. 2020, 2021). CMC research in Belize documents how coral species distribution is influenced by factors like sediment, water flow, and other environmental variables, with findings of high clonal reproduction in mangrove coral populations.

Farnsworth and Ellison (1996) investigated the spatial and temporal variability, distribution, and abundance of marine epibionts on mangrove roots at different cays throughout Belize (Peter Douglas Cay, Spruce Cay, Wee Wee Cay, and Twin Cays). The authors mention the presence of corals in mangrove root communities. Specifically, it notes that corals (*P. strigosa, P. astreoides*), along with other marine organisms, colonize the subtidal portions of mangrove (*R. mangle*) prop roots.

Across the four mangrove islands that make up the Twin Cayes, Rützler et al. (2004) identified corals growing on the fringing red mangrove, *R. mangle,* prop roots on the western (West Bay) and southern (Sponge Haven) portions of the West Island and western portion (Main Channel Southeast) of the East Island. The corals *P. porites*, *P. astreoides,* and *Millepora complanata* Lamarck, 1816 were observed in West Bay, while corals in the genus *Porites* were found in Sponge Haven (Rützler et al. 2004). *P. strigosa* was observed in Main Channel Southeast on the East Island. Environmental conditions, sediment, depth, and topography varied within locations on and across islands, which may explain variation in the presence, absence and/or diversity of corals and other marine species that were observed (Rützler et al. 2004). For example, the rich assemblage of flora and fauna in West Bay was speculated to be attributed to nutrient input from adjacent mangroves, lack of very fine sediments, and a moderate temperature-salinity regime (Rützler et al. 2004).

In the Pelican Cayes, Macintyre et al. (2000) documented a diverse assemblage of sessile marine species, including various coral species, along the rims of numerous circular ponds bordered by red mangroves. These open, low-energy ponds formed due to differential coral and mangrove growth on a Pleistocene subsurface. Several coral species were found growing on the roots and peat banks along the pond rims, including *P. divaricata*, *Agaricia tenuifolia* Dana, 1846, and *Millepora* spp. Additionally, a large colony of *S. siderea* was observed on the roots. *Acropora cervicornis* (Lamarck, 1816) and *O. annularis* were also seen within and at the entrances of the ponds. Macintyre et al. (2000) speculated on the extent to which environmental differences between these ponds including water replacement, drive variation in the abundance and diversity of marine sessile species.

Coral communities have also been found inhabiting mangrove channels on Turneffe Atoll. Located 32 km off the coast of Belize, the Turneffe Atoll is the largest coral atoll in the Mesoamerican Barrier Reef System, consisting of over 150 cays that are intersected by mangrove channels. These mangrove channels connect Turneffe's outer coral reefs to its inner lagoon, which is made up of interspersed patch reefs and seagrass beds (Bengtsson et al. 2019; Scavo Lord et al. 2021). Several coral species have been found inhabiting the submerged prop roots of *R. mangle* lining these channels (Bengtsson et al. 2019; Scavo Lord et al. 2020, 2021). Surveys in two of these channels by Bengtsson et al. (2019) revealed differences in species diversity, whereby a more diverse coral community was found in the channel, Crooked Creek, on the Western side of the atoll, including branching *Porites* spp., *P. astreoides, D. stokesii, O. faveolata* and *S. siderea, M. alcicornis*, while only branching *Porites* spp. were observed in the second channel on the eastern end of the Atoll. Differences in fine scale topography of the shorelines between the two channels was speculated to drive variation in species diversity, as Crooked Creek possessed a broad, shallow, well-lit shelf that may provide a larger settlement area for corals, especially those that are not suited to grow directly on the red mangrove prop roots (Bengtsson et al. 2019).

Furthermore, an investigation of the distribution, size, and morphology of a population of branching *Porites* spp. in one of these channels revealed interesting patterns in how corals occupy mangroves (Bengtsson et al. 2019). For example, coral colonies were only found inhabiting sections of the channel with high or moderate flow, and no corals were found within the channel's stagnant ponds (Bengtsson et al. 2019) suggesting the possible importance of water flow to coral survival. Investigations into the morphology of colonies showed variation in colony form, whereby colonies of similar sizes (ecological volume) exhibited a 10-fold difference in the number of branch tips. Variation in colony form was speculated to be due to microenvironmental differences in water flow and/or irradiance levels. Further investigation into the morphology of mangrove coral populations similarly revealed a high degree of variability of various traits (e.g., size, branch density, branch thickness, and corallite features) within mangrove individuals, as well as morphological variation between coral populations in the mangroves and adjacent patch reefs in Belize (Scavo Lord et al. 2021). Bengtsson et al. (2019) followed *Porites* spp. population and colony survival over a one-year interval. Scavo Lord et al. (2020) expanded upon this work, concluding that the population was viable and growing over a four-year time-period with long-term population growth projected forward. This population growth, however, was characterized by significant population turnover and high recruitment rates (Scavo Lord et al. 2020). Genomic investigations further revealed that the population was highly clonal, relying on asexual reproduction for the generation of new recruits (Scavo Lord et al. 2023). Asexual reproduction via parthenogenesis or the generation of asexual larvae, was thought to be the only viable mode of reproduction given that colonies are growing vertically on prop roots suspended over a deep mud substrate (Scavo Lord et al. 2023). Other known modes of asexual reproduction in corals (fragmentation, polyp bail-out) would not be viable in this context given that asexually produced propagules would sink in the mud benthos (Scavo Lord et al. 2023).

Panama

Panama has been one of the focal points of CMC research. Corals have been found growing in mangrove environments, with species diversity influenced by water clarity, proximity to freshwater inflows, and land development. CMC habitats in Panama support unique fish communities and are a refuge from light stress on corals. Long-term monitoring revealed coral growth is sustained in CMC habitats, with species like *S. siderea* thriving in specific zones.

To date, corals have only been reported growing within mangrove and seagrass habitats on the Caribbean coast of Panama. While CMC habitats have been recorded in the San Blas Archipelago, most observations and studies have taken place in the Bocas del Toro Archipelago. Bocas del Toro is the only location where canopy CMC habitats have been observed (Stewart et al. 2021, 2022). In these canopy CMC habitats, corals were observed growing in 27–74 cm of water and up to 19 m into the mangrove forest. In comparison, corals in the adjacent edge CMC habitat grew in water 56–79 cm deep. The fringing dwarf red mangrove forests surrounded by shallow reef patches and intertwined seagrass beds of Bocas del Toro make this archipelago an ideal site for studying corals thriving in both mangrove and seagrass habitats. Stewart et al. (2022) recorded 32 species of scleractinian corals within the mangroves, 29 species occurring under the mangrove canopy and 20 species on the mangrove edge, as well as two milleporid coral species found in both CMC habitats. Meanwhile, 24 scleractinian species and 1 milleporid coral species were observed on the shallow, adjacent reefs. Adjacent reef flats (<1 m depth) supported 13 scleractinian species and the reef slopes (>1.5 m depth) supported 22 scleractinian species (Stewart et al. 2022).

A presence-absence study found a positive association between coral richness and water clarity and the size of the mangrove area. Coral richness decreased with proximity to freshwater inflow (<1.2 km), land development, and pollution (Stewart et al. 2022). Increased sediment covering the mangrove roots following nearby land development caused a decrease in coral species richness at multiple CMC locations between 2017 and 2019 (Stewart et al. 2022).

A zonation pattern within the canopy CMC was reported where smaller corals (<10 cm) grow densely along the mangrove edge with the occasional large colony (>1 m diameter), followed by patches of corals with varying diversity and size further into the canopy, the coral community then transitions into a zone with the greatest coral density, dominated by large, visibly healthy *S. siderea* colonies, and ends in smaller, infrequent coral colonies (Stewart et al. 2022). Corals in CMC habitats grow on various substrates (e.g., mangrove roots, bottom between roots, on other corals) and display variable coloration and growth forms. Corals in edge CMC habitats grow in mounding and plating growth forms and coloration resembling corals on the adjacent reef. Canopy CMC corals predominantly grow in encrusting and plating growth forms that wrapped around mangrove roots and display an array of coloration within species.

Long-term monitoring of individual coral colonies revealed sustained growth within CMC habitats. The percentage of change in area was dependent on coral species and distance from the mangrove fringe (Stewart et al. 2022). In analyses comparing benthic cover between CMC habitats and adjacent reef, no difference in percent cover of live coral between habitat types was found. Furthermore, no difference in coral diversity between mangrove canopy and the reef slope was detected. The mangrove habitat had similar amounts of coral recruitment as both reef habitats, reflected by the number of coral colonies <5 cm and most coral recruits belonged to *Agaricia* or *Porites* genera.

In an attempt to understand why certain coral species thrive in mangrove habitats and to directly compare the effect living in reef or mangrove environments has on coral species richness, abundance, diversity, and condition (i.e., level of bleaching, tissue loss, and mortality), a reciprocal transplant experiment was conducted by Stewart et al. (2021). In that study, shading by mangroves was manipulated and tested to evaluate whether light intensity or habitat had a greater influence on coral survival. Corals transplanted from the reef into the mangrove canopy with branches pulled back to allow in more light displayed similar conditions to corals living on the reef under an artificial shade. Light intensity was found to be a key environmental parameter mediating coral bleaching and survival, with mangrove habitats providing a refuge from light stress. Coral condition was negatively correlated with light for *M. cavernosa*, *P. clivosa*, *P. astreoides*, *C. natans*, and *O. faveolata*. Meanwhile, there was a positive correlation between coral condition and light with *D. labyrinthiformis* (Stewart et al. 2021).

Seagrass beds and mangroves are essential nurseries for juvenile western Atlantic reef fishes and refuge for shoaling fishes, due to protection from predators and food availability (Nagelkerken et al. 2001). Many fish species migrate to the reef once reaching a larger size or different life stage. Fish assemblages within CMC habitats are distinct from mangrove habitats without coral (Wright et al. 2023). CMC habitats have greater fish species richness and greater abundance of herbivores (e.g., *Scarus* spp., *Stegastes* spp.) and *Halichoeres* spp. And a lower abundance of *Sphyraena*. In addition to the complex structure that submerged mangrove roots provide, coral presence increases rugosity and shelter holes within the habitat. Regardless of whether corals are present, over 80% of fish surveyed in mangrove habitats were juveniles.

In San Blas Point, along the northeastern coast of Panama, *M. areolata* were observed to attach to solid substrata (e.g., mangrove roots, coral rubble). Attaching to mangrove roots allows the free-living coral to colonize sediment-rich and dynamic habitats without being smothered, as would happen if settling on the sea floor. Johnson (1992) reported that the coral can break free of the substratum later in life to live unattached and diffuse out into more turbid environments away from the mangrove canopy.

Colombia

Only one report of CMCs exists for Colombia (Reyes and Hernando Campos 1992). The study is not focused on CMCs, but rather it aimed to investigate the colonization of macroinvertebrates on the roots of *R. mangle*. However, the authors reported the existence of corals among the fauna associated with mangrove roots at Bahía de Chengue. Specifically, the coral species *P. porites* was identified in the roots of *R. mangle* both inside a lagoon and at sea-exposed locations. This CMC is located within Tayrona National Natural Park, an area known for its coral formations, beaches, and extensive rocky shores.

Mexico

The free-living coral *M. areolata*, seagrass *T. testudinum*, and coralline algae *Neogoniolithon* sp. coexist in the reef lagoon of Xahuayxol, Quintana Roo. Ruiz-Zárate et al. (2000) studied the distribution and abundance of these organisms, focusing on why *M. areolata* was only found in medium-density seagrass areas. Transplant experiments showed that the coral suffered high mortality in low-density zones due to siltation and predation, and experienced stress in high-density areas. *M. areolata* selectively recruited onto

Neogoniolithon sp., which provided ecological benefits, increasing coral survivorship. However, as the coral grew, it eventually separated from the algae due to its weight and the algae's fragility.

Red Sea

The Red Sea is an extreme location for the presence of mangroves and corals. The area experiences high temperatures and high salinity (high evaporation rates and the lack of significant freshwater inflows). Mangroves are in one of their northernmost limits of geographical distribution (Leal and Spalding 2024). The coral reefs are among the most unique and diverse coral reefs in the world. These reefs are known for their high levels of endemism, with many species found nowhere else (Khalil 2015). *Avicennia marina* (Forssk.) Vierh. is the most abundant mangrove species in the Red Sea, with *Ceriops tagal* (Perr.) C.B.Rob and *Rhizophora mucronata* Lam. also present on Musseri Island. In the Gulf of Aqaba, colonies of the hood coral *S. pistillata* have been observed attached to the roots of *A. marina*, marking the northernmost distribution of this mangrove species in the Red Sea (Fishelson 1971). Por and Dor (1975) described a similar association between *A. marina* and *S. pistillata* for the Red Sea. The authors observed corals growing attached to mangrove roots along with sponges and the sea urchin *Echinometra mathaei* (Blainville, 1825). These mangrove forests, mostly located on the seaward side of their sampling locations, host a diverse community of species, including cyanobacteria (blue-green algae), the brown algae *Hydroclathrus clathratus* (C.Agardh) M.Howe and *Hormophysa cuneiformis* (J.F.Gmelin) P.C.Silva, the red alga *Laurencia*, and invertebrates such as *Amphibalanus amphitrite* (Darwin, 1854), *Magallana angulata* (Lamarck, 1819), *Littorina scabra* (Linnaeus, 1758), and the urchin *E. mathaei*. Loya (1976) likewise reported *S. pistillata* growing in shallow lagoons (Eilat) attached to the roots of *A. marina* in situations in which water temperature reached 36 °C and salinity 48 ppt. Loya attributes this to the capacity of *S. pistillata* to behave like an *r*-strategist species (the species lives in fluctuating environments with no crowding, therefore it is advantageous to increase the size of the population) since in addition to mangrove roots, the coral also attaches to shells of the gastropod *Tectus dentatus* (Forsskål, 1775) (*see* Fig. 1 in Loya 1976).

Indian Ocean

Reports of CMCs in the Indian Ocean have provided important information to aid in our understanding of mangrove-coral associations. These studies have illuminated how low-energy environments are critical for CMCs (Sesh Serebiah et al. 2011), explored how mangroves and seagrass habitats can mitigate the effects of climate change on coral reefs, and documented how corals in CMCs are preconditioned to low pH conditions, aiding in their tolerance to ocean acidification, thought at a metabolic cost (Camp et al. 2016a; Greenwood 2021). These findings underscore the importance of mangrove habitats in coral resilience and adaptation. Just as importantly, studies are being conducted in areas where CMCs are likely to further our understanding to the conditions they need (Louis et al. 2025).

India

Working in the northwestern Gulf of Kachchh (Gujarat), Sesh Serebiah et al. (2011) found mangroves and corals coexisting, which induced "curiosity to the authors". This resulted in the analysis of physicochemical parameters on sites to elucidate what conditions favored the coexistence of these assemblages. The study found that mangroves and corals coexisted particularly in areas with low-energy environments provided by the intricate mesh of creeks and sea incursions. Mangroves occupy inner creeks with either sandy or muddy substrates, and colonies of *Favia* sp. are reported associated with them. The study mentions the presence of mangroves in the region, particularly in the inner creeks of Jakhau, but does not provide information on the specific mangrove species. Temperature, salinity and pH are reported for the area of study, but it is difficult to elucidate what portions [of the data] apply to mangrove-coral locations only or, more generally, to the whole area. However, the authors concluded that the presence of mangroves appears to support the proliferation of corals by providing a suitable habitat. The article is the first to report live corals in Jakhau, and as the area is being developed due to its proximity to oil-exporting countries in the Middle East, these findings might be critical for regional ecosystem management and conservation.

A study conducted in the coral reef ecosystems of the Gulf of Mannar identified 13 species of seagrass, with *O. serrulata* being the dominant species by percent cover (37%). During the survey, the authors reported associated fauna and flora including "scleractinian corals," but species are not mentioned (Manikandan et al. 2011).

Seychelles

Working across three biogeographical regions (Atlantic Ocean, Indian Ocean, and Pacific Ocean), Camp et al. (2016a) investigated how mangroves and seagrass habitats provide different biochemical services that could potentially mitigate the effects of climate change on coral reefs. The study examines whether these habitats can act as refugia by maintaining favorable chemical conditions and supporting metabolic functions of reef-building corals. The Indian Ocean location, around the island of Curieuse within the Seychelles Archipelago, contained a mangrove-dominated habitat located within Baie La Raie. This site, however, was not directly under the mangrove canopy but rather near the leeward site. The dominant mangrove species found were *R.*

mucronata, Lumnitzera racemosa Willd., *Bruguiera gymnorhiza* (L.) Lam., and *A. marina*. As for corals, *A. austera, A. muricata*, and *A. gemmifera* were reported alongside *P. damicornis*, *P. verrucosa*, and *P. attenuata* and *P. lutea*. The dominant seagrass species was *T. hemprichii*. In general, Camp et al. (2016a) found that corals growing in mangrove habitats are preconditioned to low pH conditions, which might help them tolerate increasing ocean acidification. Moreover, mangroves could act as genetic reservoirs for corals that are more tolerant to pH conditions. However, living among mangroves comes with metabolic costs to corals as the study also found that these species incur higher metabolic costs to maintain their physiological functions, which could limit their growth and calcification.

Locations within Curieuse Marine National Park were used in the dissertation work of Greenwood (2021). Greenwood set out to explore coral adaptation and acclimatization based on changes in their holobiont community composition. The model organism used for the study was *P. lutea*. The coral was used in thermal-tolerance experiments to detect temperature tolerance limits in both mangrove and reef habitats. In general, heating corals beyond their known or regional thermal maxima decreased their productivity, but "mangrove corals" (as they are named in the dissertation) suffered less thermally induced bleaching. Coral microbiome compositions changed between reef and mangrove habitats in Indonesia and the Seychelles. Under experimental manipulation, these shifts in microbial composition occurred rapidly (within 36 h) which led Greenwood (2021) to conclude that the rapid habitat-dependent flexibility of the coral microbiome is crucial for coral resilience and adaptation.

North Pacific

Japan

In their study of settlement and habitat association of juvenile humpback red snapper (*Lutjanus gibbus* (Forsskål, 1775)) in Okinawa, Nanami and Yamada (2009) described the occurrence of several species of massive (*Porites* sp. and Faviidae), and sometimes branching, and tabular corals on seagrass beds of *O. serrulata* and *T. hemprichii*. These corals were observed in areas in which water depth was about 2 m at high tide. This is like many cases of CMC reports, in which the association was not the main objective of the study but is reported as an observation. However, in Higuchi et al. (2014), the authors directly examine the biogeochemical relationship between the coral *M. digitata* and seagrass *T. hemprichii* in the reef at Bise, Okinawa. Using benthic chambers, carbon metabolism, and nitrogen flux rates, authors measured across various habitats, including coral, seagrass, mixed coral-seagrass communities, and sandy areas. Results indicate that coastal groundwater inflow contributed to high nitrate and nitrite levels in all habitats. The coral-seagrass habitat had the highest NO_x uptake rate, indicating a mutual benefit where seagrass thrives from coral presence, while carbonate dissolution in the coral-seagrass habitat promotes coral fragmentation and dispersal, aiding habitat expansion. Their findings are important in the sense that, although no synergistic effects on photosynthesis, respiration, or calcification processes were found, seagrass-driven $CaCO_3$ dissolution may facilitate coral fragmentation, benefitting habitat development for *M. digitata*. The synergistic effect found concerning nitrogen uptake, with a higher NO_x uptake rate in the mixed habitat compared to coral or seagrass habitats alone, suggests that seagrass benefits from coral presence in terms of NO_x availability.

Also working at Bise, Meekaew et al. (2014) noted increased primary production in the water column of seagrass incubations, which led to increased dissolved oxygen levels. Dissolved oxygen was, in turn, positively correlated with dissolved organic carbon concentration. As with Higuchi et al. (2014) the species used were the coral *M. digitata* and seagrass *T. hemprichii*. Notably, corals increased carbon assimilation when coexisting with seagrass, utilizing seagrass-exuded DOC via microbial activity. Likewise, seagrass benefited by assimilating nutrients and microbially degraded DOC from coral metabolism.

South China Sea

Taiwan

Pezner et al. (2024) described the coexistence of corals and seagrasses in Taiwan, specifically at Dongsha Atoll. At Dongsha Island, extensive seagrass meadows that extend up to 1 km offshore and cover between 8.2 and 11.85 km^2, contain patches of scleractinian corals, mainly massive *Porites* and some *Acropora* colonies. The seagrass meadows contain seven species of seagrass but are dominated by *T. hemprichii, Cymodocea rotundata* Asch. & Schweinf., and *Cymodocea serrulata* (R. Br.) Asch. & Magnus. Authors note how these occurrences have been documented since at least 1994 (Dai and Fan 1995). Pezner et al.'s (2024) study suggests that the seagrass beds modify local seawater chemistry, elevating pH and dissolved oxygen during daytime but also causing acidic and hypoxic conditions at night. This dynamic chemical environment may influence coral growth patterns, with no significant differences in calcification rates between sites but variations in skeletal density and extension rates.

South Pacific

Research on coral-mangrove interactions in the South Pacific reveals fascinating insights into coral adaptability and resilience. Studies across diverse regions, from Indonesia's Raja Ampat (Becking et al. 2024; Donnelly et al. 2003) and South Sulawesi (Polapa et al. 2021) to New Caledonia (Camp et al. 2017) and the Great Barrier Reef (Stephenson et al. 1931; Ferns 1994; Billyard 1995; Camp et al. 2019), highlight how corals in mangrove and marginal habitats exhibit unique physiological and biochemical adaptations to extreme environmental conditions (*see* Chap. 18). These habitats, characterized by factors such as low light, high temperatures, and low oxygen levels, challenge coral species to adapt in remarkable ways. Given these conditions, research from this region also focuses on the resilience of coral holobionts and the role of Symbiodiniaceae in these ecosystems (Camp et al. 2019; Greenwood 2021; Haydon et al. 2021; Haydon 2022) (*see* Chap. 3). Of particular interest are studies from Singapore highlighting the "accidental" recruitment of corals and seagrasses that has occurred as an unintended consequence of constructing seawalls to reinforce the coastline (Tan et al. 2025). Findings from these studies underscore the importance of understanding these extreme environments, not only to comprehend coral resilience but also to inform conservation strategies amidst ongoing climate change and anthropogenic pressures.

Indonesia

As has frequently happened, a survey of Indonesia's Raja Ampat that was not centered on mangroves and corals managed to describe such habitat. While conducting a rapid ecological assessment carried out by The Nature Conservancy and WWF, mangroves and coral reefs were seen intermingling in bays and channels where water was clear from constant flushing by tidal currents (Donnelly et al. 2003). There is no specific mention of which species of mangroves and corals were coexisting, but mangroves in Raja Ampat were described as "markedly depauperate except in a few places where estuarine flats and tidal rivers have provided ample habitat for the *Bruguiera-Rhizophora* associations." Raja Ampat biodiversity surveys in 2016, 2020, and 2023 resulted in 37 species of scleractinian corals observed in 12 marine lakes. These marine lakes have frequent flushing of seawater but low visibility due to terrigenous runoff (Becking et al. 2024). Most corals were present at depths ranging 0.5–5 m and hard coral composed 10–20% of total benthic cover. Mangroves did not grow along the shoreline of all marine lakes, but when they did, corals were observed growing on and between the roots. The authors did not specify which coral species were found within and outside of the mangrove habitat. These marine lakes fall within the Lagoon CMC habitat type.

Working in South Sulawesi, Polapa et al. (2021) report on how different coral species adapt to extreme environmental conditions by focusing on physiological differences in corals found in reef habitats and mangrove ecosystems, which in this context are considered "marginal habitats." The study finds that corals in mangroves tend to be more heterotrophic (they rely more on consuming organic matter rather than on photosynthesis). The coral genus *Dipsastraea* showed higher densities of symbiotic algae in mangrove habitats, while *Porites* had similar densities in both habitats. Moreover, *Porites* consumed more oxygen in the mangrove habitat, and this affected their growth. This research provides valuable insights into how corals can survive and adapt to less-than-ideal conditions, which is crucial for understanding their resilience in the face of environmental changes. Unfortunately, while the article does focus on the physiological responses of the corals in these marginal habitats, it fails to specify the exact species of mangroves found in the study area.

In studies by Camp et al. (2016a) and Greenwood (2021), Wakatobi Marine National Park, Southeast Sulawesi, was compared with Seychelles locations to explore metabolic functions in corals from mangrove and seagrass habitats across different biogeographical regions. The dominant seagrass species was *T. hemprichii*, and the dominant mangrove species, *Rhizophora stylosa* Griifith. Camp et al. (2016a) found that while mangroves can help corals tolerate low pH and provide genetic reservoirs for more resilient corals, they also increase metabolic costs, potentially limiting coral growth. Greenwood's (2021) research focused on coral adaptation, revealing that corals (*P. lutea*) in mangrove habitats showed less bleaching under thermal stress compared to reef corals, and their microbiomes rapidly adapted to habitat changes, which is key to their resilience.

Singapore

In Singapore, recent studies have revealed that natural coral recruitment and colonization on seawalls have occurred spontaneously, without deliberate planning or intervention (Tan et al. 2025). These coral recruits coexist alongside seagrasses, such as *Enhalus acoroides* (L.f.) Royle and *C. rotundata*, which have also naturally established themselves due to the presence of seawalls. For instance, at one of such locations (Tanah Merah), the combination of hard coastal protection structures and regenerated seagrass habitats supports a diverse ecosystem, including 64 species of hard corals (Wong et al. 2018), 7 species of seagrass, 23 species of macroalgae, and 357 species of fauna, excluding corals (Lim et al. 2020). However, these studies (Wong et al. 2018; Lim et al. 2020) do not specifically report the coexistence of corals and seagrasses, focusing instead on the diversity and distribution of species and highlighting the potential that these unplanned reefs have for restoring coral and seagrasses within an urban context (Tan et al. 2025).

New Caledonia

In New Caledonia, Camp et al. (2017) investigated coral communities in the semienclosed mangrove lagoons of Bouraké. These lagoons are characterized by extreme conditions, including elevated temperatures, low dissolved oxygen, and low pH compared to nearby reef habitats. The study highlights diverse coral communities that thrive under "extreme" conditions and provides insights into how corals acclimatize and adapt to these stressors and environmental changes (Camp et al. 2017).

Maggioni et al. (2021) further explored these extreme environments focusing on how these unique settings serve as natural laboratories for studying coral adaptation. Their research demonstrated how the lagoon's harsh conditions, such as high temperatures, low oxygen levels, and low pH, challenge corals in ways that reveal their remarkable adaptability. The study revealed that corals in Bouraké exhibit specific physiological and biochemical adaptations to cope with these extreme conditions, including alterations in metabolic pathways and enhanced stress response mechanisms. These adaptations are critical for understanding how coral reefs might endure future environmental changes (Maggioni et al. 2021). The findings underscore the significance of studying such extreme habitats to gain insights into coral resilience and to develop strategies for conserving coral ecosystems facing climate change and other anthropogenic pressures.

The semienclosed lagoon at Bouraké (site of previous CMC research by Camp et al. 2019 and Maggioni et al. 2021) also provided the ideal setting for the investigations into the environmental variability on the spawning synchrony and reproductive output of *Acropora tenuis* (Dana, 1846), *M. digitata*, and *M. stellata* (Alessi et al. 2024). The hypothesis was that corals at Bouraké would have reduced fecundity due to the stress of living in a highly variable environment. However, the study found that the three coral species did not show consistent decreases in fecundity, although egg size was smaller in *A. tenuis* and *M. digitata*. Differences in spawning timing were consistent among geographic areas and were largely driven by seawater temperature variations. This geographic spawning asynchrony could lead to reproductive isolation and local adaptation, promoting traits that increase fitness under future climate scenarios (Alessi et al. 2024).

Australia

The Great Barrier Reef Expedition of 1928–29 resulted in a report detailing the structure and ecology of the Low Isles, a group of small coral islands located on the Great Barrier Reef, near Port Douglas, Queensland (Stephenson et al. 1931). The study detailed the physical structure of the reefs, including the composition and distribution of coral species. It highlighted the zonation patterns of different coral types and the geomorphological features of the reef. The authors also documented the diverse marine life associated with the reefs, mostly fish, invertebrates, and algae, and noted the interactions between these organisms and their environment. Mangroves are reported multiple times, with *R. mucronata* being the dominant species. The area where most mangroves were found is described as a "tidal woodland of well-grown trees" where mangroves are found alongside other species common to littoral areas. Peculiarly, the authors report that the mangroves are "invaded" to a certain extent by corals and by other typical reef species. In a small pool, the authors describe an area of "most interesting nature" that lies between an inner rampart and a dense woodland. In this location, the bottom is sand, partly covered by *Thalassia*, and mangrove roots are covered by *Pocillopora acuta* (Lamarck, 1816), which flourishes, and other corals *Acropora*, *Montipora*, *Millepora,* and massive *Porites*.

On the Cobourg Peninsula in the Northern Territory, Ferns (1994) characterized the zonation patterns of intertidal mangle coral communities in the Popham Creek. Popham Creek, a 2 km tidal channel fringed by mangrove forest, consists of approximately 265 ha of intertidal mangrove and coral communities. Across the creek, coral species abundance and diversity were observed to be closely linked to differences in light availability, tidal velocity, subaerial exposure, and shelter, which varied across the five distinct zones. Nutrient levels and emersion time were similar throughout the system and not a determinant of zonation. Ferns (1994) characterized these zones as a shaded mangrove zone (MS), open mangrove zone (MO), reef flat (RF), deep-channel zone (DC), and soft coral zone dominated by the species, *Heliofungia actiniformis* (Quoy & Gaimard, 1833) (HA). Variation in light availability was found to be a dominant driver in coral community structure (Ferns 1994). The MS zone, characterized by low light attenuation, exhibited an overall lower coral cover compared to other well-illuminated zones, such as the MO zone (a tidal channel not covered by a mangrove canopy), RF zone, and DC zone. Despite low coral cover, species diversity was relatively high in the MS zone. Coral species displayed flattened, plate-like, and encrusting growth forms, which may represent an adaptive phenotypic response to low light availability. Coral colonies were more abundant in areas dappled in shade compared to completely shaded areas. While low light availability may present a challenge to some coral species, Ferns (1994) hypothesized about the potential importance of the mangrove canopy in reducing stress from solar radiation. Ferns (1994) also identified tidal velocity, subaerial exposure, and shelter as important parameters influencing coral community structure in Popham Creek. Tidal velocity plays a key role in removing excessive nutrients and fine sediments, replenishing oxygen levels, and maintaining stable temperature and salinity conditions, which collectively influence coral health and distribution. Low coral diversity in the HA zone was presumed to be associated with low tidal flushing rates. The proliferation of one species in this zone may be explained by its ability to remove sediment loads by its long-extended polyps. The regular flushing of sediments generally within Popham Creek may explain the proliferation of corals in this suboptimal environment. Subaerial exposure and shelter were also observed to structure coral communities among the zones. Higher coral diversity and cover in the MO zone compared to the RF zone were thought to be related to reduced storm energy and improved humidity levels provided by the adjacent mangrove forest. This protective effect likely contributes to the increased coral health and diversity observed in the MO zone.

Also working at Popham Creek, Billyard (1995) highlights a rich diversity (64 species) of corals in the area distributed in an array of different communities along the reef. Of these, exists what the authors call a central community (MS Zone) found in low light conditions under the canopy of the mangrove *R. stylosa*. Here, coral cover appeared to be linked to light availability: fewer colonies in complete shade, more appearing where canopy allowed some light to penetrate. Species recorded include *Cyphastrea serailia* (Forsskål, 1775), *F. abdita*, and *Goniastrea pectinata* (Ehrenberg, 1834). Average coral cover was 12%.

Mangrove lagoons in the Great Barrier Reef provided the setting for Camp et al. (2019) to investigate stress-tolerant coral populations to understand the biology of corals in the face of global coral degradation. Camp et al. (2019) described two extreme mangrove lagoons that condition corals (34 species) to stress while also providing refuge from more challenging locations. Their findings showed that these corals developed unique strategies to cope with stressors, such as altered metabolic processes and shifts in symbiotic relationships. The species of mangroves in the lagoons are not specified, but the ecosystem services they provide to the corals are highlighted including serving as natural laboratories for studying stressors, providing refuge, aiding climate change preconditioning, and housing stress-tolerant corals.

Complementing the Camp et al. (2019) study, Haydon et al. (2021) investigated how microbial communities and dinoflagellates (Symbiodiniaceae) in *P. acuta* corals adapt when transplanted between reef and mangrove environments. Their research highlighted that both bacterial communities and algal symbionts undergo significant adjustments in response to different habitats. The study revealed that the adaptability of coral holobionts is key to managing stress and maintaining coral health across varied environmental conditions (Haydon et al. 2021). Further exploration by Haydon (2022) focused on how extreme environments influence important microbial symbionts of corals, adding depth to our understanding of microbial and symbiont flexibility in supporting coral survival in diverse and extreme habitats (*see* Chap. 3).

Working in a mangrove lagoon in Woody Isles, Great Barrier Reef, Haydon et al. (2023) analyzed the metabolic profile of corals and found that corals in mangrove habitats have elevated metabolism through energy-generating and biosynthesis pathways compared to those in reef habitats. Under ambient conditions (~28 °C), *P. acuta* in reefs predominantly associated with *Cladocopium* endosymbionts, while those in the mangroves associated with *Durusdinium*. These associations remained stable even under heat stress. Most importantly, authors report that reef corals bleached and showed extensive shifts in symbiont metabolic profiles under heat stress, while mangrove corals did not bleach. Instead, only the host metabolite profiles were altered. Apparently, this is an indication of cellular responses in interpartner signaling, antioxidant capacity, and energy storage, and it provides valuable insights into the mechanisms that allow corals to thrive in extreme conditions, which is crucial for understanding their resilience to climate change.

Waller (2024) investigated the biogeochemical and ecological dynamics of CMC habitats, focusing on how proximity to mangroves influences seawater chemistry, coral diversity, and morphology. A global review identified significant knowledge gaps, particularly in edge habitats like the Great Barrier Reef. Fieldwork at Pioneer Bay, Orpheus Island, revealed that proximity to mangroves drives stark gradients in seawater chemistry, coral composition, and biodiversity. Corals near mangroves experience greater environmental fluctuations, favoring resilient species like *Porites*, while moderate variability at intermediate sites supports higher diversity. These findings highlight the potential of CMC edge habitats to foster coral adaptation to climate change, emphasizing the need for long-term monitoring and integrated management of these vital ecosystems.

11.7 Conservation and Management Strategies

Although it is widely accepted that thriving coastal habitats rely on connected healthy mangroves, coral reefs, and seagrass, management and conservation efforts have traditionally focused on habitat types separately. As more global intergovernmental agreements, including the United Nations Decade on Ecosystem Restoration and the Kunming-Montreal Global Biodiversity Framework, focus on nature-based solutions and ecosystem-based adaptation to combat impacts of climate change, more support is being gained for simultaneous restoration of coastal and marine habitats (Saunders et al. 2024; Vozzo et al. 2023). Carbon storage (e.g., blue carbon) and coastal protection have been two nature-based solutions frequently cited as justification for financing restoration projects and investing in conservation efforts for these ecosystems (Vanderklift et al. 2022; Lovelock et al. 2024). Essentially, conservation and management strategies are a two-prong approach, the first, to halt further harm to ecosystems and the second, to recover degraded ecosystems. Below we highlight some global efforts specific to mangroves, corals, and seagrass and provide examples of how these initiatives can be combined to incorporate multiple habitat types and nested community assemblages (e.g., CMC).

Global assessments and partnerships

The International Union for the Conservation of Nature (IUCN) Red List of Mangrove Ecosystems, completed in 2024, was the first global assessment of its kind. The assessment took 3 years to complete and found that 50% of mangrove ecosystems are at risk of collapse while 20% are endangered or critically endangered. Sea-level rise was identified as the main threat affecting mangrove ecosystems with 25% of the global mangrove area predicted to be submerged by 2050 (IUCN 2024b). Only 40% of the world's remaining mangrove forests are designated as protected (Leal and Spalding 2024). Globally, average hard coral cover has decreased from 36 ± (SE) 1.4% in 1998 to 19 ± (SE) 0.4% in 2018 (Tebbett et al. 2023). Some of the most at-risk corals are the shallow nearshore reefs, which are more directly impacted by anthropogenic factors such as coastal development, exploitation of resources, nutrient runoff, pollution, and overfishing, as well as bleaching events, and algal competition (Biscéré et al. 2015; Tebbett et al. 2023). These shallow nearshore corals are the most likely to extend into neighboring mangrove forests and seagrass beds. Thus, successful conservation and management strategies must maintain a healthy connection among these habitats. Conservation and restoration of coexisting mangrove-coral or seagrass-coral habitats requires proactive efforts and novel techniques.

The Global Mangrove Alliance (GMA - https://www.mangrovealliance.org/) launched at the World Ocean Summit in 2018 and is coordinated by Conservation International, IUCN, The Nature Conservancy, Wetlands International, and the World Wildlife Fund. The alliance grew out of the urgency of conserving and restoring mangroves. Founded on the principles of collaboration and collective action, the GMA unites diverse stakeholders from government, academia, nonprofit organizations, and industry to tackle challenges facing mangrove ecosystems. The alliance's central roles are to coordinate partnerships, develop strategies to ensure successful conservation efforts and mobilize science-based resources and tools. The Global Mangrove Alliance (GMA) set three critical goals to be achieved by 2030: (1) to halt all mangrove loss due to human action, (2) to reach 80% global mangrove protection, and (3) to restore half of the world's mangroves (Leal and Spalding 2024). Although the focus of the GMA is to promote the conservation of mangrove ecosystems, the alliance also works to protect seagrass meadows, tidal flats, and coral reefs as habitat connectivity is essential for maintaining coastal resilience and ecological functioning.

Monitoring and mapping initiatives

Mangrove mapping efforts, such as the Global Mangrove Watch (GMW - www.globalmangrovewatch.org), provide open-access annual maps of mangrove distribution. The Global Mangrove Watch project uses both optical and radar satellite data to map mangrove extent and optimized normalized difference vegetation index thresholds to identify losses and filter false positives with a temporal scoring system (Bunting et al. 2023). Mangrove loss alerts are then published monthly on the Global Mangrove Watch portal. The GMW portal is an important conservation and management tool as it enables users without technical geographic information system (GIS) or remote sensing skills to access and analyze mangrove loss in their region. Utilizing this platform enables practitioners with limited resources to monitor changes and highlight regions of concern to prioritize restoration. The GMW portal estimated global mangrove loss to be 5245.24 km^2 between 1996 and 2020 (Bunting et al. 2024). The GMA utilizes the GMW disturbance alerts to increase awareness and mangrove protection.

The Allen Coral Atlas (https://allencoralatlas.org/) is another conservation tool developed out of partnerships between coral reef scientists, NGOs, universities and private entities. The goal of the coral atlas is to assist stakeholders in reaching their coral reef conservation goals. Like the GMW portal, the coral atlas allows users access to high-resolution satellite imagery and advanced analytics. One of the global products on the Allen Coral Atlas is the Coral Reef Watch (https://coralreefwatch.noaa.gov/), which provides near real-time 5 km bleaching alerts. The Allen Coral Atlas Field Engagement team, based at the National Geographic Society, has three primary goals: (1) to raise awareness about the Atlas maps and products, (2) for the Atlas to serve as a tool to improve conservation efforts and management strategies, and (3) to facilitate collaborations.

Global distribution databases can be combined to predict the overlap of corals with mangroves and seagrass and identify locations in possible need for special protection or alternative management practices. The presence of CMC habitats has served as justification for areas to receive international protection. For instance, the occurrence of corals in mangrove channels beneath the *R. stylosa* canopy of Popham Creek was cited as national significance and a criterion for Cobourg Peninsula to be designated as a Ramsar wetland of international importance (Billyard 1995). This is a conservation tool that more natural resource management agencies could be utilizing.

Mangrove forests and coral reefs were previously predicted to overlap in a global area of 805 km^2 (Stewart et al. 2022). This estimation was produced using global mangrove (25 m spatial resolution; Spalding et al. 2010; Bunting et al. 2018) and coral reef (30 m spatial resolution; UNEP-WCMC et al. 2021) habitat extents from 1999 to 2003. Since the publication of the work by Stewart et al. (2022), updated, higher resolution global extents for both mangroves (10 m spatial resolution; Bunting et al. 2024) and coral reefs (5 m spatial resolution; Lyons et al. 2022) have been released. Repeated global mapping of CMC habitats based on these updated mangrove and coral reef extents resulted in an increased overlapping global area of 963 km^2 (Fig. 11.5). This increase highlights the impact of resolution on habitat mapping: finer resolution imagery allows for more precise delineation of habitat boundaries and a more accurate assessment of overlaps. The increased resolution can help differentiate between mangrove species and other similar vegetation, which can commonly be conflated (Pham et al. 2019). Interestingly, the use of the finer-scale habitat extents resulted in a notable increase in potential CMC habitat extent for the Caribbean and the Indian Ocean and a reduction for the Red Sea and the South China Sea. The updated analysis also reveals a general shift in the distribution of CMC habitats. The Caribbean now accounts for 33% of the global potential CMC habitat, followed by the South Pacific (30%), the Indian Ocean (17%), and the South China Sea (3%; Table 11.2). These changes underscore how advancements in technology can refine our understanding of habitat distribution and potentially inform conservation priorities.

Coastal habitats are difficult to capture with moderate-resolution satellite data, particularly small, fringing, scrub mangrove forests (Doughty et al. 2024). As very high-resolution (VHR) imagery (<5 m) becomes increasingly available and smaller pixels better match the size of wetland features, mapping accuracy among wetland classes dramatically improves (Doughty et al. 2024). The same is true for accurate mapping of CMC habitats. Technological advances are crucial to the precision of remote sensing for CMC habitats. Remote sensing aids in identifying potential CMC habitats to ground truth and monitor. With VHR mapping, we can improve our ability to globally track shifts in habitat ranges and extents, which are increasingly influenced by human activities and climate change (Nagendra et al. 2013). As advancements in remote sensing technology continue to enhance our ability to map and monitor habitat extents remotely, ground truthing will remain essential to validate these findings, particularly in challenging environments like mangrove canopies (Nagai et al. 2020). This synergy between high-resolution mapping and field verification is crucial for effective conservation and management, ensuring that targeted protection efforts are both precise and responsive to the dynamic changes occurring in these vital ecosystems.

Restoration frameworks and tracking

Ecological Mangrove Restoration (EMR), was developed in 1997 as a response to low success rates of mangrove restoration projects (Lewis and Marshall 1997). EMR is a robust scientific framework that incorporates physical and ecological attributes into restoration planning. This restoration framework mimics natural processes to reach a point where ecosystems can once again self-regulate. The EMR framework has been adapted and modified to incorporate community involvement (e.g., Community-based Ecological Mangrove Restoration) and to guide mangrove recovery following hurricane disturbance (Krauss et al. 2023). EMR has been well refined and provides the basis of best management practice guidelines around the world (Agraz Hernandez et al. 2024; Beeston et al. 2023; Teutli-Hernández et al. 2021; UNEP-Nairobi Convention/USAID/WIOMSA 2020). When properly applied, EMR and its variations, has led to great success, gaining further support. While global guidelines and resources provide a framework to aid practitioners and researchers, local ecological knowledge should be integrated in conservation and management efforts (Grimm et al. 2024). Furthermore, with climate change, restoration practices need to not only look at the needs of today, but how the environment may be different in the future. The World Wildlife Fund (WWF) designed a climate-smart mangrove tool for conservation practitioners to assess climate vulnerability and select management actions using many of the EMR steps. Below is a synopsis of EMR steps and how they can be applied for coexisting mangrove-coral or seagrass-coral habitats.

- **Co-design and collaborate:** Work at a local scale with communities and agencies to identify priorities, perceptions, and sustainable utilization. If selecting restoration sites, this should be done with a diverse stakeholder group that is knowledgeable of the system. It is important that there is mutual respect, and that knowledge is being shared effectively. An evidence-based approach is necessary for decision-making and prioritization. The group should meet frequently to evaluate the success of the project and adapt as restoration needs change. There should be clear strategies to adapt to climate change. The design of the project should include funding source to ensure there is no gap or that efforts are not abandoned prematurely. Funding is needed not just for initial restoration efforts but also for long-term monitoring and adaptive management.
- **Observe:** To understand both autoecology and community ecology of species, one must consider the physiological needs of the species they are working with, as well as and how they interact with other species in the ecosystem, which provides vital insight to potential stressors or disturbances. This helps identify suitable biophysical settings which provides a blueprint of the desired outcome. Identifying a thriving, local, natural reference ecosystem to compare restoration sites to is important for evaluating success and progress.
- **Identify:** Assess factors hampering succession. In the case of mangroves, the tidal inundation at a site might be outside of the species-specific threshold or hydrology might have been altered, no longer allowing for tidal flushing or dispersing propagules. Tidal flushing is also

important for water quality and thus impacts nearshore seagrass and coral communities. In deforested or degraded areas, the propagule source may no longer exist.

- **Remove barriers:** Once barriers have been identified, it is time to determine whether interference or mitigation is warranted. Ecosystems naturally recover and regenerate once barriers are removed. In some cases, the stressor may have been acute and no longer exists, and natural recovery is occurring but at a longer temporal scale than desired. If barriers to natural regeneration still exist, it is always best to address these barriers prior to further action (e.g., planting). Otherwise, practitioners will continue to fight these forces. For example, if the barrier is pollution, addressing water quality will have a greater positive impact than continuing to outplant mangrove propagules, seagrass plugs, or coral fragments.
- **Protect:** While an ecosystem is recovering, it has a reduced capacity to handle additional stressors. By providing protection during this time, the potential for natural recovery is increased. Protection can fall under policy, governance of management, access/closure, etc. An action, such as creating no motor traffic areas, can reduce wave energy that negatively impacts mangrove regeneration, avoid prop scars to seagrass, and prevent collisions with shallow corals.
- **Monitor:** One of the most important and overlooked steps of restoration is long-term monitoring. Monitoring is the key to evaluating the efficacy of restoration and should be incorporated into an adaptive management approach. Restoration projects should continue monitoring for no less than 5 years. Data from monitoring efforts must be standardized and accessible so that knowledge can be openly shared. Through this knowledge exchange, other practitioners in similar regions or with similar barriers can learn from one another and avoid a pattern of repeating mistakes.
- **Repeat:** Monitoring and evaluating restoration efforts help identify whether barriers have successfully been removed and whether underlying stressors are revealed. Restoration is an iterative process.

To increase the success rate of restoration efforts, the GMA developed a Mangrove Restoration Tracker Tool for mangrove practitioners to record and track their restoration projects and enable learning and information exchange between practitioners. The intent of the tool is to continue to develop best management practices globally as new challenges are faced. Sharing knowledge through open-source platforms expands the potential spatial impact of restoration and conservation efforts and limits issues of important information being inaccessible in internal agencies or governmental reports.

Climate change

Mangroves and seagrasses, labeled as blue carbon ecosystems, have been gaining the attention of countries and companies looking for means to achieve net-zero carbon emissions due to their ability to sequester and store carbon at levels far exceeding other ecosystems (Friess et al. 2022). There are two ways that natural ecosystems can help achieve net emission reductions. The first is by protecting these systems, which release greenhouse gases when they are cleared or degraded. The second is by restoring ecosystems and increasing vegetation, which can sequester carbon (Vanderklift et al. 2022). By increasing marine protected areas (e.g., 30 × 30) or expanding boundaries of marine sanctuaries, these ecosystems are buffered from direct anthropogenic pressures such as vessel groundings, prop scarring of seagrass beds, and overfishing. Some international organizations (e.g., the World Bank, the World Wildlife Fund, the United Nations Development Programme) have been advocating for alternative, resilient livelihood programs (e.g., tourism, beekeeping) to alleviate anthropogenic strain on marine resources and to compensate coastal communities for lost opportunities. However, case studies have shown livelihood interventions frequently fail to ensure economic well-being for participants (Katikiro 2016; Pham 2020). Indirect anthropogenic factors are being addressed with multistakeholder groups working to reduce effluent and enforcing stronger conservation management practices. Additionally, time and energy-intensive efforts include green-gray engineering initiatives (e.g., SEAHIVE) and massive propagation and transplanting/outplanting efforts. While short-term, large-scale planting efforts are photogenic and attractive to funders, they have been fraught with failure (Lee et al. 2019; Lovelock et al. 2022). Project scale is an important consideration for the success of restoration. Large projects can support high levels of biodiversity by reconnecting habitats and attracting more attention from investors. Meanwhile, local projects empower communities, have simpler governance, and can focus on regional specificities.

Dynamic systems

While CMC and nested seagrass-coral habitats occur globally, there are vast differences geographically. Where these habitats may be extreme environments in one location, they might be refugia in another. Additionally, the type of refugia the habitat offers may vary. For example, some CMC or CSC habitats may provide spatial refugia, mitigating abiotic stress as is observed with shading. Other CMC or CSC habitats may provide adaptive refugia where the stress tolerance of corals are enhanced (e.g., temperature or pH variability). It is important to consider the dynamics of these systems to determine the best steps forward. Here we explore the example of Caribbean reefs and mangrove forests, which have evolved under pressures of hurricane disturbance. Degraded reefs have become more resistant to severe storms, and hurricane models suggest that storms increase the probability of long-distance larval exchanges which could promote coral resilience, particularly *Agaricia* spp., *Madracis* spp., and *Porites* spp., but they can also spread diseases and nonnative or invasive species (Edmunds 2019; Mudge and Bruno 2023; Dobbelaere et al. 2024). Many Caribbean mangroves are disturbance-adapted and limit their canopy height (e.g., scrub mangrove), which are less susceptible to wind damage and storm surge attenuation (Chen et al. 2021). These features of increased long-distance coral larval exchange and reduced mangrove canopy, which are predominant in hurricane prone regions, promote CMC conditions. However, the increased frequency and magnitude of hurricanes in the western Atlantic may threaten CMC habitats in the Caribbean by not allowing enough regeneration time between storms or wiping out larval and propagule sources (Krauss et al. 2023). Furthermore, the natural zonation of mangroves makes some species more vulnerable to tropical storms based on proximity to the water. For example, mangrove species found at the fringe of mangrove forests (e.g., *R. mangle*) endure the greatest wind damage and wave attenuation. These mangroves at the fringe block winds and reduce storm surge, protecting other mangrove species further inland (e.g., *Laguncularia, Avicennia*). Following hurricane damage, *Laguncularia* and *Avicennia* have the physiological advantage of epicormic growth, which *Rhizophora* do not. These physiological differences between mangrove species can drive subsequent community composition. For example, *R. mangle* may be a dominant species prior to a hurricane, but their lack of viable dormant buds in mature trunks and branches makes them highly susceptible to mortality post-disturbance. Meanwhile, species in the genera *Avicennia* and *Laguncularia* can grow epicormic sprouts in response to canopy damage and thus regenerate and have lower mortality following storm damage (Radabaugh et al. 2020) (Fig. 11.7). Due to differences in response following repeated storms, shifts in mangrove community structure may occur. Although these shifts may not have a negative impact on the mangrove forest, the species composition change could impact whether conditions remain for CMC habitats to occur.

Prior to Hurricanes Irma and Maria, the U.S. Virgin Islands were home to some of the most specious CMC habitats. These CMC habitats were dominated by *R. mangle*, which have prop roots that branch from the trunk and hold the tree up in the sediment. In areas like Hurricane Hole, these prop roots were covered in corals. Large corals also grew at the base of roots and between roots under the canopy of the mangroves. Following the storms, the mangrove canopies were largely defoliated, and many trees suffered from snap branches and trunks. Expanses of *R. mangle* were unable to recover following physical damage from the storm and boats that crashed into the forests as well as poor water conditions for months (Rogers 2019). The roots of these trees deteriorated and the corals they held fell into the sediment as the roots decomposed. Corals that grew on rocks and sediment between the roots persisted but did suffer bleaching with the sudden destruction of the canopy. The increase in light in the area also allowed exponential algal growth. Sections of the fringing mangrove canopy were replaced with *Laguncularia.* Since *Laguncularia* does not have roots that grow into the water, they do not provide structure for corals to attach, however, they still provide shade which altered the coral distribution within the mangrove canopy (Fig. 11.8). Prior to restoration efforts, both mangroves and

Fig. 11.7 *Rhizophora*, defoliated and with limb damage from hurricane force winds, protecting *Laguncularia*. *Avicennia* epicormic growth following hurricane. (Photographs taken by H.A. Stewart)

Fig. 11.8 Mangrove fringe transitioning from *Rhizophora* dominated to a mix of *Rhizophora* and *Laguncularia* following hurricane damage. *Laguncularia* provides shade, but do not have prop roots like *Rhizophora* and thus provides less structure for corals. Corals in figure include *Orbicella annularis*, *Siderastrea siderea*, and *Colpophyllia natans*. Opening in the mangrove canopy from defoliated trees leads to increased algal growth. (Photographs taken by H.A. Stewart)

corals were naturally recovering and regenerating, but concerns of the extent of loss led to human intervention 5–6 years post-storm.

Although many novel restoration techniques are being deployed for the simultaneous restoration of mangroves, corals, and seagrass; results from these efforts have not been published or openly shared at this time. Long-term monitoring is needed to determine how these techniques may need to be altered and to refine what successful CMC restoration entails.

11.8 Conclusion

This chapter demonstrates that mangroves and seagrasses provide coral habitat globally with at least 177 species of corals in CMC habitats and 39 coral species in seagrass meadows. Poritidae, Acroporidae, and Siderastreidae were the coral families greatest represented in both mangrove and seagrass habitats. Darling et al. (2012) grouped *Acropora* species into the competitive life-history strategy, while *Porites* and *Siderastrea* species were divided among multiple groups. The specific *Porites* species observed in CMC and CSC habitats equally divide between stress-tolerant and weedy life-history strategies. Similarly the two species of Siderastrea also fall between these two groups. The difference in the life-history strategies of the most common corals further supports the nuance and complexity of these CMC and CSC habitats as more than just categories within marginal or extreme habitats (Schoepf et al. 2023). Whether this is a new novel ecosystem or a marginal habitat attracting more attention as reefs degrade worldwide, these ecosystems demand further scientific investigation to improve our understanding of facilitation cascades and ecosystem functioning within these intertwined habitats. Furthermore, as the case studies highlight, these unique habitats require specialized conservation and management techniques, which will help create stronger marine ecosystems and protect biodiversity in the face of climate change.

Acknowledgments The authors wish to thank K. Yates for her insight, expertise, and contributions to this chapter. Thank you to K. Novelo and J.R. Finnerty as well for use of their photographs.

References

Agraz Hernandez C, Andrieu J, Cormier-Salem MC et al (2024) Guide méthodologique pour la restauration des mangroves. Gabrié C, Etienne JE Fonds français pour l'environnement mondial (FFEM), p 138

Ahmad F, Azman S, Said MIM et al (2014) Metals in tropical seagrass—accumulation of mercury and lead. World Appl Sci J 32(8):1468–1473. https://doi.org/10.5829/idosi.wasj.2014.32.08.1028

Albright R, Mason B, Miller M et al (2010) Ocean acidification compromises recruitment success of the threatened Caribbean coral *Acropora palmata*. Proc Natl Acad Sci USA 107:20400–20404. https://doi.org/10.1073/pnas.1007273107

Alessi C, Randall CJ, Dumas M et al (2024) Spawning window and fecundity in three Acroporid corals from the environmentally variable semi-enclosed lagoon of Bouraké. Coral Reefs 43:1469–1482. https://doi.org/10.1007/s00338-024-02551-3

Almy CC, Carrion-Torres C (1963) Shallow-water stony corals of Puerto Rico. Institute of Caribbean Science, University of Puerto Rico and University of Miami. Available via NOAA Miami Regional Library. https://www.aoml.noaa.gov/general/lib/CREWS/Cleo/PuertoRico/prpdfs/almy-shallowwater.pdf. Accessed 9 Oct 2024

Alongi DM (2018) Climate regulation by capturing carbon in mangroves. In: Finlayson CM et al (eds) The wetland book. Springer, Dordrecht. https://doi.org/10.1007/978-90-481-9659-3_236

Alongi DM, Trott LA, Tirendi F et al (2008) Growth and development of mangrove forests overlying smothered coral reefs, Sulawesi and Sumatra, Indonesia. Mar Ecol Prog Ser 370:97–109

Anthony KR, Diaz-Pulido G, Verlinden N et al (2013) Benthic buffers and boosters of ocean acidification on coral reefs. Biogeosciences 10:4897–4909. https://doi.org/10.5194/bg-10-4897-2013

Aoki N, Weiss B, Jézéquel Y et al (2024) Soundscape enrichment increases larval settlement rates for the brooding coral *Porites astreoides*. R Soc Open Sci 11:231514. https://doi.org/10.1098/rsos.231514

Ashrafuzzaman M (2023) Mangrove is the facto nature-based solutions to tackle the climate change around the globe. In: Leal Filho W, Nagy GJ, Ayal D (eds) Handbook of nature-based solutions to mitigation and adaptation to climate change. Springer, Cham. https://doi.org/10.1007/978-3-030-98067-2_61-1

Attri K, Kerkar S (2011) Seasonal assessment of heavy metal pollution in tropical mangrove sediments (Goa, India). J Ecobiotcchnol 3(8):9–15

Barry SC, Frazer TK, Jacoby CA (2013) Production and carbonate dynamics of *Halimeda incrassata* (Ellis) Lamouroux altered by *Thalassia testudinum* banks and *Soland ex* König. J Exp Mar Biol Ecol 444:73–80. https://doi.org/10.1016/j.jembe.2013.03.012

Becking LE, Martinez SJ, Aji LP et al (2024) Stony corals and their associated fauna residing in marine lakes under extreme environmental conditions. Diversity 16(5):295. https://doi.org/10.3390/d16050295

Beeston M, Cameron C, Hagger V et al (2023) Best practice guidelines for mangrove restoration. Global mangrove alliance and blue carbon initiative. The Blue Carbon Initiative and Global Mangrove Alliance. Available via Global Mangrove Alliance. https://www.mangrovealliance.org/best-practice-guidelines-for-mangrove-restoration/. Accessed 9 Oct 2024

Bellwood DR, Pratchett MS, Morrison TH et al (2019) Coral reef conservation in the Anthropocene: confronting spatial mismatches and prioritizing functions. Biol Conserv 236:604–615

Bengtsson ZA, Kuhn KM, Battaglino AT et al (2019) Corals of the genus *Porites* are a locally abundant component of the epibiont community on mangrove prop roots at Calabash Caye, Turneffe Atoll, Belize. Caribbean Nat 67:1–16

Berumen ML, Arrigoni R, Bouwmeester J et al (2019) Corals of the Red Sea. In: Voolstra C, Berumen M (eds) Coral reefs of the Red Sea. Coral reefs of the world, vol 11. Springer, Cham, pp 123–155. https://doi.org/10.1007/978-3-030-05802-9_7

Bhowmik AK, Padmanaban R, Cabral P et al (2022) Global mangrove deforestation and its interacting social-ecological drivers: a systematic review and synthesis. Sustainability. 14:4433. https://doi.org/10.3390/su14084433

Billyard RW (1995) The cultural and ecological significance of Popham Creek, Cobourg Peninsula. Parks and Wildlife Commission of the Northern Territory, Darwin, Northen Territory

Bimrah K, Dasgupta R, Hashimoto S et al (2022) Ecosystem Services of Mangroves: a systematic review and synthesis of contemporary scientific literature. Sustainability 14(19):12051. https://doi.org/10.3390/su141912051

Biscéré T, Rodolfo-Metalpa R, Lorrain A et al (2015) Responses of two scleractinian corals to cobalt pollution and ocean acidification. PLoS One 10:e0122898. https://doi.org/10.1371/journal.pone.0122898

Blaber SJM, Blaber TG (1980) Factors affecting the distribution of juvenile estuarine and inshore fish. J Fish Biol 17:143–162

Blaber SJM, Milton DA (1990) Species composition, community structure and zoogeography of fishes of mangrove estuaries in the Solomon Islands. Mar Biol 105:259–267

Bunting P, Rosenqvist A, Lucas RM et al (2018) The global mangrove watch—a new 2010 global baseline of mangrove extent. Remote Sens 10:1–19. https://doi.org/10.3390/rs10101669

Bunting P, Hilarides L, Rosenqvist A et al (2023) Global mangrove watch: monthly alerts of mangrove loss for Africa. Remote Sens 15(8):2050. https://doi.org/10.3390/rs15082050

Bunting P et al (2024) Global Mangrove Watch: Annual Mangrove Extent (4.0.19) Zenodo. https://doi.org/10.5281/zenodo.12756047

Buob AM (2019) Characterization of red mangrove proproot epibiont communities of St. Johns USVI. Dissertation, Montclair State University

Burke S, Pottier P, Lagisz M et al (2023) The impact of rising temperatures on the prevalence of coral diseases and its predictability: a global meta-analysis. Ecol Lett 26:1466–1481. https://doi.org/10.1111/ele.14266

Camp EF, Suggett DJ, Gendron G et al (2016a) Mangrove and seagrass beds provide different biogeochemical services for corals threatened by climate change. Front Mar Sci 3:52. https://doi.org/10.3389/fmars.2016.00052

Camp EF, Smith DJ, Evenhuis C et al (2016b) Acclimatization to high-variance habitats does not enhance physiological tolerance of two key Caribbean corals to future temperature and pH. Proc R Soc B 283:1831. https://doi.org/10.1098/rspb.2016.0442

Camp EF, Nitschke MR, Rodolfo-Metalpa R et al (2017) Reef-building corals thrive within hot-acidified and deoxygenated waters. Sci Rep 7(1):2434. https://doi.org/10.1038/s41598-017-02383-y

Camp EF, Edmondson J, Doheny A et al (2019) Mangrove lagoons of the Great Barrier Reef support coral populations persisting under extreme environmental conditions. Mar Ecol Prog Ser 625:1–14. https://doi.org/10.3354/meps13073

Carlson RR, Evans LJ, Foo SA et al (2021) Synergistic benefits of conserving land-sea ecosystems. Glob Ecol Conserv 28:e01684. https://doi.org/10.1016/j.gecco.2021.e01684

Challener RC, Robbins LL, McClintock JB et al (2016) Variability of the carbonate chemistry in a shallow, seagrass-dominated

ecosystem: implications for ocean acidification experiments. Mar Freshw Res 67:163–172. https://doi.org/10.1071/MF14219
Chapman VJ (1976) Mangrove vegetation. J Cramer, Leutershausen, p 447
Chen Y, Li Y, Thompson C et al (2018) Differential sediment trapping abilities of mangrove and saltmarsh vegetation in a subtropical estuary. Geomorphology 318:270–282. https://doi.org/10.1016/j.geomorph.2018.06.018
Chen Q, Li Y, Kelly DM et al (2021) Improved modeling of the role of mangroves in storm surge attenuation. Estuar Coast Shelf Sci 260:107515. https://doi.org/10.1016/j.ecss.2021.107515
Conti-Jerpe IE, Thompson PD, Wong CWM et al (2020) Trophic strategy and bleaching resistance in reef-building corals. Sci Adv 6(15):eaaz5443. https://doi.org/10.1126/sciadv.aaz5443
Dahlgren C, Sherman K, Anderson L et al (2014) Rapid ecological assessment of the fish and benthic communities for East Grand Bahama. Bahamas National Trust, Nassau
Dai CF, Fan TY, Wu CS (1995) Coral fauna of Tungsha Tao (Pratas Islands). Acta Oceanog Taiwan 34(3):1–16
Darling ES, Alvarez-Filip L, Oliver TA et al (2012) Evaluating life-history strategies of reef corals from species traits. Ecol Lett 15:1378–1386. https://doi.org/10.1111/j.1461-0248.2012.01861.x
Davies SW, Meyer E, Guermond SM et al (2014) A cross-ocean comparison of responses to settlement cues in reef-building corals. PeerJ 2:e333
de Boer WF (2007) Seagrass–sediment interactions, positive feedbacks and critical thresholds for occurrence: a review. Hydrobiologia 591:5–24. https://doi.org/10.1007/s10750-007-0780-9
de la Guardia E, Angulo J, González-Sansón G et al (2004) Biodiversity in the diving area of the Punta Francés National Park, Isla de la Juventud, Cuba. Rev Invest Mar 25(2):90–102
den Hartog C (1970) The seagrasses of the world. North-Holland Publishing Company, Amsterdam, pp 1–276
den Hartog C, Kuo J (2006) Taxonomy and biogeography of seagrasses. In: Larkum ADW, Orth RJ, Duarte CM (eds) Seagrasses: biology, ecology and conservation. Springer, Dordrecht, pp 1–23
DeVantier L, Turak M (2017) Species richness and relative abundance of reef-building corals in the Indo-West Pacific. Diversity 9(3):25. https://doi.org/10.3390/d9030025
Dobbelaere T, Dekens A, Saint-Amand A et al (2024) Hurricanes enhance coral connectivity but also superspread coral diseases. Glob Chang Biol 30(6):e17382. https://doi.org/10.1111/gcb.17382
Donato DC, Kauffman JB, Murdiyarso D et al (2011) Mangroves among the most carbon-rich forests in the tropics. Nat Geosci 4:293–297. https://doi.org/10.1038/ngeo1123
Donnelly R, Neville D, Mous PJ (2003) Report on a rapid ecological assessment of the Raja Ampat Islands, Papua, Eastern Indonesia, held October 30—November 22, 2002 (Final Draft). The Nature Conservancy – Southeast Asia Center for Marine Protected Areas, Sanur, p 246
Doughty CL, Cavanaugh KC, Chapman S et al (2024) Uncovering mangrove range limits using very-high-resolution satellite imagery to detect fine-scale mangrove and saltmarsh habitats in dynamic coastal ecotones. Remote Sens Ecol Conserv 7(3):369–381. https://doi.org/10.1002/rse2.195
Du J, Hu W, Nagelkerken I et al (2020) Seagrass meadows provide multiple benefits to adjacent coral reefs through various microhabitat functions. Ecosyst Health Sustain 6(1):1812433. https://doi.org/10.1080/20964129.2020.1812433
Duarte CM (1991) Seagrass depth limits. Aquat Bot 40:363–377. https://doi.org/10.1016/0304-3770(91)90081-F
Duarte CM (2002) The future of seagrass meadows. Environmen conserv 29(2):192–206
Duarte CM, Middelburg JJ, Caraco N (2005) Major role of marine vegetation on the oceanic carbon cycle. Biogeosciences 2(1):1–8. https://doi.org/10.5194/bg-2-1-2005
Duarte CM, Marbà N, Gacia E et al (2010) Seagrass community metabolism: assessing the carbon sink capacity of seagrass meadows. Global Biogeochem Cycles 24(4). https://doi.org/10.1029/2010GB003793
Dubuc A, Waltham NJ, Baker R et al (2019) Patterns of fish utilization in a tropical Indo-Pacific mangrove-coral seascape, New Calendona. PLoS One 14(4):e0207168. https://doi.org/10.1371/journal.pone.0207168
Duke NC (2017) Mangrove floristics and biogeography revisited: further deductions from biodiversity hot spots, ancestral discontinuities, and common evolutionary processes. In: Rivera-Monroy V, Lee S, Kristensen E (eds) Mangrove ecosystems: a global biogeographic perspective. Springer, Cham, pp 17–53
Edmunds PJ (2019) Three decades of degradation lead to diminished impacts of severe hurricanes on Caribbean reefs. Ecology 100(3):e02587. https://doi.org/10.1002/ecy.2587
Farnsworth EJ, Ellison AM (1996) Scale-dependent spatial and temporal variability in biogeography of mangrove root epibiont communities. Ecol Monogr 66:45–66. https://www.jstor.org/stable/2963480
Ferns LW (1994) Preliminary investigation of Scleractinian coral assemblages of Popham Creek, Cobourg Peninsula: Patterns of zonation through an unusual tidal channel system. Technical report. Parks and Wildlife Commission of the Northern Territory, Darwin. https://www.researchgate.net/profile/Lawrance-Ferns/publication/312379297_Preliminary_investigation_of_scleractinian_coral_assemblages_of_Popham_Creek_Cobourg_Peninsula_Northern_Territory_Patterns_of_zonation_through_an_unusual_tidal_channel_system/links/587cb96608ae9a860ff0aadd/Preliminary-investigation-of-scleractinian-coral-assemblages-of-Popham-Creek-Cobourg-Peninsula-Northern-Territory-Patterns-of-zonation-through-an-unusual-tidal-channel-system.pdf
Fishelson L (1971) Ecology and distribution of the benthic fauna in the shallow waters of the Red Sea. Mar Biol 10:113–133
Flowers K, Golembeski P, Ross B et al (2023) Visual surveys reveal coral growth in mangrove fringe in a subtropical metropolis. Bull Mar Sci 99(3):145–146. https://doi.org/10.5343/bms.2023.0016
Foster T, Gilmour JP (2016) Seeing red: coral larvae are attracted to healthy-looking reefs. Mar Ecol Prog Ser 559:65–71
Fourqurean JW, Duarte CM, Kennedy H et al (2012) Seagrass ecosystems as a globally significant carbon stock. Nat Geosci 5:505–509. https://doi.org/10.1038/ngeo1477
Frieler K, Meinshausen M, Golly A et al (2013) Limiting global warming to 2 °C is unlikely to save most coral reefs. Nat Clim Chang 3:165–170. https://doi.org/10.1038/nclimate1674
Friess DA, Howard J, Huxham M et al (2022) Capitalizing on the global financial interest in blue carbon. PLoS Clim 1(8):e0000061. https://doi.org/10.1371/journal.pclm.0000061
Gillis LG, Bouma TJ, Jones CG (2014) Potential for landscape-scale positive interactions among tropical marine ecosystems. Mar Ecol Prog Ser 503:289–303. https://doi.org/10.3354/meps10716
Golbuu Y, Fabricius K, Victor S et al (2008) Gradients in coral reef communities exposed to muddy river discharge in Pohnpei, Micronesia. Estuar Coast Shelf Sci 76:14–20. https://doi.org/10.1016/j.ecss.2007.06.005
González-Ferrer S, Martínez-Daranas B, Cano Mallo M (2006) Manglares, pastos marinos y arrecifes coralinos. In: Menéndez Carrera L, Guzmán Menéndez JM (eds) Ecosistemas de manglar en el archipiélago Cubano. Editorial Academia, Cuba, pp 266–278
Granek EF, Compton JE, Phillips DL (2009) Mangrove-exported nutrient incorporation by sessile coral reef invertebrates. Ecosystems 12:462–472. https://doi.org/10.1007/s10021-009-9235-7

Green EP, Short FT (2003) World atlas of seagrasses. University of California Press, Berkeley, p 332. https://archive.org/details/worldatlasofseag03gree

Greenwood BN (2021) Corals at the extreme: Partitioning the response of coral holobionts to marginal habitats. Dissertation, University of Essex

Grimm K, Spalding M, Leal M et al (2024) Including local ecological knowledge (LEK) in mangrove conservation & restoration. A best-practice guide for practitioners and researchers. Global Mangrove Alliance. https://doi.org/10.5479/10088/118227

Guadayol O, Silbiger NJ, Donahue MJ et al (2014) Patterns in temporal variability of temperature, oxygen, and pH along an environmental gradient in a coral reef. PLoS One 9(1). https://doi.org/10.1371/journal.pone.0085213

Guannel G, Arkema K, Ruggiero P et al (2016) The power of three: coral reefs, seagrasses and mangroves protect coastal regions and increase their resilience. PLoS One 11:e0158094. https://doi.org/10.1371/journal.pone.0158094

Guerra-Vargas LA, Gillis LG, Mancera-Pineda JE (2020) Stronger together: do coral reefs enhance seagrass meadows "blue carbon" potential? Front Mar Sci 7:628. https://doi.org/10.3389/fmars.2020.00628

Hamilton LS, Snedaker SC (1984) Handbook for mangrove area management. Honolulu, UNEP and East West Center, Environment and Policy Institute, p 126

Harborne AR, Mumby PJ, Micheli F et al (2006) The functional value of Caribbean coral reef, seagrass and mangrove habitats to ecosystem processes. Adv Mar Biol 50:57–189

Harborne AR, Rogers A, Bozec YM et al (2017) Multiple stressors and the functioning of coral reefs. Annu Rev Mar Sci 9(1):445–468

Harrison PL, Wallace CC (1990) Reproduction, dispersal and recruitment of scleractinian corals. In: Dubinsky Z (ed) Coral reefs. Ecosystems of the world, vol 25. Elsevier, Amsterdam, pp 133–207

Haydon TD (2022) How do extreme environments influence important microbial symbionts of corals?. Dissertation, University of Technology Sydney

Haydon TD, Matthews JL, Seymour JR et al (2021) Rapid shifts in bacterial communities and homogeneity of Symbiodiniaceae in colonies of *Pocillopora acuta* transplanted between reef and mangrove environments. Front Microbiol 12:756091. https://doi.org/10.3389/fmicb.2021.756091

Haydon TD, Seymour JR, Raina JB et al (2023) Metabolomic signatures of corals thriving across extreme reef habitats reveal strategies of heat stress tolerance. Proc R Soc B 290(1992):20221877. https://doi.org/10.1098/rspb.2022.1877

Hemminga M, Duarte CM (2000) Seagrass ecology. Cambridge University Press, Cambridge

Hendriks IE, Olsen YS, Ramajo L et al (2014) Photosynthetic activity buffers ocean acidification in seagrass Meadows. Biogeosciences 11:333–346. https://doi.org/10.5194/bg-11-333-2014

Hernández-Fernández L (2015) Stony corals on submerged roots of *Rhizophora mangle* L. from the Jardines de la Reina National Park, Cuba. J Mar Res 35:17–21

Higuchi T, Takagi KK, Matoba K et al (2014) The nutrient and carbon dynamics that mutually benefit coral and seagrass in mixed habitats under the influence of groundwater at Bise Coral Reef, Okinawa, Japan. Int J Mar Sci 4(1):1–15. https://doi.org/10.5376/ijms.2014.04.0001

Hilmi N, Benitez Carranco MB, Broussard D et al (2023) Tropical blue carbon: solutions and perspectives for valuations of carbon sequestration. Front Climate 5:1169663. https://doi.org/10.3389/fclim.2023.1169663

Hofmann GE, Todgham AE (2010) Living in the now: physiological mechanisms to tolerate a rapidly changing environment. Annu Rev Physiol 72:127–145. https://doi.org/10.1146/annurev-physiol-021909-135900

Hogarth PJ (2015) The biology of mangroves and seagrasses, 3rd edn. Oxford University Press, Oxford

Hubbard JAE, Pocock YP (1972) Sediment rejection by recent scleractinian corals: a key to palaeo-environmental reconstruction. Geol Rundsch 61:598–626

Hughes TP, Rodrigues MJ, Bellwood DR et al (2007) Phase shifts, herbivory, and the resilience of coral reefs to climate change. Curr Biol 17(4):360–365. https://doi.org/10.1016/j.cub.2006.12.049

Hughes S, Yau A, Max L et al (2012) A framework to assess national level vulnerability from the perspective of food security: the case of coral reef fisheries. Environ Sci Pol 23:95–108. https://doi.org/10.1016/j.envsci.2012.07.012

Huxham M, Dencer-Brown A, Diele K et al (2017) Mangroves and people: local ecosystem services in a changing climate. In: Rivera-Monroy V, Lee S, Kristensen E, Twilley R (eds) Mangrove ecosystems: a global biogeographic perspective. Springer, Cham. https://doi.org/10.1007/978-3-319-62206-4_8

International Union for the Conservation of Nature (IUCN) (2024a) Q&A: IUCN Red List and threatened species coral assessment. Available via IUCN https://iucn.org/story/202411/qa-iucn-red-list-threatened-species-corals-assessment. Accessed 26 Jan 2025

International Union for the Conservation of Nature (IUCN) (2024b) Red List of Mangrove Ecosystems. Available via IUCN https://iucn.org/resources/conservation-tool/iucn-red-list-ecosystems/red-list-mangrove-ecosystems. Accessed 9 Oct 2024

Jackson JBC (1977) Competition on marine hard substrata: the adaptive significance of solitary and colonial strategies. Am Nat 111(980):743–767. https://doi.org/10.1086/283203

Jackson JBC (1986) Distribution and ecology of clonal and aclonal benthic invertebrates. In: Jackson JBC, Buss LW, Cook RE (eds) Population biology and evolution of clonal organisms. Yale University Press, New Haven, pp 297–356

James K, Macreadie PI, Kamenos NA (2024) It's time to broaden what we consider a 'blue carbon ecosystem'. Glob Chang Biol 30(5):e17261. https://doi.org/10.1111/gcb.17261

Jayathilake DRM, Costello MJ (2018) A modelled global distribution of the seagrass biome. Biol Conserv 226:120–126. https://doi.org/10.1016/j.biocon.2018.07.009

Jeffrey CFG, Clark R, Hile SD (2010) Spatial patterns in benthic composition of nearshore seascapes and implications for Scarid populations and fisheries in La Parguera, SW Puerto Rico. In: Proceedings of the 62nd gulf and Caribbean Fisheries Institute. Gulf and Caribbean Fisheries Institute, c/o Harbor Branch Oceanographic Institution, Inc, Fort Pierce, pp 36–53

Jin P, Zhang J, Wan J et al (2021) The combined effects of ocean acidification and heavy metals on marine organisms: a meta-analysis. Front Mar Sci 8:801889. https://doi.org/10.3389/fmars.2021.801889

Johnson KG (1992) Population dynamics of a free-living coral: recruitment growth and survivorship of *Manicina areolata* (Linnaeus) on the Caribbean coast of Panama. J Exp Mar Biol Ecol 164:171–191. https://doi.org/10.1016/0022-0981(92)90173-8

Jompa J, McCook LJ (2003) Coral-algal competition: macroalgae with different properties have different effects on corals. Mar Ecol Prog Ser 258:87–95. https://doi.org/10.3354/meps258087

Kahng SE, Hochberg EJ, Apprill A et al (2012) Efficient light harvesting in deep-water zooxanthellate corals. Mar Ecol Prog Ser 455:65–77. https://doi.org/10.3354/meps09657

Kahng SE, Akkaynak D, Shlesinger T et al (2019) Light, temperature, photosynthesis, heterotrophy, and the lower depth limits of mesophotic coral ecosystems. In: Loya Y et al (eds) Mesophotic coral ecosystems, coral reefs of the world 12. Springer Nature, Cham, pp 801–828. https://doi.org/10.1007/978-3-319-92735-0_42

Katikiro RE (2016) Improving alternative livelihood interventions in marine protected areas: a case study in Tanzania. Mar Pol 70:22–29. https://doi.org/10.1016/j.marpol.2016.04.025

Kellogg CA, Moyer RP, Jacobsen M et al (2020) Identifying mangrove-coral habitats in the Florida Keys. PeerJ 8:e9776. https://doi.org/10.7717/peerj.9776

Kennedy H, Björk M (2009) Seagrass meadows. In: Laffoley, D.d'A. & Grimsditch, G. (eds). 2009. The management of natural coastal carbon sinks. IUCN, Gland, Switzerland. 53 pp.

Khalil ASM (2015) Mangroves of the Red Sea. In: Rasul N, Stewart I (eds) The Red Sea. Springer, Berlin, pp 585–597. https://doi.org/10.1007/978-3-662-45201-1_33

Kitheka JU (1996) Water circulation and coastal trapping of brackish water in a tropical mangrove-dominated bay in Kenya. Limnol Oceanogr 41(1):169–176. https://doi.org/10.4319/lo.1996.41.1.0169

Krauss KW, Whelan KR, Kennedy JP et al (2023) Framework for facilitating mangrove recovery after hurricanes on Caribbean islands. Restor Ecol 31(7):e13885. https://doi.org/10.1111/rec.13885

Kumar S (2023) Exploring the wonders: 10 best mangrove forests in the world. HT Travel. Times. https://httraveltimes.com/article/exploring-the-wonders-10-best-mangrove-forests-in-the-world/50235

Kumari A, Rathore MS (2021) Roles of mangroves in combating the climate change. In: Rastogi RP, Phulwaria M, Gupta DK (eds) Mangroves: ecology, biodiversity and management. Springer, Singapore. https://doi.org/10.1007/978-981-16-2494-0_10

Laegdsgaard P, Johnson CR (1995) Mangrove habitats as nurseries: unique assemblages of juvenile fish in subtropical mangroves in eastern Australia. Mar Ecol Prog Ser 126:67–81

Lai S, Gillis LG, Mueller C et al (2013) First experimental evidence of corals feeding on seagrass matter. Coral Reefs 32:1061–1064. https://doi.org/10.1007/s00338-013-1062-9

Landau M, Curtis M, Reiley S (2013) A note on the distribution of some sponges and corals in a seagrass bed, Long Key, Florida. Gulf Mex Sci 1–2:67–73. https://doi.org/10.18785/goms.3101.06

Leal M, Spalding MD (eds) (2024) The state of the world's mangroves 2024. Global Mangrove Alliance. https://doi.org/10.5479/10088/119867

Lee SY (1995) Mangrove outwelling: a review. Hydrobiologia 295:203–212

Lee SY, Hamilton S, Barbier EB (2019) Better restoration policies are needed to conserve mangrove ecosystems. Nat Ecol Evol 3(6):870–872. https://doi.org/10.1038/s41559-019-0861-y

Lewis RR, Marshall MJ (1997) Principles of successful restoration of shrimp aquaculture ponds back to mangrove forests. Programa/resumenes de Marcuba 1997, Havana, Cuba September 15/20, Palacio de Convenciones de La Habana, Cuba, p 126

Li Y, Chen F, Zhou R et al (2023) A review of metal contamination in seagrasses with an emphasis on metal kinetics and detoxification. J Hazard Mater 454:131500. https://doi.org/10.1016/j.jhazmat.2023.131500

Lim LJW, Loh JBY, Lim AJS et al (2020) Diversity and distribution of intertidal marine species in Singapore. Raffles Bull Zool 68:396–403. https://doi.org/10.26107/rbz-2020-0056

Lima MAC, Bergamo TF, Ward RD et al (2023) A review of seagrass ecosystem services: providing nature-based solutions for a changing world. Hydrobiologia 850:2655–2670. https://doi.org/10.1007/s10750-023-05244-0

Lirman D, Manzello D (2009) Patterns of resistance and resilience of the stress-tolerant coral *Siderastrea radians* (Pallas) to sub-optimal salinity and sediment burial. J Exp Mar Biol Ecol 369:72–77. https://doi.org/10.1016/j.jembe.2008.10.024

Lohr KE, Smith DJ, Suggett DJ et al (2017) Coral community structure and recruitment in seagrass meadows. Front Mar Sci 4:388. https://doi.org/10.3389/fmars.2017.00388

Lonard RI, Judd FW, DeYoe HR et al (2021) Biology and ecology of the halophyte *Laguncularia racemosa* (L.) Gaertn. f.: a review. In: Grigore MN (ed) Handbook of halophytes: from molecules to ecosystems towards biosaline agriculture. Springer, Cham, pp 1–16. https://doi.org/10.1007/978-3-030-57635-6_71

Louis YD, Cerri F, Galli P (2025) Are coexisting mangrove-coral assemblages missing in the Maldives? Mar Biodivers 55:3. https://doi.org/10.1007/s12526-024-01490-0

Lovelock CE, Barbier E, Duarte CM (2022) Tackling the mangrove restoration challenge. PLoS Biol 20(10):e3001836. https://doi.org/10.1371/journal.pbio.3001836

Lovelock CE, Bennion V, de Oliveira M et al (2024) Mangrove ecology guiding the use of mangroves as nature-based solutions. J Ecol. https://doi.org/10.1111/1365-2745.14383

Loya Y (1976) The Red Sea coral *Stylophora pistillata* is an r strategist. Nature 259(5543):478–480

Lugo AE, Snedaker SC (1974) The ecology of mangroves. Annu Rev Ecol Syst 5:39–64. https://doi.org/10.1146/annurev.es.05.110174.000351

Lyons M, Larsen K, Skone M (2022) Allen Coral Atlas Imagery, maps and monitoring of the world's tropical coral reefs. V1.3 Zenodo https://doi.org/10.5281/zenodo.3833242

MacFarlane GR, Pulkownik A, Burchett M (2003) Accumulation and distribution of heavy metals in the grey mangrove, *Avicennia marina* (Forsk.) Vierh.: biological indication potential. Environ Pollut 123(1):139–151

MacFarlane GR, Koller CE, Blomberg SP (2007) Accumulation and partitioning of heavy metals in mangroves: a synthesis of field-based studies. Chemosphere 69(9):1454–1464

Macintyre IG, Goodbody I, Rützler K et al (2000) A general biological and geological survey of the rims of ponds in the major mangrove islands of the Pelican Cays, Belize. Atoll Res Bull 467:15–44. https://doi.org/10.5479/si.00775630.467.15

Mack MR, Langley JA, Feller IC et al (2024) The ecological consequences of nutrient enrichment in mangroves. Estuar Coast Shelf Sci 108690. https://doi.org/10.1016/j.ecss.2024.108690

Macreadie PI, Costa MDP, Atwood TB et al (2021) Blue carbon as a natural climate solution. Nat Rev Earth Environ 2:826–839. https://doi.org/10.1038/s43017-021-00224-1

Maggioni F, Pujo-Pay M, Aucan J et al (2021) The Bouraké semi-enclosed lagoon (New Caledonia)—a natural laboratory to study the lifelong adaptation of a coral reef ecosystem to extreme environmental conditions. Biogeosciences 18(18):5117–5140. https://doi.org/10.5194/bg-18-5117-2021

Manikandan S, Ganesapandian S, Singh M et al (2011) Seagrass diversity and associated flora and fauna in the coral reef ecosystem of the Gulf of Mannar Southeast Coast of India. Res J Environ Earth Sci 3:321–326

Manzello DP, Enochs IC, Melo N et al (2012) Ocean acidification refugia of the Florida reef tract. PLoS One 7:e41715. https://doi.org/10.1371/journal.pone.0041715

McKenzie LJ, Nordlund LM, Jones BL et al (2020) The global distribution of seagrass meadows. Environ Res Lett 15:074041. https://doi.org/10.1088/1748-9326/ab7d06

McMahon KW, Berumen ML, Thorrold SR (2012) Linking habitat mosaics and connectivity in a coral reef seascape. PNAS 109(38):15372–15376. https://doi.org/10.1073/pnas.1206378109

McManus JW (1988) Coral reefs of the ASEAN region: status and management. Ambio 17(3):189–193

Meekaew A, Casareto BE, Higuchi T et al (2014) Dissolved organic carbon cycling and the roles of the microbial community in coexistence of corals and seagrasses in Bise Okinawa Japan. Eco-Engin 26:81–88

Meyer JL, Schultz ET, Helfman GS (1983) Fish schools: an asset to corals. Science 220:1047–1049
Moberg F, Rönnbäck P (2003) Ecosystem services of the tropical seascape: interactions, substitutions, and restoration. Ocean Coastal Manag 46(1–2):27–46
Mudge L, Bruno JF (2023) Disturbance intensification is altering the trait composition of Caribbean reefs, locking them into a low functioning state. Sci Rep 13(1):14022. https://doi.org/10.1038/s41598-023-40672-x
Nagai S, Nasahara KN, Akitsu TK et al (2020) Importance of the collection of abundant ground-truth data for accurate detection of spatial and temporal variability of vegetation by satellite remote sensing. In: Dontsova K, Balogh-Brunstad Z, Le Roux G (eds) Biogeochemical cycles: ecological drivers and environmental impact. Wiley, Hoboken, pp 223–244. https://doi.org/10.1002/9781119413332.ch11
Nagelkerken I, Dorenbosch M, Verberk WCEP et al (2000) Day-night shifts of fishes between shallow-water biotopes of a Caribbean bay, with emphasis on the nocturnal feeding of Haemulidae and Lutjanidae. Mar Ecol Prog Ser 194:55–64
Nagelkerken I, Kleijnen S, Klop T et al (2001) Dependence of Caribbean reef fishes on mangroves and seagrass beds as nursery habitats: a comparison of fish faunas between bays with and without mangroves/seagrass beds. Mar Ecol Prog Ser 214:225–235
Nagendra H, Lucas R, Honrado JP et al (2013) Remote sensing for conservation monitoring: assessing protected areas, habitat extent, habitat condition, species diversity, and threats. Ecol Indic 33:45–59. https://doi.org/10.1016/j.ecolind.2012.09.014
Nalley EM, Tuttle LJ, Conklin EE et al (2023) A systematic review and meta-analysis of the direct effects of nutrients on corals. Sci Total Environ 856:159093. https://doi.org/10.1016/j.scitotenv.2022.159093
Nanami A, Yamada H (2009) Seasonality, lunar periodicity of settlement and microhabitat association of juvenile humpback red snapper *Lutjanus gibbus* (Lutjanidae) in an Okinawan coral reef. Mar Biol 156:407–414. https://doi.org/10.1007/s00227-008-1093-x
Ninomiya S, Kurahashi S, Masumoto T et al (2006) A study on comprehensive management of coral reef areas: restoration technologies focusing on a symbiotic relationship between corals and seagrasses in moats. In: Proceedings of 10th international coral reef symposium, pp 1617–1626
Nordlund LM, Unsworth RKF, Gullström M et al (2018) Global significance of seagrass fishery activity. Fish Fish 19:399–412. https://doi.org/10.1111/faf.12259
Obura D (2012) The diversity and biogeography of Western Indian Ocean reef-building corals. PLoS One 7(9):e45013. https://doi.org/10.1371/journal.pone.0045013
Odum WE, Heald EJ (1972) Trophic analyses of an estuarine mangrove community. Bull Mar Sci 22:671–738
Odum WE, Heald EJ (1975) The detritus-based food web of an estuarine mangrove community. In: Cronin LE (ed) Estuarine research. Academic, New York, pp 265–286
Ogden JC (1988) The influence of adjacent systems on the structure and function of coral reefs. In: Proceedings of the 6th international coral reef symposium, pp 29–37
Palacios S, Zimmerman RC (2007) Response of eelgrass (*Zostera marina* L.) to CO_2 enrichment: possible impacts of climate change and potential for remediation of coastal habitats. Mar Ecol Prog Ser 344:1–13. https://doi.org/10.3354/meps07084
Parrish JD (1989) Fish communities of interacting shallow-water habitats in tropical oceanic regions. Mar Ecol Prog Ser 58:143–160
Pezner AK, Courtney TA, Chou WC, Chu HC, Frable BW, Kekuewa SAH et al (2024) Coral growth along a natural gradient of seawater temperature, pH, and oxygen in a nearshore seagrass bed on Dongsha Atoll, Taiwan. PLoS ONE 19(10):e0312263. https://doi.org/10.1371/journal.pone.0312263
Pham TTT (2020) Tourism in marine protected areas: can it be considered as an alternative livelihood for local communities? Mar Policy 115:103891. https://doi.org/10.1016/j.marpol.2020.103891
Pham TD, Yokoya N, Bui DT et al (2019) Remote sensing approaches for monitoring mangrove species, structure, and biomass: opportunities and challenges. Remote Sens 11(3):230. https://doi.org/10.3390/rs11030230
Polapa FS, Werorilangi S, Ali SM et al (2021) Physiological responses of scleractinian corals in marginal habitat. Biodiversitas J Biol Divers 22(9):4011–4018. https://doi.org/10.13057/biodiv/d220949
Por FD, Dor I (1975) The hard bottom mangroves of Sinai, Red Sea. Rapp Comm Int Mer Medit 23(2):145–147
Quinn NJ, Kojis BJ (1985) Does the presence of coral reefs in proximity to a tropical estuary affect the estuarine fish assemblage? In: Proceedings of the 5th international coral reef congress, vol 5, pp 445–450
Radabaugh KR, Moyer RP, Chappel AR et al (2020) Mangrove damage, delayed mortality, and early recovery following Hurricane Irma at two landfall sites in Southwest Florida, USA. Estuar Coasts 43:1104–1118. https://doi.org/10.1007/s12237-019-00564-8
Reyes R, Hernando Campos NE (1992) Macroinvertebrados colonizadores de raíces de *Rhizophora mangle* en la Bahía de Chengue, Caribe colombiano. Bol Investig Mar Costeras 21(1):106–116
Robertson AI, Blaber SJM (1992) Plankton, epibenthos and fish communities. Coast Estuar Stud 41:173–224
Rogers C (1990) Responses of coral reefs and reef organisms to sedimentation. Mar Ecol Prog Ser 62:185–202
Rogers CS (2009) High diversity and abundance of scleractinian corals growing on and near mangrove prop roots, St. John, US Virgin Islands. Coral Reefs 28(4):909–909. https://doi.org/10.1007/s00338-009-0526-4
Rogers CS (2017) A unique coral community in the mangroves of Hurricane Hole, St. John, US Virgin Islands. Diversity 9(3):29. https://doi.org/10.3390/d9030029
Rogers CS (2019) Immediate effects of hurricanes on a diverse coral/mangrove ecosystem in the US Virgin Islands and the potential for recovery. Diversity 11(8):130. https://doi.org/10.3390/d11080130
Rogers CS, Herlan JJ (2012) Life on the edge: corals in mangroves and climate change. In: Yellowlees D, Hughes TP (eds) Proceedings of the 12th international coral reef symposium. James Cook University, Cairns
Rogers A, Mumby PJ (2019) Mangroves reduce the vulnerability of coral reef fisheries to habitat degradation. PLoS Biol 17:e3000510. https://doi.org/10.1371/journal.pbio.3000510
Rovai AS, Twilley RR, Castañeda-Moya E et al (2021) Macroecological patterns of forest structure and allometric scaling in mangrove forests. Glob Ecol Biogeogr 30:1000–1013. https://doi.org/10.1111/geb.13268
Rubin ET, Enochs IC, Foord C et al (2021) Molecular mechanisms of coral persistence within highly urbanized locations in the Port of Miami, Florida. Front Mar Sci 8:695236. https://doi.org/10.3389/fmars.2021.695236
Ruiz-Zárate MA, Espinoza-Avalos J, Carricart-Ganivet JP et al (2000) Relationships between *Manicina areolata* (Cnidaria: Scleractinia), *Thalassia testudinum* (Anthophyta), and *Neogoniolithon sp.* (Rhodophyta). Mar Ecol Prog Ser 206:135–146. https://doi.org/10.3354/meps206135
Rützler K, Macintyre IG (1982) The habitat distribution and community structure of the barrier reef complex at Carrie Bow Cay Belize. In: The Atlantic Barrier Reef Ecosystem at Carrie Bow Cay Belize I. Structure and communities. Smithsonian Institution Press, Washington DC, pp 9–45
Rützler K, Goodbody V, Diaz MC et al (2004) The aquatic environment of Twin Cays, Belize. Atoll Res Bull:512, 1
Saderne V, Fietzek P, Herman PMJ (2013) Extreme variations of pCO2 and pH in a macrophyte meadow of the Baltic Sea in summer:

evidence of the effect of photosynthesis and local upwelling. PLoS One 8:e62689. https://doi.org/10.1371/journal.pone.0062689

Sánchez-Núñez DA, Bernal G, Mancera Pineda JE (2019) The relative role of mangroves on wave erosion mitigation and sediment properties. Estuar Coast 42:2124–2138. https://doi.org/10.1007/s12237-019-00628-9

Saunders MI, Cannard T, Fischer M et al (2024) A roadmap to coastal and marine ecological restoration in Australia. Environ Sci Pol 159:103808. https://doi.org/10.1016/j.envsci.2024.103808

Scavo Lord K, Lesneski KC, Bengtsson ZA et al (2020) Multi-year viability of a reef coral population living on mangrove roots suggests an important role for mangroves in the broader habitat mosaic of corals. Front Mar Sci 377. https://doi.org/10.3389/fmars.2020.00377

Scavo Lord K, Barcala A, Aichelman HE et al (2021) Distinct phenotypes associated with mangrove and lagoon habitats in two widespread Caribbean corals, *Porites astreoides* and *Porites divaricata*. Biol Bull 240(3):169–190. https://doi.org/10.1086/714047

Scavo Lord K, Lesneski KC, Buston PM et al (2023) Rampant asexual reproduction and limited dispersal in a mangrove population of the coral *Porites divaricata*. Proc R Soc B 290. https://doi.org/10.1098/rspb.2023.1070

Schoepf V, Baumann JH, Barshis DJ et al (2023) Corals at the edge of environmental limits: a new conceptual framework to re-define marginal and extreme coral communities. Sci Total Environ 884:163688. https://doi.org/10.1016/j.scitotenv.2023.163688

Scott CP, Mach L, Lucas KM et al (2024) Whose cultural ecosystem service values matter? Exploring power inequities in diverse mangrove communities. Hum Ecol 52:81–97. https://doi.org/10.1007/s10745-023-00462-5

Sealey KS, Lennon E, Roskar AG et al (2024) Ecological shifts in lagoon patch reefs: consequences of large-scale alterations to coastal systems. Acad Biol 2(3). https://doi.org/10.20935/AcadBiol7357

Semesi IS, Beer S, Björk M (2009) Seagrass photosynthesis controls rates of calcification and photosynthesis of calcareous macroalgae in a tropical seagrass meadow. Mar Ecol Prog Ser 382:41–47. https://doi.org/10.3354/meps07973

Sesh Serebiah J, Rajkumar M, Sun J et al (2011) New discovery of coral rubbings in the north-western Gulf of Kachchh, Gujarat, Western India—GIS based evaluation. J Ocean Univ China 10:153–156

Short FT, Coles RG, Pergent-Martini C (2001) Global seagrass distribution. In: Short FT, Coles RG (eds) Global seagrass research methods, vol 5. Elsevier, Amsterdam, p 30

Short FT, Carruthers T, Dennison W et al (2007) Global seagrass distribution and diversity: a bioregional model. J Exp Mar Biol Ecol 350(1–2):3–20. https://doi.org/10.1016/j.jembe.2007.06.012

Shulman MJ (1985) Recruitment of coral reef fishes: effects of distribution of predators and shelter. Ecology 66:1056–1066

Smith SV (1978) Coral-reef area and the contribution of reefs to processes and resources of the world's oceans. Nature 273:225–226

Souter D, Planes S, Wicquart J et al (2021) Status of coral reefs of the world: 2020 report. Global Coral Reef Monitoring Network (GCRMN) and International Coral Reef Initiative (ICRI). https://doi.org/10.59387/WOTJ9184

Spalding MD, Kainuma M, Collins L (2010) World atlas of mangroves (version 3.1). A collaborative project of ITTO, ISME, FAO, UNEP-WCMC, UNESCO-MAB, UNU-INWEH and TNC. London (UK): Earthscan, London. 319 pp. Data https://doi.org/10.34892/w2ew-m835

Spalding M, McIvor A, Tonneijck FH et al (2014) Mangroves for coastal defence. Guidelines for coastal managers & policy makers. Published by Wetlands International and The Nature Conservancy, p 42

Stafford-Smith MG, Ormond RFG (1992) Sediment rejection mechanisms of 42 species of Australian scleractinian corals. Aust J Mar Freshwat Res 43:683–705. https://doi.org/10.1071/MF9920683

Stambler N, Dubinsky Z (2005) Corals as light collectors: an integrating sphere approach. Coral Reefs 24:1–9. https://doi.org/10.1007/s00338-004-0452-4

Stehlé H (1945) Forest types of the Caribbean islands. Part I. Caribbean Forester 6(Supp 1):273–414

Stephenson TA, Stephenson A, Tandy G et al (1931) The structure and ecology of low Isles and other reefs. Great Barrier Reef expedition 1928–29. Sci Rep 3:17–112

Stewart HA, Kline DI, Chapman LJ et al (2021) Caribbean mangrove forests act as coral refugia by reducing light stress and increasing coral richness. Ecosphere 12(3):e03413. https://doi.org/10.1002/ecs2.3413

Stewart HA, Wright JL, Carrigan M et al (2022) Novel coexisting mangrove-coral habitats: extensive coral communities located deep within mangrove canopies of Panama, a global classification system and predicted distributions. PLoS One 17(6):e0269181. https://doi.org/10.1371/journal.pone.0269181

Tan YHJ, Lai S, Tun KPP (2025) "Accidental" artificial reefs—lessons from Singapore for coral and seagrass restoration along urban shorelines. Bull Mar Sci 101. https://doi.org/10.5343/bms.2023.0134

Tavakoli-Kolour P, Sinniger F, Morita M et al (2024) Shallow corals acclimate to mesophotic depths while maintaining their heat tolerance against ongoing climate change. Mar Pollut Bull 209:117277

Taylor PD, Wilson MA (2003) Palaeoecology and evolution of marine hard substrate communities. Earth Sci Rev 62:1–103. https://doi.org/10.1016/S0012-8252(02)00131-9

Tebbett SB, Connolly SR, Bellwood DR (2023) Benthic composition changes on coral reefs at global scales. Nat Ecol Evol 7:71–81. https://doi.org/10.1038/s41559-022-01937-2

Teutli-Hernández C, Herrera-Silveira J, Cisneros-de la Cruz D et al (2021) Manual for the ecological restoration of mangroves in the Mesoamerican Reef System and the Wider Caribbean. In: Integrated Ridge-to-Reef Management of the Mesoamerican Reef Ecoregion Project - MAR2R, UNEP-Cartagena Convention, Mesoamerican Reef Fund. Guatemala City, Guatemala. https://www.unep.org/cep/resources/toolkits-manuals-and-guides/manual-ecological-restoration-mangroves-mesoamerican-reef

Thanopoulou Z, Patus J, Sealey KS (2022) Water quality negatively impacts coral occurrence in eutrophic nearshore environments of the Florida Keys. Front Mar Sci 9:1005036. https://doi.org/10.3389/fmars.2022.1005036

Tomlinson PB (1986) The botany of mangroves. Cambridge University Press, Cambridge

Tomlinson PB (2016) The botany of mangroves, 2nd edn. Cambridge University Press, Cambridge

Travaglione N, Evans R, Moustaka M et al (2023) Scleractinian corals rely on heterotrophy in highly turbid environments. Coral Reefs 42(5):997–1010. https://doi.org/10.1007/s00338-023-02407-2

Uhrin AV, Slade CL, Holmquist JG (2005) Self-righting in the free-living coral *Manicina areolata* (Cnidaria: Scleractinia): morphological constraints. Caribbean J Sci 41:277–282

UNEP-Nairobi Convention/USAID/WIOMSA (2020) Guidelines on mangrove ecosystem restoration for the Western Indian Ocean region. UNEP, Nairobi, p 71. https://www.nairobiconvention.org/CHM%20Documents/WIOSAP/guidelines/GuidelinesonMangroveRestorationForTheWIO.pdf

UNEP-WCMC, WorldFish Centre, WRI, TNC (2021) Global distribution of warm-water coral reefs, compiled from multiple sources including the Millennium Coral Reef Mapping Project. Version 4.1. Includes contributions from IMaRS-USF and IRD (2005), IMaRS-USF (2005) and Spalding et al. (2001). Cambridge (UK): UN Environment World Conservation Monitoring Centre. Data DOI: https://doi.org/10.34892/t2wk-5t34

Unsworth RKF, Collier CJ, Henderson GM, McKenzie LJ (2012) Tropical seagrass meadows modify seawater carbon chemistry: implications for coral reefs impacted by ocean acidification. Environ Res Lett 7:24026. https://doi.org/10.1088/1748-9326/7/2/024026

Van de Koppel J, van der Heide T, Altieri AH et al (2015) Long-distance interactions regulate the structure and resilience of coastal ecosystems. Annu Rev Mar Sci 7:139–158. https://doi.org/10.1146/annurev-marine-010814-015805

Vanderklift MA, Herr D, Lovelock CE et al (2022) A guide to international climate mitigation policy and finance frameworks relevant to the protection and restoration of blue carbon ecosystems. Front Mar Sci 9:872064. https://doi.org/10.3389/fmars.2022.872064

Victor S, Golbuu Y, Wolanski E et al (2004) Fine sediment trapping in two mangrove-fringed estuaries exposed to contrasting land-use intensity, Palau, Micronesia. Wetl Ecol Manag 12:277–283

von Hammerstein H, Fett TM, Ferse SC-A et al (2024) Individual mangrove trees provide alternative reef fish habitat on backreefs. Sci Rep 14:18574. https://doi.org/10.1038/s41598-024-69524-y

Vozzo ML, Doropoulos C, Silliman BR et al (2023) To restore coastal marine areas, we need to work across multiple habitats simultaneously. PNAS 120(26):e2300546120. https://doi.org/10.1073/pnas.2300546120

Waller A (2024) Coexisting Mangrove-Coral Habitats: Trends in Seawater Chemistry and Coral Diversity. Dissertation, The University of Sydney. 97 p

Whitfield AK (2017) The role of seagrass meadows, mangrove forests, salt marshes, and reed beds as nursery areas and food sources for fishes in estuaries. Rev Fish Biol Fish 27:75–110

Wong JSY, Chan YKS, Ng CSL et al (2018) Comparing patterns of taxonomic, functional and phylogenetic diversity in reef coral communities. Coral Reefs 47:737–750. https://doi.org/10.1007/s00338-018-1698-6

Wright JL, Stewart HA, Candanedo I et al (2023) Coexisting mangrove-coral habitat use by reef fishes in the Caribbean. Biotropica 00:1–7. https://doi.org/10.1111/btp.13200

Wulff JL (2008) Collaboration among sponge species increases sponge diversity and abundance in a seagrass meadow. Mar Ecol 29:193–204. https://doi.org/10.1111/j.1439-0485.2008.00224.x

Yates KK, Rogers CS, Herlan JJ et al (2014) Diverse coral communities in mangrove habitats suggest a novel refuge from climate change. Biogeosciences 11(16):4321–4337. https://doi.org/10.5194/bg-11-4321-2014

Yeemin T, Pengsakun S, Yucharoen M et al (2013) Long-term changes in coral communities under stress from sediment. Deep Sea Res Part II Top Stud Oceanogr 96:32–40. https://doi.org/10.1016/j.dsr2.2013.04.019

Mesophotic Coral Ecosystems as Extreme and Marginal Habitats

12

Tyler B. Smith and Gal Eyal

Abstract

Mesophotic coral ecosystems (MCEs), occurring in water depths where light is strongly attenuated, represent extensive yet understudied coral habitats characterized by low-light, blue-shifted spectral conditions and substantial variability in environmental factors such as temperature, dissolved gases, nutrients, and water column structure. Although historically perceived as refugia or minimally impacted environments, recent evidence indicates that MCEs experience significant anthropogenic pressures, including temperature-driven bleaching, pollution, fishing, and physical disturbance. This review synthesizes current understanding of MCEs as both **extreme** and **marginal** environments. Extremeness arises from predictable constraints, particularly light limitation, but also from context-dependent variation in temperature stratification, oxygen availability, acidification dynamics, and pelagic resource fluxes. Marginality, defined by deviations in ecosystem functioning and community composition relative to shallow reefs, is prominent for light-dependent scleractinian corals, which exhibit declining primary productivity, altered species richness, and shifts toward deep specialist taxa with depth. However, other groups—such as sponges, octocorals, and antipatharians—may experience peak abundance or diversity at mesophotic depths, highlighting that marginality is taxon-specific. Overall, MCEs function as complex, heterogeneous ecosystems shaped by interacting physical, biological, and anthropogenic drivers, emphasizing the need for expanded study to understand their ecological roles, vulnerabilities, and contributions to coral reef persistence.

Keywords

Mesophotic coral ecosystems (MCEs) · Depth gradients · Light attenuation · Temperature stratification · Coral bleaching

12.1 Introduction

Mesophotic coral ecosystems (MCEs) form at the deeper and dimmer margins of much better understood shallow water ecosystems. Some global estimates of their extent surpass that of shallow reefs (Pyle and Copus 2019). They are also extreme and oftentimes marginal coral habitats, suggesting MCEs may also compose the largest extent of marginal and extreme coral habitats globally. Early studies of the MCEs called these habitats deep reefs or twilight zone reefs, with pioneering studies mostly carried out by submersibles in the 1960's and 1970's in such areas as the Red Sea (Fricke and Knauer 1986; Fricke 1996), Bermuda (Fricke and Meischner 1985), Belize (James and Ginsburg 1979) and Hawaii (Brock and Chamberlain 1968). MCEs were first given the title "mesophotic" in a call to action for increased study by Dr. Robert Ginsburg in 2007 (Ginsburg 2007). This eventually led to the first codified definition of these ecosystems as "characterized by the presence of light-dependent corals and associated communities that are typically found at depths ranging from 30 to 40 m and extending to over 150 m in tropical and subtropical regions" (Hinderstein et al. 2010). The oceanographic, ecological, and physiological study of MCEs has blossomed in recent years with increased technological capability for direct and remote observation and a growing understanding of MCEs as integral components of coral ecosystems globally (Turner et al. 2017; Bongaerts et al. 2019).

T. B. Smith (✉)
University of the Virgin Islands, St. Thomas, VI, USA
e-mail: tsmith@uvi.edu

G. Eyal
Bar Ilan University, Ramat Gan, Israel

University of Queensland, Brisbane, Australia

M. O. Soares et al. (eds.), *Extreme and Marginal Coral Reefs and Communities*, Coral Reefs of the World 24,
https://doi.org/10.1007/978-3-032-28035-0_12

MCEs have been characterized as both extreme and marginal environments based on their altered environmental conditions and faunal composition relative to shallow water reefs systems (Camp et al. 2018; Schoepf et al. 2023). However, MCEs vary tremendously in conditions and characteristics geographically and with depth. Thus, the degree to which they are extreme and/or marginal can also vary spatially. There are several thoughtful reviews of MCE biodiversity, ecology, physiology, and stability (Lesser et al. 2009; Bongaerts et al. 2010; Loya et al. 2016; Lesser et al. 2018; Bongaerts and Smith 2019; Slattery et al. 2024). This review of MCEs as extreme and marginal habitats seeks to describe the extreme environmental conditions influencing these environments and how this structures the often unique sessile fauna that form the foundation of these habitats (Table 12.1).

12.2 Extreme: Environmental Conditions Found on MCEs

Light. Light always declines with increasing depth. By definition, MCEs are extreme coral habitats based on their light properties, aligning them with turbid reefs (Chap. 9). As the name would imply, mesophotic, or "middle light", habitats are defined by light availability and quality. Light is one of the main structuring forces of MCEs as it controls the potential for photosynthesis by phototrophic and mixotrophic sessile organisms. In the case of corals, the amount of photosynthetically active radiation (PAR, 400–700 nm) sets how much these mixotrophic organisms can rely on photosynthesis for the availability of carbohydrates for growth, maintenance, and reproduction, and how much must be supplemented by heterotrophic feeding.

Light reaching a coral's tissues is always less than the idealized quantity available at a given depth in an open and optically clear water column. The amount of light on any given area of seafloor and the photons available to drive photosynthesis depend on many factors including the incident light hitting the water's surface, the penetration of the light through the water's surface, the absorption and scattering of light as it travels downward due to the properties of the water and the materials dissolved or suspended in water, and characteristics of the bottom morphology (Lesser et al. 2018; Kahng et al. 2019) Fig. 12.1. Critical thresholds of light quantity have been noted at 10% of surface light (the approximate midpoint of the photic zone; <55 m) and 1% of surface light (the bottom of the euphotic zone; <102 m) (Lesser et al. 2018; Tamir et al. 2019; Laverick et al. 2020) (Figs. 12.2 and 12.3). Thus, the 10% and 1% limits are typically within mesophotic depths or shallower. In addition, the quality of light is altered as water absorbs higher wavelength (redder) light first, causing a blue shift in the light field. This provides a more constricted spectra with which to power photosynthesis. Dim, blue-shifted spectra in MCEs may also differ from the dim, but not necessarily blue-shifted spectra in turbid reefs, causing somewhat different photo-physiological acclimation and adaptation to each environment.

Light is also spatially and temporally variable. Most obviously, time of day affects light availability and the periods when there is sufficient light to drive photosynthesis above the daily colony energy demands. In MCEs this period can be restricted to a narrow period around solar noon. At higher

Table 12.1 Variables associated with the extremeness of mesophotic coral ecosystems. Estimated for each variable are their patterns with depth and time (diel variability), the consistency of these patterns among MCEs, and the factors affecting the pattern. In some cases, universal patterns and the factors influencing them are not sufficiently studied to make estimates

Variable	Magnitude with increasing depth	Variability-Diel	Consistency of pattern	Factors affecting
Light	Decreases—blue-shifted spectrum	Maximum	Highest	Season, atmospheric attenuation (clouds, aerosols), diurnal cycles, water attenuation (turbidity, CDOM, topography)
Density	Increases	Low-High	Highest	Season, salinity, temperature, internal waves, tidal bores, currents
Temperature	Decreases	Low-High	Very high	Season, vertical stratification/mixing, internal waves, tidal bores, currents
Salinity	Increases	Low-High	High	Season, vertical stratification/mixing, internal waves, tidal bores, currents, terrestrial runoff
Dissolved Oxygen	Decreases	Low-High	High	Season, vertical stratification/mixing, surface productivity, decomposition, photosynthesis
Carbon Dioxide	Increases—aragonite saturation drops	?	?	Season, vertical stratification/mixing, surface productivity, decomposition, photosynthesis
Nutrients	Increases	Low-High	Moderate	Season, vertical stratification/mixing, surface productivity, decomposition, photosynthesis, terrestrial runoff
Turbidity	Variable	Low	Low	Season, light, temperature, vertical stratification/mixing, surface productivity, nutrient availability, photosynthesis, terrestrial runoff
Pelagic Resources	Variable	Low-High	Moderate	Season, vertical stratification/mixing, surface productivity, decomposition, photosynthesis, nutrient availability, phytoplankton availability

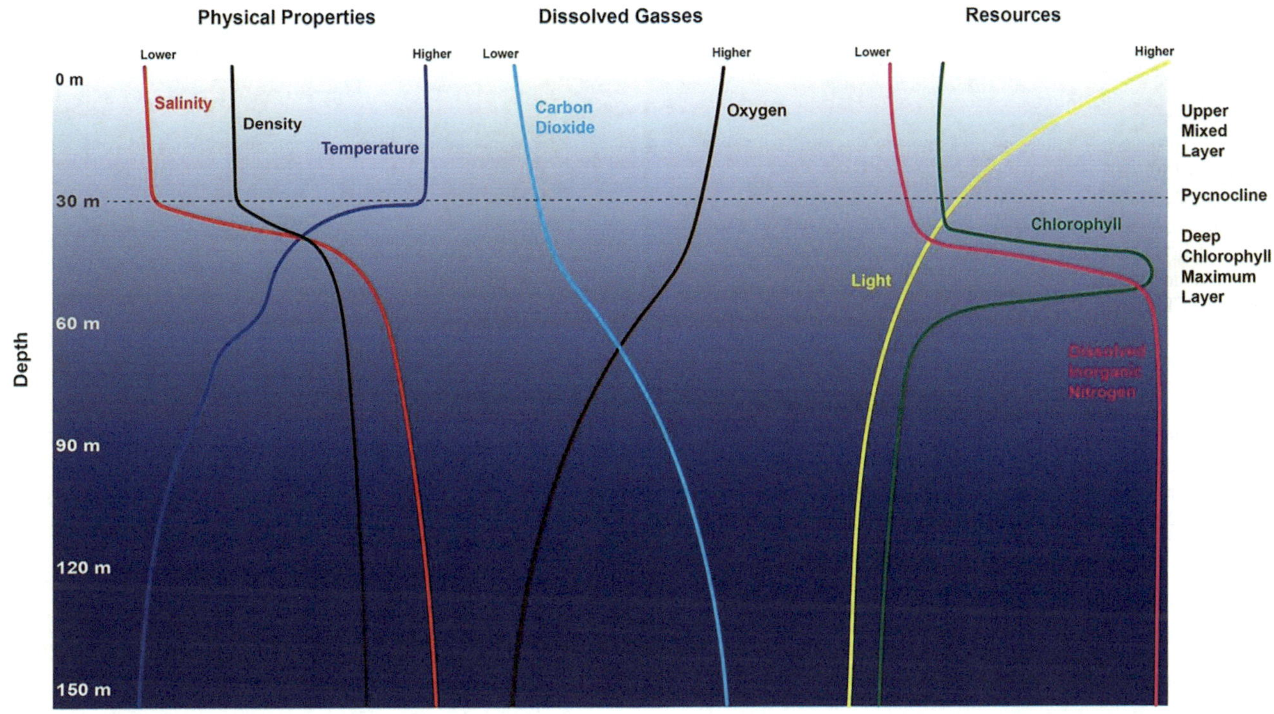

Fig. 12.1 Conceptualized vertical distribution of water quality properties and resources in global tropical oceans that can affect the extremeness of mesophotic coral ecosystems and influence the community composition. The upper ocean from the surface to about 30 m—100 m depth (depending on location and season) is relatively homogeneous in water properties due to wind and current mixing, and is referred to as the Upper Mixed Layer

latitudes the light availability in the austral and boreal winter can be limited by the seasonal angle of the incident solar light (Eilat, Bermuda, Houtman-Abrolhos; (Kahng et al. 2019)). Further, light can be limited by stochastic events in the atmosphere, such as cloud cover (Mumby et al. 2002), and in the water column, such as increased turbidity events associated with terrestrial run-off (Anthony and Connolly 2004) or seasonal algae blooms (Tamir et al. 2019). Therefore, the "extremeness" of MCEs based on light availability and quality is spatially and temporally variable.

12.3 Other Extreme Factors

Many other characteristics of water can vary with depth (Fig. 12.1), contributing to the extreme aspects of MCEs. These water characteristics include abiotic physical properties, such as temperature, density and salinity, and factors that can be related to biological processes within or external to the MCE zone, such as dissolved oxygen, pH, turbidity, nutrients, and pelagic resources, as well as phytoplankton (often indicated by water column chlorophyll concentration), zooplankton, and particulate/dissolved organic matter. Unlike light, none of these characteristics necessarily will decline or increase with depth in a totally predictable way. For example, temperature tends to decrease with increasing depth, but there can be inversions with salty, warm layers forming at depth. Or, there can be a chlorophyll maximum layer at depth, but where it is located, its shape with depth, and its strength are dependent on many factors. Most of these characteristics can be related to changing layers of water density with depth and the ways that water density influences the distribution of water constituents.

Density In a stable water column density increases with depth and can form regions of vertically homogeneous density stratification, or mixed layers, and areas of rapid change in density, or clines (Fig. 12.1). Density does not contribute directly to the extremeness of a habitat, but indirectly controls the vertical water column structure that in turn influences extremeness. Salinity and temperature contribute most consequentially to seawater density, with higher salinity (more salts per volume) and lower temperature (more molecules of water per volume) contributing to higher density. Clines associated with density, salinity, and temperature are called pycnoclines, haloclines, and thermoclines, respectively. Deeper water tends to get saltier and colder. One of the most pronounced mixed layers is the upper mixed layer (UML), a relatively consistent zone of mixed water properties from the water's surface to the first pycnocline. The upper mixed layer depth is dependent

Fig. 12.2 Cane Bay wall, St. Croix, US Virgin Islands as an example of the changing light regime for a shallow to deep coral Caribbean reef system from 10–100 m depth. All photos separated by less than 100 m horizontal distance. (**a**) 10 m: The shallow water reef with abundant light with relatively more red and yellow parts of the light spectrum and high biodiversity. (**b**) 16 m: Still abundant light along a sand channel. (**c**) 40 m: On the wall slope at the beginning of the upper mesophotic depths showing a dimmer and more blue-shifted light spectra. (**d**) 55 m: The bottom of the upper mesophotic showing dimmer light and a more heavily blue shifted light spectrum. (**e**) 70 m looking to 90 m: The beginning of the lower mesophotic depths showing a dim and intensely blue-shifted light spectra with very constricted and more specialized fauna. (**f**) 85 m looking to 100 m (location of diver): the bottom of the mesophotic zone for Caribbean coral reefs showing very dim and entirely blue-shifted light. (credit: A-E: J. Quetel, 10:30–11:30 Apr. 16, 2025; F: S. Heidmann, 14:00 Nov. 12, 2023)

on many things. At lower latitudes nearer to the equator the depth of the UML can be changed by stochastic events that cause pycnocline depth shifts, such as the passage of warm-core (UML-depressive) and cold-core (UML-shoaling) eddies (Wyatt et al. 2023), superimposed on longer oceanic cycles, such as the Pacific El Niño-Southern Oscillation (ENSO; (Wen et al. 2014; Colin 2016)), and the Indian Ocean Dipole (IOD; (Sun et al. 2022). At higher latitudes, the depth of the UML can be driven by these oceanic phenomena and a more complex set of drivers, including local wind and solar insolation and the passage of tropical storms.

Fig. 12.3 Eilat, Israel, as an example of the changing light regime for a shallow to deep Red Sea coral reef system from 5–120 m depth. All photos separated by less than 1000 m horizontal distance. (**a**) 5 m: A very shallow and well illuminated seafloor near The Interuniversity Institute for Marine Sciences. (**b**) 20 m: A deeper aggregation of corals and carbonate with losses of red and yellow parts of the light spectrum relative to very shallow water. (**c**) 35 m: An upper mesophotic coral reef showing stronger blue shifting of the light regime. (**d**) 50 m: A break in the shelf with corals. (**e**) 90 m: Intensely blue shifted and dim ambient light among carbonate outcrops in sand. (**f**) 120 m: Carbonate outcrop with small colonies of *Leptoseris fragilis* (brown spots) as taken by a remotely operated vehicle in artificial light. (Credit: A-C, E: Gal Eyal, 2014–2018, D: Tyler B. Smith, 14:00 Oct. 29, 2014, F: Oded Ezra, 12:00 Oct. 28, 2014)

Locally, the amount of mixing energy caused by wind-induced surface waves and/or currents influences the pycnocline depth, with more energy leading to a deepening of the UML. In regions where the temperature changes seasonally, such as higher latitudes, the depth of the mixed layer and associated water properties can shift (e.g., (Carlson et al. 2014; Smith et al. 2019a). In warm seasons there is often a greater contrast of the upper mixed layer temperature with that below the thermocline in the depth ranges of MCE. There also may be greater rainfall due to warmer sea surface temperature, higher rainfall/tropical storm feedbacks. Further, with the warm season there can be low-wind doldrum conditions with less wind energy for UML mixing. The warm and less saline (less dense) UML under conditions of less surface turbulence can cause warm season shoaling of the pycno−/halo−/thermo-clines. In cool seasons the difference between the UML temperature and the thermocline can be less pronounced, rain can be reduced (more salty UML), and winds can increase (more surface turbulence). This can lead to a deeper UML. In addition, other seasonal and non-seasonal factors can induce upwelling or subthermocline intrusions onto deep island or continental shelves can cause variability in the UML depth and deep vertical mixing. For example, Coral Sea water can intrude onto the shelf of the Great Barrier Reef in some regions and dramatically influence water column properties (Andrews and Gentien 1982). Or, volcanic eruptions ejecting vast amounts of ash and gas high into the atmosphere, can change upper ocean cooling and the UML depth (Genin et al. 1995). Further, on shorter timescales of less than a day internal waves along the pycnocline can cause vertical excursions of meters or tens of meters timed to tidal or other periodicities. This can create diurnal fluctuations in the depth of the UML and first pycnocline, which for a sessile organism within the pycnocline could cause large fluctuations in water properties, such as temperature and heterotrophic resources (Leichter et al. 1995; Leichter and Genovese 2006).

Temperature Temperature generally declines with increasing depth due to decreasing solar energy input with depth and atmospheric heat exchange (Fig. 12.1). Light-dependent tropical scleractinian corals are generally limited to regions where water temperatures are consistently maintained above about 18 °C, but this may be dependent on cold water anomalies below what corals normally experience in the coolest months in their local environment (González-Espinosa and Donner 2020; Bellworthy and Fine 2021). Below this temperature there can be cold-water induced coral bleaching (Foreman et al. 2024). Thus, there is the potential for the 18 °C isotherm to intersect MCE depths and limit coral formations through extreme cold conditions. This might be especially true in eastern ocean basins where the thermocline depth already tends to be shallower (Smith et al. 2017; Foreman et al. 2024) or at higher latitudes where conditions are cooler in general. On the other hand, cooler temperature at MCE depths could be helpful during surface marine heat waves and warm water coral bleaching conditions (Glynn 1996; Riegl and Piller 2003). However, the upper bleaching threshold temperature of corals is often set by local thermal conditions, with corals having a bleaching threshold temperature approximately 1 °C above the mean warmest conditions they tend to experience climatologically (Glynn and D'Croz 1990). Thus, corals can have depth(thermal)-dependent bleaching thresholds and bleach even when conditions are cooler at depth (Smith et al. 2016; Eyal et al. 2022). For example, in the US Virgin Islands Smith et al. (2016) predicted approximately a 0.25 °C decrease in the bleaching threshold for every 10 m increase in depth for star corals (*Orbicella* spp.). In addition, the depth stratification can break down or be weakened by oceanographic features, such as mesoscale eddies, and this can lead to anomalously warm conditions on MCEs and bleaching (Wyatt et al. 2023).

Dissolved Gases and Acidification Dissolved gases associated with biological processes, such as carbon dioxide and oxygen, can have contrasting patterns with depth also associated with biological processes. Near the surface gases can equilibrate with the atmosphere, but away from the surface gases are influenced by such biological processes as photosynthesis, respiration, and calcification. Physical and biological processes contribute to changing patterns of dissolved gases which can influence the potential for low oxygen stress conditions and acidification from carbon dioxide saturation. Dissolved oxygen concentrations tend to be highest near the surface with free gas exchange with the atmosphere and in association with high rates of photosynthesis. Where surface productivity is high because of high nutrient influxes, such as in areas of high terrestrial runoff and upwelling, there can be an accumulation of organic matter in deeper depths. This fuels aerobic bacterial production through the processing of organic matter, with the drawdown of dissolved oxygen and the release of carbon dioxide. Thus, in areas with high surface productivity there can be oxygen minima that coincide with deeper reef zones and the potential for oxygen levels that are potentially stressful for animals (hypoxia), such as corals and other sessile invertebrates that cannot move to avoid conditions.

Low oxygen concentrations reaching levels of near hypoxia are known for coral reefs and are predicted to increase under climate change (Pezner et al. 2023). Oxygen tends to decline with depth due to decreasing photosynthesis and increasing aerobic degradation of sinking organic material and oxygen minimum zone can form in MCE depths (Karstensen et al. 2008). This is accentuated in areas with higher productivity and thinner UML's, such as in the Eastern Tropical Pacific (Smith et al. 2014, 2017). While low oxygen stress events are not yet documented for MCEs it seems likely that there are problems with hypoxia in some regions. Hypoxia has been demonstrated in coral reefs just shallower than MCE depths (e.g., the Flower Gardens Bank; (Kealoha et al. 2020), thus, is stands to reason that hypoxia could also influence MCEs. While the extent and importance of hypoxia on reefs needs more study in general, MCEs may be more vulnerable near high productivity areas, including upwelling (Castrillón-Cifuentes et al. 2023), riverine (Kealoha et al. 2020), and human altered systems (Altieri et al. 2017).

Carbon dioxide accumulation is associated with lowering of seawater pH (acidification) and decreasing calcium carbonate (aragonite) saturation state. Declining saturation state makes it more difficult for corals to create skeletons and contributes to skeleton dissolution, which can in turn increase bioerosion, leading to more skeletal instability (Manzello et al. 2014). There is a generally declining pattern of aragonite saturation state with depth because of increasing pressure, declining temperate and increasing CO_2 from organic matter breakdown away from the surface air-sea interface (Jiang et al. 2015). The dynamics of CO_2 system with depth and how it impacts MCEs has not been well studied in MCEs (Hoegh-Guldberg et al. 2017). There is evidence that some mesophotic corals may be more resistant to lower pH (Scucchia et al. 2021), but this could still pose challenges for calcification in mesophotic corals and contribute to greater rates of dissolution and bioerosion in extant coral skeletons.

Nutrients Nutrients tend to increase with increasing depth. Elements that are needed for the growth of photoautotrophs (e.g., benthic algae, phytoplankton, bacterioplankton, stony corals) include nitrogen, phosphorous, calcium, iron, etc. Demand for these elements is strongest in well-lit upper ocean areas where photosynthesis is strongest. Here concentrations of vital limiting nutrients can be very low as they are rapidly taken up and sequestered by phototrophs, creating oligotrophic conditions that limit further photosynthesis (Furnas et al. 2005). As light and photosynthesis rates decline with depth so does nutrient demand. Due to this and the constant sinking of organic material produced in the upper ocean, higher concentrations of nutrients can develop with depth. For benthic mesophotic communities the consequences of this increase in nutrients are not totally clear, as benthic photoautotrophs are similarly light limited. However, this nutrient pattern can affect the availability of other pelagic resources that could be utilized by benthic consumers, such as mollusks, sponges, octocorals, and hexacorals, including stony corals. Notably, recent research suggests that stony corals may be able to grow their symbiotic algae and digest a portion of them to access nitrogen and phosphorus, thereby enhancing their ability to exploit dissolved inorganic nutrient pools in oligotrophic environments (Wiedenmann et al. 2023).

Pelagic Heterotrophic Resources Bacterioplankton, phytoplankton and zooplankton tend to have a peak in abundance in mid-depths (~30–80 m depth depending on location and season). This corresponds to the deep chlorophyll maximum layer created by the confluence of increased nutrient sequestered at and below the first pycnocline and the useable light downwelling from above (Fig. 12.1). Sessile benthic heterotrophs tend to feed from the water column in a variety of ways and many can utilize pelagic resources. Because the deep chlorophyll maximum layer tends to occur in mesophotic depths the associated pelagic heterotrophic resources may help offset the decline in light available for photosynthesis. While there is variability in the ability of stony corals to utilize heterotrophic resources, with even some very deep forming species being highly autotrophic (Carmignani et al. 2023), some species can benefit from pelagic resources. Leichter and Genovese (2006) found a bimodal growth pattern of a depth generalist coral, the Atlantic stony coral *Madracis auretenra* (*mirabilis*), with declining growth rates from the surface to the depth of the first pycnocline and deep chlorophyll maximum layer where growth increased to rates comparable to the surface. Variations in stony coral's abilities to photosynthesize at low light availability and utilize heterotrophic resources could contribute to species-specific patterns with depth.

Anthropogenic Impacts Given that MCEs occur in deeper water, often located farther from shore there seems to be a perception that anthropogenic disturbances are low (Schoepf et al. 2023). However, recent research challenges this view (Rocha et al. 2018). Global warming impacts from human greenhouse gas emissions are likely being felt already in MCEs, with episodes of high temperature bleaching increasingly being reported down to the depth limits of scleractinian corals (Smith et al. 2016; Eyal et al. 2022; Diaz et al. 2023; Wyatt et al. 2023). Projections from climate models suggest that MCEs of the GBR will also experience warming across a range of future emissions scenarios and could be very threatened by 2050 (McWhorter et al. 2024). Anthropogenic pollution may also be high on MCEs (see (Smith et al. 2019b)), including plastics (Pinheiro et al. 2023b), fishing gear (Rocha et al. 2018), and other trash (Smith et al. 2019a), mining discharges (Edinger 2012), submarine discharges, dredge dumping (Appeldoorn et al. 2016), and petrochemical remediation (Etnoyer et al. 2016). MCEs are also sites of industrial infrastructure, like cable laying (Smith et al. 2019b). In addition, an "out of sight, out of mind" effect could contribute to human disturbances, such as anchor damage (Smith et al. 2019a). Further, fishing can be as high or higher of MCEs than adjacent shallow water ecosystems (Sattar and Adam 2005) or at least extend into MCE depths (Kadison et al. 2017). More collection of MCE organisms for non-consumptive activities, like jewelry making, can have impacts on the populations of specific species (Tsounis et al. 2010). Thus, it seems unlikely that MCEs are under universally low anthropogenic impact, with the degree of local impacts related to the degree of impact on adjacent shallow water coral reefs.

Interacting Stressors Contributing to Extremeness Each of the factors contributing to extremeness detailed above are not experienced by MCE corals in isolation. Many can covary with each other. For example, areas or depth strata with high water column degradation of organic materials could have both low oxygen concentrations (aerobic oxygen consumption) and lower pH (release of organic acids and CO_2) because of biological processes, while also potentially having higher particulates, dissolved nutrients, and turbidity. Thus, MCE corals may experience these potential stressors as a suite of challenges, which additively or synergistically could contribute to extremeness.

Summary of MCEs as Extreme Coral Reef Habitats All MCE are extreme based on their low-light and spectrally altered conditions. However, extremeness may be impacted by the other biophysical conditions that might occur on MCEs. Schoepf et al. (2023, supplemental table S1) characterized the extremeness of MCE habitats based on their criteria of high environmental variance, shift from mean environmental conditions, susceptibility to anthropogenic stressors. It is insightful to compare their determinations to more recent literature as outlined in this section. Locations they included in their determination included the Red Sea in Israel, NW Australia, N Great Barrier Reef, the Ryukus of Japan, Pulley Ridge in the Gulf of Mexico, Discovery Bay in Jamaica, and Bermuda. Across these sites they determined that environmental variance was low, there was a shift in mean environmental conditions, and that local anthropogenic stressors were low. Compared with shallow water reefs in the same locations, MCEs clearly have shifted mean conditions for light and often for temperature, dissolved gases, nutrients, and pelagic heterotrophic resources. In addition, and in contrast to the determination of Schoepf et la. (2023) characterizations in this chapter suggest that MCEs also can have high variability, particularly in systems where the pycnocline and associated changes in water and pelagic biological properties shift in depths due to high-frequency and predictable phenomena, such as internal waves, and more stochastic phenomena, such as mesoscale oceanic eddies and longer-term oceanic cycles (e.g., the El Niño-Southern Oscillation). Also, in contrast to Schoepf et al. (2023), anthropogenic impacts may be high, may be related to shallow coral reef anthropogenic impacts, and could contribute to the extremeness of MCEs. Thus, except for light, which defines the extremeness of MCEs, other biophysical aspects of MCEs that might affect extremeness may be context dependent.

12.4 Marginal: Ecological and Biological Characteristics of MCE

While all MCEs are extreme based on light and potentially other biophysical attributes, this does not mean that they are necessarily marginal, or "limited in extent, significance, or stature" (Schoepf et al. 2023). Shallow and mesophotic habitats exist as a mosaic of conditions, or "extremeness", that meet the niche requirements of a given coral species. These conditions include the mean condition necessary for maintenance of the corals and coral reef associated organisms, such as the supply of light or heterotrophic resources, and stochastic conditions associated with stress and disturbance, such as water turbulence, thermal stress, diseases, and anthropogenic factors. For any patch of seafloor these conditions set the minimum that a MCE is shifted away from the local shallow water community.

Schoepf et al. (2023) emphasized both deviations from local or regional mean ecosystem functioning and community composition as indicative of marginality. Metrics of ecosystem functioning as indicative of marginality can be assessed by understanding key dynamic biological processes on a long-term scale (measured repeatedly) whose mean varies from the local or regional conditions. This can include such things as reef accretion and bioerosion, primary production and herbivory, secondary production and predation, and nutrient uptake and release. Community composition as indicative of marginality is typically assessed with point-in-time assessments of ecological communities. For MCEs this could include comparative metrics of local, habitat-specific (alpha) diversity and coral abundance (e.g., cover or density) relative to the total regional (gamma) diversity or abundance. In some cases, ecosystem functioning may control community composition, and this is particularly true for photosynthesis and heterotrophic resources in MCEs. Following is a brief review of the components of reef marginality for MCEs.

Primary Production and Herbivory The decline in light and change to a more blue-shifted spectrum is the primary factor affecting ecosystem functioning along a depth gradient (Lesser et al. 2009; Slattery et al. 2024). Thus, it stands to reason that its impact on primary production and herbivory might also be equally strong. Photosynthetic organisms have the capacity to adjust physiological and morphological aspects of their photosystem to adjust to declining light, and this capacity can be highly variable across species and taxa. Acclimation could allow phototrophs to maintain higher than expected rates of primary productivity than if the relationship between light and productivity were one to one (Spalding et al. 2019). Both acclimation (Kramer et al. 2022) and declining primary productivity are true for stony corals (Mass et al. 2007; Lesser et al. 2010). But this may not always be the case for macroscopic algae, with many taxa showing strong adaptations to compensate for declining light with depth (Spalding et al. 2019). Overall, there is a supposition that whole reef primary productivity declines with depth (Kramer et al. 2022; Slattery et al. 2024), but actual in-field measurements are rare.

A decline in herbivory with increasing depth seems to be a general feature globally. In some cases grazing pressure was measured directly and showed a steep decline, such as in the northern Red Sea where grazing pressure on turf algae was more than halved between shallow water and 65 m depth (Brokovich et al. 2010). This mirrored an exponential decline in herbivorous fish biomass. This same pattern can be inferred in other regions by a similar decline in abundance and biomass of reef herbivores with depth. Reef fish herbivores, including major taxa such as parrotfish and surgeonfish, show lowered abundance, biomass, and diversity in MCEs compared with shallow water reefs in all reefs examined form the Indian Ocean, Pacific, and Atlantic (Brokovich et al. 2010; Bejarano et al. 2014; Weinstein et al. 2014; Pinheiro et al. 2016, 2023a; Scott et al. 2022; Richardson et al. 2023; Heidmann et al. 2024). This may indicate a change in the strength of trophic pathways that could mirror a decline in reef primary productivity with depth.

Secondary Production and Predation Secondary production includes all sessile (invertebrate) and motile (invertebrate and vertebrate animals) that consume primary consumers, including omnivores, carnivores, and planktivores. In contrast to primary consumers (herbivores) abundance and biomass of secondary consumers can actually increase and peak in MCE depths, particularly for carnivores and planktivores (Pinheiro et al. 2023a; Heidmann et al. 2024), although this is not universally true and may reflect both changing trophic pathways and availability of habitat (Scott et al. 2022). Patterns for fishes and commercially important invertebrates may also be obscured by fishing that is particularly concentrated or more effective in shallow water (Bejarano et al. 2014; Kadison et al. 2017; Pinheiro et al. 2023a; Heidmann et al. 2024).

Nutrient Uptake and Release Nutrient cycling in MCEs remains poorly characterized, particularly in comparison to shallow reefs. While the uptake and release of nitrogen (N) and phosphorus (P) by mesophotic corals, algae, and microbes are not well understood, sponges which often increase in abundance with depth may play a central role. In shallow reefs, sponges are key mediators of nitrogen cycling through processes such as dissolved organic matter (DOM) uptake and microbial-driven nitrification and denitrification (Southwell et al. 2008; Schläppy et al. 2010; Fiore et al. 2013). If these functions persist or intensify with depth, sponges could represent a dominant pathway for nutrient transformation in MCEs. Additionally, recent studies have shown that reef-building corals can acquire inorganic N and P through their symbiotic algae, which assimilate dissolved nutrients and are partially digested by the host providing a potentially important nutrient source under oligotrophic conditions (Wiedenmann et al. 2023). Whether this mechanism functions similarly under the reduced light but higher nutrient conditions typical of MCEs remains unknown.

Diversity and Abundance Diversity and abundance can have shifting patterns with depth and variance among taxa. Diversity can include consideration of species richness and evenness, or how abundance is spread amongst taxa. Foundational species like mixotrophic corals generally see a decline in species richness with depth (Muir and Pichon 2019; Diaz et al. 2024) as taxa and species adapted to a phototrophic lifestyle in shallow, higher light environments give way to species with special adaptations to low light or higher propensity for heterotrophy (Lesser et al. 2010; Laverick et al. 2017, 2018; Tamir et al. 2019). At light levels around 1% of surface photosynthetically active radiation (typically in 60–80 m depth) there is often a transition in stony corals to deep specialist corals and very tolerant depth generalists (Lesser et al. 2010; Kahng et al. 2014; Tamir et al. 2019). This may contribute to a decline in stony coral diversity with depth represented by a transition to dominance (low evenness) of a few of these tolerant species (low species richness) in lower mesophotic MCEs (60–150 m depth). Such low diversity systems have been reported from Hawaii (Kahng et al. 2014), the Red Sea (Tamir et al. 2019), Great Barrier Reef (Englebert et al. 2017), Eastern Pacific (Pérez-Castro et al. 2023) and Caribbean (Hoeksema et al. 2016; Smith et al. 2019a) and many other locations (see (Loya et al. 2019)). Even in these lower diversity stony coral communities, abundance of coral can be quite high, but also patchy.

Abundance of sessile invertebrates who rely on planktonic resources (microbes, plankton, particulate organic matter, dissolved organic matter) and are less phototrophic may have different relationships with biodiversity with depth compared with stony corals. For sponges there is a debate about whether sponges universally increase with depth in relation to an increase in trophic resource (Lesser and Slattery 2018) or if the pattern is less generalized and shows site-to-site or regional variation based on other factors (Pawlik and Scott 2019). In addition, diversity may also be variable and have indistinct patterns with depth, increasing in some regions like the Red Sea (Raijman-Nagar et al. 2024). Regardless of the final judgement of sponge diversity and depth, it does not seem that mesophotic environments are marginal for sponge or other filter-feeders growth. Likewise, other sessile animal taxa diversity and abundance may also peak or increase in mesophotic depths, including antipatharians and octocorals (see Supplemental Table S2 in Diaz et al. 2024).

Reef Accretion Patterns of whole reef accretion (production—erosion) into mesophotic depths have not been well-studied. Given the decline in primary production with depth it may seem reasonable that the growth and calcification of autotrophs, like stony corals and coralline algae, and their contribution to carbonate production also declines. For a shallow specialist stony coral in the Caribbean, *Orbicella franksi*, skeletal calcification and linear extension growth has been shown to decline by up to half between 10 m depth and the upper mesophotic at 30–45 m depth (Weinstein et al. 2016); whereas a depth generalist, *P. astreoides*, showed level calcification and extension of the same depth gradient (Groves et al. 2018). Other depth generalist corals have also been shown to maintain relatively high rates of skeletal growth in lower mesophotic environments (Bongaerts et al. 2015). A notable mesophotic depth specialist in the Red Sea, *Fimbriaphyllia (Euphyllia) paradivisa*, exhibits light-enhanced calcification despite its occurrence in a low-light zone (Eyal et al. 2019), which suggest that mesophotic corals can also retain phototrophic plasticity and benefit from subtle increases in irradiance to support skeletal growth. A recent study from Puerto Rico showed that mesophotic ecosystems dominated by platy *Agaricia* spp. are actively accreting at slow rates (~30 cm ka^{-1}), with micritic crusts playing a major structural role (Sherman et al. 2023). Until comprehensive studies of reef carbonate budget in MCEs are conducted it is hard to say if they are more or less marginal for this factor.

In contrast to overall accretion, bioerosion tends to decrease with depth (Weinstein et al. 2019). This pattern has been attributed to a decline in the importance of autotrophic microborers with declining light and the aforementioned decline in grazers (external bioeroders) with depth. Microboring rates can decline by orders of magnitude from shallow to mesophotic depths (Vogel et al. 2000). External bioeroders, such as scraping and excavating parrotfish and sea urchins, decline in abundance with depth and, thus, their ability to erode external reef surfaces (Weinstein et al. 2014). Declining bioerosion with depth may offset some of the declining carbonate production with depth, which could help favor the creation of stable and structurally complex deeper reef environments (Weinstein et al. 2016).

Interacting Factors Contributing to Marginality As with drivers of extremeness, biological processes and ecological factors are often interdependent. For instance, from the discussion above depth specific patterns of abundance and diversity of key taxa, such as reef grazers, contribute to changes in key processes, such as herbivory and bioerosion.

Changes in light and thus energy availability can also change the ability of organisms to tolerate other conditions that change in deeper depths, such as competition or low pH.

Summary of MCEs as Marginal Coral Reef Habitats According to Schoepf et al. (2023), a coral community can be "classified as marginal if at least two of the following parameters are considered substantially different from 'normal' reefs: coral cover, community composition, species richness or ecosystem functioning". In most cases MCEs will fit the definition of marginal for light-dependent stony corals because of changes in key processes (primary production, herbivory, accretion) and community characteristics (diversity). However, they may not be marginal for other taxa, such as sponges, octocorals, and antipatharians. For many anitipatharian taxa in particular, where perhaps 75% of taxa occur deeper than 50 m (Cairns 2007), shallow water might be considered marginal.

References

Altieri AH, Harrison SB, Seemann J, Collin R, Diaz RJ, Knowlton N (2017) Tropical dead zones and mass mortalities on coral reefs. Proc Natl Acad Sci USA 114:3660–3665

Andrews JC, Gentien P (1982) Upwelling as a source of nutrients for the Great Barrier Reef ecosystems: a solution to Darwin's question? Mar Ecol Prog Ser 8:257–269

Anthony KR, Connolly SR (2004) Environmental limits to growth: physiological niche boundaries of corals along turbidity-light gradients. Oecologia 141:373–384

Appeldoorn R, Ballantine D, Bejarano I, Carlo M, Nemeth M, Otero E, Pagan F, Ruiz H, Schizas N, Sherman C, Weil E (2016) Mesophotic coral ecosystems under anthropogenic stress: a case study at Ponce, Puerto Rico. Coral Reefs 35:63–75

Bejarano I, Appeldoorn RS, Nemeth M (2014) Fishes associated with mesophotic coral ecosystems in La Parguera, Puerto Rico. Coral Reefs 33:313–328

Bellworthy J, Fine M (2021) Warming resistant corals from the Gulf of Aqaba live close to their cold-water bleaching threshold. PeerJ 9:e11100

Bongaerts P, Smith TB (2019) Beyond the "deep reef refuge" hypothesis: a conceptual framework to characterize persistence at depth. In: Loya Y, Puglise KA, Bridge TCL (eds) Mesophotic coral ecosystems. Springer, Cham, pp pp881–pp895

Bongaerts P, Ridgway T, Sampayo E, Hoegh-Guldberg O (2010) Assessing the 'deep reef refugia' hypothesis: focus on Caribbean reefs. Coral Reefs 29:309–327

Bongaerts P, Frade PR, Hay KB, Englebert N, Latijnhouwers KRW, Bak RPM, Vermeij MJA, Hoegh-Guldberg O (2015) Deep down on a Caribbean reef: lower mesophotic depths harbor a specialized coral-endosymbiont community. Sci Rep 5:art 7652

Bongaerts P, Perez-Rosales G, Radice VZ, Eyal G, Gori A, Gress E, Hammerman NM, Hernandez-Agreda A, Laverick J, Muir P, Pinheiro H, Pyle RL, Rocha L, Turner JA, Booker R (2019) Mesophotic.org: a repository for scientific information on mesophotic ecosystems. Database 2019:baz140

Brock VE, Chamberlain TC (1968) A geological and ecological reconnaissance off western Oahu, Hawaii, principally by means of the research submarine "Asherah". Pac Sci 22:373–394

Brokovich E, Ayalon I, Einbinder S, Segev N, Shaked Y, Genin A, Kark S, Kiflawi M (2010) Grazing pressure on coral reefs decreases across a wide depth gradient in the Gulf of Aqaba, Red Sea. Mar Ecol Prog Ser 399:69–80

Cairns SD (2007) Deep-water corals: an overview with special reference to diversity and distribution of deep-water scleractinian corals. Bull Mar Sci 81:311–322

Camp EF, Schoepf V, Mumby PJ, Hardtke LA, Rodolfo-Metalpa R, Smith DJ, Suggett DJ (2018) The future of coral reefs subject to rapid climate change: lessons from natural extreme environments. Front Mari Sci 5:433

Carlson DF, Fredj E, Gildor H (2014) The annual cycle of vertical mixing and restratification in the northern Gulf of Eilat/Aqaba (Red Sea) based on high temporal and vertical resolution observations. Deep-Sea Res I Oceanogr Res Pap 84:1–17

Carmignani A, Radice VZ, McMahon KM, Holman AI, Miller K, Grice K, Richards Z (2023) Levels of autotrophy and heterotrophy in mesophotic corals near the end photic zone. Front Mar Sci 10:1089746

Castrillón-Cifuentes AL, Zapata FA, Wild C (2023) Physiological responses of Pocillopora corals to upwelling events in the eastern tropical Pacific. Front Mar Sci 10:1115535

Colin PL (2016) Chapter 3.6 spotlight on the Palau Island group. In: Baker EK, Puglise KA, Harris PT (eds) Mesophotic coral ecosystems — a lifeboat for coral reefs? The United Nations Environment Programme and GRID-Arendal, Nairobi/Arendal, pp 31–36

Diaz C, Foster NL, Attrill MJ, Bolton A, Ganderton P, Howell KL, Robinson E, Hosegood P (2023) Mesophotic coral bleaching associated with changes in thermocline depth. Nat Commun 14:6528

Diaz C, Howell KL, Hosegood P, Bolton A, Foster NL (2024) Diverse and ecologically unique mesophotic coral ecosystems in the Central Indian Ocean. Coral Reefs 43:1259–1270

Edinger EN (2012) Gold mining and submarine tailings disposal: a review and case study. Oceanography 25:184–199

Englebert N, Bongaerts P, Muir PR, Hay KB, Pichon M, Hoegh-Guldberg O (2017) Lower mesophotic coral communities (60-125 m depth) of the Northern Great Barrier Reef and Coral Sea. PLoS One 12:e0170336

Etnoyer PJ, Wickes LN, Silva M, Dubick JD, Balthis L, Salgado E, MacDonald IR (2016) Decline in condition of gorgonian octocorals on mesophotic reefs in the northern Gulf of Mexico: before and after the Deepwater Horizon oil spill. Coral Reefs 35:77–90

Eyal G, Cohen I, Eyal-Shaham L, Ben-Zvi O, Tikochinski Y, Loya Y (2019) Photoacclimation and induction of light-enhanced calcification in the mesophotic coral Euphyllia paradivisa. R Soc Open Sci 6:180527

Eyal G, Laverick JH, Ben-Zvi O, Brown KT, Kramer N, Tamir R, Lindemann Y, Levy O, Pandolfi JM (2022) Selective deep water coral bleaching occurs through depth isolation. Sci Total Environ 844:157180

Fiore CL, Baker DM, Lesser MP (2013) Nitrogen biogeochemistry in the Caribbean sponge, Xestospongia muta: a source or sink of dissolved inorganic nitrogen? PLoS One 8:e72961

Foreman AD, Duprey NN, Yuval M, Dumestre M, Leichliter JN, Rohr MC, Dodwell RCA, Dodwell GAS, Clua EEG, Treibitz T, Martínez-García A (2024) Severe cold-water bleaching of a deep-water reef underscores future challenges for Mesophotic coral ecosystems. Sci Total Environ 951:175210

Fricke H (1996) Deep-water exploration of the Red Sea by submersible. In: Uiblein F, Ott J, Stachowitsch M (eds) Deep sea and extreme shallow-water habitats: affinities and adaptations. Österreichische Akad. der Wiss, pp 67–89

Fricke HW, Knauer B (1986) Diversity and spatial pattern of coral communities in the Red Sea upper twilight zone. Oecologia 71:29–37

Fricke H, Meischner D (1985) Depth limits of Bermudan scleractinian corals: a submersible survey. Mar Biol 88:175–187

Furnas M, Mitchell A, Skuza M, Brodie J (2005) In the other 90%: phytoplankton responses to enhanced nutrient availability in the Great Barrier Reef Lagoon. Mar Pollut Bull 51:235–265

Genin A, Lazar B, Brenner S (1995) Vertical mixing and coral death in the Red Sea following the eruption of Mount Pinatubo. Nature 377:507–510

Ginsburg RN (2007) Mesophotic coral reefs (30-100m) are the frontier of reef exploration and research 33rd scientific meeting of the Association of the Marine Labs of the Caribbean. US Virgin Islands, St. Thomas, p A26

Glynn PW (1996) Coral reef bleaching: facts, hypotheses and implications. Glob Chang Biol 2:495–509

Glynn PW, D'Croz L (1990) Experimental evidence for high temperature stress as the cause of El Nino-coincident coral mortality. Coral Reefs 8:181–191

González-Espinosa PC, Donner SD (2020) Predicting cold-water bleaching in corals: role of temperature, and potential integration of light exposure. Mar Ecol Prog Ser 642:133–146

Groves SH, Holstein DM, Enochs IC, Kolodzeij G, Manzello DP, Brandt ME, Smith TB (2018) Growth rates of *Porites astreoides* and *Orbicella franksi* in mesophotic habitats surrounding St. Thomas, US Virgin Islands. Coral Reefs 37:345–354

Heidmann SL, Olinger LK, Brandtneris VW, Ennis RS, Jeremiah Blondeau J, Smith TB (2024) Depth and structure as environmental drivers of fish communities across a shallow to mesophotic gradient in the northern US Virgin Islands. Bull Mar Sci 100:133–154

Hinderstein L, Marr J, Martinez F, Dowgiallo M, Puglise K, Pyle R, Zawada D, Appeldoorn R (2010) Theme section on mesophotic coral ecosystems: characterization, ecology, and management. Coral Reefs 29:247–251

Hoegh-Guldberg O, Poloczanska ES, Skirving W, Dove S (2017) Coral reef ecosystems under climate change and ocean acidification. Front Mar Sci 4:158

Hoeksema BW, Bongaerts P, Baldwin CC (2016) High coral cover at lower mesophotic depths: a dense Agaricia community at the leeward side of Curaçao, Dutch Caribbean. Marine Biodiversity 47:67–70

James NP, Ginsburg RN (1979) The seaward margin of Belize barrier and atoll reefs. Wiley, New York

Jiang L-Q, Feely RA, Carter BR, Greeley DJ, Gledhill DK, Arzayus KM (2015) Climatological distribution of aragonite saturation state in the global oceans. Glob Biogeochem Cycles 29:1656–1673

Kadison E, Brandt M, Nemeth R, Martens J, Blondeau J, Smith T (2017) Abundance of commercially important reef fish indicates different levels of over-exploitation across shelves of the U.S. Virgin Islands. PLoS One 12:e0180063

Kahng SE, Copus JM, Wagner D (2014) Recent advances in the ecology of mesophotic coral ecosystems (MCEs). Curr Opin Environ Sustain 7:72–81

Kahng SE, Akkaynak D, Shlesinger T, Hochberg EJ, Wiedenmann J, Tamir R, Tchernov D (2019) Light, temperature, photosynthesis, heterotrophy, and the lower depth limits of mesophotic coral ecosystems. In: Loya Y, Puglise KA, Bridge TCL (eds) Mesophotic coral ecosystems. Springer, Cham, pp 801–828

Karstensen J, Stramma L, Visbeck M (2008) Oxygen minimum zones in the eastern tropical Atlantic and Pacific oceans. Prog Oceanogr 77:331–350

Kealoha AK, Doyle SM, Shamberger KEF, Sylvan JB, Hetland RD, DiMarco SF (2020) Localized hypoxia may have caused coral reef mortality at the Flower Garden Banks. Coral Reefs 39:119–132

Kramer N, Tamir R, Ben-Zvi O, Jacques SL, Loya Y, Wangpraseurt D (2022) Efficient light-harvesting of mesophotic corals is facilitated by coral optical traits. Funct Ecol 36:406–418

Laverick JH, Andradi-Brown DA, Rogers AD (2017) Using light-dependent scleractinia to define the upper boundary of mesophotic coral ecosystems on the reefs of Utila, Honduras. PLoS One 12:e0183075

Laverick JH, Piango S, Andradi-Brown DA, Exton DA, Bongaerts P, Bridge TCL, Lesser MP, Pyle RL, Slattery M, Wagner D, Rogers AD (2018) To what extent do mesophotic coral ecosystems and shallow reefs share species of conservation interest? A systematic review. Environmental Evidence 7:15

Laverick JH, Tamir R, Eyal G, Loya Y (2020) A generalized light-driven model of community transitions along coral reef depth gradients. Glob Ecol Biogeogr 29:1554–1564

Leichter JJ, Genovese SJ (2006) Intermittent upwelling and subsidized growth of the scleractinian coral *Madracis mirabilis* on the deep fore-reef of Discovery Bay, Jamaica. Mar Ecol Prog Ser 316:95–103

Leichter J, Wing S, Miller S, Denny M (1995) Impact of internal tidal bores on Conch Reef, Florida Keys. Rutgers-the State University, New Brunswick

Lesser MP, Slattery M (2018) Sponge density increases with depth throughout the Caribbean. Ecosphere 9:e02525

Lesser MP, Slattery M, Leichter JJ (2009) Ecology of mesophotic coral reefs. J Exp Mar Biol Ecol 375:1–8

Lesser MP, Slattery M, Stat M, Ojimi M, Gates RD, Grottoli A (2010) Photoacclimatization by the coral *Montastraea cavernosa* in the mesophotic zone: light, food, and genetics. Ecology 91:990–1003

Lesser MP, Slattery M, Mobley CD (2018) Biodiversity and functional ecology of mesophotic coral reefs. Annu Rev Ecol Evol Syst 49:49–71

Loya Y, Eyal G, Treibitz T, Lesser MP, Appeldoorn R (2016) Theme section on mesophotic coral ecosystems: advances in knowledge and future perspectives. Coral Reefs 35:1–9

Loya Y, Puglise KA, Bridge TCL (eds) (2019) Mesophotic coral ecosystems. Springer, New York

Manzello DP, Enochs IC, Bruckner A, Renaud PG, Kolodziej G, Budd DA, Carlton R, Glynn PW (2014) Galápagos coral reef persistence after ENSO warming across an acidification gradient. Geophys Res Lett 41:9001–9008

Mass T, Einbinder S, Brokovich E, Shashar N, Vago R, Erez J, Dubinsky Z (2007) Photoacclimation of *Stylophora pistillata* to light extremes: metabolism and calcification. Mar Ecol Prog Ser 334:93–102

McWhorter JK, Halloran PR, Roff G, Mumby PJ (2024) Climate change impacts on mesophotic regions of the Great Barrier Reef. Proc Natl Acad Sci USA 121:e2303336121

Muir PR, Pichon M (2019) Biodiversity of reef-building, scleractinian corals. In: Loya Y, Puglise KA, Bridge TCL (eds) Mesophotic coral ecosystems. Springer, Cham, pp 589–620

Mumby PJ, Chisholm JRM, Edwards AJ, Andrefouet S, Jaubert J (2002) Cloudy weather may have saved Society Island reef corals during the 1998 ENSO event. Mar Ecol Prog Ser 222:209–216

Pawlik JR, Scott AR (2019) Sponge density increases with depth throughout the Caribbean: comment. Ecosphere 10:e02689

Pérez-Castro MÁ, Eyal G, Leyte-Morales GE, Hinojosa-Arango G, Enríquez S (2023) Benthic Characterization of Mesophotic Communities Based on Optical Depths in the Southern Mexican Pacific Coast (Oaxaca). Diversity 15:531

Pezner AK, Courtney TA, Barkley HC, Chou W-C, Chu H-C, Clements SM, Cyronak T, DeGrandpre MD, Kekuewa SAH, Kline DI, Liang Y-B, Martz TR, Mitarai S, Page HN, Rintoul MS, Smith JE, Soong K, Takeshita Y, Tresguerres M, Wei Y, Yates KK, Andersson AJ (2023) Increasing hypoxia on global coral reefs under ocean warming. Nat Clim Chang 13:403–409

Pinheiro HT, Goodbody-Gringley G, Jessup ME, Shepherd B, Chequer AD, Rocha LA (2016) Upper and lower mesophotic coral reef fish communities evaluated by underwater visual censuses in two Caribbean locations. Coral Reefs 35:139–151

Pinheiro HT, MacDonald C, Quimbayo JP, Shepherd B, Phelps TA, Loss AC, Teixeira JB, Rocha LA (2023a) Assembly rules of coral reef fish communities along the depth gradient. Curr Biol 33:1421–1430

Pinheiro HT, MacDonald C, Santos RG, Ali R, Bobat A, Cresswell BJ, Francini-Filho R, Freitas R, Galbraith GF, Musembi P, Phelps TA, Quimbayo JP, Quiros TEAL, Shepherd B, Stefanoudis PV, Talma S, Teixeira JB, Woodall LC, Rocha LA (2023b) Plastic pollution on the world's coral reefs. Nature 619:311–316

Pyle RL, Copus JM (2019) Mesophotic coral ecosystems: introduction and overview. In: Mesophotic coral ecosystems. Springer, pp 3–27

Raijman-Nagar L, Goren L, Shefer S, Ilan M (2024) Sponge abundance and diversity patterns in the shallow and mesophotic reefs of the northern Red Sea. Front Mar Sci 11:1385686

Richardson LE, Heenan A, Delargy AJ, Neubauer P, Lecky J, Gove JM, Green JAM, Kindinger TL, Ingeman KE, Williams GJ (2023) Local human impacts disrupt depth-dependent zonation of tropical reef fish communities. Nat Ecol Evol 7:1844–1855

Riegl B, Piller WE (2003) Possible refugia for reefs in times of environmental stress. Int J Earth Sci 92:520–531

Rocha LA, Pinheiro HT, Shepherd B, Papastamatiou YP, Luiz OJ, Pyle RL, Bongaerts P (2018) Mesophotic coral ecosystems are threatened and ecologically distinct from shallow water reefs. Science 361:281–284

Sattar SA, Adam MS (2005) Review of grouper fishery of the Maldives with additional notes on teh Faafu Atoll fishery. Marine Research Center, Malé, p 54

Schläppy M-L, Schöttner SI, Lavik G, Kuypers MMM, de Beer D, Hoffmann F (2010) Evidence of nitrification and denitrification in high and low microbial abundance sponges. Mar Biol 157:593–602

Schoepf V, Baumann JH, Barshis DJ, Browne NK, Camp EF, Comeau S, Cornwall CE, Guzmán HM, Riegl B, Rodolfo-Metalpa R, Sommer B (2023) Corals at the edge of environmental limits: a new conceptual framework to re-define marginal and extreme coral communities. Sci Total Environ 884:163688

Scott ME, Tebbett SB, Whitman KL, Thompson CA, Mancini FB, Heupel MR, Pratchett MS (2022) Variation in abundance, diversity and composition of coral reef fishes with increasing depth at a submerged shoal in the northern Great Barrier Reef. Rev Fish Biol Fish 32:941–962

Scucchia F, Malik A, Putnam HM, Mass T (2021) Genetic and physiological traits conferring tolerance to ocean acidification in mesophotic corals. Glob Chang Biol 27:5276–5294

Sherman CE, Carlo M, Tuohy E, Adey WH (2023) Styles and rates of mesophotic reef accretion on a Caribbean insular slope. Coral Reefs 42:1025–1044

Slattery M, Lesser MP, Rocha LA, Spalding HL, Smith TB (2024) Function and stability of mesophotic coral reefs. Trends Ecol Evol 39:585–598

Smith TB, Glynn PW, Maté JL, Toth LT, Gyory J (2014) A depth refugium from catastrophic coral bleaching prevents regional extinction. Ecology 95:1663–1673

Smith TB, Gyory J, Brandt ME, Miller WJ, Jossart J, Nemeth RS (2016) Caribbean mesophotic coral ecosystems are unlikely climate change refugia. Glob Chang Biol 22:2756–2765

Smith TB, Maté JL, Gyory J (2017) Thermal refuges and refugia for stony corals in the eastern tropical Pacific. In: Glynn WP, Manzello PD, Enochs CI (eds) Coral reefs of the eastern tropical Pacific: persistence and loss in a dynamic environment. Springer, Dordrecht, pp 501–515

Smith TB, Brandt ME, Brandtneris VW, Ennis RS, Groves SH, Habtes S, Holstein DM, Kadison E, Nemeth RS (2019a) The United States Virgin Islands. In: Loya Y, Puglise KA, Bridge T (eds) Coral Reefs of the world: mesophotic coral ecosystems. Springer, pp 131–147

Smith TB, Brandt ME, Brandtneris VW, Ennis RS, Groves SH, Habtes S, Holstein DM, Kadison E, Nemeth RS (2019b) Disturbance in mesophotic coral ecosystems and linkages to conservation and management. In: Loya Y, Puglise KA, Bridge T (eds) Coral reefs of the world: mesophotic coral ecosystems. Springer

Southwell MW, Weisz JB, Martens CS, Lindquist N (2008) In situ fluxes of dissolved inorganic nitrogen from the sponge community on Conch Reef, Key Largo, Florida. Limnol Oceanogr 53:986–996

Spalding HL, Amado-Filho GM, Bahia RG, Ballantine DL, Fredericq S, Leichter JJ, Nelson WA, Slattery M, Tsuda RT (2019) Macroalgae. In: Loya Y, Puglise KA, Bridge TCL (eds) Mesophotic coral ecosystems. Springer, Cham, pp 507–536

Sun S, Fang Y, Zu Y, Liu B (2022) Seasonal differences of decadal thermocline depth anomalies in the tropical Indian Ocean. J Clim 35:3947–3959

Tamir R, Eyal G, Kramer N, Laverick JH, Loya Y (2019) Light environment drives the shallow-to-mesophotic coral community transition. Ecosphere 10:e02839

Tsounis G, Rossi S, Grigg R, Santangelo G, Bramanti L, Gili J-M (2010) The exploitation and conservation of precious corals oceanography and marine biology. CRC Press, pp 161–211

Turner JA, Babcock RC, Hovey R, Kendrick GA (2017) Deep thinking: a systematic review of mesophotic coral ecosystems. ICES J Mar Sci:fsx085

Vogel K, Gektidis M, Golubic S, Kiene WE, Radtke G (2000) Experimental studies on microbial bioerosion at lee Stocking Island, Bahamas and one tree Island, great barrier reef, Australia: implications for paleoecological reconstructions. Lethaia 33:190–204

Weinstein DK, Smith TB, Klaus JS (2014) Mesophotic bioerosion: variability and structural impact on U.S. Virgin Island deep reefs. Geomorphology 222:14–24

Weinstein DK, Sharifi A, Klaus JS, Smith TB, Giri SJ, Helmle KP (2016) Coral growth, bioerosion, and secondary accretion of living orbicellid corals from mesophotic reefs in the US Virgin Islands. Mar Ecol Prog Ser 559:45–63

Weinstein DK, Maher RL, Correa AMS (2019) Bioerosion. In: Loya Y, Puglise KA, Bridge TCL (eds) Mesophotic coral ecosystems. Springer, Cham, pp 829–847

Wen C, Kumar A, Xue Y, McPhaden MJ (2014) Changes in tropical Pacific thermocline depth and their relationship to ENSO after 1999. J Clim 27:7230–7249

Wiedenmann J, D'Angelo C, Mardones ML, Moore S, Benkwitt CE, Graham NAJ, Hambach B, Wilson PA, Vanstone J, Eyal G, Ben-Zvi O, Loya Y, Genin A (2023) Reef-building corals farm and feed on their photosynthetic symbionts. Nature 620:1018–1024

Wyatt ASJ, Leichter JJ, Washburn L, Kui L, Edmunds PJ, Burgess SC (2023) Hidden heatwaves and severe coral bleaching linked to mesoscale eddies and thermocline dynamics. Nat Commun 14:25

Chillin' Out: High-Latitude and Cold-Water Coral Communities Are Cool but No Climate Change Refugia

13

Ciara Baines, James D. Reimer, Karolina M. Zarzyczny, Brigitte Sommer, Katie M. Watson, Jessica Bleuel, Margarita Brandt, Camille Burke, Fiona Chong, Charlotte Clay, Nicolás Dávalos, Stéphane De Palmas, Fabian Gösser, Wanchien Victoria Hsiao, Liberty Jackson, James Lock, Megumi Nakano, Ayaka Umeda Paul, David A. Paz-García, Sean N. Porter, Michael H. Schleyer, Daniel Velasco-Cedeño, Daito Yamaguchi, and Maria Beger

Abstract

High-latitude and other cool extreme and/or marginal coral communities host a dynamic mix of warm and cold-water species, including zooxanthellate scleractinian corals, kelp seaweeds, sponges and ascidians. They have been heralded as ecological refugia for climate change, but at the same time, resident species are often endemic and adapted to living at the limits of their environmental tolerances. Here, we characterise such "cool" temperate reefs with coral communities around the world, summarising what is known about their biology, ecology, and genetics, including adaptation to their extreme and/or marginal environments. We showcase cool reefs around the world, emphasising locally published content and local authors.

As global climate change results in regions where inter-annual ocean temperatures increase and become more variable, tropicalisation occurs. Tropicalisation is

C. Baines · C. Clay · W. V. Hsiao · L. Jackson · J. Lock
School of Biology, Faculty of Biological Sciences, University of Leeds, Leeds, UK

J. D. Reimer
Molecular Invertebrate Systematics and Ecology Laboratory, Faculty of Science, University of the Ryukyus, Nishihara, Okinawa, Japan

Tropical Biosphere Research Center, University of the Ryukyus, Nishihara, Okinawa, Japan

K. M. Zarzyczny
School of Ocean and Earth Science, National Oceanography Centre, University of Southampton, Southampton, UK

Natural History Museum, London, UK

B. Sommer
School of Life and Environmental Sciences, The University of Sydney, Sydney, NSW, Australia

K. M. Watson
Department of Botany and Zoology, Stellenbosch University, Stellenbosch, Western Cape, South Africa

School of Biological Sciences and the SWIRE Institute of Marine Science, The University of Hong Kong, Pok Fu Lam, Hong Kong

J. Bleuel
Institute of Marine Sciences (LABOMAR), Federal University of Ceará (UFC), Fortaleza, Brazil

Colegio de Ciencias Biológicas y Ambientales COCIBA, Universidad San Francisco de Quito USFQ, Quito, Ecuador

M. Brandt · D. Velasco-Cedeño
Colegio de Ciencias Biológicas y Ambientales COCIBA, Universidad San Francisco de Quito USFQ, Quito, Ecuador

Galápagos Reef Revival, Puerto Villamil, Isabela, Galápagos, Ecuador

C. Burke
Department of Earth and Ocean Sciences, University of Liverpool, Liverpool, UK

F. Chong
Molecular Invertebrate Systematics and Ecology Laboratory, Faculty of Science, University of the Ryukyus, Nishihara, Okinawa, Japan

Energy and Environment Institute, University of Hull, Hull, UK

N. Dávalos
Galápagos Reef Revival, Puerto Villamil, Isabela, Galápagos, Ecuador

M. O. Soares et al. (eds.), *Extreme and Marginal Coral Reefs and Communities*, Coral Reefs of the World 24,
https://doi.org/10.1007/978-3-032-28035-0_13

the process by which tropical species shift or expand their ranges poleward, with ecosystem transitions from algal habitat-forming species to others, including hard corals, and sometimes soft corals, zoantharia or sponges. Range shifting is both experienced and mediated by organisms that associate with kelp or coral habitat builders. For example, herbivorous fish and sea urchin species that have expanded poleward reinforce the decline in temperate macroalgae through exerting high grazing pressure. High-latitude coral communities are often hoped to act as refugia for tropical coral species that are exposed to increasingly extreme marine heatwaves in their original ranges. Here, we explain that high-latitude cool reefs are unlikely to save tropical corals, not only because the area of suitable habitat is small, but more importantly, because it is becoming clear that cool coral communities are largely comprised of high-latitude specialist and endemic coral species and their high-latitude endosymbionts that are expanding their abundances, rather than tropical coral and endosymbiont species. Cool coral communities are, however, a unique ecological system in their own right–climate warming induced changes transform local ecology, function, and ecosystem services.

Addressing global biodiversity and climate change policy targets, conservation and management efforts need to consider both the needs of native species, and those with shifting ranges. We discuss management options, including both local static and dynamic conservation strategies as well as communication amongst managers. We explore these issues in more detail in a case study of Japan.

To conclude, here we summarise the ecology, future trajectories, and knowledge gaps for cool reefs, including high-latitude subtropical reefs with coral communities and other extreme and/or marginal coral communities in cold waters. We highlight the importance, ecosystem services and beauty of these reefs, and we explain that our knowledge on range shifting and tropicalisation is very limited and might need to be re-evaluated. We argue the need for both traditional and novel research technologies to fully understand the nature of these ever-changing cool reef communities.

This chapter emphasises how marginal shallow coral communities in cool waters (mostly on high-latitude reefs) link to both their tropical and temperate neighbours and that they are stand-alone productive and beautiful ecosystems. We present the global face of "cool reefs" and outline the specific challenges they face, providing a comprehensive summary of the state of knowledge on these cool coral ecosystems

Keywords

Marginal reefs · Extreme reefs · Temperate reefs · Tropicalisation · Coral communities · High-latitude reefs

13.1 Introduction

Extreme and/or marginal coral ecosystems exposed to cool waters in tropical, subtropical and temperate regions (hereafter cool reefs, Box 13.1) and their coral communities are intertwined with kelp and algal systems to form distinct ecosystems with different habitat-building taxa (Chap. 1; Beger et al. 2014; Vergés et al. 2014a; Sommer et al. 2018). Extreme environmental conditions, such as low ocean temperature (often estimated as sea surface temperature (SST)) and light extremes, stress zooxanthellate scleractinian corals that are adapted to live in warm, low variability, high light environments (Guinotte et al. 2003). Such extreme environments are stressful for many tropical species, leading to "cool reef" coral assemblages, characterised by species with traits that allow them to persist in these extreme and/or marginal settings (Sommer et al. 2014), but often dominated by macroalgal species (Steneck et al. 2002). Most cool extreme and/or marginal reefs occur at high latitudes, but some occur in cool waters caused by upwelling or ocean currents (See Chap. 10). Coral communities present in cool extreme and/or marginal reef systems (up to ~35° latitude) also host unique genetic adaptations, endemism, and specific symbionts (e.g., Wicks et al. 2010; Thomas et al. 2017; Nakabayashi et al. 2019; Wu et al. 2021; Yu et al. 2021; Evans et al. 2021; Fifer et al. 2022). The unique conditions on these ecosystems allow them to support distinct communities, including many endemic species of fishes (e.g., *Amphiprion mccullochi* in

S. De Palmas · F. Gösser · A. U. Paul · D. Yamaguchi
Molecular Invertebrate Systematics and Ecology Laboratory, Faculty of Science, University of the Ryukyus, Nishihara, Okinawa, Japan

M. Nakano
General Incorporated Association Urashima Labo, Tokyo, Japan

D. A. Paz-García
Centro de Investigaciones Biológicas del Noroeste (CIBNOR), Laboratorio de Genética para la Conservación, La Paz, Mexico

S. N. Porter
Oceanographic Research Institute, Durban, South Africa

School of Life Sciences, University of KwaZulu-Natal, Durban, South Africa

M. H. Schleyer
Oceanographic Research Institute, Durban, South Africa

M. Beger (✉)
School of Biology, Faculty of Biological Sciences, University of Leeds, Leeds, UK

Centre for Biodiversity and Conservation Science, School of the Environment, University of Queensland, Brisbane, QLD, Australia
e-mail: M.Beger@leeds.ac.uk

Box 13.1 Defining Cool Extreme and/or Marginal Shallow Reefs with Coral Communities ('Cool Reefs')

What are extreme and/or marginal cool reefs?

In this chapter we focus on shallow coral communities that do not form accreting reefs as conditions are cool and variable, which often (but not always) occur at high latitudes up to 35°. Here, corals often occur in mosaics with kelps to form distinct ecological communities with a mix of tropical, subtropical, and temperate species. Marginal cool reefs support species with wide geographical ranges (Williams et al. 1999), and endemics (Fifer et al. 2022; Nakabayashi et al. 2019). Mostly, these species are generalists with broad niche requirements (Mizerek et al. 2016) that can tolerate sea surface temperature (SST) fluctuations from 10–25 °C.

Where do we find extreme and/or marginal cool shallow reefs?

Extreme and/or marginal cool reefs occupy transitional biogeographical zones from subtropical to temperate latitudes and are typically located beyond the tropic of cancer and Capricorn (25° to 35° latitude), where environmental conditions differ significantly from those found in tropical areas (Fig. A). For example, cool reefs occur along the east and west coasts of Australia, in New Zealand, Japan, South Africa, and Brazil. Unlike their tropical counterparts, these reefs experience greater seasonal fluctuations in SST and light availability, resulting in distinct environmental challenges for these ecosystems and their resident species.

Coral and kelp

Extreme and/or marginal cool reefs form distinct marine ecoregions (Spalding et al. 2007), where transitional communities contain tropical, subtropical, and temperate species. For example, in Australia, high-latitude subtropical shallow reefs include zooxanthellate scleractinian corals at or close to their southern range limits, which overlap with the northern range limit of the canopy-forming kelp *Ecklonia radiata* (Agardh, 1848), forming a community characterised by temperate invertebrate taxa including sea urchins, barnacles and ascidians, as well as tropical, subtropical, and temperate species of fish and molluscs (Beger et al. 2014).

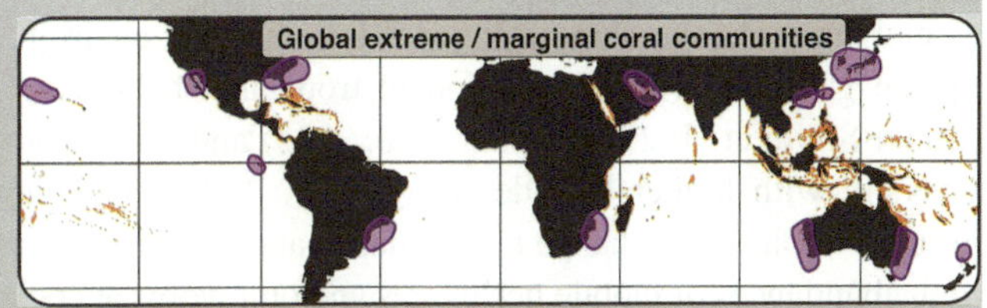

Fig. A Locations of extreme and/or marginal cool shallow reefs, most of which are discussed here in Boxes 2–10, and Riegl Jr. et al. (2026)

Australia; van der Meer et al. 2012), corals (e.g., *Pocillopora aliciae* in Australia; Schmidt-Roach et al. 2013), or *Acropora* spp. (cryptic lineage of *A. hyacinthus* in Japan; Nakabayashi et al. 2019), and molluscs (Okutani 2000).

Increases in SST are causing tropical organisms, such as fish and coral to shift or expand their ranges poleward, resulting in ecosystem transitions, a phenomenon termed tropicalisation (Vergés et al. 2014a). Tropicalisation historically refers to tropical species advancing towards temperate regions, while temperate species retreat (Vergés et al. 2014a; Pecl et al. 2017; Oliver et al. 2018; Kumagai et al. 2018; Stuart-Smith et al. 2021; Zarzyczny et al. 2024b), but here we extend the concept to native coral species expanding their cover or abundance within temperate reefs. We extend the definition because the high endemicity on extreme and/or marginal cool reefs is shaped by increases in local endemic and specialist coral communities, rather than tropical corals invading these ecosystems (Nakabayashi et al. 2019; Fifer et al. 2022). But how does tropicalisation work? Are cool reefs refugia for tropical corals and associated species? Answering these questions will require a deeper understanding of the specific ecological, evolutionary, and anthropogenic-induced processes that together drive community transitions, many of which remain uncertain.

In addition to coral communities on cool reefs, macroalgal forests such as kelps dominate many cool reef environments and are important, biodiverse ecosystems that support a wide array of associated species (Steneck et al. 2002; Wernberg et al. 2019). These forests, composed of macroalgal species usually of the order Laminariales, support a greater diversity of species than terrestrial forests, providing complex habitats for numerous sessile and mobile organisms as well as a numerous beneficial socio-ecological ecosystem services (Steneck et al. 2002; Teagle et al. 2017; Filbee-Dexter and Wernberg 2018). In addition to macroalgae, cool reef benthic communities are frequently dominated by sponges, soft corals and other invertebrates rather than zooxanthellate scleractinian corals (Beger et al. 2014; Reimer et al. 2021). Where coral communities do exist on cool reefs, they do not usually form accreting reefs and are surviving at their environmental limits (Kleypas et al. 1999; Beger et al. 2014). Over the past decade, some marginal cool reefs with coral communities have lost much of their kelp cover due to stress events such as marine heat waves, storms, and high grazer abundances (Wernberg et al. 2013; Filbee-Dexter and Wernberg 2018; Wernberg et al. 2019; Gorman et al. 2020).

Reef fish assemblages on cool reefs are dependent on the benthic community structure to provide food, shelter from predators, and nursery grounds (Holland et al. 2020). Fish assemblages change along a tropical to temperate environmental gradient, with a reduction in tropical fish biomass (Stuart-Smith et al. 2013; Miller et al. 2023) and functional diversity with higher latitude (Clay et al. 2024). Trophic groups of fishes also change in abundance and biomass, with high abundances of mobile herbivores and omnivores at cool coral communities (Ferreira et al. 2004; Kumagai et al. 2018; Holland et al. 2020). For example, roving herbivores such as kyphosids, omnivores and piscivores are highly abundant further south off the coast of Brazil (Ferreira et al. 2004), with kyphosids feeding on the highly abundant brown macroalgae that are prevalent on subtropical and temperate reefs. Beyond trophic groups, functional niches defined across multiple traits of fishes support the presence of different herbivore groups—the roving large-bodied species that drive tropicalisation (Zarco-Perello et al. 2021) and site-attached small herbivores that, apart from temperate fishes, hold only a few tropical species (Miller et al. 2023), also Chap. 3 (point to Morais et al. (2026)).

As ongoing climate change transforms many marine communities, cool extreme and/or marginal coral communities are situated in warming hotspots (Hobday and Pecl 2014; Schlegel et al. 2017) and are experiencing rapid and ongoing spatiotemporal changes (Yamano et al. 2011; Kumagai et al. 2018; Bleuel et al. 2024). In response, species range shifts and expansions due to tropicalisation are causing whole-scale community transformations in subtropical and temperate reef systems (Beger et al. 2014; Pecl et al. 2017; Smith et al. 2021). There is clear evidence that climate change and associated thermal stress events have already resulted in large-scale community shifts globally, including on the Great Barrier Reef (Hughes et al. 2018), extended the range and diversity of corals in Japan (Yamano et al. 2011), and transformed temperate kelp communities (Oliver et al. 2018) through range contractions of kelp habitat-builders (Beger et al. 2014; Vergés et al. 2014a). A shift from macroalgae-dominated to other taxa-dominated habitat structures brings profound changes to three-dimensional habitat structures, species interactions, and functional trait compositions, altering the presence and abundances of fishes and invertebrates associated with habitat-building corals and kelps (Vergés et al. 2014a), and homogenising ecological communities (Pessarrodona et al. 2021; Beger 2021; Stuart-Smith et al. 2021). Range shift dynamics and their drivers are known for certain species (e.g., the sea urchin *Centrostephanus rodgersii*, (Ling et al. 2009b), fishes (Feary et al. 2014; Clay et al. 2024), and habitat builders such as zooxanthellate scleractinian corals (Yamano et al. 2011; Sommer et al. 2014) and kelps (Smale and Wernberg 2013; Kumagai et al. 2018), but there are many unknown responses across whole ecological communities.

In this chapter, we characterise extreme and/or marginal cool reefs with coral communities and how they differ from other reef systems. We first discuss defining characteristics of these reefs such as biogeography, environmental factors, and biological communities. Second, we review the effects of climate change and tropicalisation on temperate reef species and communities and challenges to the concept of tropicalisation due to regional differences. We further highlight understudied taxa on cool reefs and synthesise evolutionary processes, including patterns of genetic diversity, phenotypic plasticity, and physiology of the local ecological communities. We discuss and compare the population dynamics of zooxanthellate scleractinian corals and resident communities in terms of the differences in their population structure and transient (short-term) population dynamics, and their potential implications under climate change. We also consider whether cool reefs can be considered novel ecosystems due to the changes in community composition resulting from climate change and the role of functional diversity in defining novel systems. Finally, we highlight management and conservation strategies and discuss potential future research avenues.

13.2 The Environment of Cool Reefs with Coral Communities: Why Are they Extreme and/or Marginal?

13.2.1 Environmental Factors

Cool coral communities and their associated taxa live in extreme and/or marginal environmental conditions for the habitat building coral taxa (Chaps. 1 and 2; Fig. 13.1). Extreme and/or marginal conditions cause slow growth and high mortality rates of corals, alongside reduced calcification rates below that necessary for accretion (e.g., Kleypas et al. 1999). Extreme environments are characterised by environmental extremes close to species' physiological tolerances and/or high seasonal variability in environmental factors (Chap. 2; Box 1; Beger et al. 2014; Vergés et al. 2014a). By virtue of their high-latitude location, most extreme and/or marginal cool reefs are found between 25° to 35° in latitude. However, latitude alone cannot predict non-accreting extreme and/or marginal coral communities, as non-accreting extreme and/or marginal reefs are found at 22°N in Hong Kong and 25°N in Taiwan due to upwelling events (Boxes 3 and 4; Chap. 12), whereas coral reefs in Okinawa at 26° N support accreting coral communities.

Several environmental factors drive the distribution of cool coral communities, such as SST, light availability and pH, generally linked to poleward flowing boundary currents (Veron and Minchin 1992; Kleypas et al. 1999; Muir et al. 2015; Mizerek et al. 2021). The main drivers of extreme and/

or marginal reef coral community structure are SST variability and reduced light availability (Veron and Minchin 1992; Halfar et al. 2005; Malcolm et al. 2011; Muir et al. 2015; Sommer et al. 2018). Compared to tropical coral reefs, organisms living on cold reefs must cope with elevated SST variability across seasons (Fig. 13.1), and often over shorter timespans, for instance, SST can fluctuate by 6 °C within a 24-hour period in the Solitary Islands, Australia (Malcolm

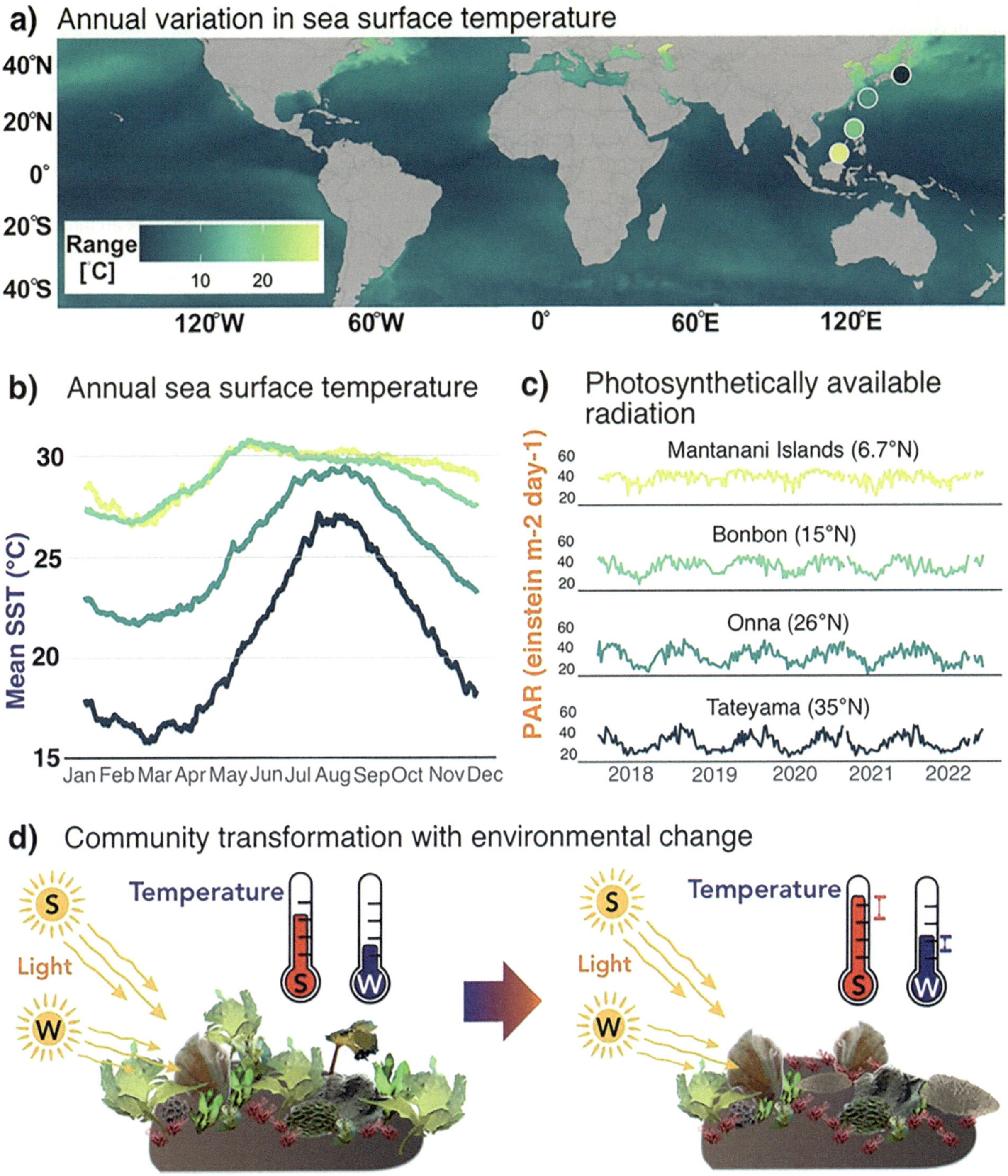

Fig. 13.1 Extreme and/or marginal cool reefs that host coral communities exist in highly variable environments, where sea surface temperature (SST) and photosynthetically available radiation (PAR) vary throughout the year, with (**a**) global annual variation in SST; (**b**) daily mean SST recorded at four sites of different latitudes (Mantanani Islands 6.7°N, Bonbon 15°N, Onna 26°N, Tateyama 35°N) between 2018 and 2022; (**c**) variation in PAR at the same four sites over the same period; and (**d**) a schematic of how changes in SST and PAR influence extreme and/or marginal cool reef communities

et al. 2011). Light availability diminishes with latitude, particularly during winter, causing some staghorn corals to live at shallower depths with increasing latitudes (Muir et al. 2015; but see Madin et al. 2016). In addition, some high-latitude coral communities are also exposed to low pH values through their proximity to CO_2 vents (Chap. 11). For example, in Shikine-jima, Japan, communities are characterised by simplification, with very few hard corals surviving (Agostini et al. 2021). Another key driver of the distribution of corals at subtropical and temperate latitudes is the presence of boundary currents that carry the larvae of corals and other tropical species poleward (Beger et al. 2011), such as those found in Australia (Booth et al. 2007) and Japan (Kumagai et al. 2018).

13.2.2 Anthropogenic Factors

Extreme and/or marginal cool reefs are subject to several local anthropogenic pressures such as overexploitation, pollution, and urbanisation, causing declines and species turnover in corals and their associated communities. For example, extensive fishing pressure is present in most cool reefs, causing significant declines in some species (e.g., the grouper *Epinephelus itajara* in Brazil) on high-latitude reefs (Giglio et al. 2017). Furthermore, overfishing of lobsters (*Jasus edwardsii*) in Australia has been shown to increase the rate of ecosystem phase shifts from kelp beds to sea urchin barrens as predators are removed (Ling et al. 2009a). The combined impact of high rainfall, forest degradation and sea level rise, will also likely instigate high soil erosion rates and wash sediment onto nearshore cold reefs as it does on tropical reefs (Brodie et al. 2012; Maina et al. 2013). Furthermore, pollution such as industrial, agricultural and urban runoffs, reaches reefs as high sediment loads smothering coral communities (Soares 2020). Marine contamination can have serious impacts on the health of marine species and populations from invertebrates to vertebrates (e.g., Baines et al. 2021). Such exposure to long-term contamination on cool reefs has the potential to lead to declines in species (e.g., links to accumulation of agricultural contamination with declines in soft coral cover (Porter et al. 2018)). Coastal urbanisation similarly reduces species diversity and have been associated with community shifts along cool reefs in Brazil (Scherner et al. 2013). Phase shifts have also been related to proximity to anthropogenic activities such as highly populated areas, urbanised surfaces and dredged ports which indicate significant local impacts of anthropogenic activities on marine community structures (Cruz et al. 2018). The cumulative impact of multiple anthropogenic stressors can seriously impact biological communities on cool reefs with synergistic, antagonistic and additive effects on marine species, communities and ecosystems (e.g., Alava et al. 2023).

13.2.3 Future Predicted Global Climatic Changes

Globally, the exposure of coral reefs to thermal stress events, or marine heatwaves (MHWs), will increase in both magnitude and frequency to a point where at 1.5 °C warming above pre-industrial levels, only 0.2% of ecological refugia are predicted to remain across all climate scenarios (Dixon et al. 2022; Kalmus et al. 2022). On high-latitude cool reefs, the minimum, mean, and maximum SST are all likely to increase (Malcolm et al. 2011), and changes in seasonal and short-term variability are likely, but unknown (Ribas-Deulofeu et al. 2023). Similarly, future thermal exposure trajectories remain unknown for cold reefs in upwelling areas. Whilst a general trend of overall warming is predicted by global climate models (Dixon et al. 2022), interannual variability and increasing instability of ocean currents may result in cold spells that will also extirpate tropical species (Leriorato and Nakamura 2019). Similarly, large-scale extreme flooding events may trigger the transformation of cool coral communities by eliminating tropicalising corals (e.g., Butler et al. 2013). In addition, altered rainfall patterns may alter light regimes through more, or less, cloud cover compared to present light availability levels (González-Espinosa and Donner 2023). Such community 'resets' have already occurred, for example during the 2011 floods in southeast Australia (Butler et al. 2013) and the 2018 cold spell in Kochi, Japan (Leriorato and Nakamura 2019).

13.3 Biology and Ecology of Cool Coral Ecosystems

13.3.1 Community Assembly on Cool Coral Ecosystems

Despite their affinity for tropical areas, zooxanthellate scleractinian corals persist on extreme and/or marginal cool reefs where the environmental conditions can be considerably cooler and more variable than in the tropics (Harriott et al. 1994). These conditions can include SST dropping below the lower temperature limit for reef formation in the tropics (e.g., below 18 °C, Johannes et al. 1983; Yamano et al. 2001; Denis et al. 2013). The species within extreme and/or marginal cool reefs often live at their environmental extremes, which gives rise to a unique assemblage of species with tropical, subtropical and temperate affinities (Hoey et al. 2011; Beger et al. 2014; Schleyer et al. 2018). The community

South Africa's Extreme and/or Marginal Cool Reefs

Sean N. Porter and Michael H. Schleyer

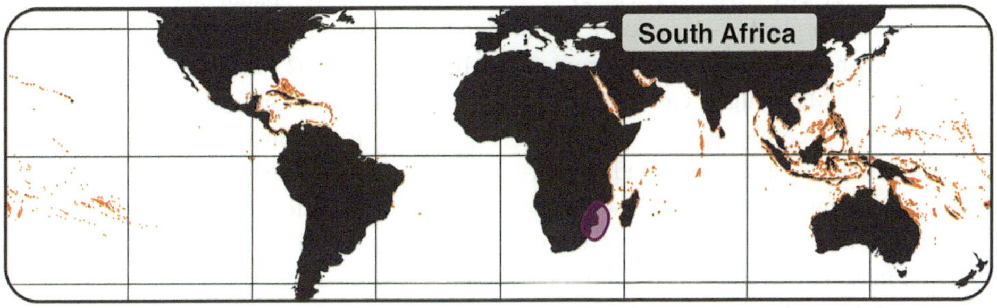

Coral communities in South Africa constitute the southernmost distribution of scleractinians on the African coast, and fall within the iSimangaliso Wetland Park, a World Heritage Site in northern KwaZulu-Natal (Schleyer and Porter 2018). The reefs support corals growing as a thin veneer on the limited Pleistocene sandstone substrata, making them patch reefs. They are clustered in a Northern, Central and Southern Complex with only some scattered reef in between and are only 40 km^2 in extent (Schleyer et al. 2018). Corals extend from a depth of 8 m down to 27 m. South Africa's extreme and/or marginal reefs are dominated by soft corals, whose cover (30%; 39 species) exceeds that of zooxanthellate scleractinian cover (23%; 93 species) (Porter and Schleyer 2019). They are comprised of tropical and subtropical Indo-Pacific species and also include endemics.

Jerker Tamelander©

Jerker Tamelander©

These reefs are deep and exposed to canyon-induced upwelling, particularly during summer. Only negligible bleaching occurred during the 1998 global bleaching event (Schleyer and Celliers 2000), and in subsequent bleaching events (<10% pale or bleached corals, Porter et al. 2021). There is some evidence of tropicalisation, with changes in coral distribution. *Parasphaerasclera aurea* (Benayahu and Schleyer 1995), which was fairly abundant at depth in the 1990s, now only occurs in the south (Schleyer *pers. obs.*). *Acropora robusta* (Dana, 1846), which is relatively abundant further north, has now made an appearance in South Africa (Floros 2012).

Jerker Tamelander©

structure and ecological dynamics of these extreme and/or marginal cool reefs differ significantly compared to their tropical counterparts (Beger et al. 2014; Schoepf et al. 2023). For example, extreme and/or marginal cool reefs are often dominated by macroalgae as opposed to corals on tropical reefs (Johannes et al. 1983). Macroalgae can potentially limit the growth and settlement of corals, thus possibly limiting coral diversity and abundance on extreme and/or marginal cool reefs through competition with macroalgae (Johannes et al. 1983). Yet, many cool reefs support both types of habitat builders, macroalgae and corals (Wernberg et al. 2013; Filbee-Dexter and Wernberg 2018; Gorman et al. 2020). Furthermore, lower coral diversity or abundance can limit the diversity of some coral-associated invertebrates such as barnacles due to a lack of hosts (Chan et al. 2018; Hammerman et al. 2022). The impacts of herbivory can be amplified on macroalgae dominant reefs, so an abundance of grazing fishes may promote coral larvae settlement by creating suitable settlement substrata (Hoey et al. 2011; Tuckett et al. 2017; Ling et al. 2018). Alternatively, some extreme and/or marginal cool reefs support high coral cover (up to 72.5%) and lack macroalgae, often as high abundances of sea urchins exert high grazing pressure (Thomson and Frisch 2010; Zarzyczny et al. 2022).

Extreme and/or marginal coral ecosystems are characterised by challenging environmental conditions, including SST, light and aragonite saturation (Kleypas et al. 1999; Muir et al. 2015) as well as salinity, water quality, the type and depth of the substratum, wave climate and siltation regime (Veron 1995; Kleypas et al. 1999). Environmental conditions also play a fundamental role in shaping the community structure and dynamics of cool coral ecosystems (Fig. 13.2; Beger et al. 2014; Bosch et al. 2022; Miller et al. 2023). Availability of resources, including space, light, and nutrients are primary drivers of variation in community structure within these ecosystems, with temperature gradients a key driver of coral biodiversity patterns along tropical-to-temperate transition zones (Veron and Minchin 1992; Mizerek et al. 2016; Sommer et al. 2018). Extreme and/or marginal cool reef communities exhibit high species turnover or rapid gradients in species richness, containing both species' range edges and endemic species, with evidence of potential adaptive genetic and phenotypic characteristics (Wernberg et al. 2016; Hughes et al. 2018; Travers et al. 2018). Understanding the complex interactions among species, their thermal niches, and ecological processes is crucial for comprehensively elucidating the functioning and resilience of extreme and/or marginal cool reef ecosystems (Fig. 13.2; Wang et al. 2020; Browne and Bauman 2023).

One defining feature of cool reef communities is the distributional overlap of species with diverse thermal tolerances (Jones et al. 2020a; Miller et al. 2023), adapted to a wide range of temperature regimes. Species in these environments exhibit complex spatial distributions along thermal, light, and exposure gradients (Ross et al. 2021; Cant et al. 2023; Miller et al. 2023). These transitional reef communities exhibit a biogeographical overlap of tropical and temperate species at their range edges, locally abundant subtropical species that are absent at lower latitudes, and endemic species with restricted geographic ranges (Crean et al. 2010; Wicks et al. 2010; Soares 2020; Favoretto et al. 2022; Schoepf et al. 2023).

Biogeographic variation in community composition along latitudinal gradients is evident across taxa, with coral cover, reef accretion, and development likely limited due to cumulative stress or range limits from physical and biological factors (Rasher et al. 2012; Sommer et al. 2014; Camp et al. 2018). Despite observed high survival of coral recruits in the subtropics (Cant et al. 2023), coral species richness and diversity generally decrease poleward (Mellin et al. 2010), with the benthos at high latitudes often dominated by macroalgae, soft corals, and other sessile invertebrates (Harriott and Banks 2002; Olbers et al. 2009; Wicks et al. 2010; Schleyer and Porter 2018). Similarly, the diversity and abundance of fish assemblages broadly declines poleward (Saeedi et al. 2019), with an observed higher diversity and abundance in the area where tropical and temperate species overlap (Miller et al. 2023). Notably, fish species traits facilitate successful settlement of tropical species into high-latitude extreme and/or marginal cool reef communities (Feary et al. 2014), typically those with generalist traits. Further, temperate reefs support higher functional diversity of fish assemblages compared to the tropics at a global scale (Stuart-Smith et al. 2013). However, regional studies suggest that possible local or fine-scale processes also support the contrasting decline of fish (Clay et al. 2024), mollusc (Floyd et al. 2020), and soft coral functional diversity (Otis et al. 2024) poleward along the tropical-to-temperate transition in Japan. Extreme and/or marginal cool reefs also exhibit faster rates of speciation (Rabosky et al. 2018), and perhaps have high incidences of endemism (Nakabayashi et al. 2019; Fifer et al. 2022). Biogeographic trends vary among invertebrates, with some taxa exhibiting reduced diversity, while others show no clear latitudinal pattern (Sawall et al. 2014; Ulate et al. 2016; Bracewell et al. 2018).

13.3.2 Structural and Functional Marginality

Marginal cool coral communities can be defined as such based on structural and functional marginality (Box 1; Chapter 1). Whilst structural marginality refers to the

Taiwan Extreme and/or Marginal Cool Reefs

Wan-Chien Hsiao

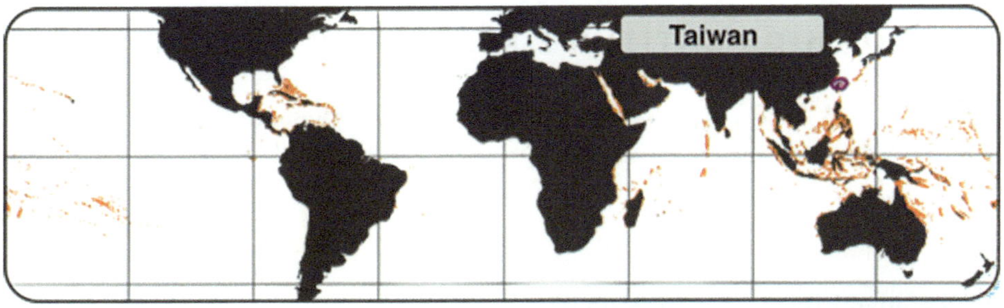

In northern Taiwan (~25°N) and the northern Penghu Archipelago (~23.5°N), coastal benthic communities along the coastline are comprised of extreme and/or marginal non-accreting coral communities (Hsiao et al. 2021). These coral ecosystems are shaped by ocean currents and strong seasonal environmental variation from the east Asian Monsoon (Kuo et al. 2023). On Taiwan's east coast, the warm northward Kuroshio current supports accreting coral reef systems, whilst the Chinese coastal current brings cool waters to northern Taiwan in the winter (Dai 2018), constraining reef development (Denis et al. 2019).

Coral communities are generally found at shallow depth <20 m due to high turbidity and sandy substrate, supporting ~100 coral species on the northern extreme and/or marginal reefs. Non-reefal coral communities in northern Taiwan are mainly composed of massive and encrusting Merulinidae corals, with a seasonal establishment of macroalgae in the winter (Lin et al. 2018).

Accreting reefs in the south are expected to expand northwards in Taiwan (Denis et al. 2020) and the accentuated warming in the winter for the northern extreme and/or marginal reefs (Lin et al. 2022), as well as intense warming of the western boundary current (Vergés et al. 2019) and observed homogenisation of community composition across the latitudinal gradient supports the idea of expansion of tropical coral species (Ribas-Deulofeu et al. 2016).

Wan-Chien Hsiao © 2024

Wan-Chien Hsiao © 2024

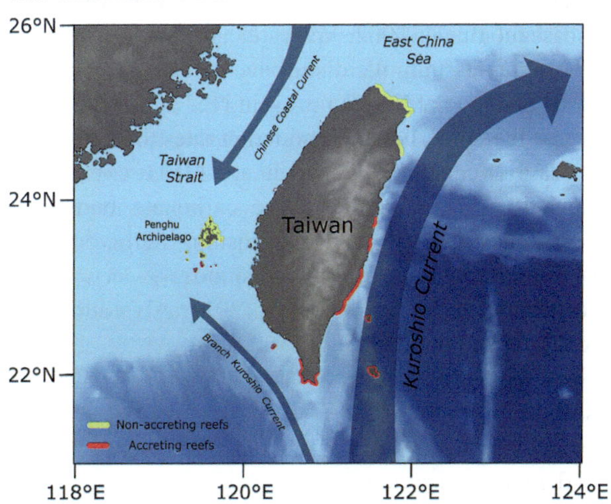

Map of Taiwan. Credit: Wan-Chien Hsiao

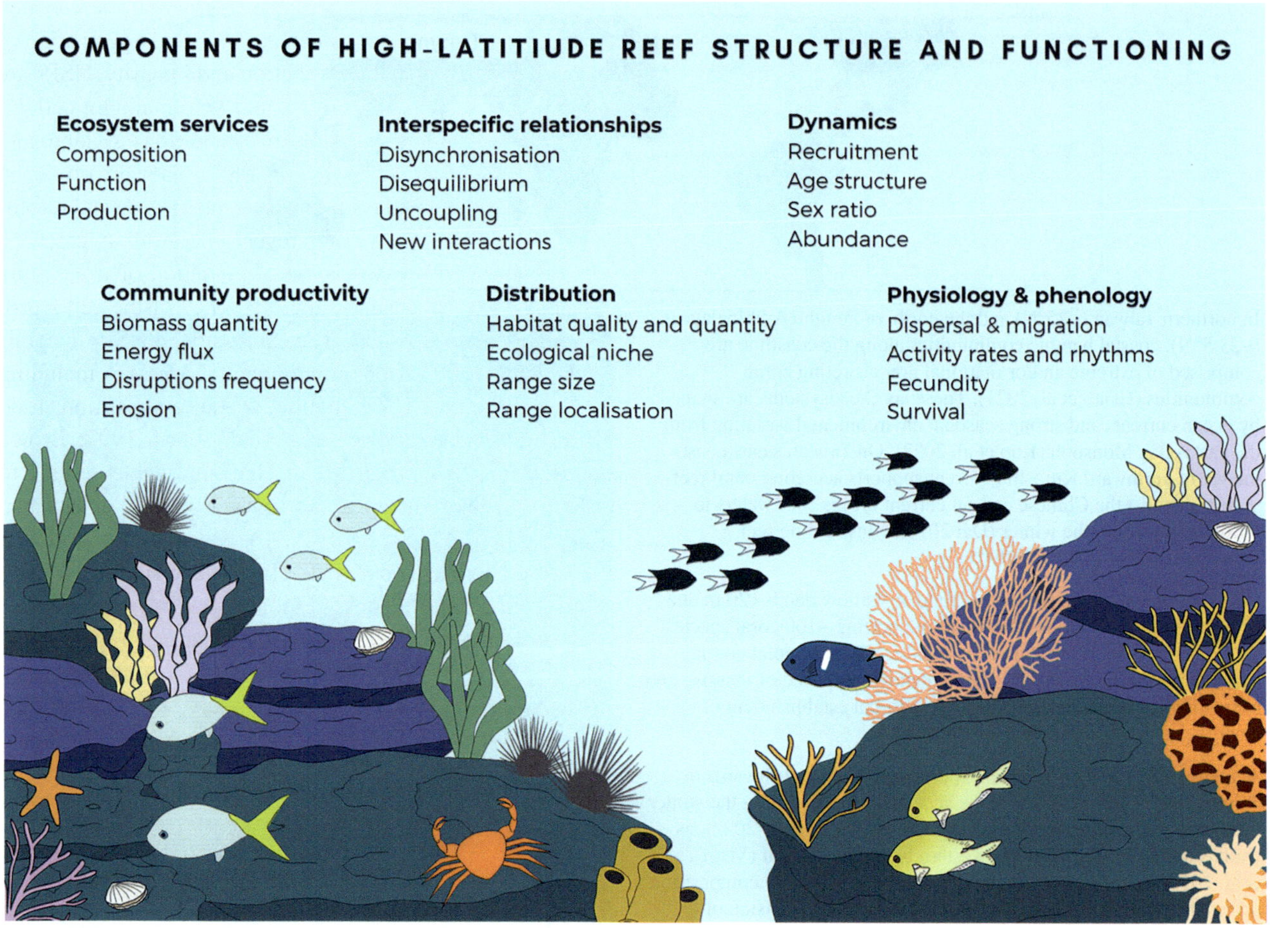

Fig. 13.2 The key diversity components within extreme and/or marginal cool reefs across distinct levels of diversity from ecosystems and communities to species and populations

community composition and functional marginality considers ecosystem functioning; both are strongly interconnected (Schoepf et al. 2023). For example, non-accreting marginal coral communities are often characterised by differences in lower scleractinian coral composition and/or species richness (structural marginality) than tropical reefs, which in turn affects ecosystem functioning such as productivity and reef growth (functional marginality) (Harriott and Banks 2002; Wicks et al. 2010; Booth and Sear 2018). Whilst abiotic disturbances such as high wave action or SST anomalies can prevent reef development by lowering the reef carbonate production rates through physical stress (Riegl 2001; Perry et al. 2013), community composition strongly impacts the carbonate budget of coral communities through bioerosion, herbivory and competition (Quan-Young and Espinoza-Avalos 2006; Rasher and Hay 2010; Perry et al. 2013; Alvarado et al. 2016; Zarzyczny et al. 2022).

Erosion is critical for healthy reef function as it maintains biochemical cycles and suspends $CaCO_3$ particles from dead coral, allowing accretion to take place (Tribollet and Golubic 2011). Whilst mechanical abrasion from wave action or storms contributes to erosion of coral reefs, the majority of erosion is driven by biotic processes, i.e., bioerosion (Martin-Garin and Montaggioni 2023). The extent of bioerosion of corals is closely associated with the feeding behaviours and living modes of important herbivorous fish (e.g., parrotfishes) and sea urchins bioeroders (Perry et al. 2013; Zarzyczny et al. 2022). As herbivores feed on algal turf, they ingest coral tissue resulting in external bioerosion (Alvarado et al. 2016). The proportion of ingested coral tissue varies and is largely affected by the method of feeding. For instance, excavating parrotfish are the most important bioeroding herbivorous group of fishes on tropical reefs (Peyrot-Clausade et al. 2000; Perry et al. 2013; Januchowski-Hartley et al. 2017), whilst main urchin bioeroders include genera of *Diadema*, *Echinometra*, *Echinostrephus* and *Eucidaris* (Perry et al. 2013). Given the community structure of herbivores displays a clear change on a latitudinal gradient (Steneck et al. 2017; McLaren et al. 2024) there are notable differences in herbivory and external bioerosion between high-latitude coral communities and tropical reefs (Longo et al. 2014; Ling et al. 2018; Zarzyczny et al. 2022). Furthermore, large herbivores have been over-fished all around the world with evident cases of "functional extinction" along extreme and/or marginal cool coral communities (Comeros-Raynal et al. 2012; Pereira et al. 2021).

Primary production also differs between 'typical' tropical coral reefs and extreme and/or marginal cool coral communities. An estimated 50–70% of primary productivity on coral reefs is attributed to zooxanthellae (family Symbiodiniaceae), the symbiotic algae living within coral tissue (Douglas 2009). Turf and crustose coralline algae account for most of the remaining primary productivity on coral reefs (Hatcher 1988; Douglas 2009; Owen et al. 2021). In contrast, key primary producers on extreme and/or marginal cool reefs are typically macroalgal beds (Filbee-Dexter et al. 2023). For example, Australian kelp forests are responsible for over a third of blue carbon sequestration in Australia, and for ~3% of global carbon sequestration (Filbee-Dexter and Wernberg 2020).

The presence and abundance of macroalgae is particularly relevant when considering extreme and/or marginal coral communities. It is widely accepted that coral and macroalgae compete for space and light (McCook et al. 2001). Direct contact between chemically rich macroalgae and coral (Rasher and Hay 2010), as well as contact through the release of hydrophobic allelochemicals, can result in coral bleaching, decreased photosynthesis, and increased coral mortality (Quan-Young and Espinoza-Avalos 2006; Rasher and Hay 2010). The release of allelochemicals by algae can inhibit coral larval growth (Morrow et al. 2016) and settlement (Birrell et al. 2005; Morrow et al. 2016). Furthermore, macroalgae can reduce coral fecundity (Foster et al. 2008) whilst algal turfs reduce coral fitness by reducing their photochemical efficiency, chlorophyll concentrations, zooxanthellae density and coral tissue thickness (Quan-Young and Espinoza-Avalos 2006).

Given the competitive interactions between coral and algae, herbivory is one of the key processes involved in shaping ecological dynamics on along biogeographic transition zones (Vergés et al. 2014a, b; Kumagai et al. 2018). Herbivores maintain low algal cover allowing coral and other invertebrates to settle (Vergés et al. 2014a), with herbivory pressure decreasing with increasing latitude (Steneck et al. 2017; Zarzyczny et al. 2022, 2024b). The presence of browsers that actively feed on macroalgae, and grazers that maintain low algal turf, promotes calcification (Jouffray et al. 2015) and therefore coral growth. In fact, extreme and/or marginal cool coral communities can exhibit very high coral cover (Thomson and Frisch 2010; Denis et al. 2013; Sommer et al. 2014), which is typically mediated by herbivory maintaining low macroalgal cover (Vergés et al. 2016; Kumagai et al. 2018). In contrast, release from herbivory pressure, for example through overfishing of herbivores, can increase algal cover (Norström et al. 2009; Vergés et al. 2014b; Kuempel and Altieri 2017; Steneck et al. 2017). Consequently, herbivory is thought to underpin ecosystem stability and increase reef resilience improving its capability of recovering from bigger environmental perturbations in the tropics (Cheal et al. 2010) but whether this is true for cool reefs remains to be determined.

13.4 Climate Change and Tropicalisation

13.4.1 Climate Change Impacts

Global anthropogenic activities are rapidly advancing the progression of ocean warming and acidification. In subtropical and temperate regions, warming and MHWs pose direct physiological stress to cold-affinity species (e.g., kelps, seaweeds; Arafeh-Dalmau et al. 2019). Previously, high-latitude communities were expected to be more resilient to thermal-stress related bleaching events than tropical reefs (van Hooidonk et al. 2014). The increasing frequency and intensity of MHWs (Yao and Wang 2022) has begun to cause bleaching even at some of the highest latitude reefs such as at Lord Howe Island in Australia (31.5°S) (Moriarty et al. 2023). Moreover, emerging evidence shows that subtropical corals and high-latitude endemics may be particularly vulnerable to heat stress (Cant et al. 2021). In southeastern Australia, subtropical endemic and specialist corals were disproportionately affected by heat-related bleaching events (Goyen et al. 2019; Kim et al. 2019; Lachs et al. 2021). Nevertheless, coral species which are typically first to bleach on tropical reefs, such as *Acropora* spp. corals, appear to bleach less at higher latitudes (Sommer et al. 2024).

Due to increased atmospheric CO_2 being absorbed by the oceans, ocean acidification is becoming a global threat to marine communities. For corals, declining aragonite saturation with increasing latitude can limit the latitudinal distribution of calcium-carbonate accreting coral reefs (Kleypas et al. 1999), and as aragonite saturation levels continue to decline, optimal conditions for calcification are expected to contract toward the equator, potentially leading to weaker skeletons and limiting the geographic distribution of many coral species to low latitudes (Orr et al. 2005). Acidification can promote the growth of algae (Kroeker et al. 2010; Connell et al. 2013) and simultaneously affect calcification rates which adversely impact calcifying organisms, including scleractinian corals (Pandolfi et al. 2011; Kurihara et al. 2021). This can lead to phase shifts from calcareous corals to soft coral, algal or other non-scleractinian dominated ecosystems (Kroeker et al. 2010; Inoue et al. 2013; Reimer et al. 2023). When considering ocean warming and acidification together, these processes are predicted to have opposing effects: warming can facilitate a poleward shift of scleractinian corals and warm-water fishes, while acidification can lead to decreased diversity of calcifying taxa including scleractinian corals and other marine invertebrates (Hall-Spencer et al. 2022). For example, at a natural analogue site at Shikine Island, in temperate Japan, ocean acidification stopped the poleward progression and tropicalisation of calcifying organisms, suggesting warming and acidification could lead to a simplification of ecosystems to turf-dominated communities (Agostini et al. 2021).

13.4.2 What Is Tropicalisation?

Tropicalisation is a marine phenomenon characterised by the poleward shifts of marine organisms (Fig. 13.3; Wernberg et al. 2013; Vergés et al. 2014a; Zarzyczny et al. 2024b). The term was first used by Bianchi and Morri (2003) to describe the increasing occurrence of warm-water species within the Mediterranean Sea, and the consequent transformation of community assemblages to more 'tropical-like'. Whilst tropicalisation of the Mediterranean Sea was, and is, largely driven by both ocean warming and anthropogenic activities (namely Lessepsian migration and transport via the Suez Canal), the term has since been adopted globally to describe changes in marine communities arising from climate-induced range shifts (Zarzyczny et al. 2024b). Specifically, tropicalisation refers to the increase in the ratio of tropical species, in relation to all species which are present within a community (Wernberg et al. 2013). Consequently, tropicalisation encompasses both poleward range expansions of (sub)tropical species, and range retractions or loss of temperate species which results in the formation of novel species assemblages (Zarzyczny et al. 2024b).

13.4.3 Ecosystem Stability during an Era of Change

As climate-induced range shifts continue, new species interactions combined with increased environmental stressors have already resulted in ecosystem phase-shifts on some extreme and/or marginal cool reefs (Figs. 13.3 and 13.4; Vergés et al. 2014a; Zarzyczny et al. 2024b). Phase-shifts are ecological phenomena where an ecosystem switches to an alternative state (Conversi et al. 2015), typically because of environmental or biological perturbations pushing an ecosystem past a critical threshold (Dakos et al. 2015). Phase-shifts are caused by a combination of interacting factors such as changes in functional diversity of reef organisms (Fig. 13.4; Eklöf et al. 2008; Steneck et al. 2017) and environmental stress caused by increased excess heat or destructive storms (Cheal et al. 2010). For example, increasing heat stress associated with an El Niño southern oscillation event, has led to a steady decline in *Ecklonia cava* kelp beds in Tosa Bay, Japan (Serisawa et al. 2004; Tanaka et al. 2012; Vergés et al. 2014a). Simultaneously, increasing SST has facilitated the range expansion of tropical herbivorous fishes into the region, leading to overgrazing of kelp and consequently preventing the kelp population from recovering (Vergés et al. 2014a). Subsequently, the barren habitat was colonised by reef building corals leading to an ecosystem phase-shift from a kelp bed to a coral community (Mezaki and Kubota 2012; Vergés et al. 2014a). Whilst climate change induced phase-shifts are an example of extreme consequences of tropicalisation on extreme and/or marginal cool reefs (Vergés et al. 2014a; Zarzyczny et al. 2024b), tropicalisation can also have species-specific and evolutionary consequences (Zarzyczny et al. 2024b).

13.4.4 Effects of Climate Warming on Species Living on Cool Reefs

The way in which species respond to climatic changes depends on their physiological tolerances, life history traits and strategies, geographic origin, and phylogeny (Fig. 13.2; Sommer et al. 2014; Feary et al. 2014; Madin et al. 2016; Sommer et al. 2017; Dahlke et al. 2020; Fujiwara et al. 2022). These traits include the age of sexual maturity, maximum population growth rate, generation time, and mobility (Fig. 13.2; Winemiller and Rose 1992; Fujiwara et al. 2022). For example, spawning fish and their resulting embryos had a higher physiological sensitivity to fluctuations in ambient temperature than other life history stages (Dahlke et al.

Hong Kong Extreme and/or Marginal Cool Reefs

Fiona Chong

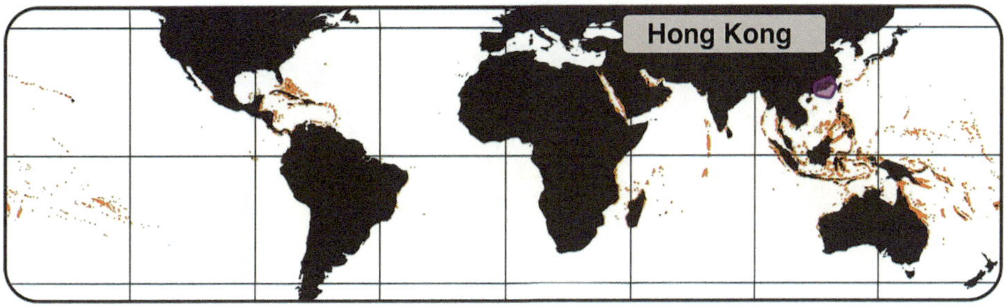

Hong Kong (22°08′-22°35'N, 113°49′-114°31′E) is a coastal city on the southern coast of China, consisting of more than 200 islands, with an extensive coastal reef environment (Lands Department 2024). Hong Kong has a humid subtropical climate with a cold, dry winter and a hot, humid summer with high precipitation (HKO 2024). In summer, the SST rise to 31 °C, while in winter, SST can be as low as 14 °C (Goodkin et al. 2011).

Due to Hong Kong's proximity to the second largest river in China, water salinity is strongly affected by seasonally variable freshwater discharge into the marine environment.

Scleractinian corals are thus mainly found in the south and east, away from the mouth of the estuary. Exposure to high anthropogenic pressures such as pollution, urban development (Wong et al. 2018), and recreational activities (Au et al. 2014) has driven the community composition and structure of the 41 fringing coral communities in Hong Kong (Yeung et al. 2021).

Coral cover ranges from 2–79%, with around half of the sites boasting >40% coral cover, with 21 coral genera recorded, amongst them the most abundant are *Pavona decussata*, massive *Porites* spp., and *Platygyra* spp. There is no evidence and limited research on the effects of tropicalisation on coral communities in Hong Kong, though both marine heatwaves and cold spells have led to bleaching events.

Gomen See©

Gomen See©

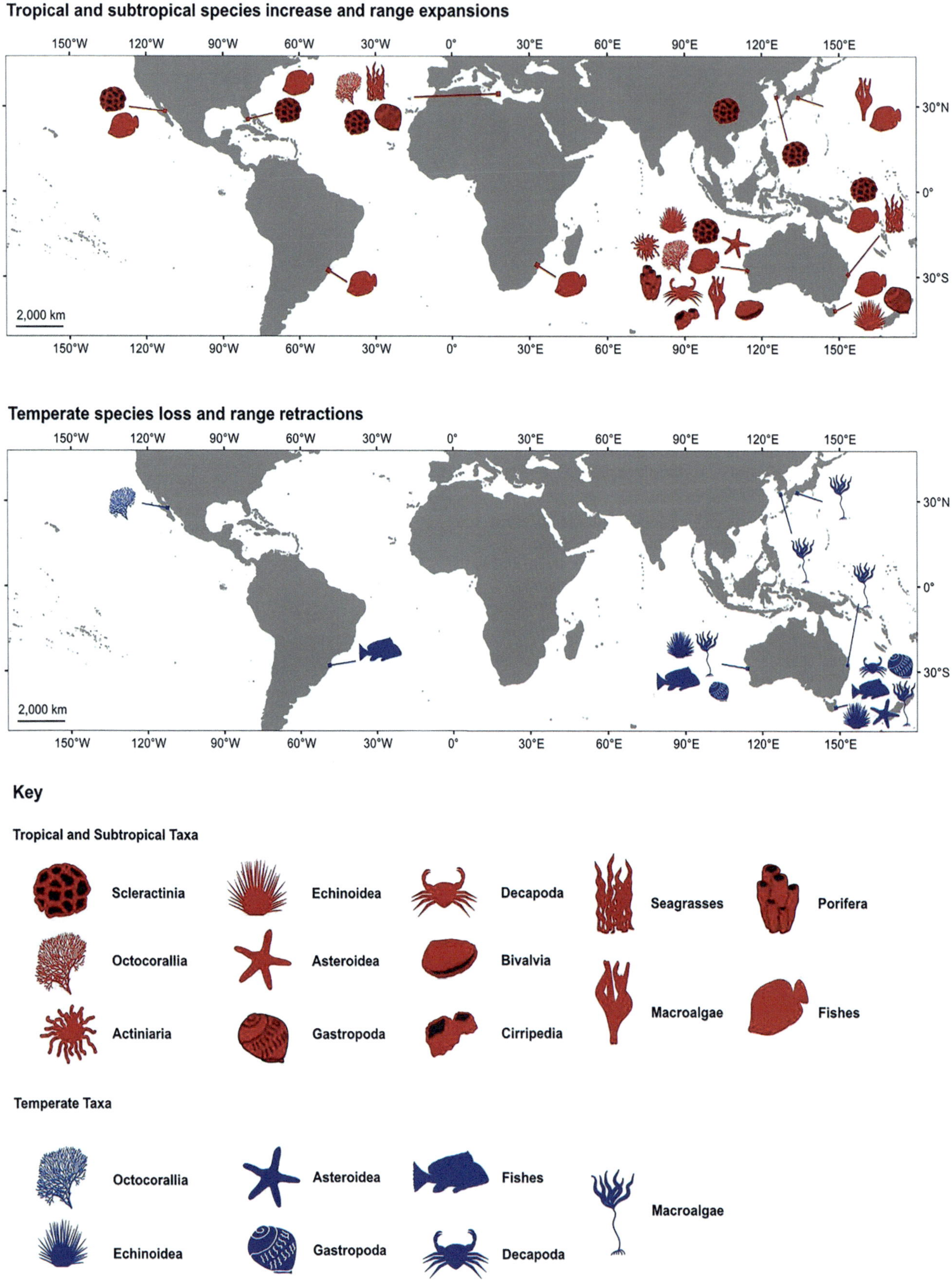

Fig. 13.3 Extreme and/or marginal cool reefs undergoing tropicalisation and the associated taxa. Data underpinning the figure was extracted from tropicalisation literature until April 2024 (Zarzyczny et al. 2024b)

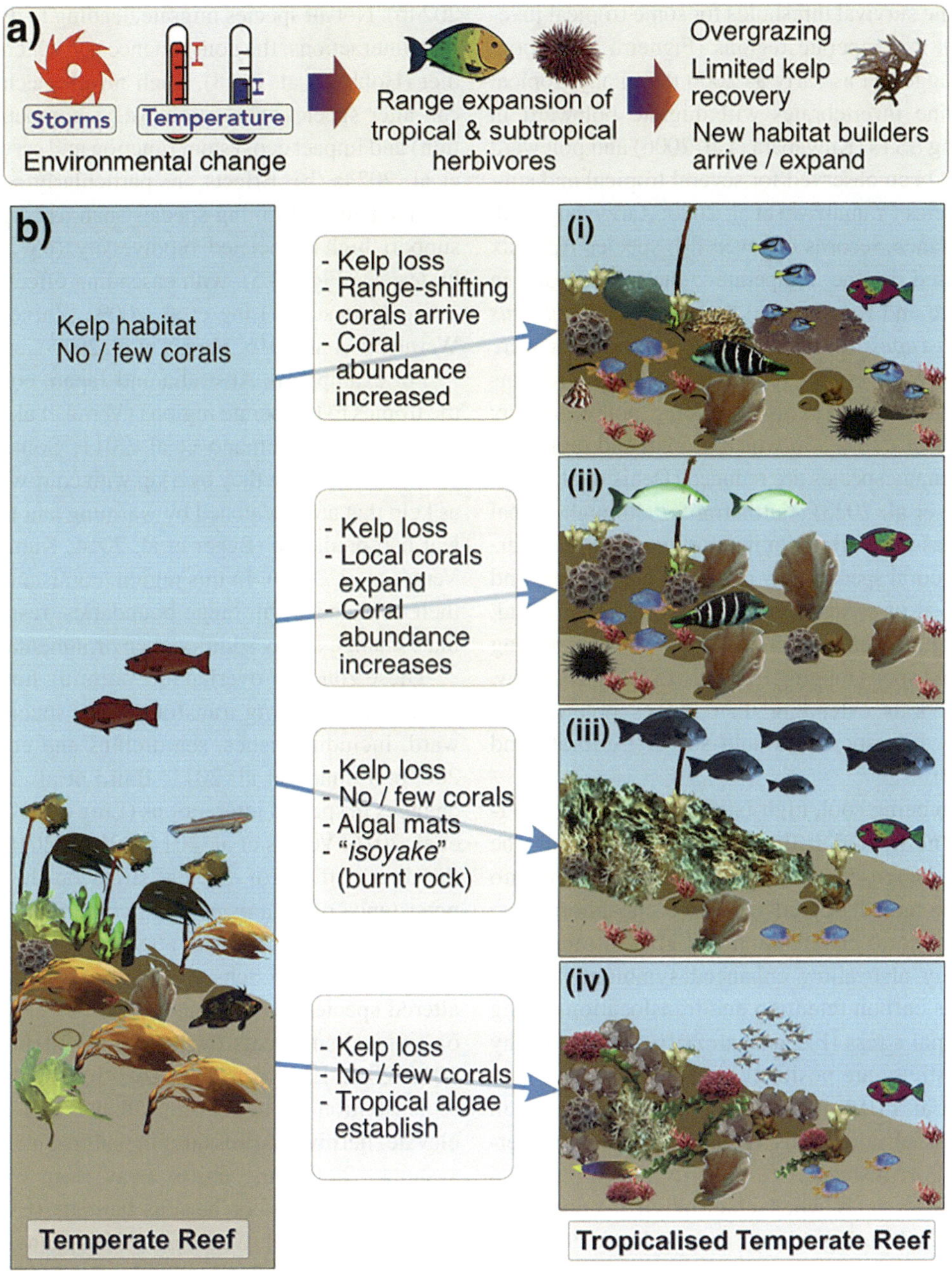

Fig. 13.4 Tropicalisation of temperate reef systems, outlining (**a**) drivers of tropicalisation, and (**b**) the processes that lead to kelp loss and different tropicalisation outcomes of (i) novel coral communities made up of local and range-shifting coral species, (ii) coral communities of local coral species with increased abundance or cover, (iii) bare rock barrens, called "isoyake" in Japan, with algal mats but few corals or other structure-building biota, and (iv) communities of tropical algae species

2020). Therefore, long-lived species adapted to warmer SST may find higher summer SSTs more suitable (Fujiwara et al. 2022), but suffer during winter. Sessile species face greater vulnerability to ambient changes because when environmental factors fluctuate outside of their tolerance range, they face an increased risk of mortality compared to motile species that can move longer distances out of their specific ecosystem (Wong et al. 2023).

The overwintering ability of many marine species is crucial to their survival, particularly for warm-affinity species near/at their poleward range limits on extreme and/or marginal cool reefs. For instance, biodiversity patterns of corals along the tropical-to-temperate transition zones in eastern Australia (Sommer et al. 2018) and Japan (Veron and Minchin 1992; Yamano et al. 2011) are closely linked to minimum temperature. Warming therefore alters SST extremes affecting the overwintering ability and survival rates of many marine species along biogeographic transition zones. For example, winter SSTs are increasing more rapidly in southeastern Australia compared to the global average and have exceeded the survival thresholds for some tropical juvenile fish species in temperate regions (Figueira and Booth 2010). It was suggested as early as 2006 that many tropical fishes and marine invertebrates will migrate poleward in response to rising SSTs (Kuwabara et al. 2006) and poleward trends have also been observed for several tropical and subtropical fish species (Yamakawa et al. 2020; Zarzyczny et al. 2024b). For instance, records of seven fish species from six families increased in the temperate zones of Japan, in Honshu, Shikoku and Kyushu, and the species *Epinephelus coioides*, *Kuhlia rupestris* and *Oxyurichthys cornutus* were thought to have overwintered in temperate regions (Yamakawa et al. 2020). Corals expanding poleward may experience SSTs of <18 °C, in which growth and reproductive abilities of many species are reduced (Denis et al. 2013; Ribas-Deulofeu et al. 2023). Furthermore, although global SSTs may increase overall, the minimum temperature tolerances of certain coral species may prevent establishment and long-term survival of corals at high latitudes (Abrego et al. 2021). For example, even though winter SSTs are increasing (by >1.5 °C), cold spells ranging from 12–15 °C in Tosa Bay, Japan, expose corals extending their ranges poleward to severe mortality in response to cold stress (Leriorato and Nakamura 2019).

In addition to being cool, high-latitude waters are generally also more nutrient rich than ocean conditions in the tropics due to increased deep-water mixing (Sarmiento et al. 2004). Though dissolved inorganic phosphorus can negatively affect coral calcification and growth rates, elevated levels may also allow enhanced symbiont maintenance as well as carbon retention and translocation during periods of thermal stress (Ezzat et al. 2016). Heterotrophy and trophic plasticity are positively linked to primary production (Fox et al. 2018) and can facilitate persistence in abiotically stressful conditions (Houlbrèque and Ferrier-Pagès 2009). For instance, *Turbinaria reniformis* had higher calcification rates when chlorophyll *a* concentrations were elevated during winter, despite lower SSTs (at 34.3°S; Ross et al. 2019), possibly resulting in a greater number of juvenile *Turbinaria* spp. corals due to higher nutrient availability (at 30°S; Sommer et al. 2024). The effect of nutrients on coral expansion and longevity in high-latitude environments remains unclear, but likely varies among species (Abrego et al. 2021).

13.4.5 Effects of Climate Change on Cool Reef Communities

Species migrations along biogeographic transition zones inevitably lead to new combinations of species with different thermal affinities occurring side by side (Zarzyczny et al. 2024b). Not all species migrate, leading to dynamic interspecific interactions, the consequences of which are hard to predict (Hobbs et al. 2018). Such novel species combinations can alter species interactions (e.g., competition and predation) and impact ecosystem function and services (Zarzyczny et al. 2024a, b). Effects are particularly severe when they involve habitat-forming species such as corals and kelp that support high associated biodiversity (e.g., mobile invertebrates, see Fig. 13.5), with cascading effects throughout the entire ecosystem (Ling et al. 2009a; Zarnetske et al. 2012; Wernberg et al. 2016; Vergés et al. 2016).

For example, in Australia and Japan, corals extend from the tropics to temperate regions (Veron et al. 1974; Veron and Minchin 1992; Yamano et al. 2011; Sommer et al. 2014, 2017, 2018) where they overlap with cold-water species such as kelp that are threatened by warming and the influx of tropical fish predation (Beger et al. 2014; Kumagai et al. 2018; Vergés et al. 2019). In this region, corals and kelp overlap at their cold and warm range boundaries, respectively, yet climate change exerts additional environmental stress.

These zones of overlap are warming hotspots (Wu et al. 2012) and are being transformed by species moving poleward, including fishes, sea urchins and corals (Ling et al. 2009b; Yamano et al. 2011; Baird et al. 2012), leading to changes in species interactions (Ling et al. 2009b; Wernberg et al. 2016; Vergés et al. 2014b, 2016, 2019). In addition to the direct effects of thermal stress on the physiology and persistence of organisms (e.g., coral bleaching, declines of habitat-forming seaweeds) (Smale and Wernberg 2013; Kim et al. 2019; Lachs et al. 2021), the indirect effects of altered species interactions affect the ecosystem dynamics on tropicalising reefs (Vergés et al. 2016; Wernberg et al. 2016, 2024; Zarzyczny et al. 2024b). Notably, the poleward redistribution of herbivorous fishes and sea urchins can elevate herbivory pressure in subtropical and temperate systems, facilitating shifts away from macroalgal-dominated areas to turf or barrens habitats (Ling et al. 2009b; Filbee-Dexter and Wernberg 2018; Kumagai et al. 2018; Vergés et al. 2019; Ling and Keane 2024) that prevent kelp recovery due to herbivory (Bennett et al. 2015; Barrientos et al. 2021; Pessarrodona et al. 2022; Ling and Keane 2024).

As corals and macroalgae compete for space, declines in cold-water seaweeds can release corals and other taxa from algae competition, facilitating increases in the abundance of tropicalising taxa (Bates et al. 2017; Tuckett et al. 2017; Ling et al. 2018). For instance, research from Tasmania highlights the potential of the range-expanding subtropical-temperate sea urchin *Centrostephanus rodgersii* in facilitating the increased abundance of corals (Ling et al. 2018) and other tropicalising species (e.g., the ascidian *Herdmania* spp; Bates et al. 2017) in urchin barrens on temperate reefs.

Fig. 13.5 Diversity of high-latitude coral-reef invertebrate communities experiencing tropicalisation across the Kuroshio Current in Okinawa and Kochi Prefectures, Japan. (Images: K.M. Watson 2019©). (**a**) Peacock mantis shrimp (*Odontodactylus scyllarus*), (**b**) Christmas tree worm (*Spirobranchus giganteus*), (**c**) Burrowing urchin (*Echinometra mathaei*), (**d**) Halloween hermit crab (*Ciliopagurus strigatus*), (**e**) Blue sea squirt (*Rhopalaea crassa*), (**f**) Strawberry conch (*Conomurex luhuanus*), (**g**) Long-spined sea urchin (*Diadema savignyi*), (**h**) Headshield slug (*Chelidonura amoena*), (**i**) Dancing shrimp/camel shrimp/hingebeak prawn (*Rhynchocinetes durbanensis*)

13.4.6 Challenges to the Tropicalisation Concept and Future Trajectories

Tropicalisation trajectories across temperate and subtropical gradients vary amongst regions and taxa, with unexpected consequences (Vergés et al. 2014a, 2019; Wernberg et al. 2016). For instance, in Japan, rocky reefs previously dominated by cold-water kelps have now been colonised by corals (Vergés et al. 2014a; Kumagai et al. 2018), causing the local abalone fishery to collapse within five years (Serisawa et al. 2004). Several studies have projected SST increases will further extend the poleward range limits of corals in Japan (Yamakawa et al. 2020; Abe et al. 2022), with the geographical limit of coral reef formation (i.e., at 18 °C) along the southern Kyushu coast predicted to migrate poleward to the northern Kyushu coast by 2100 (Yamakawa et al. 2020; Abe et al. 2022). In contrast, subtropical coral assemblages in eastern Australia have remained remarkably stable over a 20-year period (Mizerek et al. 2021), despite regional declines in kelp and tropicalisation of fish assemblages (Vergés et al. 2016; Smith et al. 2021).

The species and biogeographical differences currently seen in range expansions along biogeographic transition zones suggest that factors beyond warming likely influence these patterns, including ocean acidification (Agostini et al. 2021), different life histories and species traits (Sommer et al. 2014; Madin et al. 2016; Sommer et al. 2017), oceanographic conditions (Mizerek et al. 2021), the types of symbionts they have (La Motta et al. 2024), and the role of microbial biofilms (Cooney et al. 2024). In eastern Australia for example, coral population genetics and larval dispersal models have revealed limited connectivity between tropical and subtropical reefs (Thornhill et al. 2017; Mizerek et al. 2021), with results variable among species (Miller and Ayre 2008; Noreen et al. 2009, 2013). The offshore path of the East Australian Current at coral spawning time is thought to have so far limited tropicalisation from the Great Barrier Reef to high-latitude coral ecosystems at the Solitary Islands Marine Park (30°S; Mizerek et al. 2021), where cosmopolitan and subtropical corals make up the bulk of coral abundance (Sommer et al. 2014).

In Japan, tropicalisation is influenced by both rising SSTs and boundary current transport by the Kuroshio and Tsushima Warm Currents (Kumagai et al. 2018) and varies among coral species. For example, whilst in western Japan, *Acropora hyacinthus* has markedly expanded its distribution poleward along the Sea of Japan, *Alveopora japonica* is restricted to certain parts of the Pacific coast, and *Oulastrea crispata* can be distributed as far as the Tsugaru Straits in northern Japan (Abe et al. 2022). Additionally, several studies suggest projected SST increases will extend the range limits of corals and coral reefs poleward in Japan (Yamakawa et al. 2020; Abe et al. 2022). Notably, the northern limit of coral reef formation (i.e., at 18 °C), along the southern Kyushu coast, may migrate poleward to the northern Kyushu coast by 2100, whilst the poleward limit of high-latitude coral distribution (i.e., at 10 °C) along the Niigata and Chiba coasts may extend poleward to the Aomori and Iwate coasts (Yamakawa et al. 2020). These changes in coral community distribution could lead to a restructuring of marine communities along Japan's coastline, with some native temperate species being outcompeted by invading tropical species (Yamano et al. 2011).

Taxonomic uncertainty further complicates the prediction of future trajectories for extreme and/or marginal cool reefs. Mounting evidence shows that these reefs support higher coral endemism than previously thought (Schmidt-Roach et al. 2013; Baird et al. 2017; Fifer et al. 2022; Juszkiewicz et al. 2022; Chong et al. 2026), with several species originally considered tropical now understood to be subtropical coral specialists with distinct lineages, such as *Pocillopora aliciae* (Schmidt-Roach et al. 2013), *Cyphastrea salae* (Baird et al. 2017), and *Plesiastrea versipora* (Juszkiewicz et al. 2022) in Australia, and *Acropora hyacinthus* in Japan (Fifer et al. 2022). Although these subtropical specialists appear to be well-adapted to high-latitude conditions (González-Pech et al. 2022), they live close to their physiological limits, increasing their susceptibility to disturbance. On Australia's high-latitude reefs, endemic species (e.g., *Pocillopora aliciae*), subtropical specialists and species near their northern distributional limits (e.g., *Coscinaraea marshae*) are more vulnerable to bleaching than species typically susceptible to bleaching in the tropics (e.g., *Acropora* spp.; Thomson et al. 2011; Camp et al. 2018; Goyen et al. 2019; Kim et al. 2019; Cant et al. 2021; Lachs et al. 2021; Schoepf et al. 2023; Sommer et al. 2024). High-latitude specialist corals typically have small geographic ranges and are predicted to experience environmental conditions completely outside their physiological limits across their entire distributional ranges in the future (Kim et al. 2023). Therefore, whilst high-latitude regions may provide ecological refugia for some tropical corals facing warming SSTs, they are unlikely to become universal havens for all corals in the future (Keshavmurthy et al. 2023; Kim et al. 2023). Notably, high-latitude specialists are particularly vulnerable due to their limited capacity to avoid climate risks (Kim et al. 2023), which in addition to geographical isolation, fragmented populations, and sporadic recruitment, further limits their capacity for recovery (Fairfull and Harriott 1999; Wilson and Harrison 2005; Beger et al. 2014; Precoda et al. 2018; Lachs et al. 2021; Schoepf et al. 2023; Sommer et al. 2021, 2024).

The ecology of extreme and/or marginal cool reefs is complex and future trajectories are difficult to predict. Ecological dynamics involve many taxa that respond to a range of environmental and/or biological factors in different ways and are mediated by processes that play out at local (e.g., competition, predation) to regional (e.g., connectivity) scales. Climate change modelling that focuses on the response of a single taxon (e.g., coral) to warming alone therefore doesn't capture the complex nature of these systems and might over- or underestimate tropicalisation. For instance, field surveys and in-situ transplants at natural analogues for present and future conditions found that increased herbivory by warm-water fishes accelerated kelp forest loss in temperate Japan and that ocean acidification negated any benefits of warming for range expanding tropical corals (Agostini et al. 2021). This led to simplification rather than tropicalisation typically expected under warming alone (Agostini et al. 2021). Realistic predictions of future trajectories of extreme and/or marginal cool reefs therefore require community level approaches at local to regional scales that consider the complex interplay of climate-mediated species redistributions and associated changes in species interactions.

13.5 Little-Known and Often Ignored Taxa

Scleractinian corals and their associated fish communities have garnered much research attention, while other benthic taxa, such as sponges, zoantharians and gorgonians, remain generally overlooked in both research and conservation

Mexico's Extreme and/or Marginal Cool Reefs

David A. Paz-García

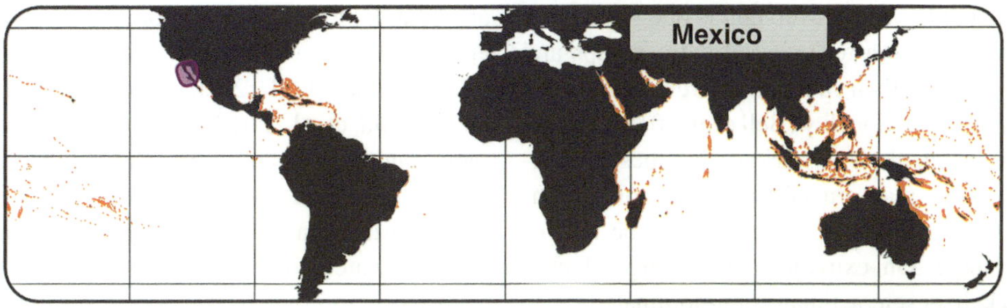

Extreme and/or marginal cool reefs in Mexico can be divided according to their geographical location: Gulf of California and western coast of the Baja California Peninsula. High-latitude reefs in the Gulf of California (28–31°N) are exposed to high turbidity caused by strong currents (10 m tidal range), high-nutrient upwelling conditions, and high annual sea surface temperature ranges (14–30 °C; Halfar et al. 2005). These extreme environmental conditions limit species distribution, promote change in the range distribution of species (Paz-García and Balart 2016; Paz-García et al. 2018), and favour those capable of tolerating them (Dennis et al. 2024). These high-latitude rocky reefs are composed by the scleractinian coral *Porites panamensis* with a vertical distribution from 2 to 10 m depth with a significant carbonate sediment production. Octocorals of the genus *Muricea* spp. are dominant taxa at depths of 5–20 m. Cool reefs of the western coast of the Baja California Peninsula (28–32°N) support a temperate ecosystem as an extension of the Southern California Bight with seasonal northwest winds that promote primary productivity and strong upwelling conditions (Beas-Luna et al. 2020; Zarzyczny et al. 2024a). The nearshore is dominated by bedrock and a mix of low rocky cover and sandy beaches, where species with temperature and subtropical affinities are distributed (Zarzyczny et al. 2024a). The rocky reef is dominated at 10–15 m depth by giant kelp (*Macrocystis pyrifera*), but it's diminution and changes in community structure have been recorded in these communities due to ENSO events and marine heatwaves (Beas-Luna et al. 2020). In the rock substrate, the only scleractinian coral found in these latitudinal communities is *Balanophyllia elegans* from high densities 10-20 m depth (Hellberg and Taylor 2002).

David A. Paz-García©

David A. Paz-García©

R. Beas-Luna©

David A. Paz-García©

efforts (Brandl et al. 2019). This oversight is due to both the lack of taxonomic resolution for many of these groups and insufficient knowledge of their functional diversity and contribution to ecosystem functionality (Bell et al. 2018a; Brandl et al. 2019; Reimer et al. 2019). However, these lesser-studied organisms may play crucial roles in shaping community dynamics and resilience, particularly in the face of environmental change. On extreme and/or marginal cool reefs, shifts from kelp or scleractinian coral-dominated states to those dominated by other benthic invertebrate, such as anemones, zoantharians, corallimorpharians, soft corals, sponges, and ascidians can occur (Bell et al. 2018b, 2021; Reimer et al. 2021). Below, we will briefly introduce some of these lesser-studied benthic groups and discuss their future perspectives under climate changes at extreme and/or marginal cool reefs.

13.5.1 Sponges

Sponges (Phylum: Porifera) are considered potential successors to scleractinian corals on tropical coral reefs in the face of changing environmental conditions (Bell et al. 2018a; Pawlik and McMurray 2020) yet, whether this will also occur on high-latitude reefs remains largely unexamined. Many sponges are noted for their ability to tolerate a range of environmental stressors, such as ocean acidification and moderate SST increases (Goodwin et al. 2014; Bennett et al. 2017). Sponges can also withstand other stressors that affect scleractinian corals, such as sedimentation and nutrient influx, though responses vary among species (Ward-Paige et al. 2005; Bell et al. 2015). However, the impact of changing environmental conditions on sponges is likely species-specific (Bell et al. 2018a), as for example, MHWs severely impacted temperate sponge communities in New Zealand (Bell et al. 2023). There is evidence suggesting sponges' resilience and transitions to sponge-dominated reefs have been reported, primarily in the Caribbean and Indo-Pacific (Pawlik and McMurray 2020; Bell et al. 2021). These transitions include both bioeroding and non-bioeroding sponge species, highlighting the expanding role of sponges in reef ecosystems beyond the Caribbean (Bell et al. 2021).

13.5.2 Soft Corals

Soft corals or octocorals, are a diverse group of marine cnidarians belonging to the class Octocorallia, and while they are commonly associated with tropical reefs, many are found on extreme and/or marginal cool reefs (Otis et al. 2024). Octocorals, including invasive species (Concepcion et al. 2010; Ofwegen and Haddad 2011; Quintanilla et al. 2017), are abundant on tropical coral reefs, and have been reported as the dominant benthos in some areas (Inoue et al. 2013; Bell et al. 2021). Octocoral communities can fill many different functional roles within shallow water communities (Otis et al. 2024), although even basic zoological data are currently missing for most species. Many soft coral species can outcompete other habitat-forming species, due to factors such as their symbiotic relationships, comparatively fast growth rates, and upright morphologies, which allow them to quickly occupy space and prevent algal overgrowth (Lasker et al. 2020). Few studies have explored the impacts of anthropogenic stressors on these organisms, particularly for extreme and/or marginal cool reefs, although some species appear to increase in numbers as zooxanthellate scleractinian corals decrease due to local stressors (Lalas et al. 2024). While some octocorals exhibit resilience to SST-related bleaching stress (Schubert et al. 2017), more research is needed to understand the responses across a wider range of species. It has also been hypothesised some octocoral species are resilient to high levels of turbidity (e.g. De'ath and Fabricius 2000). Furthermore, the potential effects of ocean acidification on soft corals, which lack massive calcareous skeletons as seen in scleractinians, requires further investigation, as for example, octocoral and zoantharian species were noted to be dominant in close proximity to a CO_2 seep on Iwotorishima Island in northern Okinawa, in contrast to nearby reefs dominated by scleractinian corals (Inoue et al. 2013; Reimer et al. 2021).

13.5.3 Zoantharians

Zooxanthellate zoantharians are common components of coral reefs and extreme and/or marginal cool reefs with coral communities. Zooxanthellate species within the genera *Zoanthus* spp. (Family: Zoanthidae) and *Palythoa* spp. (Sphenopidae) exhibit broad geographical ranges and plastic morphologies that enable adaptation to different environments. *Zoanthus* spp. and *Palythoa* spp. demonstrate fast colony growth speeds compared to scleractinian corals and can have associations with different Symbiodiniaceae symbionts (Reimer et al. 2006; Kamezaki et al. 2013; Wee et al. 2021; Zamora-Jordán et al. 2022; Fujiwara et al. 2022). There is growing evidence that some reefs may transition from coral or macroalgal dominant reefs to *Zoanthus* spp. or *Palythoa* spp., resulting in 'zoantharian barrens' (Reimer et al. 2021), with additional evidence of range expansion in the northwest Pacific (Reimer et al. 2018). However, there is currently a global lack of knowledge regarding Zoantharia dominance, barrens, and phase shifts, highlighting a need for further research (Reimer et al. 2021), as well as a need to document their northern and southern limits (e.g., López et al. 2020; Reimer et al. 2021). With increasing levels of anthropogenic degradation and ongoing climate change, the occurrence of zoantharian phase shifts and barrens is expected to increase (Reimer et al. 2021; Bell et al. 2021), with predictive modelling suggesting that higher SST's alongside decreased water quality could leave subtropical and temperate regions particularly vulnerable to new zoantharian barrens (Reimer et al. 2021).

13.5.4 Corallimorpharia and Other Cnidarians

Corallimorpharians are an order of hexacoral cnidarians closely related to scleractinian corals, however, they lack a calcareous skeleton (den Hartog 1980; Fautin 2016). Whilst more commonly associated with tropical reefs, many species of corallimorpharians occur on extreme and/or marginal cool reefs (den Hartog 1980; Chadwick 1991; Häussermann and Försterra 2005; Fautin 2016). Understanding of their ecological and functional roles on reef systems is still lacking. Corallimorpharia can compete with other coral species for space and resources (Chadwick 1991; Chadwick-Furman and Spiegel 2000; Kuguru et al. 2004). Like other organisms on extreme and/or marginal cool reefs, corallimorpharians are likely threatened by anthropogenic activities, including coastal development, pollution, overfishing, and climate change. However, studies reporting rapid expansion of Corallimorpharia at local scales show that they can also be beneficiaries of ecological disturbances such as extreme low tides, hurricanes and shipwrecks (Chadwick-Furman and Spiegel 2000; Kuguru et al. 2004; Work et al. 2008; Crane et al. 2016; Carter et al. 2019).

Other tropical and subtropical cnidarian benthos such as sea anemones (Malcolm and Scott 2016; Martín-Arjona et al. 2024), antipatharians (Ocaña Vicente et al. 2006), and hydrocorals (Clemente et al. 2011) have also been reported as potentially increasing their ranges polewards or into colder regions due to climate change, but less data exist for these groups. For example, North Solitary Island in eastern Australia hosts the highest published abundance of the anemone *Entacmaea quadricolor* (Richardson et al. 1997), with considerable abundance increases recorded between 1994 and 2008 (Scott et al. 2011).

13.5.5 Ascidians

Ascidians (Phylum: Chordata), also known as sea squirts, are marine organisms found in diverse habitats, ranging from the tropics to high latitudes and shallow waters to the deep sea, and they inhabit various substrates such as soft sediments, coral reefs, and rocky substrates (Sawada et al. 2001; Shenkar and Swalla 2011). Additionally, they can thrive on artificial structures such as ship hulls and jetties (Sawada et al. 2001; Airoldi et al. 2015). Ascidians exhibit varying degrees of tolerances to salinity and temperature changes, with some species capable of surviving in extreme conditions (Shenkar and Swalla 2011). Ecologically, ascidians play important roles, but they can also become invasive, particularly in nutrient-rich environments (Lambert 2002) and introductions into new habitats can lead to overgrowth and displacement of native species, causing ecological imbalances (Rius et al. 2014). Ascidian species richness peaks in tropical regions, where colonial species are dominant. However, as latitude increases, solitary species gradually become more significant contributors to total species richness. However, information of the ascidian fauna remains scarce in many areas worldwide (Shenkar and Swalla 2011), as well as information of the impacts of climate change on this benthic group across time (Bell et al. 2018b).

13.5.6 Microbial Communities

The astonishing diversity of microscopic organisms and their host interactions have only been explored in recent years (Dittami et al. 2021). Microbial communities likely represent another influential component of ecological dynamics along biogeographic transition zones (Messer et al. 2020; Cooney et al. 2024). Biofilms, consisting of microorganisms adhering to surfaces and secreting a matrix, play various functional roles within ecosystems, including nutrient cycling, energy flow, disease regulation, and serving as settlement cues for other organisms (Dang and Lovell 2015; Klassen 2018; Rischer et al. 2022). Investigating the structure, composition, and dynamics of microbial communities can provide valuable insights into their interactions with reef organisms and their responses to environmental change (van der Loos et al. 2019; Cooney et al. 2024). The impact of climate change stressors on benthic biofilms and how this may affect their interactions with habitat-forming organisms has not been thoroughly addressed. Like other marine organisms, biofilm communities may exhibit range shifts in response to climate change (Messer et al. 2020; Cooney et al. 2024), potentially extending into higher latitudes due to increased SST's and ocean currents. This could lead to the development of tropical biofilms in regions where they were previously absent, facilitating the settlement of tropical habitat-forming organisms. Changes in biofilms may favour the dominance of turf algae, inhibiting the settlement of kelp or corals and altering ecosystem states (Harvey et al. 2021). However, it is important to establish baselines for monitoring microbial shifts in biofilms across different environments to understand their implications for macro-organisms (Cooney et al. 2024).

As extreme and/or marginal cool reefs and their coral communities face increasing pressures from climate change and other anthropogenic disturbances, understanding how underreported taxa acclimatise and adapt to changing environmental conditions is paramount. Many of these taxa, or species within these groups, appear to have competitive advantages over many hard coral, macroalgae, and other benthic species in the face of climate change, tropicalisation, and ocean acidification.

13.6 Evolutionary Processes and Biodiversity at Different Levels of Organisation

13.6.1 Patterns of Genetic Diversity and Isolation

Under the current tropicalisation of extreme and/or marginal cool reef ecosystems, the geographic distributions of marine species are changing rapidly (Poloczanska et al. 2016; Zarzyczny et al. 2024b). The dynamic of this redistribution can generally be decomposed into leading range expansion (i.e., populations moving toward more suitable environmental conditions) and trailing edge extirpation (i.e., populations being wiped out under detrimental environmental conditions, Pinsky et al. 2020). Tropicalising taxa must be able to colonise environmentally suitable regions, demonstrate the ability to survive in new habitats, and establish stable populations (Burgess et al. 2012; Grieve et al. 2016). A population's persistence at range boundaries is favoured by three major factors: first, extreme and/or marginal populations may be genetically isolated from their core population, preserving mutations and potential local adaptation (Table 13.1; Kawecki 2008). Secondly, smaller populations at the distribution margins may also contribute to introgressive hybridisation which can in turn increase genetic diversity among populations and constitute sources of evolutionary innovation (Table 13.1; Budd and Pandolfi 2010). Finally, strong connectivity between leading edge and core populations can decrease inbreeding effects and improve individual fitness (Table 13.1; Baums 2008). For these reasons, investigating population genetics along species distributional ranges appears critical to understanding leading-edge expansion. Yet, several extreme and/or marginal cool reef populations represent groups of cryptic species with different distributions (e.g., in hard corals: *Acropora hyacinthus*, Nakabayashi et al. 2019; *Galaxea fascicularis*, Nakajima et al. 2016; *Goniopora djiboutiensis*, Yasuda et al. 2021; *Plesiastrea versipora*, Juszkiewicz et al. 2022; *Pocillopora* spp., Schmidt-Roach et al. 2013; in octocoral: *Hanabira yukibana*, Lau et al. 2019; in Zoantharia: *Zoanthus sansibaricus*, Albinsky et al. 2018). These findings call for careful re-examination of taxa through integrated systematic approaches, but also highlight the fact that cool reef taxa might represent independent genetic lineages, or new species, already adapted to high-latitude extreme and/or marginal environments.

13.6.2 Acclimatisation and Adaptation

Acclimatisation is a physiological or biochemical response of an individual as a result of exposure to naturally variable environmental conditions (Putnam 2021). Individual differences in acclimatisation are fundamental for the survival, potential reproduction, and adaptation of organisms in high-latitudinal communities (Burt et al. 2020; Abrego et al. 2021). Communities and individuals capable of tolerating changing environmental conditions are garnering attention as potential sources of genetic interventions, or rescue populations experiencing declines (Gonzalez et al. 2013; Torda and Quigley 2022). Holobiont organisms such as corals, living with a symbiotic association, have phenotypic changes that occur at different levels (Fig. 13.6a).

Firstly, change occurs at the molecular level when molecular functions are modified and production of key proteins is regulated through gene expression (Veilleux et al. 2018; Yu et al. 2023; Castillo et al. 2024). Changes in gene expression, known as transcriptomic plasticity, are crucial for cellular processes and physiological responses to environmental stressors (Drury et al. 2022). For example, *Acropora pruinosa* and *Pavona decussata* show a plastic transcriptomic response to large seasonal SST fluctuations by increasing their energy supply and maintaining a high level of immunity (Yu et al. 2023).

Secondly, symbiont association in hosts with thermally tolerant symbionts is key to surviving stressful environmental conditions resulting from latitudinal shifts in algal symbiont community and microbiome composition (Sawall et al. 2015; Howells et al. 2020; Yu et al. 2023). Corals are able to adjust their associations with Symbiodiniaceae dinoflagellates. A well-known example is that under heat stress, symbionts shuffle from heat-sensitive *Cladocopium* spp., to the more heat-tolerant *Durisdinium* spp. (Berkelmans and van Oppen 2006). When conditions return to baseline, the symbiont community reverts back to their pre-exposure state (e.g., Thornhill et al. 2006). Although symbiont shuffling has little cost for the coral host (Abbott et al. 2021), over the long-term, hosting a distinct *Cladocopium* or *Durusdinium* spp. symbiont community will change coral physiology. For

Japan & Korea's Extreme and/or Marginal Cool Reefs

James Reimer, Daito Yamaguchi

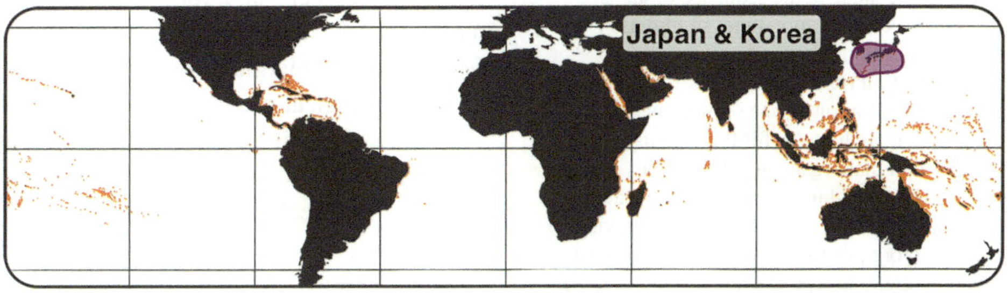

The Japanese Archipelago and Korean Peninsula host some of the worlds highest-latitude shallow-water coral communities. Coral communities occur north of Yakushima and Tanegashima islands to the Chiba Prefecture along the Pacific coast, and from Jeju Island to the Okinawa Islands in the Sea of Japan due to northward-flowing warm water currents: the Kuroshio along the Pacific, and the Tsushima in the Sea of Japan. Many of these high-latitude areas with coral communities are also composed of rapidly disappearing large macroalgal communities (e.g., Hwang et al. 2017), a term called '*isoyake*' in Japanese (Fujita 2010). Tropicalisation is well documented in Japan, for example, significant warming of winter SSTs in Amakusa, Kumamoto, Japan, between 1955 and 1995 led to increased occurrences of *Acropora solitaryensis* (Nojima and Okamoto 2008).

Furthermore, in 2007, *A. solitaryensis*, whose poleward limit was considered the Izu Peninsula (34°54′N, 138°57′E), was found at Tateyama, Chiba (34°58'N, 139°47'E; Yamano and Namizaki 2009). Similarly, fewer winter days with SSTs <14 °C and 18 °C have been noted from Jeju Island (Ribas-Deulofeu et al. 2023). In 2006 it was suggested that fishes and invertebrates would migrate poleward as SSTs rise (Kuwabara et al. 2006) and northward trends have since been observed for tropical and subtropical species in both countries (Yamakawa et al. 2020; Reimer et al. 2018).

Maria Beger ©

Maria Beger ©

Brigitte Sommer ©

Table 13.1 Definitions of genetic concepts and terminology

Term	Definition
Mutation	Mutation is a change in the sequence of an organism's DNA. Common types of mutation include substitution (change from one base to another), deletion (omission of one or more bases), and insertion (addition of one or more bases). Mutations that are harmful for the organism usually disappear with that organism's death. Yet, mutations are crucial in evolution as they can introduce new genetic variants that may provide an advantage to organisms and spread throughout a population (e.g., via reproduction).
Local adaptation	Through the selective pressure of their local environment, organisms which present new genetic variants, or an original set of those variants, may become more well-suited to survive than other members of the same species living elsewhere, and become locally adapted to their environment.
Introgressive hybridisation	Local adaptation can cause populations to diverge and eventually evolve into distinct species. However, hybridisation or repeated backcrossing can be responsible for transfer of pools of alleles, or small genomic regions, between different species.
Inbreeding	Inbreeding refers to the mating of closely related organisms (e.g., between parents and offspring). Inbreeding usually causes alleles to be identical by descent, resulting in an impoverished allelic pool, but its overall effect on populations can be positive by eliminating harmful and recessive alleles.
Individual fitness	Individual fitness (also called Darwinian fitness) is often seen as the survival and reproductive success of an individual within its population.

example, *Durusdinium* spp. are comparatively less photosynthetically efficient and translocate less photosynthetically fixed carbon to the coral host, implying a trade-off of thermal resistance for a lowered growth rate (Wang et al. 2023), unless the energy deficit is overcome by other mechanisms, such as changing their nutritional strategy (Wang et al. 2024).

In the extreme and/or marginal cool coral community of Jeju Island, several scleractinian corals harbour *Clacopium* spp. symbionts (C1 and C17; De Palmas et al. 2015). The large pan-tropical distribution of these symbionts suggests that acclimatisation abilities might heavily rely on the coral host rather than its symbionts (De Palmas et al. 2015). Alternatively, high-latitude scleractinian corals might also have evolved and harboured symbiont communities adapted to their environment. For instance, *Alveopora japonica* from Jeju Island associates with an undescribed genus of *Symbiodiniaceae* spp. (clade Fr2) and with *Effrenium voratum*, two symbionts only reported in extreme and/or marginal cool coral communities (De Palmas et al. 2015; LaJeunesse et al. 2018). Similarly, *Oulastrea crispata* from Jeju harbours *Durusdinium boreum*, another symbiont species only reported on extreme and/or marginal cool reefs (De Palmas et al. 2015). The association between *O. crispata* and its *Symbiodiniaceae* spp. has been suggested to be the result of adaptive radiation to their environmental conditions across latitudinal gradients (Lien et al. 2013). Overall, refined molecular markers and further investigation is needed to uncover the diversity of *Symbiodiniaceae* spp. symbiont in extreme and/or marginal cool reef coral communities, and their physiological performances in a changing climate.

Thirdly, physiological plasticity allows a species to tolerate a wide range of environmental conditions and adjust their physiological mechanisms to survive under stressful conditions (Sawall et al. 2015; Keshavmurthy et al. 2021). For example, elevated turbidity (increased suspended sediment concentration and/or low light) are also main sources of environmental stress for corals in extreme and/or marginal reefs (Dellisanti et al. 2021). In a laboratory setting, Jones et al. (2020b) found that *Acropora* spp., *Pocillopora* spp. and *Porites* spp. all adjusted to low-medium turbidity levels by increasing the density of their algal symbiont, presumably to increase light absorption under increasing limitation, while *Turbinaria* spp. (a turbid water specialist) symbiont density slightly decreased. At the highest turbidity with little to no light, bleaching (dissociation of the coral-symbiont relationship) and partial mortality occurred for all species (Jones et al. 2020b). Divergent coping mechanisms were found *in-situ* for *Montipora* spp., *Pocillopora* spp., *Porites* spp. and *Galaxea* spp. by Xu et al. (2020) in response to the combination of stronger radiation and higher turbidity from suspended particles in the South China Sea. This evidence points towards possible physiological adaptation to increasing environmental stress, brought on by increasing marginality, although some species are more sensitive than others (Jones et al. 2020b) and coral survival and adaptation under stress is reliant on available energy stores (Grottoli et al. 2004).

Fourthly, morphological plasticity in coral growth varies geographically and seasonally (Sawall et al. 2015). Micro and macroscopic skeletal characteristics in gonochoristic corals *Porites lobata* and *Porites panamensis* are displayed as sexual dimorphism possibly as a result of different selective pressures and energetic cost of gametes in extreme and/or marginal reefs in the Eastern Tropical Pacific (Pedraza-Pohlenz et al. 2023). Fluctuating interannual and seasonal environmental conditions from MHWs, ENSO events, light-limitation from upwelling as well as increased wave energy during storms and hurricanes, are important factors for coral survival on cool reefs (e.g., Wernberg et al. 2013; Leriorato and Nakamura 2019; Beas-Luna et al. 2020; Jones et al. 2020a; Dellisanti et al. 2021; Grove et al. 2022; Moriarty et al. 2023). These fluctuations are associated with morphological plasticity in *Pocillopora grandis,* showing variation in the thickness and branch density of the coral colonies creating many morphotypes, making taxonomic identification based on morphological features challenging (Paz-García et al. 2015a). Additionally, *Pocillopora grandis* has the capacity to alter their morphology in response to temporal shifts in environmental conditions and revert to their original

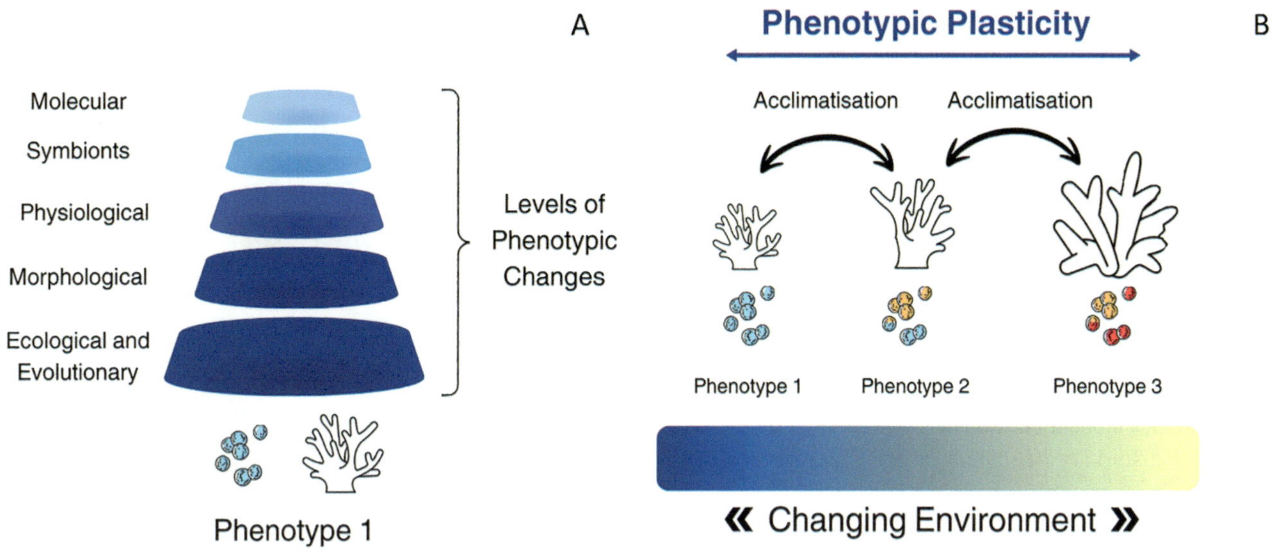

Fig. 13.6 Levels of phenotypic plasticity and acclimatisation. (**a**) An individual can display distinct levels of phenotypic changes from molecular to morphological traits and potentially, if a combination of traits is successful, this will increase its frequency in the population and have ecological and evolutionary effects. (**b**) It is through phenotypic plasticity that different phenotypes can occur as a response to the changing environment and during each period of change acclimatisation will take place. The coloured dots represent the composition of the microbiome (Bacteria or Symbiodiniaceae) that may or may not coordinate with morphological plasticity

morphology within a few months (Paz-García et al. 2015b). Finally, organisms with these levels of phenotypic plasticity and acclimatisation (for example showing a dynamic morphological plasticity and high response to fluctuating conditions) have the potential to impact ecological and evolutionary processes in high-latitudinal reefs (Fig. 13.6a).

Mechanisms for coping with a changing environment are key to the survival of marine organisms living at their physiological limits in extreme and/or marginal cool reef communities (Vázquez et al. 2017; Burt et al. 2020), and be key in the study of phenotypic plasticity, acclimatisation, and adaptation (Sawall et al. 2015; Vázquez et al. 2017; Jurriaans and Hoogenboom 2019; Armstrong et al. 2023). Phenotypic plasticity refers to the ability of the same genotype to adjust its phenotype under different environmental conditions without genetic change and occurs within a generation (Putnam 2021). In long-lived organisms such as corals, the different levels of phenotypic changes can occur simultaneously allowing them to survive fluctuating environmental conditions in cool reef communities (Fig. 13.6b). There is also increasing evidence that acclimatisation can be transgenerational through epigenetic changes in chromatin structure or DNA methylation (Putnam et al. 2016; Putnam 2021). Thus, heritable epigenetic modifications through DNA methylation can promote acclimatisation within generations, and rapid intergenerational transfer of advantageous genomic modifications (Liew et al. 2020).

Studies along these environmental gradients have shown high phenotypic plasticity despite large distances between sites (Sawall et al. 2015; Vázquez et al. 2017). For example, the coral species *Pocillopora verrucosa* at 12° latitude (28.5°N – 16.5°N) on six reefs along the Saudi Arabian Red Sea coast showed little indication of regional adaptation and high phenotypic plasticity in metabolic performance (Sawall et al. 2015). Similarly, at Weizhou Island in the South China Sea (21°N), *Acropora pruinosa* and *Pavona decussata* showed stability of Symbiodiniaceae identity (*Cladocopium*) and changes in the diversity and dominance of their bacterial communities in response to large seasonal SST fluctuations (Yu et al. 2023). In contrast, reef systems along the coastline of Western Australia showed restricted gene flow and local adaptation in *Pocillopora damicornis* (Thomas et al. 2017). A study at Kermadec Islands (New Zealand) and Lord Howe Island (Australia) found that 5 and 12 coral species respectively, (the most common were *Acropora cuneata*, *Goniastrea favulus* and *Pocillopora damicornis*) had high diversity and local endemicity of Symbiodiniaceae association with the genus *Cladocopium* spp. (Wicks et al. 2010). Furthermore, turbidity was associated with all host species and some symbionts (Wicks et al. 2010). Evidence of both local acclimatisation and genetic adaptation have been found in extreme and/or marginal cool reefs, potentially associated with tolerance to heterogenous environmental conditions as well as high levels of gene flow and genetic diversity (Wicks et al. 2010; Sawall et al. 2015; Thomas et al. 2017; Yu et al. 2023). Differences in species' performance across environmental gradients also reveal different mechanisms to cope with spatiotemporal environment variability, shedding light on how they may acclimate or adapt to changing conditions (Jurriaans and Hoogenboom 2019; Burt et al. 2020; Turnham et al. 2023).

Brazil's Extreme and/or Marginal Cool Reefs

Jessica Bleuel

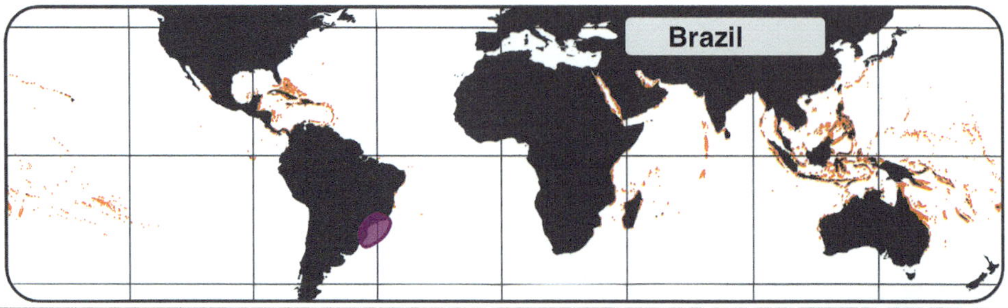

The subtropical reefs of Brazil are found in high nutrient and turbidity conditions spanning from Arraial do Cabo (~23°S) to the coast of Santa Catarina state (27°S). These reefs are composed of shallow rocky reefs with depths ranging from 1–25 m, with low zooxanthellate and hydrocoral diversity of five coral species at 23°S (*Millepora alcicornis*, *Porites branneri*, *Siderastrea stellata*, *Mussismilia hispida* and *Madracis decactis*) and only *M. hispida* and *M. decactis* remaining beyond 24°S (Bleuel et al. 2024). These reefs typically exhibit relatively low coral cover amidst algal communities (Aued et al. 2018), although exceptions exist. Recently high coral cover (<50%) was found at Queimada Grande Island (24°29'S), considered as the southern limit of biogenic reefs (Grillo et al. 2021). Similarly high coral cover of *M. decactis* exists in the southernmost rocky reefs (27°S), forming free-living colonies that aggregate into 'corallith beds', with up to 80 colonies per m^2 (Capel et al. 2012).

Grillo et al. 2021 ©

Grillo et al. 2021 ©

These high-latitude reefs have rarely been exposed to high thermal stress and show high thermal tolerance. In 2019, mass coral bleaching occurred at 23–24°S with exposure of up to 22.65 degree heating weeks, but corals showed a high tolerance (Banha et al. 2020), potentially supported by heterotrophic feeding. Such bleaching events will likely occur more frequently but warming likely will enhance more widespread coral occurrence in these regions, particularly for *M. decactis* and *M. hispida*, which are already dominant in Brazil's subtropical reefs (Martello et al. 2024).

Grillo et al. 2021©

13.7 Transient Population Dynamics

Population dynamics are governed by demographic processes that affect individuals at every stage of their life cycle, from reproduction, recruitment, growth, and eventually mortality (Berryman 2020). Collectively, these are termed vital rates, which for reef organisms in extreme and/or marginal cool reefs differ significantly from those in tropical reefs due to the distinct extreme and/or marginal environmental conditions they experience and biological characteristics (Nozawa et al. 2021; Cant et al. 2022). Demographic processes are typically linked through sharing of the same energy reservoir within an individual, therefore incurring trade-offs when allocating energy under different environmental conditions (Booth and Beretta 2021). The variable vital rates of organisms at different life stages allow them to take advantage of changing environmental conditions (Jonsson 2021).

There are fewer, but bigger, corals in extreme and/or marginal cool reefs than in tropical reefs (Chong et al. 2023). This fact suggests that extreme and/or marginal populations are characterised by low survival of small corals and/or poor recruitment (Lachs et al. 2021), but are sustained by the growth and persistence of those that survive (Cant et al. 2022; Sommer et al. 2024). For corals in extreme and/or marginal cool reefs, reproduction and successful recruitment are crucial in maintaining population viability at local scales, where populations infrequently receive coral larvae from tropical conspecifics (Fellegara et al. 2013; Mizerek et al. 2021), due to the long pelagic larval duration required to be received by downstream extreme and/or marginal reefs. Coral growth rates vary greatly among taxonomic and spatial scales (Pratchett et al. 2015), as for example, growth rates of the same species are often lower at high-latitude reefs under lower, light availability and carbonate saturation (Pratchett et al. 2015; Nozawa et al. 2021). Extreme and/or marginal cool reef organisms typically have lower survival compared to their tropical reefs counterparts, which results from coral recruits dying at the post-settlement and recruit stages (Wilson and Harrison 2005), due to several factors including a lack of adequate substrate, settlement cues and high wave energy (Willis and Oliver 1990; Harii et al. 2002; Beger et al. 2014; Hata et al. 2017). However, there is emerging evidence that many corals found in extreme and/or marginal reefs are endemic, thereby occupying a different ecological and trophic niche to their tropical counterparts, with these extreme environmental conditions selecting corals with robust or fast-growing life history strategies (Ross et al. 2018).

Fish populations are however governed by different demographic mechanisms, as they are mobile throughout most of their lifecycle, with higher latitudes producing bigger and higher trophic level fishes (Fernández-Torres et al. 2018; Coghlan et al. 2024), such as predators (Lin and Costello 2023; Clay et al. 2024). This macroecological pattern might be driven by the latitudinal gradient in SST and oxygen levels (Lin and Costello 2023). These differences in size suggest that extreme and/or marginal reef fish populations might have slower growth rates and maturity at larger body sizes (Trip et al. 2014). However, it is important to consider other ecological factors (and their trade-offs) that control fish population structure at tropical and extreme and/or marginal reefs, as trophic levels can be a confounding factor in making size-based demographic generalisations in fish communities. This is because fish are affected by (1) gape limitation, where predators are generally larger than the prey, and (2) ontogenetic diet shifts, where individuals feed at higher trophic levels as they grow (Dunic and Baum 2017). In addition to larger sizes, some traits that are beneficial to survival in variable environmental conditions are: high swimming ability, pelagic spawning behaviour, flexible habitat selection and feeding behaviours. Species with these traits might be more likely to range-expand into extreme and/or marginal reefs (Feary et al. 2014), though whether these tropicalising populations are viable or not is questionable (Chaikin et al. 2024). Nonetheless, fish populations might decline if their physiological and behavioural adaptations cannot adapt at the rate of ongoing global climatic change (Scott et al. 2019; Beger 2021; Stuart-Smith et al. 2021).

With differences in the demographic processes of extreme and/or marginal cool and tropical coral reefs in mind, it is crucial to understand how populations resist and recover from demographic alterations caused by environmental disturbances (Stott et al. 2011; Capdevila et al. 2020). Although cool reef high-latitude coral populations have lower population growth rates compared to tropical populations, they have the ability to enhance their short-term (transient) growth following disturbance events to ensure population persistence (Cant et al. 2022). The latitudinal variation in transient dynamics highlights the demographic plasticity of populations, and the need to expand our knowledge across more taxonomic groups of reef organisms. While there is a considerable amount of demographic work being done on coral populations (e.g., Cant et al. 2022; Cant et al. 2023; Chong et al. 2023), little is known about the demographic responses of coral reef fish and other benthic taxa across large-scale environmental gradients spanning different biogeographic zones.

Florida's Extreme and/or Marginal Cool Reefs

David A. Paz-García

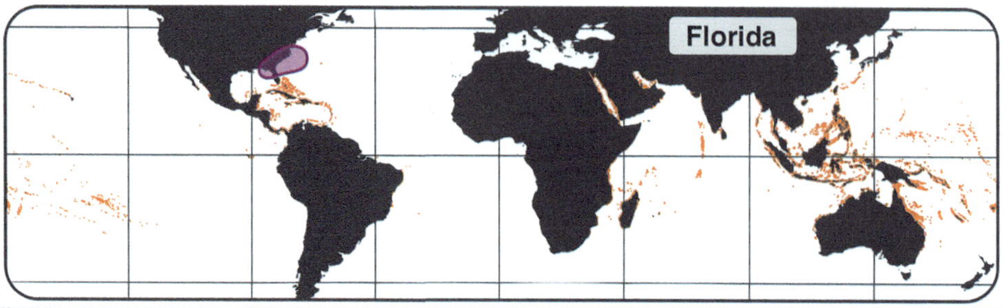

Florida's extreme and/or marginal cool reefs extend over 150 km from Martin County in the northeast (27°09′N) to Miami Beach (25°47′N). They are non-accreting with low coral cover at the northern limit of coral distribution in the western Atlantic. These reefs have been exposed to acute and frequent disturbances such as stony coral tissue loss disease, storms, and heat (>31 °C) and cold stress (<22 °C), in addition to anthropogenic stressors such as eutrophication and sedimentation from widespread coastal construction and sewage outfalls (Jones and Gilliam 2024).This has led to reduced coral cover of only 1–6% within the last decade (Grove et al. 2022; Jones et al. 2020a, b).

The most abundant species are *Siderastrea siderea*, *Porites astreoides*, *Stephanocoenia intersepta*, *Montastraea cavernosa*, *Siderastrea radians* and *Porites porites* (Enochs et al. 2023). The local extinction of *Dendrogyra cylindrus* occurred after bleaching and coral disease events between 2014 and 2020 (Jones et al. 2021). (Stony coral tissue loss disease spread across the entirity of Florida's Coral Reef system from 2014 to 2021 and continues to persist and cause extensive coral mortality (Grove et al. 2022). Recovery of stony corals has been limited, whilst there has been an increase in macroalgae and octocorals cover (Jones et al. 2020a; Jones et al. 2021).

all photos by Tanya Ramseyer©

13.8 Novel Ecosystems, Functioning and Refugia

13.8.1 Extreme and/or Marginal Cool Reefs as Novel Ecosystems

Novel ecosystems refer to ecological systems that have been significantly altered by anthropogenic activities to the extent that they no longer resemble their historical state or the state of typical natural ecosystems (Hobbs et al. 2009; Morse et al. 2014). The concept of novel ecosystems has gained increasing attention in ecological research due to the significant recent changes occurring in species distributions and environmental conditions, primarily driven by factors such as climate change and other anthropogenically driven activities (Hobbs et al. 2009; Vergés et al. 2014a). These changes can lead to the emergence of ecosystems that differ in species composition, community structure, and/or ecosystem functions (Hobbs et al. 2009; Morse et al. 2014; Graham et al. 2014).

While change is a natural characteristic of all ecosystems, the unprecedented rates of change and the breakdown of biogeographic barriers set the current era apart, leading to the increasing prevalence of novel ecosystems (Hobbs et al. 2009). Most research on novel ecosystems has focused on the terrestrial realm, where land-use change, restoration, and invasive species often contribute to the formation of novel ecosystems. Furthermore, abiotic changes and alterations in native species combinations also play significant roles. Given the substantial changes reported for a range of coastal marine ecosystems due to anthropogenic impacts (Hughes et al. 2017; Shore et al. 2021; Gissi et al. 2021; Mazzocchi et al. 2023; Daru and Rock 2023), urgent evaluations of how the concept of novel ecosystems applies to cool reefs are needed (Graham et al. 2014; Sommer 2022; Pandolfi et al. 2020).

While extreme and/or marginal cool coral communities are not novel ecosystems per se, their structure and species composition have been shown to be subject to alteration in the context of climate change in the last decades (See Sect. 13.3; Yamano et al. 2011; Baird et al. 2012; Beger et al. 2014). Such changes may involve shifts in species composition, changes in community dynamics, or modifications to ecosystem functions. For several regions globally, poleward shifts of species including scleractinian corals, fish and crabs are recorded (Figueira and Booth 2010; Yamano et al. 2011; Baird et al. 2012; Vergés et al. 2014a). Such shifts are typically coupled with the decline or complete disappearance of kelp forests in these regions (Figueira and Booth 2010; Yamano et al. 2011; Baird et al. 2012; Vergés et al. 2014a; Favoretto et al. 2022). It must be noted that it is unclear whether such observations can all be attributed to the migration of species from tropical areas, or instead are due to the increase in abundances of resident species (Keshavmurthy et al. 2023), though it is likely a combination of both mechanisms.

Anthropogenically-induced climate change is a key driver in the establishment of novel ecosystems in high-latitude communities, causing gradual increases in average SST and greater overall SST fluctuations in high-latitude regions than in the tropics (Lough and Hobday 2011; Burrows et al. 2011). In addition to this greater variability in SSTs, it is assumed that extreme weather events will become more frequent, which can damage communities at high-latitudes and thereby influence species composition (e.g. Wernberg et al. 2013; Smale and Wernberg 2013). However, these changes can be spatiotemporally and taxonomically variable, leading to different responses of species and thus reef communities. Contrary to the assumption that remote high-latitude reefs provide an ecological refuge from climate change, having traditionally been considered more resilient to elevated SSTs than their low-latitude counterparts, heat-related coral bleaching and mortality increasingly occurs on extreme and/or marginal cool reefs (Abdo et al. 2012; Kim et al. 2019; Lachs et al. 2021; Moriarty et al. 2023; Sommer et al. 2024). This altered combination of abiotic and biotic stressors is likely to increase the formation of novel marine ecosystems, such as the increase in coral communities and the decline of kelp forests at high-latitudes (Keshavmurthy et al. 2023). These ecosystems exhibit altered species compositions, ecological dynamics, functional diversities, and adaptation strategies comp-ared to traditional tropical coral reefs (Hobbs et al. 2009). Anthropogenic intervention and disturbances further influence the development and trajectory of these novel ecosystems (Bulleri et al. 2020). Understanding and managing these ecosystems requires interdisciplinary approaches that consider ecological, evolutionary and socio-economic factors (Beger et al. 2014). It is important to note that these novel configurations may still provide desirable socio-economical ecosystem goods and services, and therefore there is a need to understand the characteristics of these emerging ecosystems and determine the most appropriate management for these novel ecosystems.

13.8.2 Functioning of Marginal Cool Reefs

The consequences of functional changes in cool reefs and the services they provide are not yet fully understood. It is a common misconception that these novel coral communities will resemble tropical habitats, such as subtropical and tropical accreting coral reefs. Although there are similarities due to some shared environmental conditions, each ecosystem is unique, with coral communities on cool reefs exposed to different abiotic factors than subtropical and tropical coral reefs, such as higher variability in SST and light availability (Kleypas et al. 1999; Beger et al. 2014; Cant et al. 2023). Cold snap-related mortality events have been reported in scleractinian corals and reef fish at higher latitudes (Lirman et al. 2011; Leriorato and Nakamura 2019; Gomez et al. 2024). Furthermore, some extreme and/or marginal cool coral communities are often located close to shore (e.g., in Eastern Asia), and are therefore directly affected by both terrestrial and marine anthropogenic impacts such as runoff, coastal erosion, and pollution (Keshavmurthy et al. 2023). This environmental variability has profound effects, including disruptions in species distributions, changes in ecological interactions, and changes in ecosystem functions that can lead to entirely new ecosystem functions and interactions.

Functionality of an ecosystem is often measured by using a range of ecological traits to determine the processes regulating energy flux through an ecosystem (Laureto et al. 2015). Shifts in species ranges, as discussed in Sect. 13.3, bring about changes in trait compositions impacting ecosystem functioning (Hughes et al. 2017; Vergés et al. 2014a, 2019). Functional diversity is another commonly "controversial" used metric which describes how different species traits impact the functioning of the local environment (Petchey and Gaston 2006), and thereby describes the different niches exploited within a community. Highly diverse ecosystems, such as coral reefs, are characterised by high functional redundancy, where multiple species fulfil similar functional roles, thus buffering against the loss of species unable to adapt to rising SSTs (Yachi and Loreau 1999; Fonseca and Ganade 2001). Many studies have explored the functional diversity and redundancy of various taxa on reefs, observing declining functional diversity with latitude, and trait turnover but unchanged trait scope over time (Edwards et al. 2013; Mouillot et al. 2014; Coleman et al. 2015; Floyd et al. 2020; Cook et al. 2022; Miller et al. 2023; Clay et al. 2024).

As biological community structures change with more tropical species expanding their ranges into subtropical and temperate regions, new interactions inevitably occur (see Sect. 13.3). One such interaction is the range expansion of herbivorous tropical fishes poleward, which can lead to the decimation of kelp forests, resulting in shifts from macroalgal dominated to either turf, or turf and coral dominated areas (Zarco-Perello et al. 2021). Shifting compositions of habitat forming species has implications for ecosystem functions, as the loss of complex habitat builders such as kelp can implicate species reliant on these habitats for shelter against predation, food or nursery grounds (Vergés et al. 2019). Alternatively, increases in functional herbivory, as tropical herbivorous fishes expand their ranges, can result in shifts in energy flow from a higher intensity of feeding (Pessarrodona et al. 2022). It has been suggested that trophic pathways such as algivory and detritivory could be improved as herbivorous fish migrate poleward and turf barrens become more common (Pessarrodona et al. 2022).

The impact of tropicalising reefs on ecosystem functionality could depend on both the species that dominate the novel reef, and the interactions they have with other resident taxa (Vergés et al. 2019). For example, one study found that the loss of coral assemblages minimally affected the functional trait space on extreme and/or marginal reefs in the SW Atlantic, with only a 5.3% loss in adult fish and 1.14% loss for juvenile fish in the case of complete coral loss (Luza et al. 2022). However, on many reefs, specialist species require specific conditions to survive, and large-scale changes to an ecosystem can result in the loss of such specialists, and an increase in generalist species (Beger 2021; Clay et al. 2024). This, in turn, can lead to the homogenisation of the functional traits present, leading to a reduction in functional redundancy (Beger 2021; Clay et al. 2024).

Galápagos Extreme and/or Marginal Cool Reefs

Margarita Brandt, Daniel Velasco-Cedeño, Nicolás Dávalos

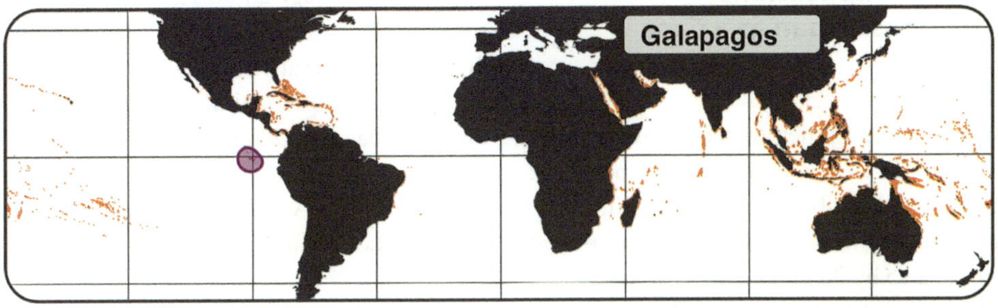

Galápagos hermatypic corals endure challenging oceanic conditions including fluctuating sea surface temperatures (SST), severe El Niño cycles, heightened nutrient influxes, and relatively acidic waters. Despite this, they survive in a patchy distribution and form communities rather than structural reefs (Hickman 2008), especially in the cooler western region of the archipelago, where SST ranges from 14–26 °C, compared to 20–28 °C in the central archipelago. Corals were once abundant in some western locations until a tectonic uplift in 1954 caused extensive mortality (Glynn and Wellington 1983). In Punta Espinosa (Isla Fernandina), a significant Pocilloporid patch was lost, as were large colonies of *Pavona* spp. and *Porites* spp. from Bahía Urvina (Isla Isabela).

Subsequent explorations in 1975–76 (Glynn and Wellington 1983) reported a continuous *Pocillopora* spp. patch of 0.5 ha, with occasional large (1-2 m in diameter) *Porites lobata* colonies in Punta Espinosa. *Porites* spp. and *Pavona* spp. were spotted at the innermost riches of Bahía Elizabeth in Isla Isabela (Glynn and Wellington 1983), and more recently in 2012, large but scarce colonies of *Porites lobata* were found at 1-4 m depth at Bahía Urvina (Manzello et al. 2014). Isla Pinzón (in the central region of the archipelago), with average temperatures of 19.96 °C during the cold season, and the shallow reefs of Puerto Villamil in Isla Isabela (mean SSTs of 21.63 °C during the cold season), seem to be the last refugia of corals adapted to cool waters in Galápagos. Over the past decade, scattered Pavonids and Pocilloporids have been recorded in Pinzón (GBIF, 2024; iNaturalist, 2016), while the densest Pocillopora aggregation is found in Puerto Villamil (Baums et al. 2014). This area contains about 3500 colonies (Dávalos et al., in press), with Pocilloporids accounting for most of them (74%). Pavonids, *Porites* spp. and Psammocorids are also present.

Puerto Villamil and Margarita Brandt©.

Puerto Villamil and Margarita Brandt©

Map of the western, cooler region of the Galápagos Archipelago. Credit: Mateo Dávila.

13.9 Management Implications and Challenges: Japan as a Case Study

The 15th Conference of the Parties to the Convention on Biological Diversity (CBD-COP), held in December 2022, adopted a new global goal on biodiversity, the Kunming-Montreal Biodiversity Framework (GBF), aiming to protect 30% of land and ocean area by 2030 (i.e., 30 × 30; Dinerstein et al. 2019). As a result, ratified countries plan to expand their marine protected areas, a goal that also affects extreme and/or marginal cool reef ecosystems. In a case study presented here for Japan, one of the most intensely tropicalising regions, we illustrate how this management process might work. It is important to note that the '30 x 30' process is subject to each countries legislative framework, and thus the Japan case study cannot necessarily be applied to other countries in all its facets.

13.9.1 Marine Protected Areas (MPAs) in Japan

The Ministry of the Environment (MoE) of Japan plans to double the area of marine parks by 2030, and establish new marine protected areas (MPAs) offshore. Currently, 72.1% of coastal areas (territorial waters with a depth of less than 200 m) and 10% of offshore areas (Japan's exclusive economic zone excluding coastal areas) are designated as MPAs. However, the definition and role of MPAs in Japan are different than in many Western countries, being defined as: "*A marine area designated with the aim of conserving biodiversity and sustainable use of ecosystem services that support the healthy structure and function of marine ecosystems...*" (Marine Biodiversity Conservation Strategy, MoE 2011).

A characteristic of MPAs in Japan's coastal areas is the multi-layered existence of multiple laws and management systems with different purposes (Makino 2017). In particular, designated areas with the purpose of protecting and cultivating marine resources account for a large proportion of Japan's MPAs; for example, of the 1161 MPAs in coastal areas that were confirmed in 2010, over 90% were restricted to no-fishing areas managed by local fishers (Yagi et al. 2010) . Japan's coastal areas have a long history of being used by coastal communities, with a high proportion of them falling within the IUCN protected area categories IV to VI (Badman and Bomhard 2008). Locally, the creation and maintainence of the natural environment through anthropogenic management is called "*satochi-satoyama*" or "*satoyama*" (MoE: https://www.env.go.jp/nature/satoyama/seibutu.html), and similarly, where marine biodiversity has been maintained through anthropogenic activities it is called "*satoumi*" (Yanagi 2006; Kakuma et al. 2018). In *satoumi*, some local communities have been actively involved in biodiversity conservation, including in surrounding land areas, with the aim of maximising the number of resources in marine areas (Nakamura et al. 2010). To conserve Japan's MPAs, it is crucial to take into account historical and cultural backgrounds, as well as promote the involvement of fishers and local communities.

13.9.2 Impact of Climate Change

In addition to the climate change impacts (see Sect. 13.3), in Japan, there are concerns about the future occurrence of large-scale disasters such as larger typhoons and an increase in storm surges (MoE 2020a). The global SST has increased at +0.61 °C/100 years, but in Japan's coastal waters, the rate of increase is +1.28 °C/100 years, over twice the global average (Japan Meteorological Agency, 2024). These rises in SSTs are bringing about changes such as coral bleaching (Hoegh-Guldberg 1999;Donner et al. 2005), expansion of the distribution area of herbivorous fishes (Vergés et al. 2014a, b), reduction and disappearance of macroalgal beds (Martínez et al. 2018; Kumagai 2020), and replacement by more tropical species (Nishida et al. 2005; Tanaka et al. 2012; Ishikuro and Misawa 2024). There are therefore concerns about the deterioration of ecosystem services (Fujii 2020; MoE 2020a).

In the fishing industry, changes in the poleward movement of yellowtail *mackerel (Seriola quinqueradiata)* and Spanish mackerel (*Scomberomorus niphonius*; Japan Ministry of Fishes (MoF) 2018), as well as changes in the habitats of spiny lobster (*Panulirus japonicus*) and abalone (*Haliotis* spp.; Takahashi 2022; Fisheries Agency 2020, 2023; Ministry of Agriculture, Forestry and Fisheries of Japan (MoAFF) 2022), have already been observed, resulting in negative impacts on local fishing industries (Fisheries Agency 2018). It is predicted that suitable habitats for sea urchins, some seaweeds, and farmed yellowtail mackerel that are currently found south of Kyushu will move poleward within 30 years (Kuwabara et al. 2006). Based on stakeholder interviews, fisheries processing businesses and seafood restaurants have not been able to keep up with changes in the fish species being caught (Nakano, unpublished data).

13.9.3 Climate Change Adaptation in Japanese MPAs

In Japan, the 'Climate Change Adaptation Act' was enacted in 2018 to intergrate the country's climate change adaptation measures into policy, with the 'Climate Change Adaptation Plan' subsequently formulated in 2019. Regarding marine ecosystems, the National Institute for Environmental Studies, Climate Change Adaptation Center (NIES CCCA) has been tasked with collecting and publishing scientific research, including increasing observations and monitoring, and restoring ecosystems. NIES CCCA provides data, develops manuals, and implements training programs to support the formulation of local climate change adaptation plans in the region, with 279 local governments having developed specific regional plans (NIES CCCA 2024).

In 2019, the MoE compiled 'Guidelines for considering adaptation measures to climate change in protected areas such as national parks' for protected areas in Japan. Specific tools and procedures for consideration have been provided for managers of protected areas, including MPAs, which are being used to evaluate climate change impacts and consider specific adaptation measures in two land based National Parks and two Marine Parks (MoE 2020b). In Ashizuri-Uwakai National Park (Shikoku, Japan), predictions of changes in the distribution of macroalgae and corals, as well as adaptation options for coral conservation and adjustment of use in fisheries, have been proposed (MoE 2020b). In Amami Islands National Park, predictions of coral distribution changes and surveys of inter-island networks were conducted, and countermeasures such as reducing runoff loads from the catchment and preserving source habitats have been proposed (MoE 2020b).

The elements required for effective marine management, such as responding to change and securing resources are often wide-ranging and complex. However, maintaining a healthy ecosystem in a sustainable condition is essential for tourism industries such as fishing and marine leisure that utilise the biota that inhabit shallow coastal waters. Public systems, such as National Parks, have built-in mechanisms for ecosystem management, such as regular monitoring, review processes by experts, and subsidy systems. However, with the exception of urban areas, coastal communities in Japan face many common challenges such as population decline, ageing, and economic stagnation, making it difficult to even implement systems and guidelines prepared by the government (MPA). It is hoped that a practical tool can be developed for uptake by small-scale fishers, local communities and non-profit organisations, while preserving small-scale MPAs and voluntary management mechanisms. Possible examples include (1) a tool for current situation analyses and future predictions that anyone can easily use, (2) a mechanism for providing experts, business partners, equipment, funds, to areas in need, (3) a scientific database that can be easily accessed and used by all, and (4) the construction of an information sharing network including experts.

An example of the formation of a global expert network is the international team of scientists and non-governmental organisations, carrying out the 'MPA Shift Project' in Japan's Kuroshio region, with support from the Lenfest Ocean Program. This project, launched in 2023, has been investigating how coral reefs and extreme and/or marginal cool reef coral communities have changed over the past decade, benchmarking abundances and biomass of the corals, fish, algae, and invertebrates inhabiting local reefs (Lenfest Ocean Program 2024). By clarifying the mechanisms by which organisms migrate poleward using ocean currents, the extent of distribution expansion of species and communities, and thus predict the future of the biota in the Kuroshio region, scientists and policymakers are enabled to make better recommendations for wide-ranging ecosystem management. Furthermore, such research and associated results underscore efforts to develop tools and build a network to help small-scale MPAs affected by climate change consider adaptation measures.

Japan's MPAs and management authorities have generally been slow to respond to changes brought about by climate change. Small-scale MPAs have not been successfully managed to date, but with the introduction of ecosystem management and climate change adaptation measures supported by scientific evidence, these MPAs will be more effectively conserved and likely lead to increased local community resilience.

Australia's Extreme and/or Marginal Cool Reefs

Brigitte Sommer, Maria Beger

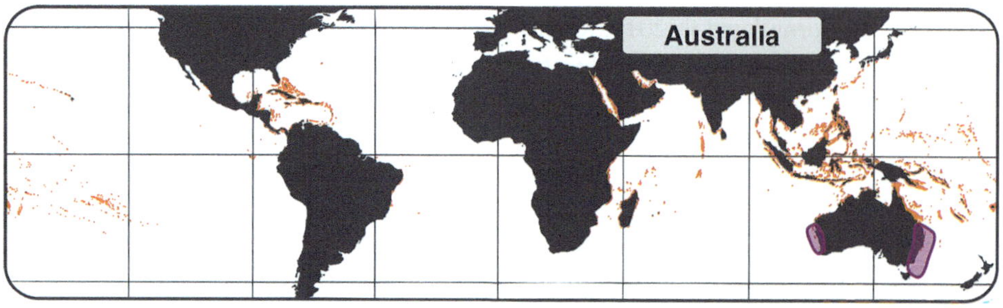

In Australia, poleward flowing currents enable corals to grow at high latitudes on both coastlines, where corals grow directly on rock, next to cold-water species (Sommer 2022). Some of these subtropical reefs, such as the Solitary Islands, are very diverse (30°S, >90 species, Harriott et al. 2014), with Lord Howe Island (31°S) the world's most southerly accreting coral reef. Australia's high-latitude reefs are typically dominated by generalist (e.g., *Turbinaria* spp.), stress-tolerant (e.g., *Paragoniastrea australensis, Coelastrea aspera*), subtropical specialist (e.g., *Acropora solitaryensis*) and endemic species with massive or flat growth (Tuckett et al. 2017; Sommer 2022).

Despite declines in macroalgae and the tropicalisation of fish assemblages, high-latitude corals have undergone limited tropicalisation (Mizerek et al. 2021), with expansion mainly restricted to high-latitude specialists and endemics such as *Plesiastrea versipora* and *Pocillopora aliciae* (Tuckett et al. 2017; Mizerek et al. 2021). These species are also sensitive to heat stress and recovery bottlenecks linked to sporadic recruitment, whereas *Turbinaria* spp. populations have remained remarkably stable despite bleaching (Sommer et al. 2024).

Although some subtropical reefs have high coral abundance, macroalgae, soft corals, and filter-feeding invertebrates are often more dominant than corals (Harriott et al. 1994). Along with subtropical and temperate sea urchins (McLaren et al. 2024) these taxa likely mediate ecological dynamics of coral communities along the tropical-to-temperate transition in eastern Australia.

all photos, Brigitte Sommer ©

13.10 Future Research Directions

Throughout this chapter, we have outlined the variety of research being conducted on extreme and/or marginal cool reefs, however, data deficiency remains one of the largest barriers to effective prediction and applied ecological measures on extreme and/or marginal reefs. As such, we encourage long-term investigations of the diversity, abundance, and demographic processes of all reef-associated organisms along the tropical-temperate coral community gradient. Capturing and investigating variations in demographic parameters controlling population resilience is important for the prediction of future reef states. Increasing the use of innovative and molecular approaches including photogrammetry and eDNA for monitoring alongside visual reef surveys will also allow larger quantities of data to be collected (Garlapati et al. 2019; Paz-García et al. 2023). Collections of orthomosaics and 3D models of latitudinal reefs and eDNA can estimate species distributions through species distribution models (Beng and Corlett 2020; Muenzel et al. 2024), and therefore with more long-term use could reveal range shift patterns, the arrival of new species into communities or the loss of species from an area.

Long-term ecological and environmental data is also essential for accurate species distribution modelling (SDM), which is becoming an increasingly popular tool to predict community change (Welch et al. 2020; Wilson et al. 2020; Brodie et al. 2022). Improving SDMs and applying them to extreme and/or marginal reefs will enable more accurate predictions of future range shifts (Davies et al. 2023), along with areas of refugia (Dixon et al. 2021), and therefore improve management efforts. Other interesting variations on SDMs are joint species distribution models that can predict community change, or hierarchical modelling of species communities that can evaluate the joint influences of environmental and biological factors, providing a promising route for predicting community composition change over time (Ovaskainen et al. 2017; Tikhonov et al. 2020a; Tikhonov et al. 2020b). Hierarchical modelling of species communities also allows for the combination of presence/absence or abundance data with trait data, phylogeny and environmental predictors, improving the predictive power (Ovaskainen and Abrego 2020). However, these techniques again require large quantities of data and reiterate the importance of continued and increasing data collection.

Another potential source of ecological data lies in the fossil record, looking back at how reefs of the past responded to environmental change has the potential to increase our understanding of coral communities (Pandolfi and Kiessling 2014; O'Dea et al. 2017) and improve predictive models (Anderson et al. 2006). Exploring the responses of coral communities to previous periods of warming, especially along latitudinal gradients, may improve our knowledge of range shifts and the potential for the tropicalisation of certain communities (Pandolfi and Jackson 2001; Abram et al. 2001; Yamamoto et al. 2006; Pandolfi and Jackson 2007; Greenstein and Pandolfi 2008). However, this methodology can only be applied to certain taxa (e.g., corals), as the fossil record is often biased. eDNA metabarcoding has the potential to improve our knowledge of ancient ecosystems and has already been applied in terrestrial systems to improve species identification from fossils (Sønstebø et al. 2010; Jørgensen et al. 2012; Ruppert et al. 2019), estimate biodiversity (Ruppert et al. 2019), and the diet of megaherbivores (Wood et al. 2012; Willerslev et al. 2014).

We have already drawn attention to the taxa often understudied in reef environments and the difficulties faced in identifying these species, let alone quantifying their ecological functions (Brandl et al. 2019). Even the most studied taxa of reef communities, such as corals and fish are often difficult to identify at the species level, especially with ongoing hybridisation and adaptation (Zarzyczny et al. 2024b). Furthermore, without correct species identification, characterising the composition of cool reef communities, and whether it is changing in response to environmental change, is impossible. This ecological knowledge is essential for effective conservation planning, therefore further research into morphological and especially genetic methods for species identification is needed.

To respond to changing variability in SST and subsequent range shifts, management and conservation strategies of extreme and/or marginal cool reefs must consider current conditions as well as future change when designing MPAs (Dixon et al. 2021), and crucially, MPAs must become adaptable. However, MPAs cannot mitigate climate-induced range shifts, neither by slowing the rate of range shifts (Bates et al. 2014; Carter et al. 2020; Davis et al. 2021), nor by preparing for future range shifts (Makino et al. 2014; Wilson et al. 2020). Combining these kinds of MPA and those mentioned earlier in the chapter with more dynamic conservation areas, which allow for shifting species ranges may improve conservation efforts (D'Aloia et al. 2019; Tittensor et al. 2019). It's important to consider future change as well as current conditions when designing MPAs, such as potential future thermal stress (Dixon et al. 2021). We would like to see more application of these dynamic and future focussed conservation strategies in practice.

13.11 Conclusions

Extreme and/or marginal cool reefs are home to a myriad of species, but as climate change continues to drive increases in SST, the migration of species poleward can have significant impacts on the community structure of these reef habitats. However, whilst some tropical corals may settle on these extreme and/or marginal cool reefs, these ecosystems are unlikely to act as effective refugia for all tropical corals escaping from increasing thermal stress in the tropics. Cool reefs are unique and valuable in their own right and although it is now reasonably well documented that the communities inhabiting these extreme and/or marginal cool reefs are changing, more work is needed to understand which species are being gained and lost, and the mechanisms driving these changes. Changes in benthic communities from macroalgal to turf or coral dominated likely have cascading impacts across all taxa, yet without first understanding how and what changes are occurring, it is difficult to create useful management strategies to protect species and communities that are struggling to survive. Spatiotemporal monitoring and rigorous data are urgently needed to address these knowledge gaps that include more taxonomic groups, particularly currently understudied taxa (e.g. sponges and soft corals) that could become dominant across many changing high-latitude reefs. In addition, extensive surveys and modelling is needed, supported by both traditional and novel technologies to fully understand the nature of these ever-changing high-latitude extreme and/or marginal ecosystems and the future implications for cool reef communities. To proactively manage coastal marine ecosystems in the Anthropocene, management efforts must consider both the needs of native and tropicalising species, which will require both local static and dynamic conservation strategies as well as communication amongst stakeholders.

References

Abbott E, Dixon G, Matz M (2021) Shuffling between *Cladocopium* and *Durusdinium* extensively modifies the physiology of each symbiont without stressing the coral host. Mol Ecol 30(24):6585–6595. https://doi.org/10.1111/mec.16190

Abdo DA, Bellchambers LM, Evans SN (2012) Turning up the heat: increasing temperature and coral bleaching at the high latitude coral reefs of the Houtman Abrolhos Islands. PLoS One 7(8):e43878. https://doi.org/10.1371/journal.pone.0043878

Abe H, Mitsui S, Yamano H (2022) Management and utilization of reef-building corals, macroalgae, and seaweed in national parks: Identifying issues for considering adaptation measures to climate change. Conserv Ecol Res 27(1):2120. (in Japanese). https://www.jstage.jst.go.jp/article/hozen/27/1/27_2120/_article/-char/ja/

Abram N, Webster J, Davies P, Dullo W (2001) Biological response of coral reefs to sea surface temperature variation: evidence from the raised Holocene reefs of Kikai-jima (Ryukyu Islands, Japan). Coral Reefs 20(3):221–234. https://doi.org/10.1007/s003380100163

Abrego D, Howells EJ, Smith SDA, Madin JS, Sommer B, Schmidt-Roach S, Cumbo VR, Thomson DP, Rosser NL, Baird AH (2021) Factors limiting the range extension of corals into high-latitude reef regions. Diversity 13(12):632. https://doi.org/10.3390/d13120632

Agostini S, Harvey BP, Milazzo M, Wada S, Kon K, Floc'h N, Komatsu K, Kuroyama M, Hall-Spencer JM (2021) Simplification, not "tropicalization", of temperate marine ecosystems under ocean warming and acidification. Glob Chang Biol 27(19):4771–4784. https://doi.org/10.1111/gcb.15749

Airoldi L, Turon X, Perkol-Finkel S, Rius M (2015) Corridors for aliens but not for natives: effects of marine urban sprawl at a regional scale. Divers Distrib 21(7):755–768. https://doi.org/10.1111/ddi.12301

Alava JJ, McMullen K, Jones J, Barragán-Paladines MJ, Hobbs C, Tirapé A, Calle P, Alarcón D, Muñoz-Pérez JP, Muñoz-Abril L, Townsend KA, Denkinger J, Uyaguari M, Domínguez GA, Espinoza E, Reyes H, Piedrahita P, Fair P, Galloway T, Grove JS, Lewis C, Schofield J (2023) Multiple anthropogenic stressors in the Galápagos Islands' complex social–ecological system: interactions of marine pollution, fishing pressure, and climate change with management recommendations. Integr Environ Assess Manag 19(4):870–895. https://doi.org/10.1002/ieam.4661

Albinsky D, Wham D, Shinzato N, Reimer JD (2018) Population connectivity in the common reef Zoantharian *Zoanthus sansibaricus* (Anthozoa: Hexacorallia) in Southern Japan. Zool Sci 35(4):321–329. https://doi.org/10.2108/zs180007

Alvarado JJ, Cortés J, Guzman H, Reyes-Bonilla H (2016) Bioerosion by the sea urchin Diadema mexicanum along Eastern Tropical Pacific coral reefs. Mar Ecol 37(5):1088–1102. https://doi.org/10.1111/maec.12372

Anderson NJ, Bugmann H, Dearing JA, Gaillard M-J (2006) Linking palaeoenvironmental data and models to understand the past and to predict the future. Trends Ecol Evol 21(12):696–704. https://doi.org/10.1016/j.tree.2006.09.005

Arafeh-Dalmau N, Montaño-Moctezuma G, Martínez JA, Beas-Luna R, Schoeman DS, Torres-Moye G (2019) Extreme marine heatwaves Alter kelp Forest Community Near its equatorward distribution limit. Front Mar Sci 6. https://doi.org/10.3389/fmars.2019.00499

Armstrong EJ, Lê-Hoang J, Carradec Q, Aury J-M, Noel B, Hume BCC, Voolstra CR, Poulain J, Belser C, Paz-García DA, Cruaud C, Labadie K, Da Silva C, Moulin C, Boissin E, Bourdin G, Iwankow G, Romac S, Agostini S, Banaigs B, Boss E, Bowler C, de Vargas C, Douville E, Flores M, Forcioli D, Furla P, Galand PE, Gilson E, Lombard F, Pesant S, Reynaud S, Sullivan MB, Sunagawa S, Thomas OP, Troublé R, Thurber RV, Zoccola D, Planes S, Allemand D, Wincker P (2023) Host transcriptomic plasticity and photosymbiotic fidelity underpin Pocillopora acclimatization across thermal regimes in the Pacific Ocean. Nat Commun 14(1):3056. https://doi.org/10.1038/s41467-023-38610-6

Au AC, Zhang L, Chung S, Qiu J-W (2014) Diving associated coral breakage in Hong Kong: differential susceptibility to damage. Mar Pollut Bull 85(2):789–796. https://doi.org/10.1016/j.marpolbul.2014.01.024

Aued AW, Smith F, Quimbayo JP, Cândido DV, Longo GO, Ferreira CEL, Witman JD, Floeter SR, Segal B (2018) Large-scale patterns of benthic marine communities in the Brazilian province. PLoS One 13(6):e0198452. https://doi.org/10.1371/journal.pone.0198452

Badman T, Bomhard B (2008) World heritage and protected areas. IUCN, Gland, Switzerland, pp 1–22. Microsoft Word – World Heritage and Protected Areas 2008.doc (iucn.org)

Baines C, Lerebours A, Thomas F, Fort J, Kreitsberg R, Gentes S, Meitern R, Saks L, Ujvari B, Giraudeau M, Sepp T (2021) Linking pollution and cancer in aqu atic environments: a review. Environ Int 149:106391. https://doi.org/10.1016/j.envint.2021.106391

Baird AH, Sommer B, Madin JS (2012) Pole-ward range expansion of *Acropora spp.* along the east coast of Australia. Coral Reefs 31(4):1063–1063. https://doi.org/10.1007/s00338-012-0928-6

Baird AH, Hoogenboom MO, Huang D (2017) *Cyphastrea salae*, a new species of hard coral from Lord Howe Island, Australia (Scleractinia, Merulinidae). ZooKeys 662:49–66. https://doi.org/10.3897/zookeys.662.11454

Banha TNS, Capel KCC, Kitahara MV, Francini-Filho RB, Francini CLB, Sumida PYG, Mies M (2020) Low coral mortality during the most intense bleaching event ever recorded in subtropical Southwestern Atlantic reefs. Coral Reefs 39(3):515–521. https://doi.org/10.1007/s00338-019-01856-y

Barrientos S, Zarco-Perello S, Piñeiro-Corbeira C, Barreiro R, Wernberg T (2021) Feeding preferences of range-shifting and native herbivorous fishes in temperate ecosystems. Mar Environ Res 172:105508. https://doi.org/10.1016/j.marenvres.2021.105508

Bates AE, Barrett NS, Stuart-Smith RD, Holbrook NJ, Thompson PA, Edgar GJ (2014) Resilience and signatures of tropicalization in protected reef fish communities. Nat Clim Change 4(1):62–67. https://doi.org/10.1038/nclimate2062

Bates AE, Stuart-Smith RD, Barrett NS, Edgar GJ (2017) Biological interactions both facilitate and resist climate-related functional change in temperate reef communities. Proc R Soc B Biol Sci 284(1856):20170484. https://doi.org/10.1098/rspb.2017.0484

Baums IB (2008) A restoration genetics guide for coral reef conservation. Mol Ecol 17(12):2796–2811. https://doi.org/10.1111/j.1365-294X.2008.03787.x

Baums IB et al (2014) Marginal coral populations: the densest known aggregation of *Pocillopora* in the Galápagos Archipelago is of asexual origin. Front Mar Sci 1:59

Beas-Luna R, Micheli F, Woodson CB, Carr M, Malone D, Torre J, Boch C, Caselle JE, Edwards M, Freiwald J, Hamilton SL, Hernandez A, Konar B, Kroeker KJ, Lorda J, Montaño-Moctezuma G, Torres-Moye G (2020) Geographic variation in responses of kelp forest communities of the California current to recent climatic changes. Glob Chang Biol 26(11):6457–6473. https://doi.org/10.1111/gcb.15273

Beger M (2021) Accepting the loss of habitat specialists in a changing world. Nat Ecol Evol 5(5):556–557. https://doi.org/10.1038/s41559-021-01402-6

Beger M, Babcock R, Booth DJ, Bucher D, Condie SA, Creese B, Cvitanovic C, Dalton SJ, Harrison P, Hoey A, Jordan A, Loder J, Malcolm H, Purcell SW, Roelfsma C, Sachs P, Smith SDA, Sommer B, Stuart-Smith R, Thomson D, Wallace CC, Zann M, Pandolfi JM (2011) Research challenges to improve the management and conservation of subtropical reefs to tackle climate change threats. Ecol Manage Restor 12(1):e7–e10. https://doi.org/10.1111/j.1442-8903.2011.00573.x

Beger M, Sommer B, Harrison PL, Smith SDA, Pandolfi JM (2014) Conserving potential coral reef refuges at high latitudes. Divers Distrib 20(3):245–257. https://doi.org/10.1111/ddi.12140

Bell JJ, McGrath E, Biggerstaff A, Bates T, Bennett H, Marlow J, Shaffer M (2015) Sediment impacts on marine sponges. Mar Pollut Bull 94(1):5–13. https://doi.org/10.1016/j.marpolbul.2015.03.030

Bell JJ, Bennett HM, Rovellini A, Webster NS (2018a) Sponges to be winners under near-future climate scenarios. Bioscience 68(12):955–968. https://doi.org/10.1093/biosci/biy142

Bell JJ, Rovellini A, Davy SK, Taylor MW, Fulton EA, Dunn MR, Bennett HM, Kandler NM, Luter HM, Webster NS (2018b) Climate change alterations to ecosystem dominance: how might sponge-dominated reefs function? Ecology 99(9):1920–1931. https://doi.org/10.1002/ecy.2446

Bell JJ, Micaroni V, Strano F (2021) Regime shifts on tropical coral reef ecosystems: future trajectories to animal-dominated states in response to anthropogenic stressors. Emerg Topics Life Sci 6(1):95–106. https://doi.org/10.1042/ETLS20210231

Bell JJ, Smith RO, Micaroni V, Strano F, Balemi CA, Caiger PE, Miller KI, Spyksma AJP, Shears NT (2023) Marine heat waves drive bleaching and necrosis of temperate sponges. Curr Biol 33(1):158–163.e2. https://doi.org/10.1016/j.cub.2022.11.013

Benayahu Y, Schleyer MH (1995) Corals of the South-West Indian Ocean. *Eleutherobia Aurea* Spec. Nov.(Cnidaria, Alcyonacea) from Deep Reefs on the KwaZulu-Natal Coast, South Africa. South African Association for Marine Biological Research, Oceanographic Research Institute

Beng KC, Corlett RT (2020) Applications of environmental DNA (eDNA) in ecology and conservation: opportunities, challenges and prospects. Biodivers Conserv 29(7):2089–2121. https://doi.org/10.1007/s10531-020-01980-0

Bennett S, Wernberg T, Harvey ES, Santana-Garcon J, Saunders BJ (2015) Tropical herbivores provide resilience to a climate-mediated phase shift on temperate reefs. Ecol Lett 18(7):714–723. https://doi.org/10.1111/ele.12450

Bennett HM, Altenrath C, Woods L, Davy SK, Webster NS, Bell JJ (2017) Interactive effects of temperature and pCO2 on sponges: from the cradle to the grave. Glob Chang Biol 23(5):2031–2046. https://doi.org/10.1111/gcb.13474

Berkelmans R, van Oppen MJH (2006) The role of zooxanthellae in the thermal tolerance of corals: a 'nugget of hope' for coral reefs in an era of climate change. Proc R Soc B Biol Sci 273(1599):2305–2312. https://doi.org/10.1098/rspb.2006.3567

Berryman AA (2020) Principles of population dynamics and their application. Garland Science, London

Bianchi CN, Morri C (eds) (2003) Global Sea warming and "tropicalization" of the Mediterranean Sea: biogeographic and ecological aspects. Biogeographia J Integrat Biogeograp 24(1). https://doi.org/10.21426/B6110129

Birrell CL, McCook LJ, Willis BL (2005) Effects of algal turfs and sediment on coral settlement. Mar Pollut Bull 51(1):408–414. https://doi.org/10.1016/j.marpolbul.2004.10.022

Bleuel J, Waechter L, Bender M, Longo GO (2024) Taxonomic and functional diversity of zooxanthellate corals and hydrocorals in Southwestern Atlantic reefs. Front Ecol Evol 12. https://doi.org/10.3389/fevo.2024.1322751

Booth DJ, Beretta GA (2021) Long-term demographics of a coral-reef fish: growth, survival and abundance at several spatial scales. Coral Reefs 40(4):1257–1266. https://doi.org/10.1007/s00338-021-02134-6

Booth DJ, Sear J (2018) Coral expansion in Sydney and associated coral-reef fishes. Coral Reefs 37(4):995–995. https://doi.org/10.1007/s00338-018-1727-5

Booth DJ, Figueira WF, Gregson MA, Brown L, Beretta G (2007) Occurrence of tropical fishes in temperate southeastern Australia: role of the East Australian current. Estuar Coast Shelf Sci 72(1):102–114. https://doi.org/10.1016/j.ecss.2006.10.003

Bosch NE, McLean M, Zarco-Perello S, Bennett S, Stuart-Smith RD, Vergés A, Pessarrodona A, Tuya F, Langlois T, Spencer C, Bell S, Saunders BJ, Harvey ES, Wernberg T (2022) Persistent thermally driven shift in the functional trait structure of herbivorous fishes: evidence of top-down control on the rebound potential of temperate seaweed forests? Glob Chang Biol 28(7):2296–2311. https://doi.org/10.1111/gcb.16070

Bracewell SA, Clark GF, Johnston EL (2018) Habitat complexity effects on diversity and abundance differ with latitude: an experimental study over 20 degrees. Ecology 99(9):1964–1974. https://doi.org/10.1002/ecy.2408

Brandl SJ, Rasher DB, Côté IM, Casey JM, Darling ES, Lefcheck JS, Duffy JE (2019) Coral reef ecosystem functioning: eight core processes and the role of biodiversity. Front Ecol Environ 17(8):445–454. https://doi.org/10.1002/fee.2088

Brodie JE, Kroon FJ, Schaffelke B, Wolanski EC, Lewis SE, Devlin MJ, Bohnet IC, Bainbridge ZT, Waterhouse J, Davis AM (2012) Terrestrial pollutant runoff to the Great Barrier Reef: an update of issues, priorities and management responses. Mar Pollut Bull 65(4):81–100. https://doi.org/10.1016/j.marpolbul.2011.12.012

Brodie S, Smith JA, Muhling BA, Barnett LAK, Carroll G, Fiedler P, Bograd SJ, Hazen EL, Jacox MG, Andrews KS, Barnes CL, Crozier LG, Fiechter J, Fredston A, Haltuch MA, Harvey CJ, Holmes E, Karp MA, Liu OR, Malick MJ, Pozo Buil M, Richerson K, Rooper CN, Samhouri J, Seary R, Selden RL, Thompson AR, Tommasi D, Ward EJ, Kaplan IC (2022) Recommendations for quantifying and reducing uncertainty in climate projections of species distribu-

tions. Glob Chang Biol 28(22):6586–6601. https://doi.org/10.1111/gcb.16371
Browne NK, Bauman AG (2023) Marginal reef systems: resilience in a rapidly changing world. Diversity 15(6):703. https://doi.org/10.3390/d15060703
Budd AF, Pandolfi JM (2010) Evolutionary novelty is concentrated at the edge of coral species distributions. Science 328(5985):1558–1561. https://doi.org/10.1126/science.1188947
Bulleri F, Batten S, Connell SD, Benedetti-Cecchi L, Gibbons M, Nugues MM, Gribben P (2020) Human pressures and the emergence of novel marine ecosystems. In: Oceanography and marine biology. CRC Press
Burgess SC, Treml EA, Marshall DJ (2012) How do dispersal costs and habitat selection influence realized population connectivity? Ecology 93(6):1378–1387. https://doi.org/10.1890/11-1656.1
Burrows MT, Schoeman DS, Buckley LB, Moore P, Poloczanska ES, Brander KM, Brown C, Bruno JF, Duarte CM, Halpern BS, Holding J, Kappel CV, Kiessling W, O'Connor MI, Pandolfi JM, Parmesan C, Schwing FB, Sydeman WJ, Richardson AJ (2011) The pace of shifting climate in marine and terrestrial ecosystems. Science 334(6056):652–655. https://doi.org/10.1126/science.1210288
Burt JA, Camp EF, Enochs IC, Johansen JL, Morgan KM, Riegl B, Hoey AS (2020) Insights from extreme coral reefs in a changing world. Coral Reefs 39(3):495–507. https://doi.org/10.1007/s00338-020-01966-y
Butler IR, Sommer B, Zann M, Zhao J-X, Pandolfi JM (2013) The impacts of flooding on the high-latitude, terrigenoclastic influenced coral reefs of Hervey Bay, Queensland, Australia. Coral Reefs 32(4):1149–1163. https://doi.org/10.1007/s00338-013-1064-7
Camp EF, Schoepf V, Mumby PJ, Hardtke LA, Rodolfo-Metalpa R, Smith DJ, Suggett DJ (2018) The future of coral reefs subject to rapid climate change: lessons from natural extreme environments. Front Mar Sci 5. https://doi.org/10.3389/fmars.2018.00004
Cant J, Salguero-Gómez R, Kim SW, Sims CA, Sommer B, Brooks M, Malcolm HA, Pandolfi JM, Beger M (2021) The projected degradation of subtropical coral assemblages by recurrent thermal stress. J Anim Ecol 90(1):233–247. https://doi.org/10.1111/1365-2656.13340
Cant J, Cook KM, Reimer JD, Mezaki T, Nakamura M, O'Flaherty C, Salguero-Gómez R, Beger M (2022) Transient amplification enhances the persistence of tropicalising coral assemblages in marginal high-latitude environments. Ecography 2022(10):e06156. https://doi.org/10.1111/ecog.06156
Cant J, Reimer JD, Sommer B, Cook KM, Kim SW, Sims CA, Mezaki T, O'Flaherty C, Brooks M, Malcolm HA, Pandolfi JM, Salguero-Gómez R, Beger M (2023) Coral assemblages at higher latitudes favor short-term potential over long-term performance. Ecology 104(9):e4138. https://doi.org/10.1002/ecy.4138
Capdevila P, Stott I, Beger M, Salguero-Gómez R (2020) Towards a comparative framework of demographic resilience. Trends Ecol Evol 35(9):776–786. https://doi.org/10.1016/j.tree.2020.05.001
Capel KCC, Segal B, Bertuol P, Lindner A (2012) Corallith beds at the edge of the tropical South Atlantic. Coral Reefs 31(1):75–75. https://doi.org/10.1007/s00338-011-0818-3
Carter AL, Edwards CB, Fox MD, Amir CG, Eynaud Y, Johnson MD, Lewis LS, Sandin SA, Smith JE (2019) Changes in benthic community composition associated with the outbreak of the corallimorph, *Rhodactis howesii*, at Palmyra Atoll. Coral Reefs 38(6):1267–1279. https://doi.org/10.1007/s00338-019-01841-5
Carter AL, Wilson AMW, Bello M, Hoyos-padilla EM, Inall ME, Ketchum JT, Schurer A, Tudhope AW (2020) Assessing opportunities to support coral reef climate change refugia in MPAs: a case study at the Revillagigedo Archipelago. Mar Policy 112:103769. https://doi.org/10.1016/j.marpol.2019.103769
Castillo KD, Bove CB, Hughes AM, Powell ME, Ries JB, Davies SW (2024) Gene expression plasticity facilitates acclimatization of a long-lived Caribbean coral across divergent reef environments. Sci Rep 14(1):7859. https://doi.org/10.1038/s41598-024-57319-0
Chadwick NE (1991) Spatial distribution and the effects of competition on some temperate Scleractinia and Corallimorpharia. Mar Ecol Prog Ser 70:39–48
Chadwick-Furman NE, Spiegel M (2000) Abundance and clonal replication in the tropical corallimorpharian *Rhodactis rhodostoma*. Invertebr Biol 119(4):351–360. https://doi.org/10.1111/j.1744-7410.2000.tb00103.x
Chaikin S, Riva F, Marshall KE, Lessard J-P, Belmaker J (2024) Marine fishes experiencing high-velocity range shifts may not be climate change winners. Nat Ecol Evol:1–11. https://doi.org/10.1038/s41559-024-02350-7
Chan BKK, Xu G, Kim HK, Park J-H, Kim W (2018) Living with marginal coral communities: diversity and host-specificity in coral-associated barnacles in the northern coral distribution limit of the East China Sea. PLoS One 13(5):e0196309. https://doi.org/10.1371/journal.pone.0196309
Cheal AJ, MacNeil MA, Cripps E, Emslie MJ, Jonker M, Schaffelke B, Sweatman H (2010) Coral–macroalgal phase shifts or reef resilience: links with diversity and functional roles of herbivorous fishes on the Great Barrier Reef. Coral Reefs 29(4):1005–1015. https://doi.org/10.1007/s00338-010-0661-y
Chong F, Sommer B, Stant G, Verano N, Cant J, Lachs L, Johnson ML, Parsons DR, Pandolfi JM, Salguero-Gómez R, Spencer M, Beger M (2023) High-latitude marginal reefs support fewer but bigger corals than their tropical counterparts. Ecography 2023:e06835. https://doi.org/10.1111/ecog.06835
Chong F, Soong GY, Hakim AA, Burke C, De Palmas S, Gösser F, Hsiao WV, Kise H, Nishijima M, Iguchi A, Sommer B, Joyce D, Beger M, Reimer JD (2026). Subtropical specialists dominate a coral range expansion front. Coral Reefs 45:5–19. https://doi.org/10.1007/s00338-024-02601-w
Clay CG, Reimer JD, Sommer B, Cook KM, Mizuyama M, Obuchi M, Kawamura I, Kise H, Beger M (2024) Variation in functional composition of reef fishes along a tropical-to-temperate gradient. J Biogeogr 51(3):454–466. https://doi.org/10.1111/jbi.14768
Clemente S, Rodríguez A, Brito A, Ramos A, Monterroso Ó, Hernández JC (2011) On the occurrence of the hydrocoral *Millepora* (Hydrozoa: Milleporidae) in the subtropical eastern Atlantic (Canary Islands): is the colonization related to climatic events? Coral Reefs 30(1):237–240. https://doi.org/10.1007/s00338-010-0681-7
Coghlan AR, Blanchard JL, Wotherspoon S, Stuart-Smith RD, Edgar GJ, Barrett N, Audzijonyte A (2024) Mean reef fish body size decreases towards warmer waters. Ecol Lett 27(2):e14375. https://doi.org/10.1111/ele.14375
Coleman MA, Bates AE, Stuart-Smith RD, Malcolm HA, Harasti D, Jordan A, Knott NA, Edgar GJ, Kelaher BP (2015) Functional traits reveal early responses in marine reserves following protection from fishing. Divers Distrib 21(8):876–887. https://doi.org/10.1111/ddi.12309
Comeros-Raynal MT, Choat JH, Polidoro BA, Clements KD, Abesamis R, Craig MT, Lazuardi ME, McIlwain J, Muljadi A, Myers RF, Nanola CL Jr (2012) The likelihood of extinction of iconic and dominant herbivores and detritivores of coral reefs: the parrotfishes and surgeonfishes. PLoS One 7(7):e39825. https://doi.org/10.1371/journal.pone.0039825
Concepcion GT, Kahng SE, Crepeau MW, Franklin EC, Coles SL, Toonen RJ (2010) Resolving natural ranges and marine invasions in a globally distributed octocoral (genus Carijoa). Mar Ecol Prog Ser 401:113–127. https://doi.org/10.3354/meps08364
Connell SD, Kroeker KJ, Fabricius KE, Kline DI, Russell BD (2013) The other ocean acidification problem: CO2 as a resource among competitors for ecosystem dominance. Philosophical Transactions of the Royal Society B: Biological Sciences 368(1627):20120442. https://doi.org/10.1098/rstb.2012.0442
Conversi A, Dakos V, Gårdmark A, Ling S, Folke C, Mumby PJ, Greene C, Edwards M, Blenckner T, Casini M, Pershing A, Möllmann C

(2015) A holistic view of marine regime shifts. Philos Trans R Soc Lond B Biol Sci 370(1659):20130279. https://doi.org/10.1098/rstb.2013.0279

Cook KM, Yamagiwa H, Beger M, Masucci GD, Ross S, Lee HYT, Stuart-Smith RD, Reimer JD (2022) A community and functional comparison of coral and reef fish assemblages between four decades of coastal urbanisation and thermal stress. Ecol Evol 12(3):e8736. https://doi.org/10.1002/ece3.8736

Cooney C, Sommer B, Marzinelli EM, Figueira WF (2024) The role of microbial biofilms in range shifts of marine habitat-forming organisms. Trends Microbiol 32(2):190–199. https://doi.org/10.1016/j.tim.2023.07.015

Crane N, Paddack M, Nelson P, Abelson A, Rulmal J, Bernardi G (2016) Corallimorph and Montipora Reefs in Ulithi Atoll, Micronesia: documenting unusual reefs. J Ocean Sci Found 21:10–17. https://doi.org/10.5281/zenodo.51289

Crean AJ, Swearer SE, Patterson HM (2010) Larval supply is a good predictor of recruitment in endemic but not non-endemic fish populations at a high latitude coral reef. Coral Reefs 29(1):137–143. https://doi.org/10.1007/s00338-009-0566-9

Cruz ICS, Waters LG, Kikuchi RKP, Leão ZMAN, Turra A (2018) Marginal coral reefs show high susceptibility to phase shift. Mar Pollut Bull 135:551–561. https://doi.org/10.1016/j.marpolbul.2018.07.043

D'Aloia CC, Naujokaitis-Lewis I, Blackford C, Chu C, Curtis JMR, Darling E, Guichard F, Leroux SJ, Martensen AC, Rayfield B, Sunday JM, Xuereb A, Fortin M-J (2019) Coupled networks of permanent protected areas and dynamic conservation areas for biodiversity conservation under climate change. Front Ecol Evol 7:27. https://doi.org/10.3389/fevo.2019.00027

Dahlke FT, Wohlrab S, Butzin M, Pörtner H-O (2020) Thermal bottlenecks in the life cycle define climate vulnerability of fish. Science 369(6499):65–70. https://doi.org/10.1126/science.aaz3658

Dai (2018) Coastal and shallow marine ecosystem. In: Regional oceanography Taiwan, 2nd edn. National Taiwan University Press

Dakos V, Carpenter SR, van Nes EH, Scheffer M (2015) Resilience indicators: prospects and limitations for early warnings of regime shifts. Philos Trans R Soc Lond B Biol Sci 370(1659):20130263. https://doi.org/10.1098/rstb.2013.0263

Dang H, Lovell CR (2015) Microbial surface colonization and biofilm development in marine environments. Microbiol Mol Biol Rev 80(1):91–138. https://doi.org/10.1128/mmbr.00037-15

Daru BH, Rock BM (2023) Reorganization of seagrass communities in a changing climate. Nat Plants 9(7):1034–1043. https://doi.org/10.1038/s41477-023-01445-6

Dávalos N, Velasco-Cedeño D, Brandt M. First assessment of the physiological performance of nursery corals in the Galápagos Islands: seasonal impacts on *Pocillopora* (Scleractinia: Pocilloporidae) growth. Revista de Biología Tropical 74(S1). In press.

Davies SC, Thompson PL, Gomez C, Nephin J, Knudby A, Park AE, Friesen SK, Pollock LJ, Rubidge EM, Anderson SC, Iacarella JC, Lyons DA, MacDonald A, McMillan A, Ward EJ, Holdsworth AM, Swart N, Price J, Hunter KL (2023) Addressing uncertainty when projecting marine species' distributions under climate change. Ecography 2023(11):e06731. https://doi.org/10.1111/ecog.06731

Davis TR, Champion C, Coleman MA (2021) Climate refugia for kelp within an ocean warming hotspot revealed by stacked species distribution modelling. Mar Environ Res 166:105267. https://doi.org/10.1016/j.marenvres.2021.105267

De Palmas S, Denis V, Ribas-Deulofeu L, Loubeyres M, Woo S, Hwang SJ, Song JI, Chen CA (2015) *Symbiodinium spp.* associated with high-latitude scleractinian corals from Jeju Island, South Korea. Coral Reefs 34(3):919–925. https://doi.org/10.1007/s00338-015-1286-y

De'ath G, Fabricius KE (2000) Classification and regression trees: a powerful yet simple technique for ecological data analysis. Ecology 81(11):3178–3192. https://doi.org/10.1890/0012-9658(2000)081[3178:CARTAP]2.0.CO;2

Dellisanti W, Chung JTH, Chow CFY, Wu J, Wells ML, Chan LL (2021) Experimental techniques to assess coral physiology in situ under global and local stressors: current approaches and novel insights. Front Physiol 12. https://doi.org/10.3389/fphys.2021.656562

den Hartog JC (1980) Caribbean shallow water Corallimorpharia. Zoologische Verhandelingen 176(1):1–83

Denis V, Mezaki T, Tanaka K, Kuo C-Y, Palmas SD, Keshavmurthy S, Chen CA (2013) Coverage, diversity, and functionality of a high-latitude Coral Community (Tatsukushi, Shikoku Island, Japan). PLoS One 8(1):e54330. https://doi.org/10.1371/journal.pone.0054330

Denis V, Soto D, De Palmas S, Lin YTV, Benayahu Y, Huang YM, Liu SL, Chen JW, Chen Q, Sturaro N, Ho MJ, Su Y, Dai CF, Chen CA (2019) Taiwan. In: Loya Y et al (eds) Mesophotic coral ecosystems. Springer International Publishing, Cham

Denis V, Fan T-Y, Hsiao WV, Hwang S-J, Lin YV, Nozawa Y (2020) Idea paper: tracking the distribution of accretive reef communities across the Kuroshio region. Ecol Res 35:595–598

Dennis, L. E., F. Favoretto, E. F. Balart, A. Munguía-Vega, C. Sánchez-Ortiz, and D. A. Paz-García. 2024. Isolation by disturbance: a pattern of genetic structure of the coral Pocillopora grandis in the Gulf of California. Marine Ecology Progress Series 733:43–57. https://doi.org/10.3354/meps14553

Dinerstein E, Vynne C, Sala E, Joshi AR, Fernando S, Lovejoy TE, Mayorga J, Olson D, Asner GP, Baillie JEM, Burgess ND, Burkart K, Noss RF, Zhang YP, Baccini A, Birch T, Hahn N, Joppa LN, Wikramanayake E (2019) A global deal for nature: guiding principles, milestones, and targets. Sci Adv 5(4):eaaw2869. https://doi.org/10.1126/sciadv.aaw2869

Dittami SM, Arboleda E, Auguet J-C, Bigalke A, Briand E, Cárdenas P, Cardini U, Decelle J, Engelen AH, Eveillard D, Gachon CMM, Griffiths SM, Harder T, Kayal E, Kazamia E, Lallier FH, Medina M, Marzinelli EM, Morganti TM, Pons LN, Prado S, Pintado J, Saha M, Selosse M-A, Skillings D, Stock W, Sunagawa S, Toulza E, Vorobev A, Leblanc C, Not F (2021) A community perspective on the concept of marine holobionts: current status, challenges, and future directions. PeerJ 9:e10911. https://doi.org/10.7717/peerj.10911

Dixon AM, Forster PM, Beger M (2021) Coral conservation requires ecological climate-change vulnerability assessments. Front Ecol Environ 19(4):243–250. https://doi.org/10.1002/fee.2312

Dixon AM, Forster PM, Heron SF, Stoner AMK, Beger M (2022) Future loss of local-scale thermal refugia in coral reef ecosystems. PLOS Clim 1(2):1–20. https://doi.org/10.1371/journal.pclm.0000004

Donner SD, Skirving WJ, Little CM, Oppenheimer M, Hoegh-Guldberg O (2005) Global assessment of coral bleaching and required rates of adaptation under climate change. Glob Chang Biol 11(12):2251–2265. https://doi.org/10.1111/j.1365-2486.2005.01073.x

Douglas A (2009) The productivity of corals. In: Oceanography Encyclopedia of Life Support Systems, Oxford, UK

Drury C, Dilworth J, Majerová E, Caruso C, Greer JB (2022) Expression plasticity regulates intraspecific variation in the acclimatization potential of a reef-building coral. Nat Commun 13(1):4790. https://doi.org/10.1038/s41467-022-32452-4

Dunic JC, Baum JK (2017) Size structuring and allometric scaling relationships in coral reef fishes. J Anim Ecol 86(3):577–589. https://doi.org/10.1111/1365-2656.12637

Edwards KF, Litchman E, Klausmeier CA (2013) Functional traits explain phytoplankton community structure and seasonal dynamics in a marine ecosystem. Ecol Lett 16(1):56–63. https://doi.org/10.1111/ele.12012

Eklöf JS, de la Torre-Castro M, Gullström M, Uku J, Muthiga N, Lyimo T, Bandeira SO (2008) Sea urchin overgrazing of seagrasses: a review of current knowledge on causes, consequences, and management. Estuar Coast Shelf Sci 79(4):569–580. https://doi.org/10.1016/j.ecss.2008.05.005

Enochs IC, Studivan MS, Kolodziej G, Foord C, Basden I, Boyd A, Formel N, Kirkland A, Rubin E, Jankulak M, Smith I, Kelble CR, Manzello DP (2023a) Coral persistence despite marginal conditions in the port of Miami. Sci Rep 13(1):6759. https://doi.org/10.1038/s41598-023-33467-7. env.go.jp/content/900489567.pdf

Evans RD, Thomas L, Kennington WJ, Ryan NM, Wilson NG, Richards Z, Lowe RJ, Tuckett C (2021) Population genetic structure of a broadcast-spawning coral across a tropical–temperate transition zone reveals regional differentiation and high-latitude reef isolation. J Biogeogr 48(12):3185–3195. https://doi.org/10.1111/jbi.14280

Ezzat L, Maguer J-F, Grover R, Ferrier-Pagès C (2016) Limited phosphorus availability is the Achilles heel of tropical reef corals in a warming ocean. Sci Rep 6(1):31768. https://doi.org/10.1038/srep31768

Fairfull SJL, Harriott VJ (1999) Succession, space and coral recruitment in a subtropical fouling community. Mar Freshw Res 50(3):235–242. https://doi.org/10.1071/mf97267

Fautin DG (2016) Catalog to families, genera, and species of orders Actiniaria and Corallimorpharia (Cnidaria: Anthozoa). Zootaxa 4145(1):1–449. https://doi.org/10.11646/zootaxa.4145.1.1

Favoretto F, Sánchez C, Aburto-Oropeza O (2022) Warming and marine heatwaves tropicalize rocky reefs communities in the Gulf of California. Prog Oceanogr 206:102838. https://doi.org/10.1016/j.pocean.2022.102838

Feary DA, Pratchett MS, Emslie MJ, Fowler AM, Figueira WF, Luiz OJ, Nakamura Y, Booth DJ (2014) Latitudinal shifts in coral reef fishes: why some species do and others do not shift. Fish Fish 15(4):593–615. https://doi.org/10.1111/faf.12036

Fellegara I, Baird AH, Ward S (2013) Coral reproduction in a high-latitude, marginal reef environment (Moreton Bay, South-East Queensland, Australia). Invertebr Reprod Dev 57(3):219–223. https://doi.org/10.1080/07924259.2012.752766

Fernández-Torres F, Martínez PA, Olalla-Tárraga MÁ (2018) Shallow water ray-finned marine fishes follow Bergmann's rule. Basic Appl Ecol 33:99–110. https://doi.org/10.1016/j.baae.2018.09.002

Ferreira CEL, Floeter SR, Gasparini JL, Ferreira BP, Joyeux JC (2004) Trophic structure patterns of Brazilian reef fishes: a latitudinal comparison. J Biogeogr 31(7):1093–1106. https://doi.org/10.1111/j.1365-2699.2004.01044.x

Fifer JE, Yasuda N, Yamakita T, Bove CB, Davies SW (2022) Genetic divergence and range expansion in a western North Pacific coral. Sci Total Environ 813:152423. https://doi.org/10.1016/j.scitotenv.2021.152423

Figueira WF, Booth DJ (2010) Increasing ocean temperatures allow tropical fishes to survive overwinter in temperate waters. Glob Chang Biol 16(2):506–516. https://doi.org/10.1111/j.1365-2486.2009.01934.x

Filbee-Dexter K, Wernberg T (2018) Rise of turfs: a new battlefront for globally declining kelp forests. Bioscience 68(2):64–76. https://doi.org/10.1093/biosci/bix147

Filbee-Dexter K, Wernberg T (2020) Substantial blue carbon in overlooked Australian kelp forests. Sci Rep 10(1):12341. https://doi.org/10.1038/s41598-020-69258-7

Filbee-Dexter K, Pessarrodona A, Duarte CM, Krause-Jensen D, Hancke K, Smale D, Wernberg T (2023) Seaweed forests are carbon sinks that may help mitigate CO2 emissions: a comment on Gallagher et al. (2022). ICES J Mar Sci 80(6):1814–1819. https://doi.org/10.1093/icesjms/fsad107

Fisheries Agency (2018) Fisheries White paper 2017. Agriculture and Forestry Statistics Association, pp 9–16. (in Japanese)

Fisheries Agency (2020) Fisheries White Paper 2019. Agriculture and Forestry Statistics Association, pp 3–32. (in Japanese)

Fisheries Agency (2023) Summary of the study group on the state of fisheries in response to changes in the marine environment. Ministry of Agriculture, Forestry and Fisheries. (in Japanese)

Floros C (2012) First observation of *Acropora robusta* (Dana, 1846) on the high-latitude coral reefs of South Africa: a species range extension. Western Indian Ocean Journal of Marine Science 11:209–209

Floyd M, Mizuyama M, Obuchi M, Sommer B, Miller MG, Kawamura I, Kise H, Reimer JD, Beger M (2020) Functional diversity of reef molluscs along a tropical-to-temperate gradient. Coral Reefs 39(5):1361–1376. https://doi.org/10.1007/s00338-020-01970-2

Fonseca CR, Ganade G (2001) Species functional redundancy, random extinctions and the stability of ecosystems. J Ecol 89(1):118–125. https://doi.org/10.1046/j.1365-2745.2001.00528.x

Foster NL, Box SJ, Mumby PJ (2008) Competitive effects of macroalgae on the fecundity of the reef-building coral Montastraea annularis. Mar Ecol Prog Ser 367:143–152. https://doi.org/10.3354/meps07594

Fox MD, Williams GJ, Johnson MD, Radice VZ, Zgliczynski BJ, Kelly ELA, Rohwer FL, Sandin SA, Smith JE (2018) Gradients in primary production predict trophic strategies of Mixotrophic corals across spatial scales. Curr Biol 28(21):3355–3363.e4. https://doi.org/10.1016/j.cub.2018.08.057

Fujii M (2020) Impacts of global warming and ocean acidification on marine ecosystems and societies along the coast of Japan. Fisheries Engineering 56(3):191–195. (in Japanese)

Fujita D (2010) Current status and problems of isoyake. Japan Bull Fish Res Agen 32:33–42

Fujiwara M, Simpson A, Torres-Ceron M, Martinez-Andrade F (2022) Life-history traits and temporal patterns in the incidence of coastal fishes experiencing tropicalization. Ecosphere 13(8):e4188. https://doi.org/10.1002/ecs2.4188

Garlapati D, Charankumar B, Ramu K, Madeswaran P, Ramana Murthy MV (2019) A review on the applications and recent advances in environmental DNA (eDNA) metagenomics. Rev Environ Sci Biotechnol 18(3):389–411. https://doi.org/10.1007/s11157-019-09501-4

GBIF.org (2024) GBIF Occurrence Download https://doi.org/10.15468/dl.jmq7p5

Giglio VJ, Bender MG, Zapelini C, Ferreira CEL (2017) The end of the line? Rapid depletion of a large-sized grouper through spearfishing in a subtropical marginal reef. Perspect Ecol Conserv 15(2):115–118. https://doi.org/10.1016/j.pecon.2017.03.006

Gissi E, Manea E, Mazaris AD, Fraschetti S, Almpanidou V, Bevilacqua S, Coll M, Guarnieri G, Lloret-Lloret E, Pascual M, Petza D (2021) A review of the combined effects of climate change and other local human stressors on the marine environment. Sci Tot Env 755:142564. https://doi.org/10.1016/j.scitotenv.2020.142564

Glynn PW, Wellington GM (1983) Corals and coral reefs of the Galápagos Islands. University of California Press, Berkeley, p 330

Gomez R, Tabalanza T, Toyama K, Herwening LK, Azuma-Malsol M, Fujiwara N, Yin X, Nakamura T (2024) Fish mortality following sudden cold snap in Okinawa-Jima Island, Japan. Ichthyolog Res 71(1):200–204. https://doi.org/10.1007/s10228-023-00914-4

Gonzalez A, Ronce O, Ferriere R, Hochberg ME (2013) Evolutionary rescue: an emerging focus at the intersection between ecology and evolution. Philos Trans R Soc Lond Ser B Biol Sci 368(1610):20120404. https://doi.org/10.1098/rstb.2012.0404

González-Espinosa PC, Donner SD (2023) Cloudiness delays projected impact of climate change on coral reefs. PLOS Clim 2(2):e0000090. https://doi.org/10.1371/journal.pclm.0000090

González-Pech RA, Hughes DJ, Strudwick P, Lewis BM, Booth DJ, Figueira WF, Sommer B, Suggett DJ, Matthews J (2022) Physiological factors facilitating the persistence of *Pocillopora aliciae* and *Plesiastrea versipora* in temperate reefs of South-Eastern Australia under ocean warming. Coral Reefs 41(12):1–15. https://doi.org/10.1007/s00338-022-02277-0

Goodkin NF, Switzer AD, McCorry D, DeVantier L, True JD, Hughen KA, Angeline N, Yang TT (2011) Coral communities of Hong

Kong: long-lived corals in a marginal reef environment. Mar Ecol Prog Ser 426:185–196. https://doi.org/10.3354/meps09019

Goodwin C, Rodolfo-Metalpa R, Picton B, Hall-Spencer JM (2014) Effects of ocean acidification on sponge communities. Mar Ecol 35(s1):41–49. https://doi.org/10.1111/maec.12093

Gorman D, Horta P, Flores AAV, Turra A, Berchez FA d S, Batista MB, Lopes Filho ES, Melo MS, Ignacio BL, Carneiro IM, Villaça RC, Széchy MTM (2020) Decadal losses of canopy-forming algae along the warm temperate coastline of Brazil. Glob Chang Biol 26(3):1446–1457. https://doi.org/10.1111/gcb.14956

Goyen S, Camp EF, Fujise L, Lloyd A, Nitschke MR, LaJeunensse T, Kahlke T, Ralph PJ, Suggett D (2019) Mass coral bleaching of P. versipora in Sydney harbour driven by the 2015–2016 heatwave. Coral Reefs 38(4):815–830. https://doi.org/10.1007/s00338-019-01797-6

Graham NA, Cinner JE, Norström AV, Nyström M (2014) Coral reefs as novel ecosystems: embracing new futures. Curr Opin Environ Sustain 7:9–14. https://doi.org/10.1016/j.cosust.2013.11.023

Greenstein BJ, Pandolfi JM (2008) Escaping the heat: range shifts of reef coral taxa in coastal Western Australia. Glob Chang Biol 14(3):513–528. https://doi.org/10.1111/j.1365-2486.2007.01506.x

Grieve BD, Curchitser EN, Rykaczewski RR (2016) Range expansion of the invasive lionfish in the Northwest Atlantic with climate change. Mar Ecol Prog Ser 546:225–237. https://doi.org/10.3354/meps11638

Grillo AC, Candido CF, Giglio VJ, Longo GO (2021) Unusual high coral cover in a Southwestern Atlantic subtropical reef. Mar Biodivers 51(5):77. https://doi.org/10.1007/s12526-021-01221-9

Grottoli AG, Rodrigues LJ, Juarez C (2004) Lipids and stable carbon isotopes in two species of Hawaiian corals, Porites compressa and Montipora verrucosa, following a bleaching event. Mar Biol 145(3):621–631. https://doi.org/10.1007/s00227-004-1337-3

Grove LJW, Blondeau J, Cain E, Davis IM, Edwards KF, Groves SH, Hile SD, Langwiser C, Siceloff L, Swanson DW, Towle EK (2022) National coral reef monitoring program biological monitoring summary Florida: 2020–2021. https://doi.org/10.25923/9jns-v916

Guinotte JM, Buddemeier RW, Kleypas JA (2003) Future coral reef habitat marginality: temporal and spatial effects of climate change in the Pacific basin. Coral Reefs 22(4):551–558. https://doi.org/10.1007/s00338-003-0331-4

Halfar J, Godinez-Orta L, Riegl B, Valdez-Holguin JE, Borges JM (2005) Living on the edge: high-latitude Porites carbonate production under temperate eutrophic conditions. Coral Reefs 24(4):582–592. https://doi.org/10.1007/s00338-005-0029-x

Hall-Spencer JM, Belfiore G, Tomatsuri M, Porzio L, Harvey BP, Agostini S, Kon K (2022) Decreased diversity and abundance of marine invertebrates at CO2 seeps in warm-temperate Japan. Zool Sci 39(1):41–51. https://doi.org/10.2108/zs210061

Hammerman NM, Roff G, Lybolt T, Eyal G, Pandolfi JM (2022) Unraveling Moreton Bay reef history: an urban high-latitude setting for coral development. Front Ecol Evol 10. https://doi.org/10.3389/fevo.2022.884850

Harii S, Kayanne H, Takigawa H, Hayashibara T, Yamamoto M (2002) Larval survivorship, competency periods and settlement of two brooding corals, *Heliopora coerulea* and *Pocillopora damicornis*. Mar Biol 141(1):39–46. https://doi.org/10.1007/s00227-002-0812-y

Harriott V, Banks S (2002) Latitudinal variation in coral communities in eastern Australia: a qualitative biophysical model of factors regulating coral reefs. Coral Reefs 21(1):83–94. https://doi.org/10.1007/s00338-001-0201-x

Harriott VJ, Smith SDA, Harrison PL (1994) Patterns of coral community structure of subtropical reefs in the Solitary Islands Marine Reserve, Eastern Australia. Marine Ecol Progr Ser 109(1):67–76

Harvey BP, Allen R, Agostini S, Hoffmann LJ, Kon K, Summerfield TC, Wada S, Hall-Spencer JM (2021) Feedback mechanisms stabilise degraded turf algal systems at a CO2 seep site. Commun Biol 4(1):1–10. https://doi.org/10.1038/s42003-021-01712-2

Hata T, Madin JS, Cumbo VR, Denny M, Figueiredo J, Harii S, Thomas CJ, Baird AH (2017) Coral larvae are poor swimmers and require fine-scale reef structure to settle. Sci Rep 7(1):2249. https://doi.org/10.1038/s41598-017-02402-y

Hatcher BG (1988) Coral reef primary productivity: a beggar's banquet. Trends Ecol Evol 3(5):106–111

Häussermann V, Försterra G (2005) Distribution patterns of Chilean shallow-water sea anemones (Cnidaria: Anthozoa: Actiniaria, Corallimorpharia); with a discussion of the taxonomic and zoogeographic relationships between the actinofauna of the south East Pacific, the South West Atlantic and the Antarctic. Sci Mar 69(S2):91–102. https://doi.org/10.3989/scimar.2005.69s291

Hellberg M, Taylor M (2002) Genetic analysis of sexual reproduction in the dendrophylliid coral *Balanophyllia elegans*. Mar Biol 141(4):629–637. https://doi.org/10.1007/s00227-002-0861-2

Hickman CP (2008) A field guide to corals and other radiates of Galápagos. Sugar Spring Press, Lexington

Hobbs RJ, Higgs E, Harris JA (2009) Novel ecosystems: implications for conservation and restoration. Trends Ecol Evol 24(11):599–605. https://doi.org/10.1016/j.tree.2009.05.012

Hobbs RJ, Valentine LE, Standish RJ, Jackson ST (2018) Movers and stayers: novel assemblages in changing environments. Trends Ecol Evol 33(2):116–128. https://doi.org/10.1016/j.tree.2017.11.001

Hobday AJ, Pecl GT (2014) Identification of global marine hotspots: sentinels for change and vanguards for adaptation action. Rev Fish Biol Fish 24(2):415–425. https://doi.org/10.1007/s11160-013-9326-6

Hoegh-Guldberg O (1999). Climate change, coral bleaching and the future of the world's coral reefs. Mar Freshw Res 50(8): 839–866. DOI: https://doi.org/10.1071/MF99078 1323-1650/99/080839

Hoey AS, Pratchett MS, Cvitanovic C (2011) High macroalgal cover and low coral recruitment undermines the potential resilience of the world's southernmost coral reef assemblages. PLoS One 6(10):e25824. https://doi.org/10.1371/journal.pone.0025824

Holland MM, Smith JA, Everett JD, Vergés A, Suthers IM (2020) Latitudinal patterns in trophic structure of temperate reef-associated fishes and predicted consequences of climate change. Fish Fish 21(6):1092–1108. https://doi.org/10.1111/faf.12488

Hong Kong Observatory (2024) Climate of Hong Kong. https://www.hko.gov.hk/en/cis/climahk.htm

Houlbrèque F, Ferrier-Pagès C (2009) Heterotrophy in tropical Scleractinian corals. Biol Rev 84(1):1–17. https://doi.org/10.1111/j.1469-185X.2008.00058.x

Howells EJ, Bauman AG, Vaughan GO, Hume BCC, Voolstra CR, Burt JA (2020) Corals in the hottest reefs in the world exhibit symbiont fidelity not flexibility. Mol Ecol 29(5):899–911. https://doi.org/10.1111/mec.15372

Hsiao WV, Lin YV, Lin H-T, Denis V (2021) Learning from differences: abiotic determinism of benthic communities in Northern Taiwan. Mar Environ Res 170:105361. https://doi.org/10.1016/j.marenvres.2021.105361

Hughes TP, Kerry JT, Álvarez-Noriega M, Álvarez-Romero JG, Anderson KD, Baird AH, Babcock RC, Beger M, Bellwood DR, Berkelmans R, Bridge TC, Butler IR, Byrne M, Cantin NE, Comeau S, Connolly SR, Cumming GS, Dalton SJ, Diaz-Pulido G, Eakin CM, Figueira WF, Gilmour JP, Harrison HB, Heron SF, Hoey AS, Hobbs J-PA, Hoogenboom MO, Kennedy EV, Kuo C, Lough JM, Lowe RJ, Liu G, McCulloch MT, Malcolm HA, McWilliam MJ, Pandolfi JM, Pears RJ, Pratchett MS, Schoepf V, Simpson T, Skirving WJ, Sommer B, Torda G, Wachenfeld DR, Willis BL, Wilson SK (2017) Global warming and recurrent mass bleaching of corals. Nature 543(7645):373–377. https://doi.org/10.1038/nature21707

Hughes TP, Kerry JT, Baird AH, Connolly SR, Dietzel A, Eakin CM, Heron SF, Hoey AS, Hoogenboom MO, Liu G, McWilliam MJ, Pears RJ, Pratchett MS, Skirving WJ, Stella JS, Torda G (2018) Global warming transforms coral reef assemblages. Nature 556(7702):492–496. https://doi.org/10.1038/s41586-018-0041-2

Hwang S-I, Kim D-K, Sung B-J, Jun S-K, Bae J-I, Jeon B-H (2017a) Effects of climate change on whitening event proliferation the coast of Jeju. Korean J Environ Ecol 31(6):529–536. https://doi.org/10.13047/KJEE.2017.31.6.529

iNaturalist. Corales Hermatípicos | Galápagos (2016) iNaturalist. https://www.inaturalist.org/projects/corales-hermatipicos-galapagos

Inoue S, Kayanne H, Yamamoto S, Kurihara H (2013) Spatial community shift from hard to soft corals in acidified water. Nat Clim Change 3(7):683–687. https://doi.org/10.1038/nclimate1855

Ishikuro T, Misawa R (2024) Northernmost record of *Pomacanthus imperator* (Pomacanthidae) from Aomori prefecture, northern Japan. Ichthy, Natural History of Fishes of Japan 42:34–37

Januchowski-Hartley FA, Graham NAJ, Wilson SK, Jennings S, Perry CT (2017) Drivers and predictions of coral reef carbonate budget trajectories. Proc R Soc B Biol Sci 284(1847):20162533. https://doi.org/10.1098/rspb.2016.2533

Japan Meteorological Agency (2024) Sea surface temperature (around Japan) (jma.go.jp). Accessed 30/07/2024

Johannes R, Wiebe W, Crossland C, Rimmer D, Smith S (1983) Latitudinal limits of coral reef growth. Marine ecology progress series Oldendorf 11(2):105–111

Jones NP, Gilliam DS (2024) Temperature and local anthropogenic pressures limit stony coral assemblage viability in Southeast Florida. Mar Pollut Bull 200:116098. https://doi.org/10.1016/j.marpolbul.2024.116098

Jones NP, Figueiredo J, Gilliam DS (2020a) Thermal stress-related spatiotemporal variations in high-latitude coral reef benthic communities. Coral Reefs 39(6):1661–1673. https://doi.org/10.1007/s00338-020-01994-8

Jones R, Giofre N, Luter HM, Neoh TL, Fisher R, Duckworth A (2020b) Responses of corals to chronic turbidity. Sci Rep 10(1):4762. https://doi.org/10.1038/s41598-020-61712-w

Jones NP, Kabay L, Semon Lunz K, Gilliam DS (2021) Temperature stress and disease drives the extirpation of the threatened pillar coral, *Dendrogyra cylindrus*, in Southeast Florida. Sci Rep 11(1):14113. https://doi.org/10.1038/s41598-021-93111-0

Jones N, Ruzicka R, Colella M, Pratchett MS, Gilliam D (2022) Frequent disturbances and chronic pressures constrain stony coral recovery on Florida's Coral Reef. Coral Reefs 41:1665–1679. https://doi.org/10.1007/s00338-022-02313-z

Jonsson A (2021) Robustness of life histories to environmental variability in complex versus simple life cycles. Theor Ecol 14(2):335–344. https://doi.org/10.1007/s12080-021-00501-1

Jørgensen T, Kjær KH, Haile J, Rasmussen M, Boessenkool S, Andersen K, Coissac E, Taberlet P, Brochmann C, Orlando L, Gilbert MTP, Willerslev E (2012) Islands in the ice: detecting past vegetation on Greenlandic nunataks using historical records and sedimentary ancient DNA meta-barcoding. Mol Ecol 21(8):1980–1988. https://doi.org/10.1111/j.1365-294X.2011.05278.x

Jouffray J-B, Nyström M, Norström AV, Williams ID, Wedding LM, Kittinger JN, Williams GJ (2015) Identifying multiple coral reef regimes and their drivers across the Hawaiian archipelago. Philos Trans R Soc B Biol Sci 370(1659):20130268. https://doi.org/10.1098/rstb.2013.0268

Jurriaans S, Hoogenboom MO (2019) Thermal performance of scleractinian corals along a latitudinal gradient on the Great Barrier Reef. Philos Trans R Soc B Biol Sci 374(1778):20180546. https://doi.org/10.1098/rstb.2018.0546

Juszkiewicz DJ, White NE, Stolarski J, Benzoni F, Arrigoni R, Hoeksema BW, Wilson NG, Bunce M, Richards ZT (2022) Phylogeography of recent *Plesiastrea* (Scleractinia: Plesiastreidae) based on an integrated taxonomic approach. Mol Phylogenet Evol 172:107469. https://doi.org/10.1016/j.ympev.2022.107469

Kakuma S, Yanagi T, Sato T (eds) (2018) An introduction to Satoumi studies. Bensei Publishing. (in Japanese)

Kalmus P, Ekanayaka A, Kang E, Baird M, Gierach M (2022) Past the precipice? Projected coral habitability under global heating. Earth's Future 10(5):e2021EF002608. https://doi.org/10.1029/2021EF002608

Kamezaki M, Higa M, Hirose M, Suda S, Reimer JD (2013) Different zooxanthellae types in populations of the zoanthid *Zoanthus sansibaricus* along depth gradients in Okinawa, Japan. Mar Biodiv 43(1):61–70. https://doi.org/10.1007/s12526-012-0119-2

Kawecki TJ (2008) Adaptation to marginal habitats. Ann Rev Ecol Evolut Syst 39:321–342. https://doi.org/10.1146/annurev.ecolsys.38.091206.095622

Keshavmurthy S, Beals M, Hsieh HJ, Choi K-S, Chen CA (2021) Physiological plasticity of corals to temperature stress in marginal coral communities. Sci Total Environ 758:143628. https://doi.org/10.1016/j.scitotenv.2020.143628

Keshavmurthy S, Mezaki T, Reimer JD, Choi K-S, Chen CA (2023) Succession and emergence of corals in high-latitude (temperate) areas of eastern Asia into the future. In: Takeuchi I, Yamashiro H (eds) Coral reefs of Eastern Asia under anthropogenic impacts. Springer International Publishing, Cham, pp 53–71

Kim S, Sampayo E, Sommer B, Sims C, Gómez-Cabrera M, Dalton S, Beger M, Malcolm H, Ferrari R, Fraser N, Figueira W, Smith S, Heron S, Baird A, Byrne M, Eakin C, Edgar R, Hughes T, Kyriacou N, Liu G, Matis P, Skirving W, Pandolfi J (2019) Refugia under threat: mass bleaching of coral assemblages in high-latitude eastern Australia. Glob Chang Biol 25:3918–3931. https://doi.org/10.1111/gcb.14772

Kim SW, Sommer B, Beger M, Pandolfi JM (2023) Regional and global climate risks for reef corals: incorporating species-specific vulnerability and exposure to climate hazards. Glob Chang Biol 29(14):4140–4151. https://doi.org/10.1111/gcb.16739

Klassen JL (2018) Defining microbiome function. Nat Microbiol 3(8):864–869. https://doi.org/10.1038/s41564-018-0189-4

Kleypas JA, McManus JW, Meñez LAB (1999) Environmental limits to coral reef development: where do we draw the line? Am Zool 39(1):146–159. https://doi.org/10.1093/icb/39.1.146

Kroeker KJ, Kordas RL, Crim RN, Singh GG (2010) Meta-analysis reveals negative yet variable effects of ocean acidification on marine organisms. Ecol Lett 13(11):1419–1434. https://doi.org/10.1111/j.1461-0248.2010.01518.x

Kuempel CD, Altieri AH (2017) The emergent role of small-bodied herbivores in pre-empting phase shifts on degraded coral reefs. Sci Rep 7:39670. https://doi.org/10.1038/srep39670

Kuguru BL, Mgaya YD, Öhman MC, Wagner GM (2004) The reef environment and competitive success in the Corallimorpharia. Mar Biol 145(5):875–884. https://doi.org/10.1007/s00227-004-1376-9

Kumagai N (2020) Changes in the geographic distribution of seaweed bed communities due to climate change. Algae Jpn J Phycol (Sôrui) 68:91–97. (in Japanese)

Kumagai NH, García Molinos J, Yamano H, Takao S, Fujii M, Yamanaka Y (2018) Ocean currents and herbivory drive macroalgae-to-coral community shift under climate warming. Proc Natl Acad Sci 115(36):8990–8995. https://doi.org/10.1073/pnas.1716826115

Kuo C-Y, Keshavmurthy S, Huang Y-Y, Ho M-J, Hsieh HJ, Hsiao A-T, Lo W-C, Hsin Y-C, Chen CA (2023) Transitional coral ecosystem of Taiwan in the era of changing climate. In: Takeuchi I, Yamashiro H (eds) Coral reefs of Eastern Asia under anthropogenic impacts. Springer International Publishing, Cham, pp 7–35

Kurihara H, Watanabe A, Tsugi A, Mimura I, Hongo C, Kawai T, Reimer JD, Kimoto K, Gouezo M, Golbuu Y (2021) Potential local adaptation of corals at acidified and warmed Nikko Bay,

Palau. Sci Rep 11(1):11192. https://doi.org/10.1038/s41598-021-90614-8

Kuwabara H, Akeda S, Kobayashi S, Takeshita A, Yamashita H, Kido K (2006) Prediction of changes in the distribution area of Japan's marine organisms due to global warming. Glob Environ 11(1):49–57. (in Japanese). https://cir.nii.ac.jp/crid/1520572357627287808

La Motta LM, Padula MP, Sommer B, Camp EF, Matthews JL (2024) Diversity of lipid profiles of Symbiodiniaceae under temperature and nutrient stress. Front Protistol 2:1320353. https://doi.org/10.3389/frpro.2024.1320353

Lachs L, Sommer B, Cant J, Hodge JM, Malcolm HA, Pandolfi JM, Beger M (2021) Linking population size structure, heat stress and bleaching responses in a subtropical endemic coral. Coral Reefs 40(3):777–790. https://doi.org/10.1007/s00338-021-02081-2

LaJeunesse TC, Parkinson JE, Gabrielson PW, Jeong HJ, Reimer JD, Voolstra CR, Santos SR (2018) Systematic revision of Symbiodiniaceae highlights the antiquity and diversity of coral endosymbionts. Curr Biol 28(16):2570–2580.e6. https://doi.org/10.1016/j.cub.2018.07.008

Lalas JAA, Jamodiong EA, Reimer JD (2024) Spatial patterns of soft coral (Octocorallia) assemblages in the shallow coral reefs of Okinawa Island, Ryukyu archipelago, Japan: dominance on highly disturbed reefs. Reg Stud Mar Sci 71:103405. https://doi.org/10.1016/j.rsma.2024.103405

Lambert G (2002) Nonindigenous ascidians in tropical Waters. Pac Sci 56(3):291–298

Lands Department (2024) Hong Kong Geographic Data (as at October 2023). www.landsd.gov.hk/en/resources/mapping-information/hk-geographic-data.html

Lasker HR, Bramanti L, Tsounis G, Edmunds PJ (2020) Chapter thirteen – the rise of octocoral forests on Caribbean reefs. In: Riegl BM (ed) Advances in marine biology. Academic, pp 361–410

Lau YW, Stokvis FR, Imahara Y, Reimer JD (2019) The stoloniferous octocoral, *Hanabira yukibana*, gen. nov., sp. nov., of the southern Ryukyus has morphological and symbiont variation. Contrib Zool 88(1):54–77. https://doi.org/10.1163/18759866-20191355

Laureto LMO, Cianciaruso MV, Samia DSM (2015) Functional diversity: an overview of its history and applicability. Nat Conserv 13(2):112–116. https://doi.org/10.1016/j.ncon.2015.11.001

Lenfest Ocean Program (2024) MPAShift: research to understand climate-driven shifts in Species and Habitats in Japan's MPAs. [Online]. https://www.lenfestocean.org

Leriorato JC, Nakamura Y (2019) Unpredictable extreme cold events: a threat to range-shifting tropical reef fishes in temperate waters. Mar Biol 166(8):110. https://doi.org/10.1007/s00227-019-3557-6

Lien Y-T, Keshavmurthy S, Nakano Y, Plathong S, Huang H, Hsu C-M, Fukami H, Yamashita Y, Hsieh HJ, Wang J-T, Chen CA (2013) Host genetics and *Symbiodinium* D diversity in a stress-tolerant scleractinian coral, *Oulastrea crispata*, in the West Pacific. Mar Ecol Prog Ser 473:163–177. https://doi.org/10.3354/meps10041

Liew YJ, Howells EJ, Wang X, Michell CT, Burt JA, Idaghdour Y, Aranda M (2020) Intergenerational epigenetic inheritance in reef-building corals. Nat Clim Chang 10(3):254–259. https://doi.org/10.1038/s41558-019-0687-2

Lin H-Y, Costello MJ (2023) Body size and trophic level increase with latitude, and decrease in the deep-sea and Antarctica, for marine fish species. PeerJ 11:e15880. https://doi.org/10.7717/peerj.15880

Lin S-M, Tseng L-C, Ang PO, Bolton J, Liu L-C (2018) Long-term study on seasonal changes in floristic composition and structure of marine macroalgal communities along the coast of Northern Taiwan, southern East China Sea. Mar Biol 165(5):83. https://doi.org/10.1007/s00227-018-3344-9

Lin YV, Hsiao WV, Chen W-J, Denis V (2022) Habitat change and its consequences on reef fish specialization in biogeographic transition zones. J Biogeogr 49(8):1549–1561. https://doi.org/10.1111/jbi.14450

Ling SD, Keane JP (2024) Climate-driven invasion and incipient warnings of kelp ecosystem collapse. Nat Commun 15(1):400. https://doi.org/10.1038/s41467-023-44543-x

Ling SD, Johnson CR, Frusher SD, Ridgway KR (2009a) Overfishing reduces resilience of kelp beds to climate-driven catastrophic phase shift. Proc Natl Acad Sci 106(52):22341–22345. https://doi.org/10.1073/pnas.0907529106

Ling SD, Johnson CR, Ridgway K, Hobday AJ, Haddon M (2009b) Climate-driven range extension of a sea urchin: inferring future trends by analysis of recent population dynamics. Glob Chang Biol 15(3):719–731. https://doi.org/10.1111/j.1365-2486.2008.01734.x

Ling SD, Barrett NS, Edgar GJ (2018) Facilitation of Australia's southernmost reef-building coral by sea urchin herbivory. Coral Reefs 37(4):1053–1073. https://doi.org/10.1007/s00338-018-1728-4

Lirman D, Schopmeyer S, Manzello D, Gramer LJ, Precht WF, Muller-Karger F, Banks K, Barnes B, Bartels E, Bourque A, Byrne J, Donahue S, Duquesnel J, Fisher L, Gilliam D, Hendee J, Johnson M, Maxwell K, McDevitt E, Monty J, Rueda D, Ruzicka R, Thanner S (2011) Severe 2010 cold-water event caused unprecedented mortality to corals of the Florida reef tract and reversed previous survivorship patterns. PLoS One 6(8):e23047. https://doi.org/10.1371/journal.pone.0023047

Longo GO, Ferreira CEL, Floeter SR (2014) Herbivory drives large-scale spatial variation in reef fish trophic interactions. Ecol Evol 4(23):4553–4566. https://doi.org/10.1002/ece3.1310

López C, Moreno S, Brito A, Clemente S (2020 Feb 5) Distribution of zooxanthellate zoantharians in the Canary Islands: potential indicators of ocean warming. Estuar Coast Shelf Sci 233:106519. https://doi.org/10.1016/j.ecss.2019.106519

Lough JM, Hobday AJ (2011) Observed climate change in Australian marine and freshwater environments. Mar Freshw Res 62(9):984–999. https://doi.org/10.1071/MF10272

Luza AL, Quimbayo JP, Ferreira CEL, Floeter SR, Francini-Filho RB, Bender MG, Longo GO (2022) Low functional vulnerability of fish assemblages to coral loss in Southwestern Atlantic marginal reefs. Sci Rep 12(1):17164. https://doi.org/10.1038/s41598-022-20919-9

Madin J, Allen A, Baird A, Pandolfi J, Sommer B (2016) Scope for latitudinal extension of reef corals is species specific. Front Biogeogr 8(1). https://doi.org/10.21425/F5FBG29328

Maina J, De Moel H, Zinke J, Madin J, McClanahan T, Vermaat JE (2013) Human deforestation outweighs future climate change impacts of sedimentation on coral reefs. Nat Commun 4(1986):1–7. https://doi.org/10.1038/ncomms2986

Makino K (2017) Japan's marine protected areas and sustainable fisheries. Fisheries Promotion 591:75. (in Japanese)

Makino A, Yamano H, Beger M, Klein CJ, Yara Y, Possingham HP (2014) Spatio-temporal marine conservation planning to support high-latitude coral range expansion under climate change. Divers Distrib 20(8):859–871. https://doi.org/10.1111/ddi.12184

Malcolm H, Scott A (2016) Range extensions in anemonefishes and host sea anemones in eastern Australia: potential constraints to tropicalisation. Mar Freshw Res 68(7):1224–1232. https://doi.org/10.1071/MF15420

Malcolm HA, Davies PL, Jordan A, Smith SDA (2011) Variation in sea temperature and the East Australian current in the Solitary Islands region between 2001–2008. Deep-Sea Res II Top Stud Oceanogr 58(5):616–627. https://doi.org/10.1016/j.dsr2.2010.09.030

Manzello DP et al (2014) Galápagos coral reef persistence after ENSO warming across an acidification gradient. Geophys Res Lett 41:9001–9008

Martello MF, Bleuel J, Pennino MG, Longo GO (2024) Projected climate-driven shifts in coral distribution indicate tropicalisation of Southwestern Atlantic reefs. Divers Distrib 30(8):e13851. https://doi.org/10.1111/ddi.13851

Martín-Arjona A, Muñoz-Caballero A, Serrano A, Díaz-Viñolas D, Urra J (2024) The thermophilic sea anemone Telmatactis cricoides

(Cnidaria, Hexacorallia) in the western Mediterranean: filling gaps in the knowledge of the distribution. J Mar Biol Assoc U K 104:e9. https://doi.org/10.1017/S0025315424000043

Martin-Garin B, Montaggioni LF (2023) The modern times. In: Martin-Garin B, Montaggioni LF (eds) Corals and reefs: from the beginning to an uncertain future. Springer International Publishing, Cham, pp 31–78

Mazzocchi MG, Di Capua I, Kokoszka F, Margiotta F, d'Alcalà MR, Sarno D, Zingone A, Licandro P (2023) Coastal mesozooplankton respond to decadal environmental changes via community restructuring. Mar Ecol 44(3):e12746. https://doi.org/10.1111/maec.12746

McCook L, Jompa J, Diaz-Pulido G (2001) Competition between corals and algae on coral reefs: a review of evidence and mechanisms. Coral Reefs 19(4):400–417. https://doi.org/10.1007/s003380000129

McLaren E, Sommer B, Pandolfi JM, Beger M, Byrne M (2024) Taxa-dependent temporal trends in the abundance and size of sea urchins in subtropical eastern Australia. Ecol Evol 14(5):e11412. https://doi.org/10.1002/ece3.11412

Mellin C, Bradshaw CJA, Meekan MG, Caley MJ (2010) Environmental and spatial predictors of species richness and abundance in coral reef fishes. Glob Ecol Biogeogr 19(2):212–222. https://doi.org/10.1111/j.1466-8238.2009.00513.x

Messer LF, Ostrowski M, Doblin MA, Petrou K, Baird ME, Ingleton T, Bissett A, Van de Kamp J, Nelson T, Paulsen I, Bodrossy L, Fuhrman JA, Seymour JR, Brown MV (2020) Microbial tropicalization driven by a strengthening western ocean boundary current. Glob Chang Biol 26(10):5613–5629. https://doi.org/10.1111/gcb.15257

Mezaki T, Kubota S (2012) Changes of hermatypic coral community in coastal sea area of Kochi, high-latitude, Japan. Aquabiology 34(4):332

Miller KJ, Ayre DJ (2008) Protection of genetic diversity and maintenance of connectivity among reef corals within marine protected areas. Conserv Biol 22(5):1245–1254. https://doi.org/10.1111/j.1523-1739.2008.00985.x

Miller MGR, Reimer JD, Sommer B, Cook KM, Pandolfi JM, Obuchi M, Beger M (2023) Temperate functional niche availability not resident-invader competition shapes tropicalisation in reef fishes. Nat Commun 14(1):2181. https://doi.org/10.1038/s41467-023-37550-5

Ministry of Agriculture, Forestry and Fisheries of Japan (MoAFF) (2022) FY2022 Trends in Fisheries index-220.pdf (maff.go.jp)

Ministry of Environment (MoE) (2011) Marine biodiversity conservation strategy

Ministry of the Environment (MoE) (2020a) Climate change impact assessment report: overview. (in Japanese)

Ministry of the Environment (MoE) (2020b) FY2019 regional adaptation consortium national management and research project commissioned Work report. Climate Change Adaptation Office, General Affairs Division, Global Environment Bureau, Ministry of the Environment. (in Japanese)

Mizerek TL, Baird AH, Beaumont LJ, Madin JS (2016) Environmental tolerance governs the presence of reef corals at latitudes beyond reef growth. Glob Ecol Biogeogr 25(8):979–987. https://doi.org/10.1111/geb.12459

Mizerek TL, Madin JS, Benzoni F, Huang D, Luiz OJ, Mera H, Schmidt-Roach S, Smith SDA, Sommer B, Baird AH (2021) No evidence for tropicalization of coral assemblages in a subtropical climate change hot spot. Coral Reefs 40(5):1451–1461. https://doi.org/10.1007/s00338-021-02167-x

Moriarty T, Leggat W, Heron SF, Steinberg R, Ainsworth TD (2023) Bleaching, mortality and lengthy recovery on the coral reefs of Lord Howe Island. The 2019 Marine heatwave suggests an uncertain future for high-latitude ecosystems. PLOS Climate 2(4):e0000080. https://doi.org/10.1371/journal.pclm.0000080

Morrow KM, Bromhall K, Motti CA, Munn CB, Bourne DG (2016) Allelochemicals produced by Brown Macroalgae of the Lobophora genus are active against coral larvae and associated bacteria, supporting pathogenic shifts to vibrio dominance. Appl Environ Microbiol 83(1):e02391–e02316. https://doi.org/10.1128/AEM.02391-16

Morse NB, Pellissier PA, Cianciola EN, Brereton RL, Sullivan MM, Shonka NK, Wheeler TB, McDowell WH (2014) Novel ecosystems in the Anthropocene: a revision of the novel ecosystem concept for pragmatic applications. Ecol Soc 19(2):1–10. http://www.jstor.org/stable/26269579

Mouillot D, Villéger S, Parravicini V, Kulbicki M, Arias-González JE, Bender M, Chabanet P, Floeter SR, Friedlander A, Vigliola L, Bellwood DR (2014) Functional over-redundancy and high functional vulnerability in global fish faunas on tropical reefs. Proc Natl Acad Sci 111(38):13757–13762. https://doi.org/10.1073/pnas.1317625111

Muenzel D, Bani A, De Brauwer M, Stewart E, Djakiman C, Halwi, Purnama R, Yusuf S, Santoso P, Hukom FD, Struebig M, Jompa J, Limmon G, Dumbrell A, Beger M (2024) Combining environmental DNA and visual surveys can inform conservation planning for coral reefs. Proc Natl Acad Sci 121(17):e2307214121. https://doi.org/10.1073/pnas.2307214121

Muir PR, Wallace CC, Done T, Aguirre JD (2015) Limited scope for latitudinal extension of reef corals. Science 348(6239):1135–1138. https://doi.org/10.1126/science.1259911

Nakabayashi A, Yamakita T, Nakamura T, Aizawa H, Kitano YF, Iguchi A, Yamano H, Nagai S, Agostini S, Teshima KM, Yasuda N (2019) The potential role of temperate Japanese regions as refugia for the coral *Acropora hyacinthus* in the face of climate change. Sci Rep 9(1):1892. https://doi.org/10.1038/s41598-018-38333-5

Nakajima Y, Zayasu Y, Shinzato C, Satoh N, Mitarai S (2016) Genetic differentiation and connectivity of morphological types of the broadcast-spawning coral Galaxea fascicularis in the Nansei Islands, Japan. Ecol Evolut 6(5):1457–1469. https://doi.org/10.1002/ece3.1981

Nakamura T, Kitazawa T, Honda Y (2010) Structure and function of Satoyama and Satoumi. Chiba Prefect Biodivers Center Res Rep 2:21–30. (in Japanese)

NIES CCCA (2024) A-plat climate change adaptation information platform. https://adaptation-platform.nies.go.jp/local/plan/list.html

Nishida T, Nakazono A, Oikawa M, Matsui S (2005) Changes in the coastal fish fauna of the Chikuzen Sea due to recent increases in seawater temperatures. Kyushu Univ Graduate School Agric Life Sci J 60(2):187–201. (in Japanese)

Nojima, Okamoto (2008) North migration and bleaching of reef-building corals. J Japanese Soc Fish Sci 74(5):884–888. (in Japanese)

Noreen AME, Harrison PL, Van Oppen MJH (2009) Genetic diversity and connectivity in a brooding reef coral at the limit of its distribution. Proc R Soc B Biol Sci 276(1675):3927–3935. https://doi.org/10.1098/rspb.2009.1050

Noreen AME, van Oppen MJH, Harrison PL (2013) Genetic diversity and differentiation among high-latitude broadcast-spawning coral populations disjunct from the core range. Mar Ecol Prog Ser 491:101–109. https://doi.org/10.3354/meps10480

Norström AV, Nyström M, Lokrantz J, Folke C (2009) Alternative states on coral reefs: beyond coral–macroalgal phase shifts. Mar Ecol Prog Ser 376:295–306. https://doi.org/10.3354/meps07815

Nozawa Y, Villanueva RD, Munasik M, Roeroe KA, Mezaki T, Kawai T, Guest J, Arakaki S, Suzuki G, Tanangonan JJB, Ang PO, Edmunds PJ (2021) Latitudinal variation in growth and survival of juvenile corals in the West and South Pacific. Coral Reefs 40(5):1463–1471. https://doi.org/10.1007/s00338-021-02169-9

O'Dea A, Dillon EM, Altieri AH, Lepore ML (2017) Look to the past for an optimistic future. Conserv Biol 31(6):1221–1222. https://doi.org/10.1111/cobi.12997

Ocaña Vicente O, Opresko DM, Brito A (2006) First record of the black coral *Antipathella wollastoni* (Anthozoa: Antipatharia) outside of Macaronesian waters. Rev Acad Can Cien 18:125–138

Ofwegen LPV, Haddad MA (2011) A probably invasive new genus and new species of soft coral (Octocorallia: Alcyonacea: Clavulariidae) from Brazil. Zootaxa 3107(1):38–46. https://doi.org/10.11646/zootaxa.3107.1.2

Okutani T (2000) Marine mollusks in Japan. Tokai University Press, Japan

Olbers JM, Celliers L, Schleyer MH (2009) Zonation of benthic communities on the subtropical Aliwal Shoal, Durban, KwaZulu-Natal, South Africa. Afr Zool 44(1):8–23. https://doi.org/10.1080/15627020.2009.11407435

Oliver ECJ, Donat MG, Burrows MT, Moore PJ, Smale DA, Alexander LV, Benthuysen JA, Feng M, Sen Gupta A, Hobday AJ, Holbrook NJ, Perkins-Kirkpatrick SE, Scannell HA, Straub SC, Wernberg T (2018) Longer and more frequent marine heatwaves over the past century. Nat Commun 9(1):1324. https://doi.org/10.1038/s41467-018-03732-9

Orr JC, Fabry VJ, Aumont O, Bopp L, Doney SC, Feely RA, Gnanadesikan A, Gruber N, Ishida A, Joos F, Key RM, Lindsay K, Maier-Reimer E, Matear R, Monfray P, Mouchet A, Najjar RG, Plattner G-K, Rodgers KB, Sabine CL, Sarmiento JL, Schlitzer R, Slater RD, Totterdell IJ, Weirig M-F, Yamanaka Y, Yool A (2005) Anthropogenic Ocean acidification over the twenty-first century and its impact on calcifying organisms. Nature 437(7059):681–686. https://doi.org/10.1038/nature04095

Otis N, Reimer JD, Kawamura I, Kise H, Mizuyama M, Obuchi M, Sommer B, McFadden CS, Beger M (2024) Variation in species and functional composition of octocorals and zoantharians across a tropical to temperate environmental gradient in the Indo-Pacific. Coral Reefs 43(3):613–626. https://doi.org/10.1007/s00338-024-02487-8

Ovaskainen O, Abrego N (2020) Joint species distribution modelling: with applications in R. Cambridge University Press, Cambridge

Ovaskainen O, Tikhonov G, Norberg A, Guillaume Blanchet F, Duan L, Dunson D, Roslin T, Abrego N (2017) How to make more out of community data? A conceptual framework and its implementation as models and software. Ecol Lett 20(5):561–576. https://doi.org/10.1111/ele.12757

Owen DP, Long MH, Fitt WK, Hopkinson BM (2021) Taxon-specific primary production rates on coral reefs in the Florida keys. Limnol Oceanogr 66(3):625–638. https://doi.org/10.1002/lno.11627

Pandolfi JM, Jackson JBC (2001) Community structure of Pleistocene coral reefs of Curaçao, Netherlands Antilles. Ecol Monogr 71(1):49–67. https://doi.org/10.1890/0012-9615(2001)071[0049:CSOPCR]2.0.CO;2

Pandolfi JM, Jackson JBC (2007) Broad-scale patterns in Pleistocene coral reef communities from the Caribbean: implications for ecology and management. In: Aronson RB (ed) Geological approaches to coral reef ecology. Springer, New York, pp 201–236

Pandolfi JM, Kiessling W (2014) Gaining insights from past reefs to inform understanding of coral reef response to global climate change. Curr Opin Environ Sustain 7:52–58. https://doi.org/10.1016/j.cosust.2013.11.020

Pandolfi JM, Connolly SR, Marshall DJ, Cohen AL (2011) Projecting coral reef futures under global warming and ocean acidification. Science 333(6041):418–422. https://doi.org/10.1126/science.1204794

Pandolfi JM, Staples TL, Kiessling W (2020) Increased extinction in the emergence of novel ecological communities. Science 370(6513):220–222. https://doi.org/10.1126/science.abb3996

Pawlik JR, McMurray SE (2020) The emerging ecological and biogeochemical importance of sponges on coral reefs. Annu Rev Mar Sci 12(1):315–337. https://doi.org/10.1146/annurev-marine-010419-010807

Paz-García DA, Balart EF (2016) New record of the endemic coral *Porites sverdrupi* (Gulf of California): do fluctuations in seawater temperature regulate its southernmost range limit? Mar Biodivers 46(2):499–502. https://doi.org/10.1007/s12526-015-0375-z

Paz-García DA, Aldana-Moreno A, Cabral-Tena RA, García-De-León FJ, Hellberg ME, Balart EF (2015a) Morphological variation and different branch modularity across contrasting flow conditions in dominant *Pocillopora* reef-building corals. Oecologia 178(1):207–218. https://doi.org/10.1007/s00442-014-3199-9

Paz-García DA, Hellberg ME, García-de-León FJ, Balart EF (2015b) Switch between Morphospecies of Pocillopora corals. Am Nat 186(3):434–440. https://doi.org/10.1086/682363

Paz-García DA, Valencia-Méndez O, Domínguez-Domínguez O, Balart EF (2018) Living on the edge: *Diadema mexicanum* in the upper Gulf of California. Mar Biodivers 48(2):1261–1264. https://doi.org/10.1007/s12526-016-0539-5

Paz-García DA, Armstrong EJ, Popovic I, González-Pech RA, Hellberg ME (2023) Editorial: innovative approaches to coral reef science by early career researchers. Front Mar Sci 10. https://doi.org/10.3389/fmars.2023.1322657

Pecl GT, Araújo MB, Bell JD, Blanchard J, Bonebrake TC, Chen I-C, Clark TD, Colwell RK, Danielsen F, Evengård B, Falconi L, Ferrier S, Frusher S, Garcia RA, Griffis RB, Hobday AJ, Janion-Scheepers C, Jarzyna MA, Jennings S, Lenoir J, Linnetved HI, Martin VY, McCormack PC, McDonald J, Mitchell NJ, Mustonen T, Pandolfi JM, Pettorelli N, Popova E, Robinson SA, Scheffers BR, Shaw JD, Sorte CJB, Strugnell JM, Sunday JM, Tuanmu M-N, Vergés A, Villanueva C, Wernberg T, Wapstra E, Williams SE (2017) Biodiversity redistribution under climate change: impacts on ecosystems and human well-being. Science 355(6332):eaai9214. https://doi.org/10.1126/science.aai9214

Pedraza-Pohlenz RM, Balart EF, Tortolero-Langarica JJA, Rodríguez-Troncoso AP, Hellberg ME, Norzagaray-López O, Cabral-Tena RA, Cupul-Magaña AL, Paz-García DA (2023) Sexual dimorphism in corallite size and modularity of a broadcast spawning coral, *Porites lobata*. Front Mar Sci 9. https://doi.org/10.3389/fmars.2022.1068391

Pereira PHC, Ternes MLF, Nunes JAC, Giglio VJ (2021) Overexploitation and behavioral changes of the largest South Atlantic parrotfish (Scarus trispinosus): evidence from fishers' knowledge. Biol Conserv 254:108940. https://doi.org/10.1016/j.biocon.2020.108940

Perry CT, Murphy GN, Kench PS, Smithers SG, Edinger EN, Steneck RS, Mumby PJ (2013) Caribbean-wide decline in carbonate production threatens coral reef growth. Nat Commun 4(1):1402. https://doi.org/10.1038/ncomms2409

Pessarrodona A, Filbee-Dexter K, Alcoverro T, Boada J, Feehan CJ, Fredriksen S, Grace SP, Nakamura Y, Narvaez CA, Norderhaug KM, Wernberg T (2021) Homogenization and miniaturization of habitat structure in temperate marine forests. Glob Chang Biol 27(20):5262–5275. https://doi.org/10.1111/gcb.15759

Pessarrodona A, Vergés A, Bosch NE, Bell S, Smith S, Sgarlatta MP, Wernberg T (2022) Tropicalization unlocks novel trophic pathways and enhances secondary productivity in temperate reefs. Funct Ecol 36(3):659–673. https://doi.org/10.1111/1365-2435.13990

Petchey OL, Gaston KJ (2006) Functional diversity: back to basics and looking forward. Ecol Lett 9(6):741–758. https://doi.org/10.1111/j.1461-0248.2006.00924.x

Peyrot-Clausade M, Chabanet P, Conand C, Fontaine MF, Letourneur Y, Harmelin-Vivien M (2000) Sea urchin and fish bioerosion on La Réunion and Moorea reefs. Bull Mar Sci 66(2):477–485

Pinsky ML, Selden RL, Kitchel ZJ (2020) Climate-driven shifts in marine species ranges: scaling from organisms to communities. Annu Rev Mar Sci 12(1):153–179. https://doi.org/10.1146/annurev-marine-010419-010916

Poloczanska ES, Burrows MT, Brown CJ, García Molinos J, Halpern BS, Hoegh-Guldberg O, Kappel CV, Moore PJ, Richardson AJ, Schoeman DS, Sydeman WJ (2016) Responses of marine organisms to climate change across oceans. Front Mar Sci 3. https://doi.org/10.3389/fmars.2016.00062

Porter SN, Schleyer MH (2019) Environmental variation and How its spatial structure influences the cross-shelf distribution of high-latitude coral communities in South Africa. Diversity 11(4):57. https://doi.org/10.3390/d11040057

Porter SN, Humphries MS, Buah-Kwofie A, Schleyer MH (2018) Accumulation of organochlorine pesticides in reef organisms from marginal coral reefs in South Africa and links with coastal groundwater. Mar Pollut Bull 137:295–305. https://doi.org/10.1016/j.marpolbul.2018.10.028

Porter SN, Sink KJ, Schleyer MH (2021) The third global coral bleaching event on the marginal coral reefs of the southwestern Indian ocean and factors that contribute to their resistance and resilience. Diversity 13(10):464

Pratchett M, King K, Hoogenboom M, Widman E, Baird A, Pandolfi J, Edmunds P, Lough J (2015) Spatial, temporal and taxonomic variation in coral growth – implications for the structure and function of coral reef ecosystems. Oceanogr Mar Biol Annu Rev 53:215–295

Precoda K, Baird AH, Madsen A, Mizerek T, Sommer B, Su SN, Madin JS (2018) How does a widespread reef coral maintain a population in an isolated environment? Mar Ecol Prog Ser 594:85–94. https://doi.org/10.3354/meps12537

Putnam HM (2021) Avenues of reef-building coral acclimatization in response to rapid environmental change. J Exp Biol 224:jeb239319. https://doi.org/10.1242/jeb.239319

Putnam HM, Davidson JM, Gates RD (2016) Ocean acidification influences host DNA methylation and phenotypic plasticity in environmentally susceptible corals. Evol Appl 9(9):1165–1178. https://doi.org/10.1111/eva.12408

Quan-Young LI, Espinoza-Avalos J (2006) Reduction of zooxanthellae density, chlorophyll a concentration, and tissue thickness of the coral *Montastraea faveolata* (Scleractinia) when competing with mixed turf algae. Limnol Oceanogr 51(2):1159–1166. https://doi.org/10.4319/lo.2006.51.2.1159

Quintanilla E, Wilke T, Ramírez-Portilla C, Sarmiento A, Sánchez JA (2017) Taking a detour: invasion of an octocoral into the Tropical Eastern Pacific. Biol Invasions 19(9):2583–2597. https://doi.org/10.1007/s10530-017-1469-2

Rabosky DL, Chang J, Title PO, Cowman PF, Sallan L, Friedman M, Kaschner K, Garilao C, Near TJ, Coll M, Alfaro ME (2018) An inverse latitudinal gradient in speciation rate for marine fishes. Nature 559(7714):392–395. https://doi.org/10.1038/s41586-018-0273-1

Rasher DB, Hay ME (2010) Chemically rich seaweeds poison corals when not controlled by herbivores. Proc Natl Acad Sci USA 107(21):9683–9688. https://doi.org/10.1073/pnas.0912095107

Rasher DB, Engel S, Bonito V, Fraser GJ, Montoya JP, Hay ME (2012) Effects of herbivory, nutrients, and reef protection on algal proliferation and coral growth on a tropical reef. Oecologia 169(1):187–198. https://doi.org/10.1007/s00442-011-2174-y

Reimer JD, Takishita K, Ono S, Maruyama T, Tsukahara J (2006) Latitudinal and intracolony ITS-rDNA sequence variation in the symbiotic dinoflagellate genus *Symbiodinium* (Dinophyceae) in *Zoanthus sansibaricus* (Anthozoa: Hexacorallia). Phycol Res 54(2):122–132. https://doi.org/10.1111/j.1440-1835.2006.00419.x

Reimer J, Kim S, Arai S, Keshavmurthy S, Choi K-S (2018) First records of zooxanthellate *Zoanthus* (Anthozoa: Hexacorallia: Zoantharia) from Korea and Japan (east) sea. Mar Biodivers 48:1269–1273

Reimer JD, Biondi P, Lau YW, Masucci GD, Nguyen XH, Santos MEA, Wee HB (2019) Marine biodiversity research in the Ryukyu Islands, Japan: current status and trends. PeerJ 7:e6532. https://doi.org/10.7717/peerj.6532

Reimer JD, Wee HB, López C, Beger M, Cruz ICS (2021) Widespread Zoanthus and Palythoa dominance, barrens, and phase shifts in shallow water subtropical and tropical marine ecosystems. In: Oceanography and marine biology. CRC Press

Reimer JD, Agostini S, Golbuu Y, Harvey BP, Izumiyama M, Jamodiong EA, Kawai E, Kayanne H, Kurihara H, Ravasi T, Wada S, Rodolfo-Metalpa R (2023) High abundances of zooxanthellate zoantharians (*Palythoa and Zoanthus*) at multiple natural analogues: potential model anthozoans? Coral Reefs 42(3):707–715. https://doi.org/10.1007/s00338-023-02381-9

Reimer JD, García-Hernández JE, Hoeksema BW. 6. xSpecies diversity of Zoantharia (Cnidaria: Anthozoa: Hexacorallia). Marine Biodiversity of Bonaire: a baseline survey.:20. Marine-Biodiversity--of-Bonaire-A-baseline-survey.pdf (researchgate.net)

Ribas-Deulofeu L, Denis V, Palmas SD, Kuo C-Y, Hsieh HJ, Chen CA (2016) Structure of benthic communities along the Taiwan latitudinal gradient. PLoS One 11(8):e0160601. https://doi.org/10.1371/journal.pone.0160601

Ribas-Deulofeu L, Loubeyres M, Denis V, De Palmas S, Hwang S-J, Woo S, Song J-I, Chen CA (2023) Jeju Island: a sentinel for tracking ocean warming impacts on high-latitude benthic communities. Coral Reefs 42(5):1097–1112. https://doi.org/10.1007/s00338-023-02400-9

Richardson DL, Harriott VJ, Harrison PL (1997) Distribution and abundance of giant sea anemones (Actiniaria) in subtropical eastern Australian waters. Mar Freshw Res 48(1):59–66. https://doi.org/10.1071/mf96020

Riegl B (2001) Inhibition of reef framework by frequent disturbance: examples from the Arabian gulf, South Africa, and the Cayman Islands. Palaeogeogr Palaeoclimatol Palaeoecol 175(1):79–101. https://doi.org/10.1016/S0031-0182(01)00387-X

Rischer M, Guo H, Beemelmanns C (2022) Signalling molecules inducing metamorphosis in marine organisms. Nat Prod Rep 39(9):1833–1855. https://doi.org/10.1039/D1NP00073J

Rius M, Clusella-Trullas S, McQuaid CD, Navarro RA, Griffiths CL, Matthee CA, von der Heyden S, Turon X (2014) Range expansions across ecoregions: interactions of climate change, physiology and genetic diversity. Glob Ecol Biogeogr 23(1):76–88. https://doi.org/10.1111/geb.12105

Ross CL, Schoepf V, DeCarlo TM, McCulloch MT (2018) Mechanisms and seasonal drivers of calcification in the temperate coral *Turbinaria reniformis* at its latitudinal limits. Proc R Soc B Biol Sci 285(1879):20180215. https://doi.org/10.1098/rspb.2018.0215

Ross CL, DeCarlo TM, McCulloch MT (2019) Environmental and physiochemical controls on coral calcification along a latitudinal temperature gradient in Western Australia. Glob Chang Biol 25(2):431–447. https://doi.org/10.1111/gcb.14488

Ross CL, French B, Lester EK, Wilson SK, Day PB, Taylor MD, Barrett N (2021) Coral communities on marginal high-latitude reefs in West Australian marine parks. Diversity 13(11):554. https://doi.org/10.3390/d13110554

Ruppert KM, Kline RJ, Rahman MS (2019) Past, present, and future perspectives of environmental DNA (eDNA) metabarcoding: a systematic review in methods, monitoring, and applications of global eDNA. Glob Ecol Conserv 17:e00547. https://doi.org/10.1016/j.gecco.2019.e00547

Saeedi H, Costello MJ, Warren D, Brandt A (2019) Latitudinal and bathymetrical species richness patterns in the NW Pacific and adjacent Arctic Ocean. Abstract Scientific Reports 9(1). https://doi.org/10.1038/s41598-019-45813-9

Sarmiento JL, Gruber N, Brzezinski MA, Dunne JP (2004) High-latitude controls of thermocline nutrients and low latitude biological productivity. Nature 427(6969):56–60. https://doi.org/10.1038/nature02127

Sawada H, Yokosawa H, Lambert CC (eds) (2001) The biology of ascidians. Springer Japan, Tokyo

Sawall Y, Al-Sofyani A, Kürten B, Al-Aidaroos A, Hoang B, Marimuthu N, Khomayis H, Sommer U, Gharbawi W, Wahl M (2014) Coral communities, in contrast to fish communities, maintain a high assembly similarity along the large latitudinal gradient along the Saudi Red Sea coast. J Ecosyst Ecogr 3(4):1–7

Sawall Y, Al-Sofyani A, Hohn S, Banguera-Hinestroza E, Voolstra CR, Wahl M (2015) Extensive phenotypic plasticity of a Red Sea coral over a strong latitudinal temperature gradient suggests limited acclimatization potential to warming. Sci Rep 5(1):8940. https://doi.org/10.1038/srep08940

Scherner F, Horta PA, de Oliveira EC, Simonassi JC, Hall-Spencer JM, Chow F, Nunes JMC, Pereira SMB (2013) Coastal urbanization leads to remarkable seaweed species loss and community shifts along the SW Atlantic. Mar Pollut Bull 76(1):106–115. https://doi.org/10.1016/j.marpolbul.2013.09.019

Schlegel RW, Oliver ECJ, Wernberg T, Smit AJ (2017) Nearshore and offshore co-occurrence of marine heatwaves and cold-spells. Prog Oceanogr 151:189–205. https://doi.org/10.1016/j.pocean.2017.01.004

Schleyer M, Celliers L (2000) The status of South African coral reefs. In: Souter D, Obura D, Linden O (eds) Coral Reef Degradation in the Indian Ocean, vol 2000. CORDIO, Stockholm, Sweden, pp 49–50

Schleyer MH, Porter SN (2018) Drivers of soft and stony coral community distribution on the high-latitude coral reefs of South Africa. Adv Mar Biol 80:1–55

Schleyer MH, Floros C, Laing SCS, Macdonald AHH, Montoya-Maya PH, Morris T, Porter SN, Seré MG (2018) What can South African reefs tell us about the future of high-latitude coral systems? Mar Pollut Bull 136:491–507. https://doi.org/10.1016/j.marpolbul.2018.09.014

Schmidt-Roach S, Miller KJ, Andreakis N (2013) *Pocillopora aliciae*: a new species of scleractinian coral (Scleractinia, Pocilloporidae) from subtropical Eastern Australia. Zootaxa 3626(4):576–582. https://doi.org/10.11646/zootaxa.3626.4.11

Schoepf V, Baumann JH, Barshis DJ, Browne NK, Camp EF, Comeau S, Cornwall CE, Guzmán HM, Riegl B, Rodolfo-Metalpa R, Sommer B (2023) Corals at the edge of environmental limits: a new conceptual framework to re-define marginal and extreme coral communities. Sci Total Environ 884:163688. https://doi.org/10.1016/j.scitotenv.2023.163688

Schubert N, Brown D, Rossi S (2017) Symbiotic versus nonsymbiotic Octocorals: physiological and ecological implications. In: Rossi S, Bramanti L, Gori A, Orejas C (eds) Marine animal forests: the ecology of benthic biodiversity hotspots. Springer International Publishing, Cham, pp 887–918

Scott A, Malcolm HA, Damiano C, Richardson DL (2011) Long-term increases in abundance of anemonefish and their host sea anemones in an Australian marine protected area. Mar Freshw Res 62(2):187–196. https://doi.org/10.1071/MF10323

Scott ME, Heupel MR, Simpfendorfer CA, Matley JK, Pratchett MS (2019) Latitudinal and seasonal variation in space use by a large, predatory reef fish, Plectropomus leopardus. Funct Ecol 33(4):670–680. https://doi.org/10.1111/1365-2435.13271

Serisawa Y, Imoto Z, Ishikawa T, Ohno M (2004) Decline of the *Ecklonia cava* population associated with increased seawater temperatures in Tosa Bay, southern Japan. Fish Sci 70(1):189–191. https://doi.org/10.1111/j.0919-9268.2004.00788.x

Shenkar N, Swalla BJ (2011) Global diversity of Ascidiacea. PLoS One 6(6):e20657. https://doi.org/10.1371/journal.pone.0020657

Shore A, Sims JA, Grimes M, Howe-Kerr LI, Grupstra CG, Doyle SM, Stadler L, Sylvan JB, Shamberger KE, Davies SW, Santiago-Vázquez LZ (2021) On a reef far, far away: anthropogenic impacts following extreme storms affect sponge health and bacterial communities. Fron Mar Sci 8:608036. https://doi.org/10.3389/fmars.2021.608036

Smale DA, Wernberg T (2013) Extreme climatic event drives range contraction of a habitat-forming species. Proc R Soc B Biol Sci 280(1754):20122829. https://doi.org/10.1098/rspb.2012.2829

Smith SM, Malcolm HA, Marzinelli EM, Schultz AL, Steinberg PD, Vergés A (2021) Tropicalization and kelp loss shift trophic composition and lead to more winners than losers in fish communities. Glob Chang Biol 27(11):2537–2548. https://doi.org/10.1111/gcb.15592

Soares M d O (2020) Marginal reef paradox: a possible refuge from environmental changes? Ocean Coastal Manag 185:105063. https://doi.org/10.1016/j.ocecoaman.2019.105063

Sommer B (2022) Marginal reefs: distinct ecosystems of extraordinarily high conservation value. In: Hamylton S, Hutchings P, Hoegh-Guldberg O (eds) The coral reefs of Australia: a dynamic life at the water surface. CSIRO Publishing, Australia, p 139

Sommer B, Harrison PL, Beger M, Pandolfi JM (2014) Trait-mediated environmental filtering drives assembly at biogeographic transition zones. Ecology 95(4):1000–1009. https://doi.org/10.1890/13-1445.1

Sommer B, Sampayo EM, Beger M, Harrison PL, Babcock RC, Pandolfi JM (2017) Local and regional controls of phylogenetic structure at the high-latitude range limits of corals. Proc R Soc B Biol Sci 284(1861):20170915. https://doi.org/10.1098/rspb.2017.0915

Sommer B, Beger M, Harrison PL, Babcock RC, Pandolfi JM (2018) Differential response to abiotic stress controls species distributions at biogeographic transition zones. Ecography 41(3):478–490. https://doi.org/10.1111/ecog.02986

Sommer B, Butler IR, Pandolfi JM (2021) Trait-based approach reveals how marginal reefs respond to acute and chronic disturbance. Coral Reefs 40(3):735–749. https://doi.org/10.1007/s00338-021-02077-y

Sommer B, Hodge JM, Lachs L, Cant J, Pandolfi JM, Beger M (2024) Decadal demographic shifts and size-dependent disturbance responses of corals in a subtropical warming hotspot. Sci Rep 14(1):6327. https://doi.org/10.1038/s41598-024-56890-w

Sønstebø JH, Gielly L, Brysting AK, Elven R, Edwards M, Haile J, Willerslev E, Coissac E, Rioux D, Sannier J, Taberlet P, Brochmann C (2010) Using next-generation sequencing for molecular reconstruction of past Arctic vegetation and climate. Mol Ecol Resour 10(6):1009–1018. https://doi.org/10.1111/j.1755-0998.2010.02855.x

Spalding MD, Fox HE, Allen GR, Davidson N, Ferdaña ZA, Finlayson M, Halpern BS, Jorge MA, Lombana A, Lourie SA, Martin KD, McManus E, Molnar J, Recchia CA, Robertson J (2007) Marine Ecoregions of the World: A Bioregionalization of Coastal and Shelf Areas. BioScience 57(7):573–583. https://doi.org/10.1641/B570707

Steneck RS, Graham MH, Bourque BJ, Corbett D, Erlandson JM, Estes JA, Tegner MJ (2002) Kelp forest ecosystems: biodiversity, stability, resilience and future. Environ Conserv 29(4):436–459. https://doi.org/10.1017/S0376892902000322

Steneck RS, Bellwood DR, Hay ME (2017) Herbivory in the marine realm. Curr Biol 27(11):R484–R489. https://doi.org/10.1016/j.cub.2017.04.021

Stott I, Townley S, Hodgson DJ (2011) A framework for studying transient dynamics of population projection matrix models. Ecol Lett 14(9):959–970. https://doi.org/10.1111/j.1461-0248.2011.01659.x

Stuart-Smith RD, Bates AE, Lefcheck JS, Duffy JE, Baker SC, Thomson RJ, Stuart-Smith JF, Hill NA, Kininmonth SJ, Airoldi L, Becerro MA, Campbell SJ, Dawson TP, Navarrete SA, Soler GA, Strain EMA, Willis TJ, Edgar GJ (2013) Integrating abundance and functional traits reveals new global hotspots of fish diversity. Nature 501(7468):539–542. https://doi.org/10.1038/nature12529

Stuart-Smith RD, Mellin C, Bates AE, Edgar GJ (2021) Habitat loss and range shifts contribute to ecological generalization among reef fishes. Nat Ecol Evol 5(5):656–662. https://doi.org/10.1038/s41559-020-01342-7

Takahashi K (2022) [Commentary] replacement of fish species in Sendai Bay and the Sanriku coast due to rising seawater temperatures. JAFIC Tech Rev 1:12

Tanaka K, Taino S, Haraguchi H, Prendergast G, Hiraoka M (2012) Warming off southwestern Japan linked to distributional shifts of subtidal canopy-forming seaweeds. Ecol Evol 2(11):2854–2865. https://doi.org/10.1002/ece3.391

Teagle H, Hawkins SJ, Moore PJ, Smale DA (2017) The role of kelp species as biogenic habitat formers in coastal marine ecosystems. J Exp Mar Biol Ecol 492:81–98. https://doi.org/10.1016/j.jembe.2017.01.017

Thomas L, Kennington WJ, Evans RD, Kendrick GA, Stat M (2017) Restricted gene flow and local adaptation highlight the vulnerability of high-latitude reefs to rapid environmental change. Glob Chang Biol 23(6):2197–2205. https://doi.org/10.1111/gcb.13639

Thomson DP, Frisch AJ (2010) Extraordinarily high coral cover on a nearshore, high-latitude reef in South-West Australia. Coral Reefs 29(4):923–927. https://doi.org/10.1007/s00338-010-0650-1

Thomson DP, Bearham D, Graham F, Eagle JV (2011) High latitude, deeper water coral bleaching at Rottnest Island, Western Australia. Coral Reefs 30(4):1107–1107. https://doi.org/10.1007/s00338-011-0811-x

Thornhill DJ, LaJeunesse TC, Kemp DW, Fitt WK, Schmidt GW (2006) Multi-year, seasonal genotypic surveys of coral-algal symbioses reveal prevalent stability or post-bleaching reversion. Mar Biol 148(4):711–722. https://doi.org/10.1007/s00227-005-0114-2

Thornhill DJ, Howells EJ, Wham DC, Steury TD, Santos SR (2017) Population genetics of reef coral endosymbionts (Symbiodinium, Dinophyceae). Mol Ecol 26(10):2640–2659. https://doi.org/10.1111/mec.14055

Tikhonov G, Duan L, Abrego N, Newell G, White M, Dunson D, Ovaskainen O (2020a) Computationally efficient joint species distribution modeling of big spatial data. Ecology 101(2):e02929. https://doi.org/10.1002/ecy.2929

Tikhonov G, Opedal ØH, Abrego N, Lehikoinen A, de Jonge MMJ, Oksanen J, Ovaskainen O (2020b) Joint species distribution modelling with the r-package Hmsc. Methods Ecol Evol 11(3):442–447. https://doi.org/10.1111/2041-210X.13345

Tittensor DP, Beger M, Boerder K, Boyce DG, Cavanagh RD, Cosandey-Godin A, Crespo GO, Dunn DC, Ghiffary W, Grant SM, Hannah L, Halpin PN, Harfoot M, Heaslip SG, Jeffery NW, Kingston N, Lotze HK, McGowan J, McLeod E, McOwen CJ, O'Leary BC, Schiller L, Stanley RRE, Westhead M, Wilson KL, Worm B (2019) Integrating climate adaptation and biodiversity conservation in the global ocean. Sci Adv 5(11):eaay9969. https://doi.org/10.1126/sciadv.aay9969

Torda G, Quigley KM (2022) Drivers of adaptive capacity in wild populations: implications for genetic interventions. Front Mar Sci 9. https://doi.org/10.3389/fmars.2022.947989

Travers MJ, Clarke KR, Newman SJ, Hall NG, Potter IC (2018) To what extents are species richness and abundance of reef fishes along a tropical coast related to latitude and other factors? Cont Shelf Res 167:99–110. https://doi.org/10.1016/j.csr.2018.08.006

Tribollet A, Golubic S (2011) Reef bioerosion: agents and processes. In: Dubinsky Z, Stambler N (eds) Coral reefs: an ecosystem in transition. Springer Netherlands, Dordrecht, pp 435–449

Trip ED, Clements KD, Raubenheimer D, Choat JH (2014) Temperature-related variation in growth rate, size, maturation and life span in a marine herbivorous fish over a latitudinal gradient. J Anim Ecol 83(4):866–875. https://doi.org/10.1111/1365-2656.12183

Tuckett CA, de Bettignies T, Fromont J, Wernberg T (2017) Expansion of corals on temperate reefs: direct and indirect effects of marine heatwaves. Coral Reefs 36(3):947–956. https://doi.org/10.1007/s00338-017-1586-5

Turnham KE, Aschaffenburg MD, Pettay DT, Paz-García DA, Reyes-Bonilla H, Pinzón J, Timmins E, Smith RT, McGinley MP, Warner ME, LaJeunesse TC (2023) High physiological function for corals with thermally tolerant, host-adapted symbionts. Proc R Soc B Biol Sci 290(2003):20231021. https://doi.org/10.1098/rspb.2023.1021

Ulate K, Sánchez C, Sánchez-Rodríguez A, Alonso D, Aburto-Oropeza O, Huato-Soberanis L (2016) Latitudinal regionalization of epibenthic macroinvertebrate communities on rocky reefs in the Gulf of California. Mar Biol Res 12(4):389–401. https://doi.org/10.1080/17451000.2016.1143105

van der Loos LM, Eriksson BK, Falcão Salles J (2019) The macroalgal Holobiont in a changing sea. Trends Microbiol 27(7):635–650. https://doi.org/10.1016/j.tim.2019.03.002

van der Meer MH, Jones GP, Hobbs J-PA, van Herwerden L (2012) Historic hybridization and introgression between two iconic Australian anemonefish and contemporary patterns of population connectivity. Ecol Evol 2(7):1592–1604. https://doi.org/10.1002/ece3.251

van Hooidonk R, Maynard JA, Manzello D, Planes S (2014) Opposite latitudinal gradients in projected ocean acidification and bleaching impacts on coral reefs. Glob Chang Biol 20(1):103–112. https://doi.org/10.1111/gcb.12394

Vázquez DP, Gianoli E, Morris WF, Bozinovic F (2017) Ecological and evolutionary impacts of changing climatic variability. Biol Rev 92(1):22–42. https://doi.org/10.1111/brv.12216

Veilleux HD, Ryu T, Donelson JM, Ravasi T, Munday PL (2018) Molecular response to extreme summer temperatures differs between two genetically differentiated populations of a coral reef fish. Front Mar Sci 5. https://doi.org/10.3389/fmars.2018.00349

Vergés A, Steinberg PD, Hay ME, Poore AGB, Campbell AH, Ballesteros E, Heck KL, Booth DJ, Coleman MA, Feary DA, Figueira W, Langlois T, Marzinelli EM, Mizerek T, Mumby PJ, Nakamura Y, Roughan M, van Sebille E, Gupta AS, Smale DA, Tomas F, Wernberg T, Wilson SK (2014a) The tropicalization of temperate marine ecosystems: climate-mediated changes in herbivory and community phase shifts. Proc R Soc B Biol Sci 281(1789):20140846. https://doi.org/10.1098/rspb.2014.0846

Vergés A, Tomas F, Cebrian E, Ballesteros E, Kizilkaya Z, Dendrinos P, Karamanlidis AA, Spiegel D, Sala E (2014b) Tropical rabbitfish and the deforestation of a warming temperate sea. J Ecol 102(6):1518–1527. https://doi.org/10.1111/1365-2745.12324

Vergés A, Doropoulos C, Malcolm HA, Skye M, Garcia-Pizá M, Marzinelli EM, Campbell AH, Ballesteros E, Hoey AS, Vila-Concejo A, Bozec Y-M, Steinberg PD (2016) Long-term empirical evidence of ocean warming leading to tropicalization of fish communities, increased herbivory, and loss of kelp. Proc Natl Acad Sci 113(48):13791–13796. https://doi.org/10.1073/pnas.1610725113

Vergés A, McCosker E, Mayer-Pinto M, Coleman MA, Wernberg T, Ainsworth T, Steinberg PD (2019) Tropicalisation of temperate reefs: implications for ecosystem functions and management actions. Funct Ecol 33(6):1000–1013. https://doi.org/10.1111/1365-2435.13310

Veron JEN (1995) Corals in space and time: the biogeography and evolution of the Scleractinia. Cornell University Press

Veron JEN, Minchin PR (1992) Correlations between sea surface temperature, circulation patterns and the distribution of hermatypic corals of Japan. Cont Shelf Res 12(7):835–857. https://doi.org/10.1016/0278-4343(92)90047-N

Veron JEN, How RA, Done TJ, Zell LD, Dodkin MJ, O'Farrell AF (1974) Corals of the Solitary Islands, New South Wales. Mar Freshwater Res 25(2):193–208. https://doi.org/10.1071/mf9740193

Wall M, Putchim L, Schmidt GM, Jantzen C, Khokiattiwong S, Richter C. 2025. Large-amplitude internal waves benefit corals during thermal stress. Proc. R. Soc. B 282, 20140650. (10.1098/RSPB.2014.0650), https://royalsocietypublishing.org/rspb/article/282/1799/20140650/77490/Large-amplitude-internal-waves-benefit-corals

Wang Y, Yu K, Chen X, Wang W, Huang X, Wang Y, Liao Z (2020) An approach for assessing ecosystem-based adaptation in coral reefs at relatively high latitudes to climate change and human pressure. Environ Monit Assess 192(9):579. https://doi.org/10.1007/s10661-020-08534-5

Wang C, Zheng X, Kvitt H, Sheng H, Sun D, Niu G, Tchernov D, Shi T (2023) Lineage-specific symbionts mediate differential coral responses to thermal stress. Microbiome 11(1):211. https://doi.org/10.1186/s40168-023-01653-4

Wang Q, Zheng X, Zhou X, Zhang H, Cai L, Leung JYS, Huang L (2024) Symbiont genus determines the trophic strategy of corals: implications for intraspecific competition for energy sources in coral reefs. Ecol Indic 158:111477. https://doi.org/10.1016/j.ecolind.2023.111477

Ward-Paige CA, Risk MJ, Sherwood OA, Jaap WC (2005) Clionid sponge surveys on the Florida Reef Tract suggest land-based nutrient inputs. Mar Pollut Bull 51(5):570–579. https://doi.org/10.1016/j.marpolbul.2005.04.006

Wee HB, Kobayashi Y, Reimer JD (2021) Effects of temperature, salinity, and depth on Symbiodiniaceae lineages hosted by *Palythoa tuberculosa* near a river mouth. Mar Ecol Prog Ser 667:43–60. https://doi.org/10.3354/meps13706

Welch H, Brodie S, Jacox MG, Bograd SJ, Hazen EL (2020) Decision-support tools for dynamic management. Conserv Biol 34(3):589–599. https://doi.org/10.1111/cobi.13417

Wernberg T, Smale DA, Tuya F, Thomsen MS, Langlois TJ, de Bettignies T, Bennett S, Rousseaux CS (2013) An extreme climatic event alters marine ecosystem structure in a global biodiversity hotspot. Nat Clim Change 3(1):78–82. https://doi.org/10.1038/nclimate1627

Wernberg T, Bennett S, Babcock RC, de Bettignies T, Cure K, Depczynski M, Dufois F, Fromont J, Fulton CJ, Hovey RK, Harvey ES, Holmes TH, Kendrick GA, Radford B, Santana-Garcon J, Saunders BJ, Smale DA, Thomsen MS, Tuckett CA, Tuya F, Vanderklift MA, Wilson S (2016) Climate-driven regime shift of a temperate marine ecosystem. Science 353(6295):169–172. https://doi.org/10.1126/science.aad8745

Wernberg T, Krumhansl K, Filbee-Dexter K, Pedersen MF (2019) Status and trends for the world's kelp forests. In: World seas: an environmental evaluation. Elsevier, pp 57–78

Wernberg T, Thomsen MS, Baum JK, Bishop MJ, Bruno JF, Coleman MA, Filbee-Dexter K, Gagnon K, He Q, Murdiyarso D, Rogers K, Silliman BR, Smale DA, Starko S, Vanderklift MA (2024) Impacts of climate change on marine foundation species. Annu Rev Mar Sci 16:247–282

Wicks LC, Sampayo E, Gardner JPA, Davy SK (2010) Local endemicity and high diversity characterise high-latitude coral–*Symbiodinium* partnerships. Coral Reefs 29(4):989–1003. https://doi.org/10.1007/s00338-010-0649-7

Willerslev E, Davison J, Moora M, Zobel M, Coissac E, Edwards ME, Lorenzen ED, Vestergård M, Gussarova G, Haile J, Craine J, Gielly L, Boessenkool S, Epp LS, Pearman PB, Cheddadi R, Murray D, Bråthen KA, Yoccoz N, Binney H, Cruaud C, Wincker P, Goslar T, Alsos IG, Bellemain E, Brysting AK, Elven R, Sønstebø JH, Murton J, Sher A, Rasmussen M, Rønn R, Mourier T, Cooper A, Austin J, Möller P, Froese D, Zazula G, Pompanon F, Rioux D, Niderkorn V, Tikhonov A, Savvinov G, Roberts RG, RDE MP, MTP G, Kjær KH, Orlando L, Brochmann C, Taberlet P (2014) Fifty thousand years of Arctic vegetation and megafaunal diet. Nature 506(7486):47–51. https://doi.org/10.1038/nature12921

Willis BL, Oliver JK (1990) Direct tracking of coral larvae: implications for dispersal studies of planktonic larvae in topographically complex environments. Ophelia 32(1–2):145–162. https://doi.org/10.1080/00785236.1990.10422029

Wilson J, Harrison P (2005) Post-settlement mortality and growth of newly settled reef corals in a subtropical environment. Coral Reefs 24(3):418–421. https://doi.org/10.1007/s00338-005-0033-1

Wilson KL, Tittensor DP, Worm B, Lotze HK (2020) Incorporating climate change adaptation into marine protected area planning. Glob Chang Biol 26(6):3251–3267. https://doi.org/10.1111/gcb.15094

Winemiller KO, Rose KA (1992) Patterns of life-history diversification in North American fishes: implications for population regulation. Can J Fish Aquat Sci 49(10):2196–2218. https://doi.org/10.1139/f92-242

Wong KT, Chui APY, Lam EKY, Ang P (2018) A 30-year monitoring of changes in coral community structure following anthropogenic disturbances in Tolo Harbour and channel, Hong Kong. Mar Pollut Bull 133:900–910. https://doi.org/10.1016/j.marpolbul.2018.06.049

Wong RHX, Monk J, Perkins NR, Barrett NS (2023) A systematic review on the anthropogenic stressors on sessile benthic mesophotic reef communities: implications for temperate reef management in Australia. Front Mar Sci 10. https://doi.org/10.3389/fmars.2023.1276072

Wood JR, Wilmshurst JM, Worthy TH, Cooper A (2012) First coprolite evidence for the diet of Anomalopteryx didiformis, an extinct forest ratite from New Zealand. N Z J Ecol:164–170. https://www.jstor.org/stable/24060842

Work TM, Aeby GS, Maragos JE (2008) Phase shift from a coral to a Corallimorph-dominated reef associated with a shipwreck on Palmyra atoll. PLoS One 3(8):e2989. https://doi.org/10.1371/journal.pone.0002989

Wu L, Cai W, Zhang L, Nakamura H, Timmermann A, Joyce T, McPhaden MJ, Alexander M, Qiu B, Visbeck M, Chang P, Giese B (2012) Enhanced warming over the global subtropical western boundary currents. Nat Clim Chang 2:161–166. https://doi.org/10.1038/nclimate1353

Wu Q, Huang W, Chen B, Yang E, Meng L, Chen Y, Li J, Huang X, Liang J, Yap T-K, Yu K (2021) Genetic structure of *Turbinaria peltata* in the northern South China Sea suggest insufficient genetic adaptability of relatively high-latitude scleractinian corals to environment stress. Sci Total Environ 775:145775. https://doi.org/10.1016/j.scitotenv.2021.145775

Xu H, Feng B, Xie M, Ren Y, Xia J, Zhang Y, Wang A, Li X (2020) Physiological characteristics and environment adaptability of reef-building corals at the Wuzhizhou Island of South China Sea. Front Physiol 11. https://doi.org/10.3389/fphys.2020.00390

Yachi S, Loreau M (1999) Biodiversity and ecosystem productivity in a fluctuating environment: the insurance hypothesis. Proc Natl Acad Sci 96(4):1463–1468. https://doi.org/10.1073/pnas.96.4.1463

Yagi N, Takagi AP, Takada Y, Kurokura H (2010) Marine protected areas in Japan: institutional background and management framework. Marine Policy 34(6):1300–1306

Yamakawa U, Mitsui S, Oda T, Morita Y, Aoki T, Maruyama T, Tanaka S, Saito H, Tsuda Y, Senou H (2020) Records of seven fish species showing a northward shift in distribution pattern in the Sagami Bay region, Japan. Nat History Rep Kanagawa 41:71–82. (in Japanese). https://www.jstage.jst.go.jp/article/nkpmnh/2020/41/2020_71/_article/-char/ja/

Yamamoto K, Iryu Y, Sato T, Chiyonobu S, Sagae K, Abe E (2006) Responses of coral reefs to increased amplitude of sea-level changes at the mid-Pleistocene climate transition. Palaeogeogr Palaeoclimatol Palaeoecol 241(1):160–175. https://doi.org/10.1016/j.palaeo.2006.06.014

Yamano H, Namizaki N (2009) Corals on the front lines: changes in the Acropora community in Tateyama, Chiba prefecture. J Coral Reef Soc Japan 11(1):71–72. (in Japanese)

Yamano H, Hori K, Yamauchi M, Yamagawa O, Ohmura A (2001) Highest-latitude coral reef at Iki Island, Japan. Coral Reefs 20:9–12. https://doi.org/10.1007/s003380100137

Yamano H, Sugihara K, Nomura K (2011) Rapid poleward range expansion of tropical reef corals in response to rising sea surface temperatures. Geophys Res Lett 38(4). https://doi.org/10.1029/2010GL046474

Yanagi T (2006) Satoumi theory. Koseisha Kouseikaku
Yao Y, Wang C (2022) Marine heatwaves and cold-spells in global coral reef zones. Prog Oceanogr 209:102920. https://doi.org/10.1016/j.pocean.2022.102920
Yasuda N, Kitano YF, Taninaka H, Nagai S, Mezaki T, Yamashita H (2021) Genetic structure of the *Goniopora lobata* and *G djiboutiensis* species complex is better explained by oceanography than by morphological characteristics. Front Mar Sci 8. https://doi.org/10.3389/fmars.2021.592608
Yeung YH, Xie JY, Kwok CK, Kei K, Ang P, Chan LL, Dellisanti W, Cheang CC, Chow WK, Qiu J-W (2021) Hong Kong's subtropical scleractinian coral communities: baseline, environmental drivers and management implications. Mar Pollut Bull 167:112289. https://doi.org/10.1016/j.marpolbul.2021.112289
Yu X, Yu K, Liao Z, Chen B, Deng C, Yu J, Yao Q, Qin Z, Liang J (2021) Seasonal fluctuations in symbiotic bacteria and their role in environmental adaptation of the scleractinian coral *Acropora pruinosa* in high-latitude coral reef area of the South China Sea. Sci Total Environ 792:148438. https://doi.org/10.1016/j.scitotenv.2021.148438
Yu X, Yu K, Liao Z, Chen B, Qin Z, Liang J, Gao X (2023) Adaptation strategies of relatively high-latitude marginal reef corals in response to severe temperature fluctuations. Sci Total Environ 903:166439. https://doi.org/10.1016/j.scitotenv.2023.166439
Zamora-Jordán N, Hernández M, López C (2022) Biogeography of endosymbionts (Symbiodiniaceae) associated with zoantharian species (Hexacorallia: Anthozoa) from the Macaronesia and Cape Verde ecoregions. Coral Reefs 41(3):511–522. https://doi.org/10.1007/s00338-022-02260-9
Zarco-Perello S, Bosch NE, Bennett S, Vanderklift MA, Wernberg T (2021) Persistence of tropical herbivores in temperate reefs constrains kelp resilience to cryptic habitats. J Ecol 109(5):2081–2094. https://doi.org/10.1111/1365-2745.13621
Zarnetske PL, Skelly DK, Urban MC (2012) Biotic multipliers of climate change. Science 336(6088):1516–1518. https://doi.org/10.1126/science.1222732
Zarzyczny KM, Watson KM, Verduyn CE, Reimer JD, Mezaki T, Beger M (2022) The role of herbivores in shaping subtropical coral communities in warming oceans. Mar Biol 169(5):62. https://doi.org/10.1007/s00227-022-04036-9
Zarzyczny KM, Hellberg ME, Lugli EB, MacLean M, Paz-García DA, Rius M, Ross EG, Treviño Balandra EX, Vanstone J, Williams ST, Fenberg PB (2024a) Opposing genetic patterns of range shifting temperate and tropical gastropods in an area undergoing tropicalisation. J Biogeogr 51(2):246–262. https://doi.org/10.1111/jbi.14744
Zarzyczny KM, Rius M, Williams ST, Fenberg PB (2024b) The ecological and evolutionary consequences of tropicalisation. Trends Ecol Evol 39(3):267–279. https://doi.org/10.1016/j.tree.2023.10.006

Trophic Ecology of Cnidarians and Sponges in Extreme and Marginal Tropical Reefs

14

Sergio Rossi

Abstract

Scleractinians (stony corals) and sponges play pivotal roles in the trophic ecology of coral reefs, shaping these ecosystems' structure, function, and resilience. As primary reef builders, scleractinians contribute to the formation of complex three-dimensional habitats that support diverse marine life. Their trophic strategies in marginal and extreme ecological conditions, including autotrophic energy acquisition through symbiotic relationships with photosynthetic microalgae and heterotrophic feeding on plankton and organic matter, enable them to survive in nutrient-poor oligotrophic waters. In contrast, sponges, as highly efficient suspension feeders, facilitate nutrient recycling by filtering large volumes of water and capturing dissolved organic matter, plankton, and microorganisms.

Together, scleractinians and sponges form a critical trophic bridge, linking primary producers to higher trophic levels and contributing to nutrient cycling also within marginal and extreme coral reef ecosystems. Sponges' capacity to process organic material and excrete bioavailable nutrients supports primary productivity, benefiting both coral and algal communities. Meanwhile, the structural complexity provided by corals and the bioactive compounds produced by sponges influence community dynamics and trophic interactions among reef inhabitants.

This complementary interplay underscores the importance of scleractinians and sponges as ecological engineers and trophic mediators, whose roles are vital for maintaining biodiversity, energy flow, and ecosystem stability in extreme coral reefs that can be considered outside the normal environmental conditions. Understanding their contributions to trophic ecology is crucial for conservation efforts, particularly in the face of climate change and anthropogenic stressors threatening reef health globally.

Keywords

Trophic ecology · Bleaching · Turbidity · Heat waves · Eutrophication · Symbiosis · Mixotrophy · Anthozoans · Porifera

14.1 Introduction

Animal forests worldwide are crucial for benthic-pelagic coupling processes, which are influenced by the feeding strategies, spatial distribution, and biomass of benthic species (Rossi et al. 2017a). This term refers to a living, three-dimensional structure similar to vegetation forests but composed mainly of animals such as sponges, cnidarians, bryozoans, ascidians, and other sessile benthic organisms (Rossi 2013; Rossi et al. 2017b). These organisms constitute a significant proportion of the biomass in benthic communities and form an essential part of the "animal forest" (Orejas et al. 2022).

The primary builders of these structures are clonal organisms of animal origin, which create complex habitats that enhance nutrient exchange. These animal forests capture and retain carbon, nitrogen, and other elements, incorporating them into their tissue and morphology. Nutrients are assimilated through various trophic strategies that differ among taxonomic groups (Gili and Coma 1998). Some benthic suspension feeding species are individual organisms, but those conforming a "forest" emulate the clonal ones gathering sev-

S. Rossi (✉)
Institute of Marine Sciences (LABOMAR), Federal University of Ceará, Fortaleza, Brazil

DiSTeBA, University of Salento, Lecce, Italy

Underwater Gardens International, Barcelona, Spain
e-mail: sergio.rossi@ufc.br

M. O. Soares et al. (eds.), *Extreme and Marginal Coral Reefs and Communities*, Coral Reefs of the World 24,
https://doi.org/10.1007/978-3-032-28035-0_14

eral animals on the same patch (e.g. mussels, oysters, polychaetes, etc.) (Orejas et al. 2022).

Studying the trophic ecology of benthic suspension feeders involves analyzing their feeding strategies, symbiotic relationships, and nutrient cycling processes (Gili and Coma 1998). These organisms also play a role in retaining organic and inorganic matter within their long-lived structures during benthic-pelagic coupling. This understanding is vital for predicting their responses to environmental changes and developing effective conservation strategies (Rossi et al. 2017a).

The diversity of trophic strategies is influenced by the environment in which these organisms shape the various types of forests. From shallow to deep waters, from the pole to the Equator, the marine animal forests conform efficient structures adapted to the food sources, being mixotrophic or heterotrophic. Marginal and/or extreme reefs are not an exception. Marine animal forests could develop on hard substrates and persist under suboptimal or extreme environmental conditions (Soares et al. 2021; Schoepf et al. 2023). These suboptimal conditions are compared with tropical shallow coral reefs considered to live in optimal conditions related to temperature, water transparency and pH. Among other suboptimal conditions, marginal and/or extreme reefs can include elevated siliciclastic sedimentation rates, turbid waters, low oxygen concentration, extreme pH fluctuations, and highly variable temperatures (Camp et al. 2018; Duarte et al. 2020; Burt et al. 2020; Soares 2020; Soares et al. 2021; Schoepf et al. 2023).

The environmental conditions of extreme and marginal habitats are typically unsuitable for most species found in "traditional" shallow-water coral reefs, which thrive in clear, shallow to medium depths with low nutrient concentrations and stable abiotic conditions (Soares et al. 2021; Schoepf et al. 2023). Similar concepts have been applied to terrestrial ecosystems, such as forests, where physiological tolerance to different conditions can lead to acclimatization. Therefore, while terms like 'optimal' and 'marginal' may not make sense from a species-level perspective, they are useful for distinguishing between different environmental sets within this framework (Soares 2020; Schoepf et al. 2023).

Among other habitat builders in the extreme and marginal reefs, cnidarians and sponges are the most abundant suspension feeding organisms (Bell et al. 2013a, b). Understanding the trophic ecology of these two phyla (especially reef-building corals for the cnidarians) in extreme and marginal habitats involves examining how these organisms acquire and utilize nutrients under different suboptimal conditions. The ability of these animal forest species to shape habitats and influence resource distribution underscores their significance in maintaining the health and biodiversity of marine ecosystems (Rossi et al. 2017a), being its trophic adaptations to the environmental conditions of the marginal reefs analysed in this book chapter.

14.2 Environmental Factors

A range of suboptimal environmental parameters such as temperature fluctuations, extreme pH levels, siltation, nutrient loads, and turbidity govern the environmental factors in which Extreme Reefs (ER) are surrounded (Kleypas et al. 1999; Schoepf et al. 2023). These factors result in different types of ERs identified worldwide, where reef communities continue to survive despite harsh conditions (Perry and Larcombe 2003; Perry and Alvarez-Filip 2019). Each type of ER provides unique insights into the resilience and adaptability of coral and associated marine communities in extreme conditions. Among the factors to be considered, temperature and turbidity are the most prominent in terms of ER extension (Kleypas et al. 1999) (Table 14.1).

It is well known that trespassing certain temperature thresholds implies bleaching in a wide range of organisms, especially cnidarians because of the loss of the Symbiodiniaceae (Hoegh-Guldberg 1999; Hughes et al. 2017). In shallow ER, the temperatures in summer may be much higher in certain areas such as the Persian-Arabian Gulf than in other tropical areas. However, in this area for example 55–60 species live in extreme temperature conditions (Riegl and Purkis 2012; Vaughan et al. 2019; Coles and Riegl 2013). To understand why there is such high water temperature regime, we must consider the limited exchange of water masses with the Indian Ocean, forcing corals and other organisms to live with temperatures above the 35 °C in summer and a seasonal temperature range greater than 20 °C (Sheppard et al. 2010; Coles and Fadlallah 1991). Formed about 15,000 years ago and shifting to its current tempera-

Table 14.1 Types of extreme and/or marginal reefs with considered extreme environmental conditions considered in this chapter

Turbid-Zone and High-Temperature Areas: Reefs in these regions withstand high sedimentation rates and elevated temperatures (Camp et al. 2018)
Macro-Tidal Reefs: Reefs subjected to large tidal ranges, which influence water movement and nutrient availability (Enochs et al. 2021)
Tide Pools: Small, isolated pools in the intertidal zone that experience fluctuating conditions (Camp et al. 2018)
Ojos (Low pH Springs): Springs discharging water with lower pH, impacting local marine life (Enochs et al. 2021)
Marine Areas with Submarine Groundwater Discharge: Regions where groundwater enters the ocean, often bringing nutrients and altering water chemistry (Camp et al. 2018)
Mangrove Systems: Coastal areas where mangroves interact with coral reefs, affecting sedimentation and nutrient dynamics (Camp et al. 2019)

ture regime 3–6000 years ago (see chapter 5 in this book), Arabian Gulf corals have adapted to these extreme conditions over a few millennia (Sheppard et al. 2010).

In this context, we must also understand that salinity, due to the high evaporation rates, exceeds easily the 40 psu, adding stress to the osmotic regulation of these symbiotic animals. Bleaching thresholds in the stony corals living in this area is around 35–36 °C, much higher than in Caribbean stony corals for example (31–32 °C). Temperature may arrive to 37–38 °C, being the responsible of massive bleaching and coral mortality (Coles and Fadlallah 1991; Vaughan et al. 2019). A fluctuation between 20 and 38 °C requires a specific thermal tolerance for all the organisms in such environmental conditions, especially when we are focused in their symbiotic relationship. For example, in the Red Sea there was a low latitudinal differentiation in the type and physiology of Symbiodiniaceae in the coral *P. verrucosa*. Only the northernmost reef, in the Gulf of Aqaba, showed differences in the dominant symbiont type and partly in its physiology, indicating that the Gulf of Aqaba is a distinct habitat from the rest of the Red Sea (Ziegler et al. 2019).

The physiological flexibility of the corals may be higher in these areas, having the same Symbiodiniaceae but a relationship with the cnidarian adapted to such drastic changes. For example, the mix of heterotrophy and autotrophy may be the key to understand this adaptability, and previous research has primarily focused on one of these energy strategies (Ribes et al. 1998; Baker et al. 2015; Pupier et al. 2019) and their seasonal variations (Bednarz et al. 2015; Rossi et al. 2020). It is less clear, however, how these extreme temperatures are affecting such heterotrophy-autotrophy balance. As previously mentioned, salinity is related with high temperatures and the hydrological balance. In the Arabian Gulf salinity may reach up to 50 ppt, derived from a low freshwater input from rivers and a high evaporation rate (Ateweberhan et al. 2006; Riegl and Purkis 2012; Vaughan et al. 2019). Corals and other marine organisms must have physiological adaptations to tolerate these extreme salinity levels, often involving specialized cellular mechanisms to regulate ion concentrations and prevent dehydration (Lesser 2006) (Fig. 14.1).

Turbidity is another important suboptimal condition that is considered to have specific adaptation for topical shallow reefs. In the Persian Gulf or in the Equatorial region of Brazil, turbidity is due to high levels of suspended sediments (Castro et al. 2018; Al-Ghadban and El-Sammak 2005). The turbidity interferes in the autotrophic input of mixotrophic benthic suspension feeders reducing light penetration, which is crucial for the microalgae living in the coenenchyma of the corals, for example (Costa-Lucas et al. 2023). But high sedimentation rates can also damage certain coral structures, clog the polyps or make an abrasive action on the living tissue of these organisms. As we will see, there are morphological, physiological and photobiological adaptations that can maintain the potential photosynthetic needs to reach the compensation point and give enough energy to the microalgae and to the host (Warner and Suggett 2016).

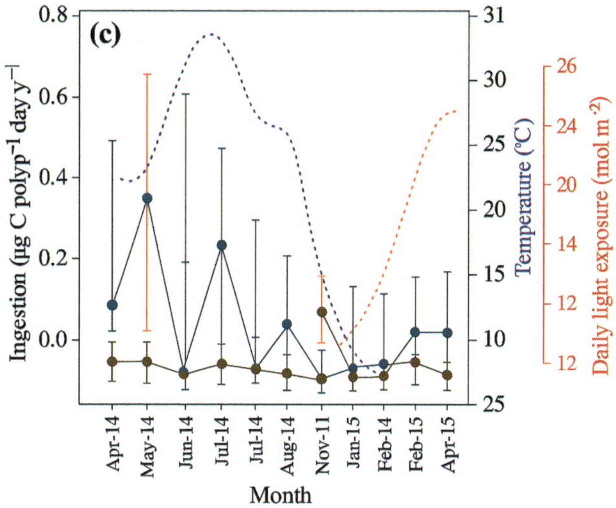

Fig. 14.1 Corresponding heterotrophic carbon ingestion rate of *Plexaurella nutans* (brown) and *Pterogorgia anceps* (green) over an annual cycle. Data represent mean ± SE ($n = 10$ per month and species), and annual variation in seawater temperature (blue) and irradiance at collection depth (orange) are indicated by dotted lines. (Transformed from Rossi et al. (2020))

14.3 Energy Sources

14.3.1 Symbiotic Relationships

Many cnidarians establish symbiotic relationships with photosynthetic algae, called Symbiodiniaceae, which reside within their tissues and supply vital organic compounds through photosynthesis. This partnership is essential for the health of corals and the stability of reef ecosystems, but it is highly susceptible to environmental stressors like elevated temperatures, low light due to turbidity and nutrient shortages (Maina et al. 2011). Algal symbionts of the family Symbiodinaceae in scleractinian corals exhibit strong responses to fluctuations in temperature, nutrients, and light (Wall et al. 2020; Silverstein 2012). These environmental factors vary considerably along different MRs, often diverging from the conditions typically deemed optimal for coral growth. Despite these challenges, different anthozoans flourish throughout these areas, indicating that microalgae have either acclimatized or adapted to the extreme conditions (Camp et al. 2018).

Physiological adjustments, such as changes in metabolic rates and energy distribution, play a crucial role in acclimatization, enhancing survival under varying environmental conditions. Typically, coral reefs do not thrive at mean summer temperatures exceeding 31 °C (Porter et al. 1999). When temperatures surpass this threshold, the photosynthetic mechanisms of microalgae (Symbiodiniaceae), the vital endosymbiotic dinoflagellates, can be compromised, jeopardizing coral health and survival (Silverstein et al. 2012). Numerous studies have demonstrated that elevated temperatures can lead to the 'inactivation' of Symbiodiniaceae, causing their expulsion and resulting in coral bleaching. While the temperature at which coral bleaching begins can differ based on localized adaptations, thriving reefs above 31–32 °C are uncommon but do exist in the Extreme Reefs where the temperature regime is higher. This rarity suggests that Symbiodiniaceae in this region may have acclimatized or evolved to withstand higher temperatures.

In addition to the anticipated rise in temperatures, nutrient loading in coastal systems is expected to increase in the coming decades due to ongoing coastal urbanization and land-use changes (Yan et al. 2019). Although nutrient enrichment is generally harmful to coral reef ecosystems, Symbiodiniaceae initially benefit from increased nutrient availability, leading to higher numbers, enhanced pigmentation, and greater protein content (Wong et al. 2021). However, this can negatively impact the coral host. For example, the coral may lose its ability to regulate nutrient supply to the Symbiodiniaceae, resulting in an uncontrolled proliferation of these symbionts and a disruption of the substrate recycling process between the host and its symbionts (Rawat et al. 2024). Additionally, an imbalance in nutrient ratios, such as excess nitrogen, can increase the stress sensitivity of the microalgae.

Over extended timescales, corals may undergo genetic adaptations that enhance their resilience to environmental stressors. Populations that survive extreme events can pass on beneficial traits to future generations, fostering evolutionary adaptation (Bay and Palumbi 2014; Fine et al. 2013). This process involves mechanisms such as natural selection, where corals with traits that provide greater tolerance to stressors like elevated temperatures, increased salinity, or low nutrient availability are more likely to survive and reproduce (Baker et al. 2004). Over time, these advantageous traits become more widespread within the population. Moreover, genetic diversity within coral populations is crucial for their adaptive potential. A broader genetic diversity offers a wider array of traits that may prove beneficial as environmental conditions change, increasing the likelihood that some individuals will be able to withstand new or intensified stressors. For example, corals in regions with high environmental variability, such as the Arabian Gulf and the southern Red Sea, have demonstrated remarkable resilience (Hume et al. 2013; Osman et al. 2018). This resilience is likely due to a combination of acclimatization and genetic adaptation, with these populations possibly harbouring unique genetic variations that allow them to survive in conditions that typically exceed the tolerance limits of most coral species (Burt et al. 2014).

Gene flow between different coral populations can introduce new genetic material, enhancing the genetic diversity and adaptive capacity of local populations (Baums et al. 2012; van Oppen and Gates 2006; Underwood et al. 2009). This gene flow often occurs through larval dispersal, where coral larvae from various regions intermingle, spreading advantageous genes across populations. The presence of stress-tolerant symbionts, such as specific strains of Symbiodiniaceae, can also result from genetic adaptation, giving corals that harbor these symbionts a survival advantage under extreme conditions (Underwood et al. 2009). Research has shown that some coral species can rapidly evolve in response to climate change, with significant changes occurring within just a few generations (van Oppen and Gates 2006). This rapid evolution is made possible by corals' high reproductive rates and large population sizes, which increase the chances of beneficial mutations arising and being passed on. Additionally, strategies like selective breeding and assisted evolution are being explored to enhance coral resilience by accelerating the process of genetic adaptation (van Oppen et al. 2015; Morikawa and Palumbi 2019; Chan et al. 2018).

With nutrient loading in coastal systems expected to increase alongside rising temperatures in the coming decades, it is crucial to understand the combined effects of elevated temperatures and nutrient concentrations on the coral holobiont, particularly on Symbiodinaceae. The coral holobiont, comprising the coral host and associated microorganisms, operates as a complex and dynamic system (LaJeunesse et al. 2018). Understanding how the microalgae respond to simultaneous increases in temperature and nutrient levels is thus vital, taking into account the whole microbiome of the organism. Research should prioritize identifying the specific thresholds and conditions under which these combined stressors become harmful. Studies need to investigate the physiological and molecular responses of Symbiodiniaceae under these circumstances, including changes in their photosynthetic efficiency, nutrient uptake, and stress tolerance mechanisms (Ziegler et al. 2017). Camp et al. (2019) suggested that physiological plasticity (photosynthetic 'strategy') and flexibility of Symbiodiniaceae taxa associations appear crucial in supporting coral capacity to thrive from reef to lagoon in the Great Coral Reef barrier. Furthermore, it is essential to assess how these stressors impact the overall health and resilience of the coral holobiont, particularly in terms of coral growth, reproduction, and disease resistance. These insights are crucial for predicting the future health of coral reef ecosystems in a changing climate and for developing effective conservation and management strategies. By understanding the synergistic effects of rising temperatures and nutrient levels, scientists and policymakers can better protect and restore coral reefs, helping to ensure their survival amid ongoing environmental changes (Vega Thurber et al. 2014).

14.3.2 Filter Feeding

Corals and sponges are filter feeders that rely on capturing plankton and organic particles from the water to meet their nutritional needs (Gili and Coma 1998; Denis et al. 2024). In extreme or marginal habitats where nutrient concentrations are low or excessive, the efficiency of filter feeding becomes vital. In these environments, corals and sponges have developed specialized adaptations to maximize their feeding efficiency and sustain their metabolic needs (Grottoli et al. 2006; Maldonado et al. 2012). For corals, the presence of symbiotic microalgae offers a significant advantage. In low-nutrient environments, however, corals still rely on their ability to capture and utilize available plankton and organic particles to fully meet their nutritional requirements. Corals enhance their feeding efficiency by extending their polyps, thereby increasing their surface area and improving their chances of capturing passing food particles (Anthony 2006). Sponges, on the other hand, are exceptionally efficient filter feeders, capable of processing large volumes of water to extract suspended particles. They use specialized cells called choanocytes, which create water currents and trap food particles. In nutrient-poor habitats, sponges can adjust their filtration rates and increase choanocyte activity to maximize nutrient intake (Maldonado et al. 2012). Additionally, some sponge species form symbiotic relationships with microorganisms, such as bacteria and cyanobacteria, which can contribute to their nutritional needs by fixing nitrogen or providing additional organic compounds. In terms of heterotrophy, sponges may be much more adapted to exploit the water column in oligotrophic conditions respect corals or other anthozoans (Maldonado et al. 2012; de Goeij et al. 2013; Bell et al. 2013a, b).

Both corals and sponges may also exhibit behavioral adaptations to enhance their feeding efficiency. For example, some corals can adjust their position or orientation to face prevailing currents, optimizing their exposure to plankton-rich water (Sebens et al. 2003; Hoogenboom et al. 2008). Sponges can modify their pumping rates and the structure of their internal canal systems to improve water flow and particle capture (Pile et al. 1997; Leys et al. 2011). In both cases, the patch diversity and structure may enhance particle retention and feeding efficiency (Nelson and Bramanti 2020). The seasonal variation of heterotrophic energy input in warm-temperate non-symbiotic octocorals has also been extensively studied to understand reproductive cycles, energy investment in respiration, trophic crises, and population dynamics based on the seasonal availability of food (Ribes et al. 1999; Tsounis et al. 2006; Coma et al. 2015; Rossi et al. 2004, 2017b, and references therein), including potential energy deficits or surpluses (Coma et al. 1998; Coppari et al. 2016). Understanding these adaptations and the efficiency of filter feeding in corals and sponges is crucial for comprehending their survival strategies in extreme or marginal habitats.

By capturing plankton and organic particles from the water, sponges and corals play a vital role in regulating nutrient cycles within reef environments. Their filter-feeding activities help remove excess particulate matter and maintain water clarity, which is essential for the photosynthetic efficiency of the microalgae and the overall health of the reef. The nutrients acquired through filter feeding support the metabolic needs and growth of these organisms, which in turn provide habitat and structural complexity to the reef ecosystem. The ability of sponges and corals to efficiently acquire nutrients in low-nutrient environments underscores their adaptability and resilience (Reiswig 1974; Grottoli et al. 2006). Their capacity to sustain themselves through filter feeding, even under challenging conditions, highlights the importance of these organisms in supporting biodiversity and ecosystem function. In nutrient-poor or extreme habitats, their presence ensures the continuation of essential ecological processes, such as primary production, nutrient recycling, and habitat formation (Hughes et al. 2007; Muscatine and Porter 1977; de Goeij et al. 2013). Moreover, the filter-feeding activities of sponges and corals have broader implications for the entire reef community. By capturing and recycling nutrients, they also indirectly support the nutritional needs of other reef organisms, including fish and invertebrates (Bell et al. 2013a, b; Graham et al. 2006). This interdependence underscores the critical role of sponges and corals in maintaining the overall productivity and resilience of reef ecosystems.

Another extreme situation may be the excess of nutrients (eutrophication). The same community, under an excess of nutrients, may drastically change the N and C equilibria (Gonzàlez-de Zayas et al. 2020). Excess nutrients, particularly nitrogen (e.g., nitrate, ammonium) and phosphorus (e.g., phosphate), has detrimental effects on coral reefs. These nutrients often enter reef ecosystems through agricultural runoff, sewage discharge, and coastal development. For example, from the trophic point of view, elevated nutrient levels exacerbate coral bleaching by increasing the susceptibility of corals to thermal stress (Wiedenmann et al. 2013). Nutrients promote the growth of symbiotic algae, but this relationship becomes unstable under thermal stress, leading to expulsion of Symbiodinaceae, as previously showed. The excess of nutrients may have an evident unbalance in the availability of some elements, which makes the symbiotic relationship more unstable. The potential bleaching effect will haves a direct consequence, the lack of autotrophic energy input on the cnidarian that may, however, substitute part of the trophic input by capturing more preys. More problematic may be the rapid growth of micro and macroalgae in the impacted area. This event shifts the reef ecosystem from coral-dominated to algae-dominated states (McCook et al. 2001). Also, taking into account the trophic balance, the excess nutrients reduces the ability of corals to calcify, leading to weaker skeletons and slower growth rates.

14.4 Adaptation to Extreme Environmental Conditions

14.4.1 Thermal Tolerance

High temperatures can disrupt the photosynthetic functions of the symbiont microalgae (Warner et al. 1999) resulting in the production of harmful reactive oxygen species (ROS, Lesser 1997). This stress may lead to programmed cell death, or apoptosis, causing the Symbiodiniaceae to be expelled from the coral, a process known as coral bleaching (Baird et al. 2009; Baker et al. 2008). The ROS can damage vital cellular components like proteins, lipids, and DNA (Lesser 1997). Different algal cell strains exhibit varying levels of thermal tolerance, and some corals host multiple types of symbionts, potentially enhancing their resilience to temperature changes (Baker et al. 2004; LaJeunesse et al. 2018). However, the overall susceptibility of corals to bleaching is often determined by the most sensitive micoralgae present in their tissues.

The loss of Symbiodiniaceae not only robs corals of vital nutrients but also subjects them to additional stress and the risk of mortality. Without their symbiotic partners, corals face challenges in meeting their energy needs, which can lead to stunted growth, weakened immune responses, and decreased reproductive success (Hume et al. 2016). If thermal stress continues and corals fail to re-establish their symbiosis with zooxanthellae, they may ultimately perish, contributing to the decline of reef ecosystems. With marine heatwaves becoming more frequent and intense due to climate change, it's essential to understand the mechanisms behind coral bleaching and the thermal tolerance of various microalgae strains for effective conservation (Cunning et al. 2015). Strategies like selective breeding, assisted evolution, and safeguarding thermally resilient coral populations are being investigated to boost corals' ability to endure higher temperatures (Berkelmans and van Oppen 2006). Addressing the impacts of thermal stress on the coral-microalgae relationship is vital for the protection and preservation of coral reef ecosystems in a warming world.

Corals and sponges in extreme environments often display thermal tolerance that allows them to endure high temperatures (Baker et al. 2004; Lesser et al. 2018; Webster and Thomas 2016; AlMealla et al. 2024). However, prolonged exposure to severe conditions, like coral bleaching events, can disrupt their symbiotic relationships and nutrient acquisition. Additionally, certain corals enhance their thermal tolerance through acclimatization, adjusting their physiology to better manage temperature fluctuations (Hume et al. 2013). Sponges have also evolved strategies to thrive in high-temperature environments, including the regulation of metabolic processes and adjustments in filtration rates to maintain homeostasis under thermal stress (Lesser et al. 2018). Moreover, some sponge species harbor symbiotic microorganisms that boost their thermal resilience by supplying essential nutrients or aiding in stress response mechanisms (Webster and Thomas 2016).

Despite their adaptations, prolonged exposure to extreme conditions can have severe consequences for corals and sponges (AlMealla et al. 2024). The loss of translocated metabolites not only hinders coral growth and reproduction but also increases their susceptibility to disease and mortality. The longer a bleaching event persists, the harder it becomes for corals to recover and re-establish their symbiotic relationships (Hughes et al. 2017). For sponges, extended thermal stress can diminish filtration efficiency and impair nutrient uptake (Maldonado and Zhang 2018). The disruption of symbiotic relationships with microorganisms can exacerbate these effects, further compromising the sponge's ability to thrive in extreme conditions, leading to weakened health, reduced reproductive output, and heightened vulnerability to environmental stressors (Gundermann et al. 2022; de Goeij et al. 2013).

As previously mentioned, corals can adapt to thermal stress by changing their symbiotic microalgae to more thermally tolerant strains, a process known as "symbiont shuffling" (Baker et al. 2004; LaJeunesse et al. 2018). This mechanism allows corals to selectively retain or acquire more resilient symbionts from their existing pool, enhancing their ability to withstand higher temperatures and mitigate bleaching effects. In addition to shuffling, corals can engage in "symbiont switching," where they form new associations with different types of zooxanthellae from their surroundings (Jones and Berkelmans 2010; Silverstein et al. 2015). This process enables the uptake of new symbionts that are better suited to current environmental conditions, further enhancing thermal tolerance and resilience. Both symbiont shuffling and switching are crucial for corals adapting to changing and stressing environments. These mechanisms help maintain their symbiotic relationships and optimize photosynthetic efficiency under stress, allowing corals to better manage the production of ROS and reduce the risk of bleaching. Research indicates that some coral species are more proficient at shuffling and switching than others, influencing their survival rates during thermal stress events (Baker et al. 2004; Silverstein et al. 2015). The potential for symbiont shuffling and switching emphasizes the dynamic nature of coral-symbiont relationships and highlights the significance of genetic and environmental diversity in enhancing coral adaptability. Conservation efforts should prioritize preserving this diversity, safeguarding thermally tolerant coral populations, and facilitating the recovery of bleached reefs. Creating conditions that support the natural processes of symbiont shuffling and switching will be essential for promoting coral resilience in the face of climate change.

For example, in the Arabian Gulf corals exhibit remarkable resilience, partly due to symbiont shuffling. In these scleractinians the host heat-tolerant symbionts are represented by genus *Symbiodinium* (formerly clade D). These symbionts are better adapted to withstand high temperatures compared to other types, such as those from clade C (Hume et al. 2016). During cooler months, corals may host a mix of symbiont types, but as temperatures rise, they increase the proportion of heat-tolerant clade D symbionts (Baker et al. 2004). In fact, the ability to shuffle symbionts allows Gulf corals to survive temperatures that would be lethal to corals in other regions. This adaptability is a key factor in their persistence in such an extreme environment (Riegl et al. 2011).

Increased production of antioxidants plays a crucial role in mitigating the damage caused by ROS generated under thermal stress. To combat the harmful effects of ROS, both corals and their symbionts increase antioxidant production. Antioxidants are molecules that neutralize ROS, helping to prevent oxidative stress and protect cellular integrity (Lesser 2006; Weis 2008). Key antioxidants include superoxide dismutase (SOD), catalase, and various peroxidases, which work synergistically to detoxify ROS and repair oxidative damage (Valko et al. 2007). This enhanced antioxidant production is a critical aspect of the coral stress response, enabling them to maintain cellular function and survivability during thermal stress. The capacity for increased antioxidant production can be an innate trait of specific coral species or zooxanthellae strains, contributing to their thermal tolerance. Some corals can also upregulate antioxidant production in response to environmental cues, providing a rapid defense against sudden temperature spikes. Research indicates that the antioxidant response is not uniform across all coral species or even within populations, highlighting the role of genetic variation in determining the effectiveness of this defense mechanism (Weis 2008). Antioxidant production is often linked to the overall health and nutritional status of corals. Adequate nutrient availability, particularly of elements like zinc and manganese—essential cofactors for antioxidant enzymes—can enhance the coral's ability to produce antioxidants (Weis 2008). his underscores the importance of maintaining healthy reef ecosystems with balanced nutrient levels to bolster coral resilience.

Heat Shock Proteins (HSP) also play a vital role in protecting cellular structures from damage caused by high temperatures. When corals experience thermal stress, elevated temperatures can lead to the denaturation and aggregation of proteins, disrupting cellular function and integrity. HSPs are a class of molecular chaperones that help maintain cellular homeostasis under these stressful conditions (Kiang and Tsokos 1998; Feder and Hofmann 1999). They are produced in response to heat and other stressors, stabilizing and refolding damaged proteins to prevent harmful aggregates. Additionally, HSPs assist in the degradation of irreparably damaged proteins, ensuring the efficiency of the cell's protein quality control system. The expression of HSPs in corals can be induced by various environmental stressors, including high temperatures, UV radiation, spatial competition, food deprivation and oxidative stress (Fang et al. 1997; Brown et al. 2002). This protective mechanism enables corals to maintain cellular function and survive periods of elevated temperatures (Downs et al. 2000). Several types of heat shock proteins, such as HSP70, HSP90, and small HSPs, are involved in the coral stress response. HSP70, for example, binds to unfolded proteins, preventing their aggregation and facilitating proper refolding (Feder and Hofmann 1999). HSP90 is crucial for the maturation and stabilization of a wide array of proteins, including those essential for signal transduction and stress responses (Taipale et al. 2010; Pratt and Toft 2003). Small HSPs act as the first line of defence by binding to partially denatured proteins, preventing aggregation and aiding their interaction with other chaperones for refolding or degradation. This rapid response is highly conserved, allowing corals to quickly adapt to sudden changes in their environment. The efficiency of HSP expression and function varies among different coral species and populations, contributing to the observed variability in thermal tolerance across coral reefs (Sharp et al. 1997). Research into the role of HSPs in coral thermal tolerance has underscored their importance in enhancing resilience to climate change. Corals capable of effectively upregulating HSPs in response to thermal stress are better equipped to survive bleaching events and other environmental challenges (Richier et al. 2006; Weis 2008).

It has been shown that Arabian Gulf corals (which may be exposed to up to 38 °C, AlMealla et al. 2024) exhibit elevated expression of HSPs, particularly HSP70 and HSP90, which help them withstand thermal stress (D'Angelo et al. 2015). Corals in the Red Sea also experience high temperatures and salinity, and it has demonstrated that they upregulate HSP genes as part of their stress response, enabling them to maintain cellular function under extreme conditions (Osman et al. 2018). Corals in extreme environments often allocate significant energy to stress responses, which may reduce growth and reproduction rates (Kenkel et al. 2013). The HSP expression has a high energetic cost in terms of nutrients dedicated to this cellular activity, so in extreme environments the quantity of energetic effort may compromise the performance of the species (Snyder and Rossi 2004).

Thermal stress also significantly impacts the metabolic rates of sponges. Elevated temperatures increase their metabolic demands, resulting in higher energy expenditure for maintenance and repair processes (Webster et al. 2008; Bell et al. 2013a, b; Lemoine et al. 2007). This added metabolic burden can diminish the energy available for growth, reproduction, and other essential functions, ultimately weakening the sponge over time. Additionally, thermal stress can compromise the structural integrity of sponges. High temperatures may weaken their skeletal components, making them more susceptible to physical damage from waves, currents, and bioerosion (Schmidt 2014). This structural vulnerability can lead to increased mortality rates and a diminished capacity to recover from environmental disturbances. The cumulative effects of prolonged thermal stress can greatly reduce the resilience of sponges to other environmental stressors, such as pollution, habitat degradation, and predation. Compromised sponges are less equipped to handle these additional challenges, resulting in higher mortality rates and potentially altering the composition and function of sponge communities within reef ecosystems (Bennett et al. 2018).

14.4.2 Turbidity

Turbid waters, marked by elevated levels of suspended sediments, have notable trophic implications for corals, affecting their growth, reproduction, and overall health, which in turn influences the broader reef ecosystem (Fabricius 2005). Increased turbidity reduces light availability, directly impacting the photosynthetic efficiency of microalgae. With diminished light penetration, the photosynthetic activity declines, leading to lower production of vital organic compounds that corals need for energy and growth (Anthony and Fabricius 2000). In these turbid conditions, corals may increasingly depend on heterotrophic feeding to offset the decrease in photosynthesis. Turbidity can enhance the availability of particulate organic matter (POM), which corals can capture and consume. This shift toward heterotrophy helps sustain corals during times of limited light availability (Grottoli et al. 2006; Houlbrèque and Ferrier-Pagès 2009). The effectiveness of corals in utilizing POM is influenced by their morphological adaptations and feeding mechanisms, with some species better equipped to take advantage of these resources (Anthony 2000). Sediment accumulation on coral surfaces can cause other challenges. The first one is the physical damage, because accumulated sediments can physically harm coral tissues, hindering their ability to capture food and perform essential functions (Fabricius et al. 2013). This has a direct consequence, a potentially higher energy expenditure. In fact, corals expend additional energy to remove sediments, often through mucus secretion and ciliary action. This extra energy expenditure can divert resources away from growth and reproduction (Anthony 2000).

Emerging evidence indicates that coral reefs in turbid waters may exhibit increased resilience to climate change impacts, potentially serving as vital conservation hotspots (Costa-Lucas et al. 2023; Denis et al. 2024). The shading effect of turbid waters can alleviate thermal stress, thereby reducing the severity of coral bleaching during heatwaves (Costa-Lucas et al. 2023). Additionally, suspended particles can offer protection from harmful UV radiation and promote enhanced heterotrophic feeding, compensating for the decline in photosynthetic activity. This unique environment may enable certain coral species to thrive, even in conditions typically considered marginal for coral growth (Anthony and Fabricius 2000; Fabricius et al. 2013). Despite these potential advantages, scientific understanding of turbid reef ecosystems remains limited. Logistical challenges in studying these environments—such as poor visibility, difficult diving conditions, and the need for specialized equipment—have led to their underrepresentation in scientific literature (Soares 2020; Soares et al. 2017). Moreover, studies focusing on turbid reefs often employ different methodologies and definitions to characterize turbidity, complicating comparisons and comprehensive conclusions. The lack of standardized protocols for measuring and defining turbidity has resulted in inconsistencies in research outcomes. Some studies may prioritize suspended sediment concentrations, while others might assess water clarity or light attenuation, influencing data interpretation and hindering a cohesive understanding of how turbidity impacts coral reef resilience and ecosystem dynamics (Costa-Lucas et al. 2023) (Fig. 14.2).

To fill these gaps, coordinated research efforts and standardized methodologies for studying turbid reefs are essential. By harmonizing definitions and measurement techniques, scientists can more effectively compare findings across diverse regions and environmental conditions. This will enhance our comprehension of the ecological roles and conservation potential of turbid reefs, providing valuable insights for reef management and protection strategies in the face of climate change. It has been proposed that turbid reefs can be more accurately described using new classifications that focus on their turbidity regimes and sediment sources. These classifications include persistent, fluctuating, and transitional turbidity regimes, as well as distinguishing between natural and anthropogenic sediment sources (Larcombe and Woolfe 1999; McLaughlin et al. 2003). This enhanced categorization aids in understanding the distinct ecological

Fig. 14.2 In July 2020, colonies of the coral *Siderastrea stellata* experienced bleaching along the coast of Ceará in the Equatorial Southwestern Atlantic, Brazil. (**a**, **b**) Encrusting colonies of *S. stellata* observed at a depth of 18 m in the Cabeço do Balanço reef. (**c**) Numerous fully bleached colonies of the key reef-building species, *S. stellata*. (Figure from Costa-Lucas et al. (2023), with the permission of the authors)

dynamics and conservation requirements of various turbid reef systems.

The resilience of certain turbid reefs to thermal stress can be attributed to several key factors. Notably, the presence of stress-tolerant species plays a significant role, as these organisms are often pre-adapted to withstand high and variable temperature conditions due to chronic exposure. Furthermore, light attenuation caused by suspended sediments reduces solar irradiance, alleviating stress on corals during high-temperature events (van Woesik and McCaffrey 2017; Teixeira et al. 2019). Recent modeling studies indicate that turbidity may help mitigate high-temperature bleaching for approximately 12% of the world's reefs, with 30% of these located within the biodiversity-rich Coral Triangle (Sully and van Woesik 2020). Additionally, corals inhabiting turbid waters have shown faster recovery rates and lower mortality following bleaching events, likely due to their enhanced heterotrophic feeding capacity (Guest et al. 2016; Banha et al. 2019). This increased capacity not only supports recovery from bleaching but also provides resilience against ocean acidification (Albright and Cooley 2019) (Table 14.2).

The Paluma Shoals Reef Complex in Australia represents a prime example of a naturally turbid ecosystem (Larcombe

Table 14.2 Turbidity regimes description

Persistent turbidity	Reefs that experience consistently high levels of turbidity throughout the year
Fluctuating turbidity	Reefs where turbidity levels vary significantly, often influenced by seasonal changes or episodic events
Transitional turbidity	Reefs that occasionally experience high turbidity but generally maintain clearer water conditions
Natural turbidity	Sediment input primarily from natural sources such as river outflows, coastal erosion, and resuspension of seabed sediments due to wave action
Anthropogenic turbidity	Sediment input predominantly from human activities such as dredging, land reclamation, coastal development, and agricultural runoff

and Woolfe 1999). This reef complex experiences natural turbidity largely due to river discharge and the resuspension of sediments caused by tidal and wave action. The coral and benthic communities in this area have adapted to these naturally fluctuating light conditions and sediment deposition (Browne et al. 2012). This natural sediment regime fosters a dynamic and resilient ecosystem, with species that are well-equipped to thrive under such conditions. As a result, the reef exhibits high resilience, having developed long-term adaptations that enable it to withstand and recover from various environmental stressors. Although generally healthier and less affected by human activities, the complex remains vulnerable to climate change and extreme weather events.

In contrast, the reefs of Singapore face significant challenges from anthropogenic turbidity (Huang et al. 2009; Dikou and van Woesik 2006). These reefs are subjected to high levels of sedimentation resulting from extensive coastal development, land reclamation, and port activities. As a consequence, the coral and benthic communities here are greatly impacted by increased sedimentation and reduced light availability (Heery et al. 2018). This anthropogenic sediment regime has altered community structures, leading to potential decreases in biodiversity and shifts in ecological functions. Compared to their naturally turbid counterparts, Singapore's reefs exhibit lower resilience, primarily due to the compounded stresses from pollution, habitat destruction, and ongoing human disturbances (Chou 2006).

By comparing the composition, function, and resilience of the Paluma Shoals Reef Complex and the Singapore reefs, it becomes clear that these two turbid reef systems represent distinct environments with varying conservation statuses (see Chapter 9 in this book). Understanding these differences is essential for developing more precise conservation strategies. Natural turbid reefs like Paluma Shoals benefit from inherent resilience mechanisms, allowing them to adapt and thrive despite fluctuating conditions. In contrast, anthropogenically impacted reefs such as those in Singapore require targeted interventions to mitigate the effects of human activities and enhance ecosystem recovery. This disparity underscores the need for incorporating specific descriptors for turbidity regimes and sediment sources in reef studies. By doing so, conservation efforts can be better informed, leading to improved management practices that address the unique challenges and needs of different turbid reef ecosystems.

14.4.3 Nutrient Cycling Efficiency

As previously explained, nutrient availability significantly impacts the coral-symbiotic algae relationship, where moderate nutrient levels can enhance zooxanthellae function and decrease susceptibility to bleaching (Vega Thurber et al. 2014). This balance is particularly delicate in regions like the Red Sea, which serves as a natural laboratory for studying the interactions between nutrient gradients and high temperatures (Sawall et al. 2014). While studies indicate that nutrient enrichment can benefit coral health to some extent, the overall impact is complex and warrants further investigation to identify optimal conditions for coral resilience (Zhao et al. 2021). Some corals form associations with nitrogen-fixing bacteria that convert atmospheric nitrogen into forms that corals can utilize, thus compensating for low nitrogen availability (Cara et al. 2010). Corals have also developed specialized cellular mechanisms to enhance nutrient uptake, such as increasing the surface area of their gastrodermal cells for improved absorption efficiency. They employ flexible feeding strategies, capturing plankton and organic particles from the water to supplement their nutrient intake, allowing them to survive periods of low nutrient availability (Muller-Parker et al. 2015).

In nutrient-poor environments, sponges play a vital role through their ability to effectively filter-feed. Research has shown that sponges can adapt to varying levels of picoplankton availability, with different species demonstrating preferences for specific types of picoplankton (Pawlik et al. 2015, 2018). For example, species like *Amphimedon compressa* and *Niphates erecta* excel at removing cyanobacteria and heterotrophic bacteria, essential for their nutrition in nutrient-limited habitats (Poppell et al. 2013). Additionally, sponges can reduce their metabolic rates to conserve energy and minimize nutrient requirements during times of scarcity, allowing them to endure fluctuating nutrient conditions. Their ability to adjust symbiotic associations based on environmental factors further optimizes their nutrient acquisition strategies, ensuring survival in challenging habitats (Freeman et al. 2021; Thacker and Freeman 2012).

14.5 Combined Impact on Ecosystems

The nutrient efficiency of corals and sponges in extreme and marginal habitats significantly impacts the health and stability of marine ecosystems (Soares et al. 2021). These adaptations not only support the survival of individual organisms but also enhance the overall productivity and resilience of reef systems. By effectively utilizing available nutrients, corals and sponges help maintain water quality and promote biodiversity, even in challenging conditions (Glasl et al. 2017). Extreme and marginal habitats may have limited resources, leading to increased competition among corals, sponges, and other organisms for available nutrients (AlMealla et al. 2024) (Fig. 14.3).

Space competition between corals and sponges is a crucial factor on the reef, as both groups vie for limited space, with the ability to overgrow or outcompete neighbors often

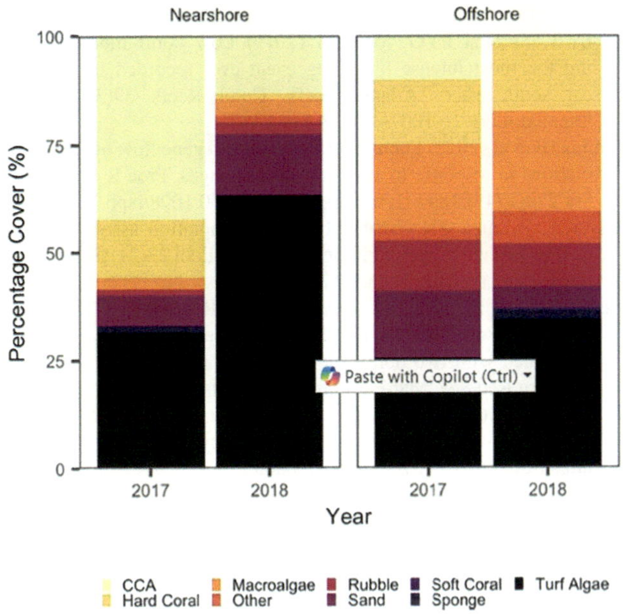

Fig. 14.3 Mean percentage cover (%) of the reef benthic community structure pre (2017) and post (2018) bleaching in nearshore and offshore sites in Bahrain. (From AlMealla et al. (2024))

determining species success (López-Victoria et al. 2006; Porter and Targett 1988). In some environments, sponges can outcompete corals due to their rapid growth rates and ability to thrive in low-light conditions. Additionally, nutrient competition plays a significant role; corals obtain nutrition through both autotrophic means, via their symbiotic algae, and heterotrophic means, by capturing plankton. However, sponges, as efficient filter feeders, can deplete the water of plankton and organic particles, potentially limiting the resources available to corals (Pawlik and McMurray 2020). Within coral species, competition for light, space, and nutrients influences growth forms and colony structures, with faster-growing or more aggressive species often dominating over slower-growing ones. Similarly, sponges compete within their own groups for space and resources, with species that possess more efficient filter-feeding capabilities or form symbiotic relationships with nutrient-cycling microorganisms gaining a competitive edge (Ruetzler 2004; Kiran et al. 2018).

Corals and sponges are vital to nutrient cycling within reef ecosystems, with sponges particularly important for recycling nutrients by breaking down organic matter and making nutrients accessible to other organisms. Resilient species that can adapt to shifting conditions will play a significant role in shaping future reef ecosystems (Anthony et al. 2015; Camp et al. 2018). Additionally, turbid conditions can alter competitive dynamics within these ecosystems; species that are better adapted to such environments—those with enhanced heterotrophic capabilities or more effective sediment-clearing mechanisms—may become more dominant. This shift can result in changes to community structure and biodiversity, further influencing the health of reef ecosystems.

14.6 Role as Ecosystem Engineers

Corals and sponges are often regarded as ecosystem engineers due to their significant contributions to the physical structure of their habitats (Rossi et al. 2017b). In extreme environments, their ecological roles become even more pronounced, influencing resource availability and providing refuge for various organisms (Smith and Buddemeier 1992). The calcium carbonate structures they create also serve as substrates for other organisms to attach and grow, thereby boosting overall biodiversity. Similarly, sponges contribute to the physical complexity of marine habitats by forming dense aggregations on reefs and other substrates. The presence of sponges enhances habitat structural diversity, supporting higher biodiversity within these ecosystems.

Both corals and sponges in extreme environments, such as those with high temperatures, low nutrient levels, or high acidity, display remarkable adaptations that enable them to survive and thrive under challenging conditions. This resilience not only stabilizes the ecosystem but also helps maintain its functionality despite environmental stressors. Furthermore, in nutrient-poor environments, their efficient nutrient acquisition and recycling capabilities play a critical role. By capturing and processing scarce resources, corals and sponges help distribute nutrients more evenly throughout the ecosystem, thereby supporting the survival of other organisms and promoting overall ecosystem health (AlMealla et al. 2024). The ability of corals and sponges to engineer their environments and create suitable habitats under extreme conditions can drive evolutionary processes. Species that successfully exploit the niches created by these organisms are more likely to survive and adapt to changing conditions (Rossi et al. 2019). As ocean temperatures rise and conditions shift, some species may migrate to more favorable environments, which can impact local biodiversity and ecosystem dynamics (Denis et al. 2024). Therefore, monitoring and managing these habitat shifts is essential for maintaining the health and stability of marine ecosystems.

14.7 Conclusions

Corals and sponges play critical roles as ecosystem engineers, and their resilience and adaptability significantly influence the overall health and stability of marine ecosystems. Maintaining their functions is essential for the continued provision of ecosystem services, including habitat creation, nutrient cycling, and water filtration. In extreme and marginal habitats, these organisms exhibit various resilience and adaptive mechanisms that enable them to cope with environmental variability. Studying these responses is vital for predicting how they will fare under the changing conditions brought about by climate change. By understanding and supporting these natural adaptations, conservation efforts can be more effectively directed toward preserving these vital components of marine ecosystems.

Acknowledgements SR is especially grateful to the Horizon Europe project OCEAN CITIZEN (Horizon Europe-Horizon-MISS-2021-OCEAN-02-01 European Blue Parks, Contract number 101093910), as well as the Deep LIFE Project (BNP Paribas Foundation). SR is in the CNPq excellence program (Process number 314460/2023-3, Level E).

References

Albright R, Cooley S (2019) A review of interventions proposed to abate impacts of ocean acidification on coral reefs. Reg Stud Mar Sci 29:100612. https://doi.org/10.1016/j.rsma.2019.100612

Al-Ghadban AN, El-Sammak A (2005) Sources, distribution and composition of the suspended sediments, Kuwait Bay, Northern Arabian Gulf. J Arid Environ 60(4):647–661. https://doi.org/10.1016/j.jaridenv.2004.11.008

AlMealla RK, Edullantes B, Hepburn LJ (2024) Bleaching threatens positive carbonate budgets on Bahraini reefs. Mar Biol 171:39. https://doi.org/10.1007/s00227-023-04351-9

Anthony KRN (2000) Enhanced particle-feeding in turbid waters: a mechanism of resilience for coral reefs subjected to terrestrial runoff. Mar Freshw Res 51(5):569–577. https://doi.org/10.1071/MF99101

Anthony KRN (2006) Feeding strategies and metabolic flexibility in reef corals. Coral Reefs 25(4):696–702. https://doi.org/10.1007/s00338-006-0126-3

Anthony KRN, Fabricius KE (2000) Shifting trophic balance of coral reefs under turbidity. Mar Ecol Prog Ser 200:1–8. https://doi.org/10.3354/meps200001

Anthony KRN, Marshall PA, Abdulla A, Beeden R, Bergh C, Black R, Akin CM, Game ET, Gooch M, Graham NAJ, Green A, Heron SF, van Hooidonk R, Knowland C, Mangubhai S, Marshall N, Maynard JA, McGinnity P, McLeod E, Mumby PJ, Nyström M, Obura D, Oliver J, Possingham HP, Pressey RL, Rowlands GWP, Tamelander J, Wachenfeld D, Wear S (2015) Operationalizing resilience for adaptive coral reef management under global environmental change. Glob Chang Biol 21(1):48–61. https://doi.org/10.1111/gcb.12700

Ateweberhan M, Bruggemann JH, Breeman AM (2006) Effects of extreme seasonality on community structure and functional group dynamics of coral reef algae in the southern Red Sea (Eritrea). Coral Reefs 25:391–406. https://doi.org/10.1007/s00338-006-0109-6

Baird AH et al (2009) Coral bleaching: the role of the host. Trends Ecol Evol 24(1):16–20. https://doi.org/10.1016/j.tree.2008.09.005

Baker AC et al (2004) Corals' adaptive response to climate change. Nature 430(7001):741–744. https://doi.org/10.1038/nature02791

Baker AC et al (2008) Thermotolerance in corals. Coral Reefs 27(3):441–453. https://doi.org/10.1007/s00338-008-0397-0

Baker DM, Freeman CJ, Knowlton N, Thacker RW, Kim K, Fogel ML (2015) Productivity links morphology, symbiont specificity, and bleaching in the evolution of Caribbean octocoral symbioses. ISME J 9:2620–2629

Banha TNS, Capel KCC, Kitahara MV, Francini-Filho RB, Francini CLB, Sumida PYG, Mies M (2019) Low coral mortality during the most intense bleaching event ever recorded in subtropical southwestern Atlantic reefs. Coral Reefs 39(3):515–521. https://doi.org/10.1007/s00338-019-01856-y

Baums IB et al (2012) Population structure and gene flow of coral populations in response to environmental changes. Proc R Soc B Biol Sci 279(1745):1894–1903. https://doi.org/10.1098/rspb.2011.2383

Bay RA, Palumbi SR (2014) Multilocus adaptation associated with heat resistance in reef-building corals. Mol Ecol 23(21):5071–5084. https://doi.org/10.1111/mec.12844

Bednarz VN, Cardini U, van Hoytema N, Al-Rshaidat MMD, Wild C (2015) Seasonal variation in dinitrogen fixation and oxygen fluxes associated with two dominant zooxanthellate soft corals from the northern Red Sea. Mar Ecol Prog Ser 519:141–152

Bell JJ et al (2013a) Sponge contributions to coral reef nutrient cycling. Philos Trans R Soc B Biol Sci 368(1617):20130065. https://doi.org/10.1098/rstb.2013.0065

Bell JJ et al (2013b) Could some coral reefs become sponge reefs as our climate changes? Glob Chang Biol 19(10):2977–2984. https://doi.org/10.1111/gcb.12234

Bennett HM et al (2018) Thermal tolerance and ecological consequences for sponges in warming oceans. Glob Chang Biol 24(2):577–590. https://doi.org/10.1111/gcb.13900

Berkelmans R, van Oppen MJ (2006) The role of zooxanthellae in coral thermal tolerance. Coral Reefs 25(4):551–555. https://doi.org/10.1007/s00338-006-0142-7

Brown BE et al (2002) Coral responses to environmental stressors: the role of heat shock proteins. In: Stress responses in marine organisms. Elsevier. https://doi.org/10.1016/B978-012171620-9/50008-7

Browne NK et al (2012) Spatial and temporal variations in turbidity on the great barrier reef: distinguishing natural from anthropogenic changes. Mar Pollut Bull 64(5):965–975. https://doi.org/10.1016/j.marpolbul.2012.02.021

Burt JA et al (2014) Coral bleaching and mortality thresholds in the UAE. Mar Pollut Bull 88(1–2):119–123. https://doi.org/10.1016/j.marpolbul.2014.08.019

Burt JA, Camp EF, Enochs IC, Johansen JL, Morgan KM, Riegl B, Hoey AS (2020) Insights from extreme coral reefs in a changing world. Coral Reefs 39:495–507. https://doi.org/10.1007/s00338-020-01966-y

Camp EF, Schoepf V, Mumby PJ, Hardtke LA, Rodolfo-Metalpa R, Smith DJ, Suggett DJ (2018) The future of coral reefs subject to rapid climate change: lessons from natural extreme environments. Front Mar Sci 5:1–21. https://doi.org/10.3389/fmars.2018.00004

Camp EF, Edmondson J, Doheny A, Rumney J, Grima AJ, Huete A, Suggett DJ (2019) Mangrove lagoons of the great barrier reef support coral populations persisting under extreme environmental conditions. Mar Ecol Prog Ser 625:1–14. https://doi.org/10.3354/meps13073

Cara F et al (2010) Nitrogen fixation and nitrogen transformations in marine symbioses. Trends Microbiol 18(10):455–463

Castro ALC, Eschrique SA, Silveira PCA, Azevedo JWJ, Ferreira HRS, Soares LS, Monteles JS, Araújo MC, Nunes JL, Silva MHL (2018) Physicochemical properties and distribution of nutrients on the inner continental shelf adjacent to the Gulf of Maranhão (Brazil) in the equatorial Atlantic. Appl Ecol Environ Res 16(4):4829–4847

Cecilia D'Angelo, Benjamin C C Hume, John Burt, Edward G Smith, Eric P Achterberg, Jörg Wiedenmann, (2015) Local adaptation constrains the distribution potential of heat-tolerant Symbiodinium from the Persian/Arabian Gulf, The ISME Journal, Volume 9, Issue 12, 2551–2560, https://doi.org/10.1038/ismej.2015.80

Chan WY et al (2018) Evolutionary potential in the symbiotic algae of corals. Nat Commun 9:2925. https://doi.org/10.1038/s41467-018-05379-x

Chou LM (2006) Marine biodiversity in Singapore's coastal waters. Raffles Bull Zool 14:3–12

Coles SL, Fadlallah YH (1991) Reef coral survival and mortality at high temperatures in the Arabian Gulf. Coral Reefs 9:1–5. https://doi.org/10.1007/BF00299393
Coles SL, Riegl BM (2013) Thermal tolerance and the global distribution of coral reefs. Coral Reefs 32:257–270. https://doi.org/10.1007/s00338-013-1066-0
Coma R, Ribes M, Gili JM, Zabala M (1998) An energetic approach to the study of life-history traits of two modular colonial benthic invertebrates. Mar Ecol Prog Ser 162:89–103
Coma R, Llorente-Llurba E, Serrano E, Gili JM, Ribes M (2015) Natural heterotrophic feeding by a temperate octocoral with symbiotic zooxanthellae: a contribution to understanding the mechanisms of die-off events. Coral Reefs 34(2):549–560
Coppari M, Gori A, Viladrich N, Saponari L, Grinyo J, Olariaga A, Rossi S (2016) The role of sponges in the benthic-pelagic coupling process in warm temperate coastal bottoms. J Exp Mar Biol Ecol 477:57–68
Costa-Lucas CC, Teixeira CEP, Braga MDA, Júnior FC, Paiva SV, Gurgel AL, Rossi S, Soares MO (2023) Heatwaves and a decrease in turbidity drive coral bleaching in Atlantic marginal equatorial reefs. Front Mar Sci 10:1061488. https://doi.org/10.3389/fmars.2023.1061488
Cunning R et al (2015) Thermal tolerance and resilience in coral-algal symbioses. Mar Ecol Prog Ser 519:107–115. https://doi.org/10.3354/meps11046
de Goeij JM et al (2013) Sponges as key players in nutrient cycling in coral reef ecosystems. Mar Ecol Prog Ser 493:147–159. https://doi.org/10.3354/meps10513
Denis V, Ferrier-Pagès C, Schubert N, Coppari M, Baker DM, Camp E, Gori A, Grottoli A, Houlbrèque F, Maier S, Mancinelli G, Martinez S, Ozdilek SY, Radice V, Ribes M, Richter C, Viladrich N, Rossi S (2024) Heterotrophy in marine animal forests in an era of climate change. Biol Rev 99:965–978. https://doi.org/10.1111/brv.13053
Dikou A, van Woesik R (2006) Survival under chronic stress from sedimentation in Singapore's coral reefs. Mar Pollut Bull 52(9):1368–1376. https://doi.org/10.1016/j.marpolbul.2006.02.017
Downs CA et al (2000) Oxidative stress and HSP expression in coral health. Free Radic Biol Med 29(10):1623–1630. https://doi.org/10.1016/S0891-5849(00)00475-4
Duarte GAS, Villela HDM, Deocleciano M, Silva D, Barno A, Cardoso PM, Vilela CLS, Rosado P, Messias CSMA, Chacon MA, Santoro EP, Olmedo DB, Szpilman M, Rocha LA, Sweet M, Peixoto RS (2020) Heat waves are a major threat to turbid coral reefs in Brazil. Front Mar Sci 7:1–8. https://doi.org/10.3389/fmars.2020.00179
Enochs IC, Toth LT, Kirkland A, Manzello DP, Kolodziej G, Morris JT, Holstein DM, Schlenz A, Randall CJ, Maté JL, Leichter JJ, Aronson RB (2021) Upwelling and the persistence of coral-reef frameworks in the eastern tropical Pacific. Ecol Monogr 91(4):e014822
Fabricius KE (2005) Effects of terrestrial runoff on the ecology of corals and coral reefs. Mar Pollut Bull 50(2):125–137. https://doi.org/10.1016/j.marpolbul.2004.11.028
Fabricius KE et al (2013) Effects of turbidity on reef-building corals. Mar Pollut Bull 74(1):224–236. https://doi.org/10.1016/j.marpolbul.2013.07.017
Fang LS et al (1997) Coral HSP expression in response to thermal stress. Coral Reefs 16(3):151–155. https://doi.org/10.1007/s003380050046
Feder ME, Hofmann GE (1999) Heat shock proteins in stress responses and adaptation. Annu Rev Physiol 61:243–282. https://doi.org/10.1146/annurev.physiol.61.1.243
Fine M et al (2013) Adaptation of marginal reef corals in the Arabian Gulf. Coral Reefs 32(3):877–888. https://doi.org/10.1007/s00338-013-1056-1
Freeman CJ, Easson CG, Fiore CL, Thacker RW (2021) Sponge–microbe interactions on coral reefs: multiple evolutionary solutions to a complex environment. Front Mar Sci 8:705053. https://doi.org/10.3389/fmars.2021.705053
Gili JM, Coma R (1998) Benthic suspension feeders: their paramount role in littoral marine food webs. Trends Ecol Evol 13:316–321
Glasl B, Webster NS, Bourne DG (2017) Microbial indicators as a diagnostic tool for assessing water quality and climate stress in coral reef ecosystems. Mar Biol 164:91
González–De Zayas R, Rossi S, Hernández-Fernández L, Soto-Jiménez M, Soares M, Merino-Ibarra M, Castillo-Sandoval FS (2020) Stable isotopes to asses pollution impacts on coastal and marine ecosystems of the Caribbean Region. RegNal Stud Mar Sci https://doi.org/10.1016/j.rsma.2020.101413
Graham NAJ et al (2006) Coral reefs as complex adaptive systems. Coral Reefs 25(3):424–429. https://doi.org/10.1007/s00338-006-0110-6
Grottoli AG et al (2006) Heterotrophic plasticity and resilience in corals. Coral Reefs 25(4):715–724. https://doi.org/10.1007/s00338-006-0134-y
Guest JR, Tun K, Low J, Vergés A, Marzinelli EM, Campbell AH, Bauman AG, Feary DA, Chou LM, Steinberg PD (2016) 27 years of benthic and coral community dynamics on turbid, highly urbanized reefs off Singapore. Sci Rep 6:36260. https://doi.org/10.1038/Srep36260
Gundermann N et al (2022) Thermal stress and sponge physiology: implications for nutrient cycling. Mar Environ Res 177:105642. https://doi.org/10.1016/j.marenvres.2022.105642
Heery E et al (2018) Urban coral reefs: degradation and resilience of hard coral assemblages in coastal cities. Mar Pollut Bull 135:47–56. https://doi.org/10.1016/j.marpolbul.2018.07.047
Hoegh-Guldberg O (1999) Coral bleaching and climate change: impacts on coral reefs. Science 285(5433):1230–1232. https://doi.org/10.1126/science.285.5433.1230
Hoogenboom MO et al (2008) Coral morphology and feeding efficiency in response to water flow. Coral Reefs 27(2):611–617. https://doi.org/10.1007/s00338-008-0403-6
Houlbrèque F, Ferrier-Pagès C (2009) Heterotrophy in tropical scleractinian corals. Biol Rev 84(1):1–17. https://doi.org/10.1111/j.1469-185X.2008.00059.x
Huang D et al (2009) Impacts of high sedimentation on coral reefs in Singapore. Mar Pollut Bull 58(4):547–553. https://doi.org/10.1016/j.marpolbul.2008.11.019
Hughes TP et al (2007) Coral reefs in a changing world: are coral reefs resilient? Science 318(5857):1733–1735. https://doi.org/10.1126/science.1152509
Hughes TP et al (2017) Global warming and recurrent mass bleaching of corals. Nature 543(7645):373–377. https://doi.org/10.1038/nature21707
Hume BCC et al (2013) Coral resilience to environmental stress in the Arabian Gulf. Mar Pollut Bull 72(1):73–81. https://doi.org/10.1016/j.marpolbul.2013.04.022
Hume BCC et al (2016) Coral symbioses: flexibility in algal partners. Mar Ecol Prog Ser 543:65–81. https://doi.org/10.3354/meps11555
Jones AM, Berkelmans R (2010) Trade-offs in coral thermal tolerance and growth associated with symbiont shuffling. Coral Reefs 29(3):697–707. https://doi.org/10.1007/s00338-010-0654-1
Kenkel CD, Meyer E, Matz MV (2013) Gene expression under chronic heat stress in populations of the mustard hill coral (Porites astreoides) from different thermal environments. Mol Ecol 22: 4322–4334. https://doi.org/10.1111/mec.12390
Kiang JG, Tsokos GC (1998) Heat shock protein 70 kDa: molecular biology, biochemistry, and physiology. Pharmacol Ther 80(2):183–201
Kiran GS, Sekar S, Ramasamy P, Thinesh T, Hassan S, Lipton AN, Ninawe AS, Selvin J (2018) Marine sponge microbial association: towards disclosing unique symbiotic interactions. Mar Environ Res 140:169–179
Kleypas JA et al (1999) Environmental limits to coral reef development: where do we draw the line? Coral Reefs 18(4):311–318. https://doi.org/10.1007/s003380050224

LaJeunesse TC et al (2018) Symbiodiniaceae diversity in coral ecosystems. Oceanography 31(1):66–73. https://doi.org/10.5670/oceanog.2018.113

Larcombe P, Woolfe KJ (1999) Increased sediment supply to GBR coral reefs: a geological perspective and review of evidence. Mar Geol 159(1–4):55–71. https://doi.org/10.1016/S0025-3227(99)00029-9

Lemoine NP et al (2007) Energetic responses of sponges to environmental stressors. Mar Ecol Prog Ser 343:1–11. https://doi.org/10.3354/meps06984

Lesser MP (1997) Oxidative stress causes coral bleaching. Nature 388(6640):741–742. https://doi.org/10.1038/41687

Lesser MP (2006) Oxidative stress in marine environments: biochemistry and physiological ecology. Annu Rev Physiol 68:253–278

Lesser MP et al (2018) Thermal tolerance in sponges and its ecological implications. Mar Ecol Prog Ser 594:105–118. https://doi.org/10.3354/meps12554

Leys SP et al (2011) The function of sponge pumping and environmental adaptation. Mar Ecol Prog Ser 437:63–76. https://doi.org/10.3354/meps09270

López-Victoria M, Zea S, Weil E (2006) Competition for space between encrusting excavating Caribbean sponges and other coral reef organisms. Mar Ecol Prog Ser 312:113–121

Maina J, McClanahan TR, Venus V, Ateweberhan M, Madin J (2011) Global gradients of coral exposure to environmental stresses and implications for local management. PLoS One 6(8):e23064. https://doi.org/10.1371/journal.pone.0023064

Maldonado M, Zhang X (2018) Functional implications of filtration changes in thermally stressed sponges. Funct Ecol 32(8):1930–1940. https://doi.org/10.1111/1365-2435.13172

Maldonado M et al (2012) Nutrient capture and filtering efficiency in sponges. Funct Ecol 26(5):1184–1193. https://doi.org/10.1111/j.1365-2435.2012.02016.x

McLaughlin CJ et al (2003) Coral response to fluctuating turbidity: implications for management. Mar Pollut Bull 46(1):70–74. https://doi.org/10.1016/S0025-326X(02)00363-5

McCook L, Jompa J, Diaz-Pulido G (2001) Competition between corals and algae on coral reefs: a review of evidence and mechanisms. Coral Reefs 19:400–417. https://doi.org/10.1007/s003380000129

Morikawa MK, Palumbi SR (2019) Using naturally occurring climate-resilient corals to construct bleaching-resistant nurseries. Proc Natl Acad Sci 116(45):22875–22883. https://doi.org/10.1073/pnas.1906525116

Muller-Parker G, D'Elia CF, Cook CB (2015) Interactions between corals and their symbiotic algae. In: Birkeland C (ed) Coral reefs in the Anthropocene. Springer, Dordrecht, pp 65–94. https://doi.org/10.1007/978-94-017-7249-5_5

Muscatine L, Porter JW (1977) Resource partitioning by reef corals. Am Zool 17(3):487–503. https://doi.org/10.1093/icb/17.3.487

Nelson H, Bramanti L (2020) From trees to octocorals: the role of self-thinning and shading in underwater animal forests. In: Perspectives on the marine animal forests of the world. Springer-Nature, Cham, pp 5–30

Orejas C, Carreiro-Silva M, Mohn C, Reimer JD, Samaai T, Allcock AL, Rossi S (2022) Marine animal forests of the world: definition and characteristics. Res Ideas Outcomes 8:e96274

Osman EO et al (2018) Thermal refugia against coral bleaching throughout the northern Red Sea. Science 359(6375):580–583. https://doi.org/10.1126/science.aan8363

Pawlik JR, McMurray SE (2020) The emerging ecological and biogeochemical importance of sponges on coral reefs. Annu Rev Mar Sci 12:1–21

Pawlik JR, McMurray SE, Erwin P, Zea S (2015) A review of evidence for food limitation of sponges on Caribbean reefs. Mar Ecol Prog Ser 519:265–283. https://doi.org/10.3354/meps11093

Pawlik JR, Loh T, McMurray SE (2018) A review of bottom-up vs. top-down control of sponges on Caribbean fore-reefs: what's old, what's new, and future directions. PeerJ 6:e 4343. https://doi.org/10.7717/peerj.4343

Perry CT, Alvarez-Filip L (2019) Changing geo-ecological functions of coral reefs in the Anthropocene. Funct Ecol 33:976–988. https://doi.org/10.1111/1365-2435.13247

Perry CT, Larcombe P (2003) Marginal and non-reef-building coral environments. Coral Reefs 22:427–432. https://doi.org/10.1007/s00338-003-0330-5

Pile AJ et al (1997) Adaptations in sponge structure for particle capture. Invertebr Biol 116(4):324–336. https://doi.org/10.2307/3226635

Poppell E, Weisz J, Spicer L, Massaro A, Hill A, Hill M (2013) Sponge heterotrophic capacity and bacterial community structure in high- and low-microbial abundance sponges. Mar Ecol 34(4):389–401. https://doi.org/10.1111/maec.12098

Porter JW, Targett NM (1988) Allelochemical interactions between sponges and corals. Biol Bull 175(2):230–239

Porter JW, Lewis SK, Porter KG (1999) The effect of multiple stressors on the Florida keys coral reef ecosystem: a landscape hypothesis and a physiological test. Limnol Oceanogr 44(3, part 2):941–949

Pratt WB, Toft DO (2003) Regulation of signaling protein function and trafficking by the hsp90/hsp70-based chaperone machinery. Annu Rev Biochem 72:655–687. https://doi.org/10.1146/annurev.biochem.72.121801.161034

Pupier CA, Fine M, Bednarz VN, Rottier C, Grover R, Ferrier-Pagès C (2019) Productivity and carbon fluxes depend on species and symbiont density in soft coral symbioses. Sci Rep 9(1):1–10

Rawat VS, Nautiyal A, Ramlal A et al (2024) Algae-coral symbiosis: fragility owing to anthropogenic activities and adaptive response to changing climatic trends. Environ Dev Sustain 27:17993–18020. https://doi.org/10.1007/s10668-024-04748-6

Reiswig HM (1974) Water transport, respiration, and energetics in a tropical sponge. J Exp Mar Biol Ecol 14(2):231–249. https://doi.org/10.1016/0022-0981(74)90033-2

Ribes M, Coma R, Gili JM (1998) Heterotrophic feeding by gorgonian corals with symbiotic zooxanthellae. Limnol Oceanogr 43:1170–1179

Ribes M, Coma R, Gili JM (1999) Heterogeneous feeding in benthic suspension feeders: the natural diet and grazing rate of the temperate gorgonian *Paramuricea clavata* (Cnidaria: Octocorallia) over a yearly cycle. Mar Ecol Prog Ser 183:125–137

Richier S et al (2006) Heat shock protein expression as a biomarker of environmental stress in coral ecosystems. Mar Ecol Prog Ser 314:1–10. https://doi.org/10.3354/meps314001

Riegl BM, Purkis SJ, Al-Cibahy AS, Abdel-Moati MA, Hoegh-Guldberg O (2011) Present Limits to Heat-Adaptability in Corals and Population-Level Responses to Climate Extremes. PLoS ONE 6(9): e24802. https://doi.org/10.1371/journal.pone.0024802

Riegl BM, Purkis SJ (2012) Coral reefs of the Gulf: adaptation to climatic extremes in the world's hottest sea. In: Riegl B, Purkis S (eds) Coral reefs of the Gulf. Springer, pp 1–25. https://doi.org/10.1007/978-94-007-3008-3_1

Rossi S (2013) The destruction of the 'animal forests' in the oceans: towards an over-simplification of the benthic ecosystems. Ocean Coast Manag 84:77–85

Rossi S, Ribes M, Coma R, Gili JM (2004) Temporal variability in zooplankton prey capture rate of the soft-bottom passive suspension feeder *Leptogorgia sarmentosa* (Cnidaria: Octocorallia), a case study. Mar Biol 144:89–99

Rossi S, Bramanti L, Gori A, Orejas C (2017a) An overview of the animal forests of the world. In: Rossi S, Bramanti L, Gori A, Orejas

C (eds) Marine animal forests: the ecology of benthic biodiversity hotspots. Springer, Berlin/Heidelberg, pp 1–28

Rossi S, Coppari M, Viladrich N (2017b) Benthic-pelagic coupling: new perspectives in the animal forests. In: Rossi S, Bramanti L, Gori A, Orejas C (eds) Marine animal forests: the ecology of benthic biodiversity hotspots. Springer, Berlin/Heidelberg, pp 855–886

Rossi S, Isla E, Bosch-Belmar M, Galli G, Gori A, Gristina M, Ingrosso GM, Milisenda G, Orejas C, Piraino S, Rizzo L, Schubert N, Soares M, Solidoro C, Thurstan R, Viladrich N, Willis T, Ziveri C (2019) Changes of energy fluxes in the marine animal forest of the Anthropocene: factors shaping the future seascape. ICES J Mar Sci 76:2008–2019. https://doi.org/10.1093/icesjms/fsz147

Rossi S, Schubert N, Soares M, Brown D, Gómez-Posada A (2020) Trophic ecology of two Caribbean octocorals: autotrophic and heterotrophic seasonal trends. Coral Reefs 39:433–449

Ruetzler K (2004) Sponges on coral reefs: a community shaped by competitive cooperation. Invertebr Biol 123(4):304–317. https://doi.org/10.1111/j.1744-7410.2004.tb00137.x

Sawall Y, Al-Sofyani A, Banguera-Hinestroza E, Voolstra CR (2014) Spatio-temporal analyses of Symbiodinium physiology of the coral *Pocillopora verrucosa* along large-scale nutrient and temperature gradients in the Red Sea. PLoS One 9(8):e103179. https://doi.org/10.1371/journal.pone.0103179

Schmidt EW (2014) Ecological implications of sponge physiology under thermal stress. Mar Ecol Prog Ser 508:1–10. https://doi.org/10.3354/meps10837

Schoepf V, Baumann JH, Barshis DJ, Browne NK, Camp EF, Comeau S, Cornwall CE, Guzmán HM, Riegl B, Rodolfo-Metalpa R, Sommer B (2023) Corals at the edge of environmental limits: a new conceptual framework to re-define marginal and extreme coral communities. Sci Total Environ 884:163688

Sebens KP et al (2003) Adaptive feeding strategies in coral colonies facing nutrient-limited conditions. Ecol Monogr 73(4):441–464. https://doi.org/10.1890/02-4009

Sharp VA et al (1997) Impact of environmental factors on coral stress responses. Mar Environ Res 44(3):295–316. https://doi.org/10.1016/S0141-1136(97)00055-9

Sheppard C et al (2010) The Persian Gulf: a young ocean basin. In: The Persian Gulf: a natural environment. Springer, Dordrecht, pp 1–10. https://doi.org/10.1007/978-90-481-3430-9_1

Silverstein RN, Correa AMS, Baker AC (2012) Specificity is rarely absolute in coral–algal symbiosis: implications for coral response to climate change. Proc R Soc B Biol Sci 279(1734):2609–2618. https://doi.org/10.1098/rspb.2012.0055

Silverstein RN et al (2015) The ecological cost of increased thermal tolerance in coral-algal symbiosis. Proc R Soc B Biol Sci 282(1804):20151825. https://doi.org/10.1098/rspb.2015.1825

Smith SV, Buddemeier RW (1992) Global change and coral reef ecosystems. Annu Rev Ecol Syst 23:89–118. https://www.jstor.org/stable/2097283

Snyder MJ, Rossi S (2004) Stress protein (HSP 70 family) expression in intertidal benthic organisms: the example of Anthopleura elegantissima (Cnidaria: Anthozoa). Scientia Marina 68 (Suppl.1): 155–162

Soares MO (2020) Marginal reef paradox: a possible refuge from environmental changes? Ocean Coast Manag 185:105063. https://doi.org/10.1016/j.ocecoaman.2019.105063

Soares M, Rossi S, Santos Martins FA, Macêdo Carneiro PB (2017) The forgotten reefs: benthic assemblage coverage on a sandstone tropical reef (South-Western Atlantic). J Mar Biol Assoc U K 97:1585–1592

Soares M, Rossi S, Rebouças-Gurgel ALA, Costa-Lucas C, Lopes-Tavares TC, Feitosa CV, Pereira PHC, Papa de Kikuchi RK, Leão ZML, Silva-Cruz IG, Alvarez-Filip L (2021) Marginal reefs under pressure (South Atlantic, Brazil). Ocean Coast Manag. https://doi.org/10.1016/j.ocecoaman.2021.105692

Sully S, Van Woesik R (2020) Turbid reefs moderate coral bleaching under climate-related temperature stress. Glob Chang Biol 26(3):1367–1373. https://doi.org/10.1111/gcb.14948

Taipale M et al (2010) HSP90 at the hub of protein homeostasis: emerging insights from systems biology. Nat Rev Mol Cell Biol 11(7):515–528. https://doi.org/10.1038/nrm2900

Teixeira CD, Leitão RLL, Ribeiro FV, Moraes FC, Neves LM, Bastos AC, Pereira-Filho GH, Kampel M, Salomon PS, Sá JA, Falsarella LN, Amario M, Abieri ML, Pereira RC, Amado-Filho GM, Moura RL (2019) Sustained mass coral bleaching (2016–2017) in Brazilian turbid-zone reefs: taxonomic, cross-shelf and habitat-related trends. Coral Reefs 38:801–813. https://doi.org/10.1007/s00338-019-01789-6

Thacker RW, Freeman CJ (2012) Sponge-microbe symbioses: recent advances and new directions. In: Becerro MA, Uriz MJ, Maldonado M, Turon X (eds) Advances in marine biology, vol 62. Academic Press, pp 57–111

Tsounis G, Rossi S, Laudien J, Bramanti L, Fernández N, Gili JM, Arntz W (2006) Diet and seasonal prey capture rate in the Mediterranean red coral (*Corallium rubrum* L.). Mar Biol 149:313–325

Underwood JN et al (2009) Genetic diversity and connectivity among coral populations in high thermal stress regions. Mol Ecol 18(11):2645–2657. https://doi.org/10.1111/j.1365-294X.2009.04285.x

Valko M et al (2007) Free radicals and antioxidants in normal physiological functions and human disease. Int J Biochem Cell Biol 39(1):44–84. https://doi.org/10.1016/j.biocel.2006.07.001

Van Oppen MJH, Gates RD (2006) Coral adaptation to climate change. Curr Opin Environ Sustain 1(1):15–19. https://doi.org/10.1016/j.cosust.2009.09.003

Van Oppen MJH et al (2015) Building coral reef resilience through assisted evolution. Proc Natl Acad Sci USA 112(8):2283–2290. https://doi.org/10.1073/pnas.1422304112

Van Woesik R, McCaffrey KR (2017) Repeated thermal stress, shading, and directional selection in the Florida reef tract. Front Mar Sci 4:182. https://doi.org/10.3389/fmars.2017.00182

Vaughan GO et al (2019) Coral resilience to climate change in the Persian/Arabian Gulf. Glob Chang Biol 25(10):3456–3470. https://doi.org/10.1111/gcb.14765

Vega Thurber RL et al (2014) Coral reefs in crisis: the interactions of temperature and nutrient enrichment. Glob Chang Biol 20(4):1158–1168. https://doi.org/10.1111/gcb.12404

Wall CB, Kaluhiokalani M, Popp BN, Donahue MJ, Gates RD (2020) Divergent symbiont communities determine the physiology and nutrition of a reef coral across a light-availability gradient. ISME J 14(4):945–958. https://doi.org/10.1038/s41396-019-0570-1

Warner ME, Suggett DJ (2016) The photobiology of Symbiodinium spp.: linking physiological diversity to the implications of stress and resilience. In: Goffredo S, Dubinsky Z (eds) The cnidaria, past, present and future. Springer, Cham, pp 465–482. https://doi.org/10.1007/978-3-319-31305-4_30

Warner ME et al (1999) Damage to photosystem II in symbiotic dinoflagellates: a determinant of coral bleaching. Proc Natl Acad Sci USA 96(26):14542–14547. https://doi.org/10.1073/pnas.96.26.14542

Webster NS, Thomas T (2016) Microbial symbioses and sponge resilience. Invertebr Surviv J 13(2):112–120. https://doi.org/10.1007/s10101-016-0869-1

Webster NS, Hill RT, Thomas T (2008) The effects of elevated temperature on microbial symbiosis and sponge metabolism. Aquat Biol 3(1):51–60. https://doi.org/10.3354/ab00057

Wiedenmann J, D'Angelo C, Smith E et al (2013) Nutrient enrichment can increase the susceptibility of reef corals to bleaching. Nature Clim Change 3:160–164. https://doi.org/10.1038/nclimate1661

Weis VM (2008) Cellular mechanisms of coral bleaching: antioxidants and coral health. Mar Pollut Bull 56(3):594–599. https://doi.org/10.1016/j.marpolbul.2007.12.024

Wong JCY, Enríquez S, Baker DM (2021) Towards a trait-based understanding of Symbiodiniaceae nutrient acquisition strategies. Coral Reefs 40:625–639. https://doi.org/10.1007/s00338-020-02034-1

Yan H, He X, Lei Y, Wang Y, Su H, Jiang S (2019) Land use-induced change in trophic state of Shenzhen Bay (South China) over the past half-century. Mar Pollut Bull 145:208–213

Zhao H, Yuan M, Strokal M, Wu HC, Liu X, Murk AT, Kroeze C, Osinga R (2021) Impacts of nitrogen pollution on corals in the context of global climate change and potential strategies to conserve coral reefs. Sci Total Environ 774:145017

Ziegler M, Apprill A, Van de Water JA, Bourne DG, Voolstra CR, Berumen ML (2017) Coral-microbe dynamics and resilience to environmental stress. Nat Microbiol 2(12):17091. https://doi.org/10.1038/nmicrobiol.2017.91

Ziegler M et al (2019) Ecophysiology of reef-building corals in the Red Sea. In: Voolstra C, Berumen M (eds) Coral reefs of the Red Sea, vol 11. Springer, Cham, pp 29–48. https://doi.org/10.1007/978-3-030-05802-9_3

Reef Geology and Carbonate Framework of Extreme and Marginal Reefs of the World

15

Alex Cardoso Bastos, Tarcila Franco, and Maitê Noda-Zanotti

Abstract

Extreme and marginal reefs have existed throughout Earth's history, from Precambrian microbial stromatolites to Cenozoic scleractinian coral reefs. Each geological Era hosted distinct reef types and builders, reflecting the dominant biological and environmental conditions of the time. Records of marginal reefs through geological time are observed worldwide. The main reef-builders also changed over time, from stromatolites to zooxanthellate corals. This chapter presents and discuss the reef framework of extreme and marginal reefs, focusing on Quaternary reef systems. We focused on turbid, high latitude and high temperature and salinity reefs. The reef framework reflects dynamic interactions between environmental stressors and ecological responses. Notably, extreme/marginal frameworks lack uniformity; even reefs with high coral coverage may shift rapidly under stress. Framework composition depends on both local conditions (e.g., temperature extremes. and high-energy events) and biogeographic context, with allochthonous/autochthonous deposits often linked to high-energy events or mortality episodes. A marginal reef can be defined solely based on the ecology (coral assemblage), but extreme environmental (seasonal or ephemeral) conditions can also affect a marginal reef, influencing the reef builders, and consequently, the reef framework. These insights are critical for paleoecological reconstructions, particularly as modern climate change exacerbates extreme conditions. Understanding contemporary marginal and extreme reef frameworks thus provides essential analogues for interpreting past and future reef responses.

Keyword

Turbid reefs · High latitude reefs · High temperature and salinity reefs · Reef framework · Extreme conditions · Reef frame-builders

15.1 Introduction

Reefs are vital to Earth's ecosystems, primarily as carbonate factories and biodiversity hotspots. The growth and development of coral reefs depend on environmental factors (temperature, salinity, turbidity, light, hydrodynamics, seabed morphology), which makes extreme and marginal reefs of particular scientific interest on modern reefs. In general, extreme and marginal reefs thrive at the edge of environmental limits for coral reef growth, supporting distinct ecological communities and offering unique insights into adaptation, resilience, and the implications of climate change. These characteristics challenge the traditional notions of modern reef-building processes (Guinotte et al. 2003; Kleypas et al. 1999; Perry and Larcombe 2003). Examples include high-latitude reefs, turbid coastal reefs, hyper and hyposaline environments, mesophotic reefs, upwelling-affected reefs, and anthropogenically stressed reefs.

Schoepf et al. (2023) proposed a framework to redefine marginal and extreme coral reefs, considering that they are related but distinct. The authors define coral reef extremeness based on environmental conditions (i.e., suboptimal conditions) and marginality based on ecological criteria (i.e., altered community composition and/or ecosystem functioning).

From a geological perspective, marginal reefs enhance our understanding of carbonate production and reef-building processes under stress or large deviation environmental conditions. They often exhibit distinct sedimentological features, such as mixed carbonate-siliciclastic deposits, low coral diversity, or unique coral community patterns, which may provide analogs for interpreting ancient reef systems.

A. C. Bastos (✉) · T. Franco · M. Noda-Zanotti
Laboratório de Geociências Marinhas, Departamento de Oceanografia, Universidade Federal do Espirito Santo, Vitória, Brazil
e-mail: Alex.bastos@ufes.br

M. O. Soares et al. (eds.), *Extreme and Marginal Coral Reefs and Communities*, Coral Reefs of the World 24,
https://doi.org/10.1007/978-3-032-28035-0_15

Throughout the course of geological time, reef growth occurred under distinctly disparate environmental and ecological conditions when compared with modern reefs. Sea-level fluctuations, global climate, and ocean chemistry were considerably different from the contemporary time (Quaternary). Therefore, modern extreme and marginal reefs are critical for reconstructing paleoenvironments, as their growth patterns and fossil assemblages may reflect past climatic and oceanic conditions.

Reef framework refers to the three-dimensional skeletal structure of a reef, primarily built by the accumulation of calcium carbonate ($CaCO_3$) secreted by reef-building organisms such as corals, coralline algae, and other calcifying species, but also marine cement. This framework provides the physical foundation of the reef ecosystem, creating complex habitats that support high biodiversity. The framework of modern extreme and marginal reefs is primarily built by stress-tolerant coral species and other calcifying organisms, such as coralline algae, sponges and bryozoans. In high-latitude reefs (e.g., Lord Howe Island, Australia), cooler temperatures limit coral diversity, but species like *Porites* spp. and *Pocillopora* spp. dominate, forming robust, low-relief structures (Harriott 1999). In turbid reefs (e.g., Abrolhos, Bahia-Brazil), mixed carbonate-siliciclastic deposits mark the framework, as well as sedimentation-tolerant species, even endemic species like *Mussismillia braziliensis* are reef-builders adapted to low light and high sediment loads (Leão et al. 2003; Ribeiro et al. 2018). Hypersaline reefs, such as those in the Persian Gulf, are characterized by thermally resilient corals like *Acropora clathrata*, which thrive in salinities exceeding 40 ppt and temperatures up to 35 °C (Riegl et al. 2011).

This chapter aims to introduce the reef geology and carbonate framework (reef builders) of extreme and marginal reefs throughout Earth's history, focusing on reef growth in the Holocene. The chapter will briefly summarize marginal reefs' framework through geological time, provide examples of marginal and extreme reefs in the Holocene, and then give an overview of the framework and reef geology of modern marginal and extreme reefs worldwide.

15.2 Extreme and Marginal Reefs Through Geological Time

Reef systems have existed for over 3.5 billion years, and their framework has been evolving in response to changes in Earth's climate, ocean chemistry, and biota. Reef distribution has varied with sea level and temperature gradients, while reef-building organisms' trophic demands and biological interactions have evolved. The oldest reef records date to the Archean Era (e.g., Strelley Pool Formation, Western Australia, Wacey 2010). Their framework is basic microbialites (organosedimentary deposits influenced by benthic microbial communities) forming laminated structures known as stromatolites (cyanobacterial origin). Stromatolites dominate as the main reef form up to the end of the Proterozoic, when the ocean and the atmosphere started to change, and until the appearance of metazoans. By the early Cambrian (520 mya), complex reef ecosystems had emerged, with calcified heterotrophs and microbial carbonate processes providing structural stability. Metazoan-built reefs appeared in the early Paleozoic, experiencing cycles of diversification and extinction. Paleozoic reefs included sponges, corals, and bryozoans. After the great Permian mass extinction, a rise of scleractinian corals in the Triassic is observed, but the Mesozoic reef framework is dominated by sponges, corals, and rudist bivalves. After the Cretaceous-Paleogene mass extinction, Scleractinia coral dominance in reefs increased in the Cenozoic, coinciding with the rise of herbivorous predators like fish, echinoids, and limpets, which controlled macroalgal competition. Table 15.1 provides a concise overview of the primary reef-builders that are pertinent to each geological era.

Thus, reef framework underwent changes throughout the course of geological time in accordance with the Earth's climate, ocean chemistry and also biological adaptation. Reef extinctions have been linked to greenhouse gas-induced warming, ocean acidification, and anoxia, leading to biodiversity loss and slowed carbonate deposition. The collapse of photosymbiotic reef ecosystems has been observed, with the potential cause attributed to the loss of symbionts. Post-extinction recoveries were slow, with new taxa emerging rather than the same ones returning. These patterns observed through the geological record highlight the vulnerability of modern reefs to climate change. In summary, for most of the geological time, fossil records can be considered as marginal or extreme reefs when compared to modern reefs and considering the climate and ocean chemistry.

15.3 Reef Geology and Framework of Modern Marginal and Extreme Reefs

The distribution of modern marginal and extreme reefs is global, occurring in a variety of environments ranging from the cooler waters of higher latitudes to the deep, dark waters of the ocean (Fig. 15.1). Despite facing environmental challenges, these reefs have developed unique adaptations that allow them to thrive at the limits of coral tolerance. These reefs are of critical importance in understanding the resilience of corals and the potential responses of marine ecosystems to future climate change scenarios. The chapter will explore three types of extreme and marginal reefs with examples of their main environmental drivers and the frame-

Table 15.1 Summary of the main reef-builders for each geological Era

Geological time	Reef-builders
Precambrian	Microbial reefs dominated shallow marine environments. Stromatolites, built by cyanobacteria, were the primary reef builders and often thrived in extreme conditions, such as hypersaline lagoons and high-temperature settings
Paleozoic	Rise of metazoan reef builders, but extreme and marginal reefs persisted in environments with fluctuating salinity and temperature, or sedimentation. During the Ordovician, bryozoans and stromatoporoids constructed reefs in high-latitude, cooler waters, representing early examples of marginal reefs (Webby 2002). In the Devonian, stromatoporoid-coral reefs in the Canning Basin, Australia, adapted to high sedimentation and turbidity, forming resilient frameworks in marginal settings (Playford 1980). The Late Devonian extinction led to the decline of these reefs, but microbialites often filled ecological niches in post-extinction environments, forming marginal reefs during biotic stress
Mesozoic	Diversification of Scleractinia corals, but extreme and marginal reefs continued to develop in challenging environments. During the Triassic, microbialites and sponges dominated reef systems in hypersaline and low-oxygen settings (Germanic Basin, Flugel 2002). In the Cretaceous, rudist bivalves constructed reefs in shallow, high-energy environments, often in marginal settings with high sedimentation or variable salinity (Johnson et al. 2002)
Cenozoic	Scleractinian corals are dominant reef builders. Reef framework is characterized by coral builders and coralgal assemblages Reef framework changes due to tectonic activity, change in ocean circulation, eustatic sea-level fluctuations, and climatic variations (glacial and interglacial stages) (Perrin 2002) Extreme and marginal reefs persisted in environments such as high-latitude regions, turbid coastal zones, and hypersaline basins

work that can characterize or typify the extreme and marginal reefs: Turbid Reefs, High-Latitude Reefs, and High Temperature-Salinity (HTS) Reefs.

15.3.1 Turbid Reefs

Turbidity reefs are located in areas of high input of terrigenous sediments, usually in coastal regions, and are influenced by high rates of turbidity and deposition. Turbid reefs are subject to episodic high turbidity at low tides, increasing at low-water spring tides; however, insufficient sediment deposition upon the colonies causes smothering (Larcombe and Woolfe 1999). Their proximity to the coast and the significant influence of river sediment discharge also make them vulnerable to anthropogenic impacts such as increased concentrations of nutrients and contaminants.

Turbidity plays a key role in restricting the depth of light penetration, influencing the development of turbid-zone reefs, apart from accommodation space (Perry and Smithers 2010). Unlike clear-water reef growth that is mainly controlled by sea-level changes (i.e., accommodation space), turbid reefs have both the vertical window for growth and sea-level changes, which act as essential controls on post-initiation development (Perry and Smithers 2010). Therefore, it is typical for phases of coral turn-off and turn-on to occur due to short-term changes (decades to hundreds of years), as a result of local shifts in sediment transport directions and depocenter locations (Woolfe and Larcombe 1998). Larcombe and Woolfe (1999) attributed episodes of reef turn-off and migration to coastal retreat (or sea-level rise) and episodes of reef turn-on to periods of relative stasis, both throughout the post-glacial sea-level rise.

The dynamics between carbonate accretion and terrigenous sediment accumulation are a major driver in turbid reefs. Woolfe and Larcombe (1999) developed a conceptual model based on the relationship between the rates of terrigenous sedimentation and reefal carbonate accumulation. The model indicates that, in turbid environments, the growth rates for reef initiation must be greater than the rate of non-framework sediment accumulation and erosion for the reef to thrive. The authors suggest that reef accretion can occur over time until non-framework sediment accumulation and erosion reach reefal carbonate production.

Thus, reef development or "reef growth" notably differs between clear-water and turbid coral reefs. Kleypas et al. (2001) classify reefs based on the mechanisms that control net accumulation. Clear water coral reefs are "production-dominated reefs", and marginal reefs (turbid reefs) are "import-dominated reefs". On import-dominated reefs, there is a great contribution of imported material (allochthonous) to the reef net accumulation (Kleypas et al. 2001). According to the authors, import-dominated reefs will be vital in future conditions where carbonate production is likely to decline either through the removal of reef builders (increase in bleaching episodes) or depressed calcification rates (due to changes in temperature and $CaCO_3$ saturation state).

The coral communities' suites from turbid reefs differ from clear water reefs once they are more tolerant to lower levels of light, have protective mechanisms about the deposition of sediment on the surface, and can recover quickly from periodic setbacks such as flood events and cyclones (Browne et al. 2010).

Signatures in the reef framework can indicate a turbidity reef, such as the dominance of floatstone and rudstone and the presence of non-framework sedimentary packages. In the following section, we will provide information of two environments characterized by the presence of turbidity reefs in

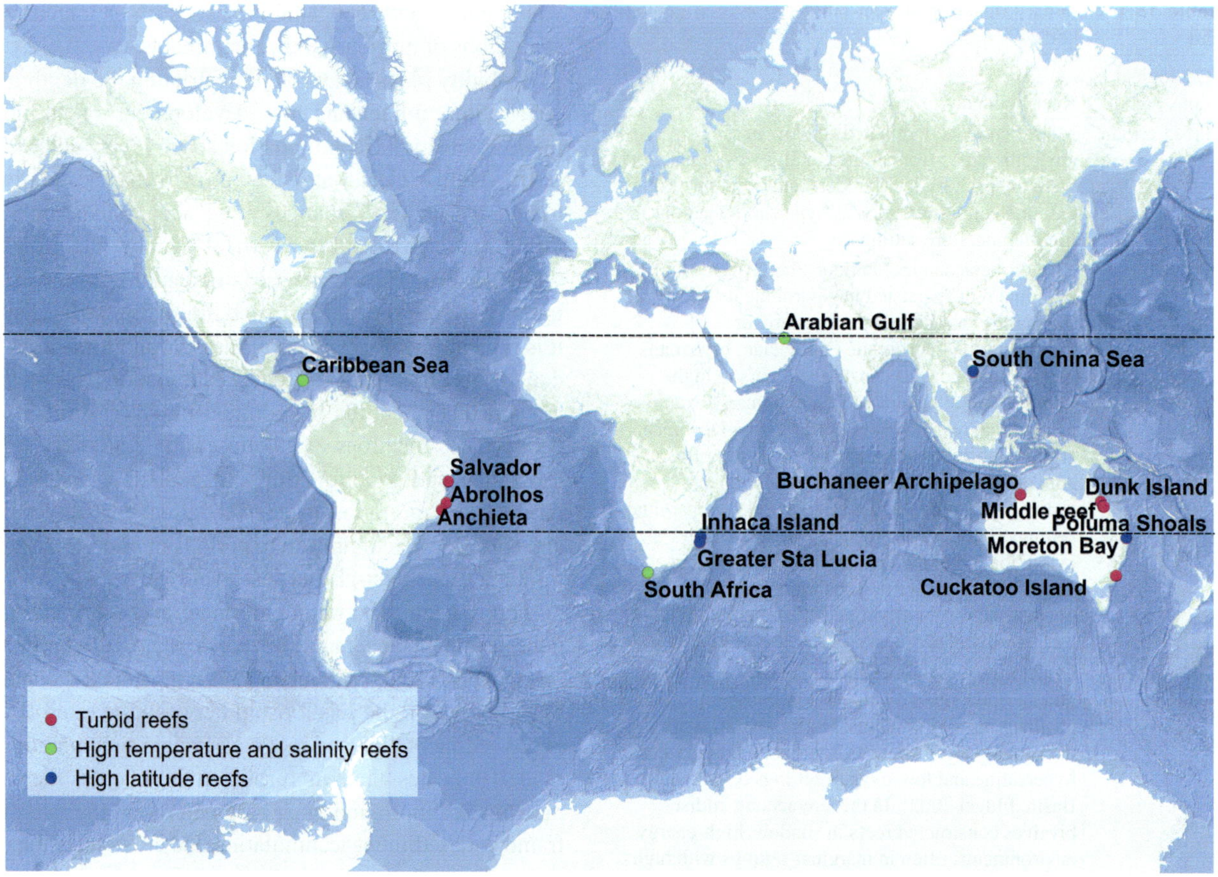

Fig. 15.1 Distribution of extreme and marginal reef types most presented and discussed in this chapter as examples

proximity to clear water reef environments: The Great Barrier Reef (Australia) and the Abrolhos Bank (Brazil). This will facilitate a comprehensive discussion on the fundamental differences between these two distinct ecological settings.

15.3.1.1 Great Barrier Reef

The Great Barrier Reef (GBR) comprises approximately 900 inshore reefs (Hopley et al. 2007), includes fringing reefs, nearshore reefs, and shoals. This represents approximately a third of the reefs on the GBR. Turbid reefs from the GBR are mostly located in Northeast and Northwest Australia. Browne et al. (2012) present an extensive review of the occurrence, composition, and growth of the coral reefs on the GBR turbid inner shelf. Webster and Davies (2003) present the contrasting framework of GBR inner and outer shelf reefs.

Within the central GBR province, turbid reefs occur close to the coast at shallow depths (<20 m) (Browne et al. 2012). Accumulation rates of terrigenous sediments vary between <0.1 and 8 mm year^{-1} in the inshore area of the GBR (Woolfe and Larcombe 1998).

Coastal turbid reefs started to grow during the post-Last Glacial Maximum (LGM) transgression, around 7500 years BP (Larcombe and Woolfe 1999). Contemporary reef sequence initiated above the underlying shallow subtidal sediments with both reef initiation and subsequent accretion progressing under presumed high turbidity conditions (Perry et al. 2009). Turbid reefs from GBR present high accretion rates (e.g., 8.3 mm/yr. at Middle Reef, Perry et al. 2012, and 3 mm/yr. at Paluma Shoal, Perry et al. 2008). Perry et al. (2012) suggest that the swift buildup of reefs is facilitated by the accumulation of fine-grained terrestrial sediments. These sediments fill the porous structure of the reef, reducing the breakdown of skeletal material after deposition. As a result, the primary reef framework is better preserved, promoting faster reef growth.

The GBR turbid reefs comprise a mixed of non-framework clast- to matrix-supported coral framework composed of entire colonies or coral rubble. The coral framework provides an important structural component to the reef. Perry et al. (2009) describe the 1.3 m thick Holocene reef from Lugger Shoal according to the variations in sediment composition (especially mud and $CaCO_3$ content) and the relative volume of sediment to coral material (i.e., clast- or matrix-supported), as: clast-supported unit (sandy matrix), clast-supported unit (sandy-clay matrix) and, matrix-supported unit (clay-rich matrix). Webster and Davies (2003) described an inner-shelf core from Boulder Reef into two primary sections: an upper sequence, defined as carbonate-dominated (0–34 m) comprising four reef units, and a mud lower sequence (34–86 m), composed of siliciclastic sediment and two thin coral-bearing units. The upper sequence comprises four reef units showing a coral framework interbedded with different layers of grainstones/packstones, rhodoliths, and mixed disaggregated coral rubble. The coral framework is dominated by a single coral assemblage (mainly *Porites* sp. and faviids), indicating low energy and turbid environments.

A comparison of turbid reefs and offshore barrier reefs reveals that the reef community types can be defined in relation to definable hydrodynamic, turbidity, and sedimentation regimes (Woolfe and Larcombe 1998). Turbid fringing reefs encompass a taxonomically diverse coral communities in the GBR (e.g., South Myall, Rykers, and Emmagen fringing reefs; Veron 1995). Webster and Davies (2003) compared the framework from Boulder Reef (turbid inshore reef) and Ribbon Reef, a outer barrier reef, both located in the northern Great Barrier Reef. The outer barrier reef framework encompasses 6 reef units with two coral assembles and coralline algae associations. The inshore reef has an upper section showing one reef unit with one coral assemblage, and a lower section characterized by siliciclastic facies with two thin coral-bearing units, showing distinct frameworks driven by the dominant environmental conditions acting in different settings of the GBR. The Queensland coast presents a high cover of living coral and long-term persistence of the dominant taxa, comprising *Acropora* sp. (including *Acropora pulchra*), *Turbinaria* sp. (including *Turbinaria frondens*), *Montipora* sp. (including *Montipora digitate* and *Montipora mollis*); *Porites* sp., Favids and *Galaxea fascicularis* (Smithers and Larcombe 2003; Perry et al. 2008, 2009, 2012; Palmer et al. 2010; Perry and Smithers 2010). The reefs' composition shows a long-term persistence throughout the reef. This is due to their potential adaptability to lower light penetration and sediment stress, where alternating dominance occurred only once the reefs reached sea level (Perry et al. 2008, 2009). The persistent dominance of a restricted coral assemblage throughout the entire reef sequence shows that reef accretion depends on the nature of the preexisting environment that facilitated reef growth. This suggests that in areas where corals are pre-adapted or exhibit rapid adaptability to these conditions, reef-building processes can persist with little impact on the composition of reef-building taxa (Perry et al. 2008, 2009).

15.3.1.2 Abrolhos Bank

The Abrolhos Bank, the largest and most diverse coral reef complex in the Southern Atlantic, is located on the Abrolhos Continental Shelf (ACS) along the Eastern coast of Brazil. This reef system is notable for its high coral species endemism, uniquely shaped reef structures, and location within a marginal high-turbidity environment and a highly anthropogenic influence (Hartt 1870; Leão et al. 2003). The Abrolhos Bank comprises two arcs parallel to the shoreline, separated by the Abrolhos Channel (Fig. 15.2). Both arcs are characterized by mushroom-shaped pinnacles known as "chapeirões" (Hartt 1870). These distinctive features have narrow bases and broad tops, ranging from 5 to 25 m in height and 5 to 50 m in width at the top (Leão et al. 2003). The coastal arc is under more anthropogenic pressure and sediment input, while the outer arc is a marine protected area, located ~80 km from the coast at the Abrolhos National Marine Park.

A combination of isolated evolutionary processes and significant terrigenous riverine input along the Brazilian coast has resulted in a coral fauna composition with distinctive characteristics compared to the Indo-Pacific and Caribbean (Leão et al. 2003). This fauna is characterized by low species diversity (eighteen species) and a high level of endemism (six species). In the Abrolhos Bank, seventeen species of Scleractinia have been identified, including six endemics: *Mussismilia braziliensis* (Verrill, 1868), *Mussismilia hispida* (Verrill, 1901), *Mussismilia harttii* (Verrill, 1868), *Siderastrea stellata* (Verrill, 1868), *Favia gravida* (Verrill, 1868), and *Favia leptophylla* (Verrill, 1868) (Leão et al. 2003). Cosmopolitan species (e.g., *Porites* sp. and *Agaricia* sp.) are also present, but to a lesser extent. The presence of a rich reef complex in this turbid marginal environment is primarily attributed to its endemic-adapted coral fauna. The larger polyp species and brain-shaped corals characteristic of this fauna are believed to be more resistant to environmental stressors such as turbidity and elevated temperatures (Leão et al. 2003).

The Abrolhos's Coastal Arc reefs' framework indicates a dominant shallow water benthic community characterized by a complex coralgal assemblage. Fauna composition is dominated by endemic scleractinian corals, with a contribution of cosmopolitan species intercalated with crustose coralline algae (CCA). Contrastingly, in the Outer Arc reefs, corals and hydrocorals compose only a small portion of the framework (11%–23%; 5 taxa) (Bastos et al. 2018). Notably, bryozoans are the major components of the reef framework (36%–44%; 20 taxa) in the Parcel dos Abrolhos reef (Outer Arc), at least from the past 2000 years BP, and are frequently intergrown with CCA, the second major reef framework component (15%–28%; 12 taxa) (Bastos et al. 2018). The authors emphasize the unique dominance of the Bryozoan framework, particularly when compared to Indo-Pacific clear-water reefs, where scleractinian corals are the primary reef builders.

Regarding the Abrolhos's reef Holocene evolution, Leão et al. (2003) proposed a general model for reef development in the Brazilian margin, controlled mainly by sea-level fluctuations and stages. The initiation of the Abrolhos reef system (Stage 1) occurred approximately 7000 years BP,

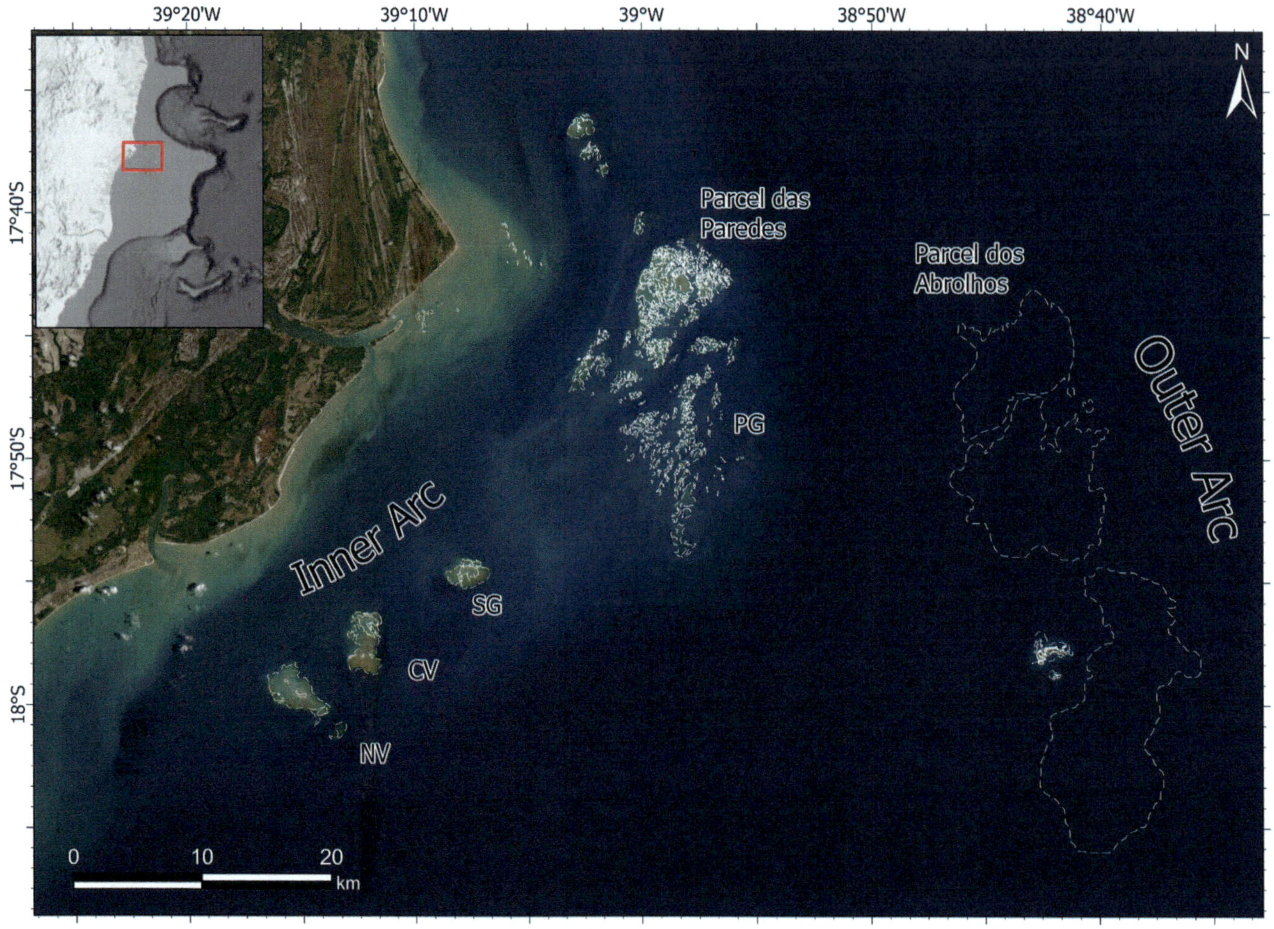

Fig. 15.2 Satellite image of the Abrolhos reefs showing the inshore/coastal reef arc and the offshore/outer reef arc

following the flooding of the Brazilian continental shelf around 8000–7000 years BP (Leão et al. 2003; Vieira et al. 2022). The oldest reef dates back to 6.66 ky BP. After initial establishment, the reef underwent a phase of rapid upward accretion (Stage 2) from 5000 to 4000 years BP, characterized by a catch-up mode of growth with accumulation rates of 7 mm/yr. This occurred during the Holocene highstand, when sea level were 2–5 m above present level. The final stages were marked by the regressive phase of the mid-late Holocene, which limited vertical accretion due to the lack of accommodation space. Vertical accretion ceased when the reefs reached sea level, and from 4000 to 2000 years BP, lateral progradation of the reefs surpassed their vertical growth (Stage 3). From around 2000 years BP to now, the reefs have experienced significant degradation and reef turn-off. As sea-level fell, coastal reefs were subjected to higher rates of terrigenous sedimentation and increased solar exposure of intertidal communities.

Leão et al. (2003) propose that those unique mushroom pinnacles were a result of limited upward accretion due to sea level regression during the last 4000 years, which would lead to lateral progradation at the pinnacle's top after the phase of dominant vertical accretion. Bastos et al. (2018) go further to state that the low disturbance regime of ACS, with no regular sub-tropical cyclone or hurricane events, does not constrain horizontal progradation of carbonate framework (e.g., physical erosion), allowing significant lateral accretion as found in the pinnacles.

Detailed evolution and framework of Coastal Arc reefs were described by Dechnik et al. (2019, 2021), Vieira (2020) and Vieira et al. (2022) based on the description and radiocarbon dating of shallow (~1.5 m) and deep (~25 m) reef cores. Coroa Vermelha (CV; Fig. 15.3), the oldest Holocene-dated reef sample, was obtained at 10.5 m deep and dated 6406 ± 205 cal yrs. BP. This reef reached mean sea level around 4760–4442 ± 220 years BP and turned off. Pedra Grande (PG; Fig. 15.3) started growing approximately 18 m deep around 6785 ± 200 years BP and cached-up mean sea level at around 2420–1145 years BP, but still presents living corals at the top. Sebastião Gomes (SG) reef approached mean sea level around 2420 years BP, similar to PG. CV approached sea level first due to a combination of environmental conditions during the reef's evolution (Dechnik et al. 2019, 2021). Alternatively, it may simply be due to the dif-

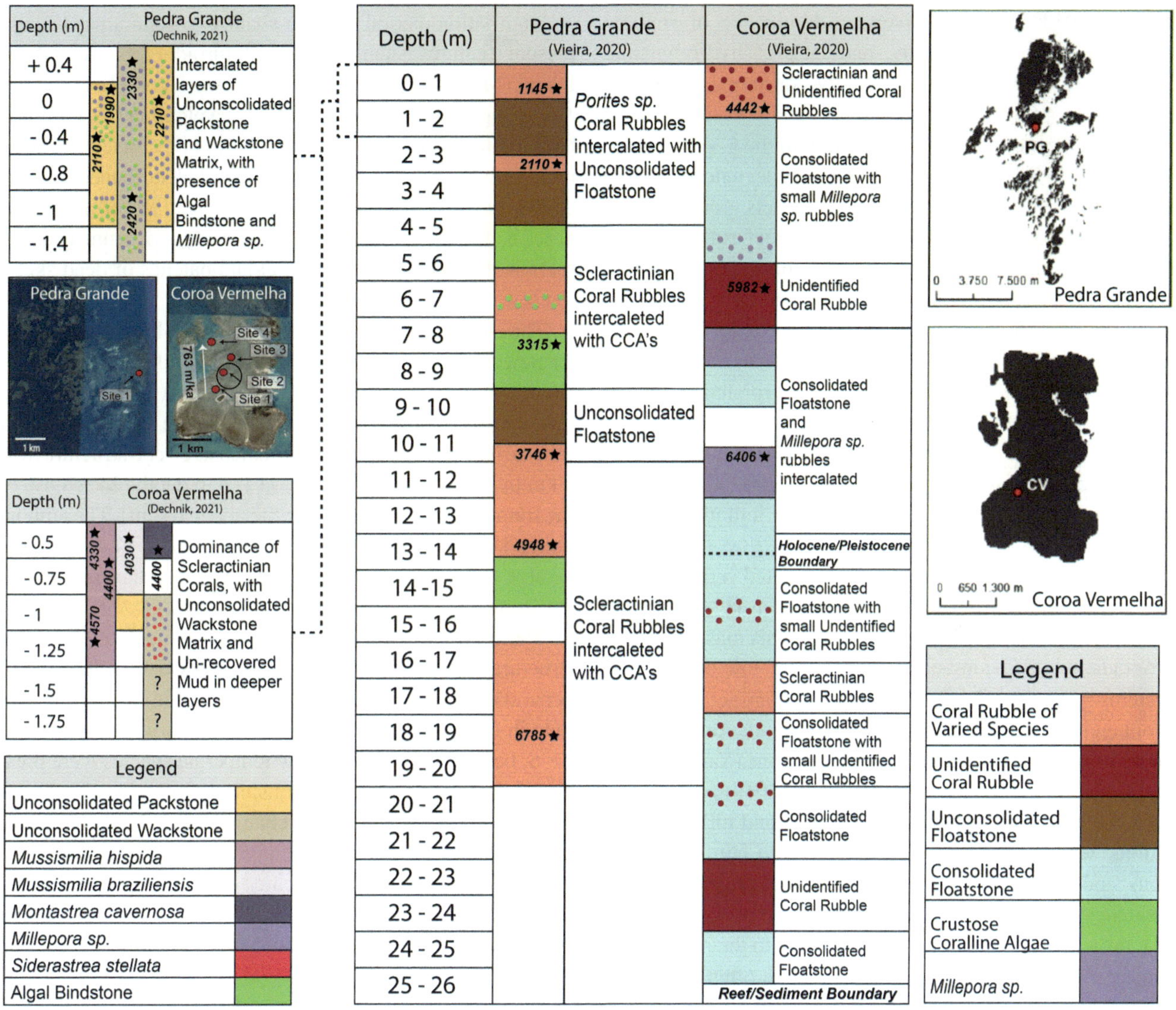

Fig. 15.3 Schematic representation and description of reef cores from the Pedra Grande and Coroa Vermelha reefs. The deepest cores (right), reaching up to 20 m, correspond to the study by Vieira (2020), whereas the shallow-water cores (left) correspond to the study by Dechnik et al. (2021). The location of each core is shown on the respective maps. Star icons indicate the dated levels of the reefs, with ages expressed in years before present (yr BP)

ference in initiation depth of these reefs, leading to a smaller accommodation space in CV because of its shallower initiation (Vieira 2020).

The framework of these Coastal Arc reefs showed intercalated facies of boundstone, floatstone, and sometimes rudstone. The predominance of one facies over the other changed spatially and throughout time, depending on sedimentation rates (Vieira et al. 2022). PG reef core presents typical boundstone facies intercalated with floatstone. Coralgal framework dominates (*M. cavernosa*, *M. harttii*, *Porites* sp. and *S. stellata*, *Millepora* sp.), with the occurrence of small bryozoans. Sponges and mollusks boring organisms are present with carbonate sediment infilling (Vieira et al. 2022). Shallow cores of PG demonstrate a packstone and wackestone with a significant amount of carbonate sediment (Dechnik et al. 2019, 2021). CV reef core is dominated by coral reef rubbles (*F. gravida*, *M. cavernosa*, *M. harttii*, *M.braziliensis*, *M. hispida*, *Porites* sp., and *S. stellata*) with infilling sediments. Floatstone and rudstone facies occur along the cores. SG shallow cores resemble some facies of PG, with packstone and wackestone matrix intercalated with *Millepora* sp. and *S. stellata* rubbles.

A shallowing upward reef assemblage sequence was not observed on both deep cores (PG and CV; Vieira 2020). Typically, as reefs approach the intertidal sea level, the coral assemblage shifts to shallow water species, as observed in Indo-Pacific and Caribbean Holocene reefs (Hopley et al. 2007; Montaggioni and Braithwaite 2009; Woodroffe and

Webster 2014). The low diversity, dominance of shallow water species, high turbidity, and lack of hydrodynamic energy may contribute to community constancy, as assemblages are not forced to change to persist.

The Holocene/Pleistocene boundary in the CV reef cores is marked at a depth of 12–13.5 m by a higher calcite content in reef fragments (98%–99%). Holocene reefs grew on top of Pleistocene boundstone, floatstone, and reef rubble facies. Interestingly, the coral fauna did not change considerably between the Holocene and Pleistocene periods of the CV reef. The predominant occurrence of floatstone facies with dark mud laminations and low carbonate content (<5.2%) at the base of the Pleistocene reef unit is interpreted as indicative of a higher turbidity environment during reef initiation (Vieira et al. 2022). Both 50-meter-deep CV reef cores show intercalated grainstone and floatstone facies with varying carbonate content before Pleistocene reef initiation (Vieira 2020).

During the Holocene, these reefs presented vertical accretion rates ranging from 2.1 m ky^{-1} to 5.3 m ky^{-1} and rapid lateral progradation (762.5–1909 m/ky). This massive lateral progradation is inconsistent with similar low-energy reef systems globally (24–300 m/ky) (Yamano et al. 2003; Montaggioni 2005). Greater reef flat progradation is usually under high-energy wind and wave conditions (Yamano et al. 2003; Montaggioni 2005). Dechnik et al. (2019) propose that ACS reefs evolved in high sedimentation and turbidity conditions, which promoted stress tolerance to inimical conditions and allowed further progradation. Longer sediment residency time would produce a rapid infill of the reef flats. The progradation direction is variable due to the mushroom-shaped morphology of ACS reefs, which is often limited by the reef slope compared to Indo-Pacific reefs.

15.3.2 High-Latitude Reefs

The major environmental variables that correlate with the biogeographical pattern of reef distribution are temperature, light, and carbonate saturation (Kleypas et al. 1999; Abrego et al. 2021), all of which limit reef occurrence on high latitudes. High-latitude reefs are subject to reduced temperatures and seasonal changes, with most coral communities occurring below 18 °C (Kleypas et al. 1999). High-latitude coral reefs are also subject to low levels of aragonite saturation, as the degree of saturation varies with latitude (4.1 at the Equator to 1.5 at the Poles) (Kleypas et al. 1999). Moreover, in high-latitudes, the angle of light incidence contributes to the loss of light penetration, and the average day length is shorter and light is less intense, so active reef growth at high latitudes is restricted to relatively shallow depths (Kleypas et al. 1999; Harriott and Banks 2002).

High-latitude coral reefs in shallow coastal areas usually present features such as tropical turbid reefs. At Inhaca Island (Mozambique), the reef framework comprises mainly *Porites* sp. bommies, and the reef development is inhibited by high turbidity conditions, highly mobile siliciclastic sediment substrates, and aperiodic fluvial disturbances (Perry 2003). On Moreton Bay (Australia), Wallace et al. (2009) described characteristics of coral framework from fringing reefs with high terrigenous input and patches of dead skeleton followed by regrowth.

The environmental limitations led to the high-latitude reef framework being constrained spatially and bathymetrically (Yamano et al. 2001; Perry 2003). Perry (2003) evidenced this restriction on Barreira Vermelha on the western shore of Inhaca Island, where the most extensive reef structure was 1 km in length and distributed on 1–6 m depth (24 ± 8.4% of the framework development in the upper 1–2 m). The authors pointed out that intense bioerosion by sponge *Cliona* sp. and bivalve *Lithophaga* sp. is likely to inhibit further framework development.

The high-latitude reefs framework also exhibits reduced structural complexity. Harriott and Banks (2002) described a general decline in coral species richness with increasing latitude in coastal eastern Australia and a sharp drop at around 31°S, beyond which only a few temperate species were present. However, the same relationship was not found between coral cover and the latitudinal change, where fewer but larger coral colonies are found (Chong et al. 2023). High-latitude marginal reefs experience greater environmental stress and temporal variability, which suggests low survival of smaller corals, slow growth, and/or poor recruitment (Chong et al. 2023).

Wallace et al. (2009) described several living corals on Moreton Bay (Australia) and coral suite persistence demonstrated by the unchanged Holocene coral fossil suite through the present, indicating that many species were continuously or intermittently present in the bay. On Moreton Bay, the Holocene assemblages studied are dominated by *Faviidae* and *Acropora*, with *Faviidae*, especially the genus *Favia*, being the dominant coral assemblage in most parts of the bay (Wallace et al. 2009). On Iki Island (Japan), Yamano et al. (2001) described the coral reef as dominated by *faviid* corals (mostly *Favia speciosa* and *Cyphastrea* sp.), suggesting that *Faviidae* corals can establish in regions beyond the environmental limits of *Acropora*.

Subtropical coral-dominated reef frameworks rarely form true reefs, with corals frequently growing directly attached to rocky substrates (Harriott and Banks 2002). Harriott and Banks (2002) attribute the shift to non-accreting reefs in eastern Australia to changes in the representation of branching *Acropora* species and massive corals, such as *Porites* colonies, along with an increase in macroalgal abundance at high-latitude sites. Kleypas et al. (1999) suggest that the decrease in aragonite saturation on high latitudes could induce a lower carbonate production; however, the covariance of temperature and light with saturation state and lack of field-measured calcification rates make this affirmation difficult.

15.3.3 High Temperature and Salinity Reefs

High-temperature and salinity (HTS) reefs are specialized coral reef ecosystems that thrive in extreme environments with elevated sea surface temperatures (above 30 °C) and high or very low salinity levels. These reefs are typically found in marginal marine environments where temperature and salinity exceed the tolerance limits of most coral species. Key global occurrences of HTS reefs include the Persian/Arabian Gulf, the Red Sea, and certain regions of the Indo-Pacific.

The Arabian Gulf is one of the best examples of HTS reefs occurrence and research. The Gulf geomorphological and hydrological settings generate extreme oceanographic conditions, especially for coral growth. The Gulf is the hottest sea on the planet during the summer (>35 °C), showing high evaporation rates and limited freshwater input, causing extreme salinity, with average values of 42 ppt, increasing to >50 ppt in the Southern bays, and reaching >70 ppt in restricted lagoons (John et al. 1990).

Sandy and muddy sediments dominate the sea bottom, but emergent cap rocks and hard-bottom salt domes occur regularly, which support coral reefs, non-accreting coral communities, and algal beds (Sheppard et al. 2010). Despite these extreme environmental conditions, this region presents 40 species of Scleractinian corals and 35 Alcyonacean corals (Riegl and Purkis 2012a; Sheppard et al. 2010). This fauna is well adapted to high ranges of temperatures and salinities, which results from a thermal adaptation of corals and their symbiotic Zooxanthellae (Howells et al. 2016; Smith et al. 2017). The most predominant coral genus in the Gulf is *Acropora*, especially among more suitable environments, such as upwind fringes of islands and exposed coastlines.

Although the Persian/Arabian Gulf is home to dense coral communities, in situ coral reef frameworks are relatively rare (Purkis and Riegl 2005). Major framework types are patch reefs, small fringing reefs, and stringer (elongated structures) reefs. Biostromes, or coral carpets, are also observed forming a thin and continuous framework of corals, but do not form reefs (Riegl and Purkis 2011). The reef framework in HTS environments is primarily built by heat- and salinity-tolerant or adapted coral species, such as *Porites* and *Acropora*, along with microbial communities and endolithic algae that contribute to reef accretion and stability. In the Arabian Gulf, coral assemblages are dominated by *Acropora clathrata* or *Acropora downingi* (Purkis and Riegl 2005). During extreme events, *Porites harrisoni* and *Porites nodifera* had lower mortality than Acropora, so their presence in a framework can be indicative of reef mortality (Riegl and Purkis 2011).

Although the coral fauna in this region is less diverse compared to Indo-Pacific reefs, the morphological variability of these reefs is notable. Rather than being shaped primarily by intricate limestone structures, this diversity stems from the underlying seafloor's structural complexity (Purkis and Riegl 2005; Riegl and Purkis 2011). Within the Gulf, a significant portion of reef volume is contributed by calcifying organisms other than corals. In areas with dense coral growth, interstitial spaces between colonies are filled by smaller carbonate-producing organisms, including mollusk shells, sea urchin tests, foraminifera, and calcareous algae. Due to the Gulf's supersaturated waters, this loose sediment undergoes rapid lithification—transforming into consolidated boundstone facies—when wave and current action provides adequate sediment ventilation (Riegl and Purkis 2012a, b). This hardened substrate then serves as an ideal foundation for subsequent coral settlement. Conversely, in zones with sparse coral growth, where carbonate debris is not retained, the same water chemistry and hydrodynamic conditions lead to the formation of hardgrounds (Riegl and Purkis 2012a, b). These surfaces eventually support dispersed, solitary coral colonies. The underdeveloped state of communities involved in frame-building in the final setting is attributable to the high mobility of sand in the flat topography, which causes the periodic smoothness of corals.

15.4 Final Remarks: What Else Can Influence Extreme and Marginal Reef Framework?

Considering the conceptual framework proposed by Schoepf et al. (2023), extreme coral reefs are related to regions that show large deviations from optimal conditions compared with classic clear-water coral reefs. Coastal and marine areas influenced by physical processes such as upwelling (high-nutrient and low temperature), river plumes, mesophotic condition (water depth), and storminess, could be considered extreme reefs. Moreover, biogeographic regions, such as the South Atlantic, could be considered as a marginal reef area due to its low coral richness (Bastos et al. 2018). Oceanographic processes will vary spatially and temporarily, so in terms of reef framework and coral or biogenic reef-builders, we may expect a reef framework resembling a combination of environmental and ecological processes through time. For example, a mesophotic reef framework usually shows a change in reef-builder composition, shifting from a shallow coral-dominated assemblage to a calcareous algae and/or heterotrophic-dominated builders. This indicates a give-up or a drowned reef due to a sea-level rise (Bridge et al. 2011; Abbey et al. 2013).

The Amazon reefs are also an example of extreme and marginal reefs (Moura et al. 2016). In this case, the Amazon river discharge provides the suboptimal environmental conditions, including sediment river discharge (suspended sediment plume—turbid) and freshwater river discharge (lower salinity), so the river and sediment plume are major environmental drivers for the extreme classification, combined with the water depth (mesophotic). The Amazon reefs are not coral reefs, but a complex mosaic of mesophotic environments that are driven, mostly, by the Amazon river plume seasonal variation. The mosaic is formed by sandstone reefs with associate mesophotic benthic community (including black coral, bryozoans, calcareous algae and sponges) and a sponge/calcareous algae reef (Moura et al. 2016; Vale et al. 2022).

Coastal or shelf upwelling environments could also be considered as extreme regions that may influence coastal reefs in positive and negative ways (Randall et al. 2020). Seasonal upwelling processes change water temperature and nutrient levels on continental shelves. This can impact coral assemblages at the same time that cool water can prevent coral bleaching. Upwelling may also influence biological and ecological competition that could be recorded in the framework. In a marginal system, such as the western South Atlantic, the occurrence of upwelling areas along the Brazilian Eastern shelf are associated with extensive rhodolith beds, with high crustose coralline algae richness (Amado-Filho et al. 2017). Rhodolith internal structure records the influence of terrigenous sedimentation, phases of rhodolith growth (turn-on and turn-off), fossil criptofauna and endofauna, and taphonomic phases of bioerosion, boring filling, and cement precipitation (Brasileiro et al. 2018; Franco et al. 2024).

Taking the South Atlantic as a marginal reef province, Bastos et al. (2018) showed that the most conspicuous reef type of the offshore reef arc on the Abrolhos Bank (mushroom-shaped pinnacle) has a complex framework-building system dominated by bryozoans, with a very low presence of the dominant coral assemblage of Abrolhos. The Abrolhos bank is the largest coral province of the South Atlantic, but one of the world's smallest provinces, with low coral richness and high endemism levels of up to 50% (Leão et al. 2003; Bastos et al. 2018).

Riegl (2001) points out that extreme conditions can also prevent *in situ* reef framework development. Temperature anomalies and high-energy waves or hurricanes/typhoons can be responsible for mass mortality and framework fragmentation, inhibiting *in situ* regrowth of reef framework. Extreme reef framework can be characterized by coral rubble forming rudstones and floatstones, meaning that high coral richness/cover and absence of framebuilding can co-occur (Riegl et al. 2011). Although high-latitude and high-temperature/salinity reefs can produce *in situ* framebuilding, extreme conditions can also inhibit their regrowth. It might be the same situation for marginal reefs of biogeographically isolated areas, such as the South and Equatorial Atlantic. South Atlantic reef framework are characterized by a coralgal framework interbedded with floatstones and rudstones (Leão and Ginsburg 1997; Vieira et al. 2022), bryozoan and coralgal framework (Bastos et al. 2018) on the Abrolhos Bank and coralline bindstone/framestone with rudstone (Gherardi and Bosence 2005) on the Rocas Atoll.

In summary, it is important to note that extreme and marginal reef frameworks are not uniform, and there is no specific framework model for each environmental condition or marginal area. Marginal reefs could show a dense coral growth with high live coverage, and it be reflected in their framework, but extreme conditions can change that. The reef framework will reflect the coral community when considering extreme conditions (e.g., high temperatures) and marginal biogeographic area, but allochthonous or autochthonous frameworks are related, mainly, to exposure to high-energy events or mass mortality conditions. All of these considerations are crucial for paleoecological interpretations, meaning that understanding the framework of modern extreme and marginal reefs is becoming more imperative, considering that the planet is facing even more extreme weather and climate conditions.

References

Abbey E et al (2013) Deglacial mesophotic reef demise on the great barrier reef. Palaeogeogr Palaeoclimatol Palaeoecol 392:473–494

Abrego D et al (2021) Factors limiting the range extension of corals into high-latitude reef regions. Diversity 13(12):632

Amado-Filho GM et al (2017) South Atlantic rhodolith beds: latitudinal distribution, species composition, structure and ecosystem functions, threats and conservation status. In: Rhodolith/Maërl beds: a global perspective. Springer, pp 299–317

Bastos AC et al (2018) Bryozoans are major modern builders of South Atlantic oddly shaped reefs. Sci Rep 8(1):9638

Brasileiro PS et al (2018) Burial rate determines Holocene rhodolith development on the Brazilian shelf. Palaios 33(10):464–477

Bridge TCL et al (2011) Topography, substratum and benthic macrofaunal relationships on a tropical mesophotic shelf margin, central Great Barrier Reef, Australia. Coral Reefs 30:143–153

Browne NK, Smithers SG, Perry CT (2010) Geomorphology and community structure of middle reef, central Great Barrier Reef, Australia: an inner-shelf turbid zone reef subject to episodic mortality events. Coral Reefs 29:683–689

Browne NK, Smithers SG, Perry CT (2012) Coral reefs of the turbid inner-shelf of the great barrier reef, Australia: an environmental and geomorphic perspective on their occurrence, composition and growth. Earth Sci Rev 115(1–2):1–20

Chong F et al (2023) High-latitude marginal reefs support fewer but bigger corals than their tropical counterparts. Ecography 2023(12):e06835

Dechnik B et al (2019) Holocene reef growth in the tropical southwestern Atlantic: evidence for sea level and climate instability. Quat Sci Rev 218:365–377

Dechnik B et al (2021) Environmental controls on holocene reef development along the eastern Brazilian margin. Coral Reefs 40:1321–1337

Flugel E (2002) Triassic reef patterns. In: Kiessling W, Flügel E, Golonka J (eds) Phanerozoic reef patterns, vol 72. SEPM Special Publication, pp 391–463. https://doi.org/10.2110/pec.02.72.0391

Franco T et al (2024) Highly bioeroded rhodoliths from the Vitória-Trindade seamount chain (Brazil, southwestern Atlantic). Palaios 39(5):161–173

Gherardi DFM, Bosence DWJ (2005) Late Holocene reef growth and relative sea-level changes in Atol das Rocas, equatorial South Atlantic. Coral Reefs 24:264–272

Guinotte JM, Buddemeier RW, Kleypas JA (2003) Future coral reef habitat marginality: temporal and spatial effects of climate change in the Pacific basin. Coral Reefs 22:551–558

Harriott VJ (1999) Coral growth in subtropical eastern Australia. Coral Reefs 18:281–291

Harriott V, Banks S (2002) Latitudinal variation in coral communities in eastern Australia: a qualitative biophysical model of factors regulating coral reefs. Coral Reefs 21:83–94

Hartt CF (1870) Geology and physical geography of Brazil. Boston, Fields, Osgood and Co, Boston. 620p

Hopley D, Smithers SG, Parnell K (2007) The geomorphology of the Great Barrier Reef: development, diversity and change. Cambridge University Press

Howells EJ et al (2016) Host adaptation and unexpected symbiont partners enable reef-building corals to tolerate extreme temperatures. Glob Chang Biol 22(8):2702–2714

John VC, Coles SL, Abozed AI (1990) Seasonal cycles of temperature, salinity and water masses of the western Arabian gulf. Oceanol Acta 13(3):273–281

Johnson CC et al (2002) Patterns and processes influencing upper cretaceous reefs. SEPM Society for Sedimentary Geology

Kleypas JA, McManus JW, Meñez LAB (1999) Environmental limits to coral reef development: where do we draw the line? Am Zool 39(1):146–159

Kleypas JA, Buddemeier RW, Gattuso J-P (2001) The future of coral reefs in an age of global change. Int J Earth Sci 90:426–437

Larcombe P, Woolfe KJ (1999) Increased sediment supply to the Great Barrier Reef will not increase sediment accumulation at most coral reefs. Coral Reefs 18:163–169

Leão ZMAN, Ginsburg RN (1997) Living reefs surrounded by siliciclastic sediments: the Abrolhos coastal reefs, Bahia, Brazil. In: Proc 8th Int Coral Reef Symp., pp 1767–1772

Leão ZMAN, Kikuchi RKP, Testa V (2003) Corals and coral reefs of Brazil. In: Latin American coral reefs. Elsevier Science, pp 9–52

Montaggioni LF (2005) History of Indo-Pacific coral reef systems since the last glaciation: development patterns and controlling factors. Earth Sci Rev 71(1–2):1–75

Montaggioni LF, Braithwaite CJR (2009) Chapter nine corals and coral reefs as Records of Climatic Change. Dev Mar Geol 5:373–428

Moura RL et al (2016) An extensive reef system at the Amazon River mouth. Sci Adv 2(4):e1501252

Palmer SE et al (2010) Internal structure and accretionary history of a nearshore, turbid-zone coral reef: Paluma shoals, central Great Barrier Reef, Australia. Mar Geol 276(1–4):14–29

Perrin C (2002) Tertiary: the emergence of modern reef ecosystems. SEPM Society for Sedimentary Geology

Perry CT (2003) Coral reefs in a high-latitude, siliciclastic barrier Island setting: reef framework and sediment production at Inhaca Island, southern Mozambique. Coral Reefs 22(4):485–497

Perry CT, Larcombe P (2003) Marginal and non-reef-building coral environments. Coral Reefs 22:427–432

Perry CT, Smithers SG (2010) Evidence for the episodic "turn on" and "turn off" of turbid-zone coral reefs during the late Holocene Sea-level highstand. Geology 38(2):119–122

Perry CT et al (2008) 1200 year paleoecological record of coral community development from the terrigenous inner shelf of the great barrier reef. Geology 36(9):691–694

Perry CT, Smithers SG, Johnson KG (2009) Long-term coral community records from lugger shoal on the terrigenous inner-shelf of the central Great Barrier Reef, Australia. Coral Reefs 28:941–948

Perry CT et al (2012) Evidence of very rapid reef accretion and reef growth under high turbidity and terrigenous sedimentation. Geology 40(8):719–722

Playford PE (1980) Devonian "Great Barrier Reef" of Canning Basin, Western Australia. AAPG Bull 64(6):814–840

Purkis SJ, Riegl BM (2005) Spatial and temporal dynamics of Arabian Gulf coral assemblages quantified from remote-sensing and in situ monitoring data. Mar Ecol Prog Ser 287(99–113):2005

Randall CJ et al (2020) Upwelling buffers climate change impacts on coral reefs of the eastern tropical Pacific. Ecology 101(2):e02918

Ribeiro FV et al (2018) Long-term effects of competition and environmental drivers on the growth of the endangered coral Mussismilia braziliensis (Verril, 1867). PeerJ 6:e5419

Riegl B (2001) Inhibition of reef framework by frequent disturbance: examples from the Arabian Gulf, South Africa, and the Cayman Islands. Palaeogeogr Palaeoclimatol Palaeoecol 175(1–4):79–101

Riegl B, Purkis SJ (2011) Persian/Arabian gulf coral reefs. In: Encyclopedia of modern coral reefs: structure, form and process. Springer, pp 790–798

Riegl BM, Purkis SJ (2012a) Coral reefs of the Gulf: adaptation to climatic extremes in the world's hottest sea. In: Coral reefs of the Gulf: adaptation to climatic extremes. Springer Netherlands, Dordrecht, pp 1–4

Riegl BM, Purkis SJ (2012b) Dynamics of gulf coral communities: observations and models from the world's hottest coral sea. In: Coral reefs of the Gulf: adaptation to climatic extremes. Springer Netherlands, Dordrecht, pp 71–93

Riegl BM et al (2011) Present limits to heat-adaptability in corals and population-level responses to climate extremes. PLoS One 6(9):e24802

Schoepf V et al (2023) Corals at the edge of environmental limits: a new conceptual framework to re-define marginal and extreme coral communities. Sci Total Environ 884:163688

Sheppard C et al (2010) The Gulf: a young sea in decline. Mar Pollut Bull 60(1):13–38

Smith EG et al (2017) Symbiont community stability through severe coral bleaching in a thermally extreme lagoon. Sci Rep 7(1):2428

Smithers S, Larcombe P (2003) Late Holocene initiation and growth of a nearshore turbid-zone coral reef: Paluma Shoals, central Great Barrier Reef, Australia. Coral Reefs 22:499–505

Vale NF et al (2022) Distribution, morphology and composition of mesophotic 'reefs' on the Amazon Continental Margin. Mar Geol 447:106779

Veron JEN (1995) Corals in space and time: the biogeography and evolution of the Scleractinia. Cornell University Press

Vieira LSV (2020) Plataforma de Abrolhos: desenvolvimento recifal durante o quaternário superior. PhD thesis, Universidade Federal Fluminense

Vieira LSV et al (2022) New insights on the Late Quaternary evolution of Abrolhos reefs, Eastern Brazilian Shelf. In: Coral Reefs and sea-level change: quaternary records and modelling. International Association of Sedimentologists (IAS), pp 69–100. https://doi.org/10.54780/IASSP49/03

Wacey D (2010) Stromatolites in the ~ 3400 Ma Strelley Pool formation, Western Australia: examining biogenicity from the macro-to the nano-scale. Astrobiology 10(4):381–395

Wallace CC et al (2009) The scleractinian corals of Moreton Bay, eastern Australia: high latitude, marginal assemblages with increasing species richness. In: Davie PJF, Phillips JA (eds) Proceedings of the thirteenth International Marine Biological Workshop, The Marine Fauna and Flora of Moreton Bay, Queensland. Memoirs of the Queensland Museum — Nature, vol 54(2), pp 1–118. Brisbane. ISSN 0079-8835

Webby BD (2002) Patterns of Ordovician reef development W. In: Kiessling E, Flügel JG (eds) Phanerozoic reef patterns, vol 72. SEPM, Tulsa, pp 129–180

Webster JM, Davies PJ (2003) Coral variation in two deep drill cores: significance for the Pleistocene development of the Great Barrier Reef. Sediment Geol 159(1–2):61–80

Woodroffe CD, Webster JM (2014) Coral reefs and sea-level change. Mar Geol 352:248–267

Woolfe KJ, Larcombe P (1998) Terrigenous sediment accumulation as a regional control on the distribution of reef carbonates. In: Reefs and carbonate platforms in the Pacific and Indian Oceans. Blackwell Science, pp 295–310

Woolfe KJ, Larcombe P (1999) Terrigenous sedimentation and coral reef growth: a conceptual framework. Mar Geol 155(3–4):331–345

Yamano H et al (2001) Highest-latitude coral reef at Iki Island, Japan. Coral Reefs 20:9–12

Yamano H et al (2003) Influence of wave energy on Holocene coral reef development: an example from Ishigaki Island, Ryukyu Islands, Japan. Sediment Geol 159(1–2):27–41

Carbonate Budgets of Marginal and Extreme Coral Reefs

16

Nicola Browne, Kyle Morgan, and Lorenzo Alvarez-Filip

Abstract

A carbonate budget measures the balance between rates of calcium carbonate production and erosion by reef communities. To date, 57 census-based studies have been conducted at >200 reef sites, with 40% of these on extreme or marginal reefs (i.e., non-traditional). Here we review 37 (19 traditional and 18 non-traditional reefs) carbonate budget studies and examine how coral communities within different environmental settings influence reef structural complexity and sediment properties in response to local conditions. Net carbonate production by turbid, urban and low pH/high nutrient reefs was greater than those rates reported for traditional reefs. These data highlight that despite suboptimal environmental conditions, many extreme reefs have a greater capacity to maintain critical reef processes that support reef accretion. Limited net carbonate production was observed on hot and marginal reefs, with many marginal reefs in a negative budget state. Critical knowledge gaps were identified relating to limited data available for calculating budgets. Specifically, future research should focus efforts towards (1) generating coral growth data for non-traditional reefs, (2) assessing colony and habitat scale rugosity, and (3) collecting environmental data in tandem with biological data to improve definitions of marginal and extreme reefs.

N. Browne (✉)
School of the Environment, University of Queensland, Brisbane, QLD, Australia
e-mail: n.browne@uq.edu.au

K. Morgan
Asian School of the Environment, Nanyang Technological University, Singapore, Singapore

Earth Observatory of Singapore, Nanyang Technological University, Singapore, Singapore

L. Alvarez-Filip
Unidad Académica de Sistemas Arrecifales, Universidad Nacional Autónoma de México (UNAM),
Puerto Morelos, Quintana Roo, Mexico

Keywords

Turbid reefs · Urban reefs · Hot reefs · Low pH reefs · Degraded reefs · Reef structural complexity · Carbonate sediment budgets · Coral growth · Bioerosion

16.1 Introduction

Coral reef ecosystems are complex three-dimensional habitats that are formed through the interaction between calcium carbonate ($CaCO_3$) production by benthic reef organisms (e.g., scleractinian corals, crustose coralline algae), which accumulate on reefs, and erosion processes (biological, physical and chemical) that breakdown or dissolve existing carbonate structures and export material away from reefs (Lange et al. 2020; Morgan and Kench 2014). Over decades to centuries, reef communities create vast physical structures that support immense marine biodiversity and biomass through habitat creation, form barriers that protect coasts from incident wave energy (Ferrario et al. 2014), and nourish adjacent shorelines with a supply of sediment (Cuttler et al. 2019; Morgan and Kench 2016a). The physical complexity (here considered as rugosity) of most contemporary reefscapes, as well as their ability to sustain these critical ecosystem services, is attributed to their carbonate budget, and the greater dominance of either accretionary (i.e., constructive) or erosional (i.e., destructive) processes on reefs (Perry et al. 2018a).

16.1.1 Carbonate Budgets

A coral reef carbonate budget is a method used to quantify the net rate of calcium carbonate production by reef communities. Early studies of reef carbonate budgets utilized hydrochemical methodologies to define net community productivity and calcification (Muehllehner et al. 2016). However, geoscientists (and more recently ecologists) have adopted a census-based approach to quantify taxa-level

M. O. Soares et al. (eds.), *Extreme and Marginal Coral Reefs and Communities*, Coral Reefs of the World 24,
https://doi.org/10.1007/978-3-032-28035-0_16

contributions to both reef-scale carbonate production and biological erosion (see review by Browne et al. 2021). The growing number of carbonate budgets within coral reef science is largely because the methods used are holistic, and reflect the local ecology, reef sedimentary processes and environmental conditions. It is therefore a useful framework for assessing overall coral reef health and geo-ecological functions. The concept can also be applied to understand the impact of disturbance periodicity on reefs, as repeat measurements at the same location through time can identify and quantify alterations in the reef budgetary state (e.g., Perry and Morgan 2017), as well as perturbations in the dominant drivers of calcification and erosion (e.g., Cornwall et al. 2021).

A **positive carbonate budget**, where the rate of reef calcification exceeds the rate of erosion, is indicative of a healthy coral reef system. These reefs are typically characterised by high biological productivity and reef growth potential, which is crucial for the development and maintenance of reef framework (e.g., Maldives; Morgan 2014). Coral reefs in net positive budgetary states are therefore typically assumed to have high coral cover and structurally complex reef communities (Perry et al. 2013). In contrast, a **negative carbonate budget** occurs where erosion processes exceed calcium carbonate accumulation, and are commonly associated with degraded reef systems, where the relative abundance of benthic calcifiers (e.g., corals) are low, leading to low relief reefscapes (Alvarez-Filip et al. 2009). Disturbances such as coral bleaching, poor water quality, or physical damage from storm activity can cause dramatic and relatively rapid shifts in a reef carbonate budget state, which may transition reefs from net positive to a net negative state as the reef ecology is altered (Perry et al. 2008; Perry and Morgan 2017). However, at some sites it has also been shown that carbonate production can rapidly recover (within a few years) and return to positive reef budget states following disturbances such as bleaching events (Lange et al. 2023b), if the post-disturbance conditions improve and there are no further pressures compromising reef condition and stability.

Within the context of marginal and extreme reefs, the conceptual framework of a carbonate budget remains relevant, although the process driving carbonate production and erosion are less well understood. Key ecological and environmental parameters that influence budget processes, and the overall budgetary states of reefs, likely differ between marginal and extreme reef types, as well as when compared to non-extreme or marginal i.e. traditional reefs (see review by Schoepf et al. 2023). This is due to the unique environmental conditions under which these reef ecosystems thrive, and as a response to the altered benthic community composition (both of carbonate producers and eroders), as specific taxa adapt to tolerate the stressful conditions. Significantly less is known about the carbonate budgets of marginal and extreme reef types, specifically how changes in reef ecology and environment manifest within aspects of the budget processes and how these alterations may influence the construction of reef framework and sediment generation on reefs.

16.1.2 Carbonate Sediment Budgets

The carbonate budget of a coral reef ecosystem provides a comprehensive measurement of net reef carbonate production as a function of biological productivity and reef erosion. However, benthic reef communities are also important generators and reservoirs of biogenically-produced carbonate sediment. These sediments contribute to the development of the internal fabric of the reef, and influence rates of vertical reef growth (Hubbard et al. 1990; Milliman 1993). The sediment budget of a coral reef is a combination of biological, physical, and chemical processes that produce calcium carbonate (e.g., coral calcification), and subsequently, the breakdown of this standing stock of material (e.g., physical abrasion, biological erosion) to boulder-gravel-sand-silt sized particles. The conversion of the calcium carbonate framework into sediment result from the gradual erosion of carbonate structures, for example, as a by-product of feeding activity, and have a profound impact on the sediment produced on reefs (Perry et al. 2016, 2019; Morgan and Kench 2016a, b). In addition to typical coral framework producers, such as corals and crustose coralline algae, carbonate sediments are also formed by direct sediment producers on reefs (e.g., formainifera, molluscs, *Halimeda* spp.), whereby an organism's skeleton is immediately converted to sediment post-mortem without the need for a secondary breakdown process (Freile 2004; Multer and Clavijo 2004). Direct sediment producers may not be highly abundant on the reef surface, or well represented in traditional benthic ecological surveys, but they can disproportionately contribute to the local detrital sediment pool. Within sediment budgets, the benthic community composition, life history and growth morphology of organisms, turnover rates of specific calcifying reef taxa, and the mechanism of breakdown and grain alteration (e.g., physical versus biological), all contribute to the type, rate and magnitude of the carbonate sediment produced by reefs (see review by Browne et al. 2021).

Within marginal and extreme reef settings, where benthic reef community structure has often been majorly altered from those commonly found on traditional reefs, or where there are significant inputs of external terrigenous sedimentation (e.g., turbid reefs), the local reef detrital sediment system can differ in terms of sediment grain size, constituent composition and mineralogy (e.g. Browne et al. 2013; Morgan et al. 2020). This is because both the source of the sediment (i.e., ecological standing stock) and the destructive processes for sediment generation also differ. Despite several attempts to provide a methodology for quantifying reef sediment budgets, it is only relatively recently that the first detailed methodology was published, termed the SedBudget (Lange et al. 2023a). This new method may provide the first critical step to improving our current understanding of coral reef geo-ecological functions for a range of reef types including marginal and extreme reefs.

16.1.3 Quantification of Carbonate Budgets

The carbonate budget of a coral reef can be measured using two main approaches. The first method is termed the **total alkalinity anomaly method** (or hydrochemical method), which is calculated by measuring changes in total alkalinity of the seawater surrounding a section of coral reef (Muehllehner et al. 2016). As corals and other reef taxa calcify, they consume alkalinity at a rate that can be measured through the use of incubation chambers (*in-situ* or *ex-situ* e.g., Comeau et al. 2018). It is therefore possible to establish the rate that these reef organisms extract calcium and carbonate ions to build their skeletons, giving precise measurements of calcification for an individual colony or section of reef. Although this technique provides useful indirect information of coral and/or community metabolism and productivity with regards to calcification, and can be applied consistently over large spatial scales, it provides no direct information on the specific taxa contributing to net calcification. It also omits important ecological data that may contribute to a better understanding of reef geo-ecological functions, including an assessment of reef rugosity and reef growth capacity. Furthermore, erosion processes from various bioeroders (e.g., parrotfish, urchins, macroboring organisms) are not captured, and therefore relating net calcification values to address ecological and geoscience research questions is limited (see review by Browne et al. 2021).

The second method for quantifying reef carbonate budgets, and one which is more widely used by ecologists, is the **census-based method**. The census-based method involves using field-derived benthic cover surveys and applying rates of taxa-specific calcification to calculate gross carbonate production for a section of reef (e.g. Browne et al. 2013; Morgan 2014). For many reef types, site-specific rates of organism calcification are not available, and global averaged data are often substituted to estimate rates of gross productivity (e.g., ReefBudget by Perry et al. 2012). Biological erosion of reefs is then calculated by conducting census surveys of key excavating and boring taxa, which include urchins, parrotfish and internal micro- and macroboring organisms. Reef bioerosion is calculated using the relative abundance of these bioeroders per unit of reef area, their population size structure, and associated erosion rates derived mostly from published empirical relationships (Perry et al. 2012). Chemical dissolution of the reef carbonate is then calculated, or in many cases a uniform rate is applied from existing scientific literature (e.g., Brown et al. 2020), which are then combined with biological erosion data to calculate gross erosion rates. The net calcification rate for a reef is calculated by subtracting the rate of reef erosion from the rate of gross carbonate production. A positive budget value indicates a net gain of calcium carbonate (expressed as kg m^{-2} y^{-1}), while a negative value suggests a net loss (Perry et al. 2008). Reefs that exhibit budget values of around zero are commonly termed as neutral budget states.

Accurate measurements of gross carbonate production therefore rely on obtaining high-quality data on coral growth, skeletal density and calcification, as well as calcification rates of other encrusting and cryptic reef taxa across a range of environmental settings (Browne et al. 2021). This is particularly important as data generated for small areas of reef (at the transect level) are often extrapolated over larger spatial reef extents, such as entire eco-geomorphic zones or whole reef archipelagos. The census-based approach therefore creates some fundamental challenges for calculating carbonate budgets for marginal and extreme reef types, largely due to the lack of taxa-specific calcification and erosion rate data within these understudied reef types. Reefs located outside the range of traditional environmental conditions will likely exhibit variances in organism growth rates, bulk skeletal density, or feeding behaviour, and therefore the application of global averaged datasets from traditional reef types may not be appropriate.

16.1.4 Approach

In this chapter, we review existing published global carbonate budget studies for extreme and marginal reef systems and use this meta-analysis to characterise the major carbonate budget processes for these reef types, and compare them to traditional reefs. To do this, we categorised and identified reefs into one of the following reef types: (1) traditional reefs, (2) marginal reefs, or (3) extreme reefs. Here we define a 'traditional' reef as a coral reef that does not experience large deviations from optimal abiotic conditions (e.g. see Kleypas et al. 1999), AND is characterised by >10% coral cover. Definitions for extreme and marginal reefs follow the recent review paper by Schoepf et al. (2023), whereby extreme reefs were defined as reefs characterised by large deviations from optimal in one or more abiotic conditions (e.g., temperature, light, pH). Extreme reefs include, for example, hot reefs (e.g., northern Red Sea), where mean sea water temperatures are around 31 °C (Roik et al. 2018), turbid reefs (e.g., inner Great Barrier Reef), which experience periodically high sedimentation and low-light conditions (Browne et al. 2012; Morgan et al. 2016), urban reefs (e.g.,

Singapore) that exist nearby large urban centres and are characterised by high terrestrial inputs (nutrients, sediments, pollutants; Browne et al. 2015; Heery et al. 2018; Morgan et al. 2020), high latitude reefs (e.g., Okinawa) that form at the environmental limits for coral reef development (Yamano et al. 2012), low pH reefs, which inhabit relatively acidified seawater conditions (Fabricius et al. 2011), and mangrove reefs, which coexist with coastal fringing forests and are subject to large daily fluctuations in pH, salinity, oxygen and temperature (Camp et al. 2019). In contrast, marginal reefs were defined as reefs with altered coral cover, community composition and/or ecosystem functioning, which may also include coral communities. Coral reefs with <10% coral cover AND exposed to non-optimal abiotic conditions were classified as both marginal and extreme. Lastly, we use the term 'non-traditional' when referring to marginal and extreme reefs collectively.

Carbonate budgets provide a comprehensive way to examine important questions about how reefs calcify, and how they construct framework that contributes to long-term reef development, because of the large range of data included. For example, carbonate budgets provide insights into coral reef community composition using census-based ecological surveys, as well as organism activity (e.g., bioerosion), and net rates of carbonate production that have longer-term geological consequences for the development of reef structure and local sediment reservoirs (Perry et al. 2018a, b). As such, by assessing key elements of carbonate budgets, this chapter aims to address the following questions:

1. What are the ecological drivers of carbonate production on traditional, marginal and extreme reefs?
2. How do rates of carbonate production differ between traditional and non-traditional reef types?
3. What are the processes that contribute to reef framework erosion on marginal and extreme reefs?
4. How do variations in carbonate budgets between reef types impact reef (or habitat) structural complexity?
5. Are there differences in reef sediment characteristics and generation processes between reef types?

16.1.4.1 Data Collection and Analysis

Differences in carbonate budgets between reefs were assessed by extracting all census-based studies from a recent review paper by Browne et al. (2021). This comprehensive review of global carbonate budgets provided details on all complete census-based carbonate budgets between 1977 to 2020 (37 studies; see Table 16.1 in Browne et al. 2021). Since 2020 (to 2024), an additional 17 studies were identified using a systematic approach and reviewed here. In addition, we also include three earlier studies that focus on bioerosion from marginal or extreme reefs (Table 16.1), where carbonate budget data is lacking.

To determine significant differences in key reef ecological parameters (coral cover, gross carbonate production, bioerosion, net production and reef structural complexity) between reef types, we conducted statistical analysis (R studio; version 4.2.3). Data were first tested for normality and homogeneity of variance using the Shapiro-Wilk's Test and Levene's test, respectively. Data on coral cover, and gross and net carbonate production were log transformed to meet parametric assumptions, followed by one-way Anova (using the lm function) with reef type as the fixed factor, and Tukey HSD post hoc test (if necessary) to identify reef types that differed significantly (alpha = 0.05). We also carried out a linear regression between coral cover and gross carbonate production, and bioerosion and net carbonate production to determine if there were significant relationships between these variables.

16.2 Carbonate Budgets of Traditional, Marginal and Extreme Reef Systems

Since the late 1970s, there have been 57 full (or part) census-based carbonate budget studies completed. These existing 57 carbonate budget studies have a wide geographic distribution, and include 49 locations and 237 reef sites. The largest number of studies have been conducted in the Indo-Pacific region (17), of which 71% were conducted in Australia and the remaining within Singapore, Indonesia and the South China Sea (Fig. 16.1). A number of carbonate studies (12) have been conducted in the central to Eastern Pacific region, of which 58% were conducted in the Eastern Pacific (e.g., Galapagos and Gulf of Chiriqui, Panama) and 42% distributed over the Western to Central Pacific. Of the 11 studies conducted in the Indian Ocean, eight studies were conducted in the Maldives and on the Chagos Archipelago, with other sites including the Seychelles, Zanzibar in East Africa and the Lakshadweep Archipelago in India. There have also been 11 studies conducted in the Caribbean, four in the Atlantic, largely concentrated on the coral reefs off the coast of Florida, and three studies in the Red Sea.

Of these 57 global carbonate budget studies, only 25 studies have been conducted on extreme and/or marginal reefs, and include eight turbid reefs, five urban reefs, five marginal reefs, three hot reefs, three low pH (and high nutrient) reefs and one mangrove reef (Table 16.1). Non-traditional reefs therefore account for 40% of the total global carbonate budget studies conducted to date. The majority of marginal and extreme reef carbonate budget studies (84%) have been conducted in the last 10 years, and 40% of these within the past 3 years. In contrast, carbonate budget studies for traditional reefs began over 40 years ago, with 50% conducted in the past 10 years, and 15% in the last 3 years. This demonstrates the growing scientific interest in the widespread application of carbonate budgets within all reef types, but also the

Table 16.1 Carbonate budget studies conducted on extreme and marginal reef types

No.	Author	Year	Title	Location	Reef type	Carbonate budget	Carbonate budget variables measured	Methods
1	Heiss	1995	Carbonate production by scleractinian corals at Aqaba, Gulf of Aqaba, Red-Sea	Gulf of Aqaba, Red-Sea	Hot reef	Part (gross coral production)	Live and dead coral cover, sediment cover, *Porites* growth rate, density and calcification rate	LIT, collection of live Porites and CT scans and X-radiographs
2	Edinger et al.	2000	Normal coral growth rates on dying reefs: Are coral growth rates good indicators of reef health?	Java, Indonesia	Mangrove (Lagun Marican), nutrients and sediments (Palau Panjang and Bondo)	Part (net coral production)	Live and dead coral cover, algal cover, sand and rubble, sponges, coral composition, growth rate and skeletal density, macro-boring rates, water clarity, maximum depth of coral growth, temperature, salinity, chlorophyll a, nitrate and phosphate concentrations, SPM, sediment accumulation and light intensity	LIT, collection of live Porites and coral rubble, in situ environmental data collection
3[a]	Tribollet and Golubric	2005	Cross-shelf differences in the pattern and pace of bioerosion of experimental carbonate substrates exposed for 3 years on the northern Great Barrier Reef, Australia	6 sites including 2 inshore turbid reefs (Snapper and Low Isles) on the Northern GBR	Natural turbidity gradient	Part (bioerosion)	Internal macro and micro-bioerosion rates, external grazing rates, composition of bioeroding organisms	Deployment of bioeroder blocks over 1 and 3 years in Porites blocks
4[a]	Manzello et al.	2008	Poorly cemented coral reefs of the eastern tropical Pacific: Possible insights into reef development in a high-CO_2 world	Eastern tropical Pacific—Galapagos, Gulf of Panama and Chiriqui	Low to high aragonite saturation state	Part (cementation and bioerosion)	Temperature, TCO_2, TA, seawater pCO_2, pH and aragonite saturation state, reef framework cementation, macro and microboring	Seawater sampling and collection of reef framework samples

(continued)

Table 16.1 (continued)

No.	Author	Year	Title	Location	Reef type	Carbonate budget	Carbonate budget variables measured	Methods
5	Browne et al.	2013	Carbonate and terrigenous sediment budgets for two inshore turbid reefs on the central Great Barrier Reef	Middle Reef and Paluma shoals, GBR	Naturally turbid reefs	Complete	Live and dead coral cover, macro and turf algal cover, CCA cover, encrusting community cover, composition and growth rates, sand, rubble and reef framework cover, rugosity, coral composition, growth rates and skeletal density, fish species, abundance and size, urchin abundance and size, macro-boring rates, light, suspended sediments, net sediment accumulation rates, sediment export rates, sediment depth layers, sediment particle size, carbonate content and composition	Census based approach
6[a]	Chen et al.	2013	Macrobioerosion in *Porites* corals in subtropical northern South China Sea: a limiting factor for high-latitude reef framework development	Daya Bay, South China Sea	Close (2 km) to far (19 km) from urban sites within the bay	Part (bioerosion)	Bioerosion intensity and taxa composition on a gradient with water quality	Collection of live Porites blocks present with a number of water quality parameters obtained from the Chinese ecosystem research network

7	Perry et al.	2013	Caribbean-wide decline in carbonate production threatens coral reef growth	Several sites across the Caribbean (Eleutherain the Bahamas, Grand Cayman and Bonaire)	Three reefs included from this study had <10% coral cover—marginal systems	Complete	Living coral cover, CCA cover, turf and macro-algal cover, *Halimeda* cover, sand, rubble and reef framework cover, rugosity, coral composition, growth rates and skeletal density, encrusting cover and growth rates, parrotfish species, size and abundance, parrotfish bite rates, scars and mass eroded, urchin abundance, size and species, clinoid sponge cover, micro-bioerosion rates	Census based approach (ReefBudget)
8	Herran et al	2017	Calcium carbonate production, coral cover and diversity along a distance gradient from Stone Town: A case study from Zanzibar, Tanzania	Four sites of a gradient (Chapwani Reef, Prison reef, Bawe reef and Chumbe Island) in Zanzibar, Tanzania	Urban—high nutrients (high volumes of untreated sewage) and sediments	Complete	Same as Perry et al. (2013)	Census based approach (ReefBudget)
9	Prouty et al.	2017	Vulnerability of coral reefs to bioerosion from land-based sources of pollution	7 sites on a gradient from a CO_2 seep, West Maui, Hawaii	Nutrient rich (nitrate 50x higher than ambient) and low pH (7.85–8.01)	Part (net coral production)	Coral linear extension, density, calcification, bioerosion, dissolved nutrients (ammonia, Si, phosphate, nitrate and nitrite, pH, TA, DIC, DO, aragonite saturation state, temperature and salinity	Coral core collection

(continued)

Table 16.1 (continued)

No.	Author	Year	Title	Location	Reef type	Carbonate budget	Carbonate budget variables measured	Methods
10	Roik et al.	2018	Coral reef carbonate budgets and ecological drivers in the Central Red Sea—a naturally high temperature and high total alkalinity environment	Three sites (inshore, midshore, offshore) Red Sea, Saudi Arabia	Hot reef	Complete	Living and dead coral cover, CCA cover, turf and macro-algal cover, reef framework cover, rugosity, coral composition and growth rates, CCA growth rates, parrotfish species, size and abundance, urchin abundance, size and species, net accretion/erosion rates (blocks), temperature, salinity, pH, nitrite/ate, ammonia, phosphate, TA	Census based approach (ReefBudget)
11	Manzello et al	2018	Resilience in carbonate production despite three coral bleaching events in 5 years on an inshore patch reef in the Florida keys	Cheeca rocks, Florida keys	Inshore, turbid reef	Complete	Same as Perry et al. (2012)	Census based approach (ReefBudget)
12	de Bakker et al.	2019	Extreme spatial heterogeneity in carbonate accretion potential on a Caribbean fringing reef linked to local human disturbance gradients	2 (Kralendijk-K and Saltpans-SP) out 12 sites considered to be highly impacted by anthropogenic activity, Bonnaire, Caribbean	Urban/anthropogenic reefs	Complete	Same as 2012 Perry et al. study, with the addition of maximum reef height and sponge specific erosion rates	Census based approach (ReefBudget)

13	Januchowski-Hartley et al.	2020	Accreting coral reefs in a highly urbanized environment	Seven sites along a turbidity/nutrient gradient Singapore	Anthropogenic turbid reef	Complete	Live coral cover, CCA and non-encrusting coralline algal cover, turf and fleshy macro-algal cover, *Halimeda* cover, sand, rubble and reef framework cover, rugosity, coral composition, shape and size, growth rates and skeletal density, CCA calcification rates, parrotfish species, size and abundance, urchin abundance, size and species, internal bioerosion rates, epilithic algal matrix, microbes and micro-algae	Census based approach (ReefBudget)
14	Molina-Hernandez et al.	2020	Two decades of carbonate budget change on shifted coral reef assemblages: are these reefs being locked into low net budget states?	34 reef sites in the Mexican Caribbean	Marginal low coral cover	Complete	Same as Perry et al. (2012)	Census based approach (ReefBudget)
15	Dee et al.	2020	The complexity of calculating an accurate carbonate budget	Eva and Fly Reef, Exmouth Gulf, Australia	Naturally semi turbid reef	Part (gross coral production)	Live and dead coral cover, macro and turf algal cover, CCA cover, encrusting community cover, composition and growth rates, sand, rubble and reef framework cover, rugosity, coral composition, growth rates and skeletal density, 3D reef morphology	Census based approach

(continued)

Table 16.1 (continued)

No.	Author	Year	Title	Location	Reef type	Carbonate budget	Carbonate budget variables measured	Methods
16	Randi et al.	2021	Calcium carbonate production in the southernmost subtropical Atlantic coral reef	Quimada Grande Island, Brazil	High natural turbidity and semi-tropical	Part (encruster)	Encruster cover, composition and calcification rates, rhodolith dimensions, abundance and annual growth	Deployment of encrusting plates and collection of rhodolith samples
17	Dee et al.	2021	Encrusters maintain stable carbonate production despite temperature anomalies among two inshore island reefs of the Pilbara, Western Australia	Eva and Fly Reef, Exmouth Gulf, Australia	Naturally semi turbid reef	Part (encruster)	Encrusting carbonate community and production rates, chlorophyll a, conductivity, salinity, pH, turbidity, temperature and light	Deployment of encrusting plates and collection of in environment and water quality data
18	Goh et al.	2021	Growth and carbonate production of crustose coralline algae on a degraded turbid reef system	7 sites in Singapore	Anthropogenic turbid reef	Part (encruster)	Encrusting community composition, CCA cover and production rates	Settlement cards deployed for 2 years in 6 month blocks
19	Morris et al.	2022	Low net carbonate accretion characterizes Florida's coral reef	Three biogeographic regions (Dry Tortugas-DRTO, Florida keys-FLK and Southeast Florida-SEFL) with 723 sites	A number of low coral cover (<10%) marginal reefs	Complete	Living and dead coral cover, CCA cover, turf and macro-algal cover, reef framework cover, rugosity, coral composition and growth rates, CCA growth rates, parrotfish species, size and abundance, urchin abundance, size and species, macro and micro bioerosion	Census based approach (ReefBudget)
20	Toth et al.	2022	The past, present, and future of coral reef growth in the Florida keys	46 sites in Florida keys assessed from 1996 to 2019	Marginal low coral cover	Complete	Living and dead coral cover, CCA cover, turf and macro-algal cover, reef framework cover, rugosity, coral composition and growth rates, CCA growth rates, parrotfish species, size and abundance, urchin abundance, size and species, macro and micro bioerosion	Census based approach (ReefBudget)

21	Dee et al.	2023	Low bioerosion rates on inshore turbid reefs of Western Australia	Eva and Fly Reef, Exmouth Gulf, Australia	Naturally semi turbid reef	Part (bioerosion)	Macro and micro-bioerosion rates and composition, fish and urchin grazer rates, chlorophyll a, conductivity, salinity, pH, turbidity, temperature and light	Bioeroder blocks and visual census of fish and urchins
22	Rodriguez-Ruano et al.	2023	Upwelling, climate change, and the shifting geography of coral reef development	Gulf of Panama (GoP; upwelling site) and Gulf of Chiriqui (GoC)	Upwelling site: lower temperatures and aragonite saturation state and higher nutrients	Complete	Living and dead coral cover, CCA cover, turf and macro-algal cover, rubble and sand, coral composition and growth rates, parrotfish species, size and abundance, urchin abundance, size and species, macro and micro bioerosion	Census based approach (ReefBudget)
23	Toth et al.	2023	The potential for coral reef restoration to mitigate coastal flooding as sea levels rise	54 shallow fore reef sites on Buck Island Reef, US Virgin Islands	Marginal low coral cover fringing reef	Complete	Living coral cover, CCA cover, turf and macro-algal cover, *Halimeda* cover, sand, rubble and reef framework cover, coral composition, growth rates and skeletal density, encrusting cover and growth rates, parrotfish species, size and abundance, parrotfish bite rates, scars and mass eroded, urchin abundance, size and species, clinoid sponge cover, micro-bioerosion rates	Census based approach

(continued)

Table 16.1 (continued)

No.	Author	Year	Title	Location	Reef type	Carbonate budget	Carbonate budget variables measured	Methods
24	Alemealla et al.	2024	Bleaching threatens positive carbonate budgets on Bahraini reefs	A nearshore and offshore site in Bahraini, Arabian Gulf pre (2017) and post (2018) bleaching	Hot reefs and inshore reef heavily impacted by reclamation and dredging	Complete	Live and dead coral cover, macro and turf algal cover, CCA cover, encrusting community cover, composition and growth rates, sand, rubble and reef framework cover, rugosity, coral composition, fish species, abundance and size, urchin abundance and size, macro-boring and micro-boring rates, light, temperature, salinity, and turbidity	Census based approach (ReefBudget)
25	Dee et al.	2024	The application of carbonate and sediment budgets to assess the stability of marginal reef systems	Eva and Fly Reef, Exmouth Gulf, Australia	Naturally semi turbid reef	Complete	Live and dead coral cover, macro and turf algal cover, CCA cover, encrusting community cover, composition and growth rates, sand, rubble and reef framework cover, rugosity, coral composition, growth rates and skeletal density, fish species, abundance and size, urchin abundance and size, macro-boring and micro-boring rates, light, temperature, chlorophyll a, conductivity, salinity, pH and turbidity, sediment depth layers, sediment particle size, carbonate content and composition	Census based approach

No.	Author	Year	Title	Gross productivity (kg m^{-2} y^{-1})	Erosion (kg m^{-2} y^{-1})	Net productivity (kg m^{-2} y^{-1})	Key producer and/or eroder	Coral cover (%)	Structural complexity at the habitat scale (rugosity with 1 denoting a flat surface)
1	Heiss	1995	Carbonate production by scleractinian corals at Aqaba, Gulf of Aqaba, Red-Sea	Mean = 1.6 (coral only)	N/A	N/A	Branching corals are the dominant form followed by massive corals	Mean = 35	Not measured
2	Edinger et al.	2000	Normal coral growth rates on dying reefs: Are coral growth rates good indicators of reef health?	L.Marican = 3.2, P.Panjang = 3.2, Bondo = 4.3 (coral only)	L.Marican = 0.9, P.Panjang = 10.1, Bondo = 5.1	L.Marican = 2.4, P.Panjang = −7.4, Bondo = −0.8	Only corals (stress tolerant) measured and only internal macro-borers assessed	L.Marican = 22, P.Panjang = 25, Bondo = 30	Not measured
3[a]	Tribollet and Golubric	2005	Cross-shelf differences in the pattern and pace of bioerosion of experimental carbonate substrates exposed for 3 years on the northern Great Barrier Reef, Australia	N/A	1 year: Snapper = 0.27 and Low Isles = 0.18 3 years: Snapper = 0.43 and Low Isles = 0.63[a]	N/A	Microbioerosion dominated the first year but macrobioerosion was the dominant by 3rd year. Grazing pressure very low year 1 and increased in year 3. Bivalves and worms first to recruit with higher rates of sponges by year 3	N/A	Not measured
4[a]	Manzello et al.	2008	Poorly cemented coral reefs of the eastern tropical Pacific: Possible insights into reef development in a high-CO_2 world	Cement abundance (%): Galapagos = 4.5, G.Panama =4.4, G.Chiriqui = 12.5	Galapagos = 25.4, G.Panama =26.1, G.Chiriqui = 13.6[a]	N/A	N/A	N/A	Not measured

(continued)

Table 16.1 (continued)

No.	Author	Year	Title		Location	Reef type	Carbonate budget	Carbonate budget variables measured	Methods
5	Browne et al.	2013	Carbonate and terrigenous sediment budgets for two inshore turbid reefs on the central Great Barrier Reef	Middle reef = 13.8, Paluma shoals = 9.6	Middle reef = 1.5, Paluma shoals = 1.6	Middle reef = 12.3, Paluma shoals = 6.9	Corals = 96 to 99 of carbonate, CCA = 1 to 4. Dominant corals = *Acropora*, *Montipora* and *Galaxea*. No to low external macro-bioerosion, but high internal macro-bioerosion	Middle reef = 40, Paluma shoals = 30	Range: Middle reef = 1.1 to 1.6 and Paluma shoals = 1.0 to 1.4
6[a]	Chen et al.	2013	Macrobioerosion in *Porites* corals in subtropical northern South China Sea: a limiting factor for high-latitude reef framework development	N/A	Bioerosion intensity () range: Close = 27.55 and far = 2.18	N/A	Bivalves and sponges dominated inshore whereas worms and barnacles less of gradient	N/A	Not measured
7	Perry et al.		Caribbean-wide decline in carbonate production threatens coral reef growth	Eleuthera (hole in the wall) = 1.2, Grand Cayman (Don Fosters) = 0.9, Bonaire (Cai) = 0.3 (all at 5 m)	Eleuthera (hole in the wall) = 1.7, Grand Cayman (Don Fosters) = 2.6, Bonaire (Cai) = 1.2	Eleuthera (hole in the wall) = −0.5, Grand Cayman (Don Fosters) = −1.8, Bonaire (Cai) = −1	Corals were the main contributors to carbonate production typically non-framework domed and encrusting corals. No information on bioeroders provided	Eleuthera (hole in the wall) = 4.6, Grand Cayman (Don Fosters) = 5.6, Bonaire (Cai) = 2.7	Not provided
8	Herran et al	2017	Calcium carbonate production, coral cover and diversity along a distance gradient from stone town: A case study from Zanzibar, Tanzania	Chapwani = 10.7, prison = 10.43, Bawe = 15.03, Chumb = 18.95	Chapwani (5 m) = 1.89, prison (5 and 10 m) = 1.25 and 1.06, Bawe (5 and 10 m) = 6.16 and 5.59, Chumbe (5 and 10 m) = 1.98 and 2.33	Chapwani = 8.03, prison = 10, Bawe = 9.2, Chumbe = 16.90	As conditions improved coral production went from submassives and massives to branching and plating corals. Microborers low across all sites. Poorest site dominated by urchins and low sponge cover. Fish erosion increased with distance from source	Chapwani (5 m) = 52.5, prison (5 and 10 m) = 65 and 74.5, Bawe (5 and 10 m) = 72.7 and 71.5, Chumbe (5 and 10 m) = 67.4 and 57.6	Chapwani (5 m) = 1.22, prison (5 and 10 m) = 1.47 and 1.51, Bawe (5 and 10 m) = 1.35 and 1.18, Chumbe (5 and 10 m) = 1.38 and 1.34

9	Prouty et al.	2017	Vulnerability of coral reefs to bioerosion from land-based sources of pollution	Coral core calcification rates ranged from 0.68 to 1.1 g cm^{-2} y^{-1}	Bioerosion rates ranged from 99.15 to 22.58 mg $cm^{-2}y^{-1}$	N/A	Bioerosion rates <8 × higher than expected for corals growing away from land-based sources of pollution	N/A	N/A
10	Roik et al.	2018	Coral reef carbonate budgets and ecological drivers in the Central Red Sea—a naturally high temperature and high total alkalinity environment	Inshore = 0.43, Midshore = 1.76, offshore = 2.81	Inshore = 1.59, Midshore = 0.75, offshore = 0.46	Inshore = −1.48, Midshore = 1.02, offshore = 2.44	CCA contributed 4 to 5 inshore to offshore, but total CCA productivity was approximately 2 orders of magnitude higher offshore. Massive corals contributed inshore whereas more diverse coral community with *Acropora* offshore. Only parrotfish and urchins assessed, both of which were more abundant inshore (urchins ×10, fish ×1.5 to 2)	Inshore = 10 to 20, Midshore = 30 to 40, offshore = 8 to 35	Measured but not provided
11	Manzello et al	2018	Resilience in carbonate production despite three coral bleaching events in 5 years on an inshore patch reef in the Florida keys	2012 = 4.5 to 2016 = 5.73	2012 = 0.94 to 2016 = 1.08	2012 = 3.56 to 2016 = 4.64	Carbonate production driven by O. annularis (75%). Bioerosion driven by parrotfish (>60%), microbioerosion (>15%). Urchins and sponges low	2012 = 24.8 to 2016 = 32.9	Measured but not provided
12	de Bakker et al.	2019	Extreme spatial heterogeneity in carbonate accretion potential on a Caribbean fringing reef linked to local human disturbance gradients	K = 0.85 and SP = 0.58	K = 1.85 and SP = 1.3	K = −2.20 and SP = −1.95	No CCA contributions and low coral cover (2–5%). Parrotfish were the drivers of erosion, no urchins and low sponge erosion	K = 5 and SP = 2.5	1.07–1.09

(continued)

Table 16.1 (continued)

No.	Author	Year	Title		Location	Reef type	Carbonate budget	Carbonate budget variables measured	Methods
13	Januchowski-Hartley et al.	2020	Accreting coral reefs in a highly urbanized environment	Range = 0.96 to 4.82, mean = 3.67	Range = 2.3 to 4.6, mean = 3.1	Range = −1.56–1.97, mean = 0.63	Main corals were massives and encrusting corals. CCA contribution around 5%. Internal bioerosion was 50–80% whereas fish and urchin were ~ 10–25%.	Range = 5–35, mean = 23	~2
14	Molina-Hernandez et al.	2020	Two decades of carbonate budget change on shifted coral reef assemblages: are these reefs being locked into low net budget states?	Range = 1 to 4, mean = 1.8	Range = 0.05 to 8, mean = 1.2	Range = −4.4 to 3.6, mean = 0.6	Carbonate production driven by non-framework encrusting corals and CCA. Bioerosion driven by parrotfish. Sponge and micro-borer rates low and urchins largely not present	Not provided per reef but all <17	Used species specific rugosity values so no reef rugosity values measured/provided
15	Dee et al.	2020	The complexity of calculating an accurate carbonate budget	Eva = 3.8 and Fly = 2.9 (corals only)	N/A	N/A	Massive and foliose corals dominated both reefs with <1% branching corals	Eva = 9.0 and Fly = 6.2	Eva = 1.44 and Fly = 1.5
16	Randi et al.	2021	Calcium carbonate production in the southernmost subtropical Atlantic coral reef	Encruster mean contribution = 0.13	N/A	N/A	N/A	N/A	N/A
17	Dee et al.	2021	Encrusters maintain stable carbonate production despite temperature anomalies among two inshore island reefs of the Pilbara, Western Australia	Eva = 0.42 and Fly = 0.35 (encrusters only)	N/A	N/A	N/A	Eva = 9.0 and Fly = 6.2	Eva = 1.44 and Fly = 1.5
18	Goh et al.	2021	Growth and carbonate production of crustose coralline algae on a degraded turbid reef system	Range = 0.09 to 0.52 (encrusters only)	N/A	N/A	N/A	N/A	N/A

19	Morris et al.	2022	Low net carbonate accretion characterizes Florida's coral reef	DRTO = 0.573, FLK = 0.567, SEFL = 0.092	DRTO = 0.596, FLK = 0.687, SEFL = 0.606	DRTO = −0.023, FLK = −0.12, SEFL = −0.514	Coral production by massives including *Orbicella* spp. *Siderastrea* and *Montastraea*. CCA contributed between 0.001–0.016 (~2%). Parrotfish were the main drivers of erosion with microbioerosion also be relatively high. No macrobioerosion noted and no to very low urchin erosion	DRTO = 7.81, FLK = 8.57, SEFL = 1.46	Measured but not provided
20	Toth et al.	2022	The past, present, and future of coral reef growth in the Florida keys	1996 = 1.9 and 2019 = 0.7	1996 = 0.57 and 2019 = 0.48	1996 = 1.33 and 2019 = 0.22	In 1996 coral cover was largely *Orbicella* with ~10% contribution from *Acropora* and *Siderastrea*. In 2019, *Acropora* is lost and coral production driven by both *Orbicella* and *Siderastrea*. Much less temporal variation in bioerosion driven by parrotfish and microbioerosion (>96%)	1996 = 15.5 and 2019 = 8.5	Measured but not provided
21	Dee et al.	2023	Low bioerosion rates on inshore turbid reefs of Western Australia	N/A	Eva = 0.151 and Fly = 0.175[a]	N/A	Bioerosion driven by both macro and micro borers with low grazing pressure from fish and none from urchins	Eva = 9.0 and Fly = 6.2	Eva = 1.44 and Fly = 1.5
22	Rodriguez-Ruano et al.	2023	Upwelling, climate change, and the shifting geography of coral reef development	GoP = 18, GoC = 10	GoP = 9, GoC = 7	GoP = 9, GoC = 3	*Pocillopora* provided >90% carbonate produced. Macro-borers contributed 88% to bioerosion. Sponges contributed about 4%	GoP = 78 GoC = 50	Not measured

(continued)

Table 16.1 (continued)

No.	Author	Year	Title		Location	Reef type	Carbonate budget	Carbonate budget variables measured	Methods
23	Toth et al.	2023	The potential for coral reef restoration to mitigate coastal flooding as sea levels rise	1.27	4.14	−2.87	Dominant corals included *Pseudodiploria*, *A. palmata* and *Orbicella* spp. Parrotfish accounted for >85% of the erosion with minimal input from internal borers and urchins.	6.7	Not measured
24	Alemealla et al.	2024	Bleaching threatens positive carbonate budgets on Bahraini reefs	Nearshore (pre and post) = 4.1 to 0.5 and offshore (pre and post) = 4 to 4.2	Nearshore (pre and post) = 0.5 to 0.2 and offshore (pre and post) = 0.3 to 0	Nearshore (pre and post) = 3.6 to 0.3 and offshore (pre and post) = 3.7 to 4.2	Massives (*Platygyra, Favia* and *Porites*) were the dominant corals at both sites. CCA was high in nearshore site pre bleaching. Nearshore site bioerosion due to urchins with no sponges and offshore site due to parrotfish (pre not post)	Nearshore (pre and post) = 13.6 to 1.4 and offshore (pre and post) = 14.8 to 15.5	Nearshore (pre and post) = 1.7 to 1.4 and offshore (pre and post) = 1.8 to 1.7
25	Dee et al.	2024	The application of carbonate and sediment budgets to assess the stability of marginal reef systems	Eva = 1.11 and Fly = 0.62	Eva = 0.05 and Fly = 0.06	Eva = 1.06 and Fly = 0.56	CCA rate was between 0.08 and 0.12 (~10%) of production. Coral production driven by massive and foliose corals. Bioerosion rates driven by macro-borers	Eva = 9.4 and Fly = 6.2	Eva = 1.44 and Fly = 1.5

Studies highlighted un grey have not been included in the analysis either due to incompatible data units or if the studies data was also available in a later publication
Note [a]denotes additional studies not included in Browne et al. (2021)

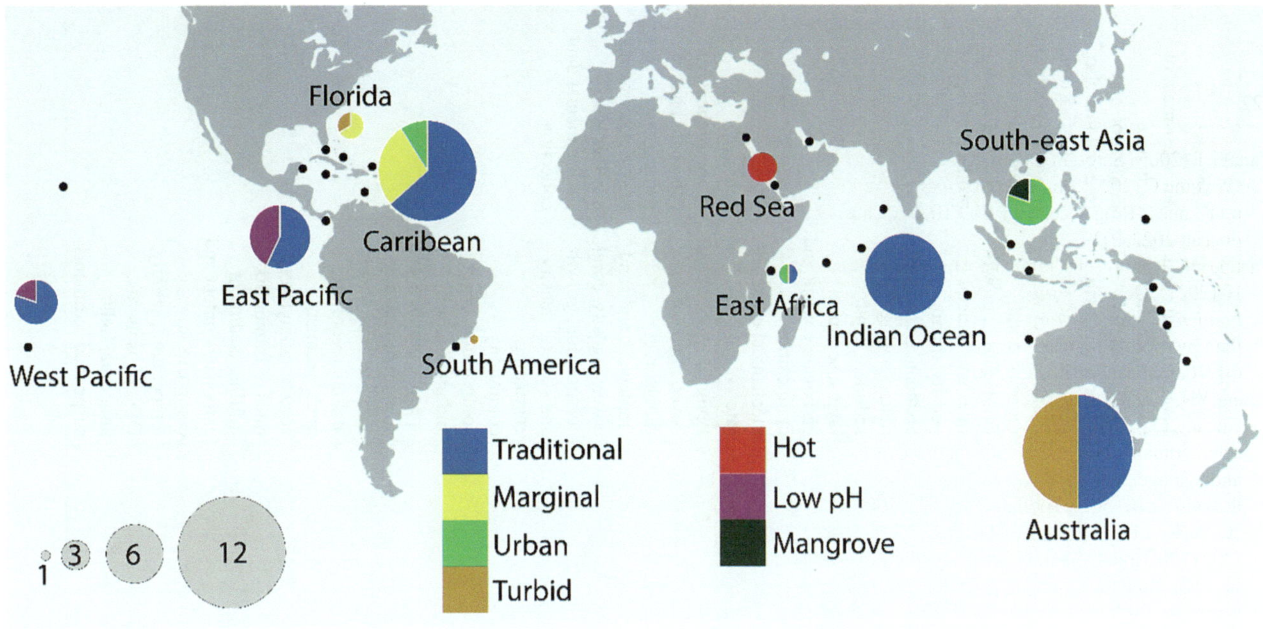

Fig. 16.1 A global map of the 57 global census-based carbonate budget studies conducted between 1977 and 2024. For each region (left to right: West Pacific, East Pacific, Florida, Caribbean, South America, Red Sea, East Africa, Indian Ocean, South-East Asia and Australia) the size of the coloured circle denotes the number of studies and colour denotes reef type. Black dots represent approximate location of studies with some dots representing many studies close together

particular recent recognition of conducting these assessments within marginal and extreme reef settings.

Studies focused on marginal and extreme reefs have a stronger geographic clustering. For example, 75% of carbonate budget studies on turbid reefs have been conducted in Australia, and 75% of studies on urban reefs have been conducted in Asia (primarily Singapore). A review of turbid reefs by Zweifler et al. (2021) found that turbid reefs have been documented in a number of regions including the Caribbean, whereas urban reefs, although less well documented and defined, have also been found in Florida (Rubin et al. 2021) and Colombia (Shantz et al. 2023). In contrast, all studies on hot reefs are located in one location, the Red Sea. These data likely reflect the concentrated research efforts in some regions, as well as the higher incidence of certain reef types in particular geographic locations. For example, the Caribbean, and Florida region have a higher concentration of marginal reef types characterised by low coral cover (<10%).

Of the 25 marginal and extreme reef carbonate budget studies, our data analysis focuses on 18 studies (Table 16.1). Seven studies were excluded because the data was in a format that was not comparable (e.g., percentage bioeroded as opposed to rates of bioerosion), or because some datasets had been included in subsequent papers (e.g., Dee et al. 2024). It is also important to note that many of these studies have included more than one reef site. For example, Molina-Hernández et al. (2020) conducted the ReefBudget method on 34 marginal reef sites (<10% coral cover) in the Mexican Caribbean, and Toth et al. (2023) used the same approach to assess coral reef restoration potential at 54 marginal reef sites on Buck Island Reef in the US Virgin Islands. Several studies also included non-marginal and/or non-extreme reef sites (i.e., traditional reefs) to demonstrate differences in carbonate budgets across environmental gradients (e.g., Edinger and Risk 2000; Chen et al. 2013). For these studies, we focused on the marginal and/or extreme reef sites for our analysis. Further, those sites where carbonate budgets were calculated over more than 1 year, we used the most recent data (e.g., AlMealla et al. 2023; Toth et al. 2022).

Of the traditional reefs where carbonate budgets have been conducted, these include remote atolls, platform reefs and mainland fringing reefs distant from anthropogenic activities (Table 16.2). A total of 32 studies have been conducted on traditional reefs since the late 1970s. Here we focus on 19 of the more recent studies, with the 11 earliest studies excluded from further comparative analysis due to highly variable methods (these studies are included in Browne et al. 2021), as well as two later studies, where the carbonate budget assessment relied heavily on offsite published data or focused on coral reef restoration (see Table 16.2). In comparison, studies since the early 2000s have used more comparable methodologies to assess budgets, with most using the ReefBudget (or modified version) approach (for more details see Browne et al. 2021). A few of these studies have been conducted before and after an acute

Table 16.2 Carbonate budget studies completed on traditional coral reefs

No.	Author	Year	Paper title	Location/s	Reef type	Carbonate budget	Carbonate budget variables measured[a]	Methods[a]
1	Harney and Fletcher	2003	A budget of carbonate framework and sediment production, Kailua Bay, Oahu, Hawaii	Kailua Bay, Oahu, Hawaii	Seaward reef platform and reef front on a nearshore fringing reef	Complete	Live and dead coral cover, macro and turf algal cover, CCA cover and growth rates, sand and reef framework cover, rugosity, coral composition, growth rates and skeletal density, urchin abundance and size, macro-boring rates, sediment depth layers, sediment carbonate content and composition, mechanical erosion rate	An assessment of carbonate production and removal that also considers variable production between zones and includes direct sediment production rates
2	Mallela and Perry	2007	Calcium carbonate budgets for two coral reefs affected by different terrestrial runoff regimes, Rio Bueno, Jamaica	Outer embayment (OE) site, Rio Bueno, Jamaica	Clearwater nearshore fore reef setting, high wave energy	Complete	Living and dead coral cover, algal cover, rubble, reef framework and sand cover, rugosity, coral composition, growth and skeletal density, encrusting community composition, cover and growth, fish species, numbers and size, urchin density and size, macroboring and micro-boring rates, sediment accumulation rates, rates of framework infilling, sediment carbonate content and composition, temperature, salinity, DIN, phosphorus, ammonia, light, suspended sediments	A comprehensive carbonate budget using transects and in situ experiments together with environmental data collection

3	Hart and Kench	2007	Carbonate production of an emergent reef platform, Warraber Island, Torres Strait, Australia	Warraber Island, Torres Strait, Australia	Remote, small cay with large reef platform surrounded by clearwater	Part (gross production)	Living coral cover and composition, CCA cover, *Halimeda* cover, sponge and brown algal cover, seagrass cover, sand and reef framework cover, rugosity, sediment depth and type, foraminifera and mollusc abundance	This study assesses gross carbonate production using quadrates and transects within ecological zones
4	Perry et al.	2012	Estimating rates of biologically driven coral reef framework production and erosion: a new census-based carbonate budget methodology and applications to the reefs of Bonaire	No dive reserve site on the NW of Bonaire	Fringing reef	Complete	Living coral cover, CCA cover, turf and macro-algal cover, *Halimeda* cover, sand, rubble and reef framework cover, rugosity, coral composition, growth rates and skeletal density, encrusting cover and growth rates, parrotfish species, size and abundance, parrotfish bite rates, scars and mass eroded, urchin abundance, size and species, clinoid sponge cover, micro-bioerosion rates	This study provides the method of a rapid census based approach termed 'ReefBudget', which focuses on quantifying the relative contributions of key calcifying organisms
5	Hamylton et al.	2013	The use of remote sensing to scale up measures of carbonate production on reef systems: a comparison of hydrochemical and census-based estimation methods	One Tree Island, Australia	Remote, planar reef with a raised rubble cay island	Part (gross production)	Live and dead coral cover, macro-algal cover and algal mat, sand, rubble and consolidated rubble cover, reef platform	Study compares rates of gross carbonate production from a census based approach with hydrochemically derived estimates, and uses remote sensing to classify reef habitat types over broad areas

(continued)

Table 16.2 (continued)

No.	Author	Year	Paper title	Location/s	Reef type	Carbonate budget	Carbonate budget variables measured[a]	Methods[a]
6	Morgan	2014	A calcium carbonate budget of a Maldivian reef platform	Vabbinfaru Reef, North Malé Atoll, Maldives	Clearwater reef platform	Complete	Live and dead coral cover, CCA and *Halimeda* cover, sand, rubble and reef framework cover, rugosity, coral composition, growth rates and skeletal density, encruster composition and calcification rate, CCA carbonate production, fish species, abundance and size, urchin abundance and size, macro-boring rates, light, suspended sediments, net sediment accumulation rates, sediment export and depth layers, particle size, carbonate content and composition, reef island planform morphology, wave energy	This study provides a detailed census based carbonate budget of different reef eco-geomorphic zones which also includes an assessment of sediment flux and a sediment budget

7	Perry et al.	2015b	Remote coral reefs can sustain high growth potential and may match future sea-level trends	Great Chagos Bank, Peros Banhos, Salomon	Clearwater, remote atoll reefs	Complete	Living coral cover, CCA and non-encrusting coralline algal cover, turf and fleshy macro-algal cover, *Halimeda* cover, sand, rubble and reef framework cover, rugosity, coral composition, shape and size, growth rates and skeletal density, parrotfish species, size and abundance, parrotfish bite rates, scars and mass eroded, urchin abundance, size and species, micro and macrobioerosion rates, sediment production rates by macro-borers	This study presents a modified version of the ReefBudget method for the Indo-Pacific region. Sites were also classified based on wave exposure for modelling purposes
8	De Bakker et al. (2016)	2016	Long term shifts in coral communities on shallow and deep reef slopes of Curacao and Bonaire: Are there any winners?	Curacao and Bonaire, Caribbean	Fringing reef	Coral community characteristics only	Live coral cover, coral composition and size distributions, dead coral cover	Study focusing on the coral community and applies growth rates from the ReefBudget method
9	Januchowski-Hartley et al.	2017	Drivers and predictions of coral reef carbonate budget trajectories	21 reefs, Seychelles Islands	Includes carbonate fringing reefs, granite rocky reefs and patch reefs—all considered to be clearwater non-marginal reefs	Complete	Same a Perry et al. (2015b)	This study was carried out using 5 years of data (1994 to 2014) along the reef front to track changes in net carbonate production over time and following the impact of the 1998 bleaching event

(continued)

Table 16.2 (continued)

No.	Author	Year	Paper title	Location/s	Reef type	Carbonate budget	Carbonate budget variables measured[a]	Methods[a]
10	Perry and Morgan	2017	Bleaching drives collapse in reef carbonate budgets and reef growth potential on southern Maldives reefs	Five sites in Gaafu Dhalu Atoll, Southern Maldives	Clearwater remote atoll	Complete	Same a Perry et al. (2015b)	This study uses the modified ReefBudget approach for the Indo-Pacific to assess changes in carbonate budgets due to a coral bleaching event in 2016
11	Perry et al.	2017	Reef habitat type and spatial extent as interacting controls on platform-scale carbonate budgets	Lhaviyani Atoll, Central Maldives	Clearwater remote atoll	Complete	Same as (2015b) Perry et al. study, with the addition of *Halimeda* plant volume, number of non-*Halimeda* calcifying green algae, number of articulated red coralline algae, number of epifaunal gastropods, sediment compositional analysis	This study uses a census based approach to demonstrate spatial differences in carbonate production at the inter-reef habitat scale
12	Perry et al.	2018b'	Loss of coral reef growth capacity to track future increases in sea level	86 sites across the Caribbean and 68 sites in the Indo-Pacific region. Here we focus on Ningaloo, WA (as many of the other reefs have already been included in previous publications)	Ningaloo WA—clearwater, large fringing reef system	Complete	Same as Perry et al. (2012)	A meta-analysis of carbonate budgets studies, most using the census based ReefBudget method or adaptations of this method

13	Van Woesik and Cacciapaglia	2018	Keeping up with sea-level rise: Carbonate production rates in Palau and Yap, western Pacific Ocean	Palau and Yap, Western Pacific Ocean	Clear water patch, outer and inner reefs	Complete	Living and dead coral cover, CCA cover, macroalgal and sponge cover, coral composition, growth rates and skeletal density, rugosity, fish species, abundance and size, urchin species, abundance and size	This study combines in situ cover of carbonate producing organisms with published production/removal rates to produce a spatial visualisation of net carbonate production over the reef
14	Hamylton and Mallela	2019	Reef development on a remote coral atoll before and after coral bleaching: A geospatial assessment	Cocos keeling	Remote, clearwater atoll—focus on the northern lagoon	Complete	Living and dead coral cover, CCA cover, turf cover, encrusters cover, seagrass cover, sand, rubble and reef framework cover, rugosity, coral composition, growth rates and skeletal density, sediment composition, sediment export rates, macro-borer erosion rate, parrotfish species, size and abundance, urchin abundance and size	A geo-spatial assessment of carbonate budget using in situ census based data together with remote sensing to provide a carbonate budget that covers a considerable spatial area (225 km^2)
15	Lange and Perry	2019	Bleaching impacts on carbonate production in the Chagos Archipelago: influence of functional coral groups on carbonate budget trajectories	Great Chagos Bank, Peros Banhos, Salomon in the Chagos Archipelago	Remote, clearwater atoll	Part (gross production)	Living coral cover, CCA and non-encrusting algal cover, turf and fleshy macro-algal cover, *Halimeda* cover, sand, rubble and reef framework cover, rugosity, coral composition, shape and size, growth rates and skeletal density	This study used the modified ReefBudget approach for the Indo-Pacific to assess the impacts of a warming event on gross carbonate production rates—2016 and 2018

(continued)

Table 16.2 (continued)

No.	Author	Year	Paper title	Location/s	Reef type	Carbonate budget	Carbonate budget variables measured[a]	Methods[a]
16	Ryan et al.	2019	Massive corals maintain a positive carbonate budget of a Maldivian upper reef platform despite major bleaching event	Reef flat on the Mahutigala reef, Gaafu Dhalu Atoll, Maldives	Remote, clearwater atoll	Complete	Live and dead coral cover, CCA and algae cover, rubble and reef framework cover, rugosity, coral composition, growth rates and skeletal density, CCA growth rates, parrotfish bioerosion rates, urchin genera, abundance and erosion rates, micro and macro-bioerosion rates	This study uses the modified ReefBudget approach for the Indo-Pacific to assess the impact of a warm SST anomaly event on net carbonate production for a reef platform
17	Brown et al.	2020	Habitat-specific biogenic production and erosion influences net framework and sediment coral reef carbonate budgets	Heron Island, Southern Great Barrier Reef, Australia	Remote clear water platform reef	Complete	Living coral cover, CCA cover, turf and macro-algal cover, *Halimeda* cover, sand, rubble and reef framework cover, rugosity, coral composition, growth rates and skeletal density, encrusting cover and growth rates, parrotfish species, size, abundance, bite rates, scars and mass eroded, urchin abundance, size and species, clinoid sponge cover, macro and micro-bioerosion rates, sediment composition and production rates, sediment dissolution	This study uses the modified ReefBudget approach for the Indo-Pacific to assess differences in net carbonate production across a number of different reef geomorphic zones

18	Courtney et al.	2022	Rapid assessment of Pacific Ocean net coral reef carbonate budgets and net calcification following the 2014–2017 global coral bleaching event	Of the 56 reef sites in the Pacific, have selected 14 sites around American Samoa	Clearwater reefs	Complete (mostly)	Living coral cover, CCA cover, turf and macro-algal cover, *Halimeda* cover, sand, rubble and reef framework cover, rugosity, coral composition, growth rates and skeletal density, encrusting cover and growth rates, parrotfish species, size and abundance, parrotfish bite rates, scars and mass eroded, urchin abundance, size and species, clinoid sponge cover, micro-bioerosion rates	This study used a simplified version of the ReefBudget approach to retroactively assess carbonate production from benthic imagery
19	Divan Patel et al.	2023	Carbonate budgets in Lakshadweep Archipelago bear the signature of local impacts and global climate disturbances	Lakshadweep Archipelago, India	Three clearwater coral atoll, with varying degrees of human influence	Complete	Living coral cover, CCA cover, turf and macro-algal cover, *Halimeda* cover, sand, rubble and reef framework cover, coral composition, growth rates and skeletal density, encrusting cover and growth rates, parrotfish species, size and abundance, parrotfish bite rates, scars and mass eroded, urchin abundance, size and species, clinoid sponge cover, micro-bioerosion rates	This study uses the modified ReefBudget approach for the Indo-Pacific to assess net carbonate production at 10 sites

(continued)

Table 16.2 (continued)

No.	Author	Year	Paper title	Location/s	Reef type	Carbonate budget	Carbonate budget variables measured[a]	Methods[a]
20	Lange et al.	2023	Recovery trends of reef carbonate budgets at remote coral atolls 6 years post-bleaching	Great Chagos Bank, Peros Banhos, Salomon in the Chagos Archipelago	Remote, clearwater atolls	Complete	Living coral cover, CCA cover, turf and macro-algal cover, *Halimeda* cover, sand, rubble and reef framework cover, rugosity, coral composition, growth rates and skeletal density, encrusting cover and growth rates, parrotfish species, size and abundance, parrotfish bite rates, scars and mass eroded, urchin abundance, size and species, clinoid sponge cover, micro-bioerosion rates	Tracking the recovery of coral reefs following the 2016 global bleaching event. Data from 2015 and 2018 already published in Lange and Perry (2019).
21	Nunez Lendo et al.	2024	Carbonate budgets induced by coral restoration of a Great Barrier Reef site following cyclone damage	Moore Reef, central Great Barrier Reef	Clearwater remote reef	Complete	Living coral cover, CCA cover, turf and macro-algal cover, *Halimeda* cover, sand, rubble and reef framework cover, rugosity, coral composition, growth rates and skeletal density, encrusting cover and growth rates, parrotfish species, size and abundance, parrotfish bite rates, scars and mass eroded, urchin abundance, size and species, clinoid sponge cover, micro-bioerosion rates	This study uses the modified ReefBudget approach for the Indo-Pacific to assess the impact of coral restoration efforts following cyclone damage

No.	Author	Year	Paper title	Gross productivity (kg m^{-2} y^{-1})	Erosion (kg m^{-2} y^{-1})	Net productivity (kg m^{-2} y^{-1})	Key producer and/or eroder	Coral cover (%)	Structural complexity at the habitat scale (rugosity with 1 denoting a flat surface)
1	Harney and Fletcher	2003	A budget of carbonate framework and sediment production, Kailua Bay, Oahu, Hawaii	Reef platform = 1.90 and reef front = 3.38	Reef platform = 0.2 and reef front = 0.4	Reef platform = 0.9 and reef front = 2.98	Encrusting corals dominate (50–95%) and high 10–75% coralline algal cover with carbonate contributions from corals (58%) and CCA (42%). Framework bioerosion from urchins and internal eroders (0.1–1.15 kg m^{-2} y^{-1})	Reef platform = 55.3 and reef front = 55.2	Reef platform = 1.23 and reef front = 1.27
2	Mallela and Perry	2007	Calcium carbonate budgets for two coral reefs affected by different terrestrial runoff regimes, Rio Bueno, Jamaica	OE site = 1.37	OE site = 0.126	OE site = 1.24	Over 80% of carbonate from corals with the remaining from secondary encrusters (CCA). Dominant corals were encrusting form. Macro-borers were the dominant bioeroders accounting from ~50% followed by microborers. Fish erosion was low and urchins very low	OE site = ~16	Measured but not provided
3	Hart and Kench	2007	Carbonate production of an emergent reef platform, Warraber Island, Torres Strait, Australia	1.66	N/A	N/A	Massives corals account for ~60% of carbonate produced with coralline algae then contributing the most next	4300%	N/A

(continued)

Table 16.2 (continued)

No.	Author	Year	Paper title	Location/s	Reef type	Carbonate budget		Carbonate budget variables measured[a]	Methods[a]
5	Hamylton et al.	2013	The use of remote sensing to scale up measures of carbonate production on reef systems: a comparison of hydrochemical and census-based estimation methods	2.16 (corals only)	N/A	N/A	Not enough detail provided	Coral cover ranged from 2 to 65	Not measured
6	Morgan	2014	A calcium carbonate budget of a Maldivian reef platform	16.88	11.86	4.95	*Acropora* was the dominant coral (75%) followed by *Porites* (13%), and corals accounted for 98% of the carbonate production. Parrotfish accounted for >99% of bioerosion	51.7	1.97
7	Perry et al.	2015a, b	Remote coral reefs can sustain high growth potential and may match future sea-level trends	6.6	3.4	3.7	Coral production was generally driven by *Acropora*, *Porites* and *Pocillopora* (>70%). Information on bioeroder groups not provided	26.8	Measured but not provided
8	De Bakker et al. (2016)	2016	Long term shifts in coral communities on shallow and deep reef slopes of Curacao and Bonaire: Are there any winners?	N/A	N/A	N/A	N/A	N/A	N/A

9	Januchowski-Hartley et al.	2017	Drivers and predictions of coral reef carbonate budget trajectories	1994 = 5.48 and 2014 = 2.95	1994 = 1.83 and 2014 = 2.97	1994 = 3.65 and 2014 = −0.01	In 1994 branching corals accounted for over 60% of the coral cover and massives approximately 35%. Erosion was predominately parrotfish. Following bleaching event in 1998, corals in 2014 still branching dominated but lower coral cover.	1994 = 27.9 and 2014 = 17	1994 = 3.17 and 2014 = 2.54
10	Perry and Morgan	2017	Bleaching drives collapse in reef carbonate budgets and reef growth potential on southern Maldives reefs	Pre bleaching = 8.43 and post bleaching = 1.83	Pre bleaching = 1.76 and post bleaching = 4.15	Pre bleaching = 5.92 and post bleaching = −2.96	*Acropora* was the dominant coral and parrotfish the dominant eroder. Negative production due to loss in branching corals and increase in erosion. Post bleaching more resilient massive and sub-massive corals were dominant.	Pre bleaching = 25.6 and post bleaching = 6.3	Pre bleaching = 2.6 and post bleaching = 2.3
11	Perry et al.	2017	Reef habitat type and spatial extent as interacting controls on platform-scale carbonate budgets	Reef patch (Z1 and Z2) = 6.8	Reef patch (Z1 and Z2) = 4.4	Reef patch (Z1 and Z2) = 2.35	Coral accounted for 95% of carbonate and encrusters for 5%. Coral community dominated by branching morphologies with some *Porites*. Parrotfish were the dominant eroder (>75%) with micro and macroborers also accounting for ~10–15% of the erosion. Very low urchin numbers and erosion	Reef patch (Z1 and Z2) = 25.5	Reef patch (Z1 and Z2) = 6.8

(continued)

Table 16.2 (continued)

No.	Author	Year	Paper title	Location/s	Reef type	Carbonate budget		Carbonate budget variables measured[a]	Methods[a]
12	Perry et al.	2018b'	Loss of coral reef growth capacity to track future increases in sea level	5.23	1.35	3.88	Mixed to branching coral community with encrusters contributing <2% production. Parrotfish contributed to ~50% of erosion with macro and micro contributing the remaining. Urchin erosion was very low.	30.88	Measured but not provided
13	Van Woesik and Cacciapaglia	2018	Keeping up with sea-level rise: Carbonate production rates in Palau and yap, western Pacific Ocean	Palau = 8.12 and Yap = 11.83 (data taken from Table 16.1 in paper—note net calcification rates are greater than gross calcification rates)	Palau = 0.05 and Yap = 0.42	Palau = 8.48 and Yap = 11.62	Corals were the main contributor to framework carbonate production and parrotfish were the key bioeroders. Some erosion by urchins but low and very site specific. *Porites* and *Acropora* species were the main contributors to carbonate production. Macrobioerosion rates minimal on both islands and microbioerosion rates not measured	Palau = 45 and Yap = 38	Measured but not provided

14	Hamylton and Mallela	2019	Reef development on a remote coral atoll before and after coral bleaching: A geospatial assessment	Northern lagoon = 1.61	Northern lagoon = 0.43	Northern lagoon = 1.18	Dominated by massive corals with low branching coral cover. CCA production estimated at 0.4 kg m^{-2} y^{-1} and is therefore an important contribution to the gross production. Parrotfish removed ~75% of the total carbonate removed. Urchins assessed but data not provided and internal borers not accounted for	Northern lagoon = 28	Northern lagoon = 1.09
15	Lange and Perry	2019	Bleaching impacts on carbonate production in the Chagos Archipelago: influence of functional coral groups on carbonate budget trajectories	2016 = 13.6 and 2018 = 3.4	N/A	N/A	*Acropora* were the main coral contributor pre bleaching event (declined by 96%) with stress tolerant massive and encrusting dominating following the event. CCA contributed ~4% pre bleaching which increased to 8% post bleaching.	2016 = 27.1 and 2018 = 8.5	2016 = 2.3 and 2018 = 1.9
16	Ryan et al.	2019	Massive corals maintain a positive carbonate budget of a Maldivian upper reef platform despite major bleaching event	5.9	3.3	2.5	Dominant corals were massives including *Porites, Goniopora, Platygyra* (63%) with branching corals accounting for 13%. Parrotfish accounted for 94% of the erosion and micro and macro was 6% whereas urchins <0.1%	1790.00%	2.6

(continued)

Table 16.2 (continued)

No.	Author	Year	Paper title	Location/s	Reef type	Carbonate budget		Carbonate budget variables measured[a]	Methods[a]
17	Brown et al.	2020	Habitat-specific biogenic production and erosion influences net framework and sediment coral reef carbonate budgets	Reef slope = 17.75 and total reef area = 5.6	Reef slope = 1.65 and total reef area = 1.5	Reef slope = 16.15 and total reef area = 4.06	*Acropora* contributed between 74–89% of gross carbonate production and parrotfish accounted for between 91 to 99% of the total bioerosion. Other eroders were less than 5%	Reef slope = 59.71 and total reef area = 29.7	Reef slope = 1.25 and total reef area = 1.1
18	Courtney et al.	2022	Rapid assessment of Pacific Ocean net coral reef carbonate budgets and net calcification following the 2014–2017 global coral bleaching event	2.97	0.41	2.76	Coral community data not provided or how much CCA contributed to the budget. However, CCA cover on transects was high (average = 24%). Parrotfish likely the main bioeroders but no breakdown of rates of bioerosion per group provided	29	Not measured
19	Divan Patel et al.	2023	Carbonate budgets in Lakshadweep Archipelago bear the signature of local impacts and global climate disturbances	4.5	2.25	2.25	Assumes that corals were the main carbonate producer, and CCA only contributed about 2.1% of carbonate production. Coral community was largely composed of stress tolerant massive and sub-massive species. Of total bioerosion, parrotfish were the dominant bioeroder (47%) but urchin densities at some locations were very high (25%)	Measured but not provided	Not provided (unclear if measured)

20	Lange et al.	2023	Recovery trends of reef carbonate budgets at remote coral atolls 6 years post-bleaching	2021 = 4.1	2021 = 4.4	2021 = −0.37	Corals were the dominant carbonate producer with CCA contributing <5%. Total bioerosion was largely due to Parrotfish (>75%) with some macro and micro bioerosion (equal) and low to no urchin erosion	2021 = 18	2021 = 1.7
21	Nunez Lendo et al.	2024	Carbonate budgets induced by coral restoration of a Great Barrier Reef site following cyclone damage	N/A	N/A	N/A	N/A	N/A	N/A

Studies highlighted in grey not included in the analysis due to either limited in situ assessment of a number of key carbonate budget metrics or focused on reef restoration
Note [a]taken from Browne et al. (2021) (up till 2020 references)

disturbance (e.g., Januchowski-Hartley et al. 2017; Lange and Perry 2019) to assess the impact of the event on reef carbonate production. Here, we focus on data from these studies collected before the disturbance event.

In summary, this chapter includes a detailed assessment of 37 (19 traditional and 18 non-traditional reefs) of the 57 carbonate budget studies that have been published to date, which encompasses 37 locations and 220 reef sites. In the following section, we characterise and compare rates of gross and net carbonate production, as well as rates of erosion between traditional reefs with marginal and extreme reef types (hot, low pH/high nutrient, mangrove, marginal, turbid and urban). In addition, we discuss coral cover and identify the main drivers of carbonate production on reefs (e.g., coral morphology types) together with differences in key eroder groups. We also evaluate sediment characteristics between reef types to determine if there are important differences in local and external sediment sources. Finally, we examine reef structural complexity (rugosity) and discuss how coral community composition, as well as coral cover, has influenced differences in complexity between reef types.

16.2.1 Gross Carbonate Production

Corals are typically the main producer of calcium carbonate on tropical and semi-temperate reef systems. To determine rates of gross coral carbonate production, a key metric is coral cover (%). In addition, growth and calcification rates for the dominant coral genera are required to estimate the amount of carbonate produced by corals (kg m^{-2} y^{-1}). More generally, coral cover is considered a key reef health metric (Castro-Sanguino et al. 2021) with reefs with low coral cover (<10%) having potentially transitioned away from a coral-dominated reef state (i.e., phase shift) to a reef dominated by non-calcifying organisms, such as macroalgae and/or soft corals (Done 1992).

Mean coral cover was higher at the traditional reef sites (32.6%) when compared to non-traditional reefs (24.7%; df = 1, F = 3.08, p = 0.086, Fig. 16.2a). However, when compared to extreme and marginal reef types separately, coral cover on traditional reefs was only significantly greater than on marginal reefs (Post Hoc Tukey = p < 0.001), whose cover was also significantly lower than on turbid, urban and hot reefs (Post Hoc Tukey = p < 0.04; low pH and mangrove reefs removed from analysis due to low sample size; Fig. 16.2b). This is somewhat unsurprising given that marginal reef types are defined here as reefs with low coral cover and/or diversity (following Schoepf et al. 2023). In contrast, turbid and urban reefs (~30–40%) had comparable or higher coral cover than traditional reef types, respectively. Historically, turbid and urban reefs have often been considered 'degraded' compared to their clear-water counterparts (Zweifler et al. 2021). These data, however, suggest that this may not always be the case, highlighting both the conservation value of these often overlooked reef types, particularly given their greater levels of environmental resilience to non-optimal conditions (e.g., water quality and temperature stress) (e.g., Morgan et al. 2017). Lastly, coral cover at the low pH site was exceptionally high. This site is located in the Gulf of Panama next to an upwelling site, which brings cooler, nutrient rich waters to the surface as well as lowering the aragonite saturation state. Here, coral cover was measured at 78% in 2016/2017 (Rodriguez-Ruano et al. 2023). Other studies that have measured coral cover at low pH and aragonite saturation sites have also found high coral cover. For example, Shamberger et al. (2014) reported the existence of diverse coral-dominated reefs (~30–65% cover) in naturally acidified waters of the Palauan reef system in the Western Pacific Ocean, and Camp et al. (2017) recorded coral cover of 24–35% within a lagoonal system in New Caledonia where corals were exposed to elevated temperature and low pH. Similarly to turbid and urban reefs, corals on low pH reefs have adapted to non-optimal conditions with corals continuing to grow, and even increase their calcification rates (Tanvet et al. 2023) and provide a key source of carbonate to the reef.

Despite traditional reefs having on average ~ 25% more coral cover compared to non-traditional reefs, gross carbonate productivity was approximately 50% greater (traditional = 6.76 and non-traditional = 4.67 kg m^{-2} y^{-1}; df = 1, F = 1.91, p = 0.17; Fig. 16.2a). Given that gross carbonate production is a product of both coral cover and rates (i.e., skeletal linear extension and bulk density), this suggests that there are colony-level differences in carbonate production rates of non-traditional reefs. This is sometimes due to the reduced calcification rates of individual coral colonies. For example, calcification of *Acropora* spp. measured using the buoyant weight method on semi-turbid reefs in Exmouth Gulf, Western Australia, was ~40% less than on clear-water reefs within the same region (Dee et al. 2024). However, gross carbonate production was only significantly lower on marginal reefs compared to normal reefs (Post Hoc Tukey = p < 0.04; Fig. 16.3b). Marginal reefs also had significantly lower coral cover, which likely explains the reduced gross carbonate production. However, coral cover on hot reefs was comparable to traditional reefs whereas gross carbonate productivity on hot reefs was lower (2.16 kg m^{-2} y^{-1}) indicating lower coral growth rates on these extreme reefs. A comprehensive assessment of coral and calcareous calcifying algae in the Red Sea across varying wave exposures and seasons, found on average, *Acropora* calcification rates to be at the lower end when compared to Caribbean and Indo-Pacific region, but *Pocillopora* and *Porites* calcification rates to be comparable to other regions (Roik et al. 2015). It is also worth noting that these hot reefs

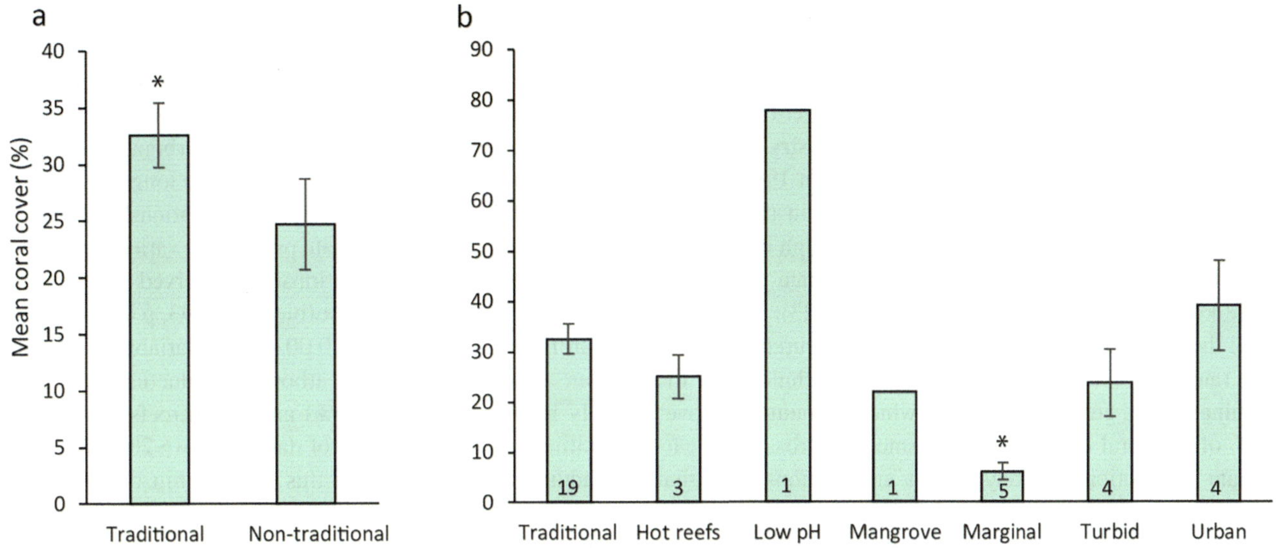

Fig. 16.2 Mean coral cover at (**a**) all traditional and non-traditional reef sites, and (**b**) at traditional, hot, low pH, mangrove, marginal, turbid and urban reefs (error bars = SE). Numbers in bars indicate number of studies included per reef type and asterisks indicate significant differences (P values for Tukey Posthoc results: Traditional:Hot = 0.95, Traditional:Marginal = <0.0001, Traditional:Turbid = 0.66, Traditional:Urban = 0.955, Hot:Marginal = 0.009, Hot:Turbid = 0.98, Hot: Urban = 0.99, Marginal:Turbid = 0.04, Marginal:Urban = 0.008, Turbid:Urban = 0.95)

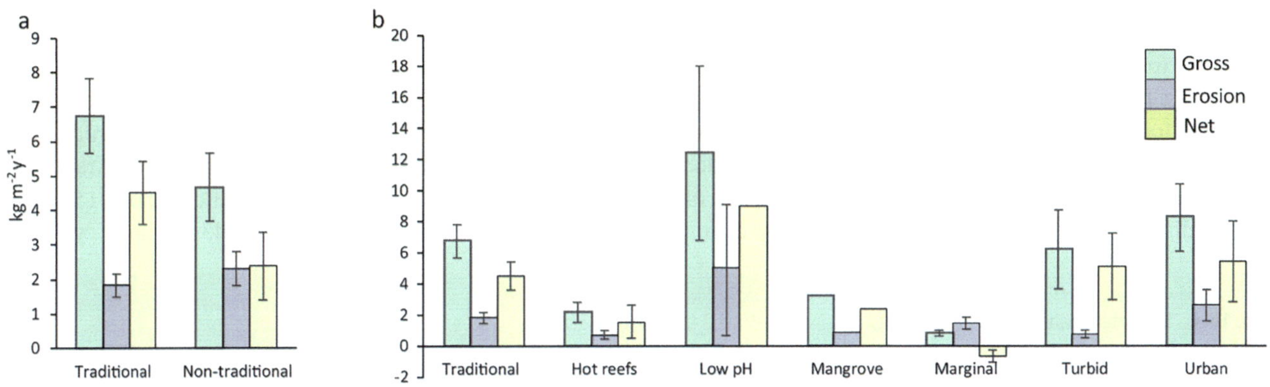

Fig. 16.3 Average rates of gross and net carbonate production (kg m^{-2} y^{-1}), and erosion (kg m^{-2} y^{-1}) for (**a**) traditional and non-traditional reefs, and for (**b**) all reef types (error bars = SE)

are also at a higher latitude and hence lower daylight hours may also be reducing coral calcification rates. This study supplied the *in-situ* calcification rates for the carbonate budget assessment in the Red Sea by Roik et al. (2018; Table 16.1) where coral cover was found to include a high cover of *Acropora,* thus potentially explaining part of the reduced gross carbonate production despite healthy coral cover.

In contrast, mean gross carbonate production on urban (8.25 ± 2.18 kg m^{-2} y^{-1}) and turbid reefs (6.17 ± 2.52 kg m^{-2} y^{-1}) was comparable to traditional reefs, although rates measured were considerably more variable. Given that carbonate production on traditional, turbid and urban reefs follows trends in coral cover, it suggests that rates of coral growth and carbonate production are also comparable. However, of the eight studies conducted on turbid and urban reefs, five applied the ReefBudget approach (see Perry et al. 2012), which relies on published global rates of coral growth as opposed to rates measured *in-situ*. Consequently, these gross production rates might be masking important differences in rates of coral calcification between taxa exposed to different environmental conditions given the lack of coral growth studies from extreme reefs. Higher gross production rates on these two reef types may also reflect biases in the methods used to quantify coral cover. These methods typically rely on two-dimensional metrics (e.g., photo-quadrats or transects) that can overestimate the relative contribution of corals on turbid and urban reefs where communities are dominated by flat and encrusting growth forms.

There is an assumption that coral growth rates on low pH and low aragonite saturation reefs may be lower than on traditional reefs due to reduced carbonate chemistry to support skeletal development (Marubini and Atkinson 1999). Coral reefs exposed to low pH due to upwellings in the Gulf of Panama, are characterised by exceptionally high coral cover and also exhibited high rates of gross carbonate production at 18 kg m^{-2} y^{-1} (Rodriguez-Ruano et al. 2023). This study used the ReefBudget approach, but also measured calcification rates *in-situ* using the buoyant weight technique for the dominant coral genus *Pocillopora*, which accounted for over 90% of the coral cover. This assessment of gross coral carbonate production is likely robust and demonstrates that some extreme reefs are capable of maintaining key ecosystem functions, such as coral growth, contrary to broader expectations. Yet, with only one comprehensive study on carbonate production for low pH reefs, more data on these extreme reefs is required to fully understand the carbonate production rates on these reef types and related coral community drivers.

Given that coral cover is a key metric for a reef gross carbonate production, it is unsurprising there is a strong positive relationship between coral cover and gross carbonate productivity for all reef types classified together ($r^2 = 0.59$, $p < 0.001$; Fig. 16.4). Other organisms such as CCA, molluscs and foraminifera, also calcify and produce calcium carbonate, but have only been identified as dominant producers at few reef sites and associated reef island systems, and often within discrete reef zones (e.g., reef flats). For example, Raine Island on the northern Great Barrier Reef, is largely composed of large benthic foraminifera, which are produced on the reef flat and are then transported onto the reef island (Dawson et al. 2014). For all carbonate budget studies included here, corals were the main carbonate producer highlighting the critical role of coral in the longer-term geological stability of reefs. However, the relationship between coral cover and gross carbonate production varied among the reef types with stronger relationships observed for marginal reefs ($r^2 = 0.38$, $p = 0.045$), turbid ($r^2 = 0.93$, $p = 0.004$) and urban reefs ($r^2 = 0.91$, $p < 0.001$). The variability in coral cover, and therefore gross carbonate production, was relatively high on turbid (6–40%) and urban reefs (2.5–73%), resulting in a greater spread of data (Figs. 16.2b and 16.3b). In addition, the strength of this relationship on these two extreme reef types may also be due to comparable coral community compositions. These reef types are largely dominated by massive and foliose (i.e., stress tolerant) corals which typically calcify at slower, but also less variable rates (Browne 2012; Morgan and Kench 2012; Lough and Barnes 1992, 2000), resulting in similar gross production rates with coral cover. In contrast, traditional reefs, where coral cover ranged from 18% to 62% ($r^2 = 0.07$, $p = 0.13$), typically have more mixed coral communities (e.g., branching to mixed) with more variable coral growth rates and, therefore, more variable gross carbonate productivity (Fig. 16.4).

In contrast, hot reefs displayed no positive linear trend between coral cover and gross carbonate production. Specifically, both inshore Red Sea reefs in Saudi Arabia and the offshore reefs in Bahraini had coral cover of ~15%. Yet, gross carbonate production was 0.43 kg m^{-2} y^{-1} and

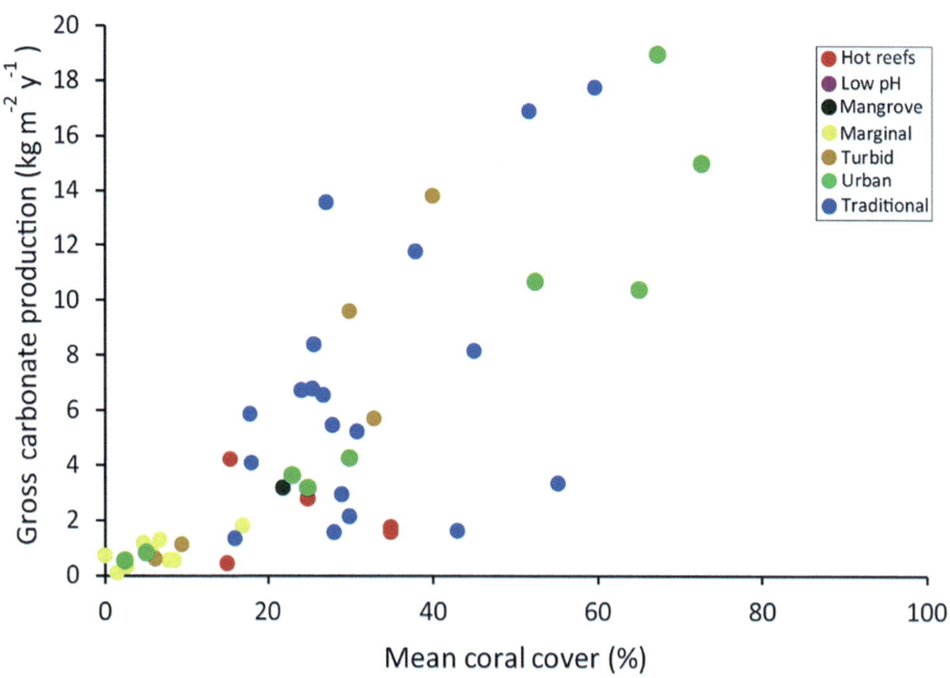

Fig. 16.4 The relationship between mean coral cover (%) and gross carbonate productivity (kg m^{-2} y^{-1}) for all reef types

4.2 kg m^{-2} y^{-1}, respectively. Importantly, both sites were also dominated by massive coral communities. However, there were important differences in how the carbonate budget was calculated, where the Saudi Arabia study used site specific coral calcification rates (Roik et al. 2018), and the Bahraini study used published rates for the Indo-Pacific region as per the ReefBudget methodology (AlMealla et al. 2023). The latter study resulted in higher rates of gross carbonate production. These differences once again highlight the importance of evaluating the source of key data used to calculate rates of gross and net carbonate production, which may in turn be masking important differences between reefs over time and space.

Despite a wide range of mean coral cover values observed on traditional reefs, there was no significant relationship between mean coral cover and gross carbonate production ($r^2 = 0.07$, $p = 0.14$). Data suggests that there is greater variability in gross carbonate production as coral cover increases. Reefs that had a higher-than-expected gross carbonate production rate compared to their coral cover, may be reefs where other sources of carbonate (e.g., encrusters) play a greater role in carbonate production, or possibly due to a higher composition of faster growing corals. For example, Great Chagos Bank in the Chagos Archipelago in 2016 displayed a higher rate of gross carbonate production (13.6 kg m^{-2} y^{-1}) than expected for a coral cover of 27.1% (Lange and Perry 2019). However, the main contributor to coral cover at this site was fast growing *Acropora*. In contrast, Kailua Bay in Oahu, Hawaii had a much lower rate of gross carbonate production (3.38 kg m^{-2} y^{-1}) than expected based on a coral cover of 55.2% (Harney and Fletcher 2003). Coral at this site was dominated by slower growing encrusting corals, which contributed 58% of gross carbonate production, and CCA contributed the remaining 42%. Similarly, gross carbonate productivity was lower than expected at Warraber Island, Torres Strait in northern Australia (coral cover = 43% and gross productivity = 1.66 kg m^{-2} $year^{-1}$; Hart and Kench 2007). The coral community surrounding Warraber Island is characterised by massive corals (Table 16.2), which account for over 60% of the carbonate produced with coralline algae being the second major contributor. These examples highlight the importance of not only coral cover, but also the life-history traits and composition of carbonate producers, for sustaining reef gross carbonate production in a range of settings (e.g., Kench et al. 2022).

Traditional reefs had the most diversity in coral community assemblages and, importantly, were the only reef type that had *Acropora* dominated reefs (Fig. 16.5). In contrast, urban reefs were dominated by massive corals, turbid reefs were dominated by massive and foliose communities, and marginal reefs by massive and encrusting corals. These differences reflect the advantages and disadvantages of specific coral morphologies within different environmental settings, and how such preferences can influence reef carbonate budget states. For example, branching corals grow fast and drive rapid reef recovery following acute disturbance events (Gilmour et al. 2013), but are more susceptible to higher temperatures (McCowan et al. 2012) and poor water quality (Browne et al. 2015; Guest et al. 2016). In contrast, slow growing massive corals are less sensitive to higher temperatures and can more effectively switch to heterotrophy under low light, high nutrient conditions (Houlbreque and Ferrier-Pages 2009). There are also a number of coral genera and species that can alter their coral morphology to acclimate to local environmental conditions (Todd 2008). Given that there were reefs dominated by massive corals in all reef types categories, with the exception of low pH reefs, this further suggests that these coral morphologies are potentially the most resilient to extreme environmental conditions. In contrast, branching morphologies appear less adaptable to non-optimal conditions, which can have clear implications for the reef carbonate budget status between reef types.

Differences in coral community composition also reflected in the gross carbonate production of a reef. Reefs that were dominated by massive corals included a range of coral cover values (1–55%) and associated gross productivity rates (<1–11 kg m^{-2} y^{-1}), most likely reflecting the diversity of morphologies (i.e., massive to submassive) and species classed together in this morphological group. Yet, most reefs dominated by massive, encrusting, foliose (or combinations) typically had both lower coral cover and productivity (Fig. 16.6). This is not surprising given that these corals calcify and grow more slowly than branching corals (Browne 2012; Morgan and Kench 2012; Guest et al. 2023). In contrast, those reefs dominated by *Acropora* spp., branching or mixed coral communities, regardless of reef type, had higher coral cover and gross productivity.

16.2.2 Bioerosion

Bioerosion is a destructive process on coral reefs and is a dominant control on reef accretion and carbonate sediment generation (Glynn 1997; Hutchings 1986). The relative importance of bioerosion on reefs is expected to increase in future years due to increased nutrient input into coastal ecosystems, which support filter feeding and bioeroding organisms (Perry and Harborne 2016). Accurately assessing and tracking rates of erosional change over time will be key to improving our understanding of the longer-term accretionary capacity and stability of coral reefs, particularly for extreme and marginal reef types which already experience high nutrient inputs, or reduced cover of reef calcifiers.

Erosion rates were comparable between traditional and non-traditional reefs (~2 kg m^{-2} $year^{-1}$; df = 1, F = 0.15, p = 0.70; Fig. 16.3a). However, erosion rates at one of the

Fig. 16.5 The dominant coral morphological groups for traditional and non-traditional reefs (Hot reefs, Low pH, Mangrove, Marginal, Turbid, Urban). Where detailed data sets were available for more than one reef site per study, each reef site was included separately. However, most studies are represented as a single data point for a reef or reef complex. In addition, five studies out of 37the coral composition data was not available (see Tables 16.1 and 16.2)

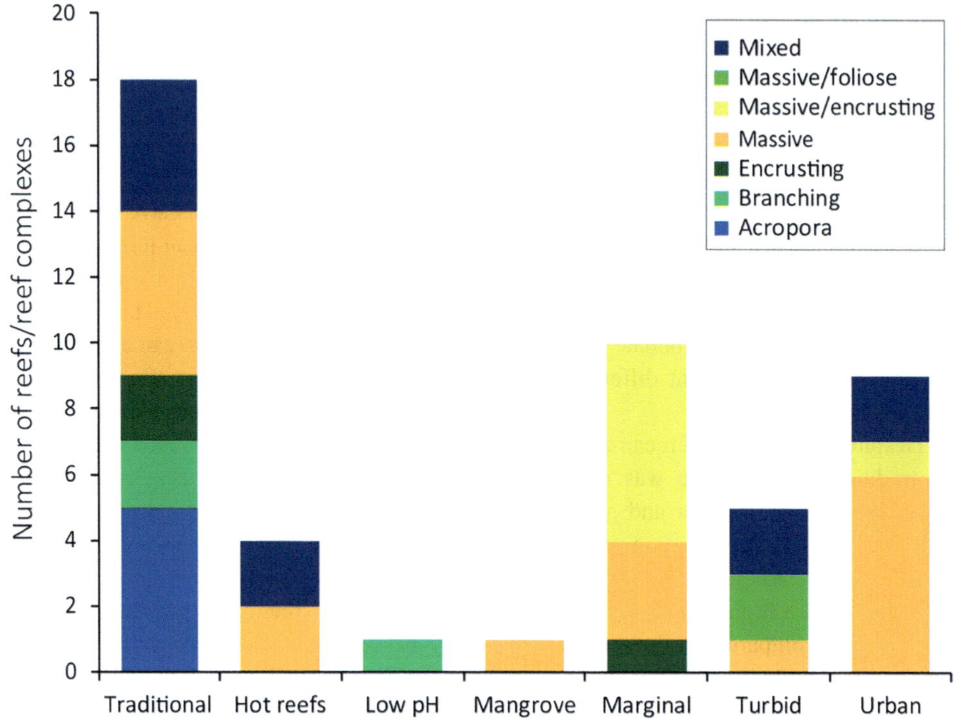

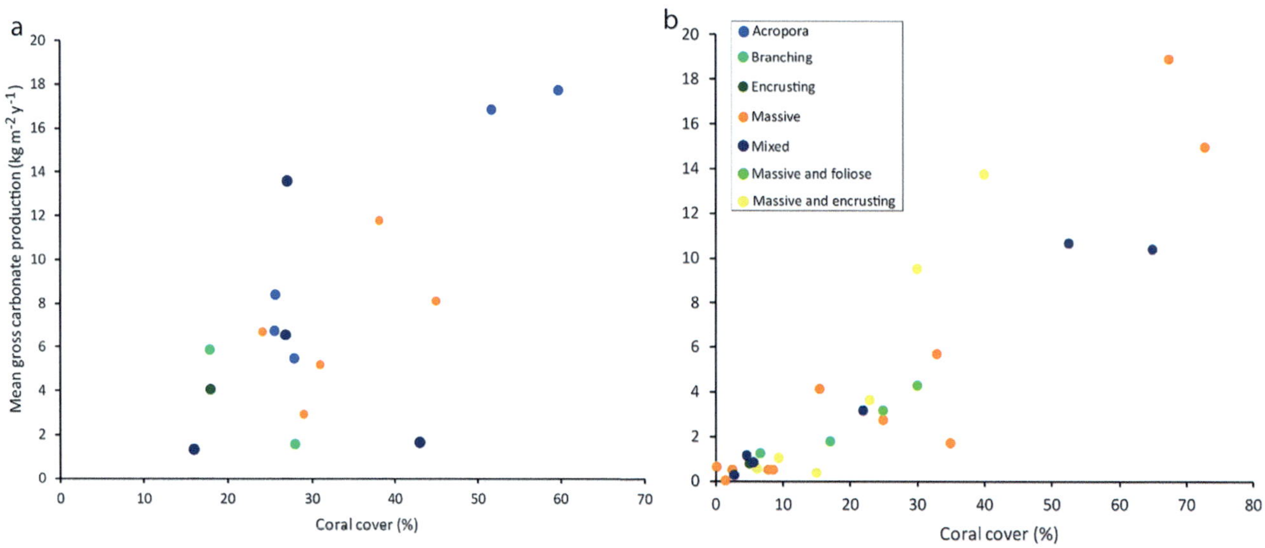

Fig. 16.6 The relationship between mean coral cover and gross carbonate productivity for the dominant coral morphologies in (**a**) traditional and (**b**) non-traditional reefs

low pH reefs were considerably higher than all other reefs (9.9 kg m^{-2} year^{-1}). This reef was located at an upwelling site in the Gulf of Panama (Rodriguez-Ruano et al. 2023), where the assessment of bioerosion rates was conducted using the ReefBudget approach. In contrast, the bioerosion rate at the second low pH site had a lower rate of erosion (0.9 kg m^{-2} year^{-1}) and had been determined using coral cores close to a CO_2 seep in west of Maui, Hawaii (Prouty et al. 2017). Reefs exposed to low pH potentially experience greater levels of bioerosion due to the weakening of the reef structure, which enables macro- and micro-borers to more easily remove carbonate material (Enochs et al. 2016; Wisshak et al. 2012). The use of the different techniques potentially explains the vastly different bioerosion rates, but also highlights the need for more data on these reef types. Given that these reefs had high to very high rates of gross carbonate production (6.8–18 kg m^{-2}y^{-1}), they are likely to continue to grow under current environmental conditions, even at high bioerosion rates. However, given future predictions on rising sea surface temperatures and declining ocean pH, these reefs may become more vulnerable to mortality events followed by rapid reef framework erosion of boring organisms on newly available substrate.

Erosion rates on urban reefs were also higher (mean = 2.61 kg m^{-2} $year^{-1}$) compared to most other reef types including turbid reefs (mean = 0.76 kg m^{-2} y^{-1}) despite both these reef types sharing similar environmental characteristics. Urban reefs are typically exposed to elevated nutrient and sediment loads (Heery et al. 2018). Several studies have observed higher rates of bioerosion on reefs subjected to higher nutrient inputs (Hallock 1988; Herrán et al. 2017; Januchowski-Hartley et al. 2020), as higher nutrient loading likely supports a more diverse and active internal boring community (e.g., sponges, molluscs, polychaetes) resulting in higher rates of internal erosion. However, there are also contrasting reports that nutrients may also suppress internal borer abundance (Chazottes et al. 2002; DeCarlo et al. 2015). Further, urban reefs are exposed to other factors that likely influence both the abundance and activity of borers, such as trace metals, sediments and low light availability (Chazottes et al. 1995; Tribollet and Golubic 2005; Morgan et al. 2020). These water quality variables often co-vary together, and therefore, it has been difficult to disentangle the individual influences of each. Yet, a recent study by Dee et al. (2023) found very low rates of bioerosion on two remote semi-turbid reef systems in Western Australia (0.163 kg m^{-2} y^{-1}). Unlike urban reefs, which are located close to several potential sources of nutrients (e.g., sewage, industrial discharge), remote and naturally turbid reefs may have high sediment loads, but not high nutrient loads. The low bioerosion rates at these sites were attributed to both low light and low nutrients. Low light due to sediment resuspension has previously been linked to lower micro-bioerosion (sponges, parasites, fungi) rates (Silbiger et al. 2017), and when taken together, these data highlight the different responses of boring organisms to declines in light and increases in nutrients, which can occur simultaneously on urban reefs. More work needs to be conducted to fully understand how bioerosion rates, particularly on reefs close to urban centers, will change with both increased local threats as well as climate change (Browne et al. 2021).

Unlike urban and turbid reefs, where bioerosion is largely driven by internal borers (>55% of studies), bioerosion on traditional reefs is typically driven by external excavators (>65% of studies; Fig. 16.7). External bioerosion includes mainly parrotfish and urchins (Tribollet and Golubic 2011). These larger, more mobile eroding organisms graze on the reef benthos, and in doing so, excavate the underlying carbonate framework (Bellwood 1995; Perry et al. 2015a, b; Morgan and Kench 2016b). As these organisms are larger and more visible on reefs, there is considerably more data (e.g., rates of carbonate removal) relating to their role in carbonate budgets when compared to internal boring taxa (Browne et al. 2021). Previous studies have demonstrated that sediments on inner shelf reefs can deter grazing by parrotfish (Gordon et al. 2016), which suggests that these fish prefer clearwater reefs where they are able to graze on the substrate more effectively. This also explains why bioerosion rates on marginal reefs were also dominated by parrotfish. These reef types are not characterised by extremes in one or more environmental variables, such as sediments or nutrients, and are therefore still attractive to parrotfish, despite declines in coral cover. Conversely, turbid reef environments, where the benthos can be covered in sediment, do not attract these fish (Dee et al. 2023). Interestingly, four of the seven urban reef studies also identified parrotfish as the main eroding agent within their carbonate budgets. Two of these sites were located in the Caribbean (De Bakker et al. 2019), and two were offshore sites in Zanzibar (Herrán et al. 2017). The remaining urban reefs, where internal bioerosion was dominant, were found in the Indo-Pacific region and two inshore reefs in Zanzibar (Table 16.1). These spatial differences may relate to variable nutrient and sediment levels between the urban settings, as well as human pressures such as fishing, resulting in variable parrotfish abundance, and therefore erosional pressure, decreasing as water quality declines.

Interestingly, urchins were only identified as a key eroder on the traditional reef of Kailua Bay Hawaii, which had both urchins and internal borers (Harney and Fletcher 2003) and on the hot inshore reef site in the Red Sea (Roik et al. 2018). In the 1980's, a number of reefs in the Caribbean suffered a significant decline in live coral cover (Hughes 1994; Jackson et al. 2001). This was largely attributed to overfishing, which removed the natural predators of urchins, as well as herbivorous fish who would compete with urchins for algal resources (Hughes 1994). This example highlights both a direct (predators) and indirect pressure (food competition) on urchin populations, and therefore it is somewhat surprising that urchins are not more prevalent in urban and turbid reef environments where fish abundance and diversity can be low (Moustaka et al. 2018; Wenger et al. 2016), and algal cover high (Browne et al. 2010; Cartwright et al. 2023). Despite less fish on these extreme reefs, key functional fish groups, such as herbivores, may still be present thereby competing with urchins for food source. In addition, many fleshy algal types do not grow well in low light, heavily sediment laden environments (Tebbett et al. 2018), which may further explain both low fish abundance as well as low to no urchins on these reefs.

The mechanism of reef erosion (i.e., internal vs. external) has important consequences for not only the reef carbonate budget, but also the geo-ecological functioning of reefs. Internal boring and removal of carbonate from the reef structure can significantly reduce the stability of reef framework, making it more prone to collapse during storm activity (De

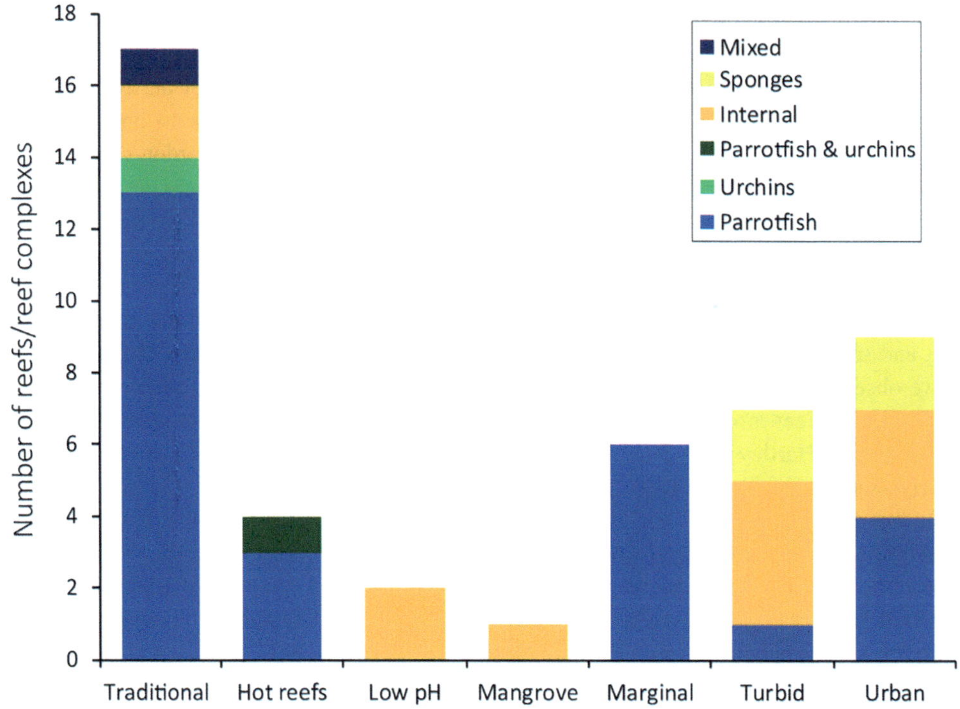

Fig. 16.7 The dominant bioeroder groups on traditional and non-traditional reefs. Where detailed data was available for more than one reef site per study, these were included as separate data points. However, most studies were either represented as a single data point for a reef or reef complex. Note that there were five marginal/extreme and four traditional reef studies that either did not measure erosion or did not provide details on the bioeroder community (see Tables 16.1 and 16.2 for further details)

Bakker et al. 2018; Enochs et al. 2015). In contrast, high rates of external bioerosion removes significant quantities of carbonate causing flattening of existing reef surfaces, however, the internal integrity of the reef fabric remains intact (Molina-Hernández et al. 2022). When considered over geological timescales, such variability in reef bioerosion processes between reef types becomes relevant to understanding long-term physical reef development.

16.2.3 Net Carbonate Production, Framework Building and Reef Accretion

Net carbonate production is the annual addition of carbonate to reef framework following erosive processes. This metric provides some insights into the longer-term reef accretionary capacity with values greater than ~1 kg m^{-2} y^{-1} considered to indicate reefs that are actively accreting (Perry et al. 2008). In contrast, negative values indicate that the reef framework is eroding more than its accreting, and that the reef is likely undergoing widespread reef framework degradation. Net carbonate production values between zero and 1 kg m^{-2} y^{-1} indicate that the reef is currently in a state of stasis and may, therefore, be more vulnerable to future disturbance events (Perry et al. 2008). Net carbonate production values can be converted to reef accretionary rates (mm y^{-1}) by considering the porosity of the reef framework (ranging from 20% to 80% depending on the dominant coral assemblages; Perry et al. 2018b) and the density of calcium carbonate (2.9 g cm^{-3}; Palmer et al. 2010). However, calculated reef accretion rates using this approach should be interpreted cautiously given that reef porosity are difficult to estimate on a reef-by-reef basis, and other critical sedimentary processes that contribute to framework-building on reefs are poorly quantified.

Here net carbonate production values ranged from −0.67 kg m^{-1} y^{-1} on marginal reefs to 9 kg m^{-2} y^{-1} on low pH reefs. Net carbonate production rates for marginal reefs were significantly lower than on traditional reefs and urban reefs (Post Hoc Tukey = $p < 0.02$). Marginal reefs had the lowest rates largely due to very low gross carbonate production values as opposed to high rates of bioerosion (Fig. 16.3b). This was the only reef type examined here that exhibited an erosive reef state. All other reef types had values greater than 1.5 kg m^{-2} y^{-1}. Reef accretion rates, calculated by converting these positive net production rates (1.5–9 kg m^{-2} y^{-1}) using the protocol as described by Browne et al. (2013; see Table 16.1 in reference for equation) that takes into account carbonate density and reef porosity, range from ~1 mm y^{-1} to 6 mm y^{-1}. Here we used 50% porosity for mixed coral communities (as per Perry et al. 2018b) given the range of coral community assemblage present at these reef types. This suggests that, on average, traditional and extreme reefs have the capacity to accrete at comparable rates despite environmental (and biological) differences.

Given that most reefs will experience over 0.5 m increase in water depth by 2100 (Perry et al. 2018b), coral reefs need to be accreting at rates >5 mm y^{-1} to keep up with sea level rise, which equates to a net carbonate production rate of approximately ~7 kg m^{-2} y^{-1}. Based on this estimation, none of the reefs included here, apart from low pH reefs, currently have this accretionary capacity. Further, due to continued increases in ocean temperatures and declines in pH, it is likely that coral reefs capacity to accrete will decline. However, these data highlight that we should not underestimate the capabilities of extreme reefs, such as low pH, urban and turbid reefs, to continue to provide key coral reef ecosystem services that arise from the reef-framework (i.e., habitat provision and coastal protection) into the future given potentially greater rates of net carbonate production than a number of traditional reef systems.

It may also be expected that as the bioerosion rates increase, net carbonate production would decline. To investigate this assumption, we compared rates of erosion (as well as dominant eroders) with net carbonate production rates. We found no significant relationship between erosion rate and net productivity for traditional (r^2 = 0.02, p = 0.26), extreme and marginal reef types (r^2 = −0.03, p = 0.81; Fig. 16.8). This was also regardless of the main driver of erosion, in this case parrotfish or internal borers. However, three traditional and seven non-traditional reef studies had both low erosion (<1 kg m^{-2} y^{-1}) and low net productivity (<2 kg m^{-2} y^{-1}) that could possibly be masking a negative relationship between erosion and net carbonate productivity. These reef sites have low coral cover and, as such, it is important that bioerosion rates remain low to prevent these reefs from tipping into an erosive reef state, which is potentially more difficult to recover from. There is only one reef site where high erosive processes have pushed a reef into a negative reef budget state, despite a healthy gross carbonate production rate (Fig. 16.8). This reef is a coral cay island located in Palua Panjang, Java, Indonesia, which is exposed to high volumes of sewage, sediment and aquaculture runoff classified here as an urban reef (Edinger and Risk 2000). This has resulted in extremely high rates of internal boring. Overall, these data suggest that for all reef types, net productivity is mainly driven by rates of carbonate production and not carbonate loss.

Importantly, there were also sites that had high net productivity and low erosion rates. These included both traditional and extreme reef sites. For example, this included two clear-water patch reefs (Yap and Palua; ~40% cover cover) in the Western Pacific Ocean, where parrotfish erosion was dominant but comparatively low (van Woesik and Cacciapaglia 2018), a turbid inshore patch reef on the central Great Barrier Reef where bioerosion was driven by internal macro-borers and coral cover was high at 40% (Browne et al. 2013), and an urban reef in Zanzibar with exceptionally high coral cover of 67% and medium rates of erosion due to parrotfish (Herrán et al. 2017). These reefs are currently in a healthy reef accretionary state and further demonstrate that extreme reefs should be accounted for in long-term coral reef conservation plans.

16.2.4 Structural Complexity

Structural complexity (rugosity) of a coral reef is a major indicator of reef geo-ecological function. For example, reefs with high structural complexity promote species richness and is therefore an important characteristic of reef resilience (Graham and Nash 2013). However, rugosity can be measured at different scales from reef geomorphic scales (e.g., Harris et al. 2023), to reef habitat scales, down to the coral colony (Husband et al. 2022). The scale at which rugosity is measured will depend (mostly) on the objective of the study. For example, studies assessing the ability of a

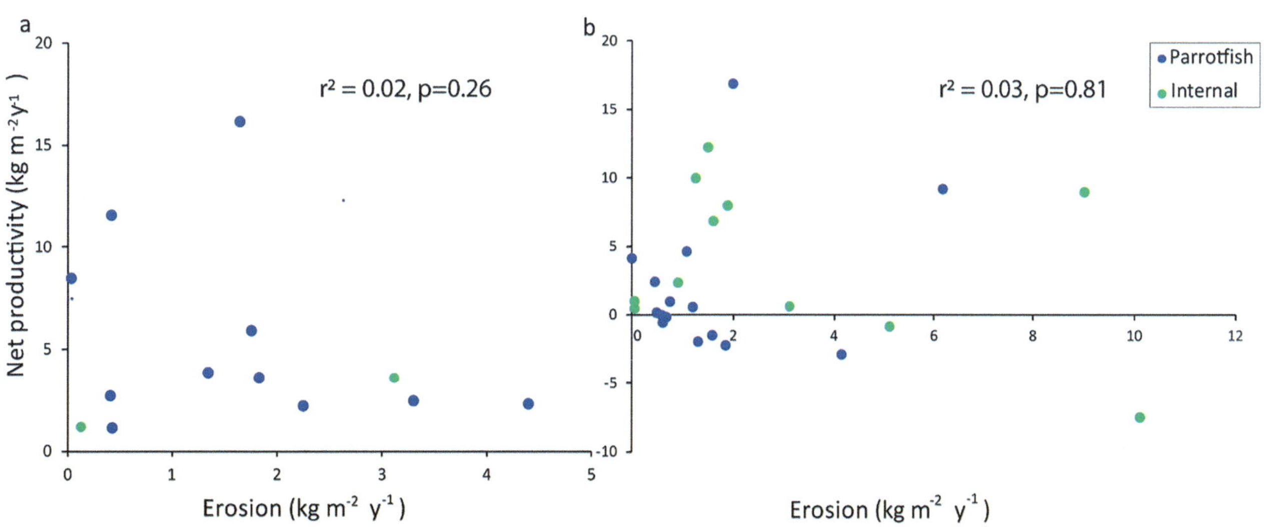

Fig. 16.8 Relationship between erosion rate and net productivity with dominant bioeroder (parrotfish or internal macro and/or microborers) for (**a**) traditional reefs and (**b**) non-traditional reefs

coral reef to protect shorelines from coastal erosion will likely focus on reef rugosity where the underlying geomorphic structure (e.g., spur and groove formations) will have a significant influence on wave energy as it passes over the reef (Harris et al. 2018). Rugosity at this scale is less related to biological productivity of the living coral cover (and carbonate production) than rugosity measured at the reef habitat scale. Here the 'chain and tape' method is often employed to quantify rugosity within reef habitats (Risk 1972), and has been used in several carbonate budget studies (Browne et al. 2013; Dee et al. 2020; Morgan 2014). Rugosity measured using this approach has been used as a multiplier in carbonate budget calculations, including the ReefBudget approach, as a means of considering coral growth in three dimensions. Measurements at this scale can be a useful metric for comparing structural complexity between reefs and over time. However, more recently, the ReefBudget approach has moved towards measuring coral colony rugosity to account for differences in benthic community diversity.

Of the 57 studies included here, 23 measured and reported rugosity values at the habitat scale. Of these 23, 11 were on traditional reefs and 12 were on extreme reefs. Of the 12 on extreme reefs, this included one hot reef, four turbid reefs and seven urban reefs (Fig. 16.9). This highlights the paucity of structural complexity data collected all on reef types, including traditional reefs. It should, however, be noted that there were several more recent studies that stated that rugosity was measured but not reported (five extreme/marginal reefs and five traditional reefs; Tables 16.1 and 16.2). These studies may have measured rugosity at the coral colony level, which is argued to provide a more accurate representation of the substrate cover (González-Barrios et al. 2021), but it does not provide a reef rugosity measurement that can be tracked over time or used to compare to other reefs. Recent technological advances have resulted in an increase of studies that are now evaluating and reporting reef rugosity, which include high-resolution Lidar (Dee et al. 2020; Harris et al. 2023) and close-range underwater photogrammetry (Storlazzi et al. 2016). We recommend that future carbonate budgets studies on all reef types include an assessment of reef rugosity at the habitat scale given the importance of this reef metric for both carbonate budgets and reef health assessments.

It would be expected that marginal reefs with low coral cover and/or diversity would also have low reef structural complexity. However, as no rugosity measurements were provided for marginal reefs where carbonate budgets had been conducted, our focus shifts to comparing structural complexity between traditional and extreme reef types. The key question is whether we would expect to see differences in structural complexity between reefs with high coral cover and diversity yet different abiotic conditions.

Here we found that there were significant differences in rugosity values between traditional (rugosity = 1.97) and extreme reefs (1.4; df = 1, F = 11.45, p = 0.003 Fig. 16.9a). This suggests that despite high coral cover, the structural complexity of extreme reefs is reduced. Typically, as coral cover increases, the structural complexity of the reef is expected to increase as each coral colony increases the surface area of the reef (Alvarez-Filip et al. 2011; Bozec et al. 2015). However, as different coral species have variable three-dimensional complexity, the rate of increase in habitat complexity with coral cover varies depending on the mix of corals within the community. Therefore, for extreme reefs such as turbid and urban reefs, which are typically dominated by low profile encrusting and massive corals, the overall complexity of the reef is less than when compared to reefs with comparable coral cover but dominated by structurally complex branching corals (Alvarez-Filip et al. 2011; González-Barrios et al. 2021). Interestingly, we also see a wide range of rugosity values at comparable coral cover levels for traditional reefs as well as a limited range of rugosity values across a wide range of coral covers for extreme reefs. This may again be explained by important differences in the coral community composition with more variable coral community structure on traditional reefs (as seen in Fig. 16.5) and more consistent low profile coral communities on turbid and urban reefs.

The relationship between coral cover and habitat scale reef rugosity is also influenced by the rate of breakdown of dead coral cover following a mortality event, as non-living benthic components also contribute to the total reef rugosity. Coral mortality is not translated into a loss of complexity at the same rate for all coral communities and therefore for all reef types. For all reefs, there is a temporal lag between a disturbance event (e.g., bleaching) and the loss of structural complexity, as the dead coral still provides complexity until it is eroded and broken down (Alverez-Filip et al. 2011). However, the length of the time lag (i.e., weeks, months to years) depends heavily on the composition of corals. For example, dead branching corals will likely be broken more quickly than large massive corals, which can remain *in situ* for several years (Aronson and Precht 2006). Given that extreme reefs such as turbid and urban reefs are typically dominated by massive corals, it suggests that the lag time between a mortality event and declines in structural complexity will be greater than on traditional reefs characterised by branching to mixed coral communities. This may, in part, explain why we do not see a decline in rugosity with lower coral cover for turbid and urban reefs (Fig. 16.9b). As such, carbonate budget calculations conducted on reefs following a mortality event, should also assess coral colony rugosity to assess rates of gross coral production given that habitat level rugosity, which includes live and dead coral cover, may overestimate carbonate production.

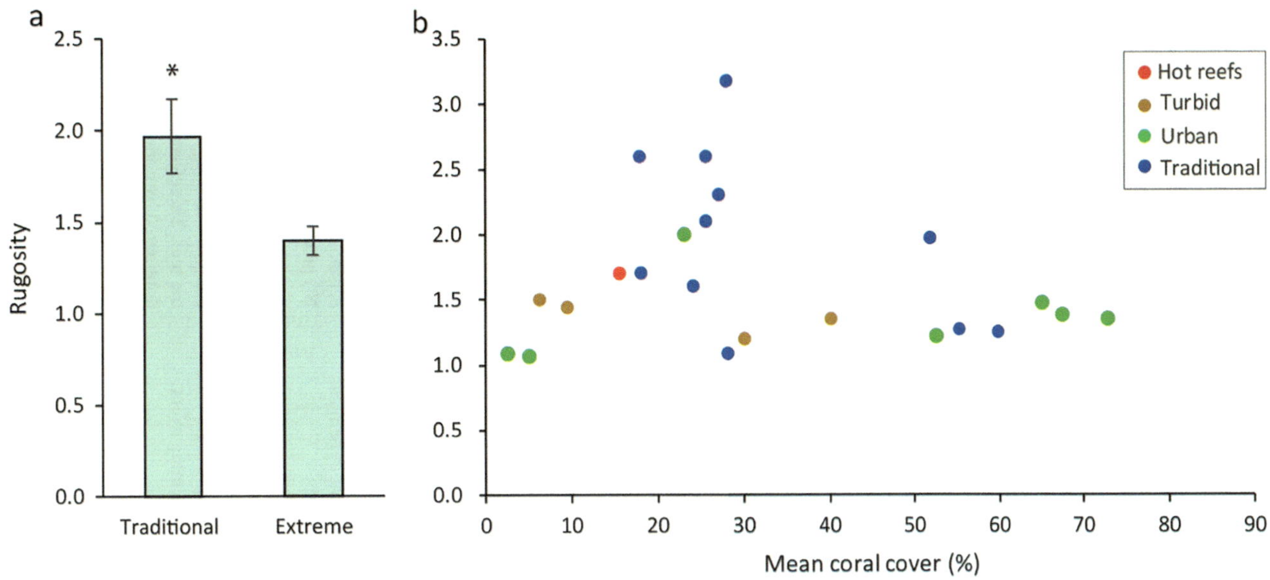

Fig. 16.9 Mean rugosity on (**a**) traditional and non-traditional reefs, and (**b**) the relationship between mean coral cover (%) and rugosity for traditional and extreme reefs. Asterisks indicate significant differences

16.2.5 Reef Sediment Budgets

Although the use of carbonate budgets has become more popular since the early 2000s, the study of carbonate production was first developed and applied by geologists who were interested in understanding sources and sinks of carbonate sediments on reefs. For example, Land (1979) provided a detailed assessment of coral carbonate production together with sediment accumulation profiles using seismic scanners and sediment cores in Discovery Bay, Jamaica. This also involved data collection on sediment carbonate content, size fraction and depth layers to determine volumes and direction of movement of reef-derived sediment. Similarly, Sadd (1984) measured sediment transport and a gross carbonate budget on a fringing reef in Cane Bay, U.S Virgin Islands. Here sediment tagging was used to assess rates of sediment transport and accumulation rates under varying wave energy conditions to determine what proportion of sediment produced on the reef by biological organisms were removed annually. As such, eight out of 10 carbonate budget studies conducted prior to 2000 (exceptions being Grigg 1982; Heiss 1995), included some aspect of a carbonate sediment budget, whereas only 10 (out of the remaining 47) of the carbonate budget studies conducted since 2000 have included (or part of) an assessment of carbonate sediments (Table 16.3).

Of the 18 studies with a carbonate sediment assessment, 16 were conducted on traditional clear-water reefs. The two studies conducted on extreme reefs were both in naturally turbid water settings off the coast of Australia (Table 16.3). The first study was conducted on two inshore patch coral reefs on the GBR in northern Queensland. Sediment composition on both reefs were characterised by a high mollusc content (~25%), with the organism producing between 0.71 and 0.83 kg m^{-2} y^{-1} carbonate (Browne et al. 2013). *Halimeda* and foraminifera production rates were comparatively low (<0.02 kg m^{-2} y^{-1}). Corals contributed the most to the sediment profile (~30%) with other contributions from CCA (10%), sponge spicules (10%) and terrigenous material 5–15% (Browne 2011). Similarly, a study conducted more recently off the coast of central Western Australia on turbid reefs, found comparable contributions of molluscs (35%) and corals (36%) to reef sediments, with mollusc carbonate production rates between 0.95–2.54 kg m^{-2} y^{-1}; Dee et al. 2024). The high mollusc content is unusual when compared to clear-water reefs, where mollusc content is typically <10% (Table 16.3), but previous (non-carbonate budget) studies have also found high mollusc content in reef habitats where muds are deposited (e.g., Masse et al. 1989). This suggests that carbonate sediment production within turbid settings may rely on a healthy mollusc population to provide carbonate sediments for reef infilling as well as reef associated island development (Bonesso et al. 2022).

In contrast, the sediment profile on traditional reefs is often more mixed with the dominant sediment producers typically being coral, and additional major constituents including CCA and/or *Halimeda* fragments (Table 16.3). One exception is Green Island on the northern GBR, where foraminifera are the dominant sediment constituent in both reef and island sediments (Yamano et al. 2000). Coral sediments are produced by erosive processes that breakdown the reef framework, and could include both biological and physical processes. The reef framework can also be chemically eroded, however, this does not produce sediment sized particles that contribute to the reef sediment profile (Lange et al.

Table 16.3 Carbonate budget studies that have included carbonate sediments

No.	Author	Date	Title	Location	Reef type	Carbonate budget variables measured	Methods	Carbonate sediment production
1	Stearn et al. (1977) and Scoffin et al. (1980)	1977 and 1980	Calcium carbonate budget of a fringing reef of the west coast of Barbados (Part 1: Zonation and productivity, AND Part 2: Erosion, sediments and internal structure)	Bellairs Reef, Barbados	Clear-water reef	Living and dead coral cover, CCA cover, sand and rubble cover, rugosity, coral growth rates and skeletal density, coral composition and community size distributions, CCA growth rates (Stearn et al. 1977), fish species, size and abundance, parrotfish gut content analysis and feeding activity, urchin abundance, test size and gut content analysis, macro-boring rates, sediment particle size and composition, sediment accumulation rates, dead reef erosion/lowering rate, diagenetic constituents and reef cementation rates (Scoffin et al. 1980)	A detailed survey of carbonate production and loss within ecological zones together with a comprehensive assessment of habitat complexity	Sediment composition was 58% coral, 13% coralline algae, 6% molluscs, 4% foraminifera, 2% echinoderms
2	Land (1979)	1979	The fate of reef-derived sediment on the north Jamaican Island slope	Discovery Bay, Jamaica	Clear-water reef	Living coral cover and composition, coral growth rates, sand cover, sediment carbonate content, sediment loss and accumulation rates, sediment size fraction, sediment depth layer, mass of reef derived sediment	This study provides an assessment of coral carbonate production with detailed sediment accumulation profiles from seismic scanners and sediment cores	No details on reef surficial sediment samples provided. All sediment analysis was conducted on deeper water, reef slope sediments.
3	Glynn et al. (1979)	1979	Coral reef growth in the Galapagos: Limitation by sea urchins	Onslow Island, Galapagos	Clear-water reef	Live and dead coral cover, CCA cover, coral composition, urchin species, abundance, size and grazing rates, urchin gut content analysis, sediment composition and particle size	This study provides an assessment of coral cover and urchin activity to estimate a net carbonate budget	Coral and CCA combined varied between 26 to 43%, with barnacle remains dominating (44 to 54%).
4	Sadd (1984)	1984	Sediment transport and $CaCO_3$ budget on a fringing reef, Cane Bay, St. Croix, U.S. Virgin Islands	Cane Bay, St. Croix	Clear-water reef	Living and dead coral cover, sand cover, CCA cover, rugosity, coral composition, skeletal growth and density, sediment transport loss and accumulation rates, sediment carbonate content	This study uses transects and sediment tagging to assess gross carbonate production and sediment dynamics, respectively	Sediment composition was 45% coral, 16% coralline algae, 13% foraminifera, 8% mollusc, 6% echinoderm, and 3% *Halimeda*. No production rates.
5	Glynn (1988)	1988	El Nino warming, coral mortality and reef framework destruction by echinoid bioerosion in the eastern Pacific	Uva Island and Onslow Reef, Eastern Pacific	Clear-water reef	Living coral cover, coral colony size and shape, rugosity, urchin species, abundance, size and grazing rates, damselfish territories, macro-boring rates, bioeroder sediment production, sediment composition	A detailed study of urchin bioerosion and sediment production pre and post an El Nino warming event	Focused on sediments eroded by urchins as opposed to a complete sediment sample from the reef. Therefore, heavily dominated by coral and CCA fragments.

6	Bak (1990)	1990	Patterns of echinoid bioerosion in two Pacific coral reef lagoons	Tiahura, Moorea	Clear-water reef	Living and dead coral cover, coral composition and condition, rubble and sand cover, urchin species, size and abundance, urchin gut content	A detailed study on urchin densities and size distributions with changing substrate	Focused on urchin sediment generation rates
7	Hubbard (1990)	1990	Production and cycling of calcium carbonate in a shelf-edge reef system (St Croix, U.S. Virgin Islands): Applications to the nature of the reef systems in the fossil record	St Croix, US Virgin Islands	Clear-water reef	Living coral cover, CCA cover, coral composition, growth rates and bulk density, sediment composition, rugosity, sediment import, export and accumulation rate, reef accretion rates	This study provides an assessment of carbonate production with an estimation of sediment import, export and accumulation rates	Corals contribute the most to sediment profile and contribute 0.22 kg m^{-2} y^{-1}. CCA contribute 0.02, molluscs, echinoderms contribute 0.06 kg m^{-2} y^{-1}
8	Eakin (1996)	1996	Where have all the carbonates gone? A model comparison of calcium carbonate budgets before and after the 1982–1983 El Nino at Uva Island in the eastern Pacific	Uva Island, Eastern Pacific	Clear-water reef	Live and dead coral cover, CCA cover, damselfish lawn cover, coral composition, rugosity, urchin density, sediment size distribution and hydrodynamic properties	This study uses in situ census based data together with published carbonate production rates to develop a carbonate budget model	Focused on sediment production from bioerosion (coral and CCA sediment production) and stated that other sources of carbonate sediment were comparatively negligible.
9	Yamano et al. (2000)	2000	The importance of foraminifera for the formation and maintenance of a coral sand cay: Green Island, Australia	Green Island, Australia	Clear-water reef	Living coral cover, algal and turf cover, *Halimeda* cover, calcareous algal cover, sand cover, coral composition, sediment trap content weight, sediment composition and particle size, foraminifera density and species composition, wind and current data	This study focuses on the contribution of foraminifera to the carbonate budget using a census based approach	Sediment were dominated by foraminifera (50–70%), corals contributed between 5–20% over the reef, molluscs 10%, and *Halimeda* 5–10%.
10	Eakin (2001)	2001	A tale of two ENSO events: Carbonate budgets and the influence of two warming disturbances and intervening variability, Uva Island, Panama	Uva Island, Eastern Pacific	Clear-water reef	Live and dead coral cover, CCA cover, damselfish lawn cover, coral composition, rugosity, urchin density, sediment size distribution and hydrodynamic properties	Study combines in situ census based data with published carbonate production rates to develop a carbonate budget model	No details on sediment production rates or composition provided.
11	Harney and Fletcher	2003	A budget of carbonate framework and sediment production, Kailua Bay, Oahu, Hawaii	Kailua Bay, Oahu, Hawaii	Clear-water reef	Live and dead coral cover, macro and turf algal cover, CCA cover and growth rates, sand and reef framework cover, rugosity, coral composition, growth rates and skeletal density, urchin abundance and size, macro-boring rates, sediment depth layers, sediment carbonate content and composition, mechanical erosion rate	An assessment of carbonate production and removal that also considers variable production between zones and includes direct sediment production rates	Corals (35%) and CCA (27%) contributions to the sediments, with *Halimeda* (19%) contributing the most for the direct sediment producers

(continued)

Table 16.3 (continued)

No.	Author	Date	Title	Location	Reef type	Carbonate budget variables measured	Methods	Carbonate sediment production
12	Mallela and Perry	2007	Calcium carbonate budgets for two coral reefs affected by different terrestrial runoff regimes, Rio Bueno, Jamaica	Outer embayment site, Rio Bueno, Jamaica	Clear-water reef	Living and dead coral cover, algal cover, rubble, reef framework and sand cover, rugosity, coral composition, growth and skeletal density, encrusting community composition, cover and growth, fish species, numbers and size, urchin density and size, macroboring and micro-boring rates, sediment accumulation rates, rates of framework infilling, sediment carbonate content and composition, temperature, salinity, DIN, phosphorus, ammonia, light, suspended sediments	A comprehensive carbonate budget using transects and in situ experiments together with environmental data collection	Dominant carbonate grains were *Halimeda*, CCA and molluscs
13	Hart and Kench	2007	Carbonate production of an emergent reef platform, Warraber Island, Torres Strait, Australia	Warraber Island, Torres Strait, Australia	Clear-water reef	Living coral cover and composition, CCA cover, *Halimeda* cover, sponge and brown algal cover, seagrass cover, sand and reef framework cover, rugosity, sediment depth and type, foraminifera and mollusc abundance	This study assesses gross carbonate production using quadrates and transects within ecological zones	Coral was the main contributor to reef sediments (74%) with a small contribution of mollusc (4%). CCA made up 19% and foraminifera 2%
14	Browne et al.	2013	Carbonate and terrigenous sediment budgets for two inshore turbid reefs on the central Great Barrier Reef	Middle Reef and Paluma shoals, Great Barrier Reef	Naturally turbid reefs	Live and dead coral cover, macro and turf algal cover, CCA cover, encrusting community cover, composition and growth rates, sand, rubble and reef framework cover, rugosity, coral composition, growth rates and skeletal density, fish species, abundance and size, urchin abundance and size, macro-boring rates, light, suspended sediments, net sediment accumulation rates, sediment export rates, sediment depth layers, sediment particle size, carbonate content and composition	Census based approach	Mollusc were a key sediment producer contributing between 0.17 and 1.48 kg m^{-2} y^{-1}, and average rates of 0.71 and 0.83 kg m^{-2} y^{-1} per reef

15	Morgan	2014	A calcium carbonate budget of a Maldivian reef platform	Vabbinfaru Reef, North Malé Atoll, Maldives	Clear-water reef	Live and dead coral cover, CCA and *Halimeda* cover, sand, rubble and reef framework cover, rugosity, coral composition, growth rates and skeletal density, encruster community composition and calcification rate, CCA carbonate production, fish species, abundance and size, urchin abundance and size, macro-boring rates, light, suspended sediments, net sediment accumulation rates, sediment export rates, sediment depth layers, sediment particle size, carbonate content and composition, reef island planform morphology, wave energy	This study provides a detailed census based carbonate budget of different reef eco-geomorphic zones which also includes an assessment of sediment flux and a sediment budget	Over 97% of sediment was generated by parrotfish (coral in origin). Direct sediment producers included *Halimeda* and foraminifera (<1.5%)
16	Perry et al.	2017	Reef habitat type and spatial extent as interacting controls on platform-scale carbonate budgets	Lhaviyani Atoll, Central Maldives	Clear-water reef	Same as 2015b Perry et al. study, with the addition of *Halimeda* plant volume, number of non-Halimeda calcifying green algae, number of articulated red coralline algae, number of epifaunal gastropods, sediment compositional analysis	This study uses a census based approach to demonstrate spatial differences in carbonate production at the inter-reef habitat scale	Most carbonate sediments on the reef were reworked carbonates from Parrotfish and from macroborers. Low *Halimeda* sediment production and no mollusc contributions
17	Brown et al.	2020	Habitat-specific biogenic production and erosion influences net framework and sediment coral reef carbonate budgets	Heron Island, Southern Great Barrier Reef, Australia	Clear-water reef	Living coral cover, CCA cover, turf and macro-algal cover, *Halimeda* cover, sand, rubble and reef framework cover, rugosity, coral composition, growth rates and skeletal density, encrusting cover and growth rates, parrotfish species, size and abundance, parrotfish bite rates, scars and mass eroded, urchin abundance, size and species, clinoid sponge cover, macro and micro-bioerosion rates, sediment composition and production rates, sediment dissolution	Census based approach	Sediment production rates were high. Sediments on the reef slopes and crest were reworked sediments from parrotfish (coral) whereas in the more sheltered lagoon sites, *Halimeda* was the main sediment component

(continued)

Table 16.3 (continued)

No.	Author	Date	Title	Location	Reef type	Carbonate budget variables measured	Methods	Carbonate sediment production
18	Dee et al.	2024	The application of carbonate and sediment budgets to assess the stability of marginal reef systems	Eva and Fly Reef, Exmouth Gulf, Australia	Naturally semi turbid reef	Live and dead coral cover, macro and turf algal cover, CCA cover, encrusting community cover, composition and growth rates, sand, rubble and reef framework cover, rugosity, coral composition, growth rates and skeletal density, fish species, abundance and size, urchin abundance and size, macro-boring and micro-boring rates, light, temperature, chlorophyll a, conductivity, salinity, pH and turbidity, sediment depth layers, sediment particle size, carbonate content and composition	Census based approach	Coral (36%) and molluscs (35%) were the main components of the carbonate sediments. Molluscs contributed between 0.95 and 2.54 kg m^{-2} y^{-1}

Note [a]also appear in Tables 16.1 or 16.2

2023a, b). Consequently, a number of studies have found that rates of bioerosion from fish and urchins are important drivers of carbonate sediments. For example, a carbonate and sediment budget study by Morgan (2014) in the Maldives found that 97% of carbonate sediments were generated by Parrotfish. Similarly, Brown et al. (2020) found that sediments on the reef slope and reef crest on Heron Island, central GBR, were reworked sediments also produced by Parrotfish. This contrasts with sediments on turbid reefs where Parrotfish are typically either absent, or in low abundance (Browne et al. 2013; Dee et al. 2024). Here coral sediments are produced from internal macroborers such as sponges, molluscs and polychaete worms, further illustrating important differences in drivers of not only bioerosion and carbonate reef framework removal, but also carbonate sediment generation.

16.2.6 Differences Between Traditional and Non-traditional Reefs

A comparison of carbonate budget studies from traditional reefs with all non-traditional reef types found that traditional reefs, on average, had greater coral cover, gross productivity and rugosity, while bioerosion rates were lower. These reefs also consisted of more mixed coral communities and importantly, communities dominated by more rugose, faster growing branching corals. However, these average trends were largely influenced by marginal reefs driving down gross production rates and low pH/high nutrient reefs driving up bioerosion rates. When placed on a carbonate budget continuum, traditional reefs lie in the relative centre in all key elements (coral cover, gross and net carbonate productivity, and bioerosion) except rugosity, which is likely greater than on marginal and extreme reefs due to a combination of healthy coral cover and more structurally complex coral assemblages (Fig. 16.10). Within the extreme reef types, hot and marginal reefs had the most comparable carbonate budget characteristics, as both reef types exhibited low gross production (1–2 kg m^{-1} y^{-1}) and bioerosion rates (1–2 kg m^{-1} y^{-1}), massive to massive and mixed coral communities, and parrotfish were the key drivers of bioerosion. Unsurprisingly, turbid and urban reefs were relatively similar in carbonate budget characteristics including coral cover (30–40%), gross productivity (6–8 kg m^{-1} y^{-1}) and coral community, as they inhabit similar environmental settings (e.g., low light, high sedimentation) and often support common coral assemblage types. Drivers of bioerosion on these reefs were internal borers, although boring rates were greater on urban reefs, likely due to elevated nutrient inputs. Highest nutrient and bioerosion rates were, however, observed on low pH/high nutrient reefs, but these reefs were also characterised by high gross carbonate production due to high coral cover of branching corals. However, given that these data have been extracted from two sites (Gulf of Panama and Maui, Hawaii), these data should be interpreted with caution. Lastly, all reefs except marginal reefs had a positive carbonate budget. These reef types have comparable abiotic conditions to traditional reefs and were largely located in the Caribbean where coral cover used to be significantly higher than present day. As such, these reefs could be classified as degraded reefs.

16.3 Knowledge Gaps

Our review of census-based carbonate budgets on marginal and extreme reefs identified key knowledge gaps broadly related to data availability and the standardisation of methodologies. Despite 57 carbonate budget studies having been conducted on coral reefs since the 1970's, comparisons between datasets focused on only 37 studies. A total of 20 studies were excluded from the analysis either due to incompatible datasets (i.e., different unit of measurement), or methodological differences typically associated with early studies (pre-2000). Approximately half of the 37 studies included were on traditional reefs, with the other half conducted on non-traditional reef types that included both marginal and extreme reefs (turbid, urban, low pH/high nutrient reefs, marginal and hot reefs). Yet, within each of the non-traditional reef types, the number of studies ranged from one to five, indicating not only limited carbonate budget datasets for these reef types, but also major gaps in our understanding of how these ecosystems function. In addition, no comprehensive carbonate budget assessment has been conducted on high latitude reefs, or reefs exposed to low pH (as opposed to low pH and high nutrients experienced by reefs next to upwellings). However, census-based carbonate budget studies have grown in number over recent years, particularly for non-traditional reef types, and these studies have applied comparable methodologies facilitating direct assessments between locations. In addition, we further encourage future research to not only target a wide range of reef types, but also reef locations with a greater geographic spread, given the current bias towards Australia, the central Indian Ocean and the Caribbean. For example, within Southeast Asia, which supports extensive and diverse coral reefs (both in terms of species and environments), only five carbonate budgets have been conducted to date.

Since the publication of the ReefBudget approach in 2012, the consistency of methods used to collect data for the census-based carbonate budget has improved. This rapid and standardised methodology provides clear instructions for data collection that can be repeated over time to assess both spatial and temporal changes in net carbonate production, as well as global reference datasets of existing budget process data (e.g., coral calcification). Those that applied this method

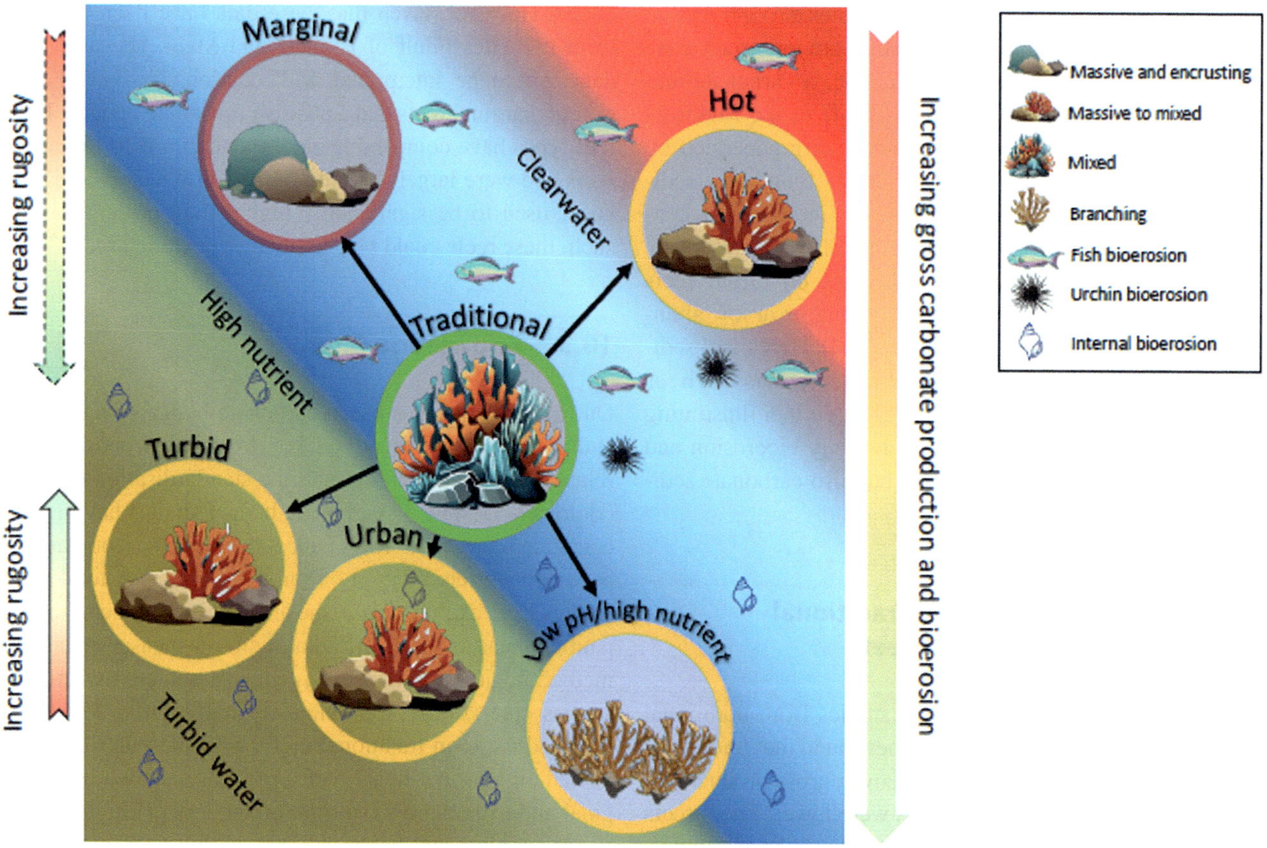

Fig. 16.10 A conceptual diagram summarising differences in key elements of the carbonate budget between traditional, extreme and marginal reefs. Red circles indicate a negative budget state, with yellow circles indicating extreme reefs and green circles stable traditional reef types. The dashed line on the top rugosity arrow indicates that this increase in rugosity from marginal to hot to traditional reef types is an assumption based on low coral cover. The coloured arrows also indicate the direction of impact with red to green indicating positive outcomes for the reef

would typically collect site specific data on benthic cover, fish and urchin abundance, and habitat or coral colony scale rugosity, and then apply production rates (e.g., coral growth rates) from previous studies. Yet, rates of production and removal are a function of the environment and, as such, we would expect different rates on different reef types as well as over time in response to seasonal changes. Two main barriers to collecting data on rates of carbonate production and removal are the investment of time and the associated cost. For example, a comprehensive assessment of coral growth rates requires a minimum of two field trips (spaced over several months) and extended time in the field per trip. Existing methodologies are also often destructive (e.g., alizarin red staining), and involve the removal of corals to quantify coral growth, which is particularly problematic for marginal reefs where coral cover is low. Similarly, an assessment of bioerosion requires multiple field excursions, but over even longer time periods (several years) to capture non-linear changes in bioerosion rates (see review by Browne et al. 2021). Yet, if we are to better understand how different reef types construct their framework in response to changes in their reef ecology and environment, it is critical that we apply more relevant production and removal rates for these reefs that span geographic ranges, environment types, as well as depth profiles.

Our meta-analysis of all reef types included here shows that net productivity is largely driven by gross carbonate production as opposed to bioerosion. This may not stand true in future years particularly for marginal reef systems with low gross carbonate production or for extreme reefs in inshore coastal waters where future declines in water quality may alter this balance. Given that corals are typically the main driver of gross carbonate production, it stands that coral growth rates are a key metric for carbonate budget assessments. To date, most coral growth studies have been conducted on a handful of coral genera (e.g., *Orbicella, Porites* and *Acropora*; Lange et al. 2020), and yet we see that for most reef types, the coral community on reefs is mixed with branching, massive, encrusting and foliose corals. Hence, to improve our carbonate budget assessments, particularly on extreme and marginal reef types, we need to not only broaden our assessment of coral growth rates within different environments but also incorporate those coral morphologies and genus with high relative abundances on non-traditional reef types.

Reef structural complexity is also a key component for calculating carbonate budgets. Here we observed significant differences in structural complexity between traditional coral reefs and extreme reefs (urban, turbid and hot), most likely due to differences in coral community composition. However, there is once again a paucity of structural complexity data on all reef types, but especially on extreme reef types, where there were only seven on urban reefs, four on turbid reefs, one on hot reefs and none on low pH/high nutrient, mangrove, and marginal reefs. It is important to note that we only assessed carbonate budget studies, not studies that may have measured rugosity for other purposes because for coral cover data to match rugosity data, these need to be measured at the same time. Another issue relating to structural complexity was the scale at which it was being assessed (and associated methods). Measures of structural complexity at all three scales (coral, habitat, reef) are useful depending on the context of the research. For the purposes of carbonate budget assessments on traditional and non-traditional reef types, we would encourage an assessment of both the habitat scale level rugosity to improve our ability to compare changes in complexity over space and time, as well as coral colony rugosity to prevent an over-estimation of coral carbonate production.

Given increasing pressures from both climate change and coastal development, it is likely that many traditional reefs will gradually transition (or have certain characteristics) to either extreme, marginal, or extreme and marginal reefs. The recent review by Schoepf et al. (2023) provided an important step towards recognising important differences between these reef types, but they also highlighted the current issues with categorising a reef into one these groups due to limited ecological AND environmental data. To improve our understanding on how these reef types, particularly extreme reefs exposed to non-optimal environmental conditions, are functioning (including carbonate production and removal) we need to be collecting environmental data in tandem with biological data. These data sets would improve our ability to define important parameters for marginal and extreme reefs as well as identify reefs at risk. This could be potentially very useful for reefs close to expanding urban centres where rapid declines in water quality could drastically impact coral cover, rates of production and bioeroder abundances, resulting in reef degradation.

16.4 Conclusions

A review of carbonate budgets on extreme and marginal reef types found that there were significant differences in carbonate production and erosion as well as structural complexity compared to traditional reef types. Specifically, we observed that although traditional reefs were characterised by healthy coral cover and mixed to branching coral communities, which resulted in higher structural complexity, net carbonate production rates were on average greater on turbid, urban and low pH/high nutrient reefs. Bioerosion rates on these reef types were either comparable or greater than on traditional reefs, but due to higher coral cover, gross production rates were also greater. In contrast, carbonate budgets on hot reefs were more comparable to marginal reefs as these reef types had the lowest average coral cover (and as such gross carbonate production) and low bioerosion rates driven by external bioeroders (e.g., parrotfish). Taken together these data highlight that despite the exposure of extreme reefs to environmental conditions outside of coral reef optimal growing conditions, many of these reef types have the capacity to maintain critical reef processes, which in relation to carbonate production and longer-term geological reef functions (i.e., reef accretion), even surpasses many traditional reefs. Lastly, we recommend the application of carbonate budgets as a means of assessing reef marginality. Here we applied a 10% coral cover threshold to categorize marginal reef systems, but perhaps a better approach would be to calculate net carbonate production. Rates that fall under 1 kg m^{-2} y^{-1}, could provide a more rigorous threshold as a means of identifying reefs that could potentially be more vulnerable to future environmental disturbance events, and therefore require active management. In contrast, extreme reefs with net production rates >2 kg m^{-2} y^{-1}, may represent important coral reef refugia in future years.

References

AlMealla RK, Edullantes B, Hepburn LJ (2023) Bleaching threatens positive carbonate budgets on Bahraini reefs. Mar Biol 171(1):39. https://doi.org/10.1007/s00227-023-04351-9

Alvarez-Filip L, Dulvy NK, Gill JA, Cote IM, Watkinson AR (2009) Flattening of Caribbean coral reefs: region-wide declines in architectural complexity. Proc Biol Sci 276(1669):3019–3025. https://doi.org/10.1098/rspb.2009.0339

Alvarez-Filip L, Dulvy NK, Côté IM, Watkinson AR, Gill JA (2011) Coral identity underpins architectural complexity on Caribbean reefs. Ecol Appl 21(6):2223–2231. https://doi.org/10.1890/10-1563.1

Alverez-Filip, L., Cote, I.,, Gill, J., Watkinson, A., & Dulvy, N. (2011). Region-wide temporal and spatial variation in Caribbean reef architecture: is coral cover the whole story? Glob Chang Biol, 17(7), 2470–2477. https://doi.org/10.1111/j.1365-2486.2010.02385.x

Aronson RB, Precht WF (2006) Conservation, precaution, and Caribbean reefs. Coral Reefs 25(3):441–450. https://doi.org/10.1007/s00338-006-0122-9

Bak R (1990) Patterns of echinoid bioerosion in two Pacific coral reef lagoons. Mar Ecol Prog Ser 66(26):272

Bellwood DR (1995) Carbonate transport and within-reef patterns of bioerosion and sediment release by parrotfishes (family Scaridae) on the Great Barrier Reef. Mar Ecol Prog Ser 117:127–136. http://www.jstor.org/stable/44634825

Bonesso JL, Browne NK, Murley M, Dee S, Cuttler MVW, Paumard V, Benson D, O'Leary M (2022) Reef to Island sediment connections within an inshore turbid reef Island system of the eastern Indian Ocean. Sediment Geol 436:106177. https://doi.org/10.1016/j.sedgeo.2022.106177

Bozec YM, Alvarez-Filip L, Mumby PJ (2015) The dynamics of architectural complexity on coral reefs under climate change. Glob Chang Biol 21(1):223–235. https://doi.org/10.1111/gcb.12698

Brown KT, Bender-Champ D, Achlatis M, van Der Zande RM, Kubicek A, Martin SB, Castro-Sanguino C, Dove SG, Hoegh-Guldberg O (2020) Habitat-specific biogenic production and erosion influences net framework and sediment coral reef carbonate budgets. Limnol Oceanogr 9999:1–17. https://doi.org/10.1002/lno.11609

Browne NK (2011) Carbonate and terrigenous sediment budgets for inshore turbid reefs on the Great Barrier Reef. Ph.D. thesis, James Cook University, Townsville, pp 83–84

Browne NK (2012) Spatial and temporal variations in coral growth on an inshore turbid reef subjected to multiple disturbances. Mar Environ Res 77:71–83. https://doi.org/10.1016/j.marenvres.2012.02.005

Browne NK, Smithers SG, Perry CT (2010) Geomorphology and community structure of Middle Reef, central Great Barrier Reef, Australia: an inner-shelf turbid zone reef subject to episodic mortality events. Coral Reefs 29(3):683–689. https://doi.org/10.1007/s00338-010-0640-3

Browne NK, Smithers SG, Perry CT (2012) Spatial and temporal variations in turbidity on two inshore turbid reefs on the Great Barrier Reef, Australia. Coral Reefs 32(1):195–210. https://doi.org/10.1007/s00338-012-0965-1

Browne NK, Smithers SG, Perry CT (2013) Carbonate and terrigenous sediment budgets for two inshore turbid reefs on the central Great Barrier Reef. Mar Geol 346:101–123. https://doi.org/10.1016/j.margeo.2013.08.011

Browne NK, Tay JK, Low J, Larson O, Todd PA (2015) Fluctuations in coral health of four common inshore reef corals in response to seasonal and anthropogenic changes in water quality. Mar Environ Res 105:39–52. https://doi.org/10.1016/j.marenvres.2015.02.002

Browne NK, Cuttler M, Moon K, Morgan K, Ross C, Castro-Sanguino C, Kennedy E, Harris D, Barnes P, Bauman A, Beetham E, Bonesso J, Bozec YM, Cornwall C, Dee S, DeCarlo T, D'Olivo J, Doropoulos C, Evans R, Eyre B, Gatenby P, Gonzalez M, Hamylton S, Hansen J, Lowe R, Mallela J, O'Leary M, Roff G, Saunders B, Zweifler A (2021) Predicting responses of geo-ecological carbonate reef systems to climate change: a conceptual model and review. Oceanogr Mar Biol Annu Rev 59:229–370

Camp EF, Nitschke MR, Rodolfo-Metalpa R, Houlbreque F, Gardner SG, Smith DJ, Zampighi M, Suggett DJ (2017) Reef-building corals thrive within hot-acidified and deoxygenated waters. Sci Rep 7(1):2434. https://doi.org/10.1038/s41598-017-02383-y

Camp EF, Edmondson J, Doheny A, Rumney J, Grima AJ, Huete A, Suggett DJ (2019) Mangrove lagoons of the Great Barrier Reef support coral populations persisting under extreme environmental conditions. Mar Ecol Prog Ser 625:1–14. https://www.int-res.com/abstracts/meps/v625/p1-14

Cartwright PJ, Browne NK, Belton D, Parnum I, O'Leary M, Valckenaere J, Fearns P, Lowe R (2023) Long-term spatial variations in turbidity and temperature provide new insights into coral-algal states on extreme/marginal reefs. Coral Reefs 42(4):859–872. https://doi.org/10.1007/s00338-023-02393-5

Castro-Sanguino C, Ortiz JC, Thompson A, Wolff NH, Ferrari R, Robson B, Magno-Canto MM, Puotinen M, Fabricius KE, Uthicke S (2021) Reef state and performance as indicators of cumulative impacts on coral reefs. Ecol Indic 123:107335. https://doi.org/10.1016/j.ecolind.2020.107335

Chazottes V, Campion-Alsumard TL, Peyrot-Clausade M (1995) Bioerosion rates on coral reefs: interactions between macroborers, microborers and grazers (Moorea, French Polynesia). Palaeogeogr Palaeoclimatol Palaeoecol 113(2–4):189–198. http://www.sciencedirect.com/science/article/B6V6R-3XWRS81-16/2/5377d5962b9606752c2ed27e60f8db24

Chazottes V, Le Campion-Alsumard T, Peyrot-Clausade M, Cuet P (2002) The effects of eutrophication-related alterations to coral reef communities on agents and rates of bioerosion (Reunion Island, Indian Ocean). Coral Reefs 21(4):375–390. https://doi.org/10.1007/s00338-002-0259-0

Chen T, Li S, Yu K (2013) Macrobioerosion in Porites corals in subtropical northern South China Sea: a limiting factor for high-latitude reef framework development. Coral Reefs 32(1):101–108. https://doi.org/10.1007/s00338-012-0946-4

Comeau S, Cornwall CE, DeCarlo TM, Krieger E, McCulloch MT (2018) Similar controls on calcification under ocean acidification across unrelated coral reef taxa. Glob Chang Biol 24(10):4857–4868. https://doi.org/10.1111/gcb.14379

Cornwall CE, Comeau S, Kornder NA, Perry CT, van Hooidonk R, DeCarlo TM, Pratchett MS, Anderson KD, Browne N, Carpenter R, Diaz-Pulido G, D'Olivo JP, Doo SS, Figueiredo J, Fortunato SAV, Kennedy E, Lantz CA, McCulloch MT, González-Rivero M, Schoepf V, Smithers SG, Lowe RJ (2021) Global declines in coral reef calcium carbonate production under ocean acidification and warming. Proc Natl Acad Sci 118(21):e2015265118. https://doi.org/10.1073/pnas.2015265118

Courtney TA, Barkley HC, Chan S, Couch CS, Kindinger TL, Oliver TA, Kriegman DJ, Andersson AJ (2022) Rapid assessments of Pacific Ocean net coral reef carbonate budgets and net calcification following the 2014–2017 global coral bleaching event. Limnol Oceanogr 67(8):1687–1700. https://doi.org/10.1002/lno.12159

Cuttler MV, Hansen JE, Lowe RJ, Trotter JA, McCulloch MT (2019) Source and supply of sediment to a shoreline salient in a fringing reef environment. Earth Surf Process Landf 44(2):552–564. https://doi.org/10.1002/esp.4516

Dawson JL, Smithers SG, Hua Q (2014) The importance of large benthic foraminifera to reef Island sediment budget and dynamics at Raine Island, northern Great Barrier Reef. Geomorphology 222:68–81. https://doi.org/10.1016/j.geomorph.2014.03.023

De Bakker DM, Meesters EH, Bak RPM, Nieuwland G, Van Duyl FC (2016) Long-term shifts in coral communities on shallow to deep reef slopes of Curaçao and Bonaire: are there any winners? [Original Research]. Front Mar Sci 3:247. https://doi.org/10.3389/fmars.2016.00247

De Bakker DM, Webb AE, van den Bogaart LA, van Heuven SM, Meesters EH, van Duyl FC (2018) Quantification of chemical and mechanical bioerosion rates of six Caribbean excavating sponge species found on the coral reefs of Curaçao. PLoS One 13(5):e0197824. https://doi.org/10.1371/journal.pone.0197824

De Bakker DM, van Duyl FC, Perry CT, Meesters EH (2019) Extreme spatial heterogeneity in carbonate accretion potential on a Caribbean fringing reef linked to local human disturbance gradients. Glob Chang Biol 25(12):4092–4104. https://doi.org/10.1111/gcb.14800

DeCarlo TM, Cohen AL, Barkley HC, Cobban Q, Young C, Shamberger KE, Brainard RE, Golbuu Y (2015) Coral macrobioerosion is accelerated by ocean acidification and nutrients. Geology 43(1):7–10. https://doi.org/10.1130/g36147.1

Dee S, Cuttler MVW, O'Leary M, Hacker J, Browne NK (2020) The complexity of calculating an accurate carbonate budget. Coral Reefs 39:1525–1534. https://doi.org/10.1007/s00338-020-01982-y

Dee S, Cuttler M, Cartwright P, McIlwain J, Browne N (2021) Encrusters maintain stable carbonate production despite temperature anomalies among two inshore Island reefs of the Pilbara, Western Australia. Mar Environ Res 169:105386. https://doi.org/10.1016/j.marenvres.2021.105386

Dee S, DeCarlo T, Lozić I, Nilsen J, Browne NK (2023) Low bioerosion rates on inshore turbid reefs of Western Australia. Diversity 15(1):62. https://www.mdpi.com/1424-2818/15/1/62

Dee S, Zwifler A, Cuttler M, Nilsen J, O'Leary M, Browne NK (2024) The application of carbonate and sediment budgets to assess the stability of marginal reef systems. Mar Geol 473:107324. https://doi.org/10.1016/j.margeo.2024.107324

Divan Patel F, Pinto W, Dey M, Alcoverro T, Arthur R (2023) Carbonate budgets in Lakshadweep Archipelago bear the signature of local impacts and global climate disturbances. Coral Reefs 42(3):729–742. https://doi.org/10.1007/s00338-023-02374-8

Done TJ (1992) Phase shifts in coral reef communities and their ecological significance. Hydrobiologia 247:121–132. https://doi.org/10.1007/BF00008211

Eakin CM (1996) Where have all the carbonates gone? A model comparison of calcium carbonate budgets before and after the 1982–1983 El Nino at Uva Island in the eastern Pacific. Coral Reefs, 15(2):109–119

Eakin CM (2001) Tale of two ENSO events: carbonate budgets and the influence of two warming disturbances and intervening variability, Uva Island. Bull Mar Sci 69(1):171–186

Edinger EN, Risk MJ (2000) Effects of land-based pollution on Central Java coral reefs. J Coast Res 3:593–613

Enochs IC, Manzello DP, Carlton RD, Graham DM, Ruzicka R, Colella MA (2015) Ocean acidification enhances the bioerosion of a common coral reef sponge: implications for the persistence of the Florida Reef Tract. Bull Mar Sci 91(2):271–290. https://doi.org/10.5343/bms.2014.1045

Enochs IC, Manzello DP, Kolodziej G, Noonan SHC, Valentino L, Fabricius KE (2016) Enhanced macroboring and depressed calcification drive net dissolution at high-CO_2 coral reefs. Proc R Soc B Biol Sci 283(1842):20161742. https://doi.org/10.1098/rspb.2016.1742

Fabricius KE, Langdon C, Uthicke S, Humphrey C, Noonan S, De'ath G, Okazaki R, Muehllehner N, Glas MS, Lough JM (2011) Losers and winners in coral reefs acclimatized to elevated carbon dioxide concentrations. Nat Clim Chang 1(3):165–169. https://doi.org/10.1038/nclimate1122

Ferrario F, Beck MW, Storlazzi CD, Micheli F, Shepard CC, Airoldi L (2014) The effectiveness of coral reefs for coastal hazard risk reduction and adaptation. Nat Commun 5(1):1–9

Freile, D., Lewis, R. D., & Panuska, B. C. (2004) Carbonate productivity rates of Halimeda in two different locations, San Salvador Island, Bahamas. In Proceedings of the 11th Symposium on the Geology of the Bahamas and Other Carbonate Regions (pp. 95–106). Gerace Research Centre, Auburn University

Gilmour JP, Smith LD, Heyward AJ, Baird AH, Pratchett MS (2013) Recovery of an isolated coral reef system following severe disturbance. Science 340(6128):69–71. https://doi.org/10.1126/science.1232310

Glynn PW (1997) Bioerosion and coral-reef growth: a dynamic balance. In: Birkeland C (ed) Life and death of coral reefs. Chapman & Hall, pp 68–95

Glynn PW, Wellington GM, Birkeland C (1979) Coral reef growth in the Galapagos: limitation by sea urchins. Science 203(4375):47–49. https://doi.org/10.1126/science.203.4375.47

Glynn, P. W. (1988) El Niño warming, coral mortality and reef framework destruction by echinoid bioerosion in the eastern Pacific. Galaxea, 7(2):129–160

Goh TZY, Bauman AG, Januchowski-Hartley FA, Morgan KM, Seah JCL, Todd PA (2021) Growth and carbonate production of crustose coralline algae on a degraded turbid reef system. Mar Pollut Bull 173:113135. https://doi.org/10.1016/j.marpolbul.2021.113135

González-Barrios FJ, Cabral-Tena RA, Alvarez-Filip L (2021) Recovery disparity between coral cover and the physical functionality of reefs with impaired coral assemblages. Glob Chang Biol 27(3):640–651. https://doi.org/10.1111/gcb.15431

Gordon SE, Goatley CHR, Bellwood DR (2016) Low-quality sediments deter grazing by the parrotfish *Scarus rivulatus* on inner-shelf reefs. Coral Reefs 35(1):285–291. https://doi.org/10.1007/s00338-015-1374-z

Graham NA, Nash KL (2013) The importance of structural complexity in coral reef ecosystems. Coral Reefs 32(2):315–326. https://doi.org/10.1007/s00338-012-0984-y

Grigg RW (1982) Darwin point: a threshold for atoll formation. Coral Reefs 1(1):29–34. https://doi.org/10.1007/BF00286537

Guest JR, Tun K, Low J, Vergés A, Marzinelli EM, Campbell AH et al (2016) 27 years of benthic and coral community dynamics on turbid, highly urbanised reefs off Singapore. Sci Rep 6(1):36260. https://doi.org/10.1038/srep36260

Guest J, Baria-Rodriguez MV, Toh TC, dela Cruz D, Vicentuan K, Gomez E, Villanueva R, Steinberg P, Edwards A (2023) Live slow, die old: larval propagation of slow-growing, stress-tolerant corals for reef restoration. Coral Reefs 42(6):1365–1377. https://doi.org/10.1007/s00338-023-02440-1

Hallock P (1988) The role of nutrient availability in bioerosion: consequences to carbonate buildups. Palaeogeogr Palaeoclimatol Palaeoecol 63(1–3):275–291. http://www.sciencedirect.com/science/article/B6V6R-48F01MR-11C/2/8de58dc041b0d24c0db010da6486e52e

Hamylton SM, Mallela J (2019) Reef development on a remote coral atoll before and after coral bleaching: a geospatial assessment. Mar Geol 418. https://doi.org/10.1016/j.margeo.2019.106041

Hamylton SM, Silverman J, Shaw E (2013) The use of remote sensing to scale up measures of carbonate production on reef systems: a comparison of hydrochemical and census-based estimation methods. Int J Remote Sens 34(18):6451–6465. https://doi.org/10.1080/01431161.2013.800654

Harney JN, Fletcher CH (2003) A budget of carbonate framework and sediment production, Kailua Bay, Oahu, Hawaii. J Sediment Res 73(6):856–868. https://doi.org/10.1306/051503730856

Harris DL, Rovere A, Casella E, Power H, Canavesio R, Collin A, Pomeroy A, Webster JM, Parravicini V (2018) Coral reef structural complexity provides important coastal protection from waves under rising sea levels. Sci Adv 4(2):eaao4350. https://doi.org/10.1126/sciadv.aao4350

Harris DL, Webster JM, Vila-Concejo A, Duce S, Leon JX, Hacker J (2023) Defining multi-scale surface roughness of a coral reef using a high-resolution LiDAR digital elevation model. Geomorphology 439:108852. https://doi.org/10.1016/j.geomorph.2023.108852

Hart DE, Kench PS (2007) Carbonate production of an emergent reef platform, Warraber Island, Torres Strait, Australia. Coral Reefs 26(1):53–68. https://doi.org/10.1007/s00338-006-0168-8

Heery EC, Hoeksema BW, Browne NK, Reimer JD, Ang PO, Huang D, Friess DA, Chou LM, Loke LHL, Saksena-Taylor P, Alsagoff N, Yeemin T, Sutthacheep M, Vo ST, Bos AR, Gumanao GS, Syed Hussein MA, Waheed Z, Lane DJW, Johan O, Kunzmann A, Jompa J, Suharsono, Taira D, Bauman AG, Todd PA (2018) Urban coral reefs: degradation and resilience of hard coral assemblages in coastal cities of East and Southeast Asia. Mar Pollut Bull 135:654–681. https://doi.org/10.1016/j.marpolbul.2018.07.041

Heiss GA (1995) Carbonate production by scleractinian corals at Aqaba, Gulf of Aqaba, Red Sea. Facies 33(1):19–34. https://doi.org/10.1007/BF02537443

Herrán N, Narayan GR, Reymond CE, Westphal H (2017) Calcium carbonate production, coral cover and diversity along a distance gradient from stone town: a case study from Zanzibar, Tanzania. Front Mar Sci 4. https://doi.org/10.3389/fmars.2017.00412

Houlbreque F, Ferrier-Pages C (2009) Heterotrophy in tropical scleractinian corals. Biol Rev Camb Philos Soc 84(1):1–17. https://doi.org/10.1111/j.1469-185X.2008.00058.x

Hubbard DK, Miller AI, Scaturo D (1990) Production and cycling of calcium carbonate budgets in a shelf-edge reef ecosystem (St Croix, U.S Virgin Islands): applications to the nature of reef systems in the fossil record. J Sediment Petrol 60(3):336–360

Hughes TP (1994) Catastrophes phase shifts and large scale degradation of a Caribbean coral reefs. Science 265(5178):1547–1551. https://doi.org/10.1126/science.265.5178.1547

Husband E, Perry CT, Lange ID (2022) Estimating rates of coral carbonate production from aerial and archive imagery by applying colony scale conversion metrics. Coral Reefs 41(4):1199–1209. https://doi.org/10.1007/s00338-022-02247-6

Hutchings P (1986) Biological destruction of coral reefs. Coral Reefs 4:239–252. https://doi.org/10.1007/BF00298083

Jackson JBC, Kirby MX, Berger WH, Bjorndal KA, Botsford LW, Bourque BJ, Bradbury RH, Cooke R, Jon E, Estes JA, Hughes TP, Kidwell S, Lange CB, Lenihan HS, Pandolfi JM, Peterson CH, Steneck RS, Tegner MJ, Warner RR (2001) Historical overfishing and the recent collapse of coastal ecosystems. Science 293(5530):629–638. http://www.jstor.org/stable/3084305

Januchowski-Hartley FA, Graham NAJ, Wilson SK, Jennings S, T. PC (2017) Drivers and predictions of coral reef carbonate budget trajectories. Proc R Soc B 284:2016253320162533. https://doi.org/10.1098/rspb.2016.2533

Januchowski-Hartley FA, Bauman AG, Morgan KM, Seah JCL, Huang D, Todd PA (2020) Accreting coral reefs in a highly urbanized environment. Coral Reefs 39(3):717–731. https://doi.org/10.1007/s00338-020-01953-3

Kench PS, Beetham EP, Turner T, Morgan KM, Owen SD, McLean RF (2022) Sustained coral reef growth in the critical wave dissipation zone of a Maldivian atoll. Commun Earth Environ 3(1):9. https://doi.org/10.1038/s43247-021-00338-w

Kleypas JA, McManus JW, Meñez LA (1999) Environmental limits to coral reef development: where do we draw the line? Am Zool 39(1):146–159. https://doi.org/10.1093/icb/39.1.146

Land LS (1979) The fate of reef-derived sediment on the north Jamaican island slope. Mar Geol 29(1–4):55–71. https://doi.org/10.1016/0025-3227(79)90102-6. Get rights and content

Lange ID, Perry CT (2019) Bleaching impacts on carbonate production in the Chagos Archipelago: influence of functional coral groups on carbonate budget trajectories. Coral Reefs 38(4):619–624. https://doi.org/10.1007/s00338-019-01784-x

Lange ID, Perry CT, Alvarez-Filip L (2020) Carbonate budgets as indicators of functional reef "health": a critical review of data underpinning census-based methods and current knowledge gaps. Ecol Indic 110. https://doi.org/10.1016/j.ecolind.2019.105857

Lange ID, Perry CT, Stuhr M (2023a) Quantifying reef-derived sediment generation: introducing the SedBudget methodology to support tropical coastline and Island vulnerability studies. Camb Prism Coast Futur 1:e26. Article e26. https://doi.org/10.1017/cft.2023.14

Lange ID, Perry CT, Stuhr M (2023b) Recovery trends of reef carbonate budgets at remote coral atolls 6 years post-bleaching. Limnol Oceanogr 68(S1):S8–S22. https://doi.org/10.1002/lno.12066

Lough JM, Barnes DJ (1992) Comparisons of skeletal density in Porites from the central Great Barrier Reef. J Exp Mar Biol Ecol 155:1–25. https://doi.org/10.1016/0022-0981(92)90024-5

Lough JM, Barnes DJ (2000) Environmental controls on growth of the massive coral Porites. J Exp Mar Biol Ecol 245(2):225–243. http://www.sciencedirect.com/science/article/B6T8F-3YN92RJ-6/2/a481dd35ee646891a865ac642988bad3

Mallela J, Perry CT (2007) Calcium carbonate budgets for two coral reefs affected by different terrestrial runoff regimes, Rio Bueno, Jamaica. Coral Reefs 26(1):129–145. https://doi.org/10.1007/s00338-006-0169-7

Manzello DP, Kleypas JA, Budd DA, Eakin CM, Glynn PW, Langdon C (2008) Poorly cemented coral reefs of the eastern tropical Pacific: possible insights into reef development in a high-CO2 world. Proc Natl Acad Sci 105(30):10450–10455. https://doi.org/10.1073/pnas.0712167105

Manzello DP, Enochs IC, Kolodziej G, Carlton R, Valentino L (2018) Resilience in carbonate production despite three coral bleaching events in 5 years on an inshore patch reef in the Florida keys. Mar Biol 165(6):99. https://doi.org/10.1007/s00227-018-3354-7

Marubini F, Atkinson MJ (1999) Effects of lowered pH and elevated nitrate on coral calcification. Mar Ecol Prog Ser 188:117–121. https://www.int-res.com/abstracts/meps/v188/p117-121

Masse JP, Thomassin BA, Acquaviva M (1989) Bioclastic sedimentary environments of coral reefs and lagoon around Mayotte island (Comoro archipelago, Mozambique channel, SW Indian Ocean). J Coast Res 419–432

McCowan DM, Pratchett MS, Baird AH (2012) Bleaching susceptibility and mortality among corals with differing growth forms. In: Proceedings of the 12th International Coral Reef Symposium, 9–13. Cairns, QLD, Australia, pp. 1–6. [1, 2]

Milliman JD (1993) Production and accumulation of calcium carbonate in the ocean: budget of a nonsteady state. Glob Biogeochem Cycles 7(4):927–957. https://doi.org/10.1029/93GB02524

Molina-Hernández A, González-Barrios FJ, Perry CT, Álvarez-Filip L (2020) Two decades of carbonate budget change on shifted coral reef assemblages: are these reefs being locked into low net budget states? Proc R Soc B Biol Sci 287:1940. https://doi.org/10.1098/rspb.2020.2305

Molina-Hernández A, Medellín-Maldonado F, Lange ID, Perry CT, Álvarez-Filip L (2022) Coral reef erosion: in situ measurement on different dead coral substrates on a Caribbean reef. Limnol Oceanogr 67(12):2734–2749. https://doi.org/10.1002/lno.12234

Morgan KM (2014) A calcium carbonate budget of a Maldivian reef platform. PhD thesis, University of Auckland

Morgan KM, Kench PS (2012) Skeletal extension and calcification of reef-building corals in the Central Indian Ocean. Mar Environ Res 81:78–82. https://doi.org/10.1016/j.marenvres.2012.08.001

Morgan KM, Kench PS (2014) A detrital sediment budget of a Maldivian reef platform. Geomorphology 222:122–131. https://doi.org/10.1016/j.geomorph.2014.02.013

Morgan KM, Kench PS (2016a) Reef to Island sediment connections on a Maldivian carbonate platform: using benthic ecology and biosedimentary depositional facies to examine Island building potential. Earth Surf Process Landf 41(13):18151825. https://doi.org/10.1002/esp.3946

Morgan KM, Kench PS (2016b) Parrotfish erosion underpins reef growth, sand talus development and Island building in the Maldives. Sediment Geol 341:50–57. https://doi.org/10.1016/j.sedgeo.2016.05.011

Morgan KM, Perry CT, Smithers SG, Johnson JA, Daniell JJ (2016) Evidence of extensive reef development and high coral cover in nearshore environments: implications for understanding coral adaptation in turbid settings. Sci Rep 6(1):29616. https://doi.org/10.1038/srep29616

Morgan KM, Perry CT, Johnson JA, Smithers SG (2017) Nearshore turbid-zone corals exhibit high bleaching tolerance on the Great Barrier Reef following the 2016 ocean warming event. Front Mar Sci 4:224. https://doi.org/10.3389/fmars.2017.00224

Morgan KM, Moynihan MA, Sanwlani N, Switzer AD (2020) Light limitation and depth variable sedimentation drives vertical reef compression on turbid coral reefs. Front Mar Sci 7:71256. https://doi.org/10.3389/fmars.2020.571256

Morris JT, Enochs IC, Besemer N, Viehman TS, Groves SH, Blondeau J, Ames C, Towle EK, Grove LJW, Manzello DP (2022) Low net carbonate accretion characterizes Florida's coral reef. Sci Rep 12(1):19582. https://doi.org/10.1038/s41598-022-23394-4

Moustaka M, Langlois TJ, McLean D, Bond T, Fisher R, Fearns P, Dorji P, Evans RD (2018) The effects of suspended sediment on coral reef fish assemblages and feeding guilds of north-west Australia. Coral Reefs 37(3):659–673. https://doi.org/10.1007/s00338-018-1690-1

Muehllehner N, Langdon C, Venti A, Kadko D (2016) Dynamics of carbonate chemistry, production, and calcification of the Florida Reef Tract (2009–2010): evidence for seasonal dissolution. Glob Biogeochem Cycles 30(5):661–688. https://doi.org/10.1002/2015GB005327

Multer HG, Clavijo I (2004) Halimeda investigations: progress and problems. NOAA/RSMAS 2004:116–127

Nuñez Lendo CI, Suggett DJ, Boote C, McArdle A, Nicholson F, Fisher EE, Smith D, Camp EF (2024) Carbonate budgets induced by coral restoration of a Great Barrier Reef site following cyclone damage [Original Research]. Front Mar Sci 10. https://doi.org/10.3389/fmars.2023.1298411

Palmer SE, Perry CT, Smithers SG, Gulliver P (2010) Internal structure and accretionary history of a nearshore, turbid-zone coral reef: Paluma Shoals, central Great Barrier Reef, Australia. Mar Geol 276(1–4):14–29. https://doi.org/10.1016/j.margeo.2010.07.002

Perry CT, Harborne AR (2016) Bioerosion on modern reefs: impacts and responses under changing ecological and environmental conditions. In: Hubbard D, Rogers C, Lipps J, Stanley JG (eds) Coral reefs at the crossroads. Coral reefs of the world, vol 6. Springer, Dordrecht, pp 69–101

Perry CT, Morgan KM (2017) Bleaching drives collapse in reef carbonate budgets and reef growth potential on southern Maldives reefs. Sci Rep 7:40581. https://doi.org/10.1038/srep40581

Perry CT, Spencer T, Kench PS (2008) Carbonate budgets and reef production states: a geomorphic perspective on the ecological phase shift concept. Coral Reefs 27(4):853–866. https://doi.org/10.1007/s00338-008-0418-z

Perry CT, Edinger EN, Kench PS, Murphy GN, Smithers SG, Steneck RS, Mumby PJ (2012) Estimating rates of biologically driven coral reef framework production and erosion: a new census-based carbonate budget methodology and applications to the reefs of Bonaire. Coral Reefs 31(3):853–868. https://doi.org/10.1007/s00338-012-0901-4

Perry CT, Murphy GN, Kench PS, Smithers SG, Edinger EN, Steneck RS, Mumby PJ (2013) Caribbean-wide decline in carbonate production threatens coral reef growth. Nat Commun 4:1402. https://doi.org/10.1038/ncomms2409

Perry CT, Kench PS, O'Leary MJ, Morgan KM, Januchowski-Hartley F (2015a) Linking reef ecology to Island building: parrotfish identified as major producers of Island-building sediment in the Maldives. Geology 43(6):503–506. https://doi.org/10.1130/G36623.1

Perry CT, Murphy GN, Graham NA, Wilson SK, Januchowski-Hartley FA, East HK (2015b) Remote coral reefs can sustain high growth potential and may match future sea-level trends. Sci Rep 5:18289. https://doi.org/10.1038/srep18289

Perry CT, Morgan KM, Salter MA (2016) Sediment generation by *Halimeda* on atoll interior coral reefs of the southern Maldives: a census-based approach for estimating carbonate production by calcareous green algae. Sediment Geol 346:17–24. https://doi.org/10.1016/j.sedgeo.2016.10.005

Perry CT, Morgan KM, Yarlett RT (2017) Reef habitat type and spatial extent as interacting controls on platform-scale carbonate budgets. Front Mar Sci 4:185. https://doi.org/10.3389/fmars.2017.00185

Perry CT, Alvarez-Filip L, Graham N (2018a) Changing geo-ecological functions of coral reefs in the Anthropocene. Funct Ecol 33:976–988. https://doi.org/10.1111/1365-2435.13247

Perry CT, Alvarez-Filip L, Graham NAJ, Mumby PJ, Wilson SK, Kench PS, Manzello DP, Morgan KM, Slangen ABA, Thomson DP, Januchowski-Hartley F, Smithers SG, Steneck RS, Carlton R, Edinger EN, Enochs IC, Estrada-Saldivar N, Haywood MDE, Kolodziej G, Murphy GN, Perez-Cervantes E, Suchley A, Valentino L, Boenish R, Wilson M, Macdonald C (2018b) Loss of coral reef growth capacity to track future increases in sea level. Nature 558(7710):396–400. https://doi.org/10.1038/s41586-018-0194-z

Perry CT, Salter MA, Morgan KM, Harborne AR (2019) Census estimates of algal and epiphytic carbonate production highlight tropical seagrass meadows as sediment production hotspots. Front Mar Sci 6:120. https://doi.org/10.3389/fmars.2019.00120

Prouty NG, Cohen A, Yates KK, Storlazzi CD, Swarzenski PW, White D (2017) Vulnerability of coral reefs to bioerosion from land-based sources of pollution. J Geophys Res Oceans 122(12):9319–9331. https://doi.org/10.1002/2017jc013264

Randi CB, Becker AC, Willemes MJ, Perry CT, Salgado LT, Tomazetto de Carvalho R, Motta FS, Leão de Moura R, Coreixas de Moraes F, Pereira-Filho GH (2021) Calcium carbonate production in the southernmost subtropical Atlantic coral reef. Mar Environ Res 172:105490. https://doi.org/10.1016/j.marenvres.2021.105490

Risk MJ (1972) Fish diversity on a coral reef in the Virgin Islands. Atoll Res Bull 153:1–4

Rodriguez-Ruano V, Toth LT, Enochs IC, Randall CJ, Aronson RB (2023) Upwelling, climate change, and the shifting geography of coral reef development. Sci Rep 13(1):1770. https://doi.org/10.1038/s41598-023-28489-0

Roik A, Roder C, Röthig T, Voolstra CR (2015) Spatial and seasonal reef calcification in corals and calcareous crusts in the central Red Sea. Coral Reefs 35(2):681–693. https://doi.org/10.1007/s00338-015-1383-y

Roik A, Röthig T, Pogoreutz C, Saderne V, Voolstra CR (2018) Coral reef carbonate budgets and ecological drivers in the central Red Sea – a naturally high temperature and high total alkalinity environment. Biogeosciences 15(20):6277–6296. https://doi.org/10.5194/bg-15-6277-2018

Rubin ET, Enochs IC, Foord C, Mayfield AB, Kolodziej G, Basden I, Manzello DP (2021) Molecular mechanisms of coral persistence within highly urbanized locations in the Port of Miami, Florida. Front Mar Sci 8:695236. https://doi.org/10.3389/fmars.2021.695236

Ryan EJ, Hanmer K, Kench PS (2019) Massive corals maintain a positive carbonate budget of a Maldivian upper reef platform despite major bleaching event. Sci Rep 9(1):6515. https://doi.org/10.1038/s41598-019-42985-2

Sadd JL (1984) Sediment transport and $CaCO_3$ budget on a fringing reef, Cane Bay, St. Croix, US Virgin Islands. Bull Mar Sci 35(2):221–238

Schoepf V, Baumann JH, Barshis DJ, Browne NK, Camp EF, Comeau S, Cornwall CE, Guzmán HM, Riegl B, Rodolfo-Metalpa R, Sommer B (2023) Corals at the edge of environmental limits: a new conceptual framework to re-define marginal and extreme coral communities. Sci Total Environ 884:163688. https://doi.org/10.1016/j.scitotenv.2023.163688

Scoffin TP, Stearn CW, Boucher D, Frydl P, Hawkins CM, Hunter IG, MacGeachy JK (1980) Calcium carbonate budget of a fringing reef on the west coast of Barbados. Bull Mar Sci 30(2):475–508

Shamberger KEF, Cohen AL, Golbuu Y, McCorkle DC, Lentz SJ, Barkley HC (2014) Diverse coral communities in naturally acidified waters of a Western Pacific reef. Geophys Res Lett 41(2):499–504. https://doi.org/10.1002/2013gl058489

Shantz AA, López-Londoño T, Gomez-Campo K, Iglesias-Prieto R, López-Victoria M, Medina M (2023) Reassessing the role of herbivores on urban coral reefs: a case study from a heavily impacted reef near Cartagena Bay, Colombia. Urban Ecosyst:1–9. https://doi.org/10.1007/s11252-023-01463-4

Silbiger NJ, Donahue MJ, Brainard RE (2017) Environmental drivers of coral reef carbonate production and bioerosion: a multi-scale analysis. Ecology 98(10):2547–2560. https://doi.org/10.1002/ecy.1946

Stearn, C. W., Scoffin, T. P., & Martindale, W. (1977) Calcium carbonate budget of a fringing reef on the west coast of Barbados—Part I: Zonation and productivity. Bulletin of Marine Science, 27(3), 479–510

Storlazzi CD, Dartnell P, Hatcher GA, Gibbs AE (2016) End of the chain? Rugosity and fine-scale bathymetry from existing underwater digital imagery using structure-from-motion (SfM) technology. Coral Reefs 35(3):889–894. https://doi.org/10.1007/s00338-016-1462-8

Tanvet C, Camp EF, Sutton J, Houlbrèque F, Thouzeau G, Rodolfo-Metalpa R (2023) Corals adapted to extreme and fluctuating seawater pH increase calcification rates and have unique symbiont communities. Ecol Evol 13(5):e10099. https://doi.org/10.1002/ece3.10099

Tebbett SB, Bellwood DR, Purcell SW (2018) Sediment addition drives declines in algal turf yield to herbivorous coral reef fishes: implications for reefs and reef fisheries. Coral Reefs 37(3):929–937. https://doi.org/10.1007/s00338-018-1718-6

Todd PA (2008) Morphological plasticity in scleractinian corals. Biol Rev Camb Philos Soc 83(3):315–337. https://doi.org/10.1111/j.1469-185x.2008.00045.x

Toth LT, Courtney TA, Colella MA, Kupfner Johnson SA, Ruzicka RR (2022) The past, present, and future of coral reef growth in the Florida keys. Glob Chang Biol 28(17):5294–5309. https://doi.org/10.1111/gcb.16295

Toth LT, Storlazzi CD, Kuffner IB, Quataert E, Reyns J, McCall R, Stathakopoulos A, Hillis-Starr Z, Holloway NH, Ewen KA, Pollock CG, Code T, Aronson RB (2023) The potential for coral reef restoration to mitigate coastal flooding as sea levels rise. Nat Commun 14(1):2313. https://doi.org/10.1038/s41467-023-37858-2

Tribollet A, Golubic S (2005) Cross-shelf differences in the pattern and pace of bioerosion of experimental carbonate substrates exposed for 3 years on the northern Great Barrier Reef, Australia. Coral Reefs 24(3):422–434. https://doi.org/10.1007/s00338-005-0003-7

Tribollet A, Golubic S (2011) Reef bioerosion: agents and processes. In: Dubinsky Z, Stambler N (eds) Coral reefs: an ecosystem in transition. Springer, pp 435–449. https://doi.org/10.1007/978-94-007-0114-4_25

van Woesik R, Cacciapaglia CW (2018) Keeping up with sea-level rise: carbonate production rates in Palau and Yap, western Pacific Ocean. PLoS One 13(5):e0197077. https://doi.org/10.1371/journal.pone.0197077

Wenger AS, Williamson DH, da Silva ET, Ceccarelli DM, Browne NK, Petus C, Devlin MJ (2016) Effects of reduced water quality on coral reefs in and out of no-take marine reserves. Conserv Biol 30(1):142–153. https://doi.org/10.1111/cobi.12576

Wisshak M, Schönberg CHL, Form A, Freiwald A (2012) Ocean acidification accelerates reef bioerosion. PLoS One 7(9):e45124–e45124. https://doi.org/10.1371/journal.pone.0045124

Yamano H, Miyajima T, Koike I (2000) Importance of foraminifera for the formation and maintenance of a coral sand cay: Green Island, Australia. Coral Reefs 19(1):51–58. https://doi.org/10.1007/s003380050226

Yamano H, Sugihara K, Watanabe T, Shimamura M, Hyeong K (2012) Coral reefs at 34 N, Japan: exploring the end of environmental gradients. Geology 40(9):835–838. https://doi.org/10.1130/G33293.1

Zweifler A, O'Leary M, Morgan K, Browne NK (2021) Turbid coral reefs: past, present and future—a review. Diversity 13(6):251. https://doi.org/10.3390/d13060251

Mechanisms of Coral Persistence in Extreme and Marginal Coral Environments and the Implications for Conservation

17

Emma F. Camp, Ariel K. Pezner, and Ian Enochs

Abstract

Tropical corals are typically found within a narrow range of environmental conditions, explaining their geographically constrained location around the equator. However, corals can be found persisting, and sometimes thriving, in abiotic conditions typically considered unfavorable for coral growth and survival. Examples include the warm waters of the Persian Arabian Gulf, cooler low light environments of deeper reefs, acidic waters of carbon dioxide vents, warm, acidic, and low oxygen waters found in mangrove lagoons, and the urban waterways of major cities. These suboptimal environments can be considered extreme and/or marginal coral systems, which provide unique natural laboratories to study the mechanisms that support coral survival in hostile environmental conditions. Interest in understanding coral tolerance is accelerating as corals have to adapt to anthropogenic environmental climate change. In this Chapter we consider key environmental influences that can impact coral survival in extreme and marginal environments. We next examine ecological, physiological, and morphological mechanisms that can support coral survival in suboptimal environments. We conclude the Chapter by considering how the knowledge on which mechanisms support coral persistence in extreme and/or marginal coral environments can be utilised to support the broader conservation and management of coral reefs.

E. F. Camp (✉) · A. K. Pezner
Climate Change Cluster, University of Technology Sydney, Sydney, NSW, Australia
e-mail: emma.camp@uts.edu.au

I. Enochs
Atlantic Oceanographic and Meteorological Laboratory, National Oceanic and Atmospheric Administration, Miami, FL, USA

Keywords

Adaptation · Acclimation · Ecology · Morphology · Physiology · Reef

17.1 Introduction

Extreme and marginal coral environments house corals living at the edge of their environmental tolerance limits (Kleypas et al. 1999; Perry and Larcombe 2003; Schoepf et al. 2023). These unique systems create a natural laboratory to assess coral performance under suboptimal environmental conditions (Camp et al. 2016a, b; Maggioni et al. 2021). Some extreme and marginal habitats have been described as coral refugia (Randall et al. 2020; Yates et al. 2014) (see previous chapters), maintaining favorable conditions being lost elsewhere (*sensu* Keppel et al. 2012). Whilst other extreme and marginal coral habitats can house stress-hardened coral populations that are potentially valuable for active reef interventions (Camp 2022). Resolving the mechanisms that facilitate coral survival in suboptimal environmental conditions is required to best integrate extreme and marginal coral environments into effective reef management practices (Camp et al. 2018b). Such knowledge has never been more critical as the impacts of anthropogenic climate change continue to intensify, resulting in more frequent, long-lasting, and severe stress events on reefs globally (Pezner et al. 2023; Reimer et al. 2024).

Stress disrupts the homeostasis of the coral holobiont (animal host, their algal symbiont Symbiodiniaceae, and other microorganisms), leading to damage, decreased fitness, and potential death (Oakley and Davy 2018). Stress is ubiquitous across reef systems and corals and their symbiotic partners have developed a suite of sophisticated mechanisms to manage stress. Corals residing in extreme and marginal habitats are presented with distinct challenges, as they can often be exposed to stress for long durations (e.g., exposure to low pH at carbon dioxide vents; Chaps. 7 and 8) and/or periods of extreme stress (e.g., extreme low oxygen levels at

M. O. Soares et al. (eds.), *Extreme and Marginal Coral Reefs and Communities*, Coral Reefs of the World 24,
https://doi.org/10.1007/978-3-032-28035-0_17

mangrove systems; Chap. 11). In this Chapter, we briefly discuss key environmental influences that can impact coral survival in extreme and marginal environments. We next examine ecological, physiological, and morphological mechanisms that can support coral survival in these suboptimal environments. We consider known molecular changes underpinning these mechanisms to shed light on whether processes are acclimation and/or adaptation. We conclude the Chapter by considering how the knowledge on which mechanisms support coral persistence in these suboptimal environments can impact current and future conservation and management of reef and reef associated systems.

17.2 Environmental Influence on Resident Taxa

Extreme and marginal coral environments are heterogeneous (Camp et al. 2018a, b). No two habitats are the same, and a single system is typically dynamic over both space and time. Consequently, resident taxa have to be able to deal with both the absolute environmental conditions as well as the dynamic range in abiotic conditions, including periodic exposure to extremes (Fig. 17.1). Capturing the dynamic nature of these environments is therefore essential to understand the mechanisms that support coral survival within these extreme systems (Camp et al. 2018b). In addition, periods of time when abiotic conditions are not considered extreme may provide respite for resident taxa, ultimately facilitating their survival and persistence. Furthermore, the number and co-occurrence of other co-varying natural and/or human-related stressors will impact the level of stress resident taxa experience.

For example, in the shallow mangrove system of Low Isles on the Great Barrier Reef, salinity frequently drops due to freshwater input during storm events (Camp et al. 2019). In most instances, resident stress-tolerant corals can survive this input, however, when a low salinity event coincided with an extreme neap tide, freshwater bleaching occurred (Bartels et al. 2023). In other systems, co-varying factors could be beneficial to resident corals and may explain why they are able to survive in suboptimal environmental conditions. For example, elevated nutrients in mangrove lagoons have been proposed to support resident coral by providing additional resources that could be accessed to offset the higher energetic demand of survival in this system (Camp et al. 2017). Desert dust has similarly been proposed to provide essential bioelements that sustain productivity of Red Sea corals (Blanckaert et al. 2022). Upwelling can also provide additional nutritional provisions supporting suspension feeders (Enochs et al. 2021) and if upwelling occurs at the time of a thermal stress event (Chap. 10), it could provide local relief from the marine heatwave, providing a temporary refuge for local taxa (Chollett et al. 2010). Furthermore, in the seagrass-coral system of Dongsha Atoll, Taiwan, despite nightly acidification and hypoxia, daytime productivity elevates local dissolved oxygen and pH, increasing coral calcification rates (Pezner et al. 2024).

Turbid sites have similarly been described as potential refugia, reducing light stress while concurrently providing greater food availability to resident taxa (Cacciapaglia and van Woesik 2016; Morgan et al. 2017). Deeper reef sites also typically experience reduced light stress and have been similarly proposed as potential refugia (Riegl and Piller 2003), although coral bleaching at depth has been recently noted in the literature (Eyal et al. 2022; Smith et al. 2016). Reduced light stress can also occur due to the presence of both natural and artificial structures. The vertical relief of these structures may provide shade, reducing the potential for bleaching under thermal stress. For example, shade can be provided to resident organisms by mangrove plants (Yates et al. 2014) and in urban waterways and navigational inlets by structures such as seawalls, riprap, and pylons (Enochs et al. 2023). The addition of anthropogenic structures can also create an environment conducive for settlement for sessile organisms that may not have had substrate to attach to otherwise. Sessile organisms that recruit to their surfaces are elevated above the benthos and oriented in the water column so as to more easily clear sediment settling on their surfaces. As such they are less susceptible to the high turbidity and sedimentation that can otherwise be a limiting factor to settlement and survival in these environments (Enochs et al. 2023).

The distinct environmental conditions of extreme and marginal coral habitats also have the potential to impact coral health when key processes, such as coral spawning, occur. Examination of *Seriatopora hystrix* in a shallow and mesophotic reef revealed the deeper corals had a shorter reproductive season and smaller planula size (Prasetia et al. 2017), potentially due to reduced energy availability at depth. Changes in spawning phenology could also pose an extinction risk to coral populations in extreme and marginal coral habitats that have a low population size, and limited connectivity to adjacent reefs, as reproductive shortfalls could occur (Mora et al. 2016). While sessile brooding species may be most at risk of local extinction due to limited connectivity, they are also more likely to adapt to local environmental conditions over multiple generations, compared to broadcast spawning species that may recruit into the environment from a less stressful habitat (Camp et al. 2018a, b).

Under suboptimal conditions, settlement substrates can also be altered, impacting rates of larval settlement (Fabricius et al. 2017) and settlement behavior (Doropoulos et al. 2012). Furthermore, the immediate and surrounding environmental conditions determine how connected extreme and marginal coral habitats are with adjacent reef systems. This in turn impacts the population genetics and determines whether there is locally distinct genetic structure. In Singapore's marginal inshore reefs, for example, subtle fine-scale genetic

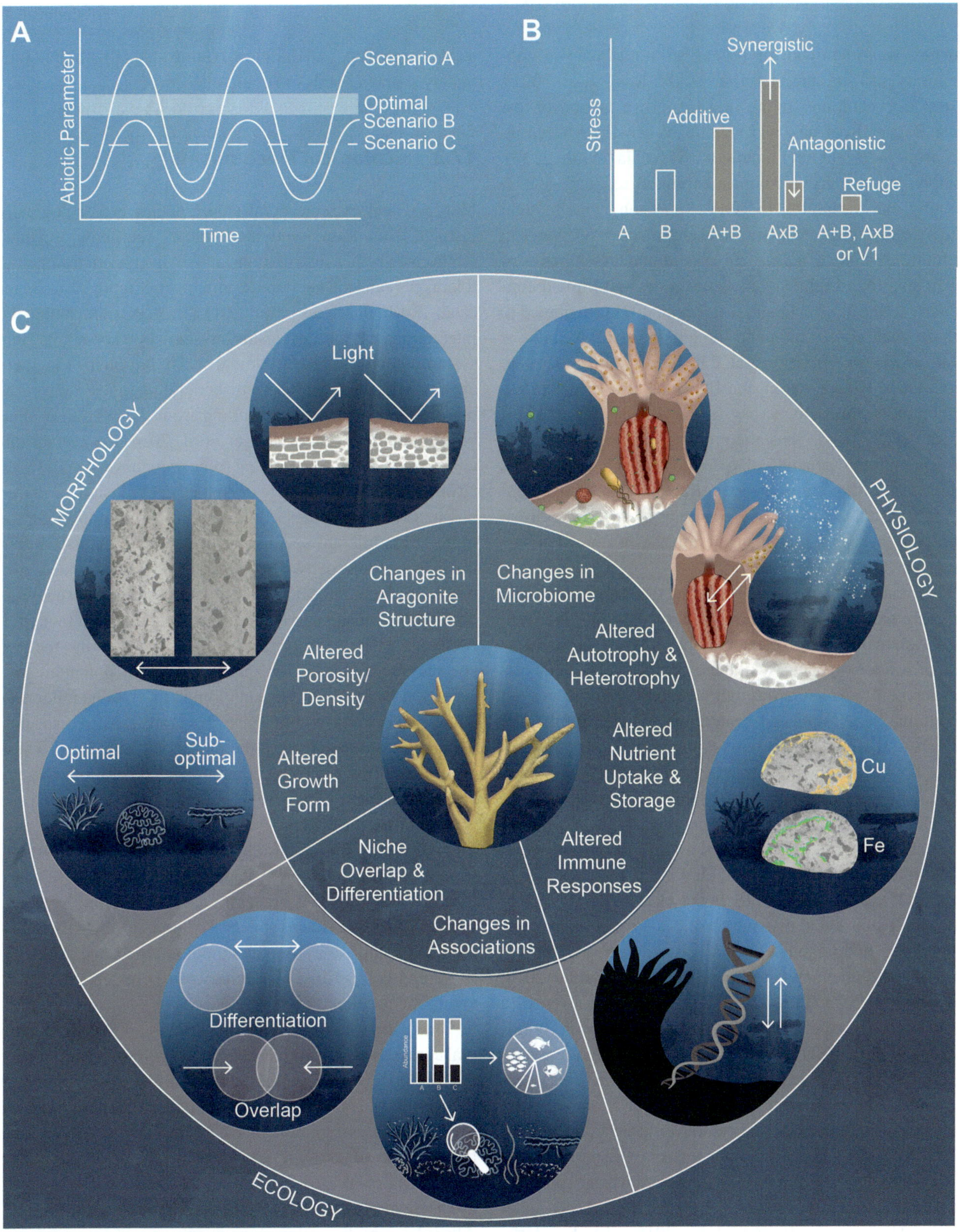

Fig. 17.1 Summary schematic of key mechanisms that support survival of corals in extreme and marginal environments. (**a**) Summary graph showing the different ways abiotic conditions can vary in suboptimal environments. In scenario **a**, high-variance in abiotic conditions can result in periods of time when conditions are above optimal, as well as periods of time when the conditions are below optimal. In scenario **b**, there is high variation again but with conditions consistently remaining below optimal. In scenario C there are static conditions that remain Fig. 17.1 (continued) below optimal. Each of these scenarios have the potential to differentially impact the corals' ability to persist in suboptimal environmental conditions. Notably each scenario can exist for single or multiple abiotic parameters. (**b**) (Adapted from Darling and Côté 2013). Multiple abiotic conditions (here, **a** and **b**) can occur in isolation or in combination, and can act as stressors in an additive, synergistic or antagonistic manner. It is also possible that combined abiotic conditions ((A + B), (A × B) or a different variable (V1)) do not result in stress, and instead create a refuge. (**c**) Summary schematic of key morphological, physiological, and ecological mechanisms that can support the survival of corals in extreme and/or marginal coral environments

structure was identified for *Porites* sp.—attributed to contemporary surface current regimes (Afiq-Rosli et al. 2021). *Porites lutea* colonies living in a mangrove lagoon on the Great Barrier Reef also had significant genetic differentiation from adjacent reef colonies (Scucchia et al. 2023). Notably, the mangrove colonies exhibited reduced genetic diversity and variability in gene expression, which might pose a disadvantage under more severe selective pressures in the future. A study of *Porites divaricata* in the mangroves of Belize also found limited genetic diversity and limited connectivity to the adjacent reefs (Lord et al. 2023), attributed to high asexual reproduction and limited dispersal. A study of *Montastraea cavernosa* separated by tens of meters depth in the Cayman Islands and in the Bahamas also revealed low population connectivity across depths (Brazeau et al. 2013; Serrano et al. 2014). Thus, the growing body of research on coral population genetics of extreme and marginal corals suggests they have unique genetic structure with limited connectivity to adjacent reefs, highlighting the need for broad conservation efforts that span the mosaic of coral habitats that exist.

Pertinent questions remain regarding the future of extreme and marginal coral environments in the Anthropocene. Will these systems become increasingly hostile, limiting the survival of resident taxa? Or, will these systems help corals better manage future stress through preconditioning to current suboptimal environmental conditions taxa? While the answers to these questions remain unknown, studies investigating how extreme and marginal corals tolerate simulated and natural stress events have revealed mixed results. In some locations, such as shallow mangrove lagoons (Barker 2018; Haydon et al. 2023; Tanvet et al. 2023 ; Roper et al. 2025), rock pools (Barker 2018; Haydon et al. 2023; Tanvet et al. 2023), and tidal flats (Schoepf et al. 2015), frequent exposure to non-lethal extreme abiotic conditions has results in enhanced fitness and resilience relative to corals in more benign environments when exposed to stress (Barker 2018; Haydon et al. 2023; Tanvet et al. 2023). However, even with the enhanced resilience, it does not leave extreme corals immune to stress (Schoepf et al. 2015) and there are examples where prior exposure to extreme abiotic conditions does not confer fitness benefits to the coral. For example, daily prior exposure to extreme temperature and pH conditions in a Caribbean seagrass bed did not improve the performance of *Acropora palmata* and *Porites astreoides* when exposed to temperature and pH conditions predicted in 2100 under IPCC A1B scenario (Camp et al. 2016a, b). Similarly, *Goniastrea retiformis* and *Porites lobata* showed no enhanced tolerance to heat stress despite living in highly dynamic rock pools that frequently experienced elevated temperatures. In fact, *P. lobata* had reduced thermal tolerance (Klepac and Barshis 2020), highlighting the complex interplay between environmental history, species, and stress exposure (both in terms of intensity and duration; (Camp et al. 2018b) that ultimately shape a corals emergent phenotype.

17.3 Ecological Mechanisms

Marginal and extreme coral environments are ecologically distinct from more benign, "typical" reef environments. Restricted species distributions due to diverse environmental tolerances can lead to altered community composition (Beger et al. 2014; Fabricius et al. 2011), in turn influencing the interactions of the organisms present. For some marginal systems, such as high-latitudinal reefs (see Chap. 13), there are relatively high numbers of endemic species, as well as more cosmopolitan taxa, resulting from the overlap in their biogeographical ranges (Beger et al. 2014). In certain circumstances, unique ecological roles or niches may confer an advantage and contribute to persistence. For instance, altered competitive hierarchies could lead to proliferation of the more stress-tolerant species. Consider the simplistic hypothetical example of two species competing for space on a reef with equal competitive ability. One of the two species is particularly sensitive to osmotic stress. In a typical reef environment, where salinity does not fluctuate wildly, the two species may compete for space with equal potential. In a coastal, urban, or mangrove environment, riverine influences could osmotically compromise the more sensitive species, leading to the advantage of the more tolerant species. These general principles can alter community composition whereby stress-sensitive corals are replaced by stress tolerant taxa and opportunistic, fast growing species that are capable of rapidly colonising space (Darling and Côté 2013). For example, at a carbon dioxide seep in Papua New Guinea, there was a loss of stress-sensitive branching corals, resulting in dominance by massive *Porites* spp. (Fabricius et al. 2014). Further, as corals comprise a complex holobiont, the same principles of altered competitive hierarchies can be extended to their symbiotic partners.

The distinct environmental conditions present in suboptimal coral environments can shape the symbiotic partnerships that ultimately impact the fitness of the coral holobiont (Mieog et al. 2009). For example, unique Symbiodiniaceae and bacterial communities have been documented within a broad range of extreme and marginal coral environments (Bongaerts et al. 2011; Camp et al. 2020; Cunning et al. 2015; Haydon et al. 2021; Hume et al. 2015; Ros et al. 2021). Altered symbiotic partnerships in suboptimal environmental conditions may reflect host-selection or symbiont competition, whereby environmental conditions shape the optimal niche space for certain symbiotic partners, allowing them to proliferate over less tolerant individuals. A recent review on the diversity and ecology of Symbiodiniaceae by Nitschke et al. (2022) highlights how environmental variables such as light, nutrients and temperature can shape the niche space occupied by Symbiodiniaceae through selection of beneficial traits.

As environmental variation in abiotic traits such as temperature can occur at regional, ecosystem, habitat, reef, colony, and within-colony scales, there are numerous opportunities for ecological mechanisms (e.g., competition) to shape coral fitness through altering symbiotic partnerships. Research on the impact of elevated temperatures on coral-associated Symbiodiniaceae highlights this point. At the regional scale, in the warm waters of the Persian Arabian Gulf (see Chap. 5), a putatively thermally-tolerant symbiont, *Cladocopium thermophilum* dominates, which has been found to tolerate temperatures exceeding 36 °C (Hume et al. 2015). At the habitat scale, *Acropora millepora* colonies living within a highly variable temperature environment have been found to associate mainly with *Durusdinium* symbionts (ITS2 type profile; D2-D1ak-D1- D2.2), whereas colonies living in a more stable reef habitat tend to associate more with *Cladocopium* (ITS2 type profile C3-C50c-C50aC3b; Camp et al. 2019). At the colony scale, within-colony zonation in Symbiodiniaceae has also been attributed to support coral fitness during a thermal stress event (Rowan et al. 1997). These temperature examples further demonstrate the heterogeneity found in partnerships within extreme coral environments, with no single "winning" host-symbiont partnerships found consistently across extreme coral systems.

Moreover, coral reefs and reef organisms do not live in isolation, and what constitutes an extreme environment for corals may not be an extreme environment for other reef species that occupy those environments. It is therefore likely that competitive potential would be altered at the detriment of the reef species. This has been observed to occur in higher productivity environments where heterotrophic calcifiers, such as barnacles and molluscs are favored over corals; or in naturally acidified environments which favor fleshy algae vs. calcifiers (e.g., (Enochs et al. 2021).

The unique ecological considerations of extreme and marginal coral reef environments also extends to interactions involving direct consumption, including predators as well as carnivorous and herbivorous grazers. This is particularly true for motile taxa, with ranges that can encompass multiple habitats. These taxa may favor more natural reef environments, whether it be due to the abundance of prey, their particular environmental tolerances, life history (e.g., recruitment), or behavior. Parrotfish, for example, are voracious grazers, and are important herbivores in natural reef environments (O'Farrell 2011). While scraping, biting, and breaking the surface of the reef, many species remove large volumes of skeletal carbonates and reef framework, acting as the greatest agent of bioerosion in many regions (Mallela and Bathgate 2018). Marginal coral environments, such as urbanised waterways, may attract fewer parrotfish, leading to reduced grazing, and ultimately, more persistent carbonates. In the highly urbanised Singapore reefs, for example, smaller fish, such as those belonging to the families Pomacentridae and Labridae, are dominant (Taira et al. 2018).

Changes in community composition can also result in more or fewer corallivores that consequently influence the level of risk to resident corals, particularly at vulnerable life-stages such as recruitment (Edmunds 2023). In other systems such as mangrove and seagrass habitats, the community as well as the ontogeny of resident taxa could also impact coral communities. Mangroves and seagrass meadows, for example, are considered important nursery habitats for fish (Whitfield 2017). How juvenile fish interact with coral in these systems could be different to how adults interact with coral on the reef. Furthermore, if the benthic community composition is different, whether there are more or less corallivores could impact coral fitness, particularly for vulnerable size classes, such as small coral recruits. These differences could have positive (e.g., nutrient or microbiota from feces; (Schiettekatte et al. 2023; Smriga et al. 2010) or negative (e.g., corallivore behavior; (Renzi et al. 2022) consequences on resident corals as well as the broader ecosystem. Understanding how different species interact within suboptimal coral environments is crucial to assess the functional roles occupied by each species. Typically (e.g., Camp et al. 2017; Fabricius et al. 2011), but not always (e.g., Maggioni et al. 2021), extreme and marginal coral habitats have reduced genetic and functional diversity, which could leave them more vulnerable to future disturbances due to the lack of functional redundancy within the system (Mora et al. 2016). It is, however, possible that the unique physiological adaptations of corals within extreme coral systems will facilitate their future survival.

17.4 Physiological Mechanisms

A broad suite of physiological mechanisms have been documented for the coral host and their symbiotic partners to support the holobiont's survival in suboptimal environmental conditions (Fig. 17.1). Perhaps the most well-studied of these resilience-support mechanisms is the microbial composition (Symbiodiniaceae, bacteria and other microorganisms) of corals that has been studied across numerous marginal and extreme environments (***see*** Chap. 4). The microorganisms associated with corals in extreme and marginal systems are often different compared to more benign reef systems (e.g. Camp et al. 2016a, b; Goulet et al. 2019; Leite et al. 2018) and appear to be coral-host niche adapted (***see*** Chap. 2). Some of these unique associations are postulated to support coral survival within extreme environmental conditions, for example association of corals in the Arabian Gulf with *Cladocopium thermophilum* (Hume et al. 2013).

Of note, some marginal reef conditions that may be relevant to other aspects of the coral holobiont may not exert strong control over the microorganism composition. For example, Noonan et al. (2013) looked at Symbiodiniacaeae communities from six species of corals collected across volcanically acidified vent sites in Papua New Guinea. Despite examining sites that ranged from contemporary acidification conditions to in excess of the end of the century projections for ocean acidification, there were no Symbiodiniacaeae community differences that were attributable to acidification. Rather, Symbiodiniacaeae community composition was largely related to host species and site, independent of pH. These data demonstrate that symbiont shuffling may only be a viable acclimatisation technique in certain types of extreme environments. Interestingly, the number of Symbiodiniaceae cells can also vary for extreme and marginal corals, for example, multiple species of coral in a mangrove lagoon on the Great Barrier Reef were found to have more Symbiodiniaceae cells than conspecifics on an adjacent reef (Camp et al. 2019; Ros et al. 2021). A higher number of Symbiodiniaceae cells may support the holobiont by helping to maintain their energetic requirements by increasing or sustaining rates of photosynthesis. However, it is also possible that the higher cell densities reflect competitive advantage of the symbiont cells within the holobiont and could challenge resource sharing that leads to compromised holobiont health (Camp et al. in review).

As the microorganisms associated with corals are increasingly studied across different geographic locations, some interesting patterns are starting to emerge. For example, corals in mangrove lagoons that are characterised by low pH, highly variable temperature and low oxygen often have homogenisation of Symbiodiniaceae communities compared to conspecifics on adjacent reefs (Assis et al. 2020; Camp et al. 2019, 2020), which could suggest species selection or environmental filtering. Furthermore, altered Symbiodiniaceae communities of corals in extreme and marginal coral systems can often be associated with potential trade-offs (e.g., reduced calcification rates; Camp et al. 2016a, b; Jones and Berkelmans 2010). Further work exploring geographic patterns in microorganism-coral associations across marginal and extreme coral habitats could help resolve conserved patterns in measures such as evenness and diversity, as well as any potential trade-offs with unique microbial associations.

Unique photosynthetic strategies have also been documented in corals living in extreme and marginal systems. For example, mesophotic corals living at depths of greater than 30 meters, face not only changes in light intensity, but also differences in the light spectrum and angular distribution (Kahng et al. 2019). Available research on these unique systems reveal diverse, and often opposing, photophysiological strategies in use by different coral species (Kahng et al. 2019), typically involving differences in the Symbiodiniaceae themselves. For example, mesophotic colonies of *Stylophora pistillata* (65 m depth) in the Red Sea have been found to associate with a unique clade of Symbiodiniaceae compared to shallower colonies (Clade C versus Clade A, respectively) and also change their light harvesting mechanism to maximise light utilisation (Einbinder et al. 2016). On mesophotic reefs in Hawaii (60–132 m), corals along this depth gradient were associated not only with different Symbiodiniaceae types, but also varied in the pigment density, pigment content, cell size, and photosynthetic yield (Padilla-Gamiño et al. 2019). In other marginal and extreme systems such as mangrove lagoons in Australia, resident *Porites lutea* and *Acropora millepora* corals have been found to have distinct excitation quenching patterns, with both species dissipating an increased proportion of absorbed excitation energy via dynamic nonphotochemical quenching [1-Q] rather than photochemical quenching [1-C] relative to reef counterparts. Furthermore, both species had significantly reduced rates of gross photosynthesis per cell compared to reef corals (Camp et al. 2019). Similarly, in the highly variable Bouraké lagoon in New Caledonia, lagoon corals had lower photosynthetic rates than reef corals, but increased chlorophyll content (Jacquemont et al. 2022). In contrast, at a volcanic carbon dioxide vent site in Papua New Guinea, net photosynthetic rates were notably higher in some species of corals at the vent site, though this varied by species (Strahl et al. 2015). While studies on this topic are relatively limited (see Chap. 14 about Trophic ecology), available data suggests that coral holobionts employ a range of strategies to maintain, or even enhance, photosynthetic activity under challenging conditions in marginal and extreme systems.

While many corals fulfill the majority of their metabolic requirements via the photosynthesis of endosymbiotic Symbiodiniaceae, they are also heterotrophs, feeding on food sources including dissolved and particulate organic matter, zooplankton, and even phytoplankton (Houlbrèque and Ferrier-Pagès 2009; Leal et al. 2014). These food sources are particularly important when autotrophy is compromised (e.g., bleaching) or light is limited (e.g., mesophotic reefs), and heterotrophy has been linked to resilience to stressors including elevated temperature (Grottoli et al. 2006) and acidification (Towle et al. 2015). As such, heterotrophy may represent an important physiological mechanism of resilience in marginal coral environments, particularly for those that have more abundant food sources than classical oligotrophic reef environments. For instance, in the Port of Miami, which has poor water quality and strong riverine inputs, corals may persist due to an abundance of suspended food sources (Enochs et al. 2023; Rubin et al. 2021). Indeed, transcriptomic analysis from *Pseudodiploria strigosa* collected within the port revealed upregulation of genes involved with digestion (Rubin et al. 2021). Similarly, mangrove lagoons in New Caledonia, which experience high temperatures, low pH and low oxygen, were characterised by high organic matter, likely contributing to higher respiration rates in the resident corals (Camp et al. 2017). While light calcification rates were still depressed in this environmentally challenging ecosystem, the energetic plasticity, or the ability to obtain more energy from heterotrophy, likely helps to offset the increased metabolic demands of sustaining physiological processes (Camp et al. 2018a, b). Similarly, studies of isotopic content of Hawaiian mesophotic corals along depth gradients of 60–132 meters revealed depth-dependent trends, suggesting an increase in heterotrophy at greater depths (Leichter and Genovese 2006; Padilla-Gamiño et al. 2019). Thus, early evidence suggests heterotrophy may play an active role in fueling coral function in marginal or extreme environments.

In addition, living in marginal and extreme reef systems can also impact a coral holobiont's use and storage of energy, key elements, and other resources that fuel basic function. For example, corals living in mangrove lagoons on the Great Barrier Reef were found to fix significantly less inorganic carbon compared to reef conspecifics, though they were still able to maintain similar proportions of inorganic carbon translocated to the coral host (Ros et al. 2021). In terms of energy storage, corals living on reefs impacted by severe nutrient runoff and enrichment in Mexico have been shown to be able to maintain high lipid contents relative to corals on healthier oligotrophic reefs, but at the expense of symbiont density (Martínez-Castillo et al. 2020). Similarly, corals growing in a marginal, high-latitude reef in China were found to have significantly higher lipid content than corals from a reef further south, especially during winter when cold stress is the highest (Yu et al. 2024). These findings suggest that corals in marginal or extreme environments acclimatise or adapt to their habitats by prioritising energy storage, perhaps as additional insurance during times of particularly high stress. While research on other uptake and storage mechanisms are limited (i.e., trace element uptake and nitrogen fixation), it is reasonable to believe that these would also differ in corals exposed to marginal and extreme conditions.

Another potential adaptation for corals living in extreme and marginal environments is increased immune response to combat reduced water quality. For the more directly anthropogenically-impacted environments such as ports, urban waterways, and other habitats near to river outputs and runoff from land, water quality is often poor, specifically in terms of nutrient, sediment, and bacterial loading. In order to thrive in these challenging conditions, studies have shown that in some cases, increased immune response may be one factor responsible for certain populations or species' ability to survive. For example, in the Port of Miami, transcriptomic analysis of *Pseudodiploria strigosa* in these waters has revealed higher expression of transcripts that correspond to innate immunity, defense, and cellular detoxification (Rubin et al. 2021). This upregulation of immune responses may allow these corals to maintain their health despite increased pathogenic activity in the surrounding seawater. While data on coral immunity from more naturally marginal or extreme systems is limited, one study suggests exposure to extreme conditions may negatively impact immune functioning. For example, a functional genomic analysis of *Acropora millepora* colonies living in a natural carbon dioxide seep system in Papua New Guinea had a downregulation of genes involved in the innate immune response, suggesting a potential impact of long-term excess pCO_2 exposure (Kenkel et al. 2018). Understanding how and to what degree coral immunity may be either boosted or compromised in marginal and extreme environments will be key in predicting how these corals may fare under increasing climate change and the spread of coral diseases.

17.5 Morphological Mechanisms

Functional attributes of coral are often linked to their morphology which can be dynamic across environments, particularly when light and water movement are variable (Todd 2008). Importantly, coral morphology is associated with coral function and resilience, noted as an important characteristic of weedy and stress-tolerant corals (Darling et al. 2012). For example, *Stylophora pistillata* was found to transition from a profusely branching structure in the shallow waters (ca. 5 m) of the Red Sea, to a planar structure at depth (ca. 50 m) (Einbinder et al. 2009). These differences in skeletal morphology result in a low surface to volume ratio that is more efficient for light trapping in low light environments (Einbinder et al. 2009). These changes in morphology may confer fitness benefits that enhance coral resilience in suboptimal environments. For example, evaluation of marginal reef systems in the Indo-Pacific found common coral traits supporting survival that included robust morphologies (Mizerek et al. 2016). Further, *S. pistillata* in the Red Sea had thinner branch diameter at depth which has been interpreted as photoacclimation to maximise light harvesting (Fricke and Schuhmacher 1983). A similar prevalence of morphologies (encrusting and plate) with a low surface to volume ratio was also documented for low light turbid reefs on the Great Barrier Reef (Woesik 1992). When corals persist in non-coral dominated systems, morphology can also be important for survival (McCook et al. 2001).

Changes in skeletal microstructure can also occur in corals living in suboptimal environments which can influence holobiont fitness. For example, *Porites lutea* in a mangrove lagoon on the Great Barrier Reef had altered micro-granular aragonite structures and reduced thickness of the skeletal walls at the single-polyp level relative to corals from a neighboring reef (Scucchia et al. 2023). The mangrove *P. lutea* also had reduced density, thickness and higher porosity in their coral skeletons. A similar observation was documented for corals at a natural carbon dioxide vent site, whereby higher porosity allowed liner extension rates to be maintained, which is potentially beneficial to reach critical size for sexual maturity (Fantazzini et al. 2015). Both macro and micro skeletal properties can influence the light availability reaching Symbiodiniaceae which can in turn contribute to coral bleaching sensitivity and overall coral resilience (Swain et al. 2018). Within a single colony, differences in light based on morphology can also occur, creating exposed and self-shaded areas (Ulstrup et al. 2006) that can result in micro-refugia that support holobiont survival under stress (Smith et al. 2017). For example, mesophotic corals have been found to have distinct coral skeletal architecture such as increased corallite marginal spacing and shorter coenosteal spines, leading to reductions in self-shading on individual polyps and promoting efficient light-capture at depth (Kramer et al. 2022).

Colony size is another important morphological feature that can be different in extreme and marginal coral environments, with suboptimal environmental conditions resulting in smaller colonies (Camp et al. 2016a, b; Woesik 1992) and sometimes unconsolidated rather than reef accreting structures (Camp et al. 2016a, b; Woesik 1992). Reduced rates of coral calcification that can impact colony size have been documented in extreme coral environments, including carbon dioxide vents (Fabricius et al. 2011), mangrove lagoons (Camp et al. 2019) and in Ojos (Crook et al. 2013). Corals from some extreme coral habitats, such as Bouraké lagoon in New Caledonia (Tanvet et al. 2023) and Palau's Rock Island (Barkley et al. 2017) have however demonstrated enhanced capacity to calcify when experimentally exposed to low pH extremes, suggesting adaptation to low pH. Systems with low accretion rates are typically considered marginal coral reef sites (Schoepf et al. 2023), and any loss of reef structure and complexity can compromise the ecological services provided by coral reefs (Rogers et al. 2015). For example, the change in structural complexity at a carbon dioxide vent site in Papua New Guinea resulted in reduced macroinvertebrate groups, demonstrating the broader ecosystem consequences resulting from altered reef complexity (Fabricius et al. 2014). In marginal coral systems where other taxon dominate, e.g., fleshy algae, coral size and morphology can also be a determinant of success (McCook et al. 2001).

Collective evidence suggests morphological mechanisms supporting coral survival in suboptimal environmental conditions are diverse, spanning from nano structural differences in aragonite deposition through to different coral morphologies (Fig. 17.1). While such morphological plasticity appears an important mechanism supporting coral survival in suboptimal environmental conditions, it could compromise coral fitness if they are moved out of their native environment. This highlights why understanding the mechanisms that support coral survival in extreme and marginal systems is necessary to inform the suitability of extreme and marginal corals in active interventions that involve cross-habitat transplantation (Scucchia et al. 2023).

17.6 Implications for Practice: Why Understanding Mechanisms for Survival in Extremes Matters

The long-term predictions for coral reefs under current rates of ocean warming remain bleak (Hoegh Guldberg et al. 2022). Consequently, habitats identified as refugia are likely to become increasingly important for the long-term persistence of coral, as they have been during past climate warming events (Dimitrijević et al. 2024). Refugia can act as a genetic reservoir for local species as well as species capable of migrating into the refugia (Fig. 17.2). Several marginal and extreme coral systems have been proposed as refugia within the academic literature (reviewed in Camp et al. 2018a, b). However, the heterogeneous nature of extreme and marginal coral systems means generalisations on refuge capacity of all of a given type of system (e.g. all deep reef locations) should be cautioned against, due to the important role of local physicochemical conditions in determining refuge capacity. Noting this caution, areas of potential climate refuge have been suggested as important locations for inclusion in proactive marine park planning, to create climate-ready marine protected area networks (outlined in Camp 2022). Within this framework, efforts to minimize local stressors can help maintain refuge potential.

Globally, efforts to buy time for reefs while climate change is being addressed are intensifying, with many science-based efforts aiming to rebuild and rehabilitate reef systems to increase their resilience under ongoing environmental change (Peixoto et al. 2024). While not a replacement for climate action, a toolbox of active interventions are being deployed for coral reefs (Suggett and van Oppen 2022). Within this toolbox, refugia can provide roles beyond housing biodiversity, including providing a blueprint for artificial refugia that can be a focus for new technological solutions. For example, increased supply of organic matter that can support enhanced heterotrophy in extreme coral systems (Camp et al. 2018a, b), could be replicated in other reef systems to support coral survival during periods of stress. Interventions could also include the managed translocation of corals into potential refugia to try and increase the biodiversity of protected species and genotypes (Fig. 17.2, Camp 2022). Furthermore, extreme and/or marginal systems that are potential refugia and also are blue carbon environments

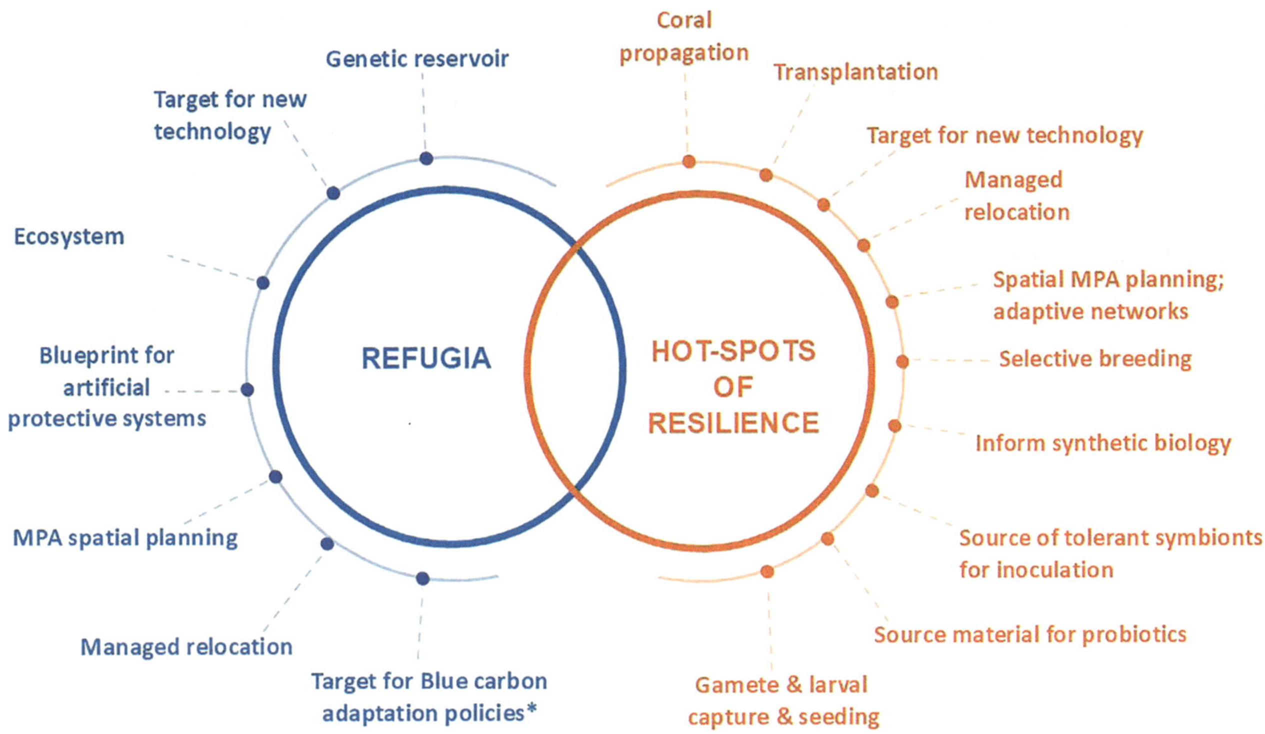

Fig. 17.2 A schematic showing the potential proactive management opportunities marginal and or extreme coral systems can provide. (Adapted from Camp 2022). MPA is a marine protected area. *Denotes that it only applies to a subset of refugia and refuge systems

(e.g., mangroves) could support coral survival while also providing carbon sequestration potential (Duarte et al. 2004).

Besides acting as potential refugia, extreme and marginal coral systems may in fact precondition corals and resident taxa to hostile abiotic conditions more representative of future climate change scenarios (Camp et al. 2016a, b). In these cases, a plethora of active interventions are possible (see Fig. 17.2), that include using these presumed hardier coral species in coral propagation, transplantation, selective breeding, managed relocation, and larval capture and re-seeding. These stress-hardened corals can also provide source microbiomes for probiotics or Symbiodiniaceae manipulations. Information on the mechanisms that ultimately support their innate tolerance can also be targeted via synthetic biology for less tolerant corals in an attempt to boost their resilience. It is important, however, to consider how well the stress hardened corals will retain their innate tolerance in new environments—whether the tolerance is in fact heritable to their offspring, and whether there are associated trade-offs with higher tolerance. As such, the unique ecology of extreme environments can have practical ramifications for restoration and species translocation. For instance, corals that have grown in areas with few corallivores and reduced predation pressure may contain fewer anti-predatory compounds, or more lipids, and may ultimately be more palatable as they lose anti-predatory traits (Harrison et al. 2024). Once moved to more typical reef environments, they may be preferentially preyed upon and therefore less effective as outplants, despite genetic or stress-hardening advantages conveyed from their upbringing in marginal environments (DeMerlis 2004). Other examples include morphological differences in extreme and marginal corals. While morphological plasticity may be an important mechanism to survive in extreme and marginal coral environments, this can raise concerns about the suitability of moving corals out of these environments, where structural changes could be a disadvantage (e.g., higher skeletal porosity and/or slower growth rates (Scucchia et al. 2023). Also, being tolerant to one stressor, e.g. temperature, does not ensure tolerance to other stressors, such as salinity stress. Consequently, integrating corals from hot-spots of resilience into new systems should be considered carefully as part of broader conservation strategies and marine management plans.

17.7 Conclusions

The heterogeneous nature of extreme and marginal coral systems has resulted in a plethora of mechanisms that contribute to coral survival and persistence within these suboptimal habitats (Fig. 17.1). In this Chapter we have summarised some key ecological, mechanical, and physiological mechanisms that have been documented to facilitate coral survival in these hostile environments. Whether future environmental change will make extreme coral systems uninhabitable remains to be seen, but there is already evidence that these systems are not immune to stress and change. Despite this uncertainty, there is compelling evidence that some corals from these systems are highly adapted to environmental stress, making them ideal candidates for active interventions. Furthermore, understanding the mechanisms that contribute to coral tolerance within extreme and marginal coral systems can act as a blueprint for designing new active interventions for other impacted reef systems. It can also help maximise conservation efforts, by guiding where limited resources should be prioritised (Schoepf et al. 2023), and reduce any associated risks of utilising extreme or marginal corals in active interventions. Ultimately, the unique nature of extreme and marginal coral habitats warrants targeted and global conservation efforts to ensure their diverse attributes are protected both now, and into the future.

Acknowledgments We thank Dr. Paige Strudwick for her contribution in making Fig. 17.1.

References

Afiq-Rosli L, Wainwright BJ, Gajanur AR et al (2021) Barriers and corridors of gene flow in an urbanized tropical reef system. Evol Appl 14(10):2502–2515

Assis JM, Abreu F, Villela HMD et al (2020) Delivering beneficial microorganisms for corals: rotifers as carriers of probiotic bacteria. Front Microbiol 11:608506

Barker V (2018) Exceptional thermal tolerance of coral reefs in American Samoa: a review. Curr Clim Change Rep 4(4) Springer Science and Business Media LLC:417–427

Barkley HC, Cohen AL, McCorkle DC et al (2017) Mechanisms and thresholds for pH tolerance in Palau corals. J Exp Mar Biol Ecol 489:7–14

Bartels N, Dilernia NJ, Howlett L et al (2023) Stress event for 'super corals' in Great Barrier Reef mangrove lagoon. Mar Biodivers 53(5):64

Beger M, Sommer B, Harrison PL et al (2014) Conserving potential coral reef refuges at high latitudes. Divers Distrib 20(3) Wiley:245–257

Blanckaert ACA, Omanović D, Fine M et al (2022) Desert dust deposition supplies essential bioelements to Red Sea corals. Glob Chang Biol 28(7) Wiley Online Library:2341–2359

Bongaerts P, Sampayo EM, Bridge TCL et al (2011) Symbiodinium diversity in mesophotic coral communities on the Great Barrier Reef: a first assessment. Mar Ecol Prog Ser 439. Inter-Research Science Center:117–126

Brazeau DA, Lesser MP, Slattery M (2013) Genetic structure in the coral, Montastraea cavernosa: assessing genetic differentiation among and within Mesophotic reefs. PLoS One 8(5):e65845

Cacciapaglia C, van Woesik R (2016) Climate-change refugia: shading reef corals by turbidity. Glob Chang Biol 22(3) Wiley:1145–1154

Camp EF (2022) Contingency planning for coral reefs in the Anthropocene; the potential of reef safe havens. Emerg Top Life Sci 6(1):107–124

Camp EF, Smith DJ, Evenhuis C et al (2016a) Acclimatization to high-variance habitats does not enhance physiological tolerance of two key Caribbean corals to future temperature and pH. Proc R Soc B Biol Sci 283(1831) Royal Society:20160442

Camp EF, Suggett DJ, Gendron G et al (2016b) Mangrove and seagrass beds provide different biogeochemical services for corals threatened by climate change. Front Mar 3(52). https://doi.org/10.3389/fmars.2016.00052

Camp EF, Nitschke MR, Rodolfo-Metalpa R et al (2017) Reef-building corals thrive within hot-acidified and deoxygenated waters. Sci Rep 7(1):2434. https://doi.org/10.1038/s41598-017-02383-y

Camp EF, Schoepf V, Suggett DJ (2018b) How can 'super corals' facilitate global coral reef survival under rapid environmental and climatic change? Glob Chang Biol 24(7):2755–2757

Camp EF, Schoepf V, Mumby PJ et al (2018a) The future of coral reefs subject to rapid climate change: lessons from natural extreme environments. Front Mar Sci 5:4

Camp EF, Edmondson J, Doheny A et al (2019) Mangrove lagoons of the Great Barrier Reef support coral populations persisting under extreme environmental conditions. Mar Ecol 625:1-14. https://doi.org/10.3354/meps13073

Camp EF, Suggett DJ, Pogoreutz C et al (2020) Corals exhibit distinct patterns of microbial reorganisation to thrive in an extreme inshore environment. Coral Reefs 39(3):701–716

Chollett I, Mumby PJ, Cortés J (2010) Upwelling areas do not guarantee refuge for coral reefs in a warming ocean. Mar Ecol Prog Ser 416. Inter-Research Science Center:47–56

Crook ED, Cohen AL, Rebolledo-Vieyra M et al (2013) Reduced calcification and lack of acclimatization by coral colonies growing in areas of persistent natural acidification. Proc Natl Acad Sci U S A 110(27) National Acad Sciences:11044–11049

Cunning R, Yost DM, Guarinello ML et al (2015) Variability of Symbiodinium communities in waters, sediments, and corals of thermally distinct reef pools in American Samoa. PLoS One 10(12) journals.plos.org:e0145099

Darling, ES and Côté IM (2013) 4.21 - Vulnerability of Coral Reefs. In: Climate Vulnerability. 4:259–270. https://doi.org/10.1016/B978-0-12-384703-4.00427-5

Darling ES, Alvarez-Filip L, Oliver TA et al (2012) Evaluating life-history strategies of reef corals from species traits. Ecol Lett 15(12) Wiley:1378–1386

DeMerlis A (2004) Molecular mechanisms of enhancing coral resilience to thermal stress. PhD Diss. University of Miami. Epub ahead of print 2004. https://paperpile.com/b/A83Odr/zbid

Dimitrijević D, Santodomingo N, Kiessling W (2024) Reef refugia in the aftermath of past episodes of global warming. Coral Reefs 43(5) Springer Science and Business Media LLC:1431–1442

Doropoulos C, Ward S, Diaz-Pulido G et al (2012) Ocean acidification reduces coral recruitment by disrupting intimate larval-algal settlement interactions. Ecol Lett 15(4):338–346

Duarte CM, Middelburg JJ, Caraco N (2004) Major role of marine vegetation on the oceanic carbon cycle. Biogeosciences 1:173

Edmunds PJ (2023) Coral recruitment: patterns and processes determining the dynamics of coral populations. Biol Rev Camb Philos Soc 98(6) Wiley Online Library:1862–1886

Einbinder S, Mass T, Brokovich E et al (2009) Changes in morphology and diet of the coral Stylophora pistillata along a depth gradient. Mar Ecol Prog Ser 381. Inter-Research Science Center:167–174

Einbinder S, Gruber DF, Salomon E et al (2016) Novel adaptive photosynthetic characteristics of mesophotic symbiotic microalgae within the reef-building coral, Stylophora pistillata. Front Mar Sci 3. Frontiers Media SA:195

Enochs IC, Toth LT, Kirkland A et al (2021) Upwelling and the persistence of coral-reef frameworks in the eastern tropical Pacific. Ecol Monogr 91(4) Wiley:e01482

Enochs IC, Studivan MS, Kolodziej G et al (2023) Coral persistence despite marginal conditions in the Port of Miami. Sci Rep 13(1):6759. https://doi.org/10.1038/s41598-023-33467-7

Eyal G, Laverick JH, Ben-Zvi O et al (2022) Selective deep water coral bleaching occurs through depth isolation. Sci Total Environ 844:157180

Fabricius KE, Langdon C, Uthicke S et al (2011) Losers and winners in coral reefs acclimatized to elevated carbon dioxide concentrations. Nat Clim Chang 1(3) Nature Publishing Group:165–169

Fabricius KE, De'ath G, Noonan S et al (2014) Ecological effects of ocean acidification and habitat complexity on reef-associated macroinvertebrate communities. Proc Biol Sci R Soc 281(1775):20132479

Fabricius KE, Noonan SHC, Abrego D et al (2017) Low recruitment due to altered settlement substrata as primary constraint for coral communities under ocean acidification. Proc Biol Sci R Soc 284(1862). https://doi.org/10.1098/rspb.2017.1536

Fantazzini P, Mengoli S, Pasquini L et al (2015) Gains and losses of coral skeletal porosity changes with ocean acidification acclimation. Nat Commun 6:7785

Fricke HW, Schuhmacher H (1983) The depth limits of Red Sea stony corals: an ecophysiological problem (a deep diving survey by submersible). Mar Ecol 4(2) Wiley:163–194

Goulet TL, Lucas MQ, Schizas NV (2019) Symbiodiniaceae genetic diversity and symbioses with hosts from shallow to mesophotic coral ecosystems. In: Coral Reefs of the World. Springer International Publishing, Cham, pp 537–551

Grottoli A, Rodrigues L, Palardy JE (2006) Heterotrophic plasticity and resilience in bleached corals. Nature 440. nature.com:1186–1189

Harrison ND, Phillips BL, Wayne AF et al (2024) Sustained predation pressure may prevent the loss of anti-predator traits from havened populations. Ecol Evol 14(7):e11668

Haydon TD, Seymour JR, Raina J-B et al (2021) Rapid shifts in bacterial communities and homogeneity of Symbiodiniaceae in colonies of Pocillopora acuta transplanted between reef and mangrove environments. Front Microbiol 12:756091

Haydon TD, Matthews JL, Seymour JR et al (2023) Metabolomic signatures of corals thriving across extreme reef habitats reveal strategies of heat stress tolerance. Proc Biol Sci R Soc 290(1992):20221877

Hoegh Guldberg O, Jacob D, Taylor M, Bindi M, Brown S, Camilloni IA, Diedhiou A, Djalante R, Ebi KL, Engelbrecht F, Guiot J, Hijioka Y, Mehrotra S, Payne A, Seneviratne SI, Thomas A, Warren R, Zhou G (2022) Impacts of 1.5°C global warming on natural and human systems. In: Global warming of 1.5°C: IPCC special report on impacts of global warming of 1.5°C above pre-industrial levels in context of strengthening response to climate change, sustainable development, and efforts to eradicate poverty. Cambridge University Press, pp 175–312

Houlbrèque F, Ferrier-Pagès C (2009) Heterotrophy in tropical scleractinian corals. Biol Rev Camb Philos Soc 84(1) Wiley:1–17

Hume B, D'Angelo C, Burt J et al (2013) Corals from the Persian/Arabian Gulf as models for thermotolerant reef-builders: prevalence of clade C3 Symbiodinium, host fluorescence and ex situ temperature tolerance. Mar Pollut Bull 72(2):313–322

Hume BCC, D'Angelo C, Smith EG et al (2015) Symbiodinium thermophilum sp. nov., a thermotolerant symbiotic alga prevalent in corals of the world's hottest sea, the Persian/Arabian Gulf. Sci Rep 5(1) Nature Publishing Group:1–8

Jacquemont J, Houlbrèque F, Tanvet C et al (2022) Long-term exposure to an extreme environment induces species-specific responses in corals' photosynthesis and respiration rates. Mar Biol 169(6):82. Springer Science and Business Media LLC

Jones A, Berkelmans R (2010) Potential costs of acclimatization to a warmer climate: growth of a reef coral with heat tolerant vs. sensitive symbiont types. PLoS One 5(5) Public Library of Science (PLoS):e10437

Kahng SE, Akkaynak D, Shlesinger T et al (2019) Light, temperature, photosynthesis, heterotrophy, and the lower depth limits of mesophotic coral ecosystems. In: Coral reefs of the world. Springer International Publishing, Cham, pp 801–828

Kenkel CD, Moya A, Strahl J et al (2018) Functional genomic analysis of corals from natural CO2 -seeps reveals core molecular responses involved in acclimatization to ocean acidification. Glob Chang Biol 24(1):158–171

Keppel G, Van Niel KP, Wardell-Johnson GW et al (2012) Refugia: identifying and understanding safe havens for biodiversity under climate change. Glob Ecol Biogeogr 21(4) Wiley:393–404

Klepac CN, Barshis DJ (2020) Reduced thermal tolerance of massive coral species in a highly variable environment. Proc Biol Sci R Soc 287(1933):20201379. https://doi.org/10.1098/rspb.2020.1379

Kleypas JA, Mcmanus JW, Meñez LAB (1999) Environmental limits to coral reef development: where do we draw the line? Integr Comp Biol 39(1) Oxford Academic:146–159

Kramer N, Guan J, Chen S et al (2022) Morpho-functional traits of the coral Stylophora pistillata enhance light capture for photosynthesis at mesophotic depths. Commun Biol 5(1) Springer Science and Business Media LLC:861

Leal MC, Nejstgaard JC, Calado R et al (2014) Molecular assessment of heterotrophy and prey digestion in zooxanthellate cnidarians. Mol Ecol 23(15) Wiley:3838–3848

Leichter JJ, Genovese SJ (2006) Intermittent upwelling and subsidized growth of the scleractinian coral Madracis mirabilis on the deep fore-reef slope of Discovery Bay, Jamaica. Mar Ecol Prog Ser 316. Inter-Research Science Center:95–103

Leite DCA, Salles JF, Calderon EN et al (2018) Coral bacterial-core abundance and network complexity as proxies for anthropogenic pollution. Front Microbiol 9:833

Lord KS, Lesneski KC, Buston PM et al (2023) Rampant asexual reproduction and limited dispersal in a mangrove population of the coral Porites divaricata. Proc Biol Sci R Soc 290(2002):20231070

Maggioni F, Pujo-Pay M, Aucan J et al (2021) The Bouraké semi-enclosed lagoon (New Caledonia)–a natural laboratory to study the lifelong adaptation of a coral reef ecosystem to extreme environmental conditions. Biogeosciences 18(18) Copernicus GmbH:5117–5140

Mallela J, Bathgate R (2018) The role of parrotfishes in the destruction and construction of coral reefs. In: Biology of parrotfishes. taylorfrancis.com, pp 161–196

Martínez-Castillo V, Rodríguez-Troncoso AP, Santiago-Valentín JD et al (2020) The influence of urban pressures on coral physiology on marginal coral reefs of the Mexican Pacific. Coral Reefs 39(3) Springer Science and Business Media LLC:625–637

McCook L, Jompa J, Diaz-Pulido G (2001) Competition between corals and algae on coral reefs: a review of evidence and mechanisms. Coral Reefs 19(4):400–417

Mieog JC, Olsen JL, Berkelmans R et al (2009) The roles and interactions of symbiont, host and environment in defining coral fitness. PLoS One 4(7):e6364

Mizerek TL, Baird AH, Beaumont LJ et al (2016) Environmental tolerance governs the presence of reef corals at latitudes beyond reef growth. Glob Ecol Biogeogr 25(8) Wiley:979–987

Mora C, Graham NAJ, Nyström M (2016) Ecological limitations to the resilience of coral reefs. Coral Reefs 35(4):1271–1280

Morgan KM, Perry CT, Johnson JA et al (2017) Nearshore turbid-zone corals exhibit high bleaching tolerance on the Great Barrier Reef following the 2016 ocean warming event. Front Mar Sci 4. frontiersin.org:224

Nitschke MR, Rosset SL, Oakley CA et al (2022) The diversity and ecology of Symbiodiniaceae: a traits-based review. Adv Mar Biol 92:55–127

Noonan SHC, Fabricius KE, Humphrey C (2013) Symbiodinium Community Composition in Scleractinian Corals Is Not Affected by Life-Long Exposure to Elevated Carbon Dioxide. PLOS ONE 8(5): e63985. https://doi.org/10.1371/journal.pone.0063985

Oakley CA, Davy SK (2018) Cell biology of coral bleaching. In: van Oppen MJH, Lough JM (eds) Coral bleaching: patterns, processes,

causes and consequences. Springer International Publishing, Cham, pp 189–211
O'Farrell S (2011) In: Mumby PJ (ed) On the dynamics of coral reef fishes: growth, senescence and mortality. University of Exeter. Available at: http://hdl.handle.net/10036/3589.
Padilla-Gamiño JL, Roth MS, Rodrigues LJ et al (2019) Ecophysiology of mesophotic reef-building corals in Hawai'i is influenced by symbiont–host associations, photoacclimatization, trophic plasticity, and adaptation. Limnol Oceanogr 64(5) Wiley:1980–1995
Peixoto RS, Voolstra CR, Baums IB et al (2024) The critical role of coral reef restoration in a changing world. Nat Clim Chang 14. Nature Publishing Group:1219–1222
Perry CT, Larcombe P (2003) Marginal and non-reef-building coral environments. Coral Reefs 22(4):427–432
Pezner AK, Courtney TA, Barkley HC et al (2023) Increasing hypoxia on global coral reefs under ocean warming. Nat Clim Chang 13(4) Nature Publishing Group:403–409
Pezner AK, Courtney TA, Chou WC, Chu HC, Frable BW, Kekuewa SAH, Soong K, Wei Y, Andersson AJ (2024) Coral growth along a natural gradient of seawater temperature, pH, and oxygen in a near-shore seagrass bed on Dongsha Atoll, Taiwan. PLOS ONE 19(10): e0312263. https://doi.org/10.1371/journal.pone.0312263
Prasetia R, Sinniger F, Hashizume K et al (2017) Reproductive biology of the deep brooding coral Seriatopora hystrix: implications for shallow reef recovery. PLoS One 12(5):e0177034
Randall CJ, Toth LT, Leichter JJ et al (2020) Upwelling buffers climate change impacts on coral reefs of the eastern tropical Pacific. Ecology 101(2) Wiley Online Library:e02918
Reimer JD, Peixoto RS, Davies SW et al (2024) The fourth global coral bleaching event: where do we go from here? Coral Reefs 43(4):1121–1125
Renzi JJ, Shaver EC, Burkepile DE et al (2022) The role of predators in coral disease dynamics. Coral Reefs 41(2) Springer:405–422
Riegl B, Piller WE (2003) Possible refugia for reefs in times of environmental stress. Int J Earth Sci 92(4):520–531
Rogers A, Harborne AR, Brown CJ et al (2015) Anticipative management for coral reef ecosystem services in the 21st century. Glob Chang Biol 21(2):504–514
Roper CD, Suggett DJ, Songsomboon K, Edmondson J, England H, HaydonTD, Goyen S, Duijser CM, Alderdice R, Voolstra CR, Camp EF (2025) Coral thermotolerance retained following year-long exposure to a novel environment. Sci Adv 8;11(32):eadu3858
Ros M, Suggett DJ, Edmondson J et al (2021) Symbiont shuffling across environmental gradients aligns with changes in carbon uptake and translocation in the reef-building coral Pocillopora acuta. Coral Reefs 40(2) Springer:595–607
Rowan R, Knowlton N, Baker A et al (1997) Landscape ecology of algal symbionts creates variation in episodes of coral bleaching. Nature 388(6639):265–269. https://doi.org/10.1038/40843
Rubin ET, Enochs IC, Foord C et al (2021) Molecular mechanisms of coral persistence within highly urbanized locations in the Port of Miami, Florida. Front Mar Sci 8. Frontiers Media SA:695236
Schiettekatte NMD, Casey JM, Brandl SJ et al (2023) The role of fish feces for nutrient cycling on coral reefs. Oikos. Wiley. Epub ahead of print 9 June 2023. https://doi.org/10.1111/oik.09914
Schoepf V, Stat M, Falter JL et al (2015) Limits to the thermal tolerance of corals adapted to a highly fluctuating, naturally extreme temperature environment. Sci Rep 5(1) Nature Publishing Group:1–14
Schoepf V, Baumann JH, Barshis DJ et al (2023) Corals at the edge of environmental limits: a new conceptual framework to re-define marginal and extreme coral communities. Sci Total Environ 884. Elsevier:163688
Scucchia F, Zaslansky P, Boote C et al (2023) The role and risks of selective adaptation in extreme coral habitats. Nat Commun 14(1):4475
Serrano X, Baums IB, O'Reilly K et al (2014) Geographic differences in vertical connectivity in the Caribbean coral Montastraea cavernosa despite high levels of horizontal connectivity at shallow depths. Mol Ecol 23(17) Wiley:4226–4240
Smith TB, Gyory J, Brandt ME et al (2016) Caribbean mesophotic coral ecosystems are unlikely climate change refugia. Glob Chang Biol 22(8):2756–2765
Smith TB, Maté JL, Gyory J (2017) Thermal refuges and refugia for stony corals in the eastern tropical Pacific. In: Glynn PW, Manzello DP, Enochs IC (eds) Coral reefs of the eastern tropical Pacific: persistence and loss in a dynamic environment. Springer Netherlands, Dordrecht, pp 501–515
Smriga S, Sandin SA, Azam F (2010) Abundance, diversity, and activity of microbial assemblages associated with coral reef fish guts and feces. FEMS Microbiol Ecol 73(1):31–42
Strahl J, Stolz I, Uthicke S et al (2015) Physiological and ecological performance differs in four coral taxa at a volcanic carbon dioxide seep. Comp Biochem Physiol A Mol Integr Physiol 184:179–186. Elsevier. Epub ahead of print 2015
Suggett DJ, van Oppen MJH (2022) Horizon scan of rapidly advancing coral restoration approaches for 21st century reef management. Emerg Top Life Sci 6(1):125–136
Swain TD, Lax S, Lake N et al (2018) Relating coral skeletal structures at different length scales to growth, light availability to symbiodinium, and thermal bleaching. Front Mar Sci 5:450
Taira D, Poquita-Du RC, Toh TC et al (2018) Spatial variability of fish communities in a highly urbanised reef system. Urban Ecosyst 21(1):85–95
Tanvet C, Camp EF, Sutton J et al (2023) Corals adapted to extreme and fluctuating seawater pH increase calcification rates and have unique symbiont communities. Ecol Evol 13(5):e10099
Todd PA (2008) Morphological plasticity in scleractinian corals. Biol Rev Camb Philos Soc 83(3):315–337
Towle EK, Enochs IC, Langdon C (2015) Threatened Caribbean coral is able to mitigate the adverse effects of ocean acidification on calcification by increasing feeding rate. PLoS One 10(4):e0123394. https://doi.org/10.1371/journal.pone.0123394
Ulstrup KE, Ralph PJ, Larkum AWD et al (2006) Intra-colonial variability in light acclimation of zooxanthellae in coral tissues of Pocillopora damicornis. Mar Biol 149(6):1325–1335
Whitfield AK (2017) The role of seagrass meadows, mangrove forests, salt marshes and reed beds as nursery areas and food sources for fishes in estuaries. Rev Fish Biol Fisher 27(1) Springer Science and Business Media LLC:75–110
Woesik V (1992) Ecology of coral assemblages on continental islands in the southern section of the Great Barrier Reef, Australia. PhD thesis, James Cook University of North Queensland. Epub ahead of print 1992
Yates KK, Rogers CS, Herlan JJ et al (2014) Diverse coral communities in mangrove habitats suggest a novel refuge from climate change. Biogeosciences 11(16) Copernicus GmbH:4321–4337
Yu X-L, Liu C-Y, Jiang L et al (2024) Local adaptation of trophic strategy determined the tolerance of coral Galaxea fascicularis to environmental fluctuations. Sci Total Environ 943:173694. Elsevier BV

GPSR Compliance

The European Union's (EU) General Product Safety Regulation (GPSR) is a set of rules that requires consumer products to be safe and our obligations to ensure this.

If you have any concerns about our products, you can contact us on ProductSafety@springernature.com

In case Publisher is established outside the EU, the EU authorized representative is:

Springer Nature Customer Service Center GmbH
Europaplatz 3
69115 Heidelberg, Germany

Batch number: 10731661

Printed by Printforce, the Netherlands